Drug-induced and Iatrogenic
Respiratory Disease

Drug-induced and Iatrogenic Respiratory Disease

Philippe Camus, MD
Professor of Pulmonary Medicine, Department of Pulmonary
and Critical Care Medicine, Hôpital du Bocage and
Université de Bourgogne, Dijon, France

Edward C Rosenow III, MD, MS, MACP,
Master FCCP
Emeritus Professor of Medicine
Former Chair Division of Pulmonary and
Critical Care Medicine
Mayo Distinguished Clinician,
Mayo Clinic School of Medicine
Rochester, Minnesota, USA

CRC Press
Taylor & Francis Group
Boca Raton London New York

CRC Press is an imprint of the
Taylor & Francis Group, an **informa** business

First published 2010 by Hodder Arnold

Published 2019 by CRC Press
Taylor & Francis Group
6000 Broken Sound Parkway NW, Suite 300
Boca Raton, FL 33487-2742

© 2010 by Taylor & Francis Group, LLC
CRC Press is an imprint of Taylor & Francis Group, an Informa business

First issued in paperback 2019

No claim to original U.S. Government works

ISBN 13: 978-0-367-45230-8 (pbk)
ISBN 13: 978-0-340-80973-0 (hbk)

Visit the Taylor & Francis Web site at
http://www.taylorandfrancis.com

and the CRC Press Web site at
http://www.crcpress.com

British Library Cataloguing in Publication Data
A catalogue record for this book is available from the British Library

Library of Congress Cataloging-in-Publication Data
A catalog record for this book is available from the Library of Congress

Cover Design: Lynda King
Index: Indexing Specialists (UK) Ltd

Typeset in 9.5/11.5 pt Minion Pro by MPS Limited, a Macmillan Company.

In respectful and loving memory of my parents, Janine and Bernard Camus.

To Clio, my affectionate and deep appreciation.

To Romain, Guigone and Augustin Camus.

To the respected physicians who shaped my interest in pulmonary medicine: Drs Julius H Comroe Jr, Robert J Fraser and JA Peter Paré, Peter T Macklem, Jere Mead, Max F Perutz and John B West. And particularly to Jacques Chrétien, Thomas V Colby, Jean-Philippe Derenne, Marc Desmeules and Harihara M Mehendale. And many others I can't name.

A great tribute goes to Dr Edward C Rosenow III, grand, inspiring academic and *connoisseur*, who pioneered and paved the way towards an understanding of drug-induced lung disease.

Philippe Camus

I want to dedicate this book to the many, many physicians who reported in the peer-reviewed literature the thousands of cases of drug-induced lung injury and disease. Without them we wouldn't have had the data to compile the material for this book. I also thank Dr Camus, whose energy and dedication made it all happen. His www.pneumotox.com website is a great asset to the many of us who need a quick resource to check the possibility of a drug-induced lung disease. I am grateful to him for his support; it has been great to work with him, even though we are 4,000 miles apart!

Edward C Rosenow III

Contents

Preface

Cherchez le médicament!
Searché le drug!

Recognizing and managing drug-induced respiratory disease (DIRD) is a real challenge to pulmonologists, critical-care medicine specialists, haematologists, oncologists, cardiologists, allergists, rheumatologists, neurologists, pathologists and primary-care physicians. Indeed, the family physician may be the first to see the patient with an adverse reaction to drugs, but may be suboptimally knowledgeable about this possibility. It is therefore necessary to disseminate knowledge about DIRD in most disciplines, with the objective of ensuring timely diagnosis and early drug stoppage should DIRD supervene.

Communication between disciplines is also of primary importance. For instance, pathologists should know about the drugs the patient is taking, since some medications can produce a distinctive lung pathology, and the information may not be made available to them at the time of biopsy. Similarly, roentgenologists should be informed of the drugs patients are being exposed to, as the radiographic characteristics of DIRD may at times enable a prompt and reliable identification of the drug aetiology. DIRD is an important topic for the patient as well, who must be educated and cognizant of the possible adverse effects of the drug he or she is taking, in order to warn us in due time should unexplained respiratory symptoms develop. Concerted efforts should concentrate on diagnosing DIRD as early and reliably as possible.

The evidence base for recognition, diagnosis and management of many drug-induced respiratory diseases is now strong. More than 430 drugs are known to cause respiratory injury (a rise from about 250 ten years ago), and the number is still increasing.

Drugs can cause major adverse respiratory reactions, depending on the target organ in the respiratory system (lung, airways, pulmonary circulation, pleura, mediastinum, lymph nodes, muscle and nerves, haemoglobin), causing entirely different clinical patterns. There are several discrete subgroups under each pattern, each corresponding to a distinct clinico-pathologic picture. Drug reactions can be a perfect mimic of naturally occurring respiratory diseases clinically, pathologically and on imaging. Drugs can cause one or more patterns of involvement with, inexplicably, little overlap between each other or in any one patient. Drugs within a pharmacological class may cause the same adverse effect (e.g. β-blockers or NSAIDs and bronchospasm, ergots and pleural effusion or

thickening, anti-TNF and opportunistic infection), suggesting a common pharmacological and/or cytopathic mechanism.

This is not just a matter of curiosity or any compulsive need to pigeon-hole this group of conditions. First, the incidence of DIRD, although wide-ranging, may be as high as 50 per cent in some oncology settings. Second, the clinical and imaging manifestations of most drug-induced respiratory diseases will remit following drug discontinuance, leaving no residuum. This represents a simple and often robust test to assess drug causality. Third, acute drug-induced interstitial lung disease (ILD), pulmonary oedema/haemorrhage, acute upper airway obstruction, bronchospasm, anaphylaxis or pleural effusion can cause severe life-threatening or fatal respiratory compromise, sometimes within minutes. This commands emergency management, early recognition of the drug aetiology, reliable identification of the culprit drug, and definitive drug removal. Fourth, some drugs produce a systemic reaction such as lupus erythematosus, vasculitis (Wegener or Churg–Strauss), polymyositis, or the drug rash and eosinophilia with systemic symptoms (DRESS), with lung and other deep-seated organ involvement. If the drug aetiology is not recognized, patients run the risk of developing irreversible organ damage or undergoing unnecessary, costly and/or invasive tests, while drug discontinuation would act as an efficient diagnostic test, abbreviating disease duration to the benefit of the patient.

Not to be forgotten, DIRD remains a disease of exclusion, in that other probable or possible causes have to be scrupulously ruled out. In the majority of patients, the causal drug cannot be diagnosed by any specific test, except with dechallenge and re-challenge, and the latter is potentially hazardous. Adding to the difficulty, some underlying diseases for which pulmotoxic drugs are exactly given can also involve the lung in a manner similar to drugs.

Traditional agents known to cause lung injury in the 1950s and 1960s were ionizing radiations, gold salts, nitrofurantoin and chemotherapy agents. The latter can produce the ill-fated 'chemotherapy lung', which remains a *bête-noire* among DIRD. Afterwards, many more drugs were shown to cause similar lung, airway and pleural injury, regardless of the route by which they were administered – including oral, inhaled, intravenous, intramuscular, ophthalmic, hepatic arterial, pleural, vaginal and topical (in the form of ointments, or in the urinary bladder). High drug dosage, prolonged *versus* episodic treatment, and combinations of chemo agents with radiation therapy and/or oxygen were shown to be more deleterious than each agent taken in isolation, or a combination of drugs could recall previous injury from previously administered therapeutic agents. Trials had to be halted for this very reason. It was surprising that the diet pill aminorex occasioned an epidemic of clinical-haemodynamic picture resembling primary pulmonary hypertension, and this was unfortunately repeated by new anorectics of the fenfluramine series in the 1980s. Both drugs had to be recalled. In the 1970s, rheumatoid arthritis patients were shown to sustain injury from disease-modifying drugs, and it appeared that drug-induced lung disease could be difficult to separate from the background rheumatoid lung. Although novel drugs including biologics are now available to treat rheumatoid arthritis in a very efficient manner, distinguishing drug-induced respiratory disease from rheumatoid lung involvement poses essentially the same problems as before, and reporting bias may still prevail as it used to do then.

A radically new pattern of pulmonary involvement emerged in the 1980s with the potentially devastating lung disease caused by the antiarrhythmic amiodarone, a distinctive pattern of drug-induced pulmonary involvement. Other novel, distinct drug-induced patterns of respiratory involvement include acute lung injury following transfusion of blood (TRALI) or administration of ATRA, drug-induced BOOP, lung injury from statins, tyrosine kinase (TKI) and mTOR inhibitors, pleural pathology from dasatinib, and opportunistic infections (mainly tuberculosis) associated with TNF-α antibody therapy, which dictates rigorous evaluation of each individual patient to be treated as regards latent tuberculosis. We can also mention the infectious and non-infectious complications that may follow stem cell or solid organ transplantation.

In addition to therapy drugs, history-taking may point to over-the-counter medications, illicit drugs of abuse and, less often, herbals or supplements, which may occasion severe pulmonary reactions, not unlike those produced by real drugs.

It remains difficult to predict drug-induced respiratory disease in the individual patient. For a limited number of drugs, however, risk assessment has identified patient-related factors (e.g. atopics and the risk of drug-induced bronchospasm, or patients with prior evidence of ILD and the risk of pulmonary adverse effects of TKI, methotrexate or amiodarone) and drug-related factors (e.g. dosage, duration of exposure, pharmacokinetics, combination of drugs, exposure to oxygen) which may expose certain subsets of patients to a higher than background incidence rate of pneumonitis. Monitoring of patients should be planned accordingly. We can imagine personalized medicine that would enable prediction of which individual patient is more likely to develop adverse effects of therapy drugs, to focus serial monitoring in those patients at risk.

In the meantime, DIRD continues to move at a swift pace, with no obvious risk of drifting into a sense of *ennui*. However, more than ever before it is difficult to master DIRD, with 13 800 literature references at the time of writing, and in some ways we seem to know less about it now than we did a few decades ago. Thus, a free web-based tool was made available as a way of keeping the busy practising physician up to date quickly about the possibility that the drug(s) his or her patient is on could explain the symptoms and imaging abnormalities (Pneumotox®, www.pneumotox.com).

In addition to that website, we thought a treatise was needed to describe DIRD in more depth. Renowned experts in the field have joined our efforts to make the endeavour possible. This is the best reward that we, as editors and co-authors, could ever imagine after several decades of combined experience and impetus to have DIRD better appreciated, classified and recognized as a meaningful clinical entity. Our deep appreciation goes to each contributor. Our acknowledgement also goes to the outstanding team of people at Hodder, who patiently and very professionally made this project possible.

Philippe Camus, Gurgy le Château, Burgundy, France
Edward C Rosenow, Rochester, Minnesota, USA

List of contributors

Lucien Abenhaim
LARISK Consultants, School of Hygiene and Tropical Medicine, London, UK

Bekele Afessa
Division of Pulmonary and Critical Care Medicine, Mayo Clinic College of Medicine, Rochester, MN, USA

James N Allen
Division of Pulmonary and Critical Care Medicine, Ohio State University, Columbus, OH, USA

Noreen Aziz
Rubin Center for Cancer Survivorship, James P Wilmot Cancer Center, University of Rochester Medical Center, Rochester, NY, USA

K Suresh Babu
Infection, Inflammation and Repair, Southampton General Hospital, Southampton, UK

Andrew D Badley
Division of Infectious Diseases, Mayo Clinic College of Medicine, Rochester, MN, USA

Francesco Bonella
Ruhrlandklinik, Department of Pneumology/Allergy, Essen, Germany

Kim Bouillon
Department of Epidemiology and Public Health, University College London, London, UK

Demosthenes Bouros
Department of Pneumonology, Medical School, Democritus University of Thrace, Alexandroupolis, Greece

Philippe Camus
Professor of Pulmonary Medicine, Department of Pulmonary and Critical Care Medicine, Hôpital du Bocage and Université de Bourgogne, Dijon, France

Jan-Wil Cohen Tervaert
Department of Internal Medicine, Cardiovascular Research Institute, Maastricht University, The Netherlands

Thomas V Colby
Department of Pathology, Mayo Clinic College of Medicine, Scousdale, AZ, USA

J Allen D Cooper Jr
Division of Pulmonary and Critical Care Medicine, University of Alabama, Birmingham, AL, USA

Ulrich Costabel
Ruhrlandklinik, Department of Pneumology/Allergy, Essen, Germany

Michiel De Vries
Department of Respiratory Medicine, ORBIS Medical Concern, Sittard/Geleen, The Netherlands

Margolein Drent
Department of Respiratory Medicine and ILD Care, Maastricht University, The Netherlands

Gary R Epler
Harvard Medical School, and Pulmonary and Critical Care Medicine, Brigham and Women's Hospital, Boston, MA, USA

Hosnieh Fathi
Castle Hill Hospital, Cottingham, Hull, UK

Marc B Feinstein
Pulmonary Division, Memorial Sloan Kettering Cancer Center and Weill Medical College of Cornell University, New York, NY, USA

Douglas B Flieder
Fox Chase Cancer Center, Philadelphia, PA, USA

Poh-Gek Forkert
Department of Anatomy and Cell Biology, Queen's University, Kingston, Ontario, Canada

Marios Froudarakis
Department of Pneumonology, Medical School, Democritus University of Thrace, Alexandroupolis, Greece

Amitava Ganguli
Department of Pharmacology and Therapeutics, University of Liverpool, Liverpool, UK

Umair A Gauhar
Division of Pulmonary and Critical Care Medicine, University of Alabama, Birmingham, AL, USA

Josune Guzman
Allgemeine und Experimentelle Pathologie, Ruhr-Universität, Bochum, Germany

Thomas E Hartman
Department of Radiology, Mayo Clinic, Rochester, MN, USA

Christopher P Holstege
Division of Medical Toxicology, Departments of Emergency Medicine and Pediatrics, University of Virginia School of Medicine, Charlottesville, VA, USA

Marc Humbert
Hôpital Antoine-Béclère, Service de Pneumologie, Assistance Publique – Hôpitaux de Paris, Université Paris-Sud 11, Clamart, France

Patricia M Kopko
Histocompatibility Laboratory, BloodSource, Sacramento, CA, USA

Robert M Kotloff
Pulmonary, Allergy and Critical Care Division, University of Pennsylvania Medical Center, Philadelphia, PA, USA

Abigail R Lara
University of Colorade Health Sciences Center, Division of Pulmonary Sciences and Critical Care, Denver, CO, USA

Deepa Lazarus
Georgetown University Hospital, Washington, DC, USA

Teofilo Lee-Chiong JR
Section of Pulmonary Medicine, Department of Medicine, National Jewish Medical and Research Center, Denver, CO, USA

Andrew H Limper
Division of Pulmonary, Critical Care and Internal Medicine, Mayo Clinic College of Medicine, Rochester, MN, USA

Michael Lippmann
Albert Einstein Medical Center, Thomas Jefferson University School of Medicine, Philadelphia, PA, USA

Demosthenes Makris
Department of Pneumonology, Medical School, Democritus University of Thrace, Alexandroupolis, Greece

Fabien Maldonado
Division of Pulmonary, Critical Care and Internal Medicine, Mayo Clinic College of Medicine, Rochester, MN, USA

Richard A Matthay
Pulmonary and Critical Care Section, Department of Medicine, Yale University School of Medicine, New Haven, CT, USA

Alyn H Morice
Head of Cardiovascular and Respiratory Studies, University of Hull, Cottingham, UK

Yola Moride
Faculty of Pharmacy, Université de Montréal, Montreal, Canada

Jaymin Morjaria
Infection, Inflammation and Repair, Southampton General Hospital, Southampton, UK

Ganesan Murali
Albert Einstein Medical Center, Thomas Jefferson University School of Medicine, Philadelphia, PA, USA

Anne O'Donnell PULM-CCM
Georgetown University Hospital, Washington, DC, USA

Steve G Peters
Division of Pulmonary and Critical Care Medicine, Mayo Clinic College of Medicine, Rochester, MN, USA

Munir Pirmohamed
Department of Pharmacology and Therapeutics, University of Liverpool, Liverpool, UK

Albert J Polito
The Lung Center, Mercy Medical Center, Baltimore, MD, USA

Mark A Popovsky
Haemonetics Corporation, Braintree, MA, USA

Tracey K Riley
Division of Medical Toxicology, Department of Emergency Medicine, University of Virginia School of Medicine, Charlottesville, VA, USA

M Patricia Rivera
Department of Medicine, Division of Pulmonary and Critical Care Medicine, University of North Carolina at Chapel Hill, Chapel Hill, NC, USA

Edward C Rosenow III
Emeritus Professor of Medicine, Former Chair Division of Pulmonary and Critical Care Medicine, Mayo Distinguished Clinician, Mayo Clinic School of Medicine, Rochester, MN, USA

Robert L Rubin
Department of Molecular Genetics and Microbiology, University of New Mexico, Albuquerque, NM, USA

Steven A Sahn
Division of Pulmonary and Critical Care, Allergy and Sleep Medicine, Medical University of South Carolina, Charleston, SC, USA

Marvin I Schwarz
University of Colorado Health Sciences Center, Division of Pulmonary Sciences and Critical Care, Denver, CO, USA

Kahoko Taki
Department of Emergency Medicine, Saga University Faculty of Medicine, Nabeshima, Saga, Japan

Lois B Travis
Rubin Center for Cancer Survivorship, James P Wilmot Cancer Center, University of Rochester Medical Center, Rochester, NY, USA

William D Travis
Department of Pathology, Memorial Sloan Kettering Cancer Center, New York, NY, USA

Esra Uzaslan
Tip Fakültesi, Uludag Universitesi, Görükle, Bursa, Turkey

Max M Weder
Department of Medicine, Division of Pulmonary and Critical Care Medicine, University of Virginia Health System, Charlottesville, VA, USA

Dorothy A White
Department of Medicine, Pulmonary Section, Memorial Sloan Kettering Cancer Center and Weill Medical College of Cornell University, New York, NY, USA

PART 1

GENERAL

1

Classification, diagnosis and management of drug-induced respiratory disease

PHILIPPE CAMUS, WITH THE INPUT AND SUPPORT OF EDWARD C ROSENOW III

INTRODUCTION TO THE TABLE ON PAGES 4–11

Only the most common patterns of drug-induced adverse respiratory effects are included. Drug-induced opportunistic infection is not covered here. For more details, see www.pneumotox.com (Pneumotox®).

Pattern	1st report	Incidence	Representative drugs 1	Risk factors	Time to onset	S&S	ARF	Chest radiograph
Interstitial – infiltrative lung disease (ILD)								
Subacute ILD*	1948	4	Methotrexate, nitrofurantoin	–	w, mo, y	d, c, f, m	–	
Acute severe ILD	1969	2	Methotrexate	–	w, mo, y	D, C, F, M	Y	
Eosinophilic pneumonia	1941	4	Minocycline, NSAIDs	–	w, mo, (y)	d, c, f, m, rash	–	
Acute eosinophilic pneumonia (w/wo the DRESS syndrome)	1991	2	Minocycline, anticonvulsants	–	w, mo, (y)	D, C, F, M	Y	
Diffuse alveolar damage (DAD) w/wo ARDS	1968	3	Chemo agents (solo or multiagent regimen), sirolimus, amiodarone	Previous lung damage, O_2	d, w, mo	D, C, F	Y	
Amiodarone pulmonary toxicity (APT)	1980	5	Amiodarone	Dosage, age	d, w, y	d, C, f, m, wl, CP	Y	
Acute amiodarone pulmonary toxicity	1983	2	Amiodarone	Dosage, age, surgery, O_2	d, w, y	D, C, F, M	Y	
Organizing pneumonia (BOOP)	1954	2	Amiodarone, mTOR-inhibitors, statins, radiation therapy	–	w, mo, y	d, c, f, m	Y	
ILD with granulomas w/wo mediastinal lymphadenopathy	1965	2	BCG-therapy, interferons, methotrexate, sirolimus, anti-TNF agents	–	mo, y	ns	–	

HRCT	Laboratory	BAL	Pathology	Pathology	Differential	Management outcome
	↑ CRP, lymphocyte stimulation test unreliable	↑ Lymphocytes w/wo neutrophils	Cellular NSIP		ILD of other causes, infection	DW (cst)/good
	↑ CRP lymphocyte stimulation test unreliable	↑ Lymphocytes	Dense cellular NSIP, occasional granulomas		ILD of other causes, infection	DW, CST/mostly good
	↑ Blood eosinophils in most patients	↑ Eosinophils	Mixed inflammatory cell infiltrate w eosinophils		EoP of other causes, infection, parasitic infestation	DW (cst)/good
	↑ Blood eosinophils in most patients	↑ Eosinophils	Dense mixed inflammatory and eosinophilic infiltrate		AEP of other causes, infection, parasitic infestation	DW, CST/mostly good
	–	↑ Neutrophils, atypical epithelial cells	Oedema, alveolar fibrin, hyaline membranes		Infection, heart failure	DW, CST/up to 50% fatality rate
	↑ ESR, ↑ globulins. KL-6 unreliable	Foam cells. ↑ lymphocytes or neutrophils in some	Interstitial inflammation and foam cells		Heart failure, infection, incidental ILD, malignancy	DW, cst/good in 85%. Fatality rate 5–10%
	–		Acute interstitial/ alveolar inflammation & foam cells		Other causes of ALI, ARDS	DW, CST/ fatality rate up to 50%
	–	↑ Lymphocytes, neutrophils and/or eosinophils	Interstitial inflammation. Ductal fibrosis		BOOP of other cause	DW, (cst)/good
	–	↑ Lymphocytes	Interstitial granulomas		Sarcoidosis and sarcoidlike reactions	DW, (cst)/good

(Contd)

Pattern	1st report	Incidence	Representative drugs 1	Risk factors	Time to onset	S&S	ARF	Chest radiograph
Acute nitrofurantoin lung	1956	2	Nitrofurantoin	Previous Rx, sensitization	d, w	D, c, F, CP	Y	
Subacute/chronic nitrofurantoin lung	1964	2	Nitrofurantoin	Prior undiagnosed acute NF lung ?	mo, y	d, c, m, wl	Y	
Exogenous lipoid pneumonia	1925	2	Liquid paraffin, mineral oil	Advanced age, aspiration	mo, y	ns	–	
Pulmonary fibrosis	1953	2	Chemo agents, radiation therapy, amiodarone	Dose, multi-agent regimens	mo, y	D, c, f, m, WL	Y	
Radiation pneumonitis	1940	2	Radiation therapy to the chest or breast	Dose, portals, concomitant chemotherapy	w, mo, y	D, C, f, m, wl	Y	
Single or multiple lung nodules or masses	1972	2	Amiodarone, bleomycin, paraffin	–	w, mo, y	d, c, f	–	
Non-cardiac pulmonary oedema	1938	3	Aspirin, chemo agents, ATRA, hydrochlorothiazide, i.v. β2 agonists, blood transfusion (TRALI)	Dose, combination therapy	min, h	D, c	Y	
ARDS	1959	2	Amiodarone, chemo agents, heroin, TACE	–	min, h, d	D, frothy sputum	Y	
Alveolar hemorrhage	1965	2	All anticoagulants, abciximab, clopidogrel, cocaine, sirolimus	Overdose, heart failure	h, d, w	D, w/wo hemoptysis	Y	

HRCT	Laboratory	BAL	Pathology	Pathology	Differential	Management outcome
	–	↑ Lymphocytes and/or eosinophils	Interstitial inflammation w/wo eosinophilia		Infection, incidental ILD	DW (cst)/good
	ANA+ in some	↑ Lymphocytes and/or neutrophils	Inflammation, fibrosis, DIP pattern		Chronic ILD of other aetiology	DW cst/may persist unchanged
	–		Lipid-laden macrophages in BAL & tissue		Aspiration pneumonia	DW; no documented effect of cst/ Will often persist
	–	↑ Neutrophils	Diffuse interstitial fibrosis. Reactive epithelial cells		Pulmonary fibrosis of other aetiology	DW, cst/ uncertain
	–	↑ Lymphocytes in both the irradiated and non-irradiated lung	Inflammation, oedema, epithelial cell sloughing. Late cases show fibrosis in irradiated area		Almost none if portals and areas of lung involvement coincide	cst/depends on volume and radiation dose delivered to the lung
	–	–	Rounded BOOP or APT		Metastatic lung disease, vasculitis	DW (cst)/good
	Identification of antibodies in donor required in suspected TRALI	↑ Neutrophils	Bland alveolar flooding		Other causes of NCPE	DW/MV/cst/ mostly good
	–	↑ Lymphocytes and/or neutrophils	Early fibrosis, alveolar fibrin, oedema		Other causes of ARDS	DW/MV/high fatality rate
	Monitor blood loss, check blood coagulation studies	Free red cells & hemosiderin-laden macrophages	DAH wo, generally, capillaritis		Other causes of DAH	DW/good

(Contd)

Pattern	1st report	Incidence	Representative drugs 1	Risk factors	Time to onset	S&S	ARF	Chest radiograph
Airway involvement								
Upper airway obstruction	1968	2	ACE-I, ARB		d, w, m, y	Asphyxia	Y	Airway narrowing
Sudden asthma attack	1933	3	NSAIDs, aspirin, β-blockers	Asthma	min, h	Sudden sever bronchospasm	Y	Hyperinflation
Cough	1985	4	ACE-I, anaesthetic agents	–	d, w, m, y	C	–	Normal
Pleural involvement								
Pleural effusion	1954	2	Ergots & several other drugs	Long-term treatment	m, y	d, c, f, m, cp	Y	
Drug *lupus*	1954	2	Lupus-inducing drugs. Novel anti-TNF agents, interferon	–	m, y	d, c, f, m, cp	–	
Pleural thickening	1966	2	Ergots	Long-term treatment	m, y	d, c, f, m, cp	–	
Hemothorax	1969	1	Oral anticoagulants	Overdosing	d, w, m, y	d, c, cp, anaemia	Y	
Pneumothorax	1973	1	Chemo agents	Subpleural tumour or masses	d, w	D, CP	Y	
Pulmonary vasculopathy								
Pulmonary microthromboembolism w/wo acute/chronic PHTn	1963	2	Illicit drugs, acrylates, contrast media, crospovidone, mercury, oral contraceptives, silicone, protamine, aprotinin, TPN	i.v. use of substances intended for oral use	h, d, w, y	D, RHF	Y	
Chronic non-embolic pulmonary hypertension	1967	2	Anorectics of the past (aminorex, fenfluramine)	Genetic background	m, y	D, PHTn, RHF	–	Similar to primary PHTn

HRCT	Laboratory	BAL	Pathology	Pathology	Differential	Management outcome
Airway narrowing	–	ND	Airway oedema		Non-drug induced UAO/ anaphylaxis	DW, cst/mostly good
–	–	ND	Overinflated lung, mucous plugging, airway oedema		Close temporality suggests causality	DW, cst, BD/ can be fatal
Normal	–	Normal	–	–	Other causes of cough may be revealed by ACE-I	DW/excellent
	Usually an exudate w/wo pleural eosinophilia	–	Pleural inflammation w/wo eosinophilia of the pleural membrane		Effusions of other cause	DW (cst?)/good
	ANA test positive	–	Inflammation		Naturally occurring *lupus*	DW, cst/good
	↑ ESR	↑ Lymphocytes and/or neutrophils	Acellular collagen w/wo inflammation		Exposure to asbestos, rheumatic conditions	DW, cst/pleural thickening may persist
	Monitor blood loss, check blood coagulation studies	–	–		Pleural infection/ malignancy	
		–	–		Spontaneous pneumothorax	–
	–	–	Microthrombi. Foreign body granulomas around pulmonary arterioles		Primary PHTn	–
	–	–	Resembles primary PHTn		Primary PHTn	DW/treat like any PHTn

(Contd)

Pattern	1st report	Incidence	Representative drugs 1	Risk factors	Time to onset	S&S	ARF	Chest radiograph
Mediastinal-subpleural lipomatosis	1967	3	Corticosteroids	Obesity	m, y	Subclinical	–	
Methemoglobinemia	1920	1	Nitrites, NO, several therapy drugs (benzocaine, dapsone), illicit substances ('poppers')	Enzymatic status	min, h	Low SpO_2, cyanosis, normal PaO_2. Symptomatic if MetHb > ~50%	Y	Pulmonary infiltrates in some cases

Key

- *Interstitial lung disease:** Other distinct patterns of interstitial lung disease have been reported as an adverse effect of drugs in a few cases. Examples are desquamative interstitial pneumonia (DIP), giant-cell interstitial pneumonia (GIP), lymphocytic interstitial pneumonia (LIP), bronchiolitis interstitial pneumonia (BIP), diffuse pulmonary calcification, and pulmonary alveolar proteinosis (PAP). See www.pneumotox.com.
- *Time to onset*: min = minutes; h = hours; d = days; w = weeks; mo = months; y = years. Parentheses indicate 'rarely so'; for example, (y) means 'rarely years'.
- *Signs and symptoms* (S&S): c = cough; cp = chest pain; d = dyspnoea; f = fever; m = malaise; MetHb = methemoglobin; ns = non-specific symptoms; PaO_2 = arterial partial pressure of oxygen; rash = cutaneous rash; RHF = right heart failure; SpO_2 = pulse saturation of oxygen; wl = bodyweight loss. When the annotation is in capitals, the sign or symptom is marked, annoying or severe.

Abbreviations

ACE-I	angiotensin converting enzyme inhibitors
AEP	acute eosinophilic pneumonia
ALI	acute lung injury
ANA	antinuclear antibody test
APT	amiodarone pulmonary toxicity
ARB	angiotensin II receptor blocker
ARDS	acute respiratory distress syndrome
ARF	acute respiratory failure
ATRA	all-*trans*-retinoic acid
BAL	bronchoalveolar lavage
BCG	Bacillus Calmette Guérin
BD	bronchodilators
BOOP	bronchiolitis obliterans organizing pneumonia
CRP	C-reactive protein
CST	high-dose corticosteroid therapy may be given according to patient status – cst: low-dose corticosteroid therapy may be given according to patient status; (cst): unclear effect of corticosteroid therapy; (cst?): usefulness of corticosteroid therapy questionable
DAD	diffuse alveolar damage
DAH	diffuse alveolar haemorrhage
DIP	desquamative interstitial pneumonia
DRESS	drug rash, eosinophilia and systemic symptoms
DW	drug withdrawal
EoP	eosinophilic pneumonia
ESR	erythrocyte sedimentation rate
HRCT	high resolution computer tomography
ILD	interstitial lung disease
IV	intravenous
KL-6	*Krebs von den Lungen* 6
mTOR	mammalian target of rapamycin
MV	mechanical ventilation
NCPE	non-cardiac/non-cardiogenic pulmonary oedema
ND	no data published (_: no personal data)
NO	nitric oxide
NSAID	non-steroidal anti-inflammatory drug
NSIP	non-specific interstitial pneumonia
O_2	oxygen
PHTn	pulmonary (arterial) hypertension
TACE	transarterial chemoembolization
TNF	tumour necrosis factor
TPN	total parenteral nutrition
TRALI	transfusion-related acute lung injury
UAO	upper airway obstruction
w, wo	with, without

HRCT	Laboratory	BAL	Pathology	Pathology	Differential	Management outcome
	–	–	Fatty mediastinal/ subpleural deposits		Obesity, hypercortisolism	
Pulmonary infiltrates in some cases	Measure MetHb *via* multi-wavelength detector	–	–	–	True cyanosis	Methylene blue/good in most

2

Mechanisms of chemically induced respiratory toxicities

POH–GEK FORKERT

INTRODUCTION

The paradigm that organ toxicities are mediated by metabolism and formation of reactive metabolites from chemically inert xenobiotic (foreign) compounds and drugs is well established. Most of the literature in this broad area referred to as 'drug metabolism' is focused on investigations in the liver. However, recognition of the susceptibility of the lung to chemically induced toxicities has produced a substantial database regarding the adverse effects of chemicals and drugs in the respiratory system, where exposure is mediated via both the circulation and inhalation. As found in the hepatic system, metabolism of chemicals or metabolic activation in the lung is usually mediated by drug-metabolizing enzymes, including the cytochrome P450 system, to yield reactive metabolites capable of binding covalently to tissue constituents including proteins, lipids and nucleic acids (DNA, RNA). The covalent binding of reactive metabolites to cellular proteins critical for maintaining normal cellular function appears to be a key event in the development of chemically induced toxicities. The pulmonary toxicities caused by a number of chemicals including 4-ipomeanol,[1] 3-methylindole,[2] naphthalene,[3] 1,1-dichloroethylene[4] and trichloroethylene[5] have been ascribed to cytochrome P450-dependent formation of reactive metabolites that bind covalently to protein.

Covalent binding of chemical or drug metabolites to proteins may also lead to the formation of chemical– or metabolite–protein adducts that act as immunogens, resulting in immune responses or hypersensitivity reactions, formation of immune complexes, or tissue toxicities.[6,7] The formation of chemical–protein adducts may be associated with some autoimmune diseases.[8] A well-recognized drug hypersensitivity reaction is that produced by the anaesthetic drug halothane; hepatocytes that contain halothane neoantigens are susceptible to attack by lymphocytes.[9] Other drugs postulated to cause immune-related toxicities include hydrazaline,

practolol, phenytoin, amodiaquine, ethynylestradiol, and propylthiouracil.[8,10,11] In contrast to the liver, data regarding chemical- or drug-related immune responses in the lung are limited.

It is more than 90 years since the first experimental tumours were produced by exposure to chemicals.[12] Since then, extensive data have accumulated from studies for a wide variety of chemicals with the potential to induce carcinogenesis. The information accrued has led to the evolution of a theory that a chemical carcinogen or its metabolite must react in a specific manner with critical cellular constituents to produce a neoplasm.[13] Upon contact, chemical carcinogens are metabolized to compounds or metabolites that react with tissue macromolecules. In the majority of cases, chemical metabolism leads to the formation of polar derivatives that are non-carcinogenic. However, in certain cases, the metabolite is more carcinogenic than the parent compound and it may react with or bind to cellular constituents including proteins and nucleic acids to initiate carcinogenesis. Early studies have focused on binding to protein as a basic mechanism leading to carcinogenesis. However, subsequent investigations have focused on binding of metabolites to DNA as a critical initiating event in the development of chemical carcinogenesis.[13,14] The importance of DNA binding was first suggested by the studies of Brookes and Lawley,[15] who found a correlation between the carcinogenicity of several polycyclic hydrocarbons in mouse skin and the covalent binding of these hydrocarbons to DNA. Further studies have shown that reactive metabolites of benzo[a] pyrene bound to DNA,[16,17] and generated deoxyguanosine as a DNA adduct, resulting in the induction of adenomas in the lungs of mice.[18]

Covalent binding is usually determined by using a radio-labelled chemical or drug, and subsequently measuring the amounts bound to proteins or other cellular constituents after extraction with agents such as solvents. Immunochemical techniques have also been used to determine covalent binding by measuring the amounts of protein-bound adducts. An immunochemical approach has been developed in the liver to characterize the protein adducts formed from the analgesic acetaminophen.[19] Interestingly, these protein adducts were detected in the serum of mice that were given toxic doses of acetaminophen.[20] Similar protein adducts were also detected in the serum of patients who had overdosed on acetaminophen and developed heptotoxicity.[21] Hence, covalent binding is a parameter that serves as a convenient index of the formation and exposure of a tissue to reactive metabolites generated from potentially cytotoxic chemicals or drugs.

Some lung toxicities are mediated by a mechanism referred to as 'oxidative stress'. In this mechanism, metabolic activation produces reactive oxygen metabolites including the superoxide anion radical ($O_2^-\cdot$), hydrogen peroxide (H_2O_2), the hydroxyl radical ($\cdot HO$), and singlet oxygen ($^1O_2$). Each of these species can act as an oxidizing agent and is capable of contributing to oxidative stress. Hydrogen peroxide may be formed from dismutation of the superoxide anion, and the hydroxyl radical may be produced by interaction of the superoxide anion or hydrogen peroxide with iron ions.[6] Reactive oxygen species may also be formed from redox cycling of compounds such as paraquat, nitrofurantoin and Adriamycin. Redox cycling is a process whereby a compound undergoes reduction by NADPH to an intermediate such as a free radical that is then oxidized by molecular oxygen to yield the superoxide anion, resulting in regeneration of the parent compound. The reactive oxygen species react with cellular constituents that can produce toxic effects including lipid peroxidation, glutathione depletion, perturbation of redox balance, and cell death.

Although covalent binding of electrophilic metabolites to nucleophilic sites on proteins, lipids and nucleic acids is linked to tissue cytotoxicities, some chemicals that produce reactive metabolites also produce reactive oxygen species. For example, the lung toxicity of cyclophosphamide, which requires metabolic activation for its therapeutic efficacy and its toxic action, is associated with the production of reactive metabolites and oxygen free radicals.[22] However, their relative contributions to cyclophosphamide toxicity have not been defined.

The specific events that occur in the period between exposure and manifestation of toxicities, including the manner as to how they contribute to the toxicities, have not been delineated. Later in this chapter we present data relating to our efforts and those of others to identify the metabolic events involved in the bronchiolar toxicity of 1,1-dichloroethylene and the carcinogenicity of vinyl carbamate.

DISTRIBUTION AND LOCALIZATION OF PULMONARY CYTOCHROME P450 ENZYMES

The cytochrome P450 monooxygenases are a superfamily of enzymes that catalyzes the oxidation of endogenous compounds such as steroids, bile acids and fatty acids, as well as a wide variety of exogenous compounds including drugs and chemicals. Cytochrome P450 is embedded in the phospholipid bilayer of the endoplasmic reticulum, thus facilitating the interactions with lipophilic xenobiotics. Although microsomal cytochrome P450 content in the lung represents only about 10 per cent of the amount in hepatic microsomes, the high concentration of the monooxygenases in a few lung cells renders them susceptible to the adverse effects of chemicals that undergo metabolic activation. This preferential localization is believed to underlie the cell selectivity of chemically induced cytotoxicities in the lung. In addition, these cells may be localized in critical regions of the lung or produce a critical substance such as surfactant, so that toxicity to a limited cell population may threaten lung integrity. Hence, the mechanisms involved in chemically induced lung cytotoxicities are linked to the localization of the different forms of cytochrome P450 in the various lung cell types. This discussion will focus on the expression and localization of the various P450 isoforms in the lung, including the substrates that they metabolize. The major cytochrome P450 enzymes that have been identified in the lungs of various species, including the human, are summarized in Table 2.1.

The nomenclature used for the cytochrome P450 superfamily is genetically based. The naming of a P450 gene

Table 2.1 Localization of cytochrome P450 enzymes in the respiratory tract

Enzyme	Species	Location	References
CYP1A1	Rabbit	Clara cells, Type II cells, endothelial cells, macrophages	24
	Rat	Clara cells, Type II cells, endothelial cells	25
	Mouse	Type II cells, endothelial cells	26
	Human	Bronchiole, Type II cells	27, 28
CYP2A5	Mouse	Olfactory mucosa	29
CYP2A10/11	Rabbit	Nasal mucosa	30, 31
CYP2A13	Human	Trachea, bronchus	32
CYP2B1	Mouse	Clara cells, Type I cells, Type II cells	33
	Rat	Clara cells, Type I cells, Type II cells	33
CYP2B4	Rabbit	Clara cells, Type II cells	34
CYP2E1	Mouse	Clara cells	40
CYP2F2	Mouse	Clara cells	43
CYP2S1	Human	Nasal mucosa, bronchi, bronchiole	50
CYP4B1	Rat	Clara cells, Type II cells	33
	Rabbit	Clara cells, Type II cells	33

includes the root symbol *CYP*, denoting *cytochrome P*450, an Arabic number designating the P450 family, a letter indicating the subfamily when two or more subfamilies are known to exist within that family, and an Arabic numeral representing the individual gene.[23] The gene product such as a protein is not italicized. The most recent updates on the P450 system are available at http://drnelson.utmem.edu/nelsonhomepage.html.

CYP1A1 is a P450 enzyme that is involved in the metabolism of a wide variety of polycyclic aromatic compounds. It is absent or present at minimal levels in the lungs of rodents, but is highly induced after treatment with agents such as 3-methylcholanthrene or 2,3,7,8-tetrachlorodibenzodioxin (dioxin).[24–26] In rabbits and rats, the induced P450 is expressed in the Clara cells, Type II cells and endothelial cells. In mice, this P450 is induced in the Type II and endothelial cells, but is not induced in the Clara cells, indicating a species difference. In humans, CYP1A1 is highly induced by cigarette smoke.[27,28]

The CYP2A subfamily has several members including CYP2A5, 2A10, 2A11 and 2A13. CYP2A5 has been localized in the olfactory mucosa.[29] CYP2A10 and CYP2A11 (formerly NMa) have been identified in the olfactory and respiratory epithelia of the nasal mucosa.[30,31] CYP2A13, which is predominantly expressed in the lung, is found in the bronchus and trachea.[32] The CYP2A enzymes are all implicated in the metabolism of aflatoxin B1 and the tobacco-specific carcinogen, 4-(methylnitrosamino)-1-(3-pyridyl)-1-butanone.[30–32] CYP2A13 is also highly efficient in the metabolism of nicotine.

Members of the CYP2B subfamily including CYP2B1 and CYP2B4 are major P450s in the lung. CYP2B1 is constitutively expressed in the bronchiolar Clara cells as well as the Type I and Type II cells (pneumocytes), and is not induced by phenobarbital, a prototypic inducer of this P450 in the liver.[33] CYP2B1 is responsible for the metabolism of butylated hydroxytoluene, O,O,S-trimethylphosphorothioate and methylcyclopentadienyl manganese tricarbonyl.[34] The toxicities are manifested in the Type I and Type II cells along with the Clara cells, which is consistent with the localization of CYP2B1 in these cell types.

CYP2E1 has been detected in the lungs of the rabbit,[35] rat,[36] hamster[37] and human.[38] CYP2E1 is found in rabbit nasal mucosa and is induced 2-fold by ethanol and 6-fold by acetone.[39] In mice, CYP2E1 is predominantly localized in the bronchiolar Clara cells, with minimal expression in the Type II cells.[40] The Clara cell necrosis induced by 1,1-dichloroethylene[41] and trichloroethylene[42] is ascribed to *in-situ* metabolic activation by CYP2E1 within the target Clara cells.

The CYP2F subfamily is distinct for its lack of gene diversity, with only a single member expressed in each of the species examined: human (2F1),[43] mouse (2F2)[44] and goat (2F3).[45] In murine lung, CYP2F2 is preferentially expressed in the Clara cells.[46] The CYP2F enzymes have a major role in the metabolism of naphthalene,[3] 3-methylindole[47] and 1,1-dichloroethylene.[48] All of these compounds induce Clara cell cytotoxicities, and support the contention that metabolic activation takes place *in situ* within the Clara cells, which possess the activating P450 enzyme.

CYP2S1 has recently been identified and localized in human lung.[49,50] Expression of CYP2S1 in the lung is highest in the epithelia of the nasal cavities, bronchi and bronchioles. In the human lung cell line A549, dioxin induces the CYP2S1

mRNA by about 2-fold. This P450 is also implicated in the metabolism of naphthalene.[49]

CYP4B1 is highly expressed in the Clara cells and Type II cells of both the rat and rabbit.[33] In rat lung, CYP4B1 plays a major role in metabolic activation of the Clara cell toxicant, 4-ipomeanol.[51] However, in rabbit lung, 4-ipomeanol, which also elicits Clara cell injury, is metabolized by CYP2B4 and CYP4B1.[52] These strain differences in 4-ipomeanol metabolism are underscored by findings showing an absence of toxicity in quail and chickens, which lack Clara cells.[53] In the context of mechanisms, covalent binding of 4-ipomeanol is 10-fold greater in isolated Clara cells than in Type II cells, which remain unaffected.[54]

This account is a brief summary of the cytochrome P450 system in the respiratory tract. More comprehensive descriptions of the expression and localization of pulmonary P450 enzymes, including their substrates, have been presented in a review by Buckpitt and Cruikshank.[55] A more detailed description of the xenobiotic metabolizing enzymes in the nasal mucosa has also been reviewed.[56] More recently, the expression, regulation and localization of P450 enzymes in human lung have been reviewed.[57,58]

GLUTATHIONE AND GLUTATHIONE S-TRANSFERASES

Conjugation of electrophilic metabolites with glutathione usually yields a product with decreased reactivity, and therefore results in detoxication by reducing or ameliorating covalent binding to cellular constituents that lead to cell cytotoxicities and/or carcinogenesis. The conjugation of electrophiles can occur non-enzymatically or be catalyzed enzymatically by the glutathione S-transferases. Accordingly, the susceptibility of cells and tissues to cytotoxicities mediated by electrophilic metabolites depends in part on the availability of both glutathione and the transferases for conjugation reactions and hence detoxication. Because of the cell-selectivity of the lung cytotoxicities, the distribution of glutathione and the transferases is of interest. Studies using histochemical labelling for glutathione revealed that the non-protein sulfhydryl is highly localized in the ciliated and Clara cells of the bronchiolar epithelium as well as the Type II cells.[59] However, the extent of labelling is greatest in the Clara cells. These findings suggested that metabolic activation and glutathione conjugation take place *in situ* in the Clara cells. Since studies involving toxic agents have shown that glutathione depletion occurs prior to cell injury,[60] it has been postulated that the susceptibility of the Clara cells to chemical cytotoxicities is the result of an imbalance between the mechanisms of activation and detoxication.

The glutathione S-transferases are a family of multifunctional proteins that have an important role in defending cells against potentially toxic compounds by catalyzing the conjugation of electrophilic metabolites to glutathione. However, detoxication can also be mediated through a non-enzymatic reaction involving binding of the transferases to electrophiles.[61,62] In some instances, transferase activity produces an intermediate that is more reactive than the parent compound or the primary metabolite, and therefore conjugation is associated with toxication rather than detoxication.[63,64] The transferases are composed of two subunits and exist as either homo- or heterodimeric proteins. They are categorized according to their primary structures into five separate families designated as the classes alpha, mu, pi, sigma and theta. On the basis of decreasing electrophoretic mobilities, three protein bands are resolved and are designated Ya, Yb and Yc.[65] Subsequent studies have shown that the Ya and Yc bands represent class alpha, while the Yb band represents class mu.[66,67]

Several studies have examined the distribution and localization of the transferases. Using 1-chloro-2,4-dinitrobenzene as a substrate, transferase activities have been found to be considerably higher in isolated Clara cells than in isolated Type II cells in both the mouse and rat.[68,69] Comparative studies have found transferase activities in human lung and rat lung to be comparable but were both less than in hamster and mouse lung.[70] Immunohistochemical studies in the lungs of mice showed that the Ya (alpha), Yp (pi) and Yb$_1$ (mu) subunits were all localized in the bronchiolar Clara cells and alveolar Type II cells.[71] In addition, Ya was localized in the alveolar Type I and endothelial cells. However, the transferase subunits were all localized to the greatest extent in the Clara cells. Parallel studies using *in-situ* hybridization and quantitative image analysis demonstrated good agreement between relative amounts of protein and mRNA transcripts. Treatment of mice with 2(3)-*tert*-butyl-4-hydroxyanisole induced the proteins for Ya and Yp as well as the corresponding mRNA transcripts in the bronchiolar epithelium. Taken together, localization of the cytochrome P450 and transferase enzymes, as well as glutathione, within the same lung cell populations likely provides optimal conditions for the detoxication of reactive metabolites formed from potential pneumotoxicants.

SUSCEPTIBILITY OF LUNG CELLS TO CHEMICALLY INDUCED CYTOTOXICITIES

Substantial evidence has accrued to establish the bronchiolar Clara cell as a cell type with enhanced susceptibility to chemically induced cytotoxicities. This is due, in part, to the localization within the Clara cells of high concentrations of cytochrome P450 enzymes that are responsible for the metabolic activation of a wide variety of chemicals. For the majority of potentially cytotoxic chemicals, cytochrome P450-dependent metabolic activation takes place *in situ* within the Clara cells. The oxidative metabolism of these chemicals yields electrophilic derivatives that are highly reactive and that bind to nucleophilic sites on tissue constituents at the site of formation, leading to necrosis of the Clara cells. Alternately, metabolic activation of chemicals within the Clara cells produces reactive metabolites that bind to DNA, leading to the formation of mutations and initiation of carcinogenicity. Thus, the Clara cell may be a cell of origin of lung tumours induced by chemicals that are metabolically activated *in situ* in this cell type. Taken together, the vulnerability of the Clara cell is linked to the high concentrations as well as the diversity of cytochrome P450 enzymes that reside in this cell population.

Information regarding the identification and distribution of the Clara cells is relevant for identifying the sites along the bronchiolar airway where the chemically induced cytotoxicities are manifested. The bronchioles are lined by a pseudostratified

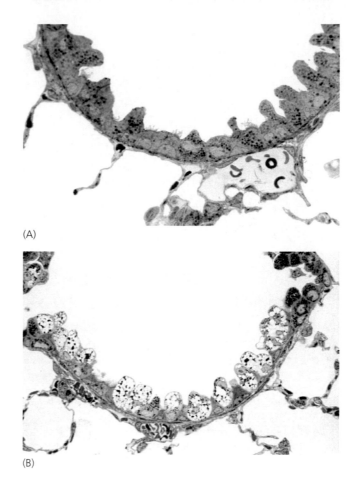

(A)

(B)

Fig. 2.1 Bronchiolar epithelium in the lungs of control and DCE-treated mice. (a) In control mice, the non–ciliated bronchiolar Clara cell is characterized by a protruding apex, while the ciliated cell is cuboidal and characterized by cilia. (b) In mice treated with DCE (100 mg/kg, i.p.), Clara cells are highly vacuolated, whereas the ciliated cells remained unaffected. *Stain: toluidine blue.*

columnar epithelium consisting of Clara cells interspersed with ciliated cells (Fig. 2.1a). There is a transition to a simple cuboidal epithelium in the distal bronchioles. The Clara cell is characterized by the presence of a bulbous protruding apex (Fig. 2.2). It contains an indented nucleus, numerous pleomorphic mitochondria, abundant smooth endoplasmic reticulum, and a small amount of rough endoplasmic reticulum. Electron-dense membrane-bound secretory granules are a unique characteristic of the Clara cells, and are located in the apical cytoplasm close to the plasma membrane. The Clara cell 10 kDa protein (CC10), a secretory protein, has been immunolocalized to these granules.[72] The Clara cells are localized predominantly in the terminal and respiratory bronchioles. The terminal bronchiole is that portion of the conducting airway that is lined by a continuous layer of epithelium, while the respiratory bronchiole represents a transitional area between the conducting portion of the lung and the respiratory portion where gas exchange occurs. The respiratory bronchiole has a conducting airway interspersed with alveoli so that the epithelium appears discontinuous. More recent studies in human lung revealed that the Clara cell, as identified by the presence of CC10, is virtually absent in the proximal airways and is restricted to the terminal and respiratory

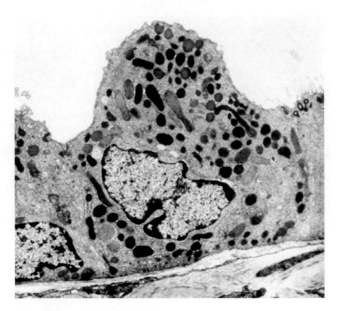

Fig. 2.2   Bronchiolar Clara cell in control murine lung. *Stains: uranyl acetate and lead citrate.* Reproduced with permission of the *Canadian Journal of Physiology and Pharmacology.*

bronchioles.[73] As the terminal and respiratory bronchioles are common sites of chemically induced lesions, the identity of the progenitor cell in epithelial cell renewal is of interest. Studies in experimental animals, mostly rodents, indicated that, under both steady-state and pathologic conditions, cell renewal in the bronchiolar region is accomplished by proliferation of the Clara cells.[74] Furthermore, the Clara cells are stem cells that give rise to both non-ciliated and ciliated cells. Hence, the Clara cell has a major role in maintaining the integrity of the bronchiolar epithelium.

Chemicals that have been reported to cause Clara cell damage include those categorized under the following classes of compounds: aromatic hydrocarbons (bromobenzene,[75] 2-methylnaphthalene,[76] naphthalene[3] and 1-nitronaphthalene[77]), chlorinated hydrocarbons (bromotrichloromethane,[78] carbon tetrachloride,[79] 1,1-dichloroethylene[4] and trichloroethylene[5]) and furans (4-ipomeanol[80] and 3-methylfuran[81]). Although by no means complete, the chemicals shown underscore the multiplicity and diversity of compounds with the capability of eliciting Clara cell necrosis. All the compounds mediate their lung toxicities through a mechanism involving metabolic activation. The isozyme-selective metabolism of these compounds by cytochrome P450 enzymes has been identified for 4-ipomeanol (CYP4B1, CYP2B4),[51,52] 3-methylindole (CYP2F),[47] naphthalene (CYP2F),[3] 1,1-dichloroethylene (CYP2E1, CYP2F)[48] and trichloroethylene (CYP2E1, CYP2F, 2B1).[5] However, it should be emphasized that changes may occur as additional P450 enzymes and their substrates are characterized. All the P450 enzymes implicated in the metabolism of the chemicals are localized in the Clara cells, consistent with the contention that xenobiotic metabolism takes place to the greatest extent in this cell type.

Although the Clara cells are a major target population for a wide variety of chemicals, other lung cells are also susceptible to chemically induced toxicities. These include the endothelial cell, alveolar Type I cell and alveolar Type II cell.

Exposure to bleomycin,[82] butylated hydroxytoluene[83] or cyclophosphamide[84] results in damage to the Type I and endothelial cells. The mechanisms by which the toxicities of these compounds are mediated are associated with the formation of reactive oxygen species, leading to oxidative stress. Monocrotaline, a member of the pyrrolizidine family of compounds, damages only the endothelial cells.[85] The mechanism of monocrotaline lung toxicity is believed to be due to its metabolism to a pyrrolic derivative that takes place in the liver and that is subsequently transported to the lung.[86] In the case of the redox cycling agent paraquat, Type I and Type II cells are both damaged.[87] Normally, repair of the epithelium subsequent to Type I cell necrosis is accomplished by proliferation and differentiation of the Type II cell to the Type I cell.[74] Since the Type II cell is a progenitor cell, damage of Type II cells produces a more severe lesion than if the Type I cell only is involved. Furthermore, reparative processes may not proceed normally if a chemical, such as bleomycin, is still present during the critical time of Type II cell division and differentiation. Cell division occurs but differentiation may be disturbed, resulting in abnormal alveolar epithelial forms. Renewal of the endothelial cells subsequent to injury is achieved by proliferation of other endothelial cells.[74] Taken together, these findings suggest that the specific lung cells that sustain cytotoxic responses are dependent, in part, on the mechanisms involved. In addition, the long-term consequences are dependent on the identities of the specific cells involved. Later in this chapter, the pulmonary lesion and the mechanism relating to paraquat toxicity are discussed in greater detail.

METABOLIC ACTIVATION OF 1,1–DICHLOROETHYLENE AND CLARA CELL NECROSIS

1,1-dichloroethylene, also known as vinylidene chloride, is used as a monomeric intermediate in the manufacture of plastics. Studies in the author's laboratory have used dichloroethylene as a model to investigate the mechanisms that mediate lung toxicity, and to identify the specific events that occur in the period intervening between exposure and cytotoxicity. Treatment of mice with dichloroethylene produces Clara cell necrosis (Figs 2.1b, 2.3 and 2.4).[4] The ciliated cells, endothelial cells as well as the alveolar Type I and Type II cells are not adversely affected by dichloroethylene at a dose that severely damages the Clara cells, indicating that Clara cells are the preferential targets of dichloroethylene in the lung.

The Clara cell injury induced by dichloroethylene is associated with cytochrome P450-dependent metabolic activation to a reactive metabolite.[4] Dose-dependent increases in covalent binding to protein and parallel decreases in glutathione levels were detected in the lungs of mice after exposure. The increases in binding and the corresponding decreases in glutathione levels correlated with the severities of Clara cell damage, suggesting that binding of a metabolite(s) to proteins mediates the cytotoxic response, and further that glutathione conjugation represents a detoxication mechanism. Conjugation of the metabolite with glutathione was achieved non-enzymatically and did not require the participation of the glutathione transferases. Subsequent studies identified the dichloroethylene epoxide as

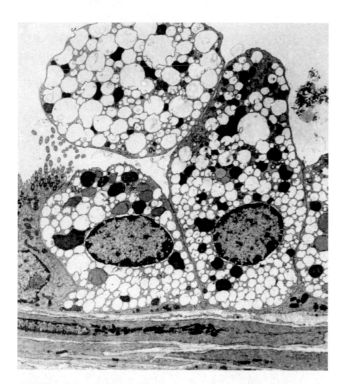

Fig. 2.3 Bronchiolar epithelium in the lung of a mouse treated with DCE (125 mg/kg, i.p.). The endoplasmic reticulum in the Clara cell is dilated, resulting in extensive vacuolization. An adjacent ciliated cell remains unaffected by DCE exposure. Reproduced with permission from [2].

(A)

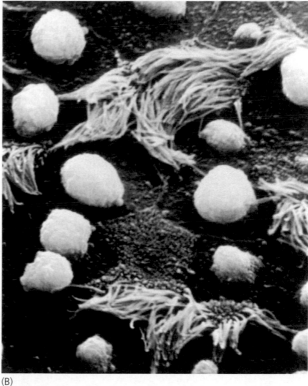

(B)

the ultimate toxic species.[88,89] The epoxide was also formed and conjugated with glutathione in incubations of human lung microsomes.[90] Two P450 isoforms were identified as being involved in the metabolism of dichloroethylene: CYP2E1 and CYP2F2.[48] Furthermore, *in-vivo* studies demonstrated that when mice were pretreated with 5-phenyl-1-pentyne, CYP2E1 and CYP2F2 were both inhibited, resulting in abrogation of epoxide formation and protection from Clara cell damage. Taken together, the data demonstrated that the Clara cell cytotoxicity induced by dichloroethylene is associated with its metabolic activation by CYP2E1 and CYP2F2, formation of the epoxide, covalent binding to protein and conjugation with glutathione.

The cell selectivity of the toxic lesion by dichloroethylene raises a question regarding whether the Clara cell damage is mediated by chemical metabolism *in situ* within this cell type. In early studies, we measured covalent binding of [14C]-dichloroethylene in isolated cell fractions enriched in Clara cells and compared the levels to that enriched in alveolar Type II cells.[68] Covalent binding in the Clara cell fraction was about 4-fold of that in the Type II cell fraction, while the levels in the mixed lung cell fraction were negligible. These results suggested that metabolic activation of dichloroethylene takes place to the greatest extent in the Clara cells. This concept is supported by findings showing that CYP2E1 and CYP2F2, which are believed to mediate the metabolism of dichloroethylene, are both preferentially localized within the Clara cells.[40,46]

It was of interest to obtain more direct evidence to demonstrate that the dichloroethylene epoxide is generated within the Clara cells. As an initial step towards this end, we have

Fig. 2.4 Scanning electron photomicrograph of the mucosal surface of the bronchiolar epithelium in murine lung. (a) In control of the lung, Clara cells with protruding apices are numerous and are distributed rather uniformly. Interspersed among the Clara cells are the ciliated cells. (b) The number of Clara cells is reduced in the lungs of mice treated with DCE (225 mg/kg, i.p.) owing to exfoliation of damaged Clara cell. The remaining Clara cells appear swollen. Reproduced with permission of the *Journal of Pathology*.

developed a polyclonal antibody specific for the dichloroethylene epoxide–glutathione conjugate.[91] Using this antibody, immunohistochemical studies in dichloroethylene-treated mice revealed specific labelling in the bronchiolar epithelium with preferential localization within the Clara cells.[92] Studies were also carried out with lung tissue from mice that were pretreated with the garlic derivative, diallyl sulfone, to inhibit CYP2E1 and CYP2F2. The combined treatment of diallyl sulfone and dichloroethylene produced diminished staining in the bronchiolar epithelium and the Clara cells, and protected from Clara cell damage. These findings have validated our working hypothesis that metabolic activation and formation of dichloroethylene epoxide takes place within the target Clara cells, an event that is responsible for mediating the cytotoxic response. These findings further suggested that the metabolic balance in the Clara cells is in favour of activation.

METABOLIC ACTIVATION OF VINYL CARBAMATE: FORMATION OF DNA ADDUCTS, MUTATIONS AND LUNG TUMOURS

Here, vinyl carbamate is used as a model compound to investigate the specific events that take place in the interval between exposure and the development of lung tumours. Vinyl carbamate is a metabolite of ethyl carbamate (urethane), a chemical formed during fermentation, and that is found in alcoholic beverages and fermented food products.[93] Ethyl carbamate was used as a co-solvent for analgesic and sedative drugs in Japan between 1950 and 1975.[94] This period represented 25 years during which millions of humans were administered 'the largest doses of a pure carcinogen that is on record'.[95] It has been estimated that the total dose of ethyl carbamate administered to a 60 kg patient was about 0.6 to 3.0 g. Ethyl carbamate has also been used as an antineoplastic agent for the treatment of chronic leukemia and multiple myeloma.[96] Today, human exposures occur inadvertently via the consumption of fermented food products and alcoholic beverages as well as tobacco use. A question has been raised regarding the potential carcinogenic risk associated with long-term or perhaps a lifetime exposure to low levels of the carbamate compounds. In this regard, regulatory agencies in Canada and the United States have set limits on the concentrations of ethyl carbamate in wines and distilled spirits.

It was first reported by Nettleship et al.[97] that ethyl carbamate produces adenomas in the lungs of mice. Lung tumours developed rapidly and were seen about 2–6 months after treatment. A similar spectrum of tumours is induced by vinyl carbamate; however, it is a more potent carcinogen than ethyl carbamate, and has been found to generate lung tumours in numbers that were 20- to 50-fold of those induced by ethyl carbamate.[98] The lung adenomas are manifested as either solid or papillary tumours. Solid tumours arise in alveolar septae and proliferate to produce a spherical, compact mass of cells with morphological characteristics of Type II cells (Fig. 2.5a). Papillary tumours arise in bronchioles, exhibit an open tubular configuration, and appear to be formed as a result of extensive and uncontrolled proliferation of columnar epithelial cells with features characteristic of non-ciliated Clara cells (Fig. 2.5b). Growth of solid tumours is restricted and regression may occur,

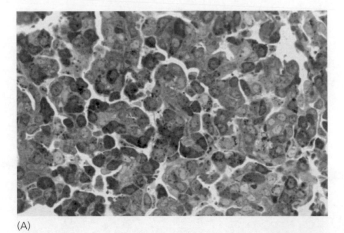

(A)

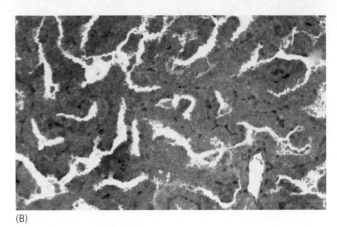

(B)

Fig. 2.5 Lung adenomas in Strain A/J mice treated with ethyl carbamate (1 mg/g body weight). Mice were sacrificed 16 weeks after carcinogen treatment. The adenomas were manifested as (a) solid or (b) papillary tumours. *Stain: toluidine blue.*

whereas papillary tumours continue to grow, are larger in size, and are more likely to progress to carcinomas.[99]

Vinyl carbamate is oxidized to form an epoxide, a metabolite postulated to be the ultimate mutagenic and carcinogenic species.[100] More recent studies confirmed a role for CYP2E1 in vinyl carbamate metabolism in murine lung.[101] Subsequent studies indicated that CYP2E1 is also involved in vinyl carbamate metabolism in human lung.[102] In-vitro studies showed that the vinyl carbamate epoxide formed adducts with DNA, and produced two guanine adducts, 7-(2′-oxoethyl)deoxyguanosine and N^2,3-ethenodeoxyguanosine, and an adenine adduct, 1,N^6-ethenodeoxyadenosine.[100,103,104] Other studies showed the formation of ethenodeoxyadenosine and 3,N^4-ethenodeoxycytidine in the lungs of mice treated with VC.[105,106] Moreover, there was a correlation between CYP2E1 levels and the production of both DNA adducts and lung tumours.[105] These findings confirmed that oxidation of vinyl carbamate by CYP2E1 leads to the formation of an epoxide that alkylates DNA to form adducts in murine lung.

Recent studies have investigated point mutations induced by vinyl carbamate in the lung, using Big Blue® transgenic mice and *cII* as a target gene.[106] Sequencing of the *cII* gene revealed that spontaneous mutations were formed in the lungs of control mice. The most common spontaneous mutation induced by vinyl carbamate in the *cII* gene is G:C→A:T transitions, which comprised about 58 per cent of the total amount of mutations formed in the lungs of control mice. The second most common mutation detected in the lung is G:C→T:A transversions (19 per cent). In mice treated with vinyl carbamate, the major mutations generated in the lung are G:C→A:T (26 per cent) and A:T→G:C (29 per cent) transitions and A:T→T:A transversions (29 per cent). These findings showed that vinyl carbamate induces a variety of mutations in the lungs of mice.

The specific mutations identified in a target tissue may be related to the specific etheno DNA adducts generated by carcinogenic compounds. The DNA adducts produced *in vivo* by vinyl carbamate in the lung are ethenodeoxyadenosine and ethenodeoxycytidine.[104,105] The A:T→G:C transitions and A:T→T:A transversions produced by VC are believed to be associated with base changes induced by ethenodeoxyadenosine.[107] On the other hand, the G:C→A:T transitions generated by vinyl carbamate is associated with the ethenodeoxycytidine adduct.[108] These findings are consistent with the theory that ethenodeoxyadenosine and ethenodeoxycytidine cause mispairing in DNA transcription.

OXIDATIVE STRESS: STUDIES WITH PARAQUAT

Paraquat, a herbicide in widespread use, produces swelling of the Type I alveolar cell early in the cytotoxic response.[109] The injury is precipitous as the Type I alveolar cells have elongated cytoplasmic processes that cover a large surface area estimated to be about 93 per cent of the epithelial surface.[110] Soon after, the Type II alveolar cells, which contribute to the remaining 7 per cent of the epithelial surface, sustain damage with loss of the contents (surfactant) of the lamellar bodies. Later in the destructive response, there is endothelial cell necrosis, and it is believed that this damage favours the development of an alveolitis characterized by oedema and infiltration of the interstitial and alveolar spaces with inflammatory cells.[111] It has been suggested that destruction of the surfactant-producing Type II cells leads to increased surface tension within the alveoli that then draws fluid from the capillaries, thus producing oedema.[112] The destructive phase is followed by a proliferative phase in which there is development of an extensive fibrosis that is rapid in onset. The alveolar spaces are invaded by immature and mature fibroblasts that deposit collagen and ground substance. Interstitial fibrosis is also observed but it is the alveolar fibrosis that is more deleterious owing to obliteration of the alveolar air spaces and obstruction of gas exchange. One of the mechanisms proposed for the pulmonary fibrosis is that it ensues when perturbations in re-epithelialization subsequent to epithelial damage occurs.[113] This mechanism is plausible in the case of paraquat since the Type I cells are damaged and normal repair of the alveolar epithelium is obviated by the concomitant destruction of the progenitor Type II cells.

Paraquat is selectively taken up in the lung but this accumulation is not associated with covalent binding, nor have any significant metabolites of paraquat been identified. However, there is general agreement that redox cycling is involved in the pulmonary toxicity of paraquat. Under aerobic conditions, the addition of paraquat to incubations containing lung microsomes and NADPH results in marked increases in NADPH oxidation and oxygen uptake.[114] Under anaerobic conditions, paraquat is reduced to a free radical metabolite by NADPH in the presence of lung microsomes[115] whereas under aerobic conditions, the oxidation of NADPH is due to the cyclic reduction and re-oxidation of paraquat.[116] As a result of these reactions, NADPH is consumed and there is a decrease in the NADPH/NADP+ ratio. The reduction is also due in part to the utilization of NADPH as a cofactor for glutathione reductase when oxidized glutathione is regenerated back to reduced glutathione. It has been suggested that glutathione oxidation is mediated through its role as a substrate in the reduction of cellular hydrogen peroxide via glutathione peroxidase, and is consistent with evidence showing paraquat-dependent formation of hydrogen peroxide in lung microsomes.[116] However, other studies have suggested that hydrogen peroxide is produced from a reactive oxygen species initially formed from the reaction of oxygen with the paraquat radical.[117] It has been demonstrated that the superoxide anion radical, a short-lived oxygen species, is formed by the reduction of oxygen by the paraquat radical. It has subsequently been confirmed that the superoxide is generated from paraquat in lung microsomal incubations.[118] The dismutation of superoxide by the enzyme superoxide dismutase can produce hydrogen peroxide, which can also be produced by reduction of superoxide by the paraquat radical. The available evidence thus indicated that several potentially toxic species could be generated from oxygen during the process of reduction and re-oxidation of paraquat.

Lipid peroxidation and membrane damage have been proposed as a possible mechanism for the toxicity of paraquat.[110] It has been shown to stimulate lipid peroxidation *in vitro*[115] and *in vivo*.[119] However, other reports have indicated that paraquat may not generate the peroxidation of lipids.[116] These findings suggested that more definitive and consistent evidence is required to demonstrate lipid peroxidation as a mechanism of paraquat toxicity. In addition to lipids, paraquat has been reported to cause modification of proteins and DNA damage.

In summary, the available evidence indicated that redox cycling of paraquat in lung cells leads to oxidative stress, consumption of NADPH, lipid peroxidation, as well as protein and DNA modification. However, it is not clear which specific events mediate the pulmonary toxicity of paraquat. It is possible that the mechanism may involve several processes that act independently or in concert with one another. The pulmonary toxicity of paraquat has recently been reviewed.[110] Ingestion of paraquat is a classic method to commit suicide in man.

CONCLUSION AND COMMENTS

The mechanisms that contribute to cytotoxic and/or carcinogenic consequences in the lung as a result of chemical exposures have been described. In the studies described herein, the compounds dichloroethylene and vinyl carbamate have been used as surrogates for chemical or drug exposures. Both are small molecules with double bonds that readily undergo oxidation by P450 enzymes to produce epoxides. The epoxides of dichloroethylene and vinyl carbamate are short-lived, are highly reactive and bind to cellular constituents including proteins and nucleic acids. The dichloroethylene epoxide targets cellular proteins at the site of formation and causes acute Clara cell necrosis. On the other hand, the epoxide formed from vinyl carbamate targets DNA, and sets into motion a cascade of events including DNA adduct formation and generation of mutations that lead subsequently to the formation of lung tumours. Hence, both dichloroethylene and vinyl carbamate require metabolic activation to exert their toxicities, and the extents to which the effects are manifested depend, in part, on the capacities of the activating enzymes to convert the chemicals to their ultimate reactive species. Of interest in this regard is the considerable variability in expression of xenobiotic-metabolizing enzymes in the general population that is due in part to the presence of genetic polymorphisms where mutations in a wild-type gene confer impaired activity. Genetic polymorphisms are linked to disease susceptibility and have been identified for P450 enzymes including CYP1A1 and CYP2E1.[120] More recently, an interethnic genetic polymorphism for CYP2F1 has been identified, and it has been proposed that it may be associated with lung cancer development. There are currently extensive investigations in this research area with the possibility of predicting risk in various population groups. In summary, the studies described herein have addressed some of the specific events evoked by chemicals, their metabolic pathways and the potential outcome of cytotoxicity and/or carcinogenicity.

KEY POINTS

- Determinants of pulmonary toxicity induced by drugs include storage, transit time, persistence, metabolism, metabolic activation and elimination of drugs and metabolites in/from lung tissue.

- Metabolic activation of drugs and chemicals may generate reactive metabolites and/or oxygen radicals that may bind to biomolecules and/or cause irreversible cell toxicity via depletion of reducing equivalents, covalent binding or oxidant stress.

- The highly diverse, heterogenous and delicate structure of the lung, including the fragile air-tissues interface, makes the lung particularly vulnerable to attack by drugs.

REFERENCES

1. Boyd MR. Role of metabolic activation in the pathogenesis of chemically induced pulmonary disease: mechanism of action of the lung-toxic furan, 4-ipomeanol. *Environ Health Perspect* 1976;**16**:127–38.
2. Skiles GL, Adams JD, Yost GS. Isolation and identification of 3-hydroxy-3-methyloxindole, the major murine metabolite of 3-methylindole. *Chem Res Toxicol* 1989;**2**:254–9.
3. Buckpitt A, Boland B, Isbell M, *et al.* Naphthalene-induced respiratory tract toxicity: metabolic mechanisms of toxicity. *Drug Metab Rev* 2002;**34**:791–820.
4. Forkert PG. Mechanisms of 1,1-dichloroethylene-induced cytotoxicity in lung and liver. *Drug Metab Rev* 2001;**33**:49–80.
5. Forkert PG, Baldwin RM, Millen B, *et al.* Pulmonary bioactivation of trichloroethylene to chloral hydrate: relative contributions of CYP2E1, CYP2F, and CYP2B1. *Drug Metab Dispos* 2005;**33**:1429–37.
6. Hinson JA, Roberts DW. Role of covalent and noncovalent interactions in cell toxicity: effects on proteins. *Ann Rev Pharmacol Toxicol* 1992;**32**:471–510.
7. Pohl LR, Sato H, Christ DD. The immunologic and metabolic basis of drug hypersensitivities. *Annu Rev Pharmacol Toxicol* 1988;**28**:367–87.
8. Park BK, Kitteringham NR. Drug-protein conjugation and its immunological consequences. *Drug Metab Rev* 1990;**22**:87–144.
9. Vergani D, Mielli-Vergani G, Alberti A, *et al.* Antibodies to the surface of halothane-altered rabbit hepatocytes in patients with severe halothane-associated hepatitis. *New Engl J Med* 1980;**303**:66–71.
10. Uetrecht JP. Idiosyncratic drug reactions: possible role of reactive metabolites generated by leukocytes. *Pharm Res* 1989;**6**:265–73.
11. Waldhauser L, Uetrecht J. Oxidation of propylthiouracil to reactive metabolites by activated neutrophils: implications for agranulocytosis. *Drug Metab Dispos* 1991;**19**:354–9.
12. Yamagiwa K, Ichikawa K. Experimental study of the pathogenesis of carcinoma. *J Cancer Res* 1918;**3**:1–29.
13. Miller EC, Miller JA. Mechanisms of chemical carcinogenesis: nature of proximate carcinogens and interactions with macromolecules. *Pharmacol Rev* 1966;**18**:805–38.
14. Belinsky SA, White CM, Devereux TR, Anderson MW. DNA adducts as a dosimeter for risk estimation. *Environ Health Perspect* 1987;**76**:3–8.
15. Brookes P, Lawley PD. Evidence for the binding of polynuclear aromatic hydrocarbons to the nucleic acids of mouse skin: relation between carcinogenic power of hydrocarbons and their binding to deoxyribonucleic acid. *Nature* 1964;**202**:781–4.
16. Gelboin HV. A microsome-dependent binding of benzo[*a*]pyrene to DNA. *Cancer Res* 1969;**29**:1272–6.
17. Grover PL, Sims P. Enzyme-catalysed reactions of polycyclic hydrocarbons with deoxyribonucleic acid and protein in vitro. *Biochem J* 1968;**110**:159–60.
18. Ross JH, Nelson GB, Wilson KH, *et al.* Adenomas induced by polycylic aromatic hydrocarbons in strain A/J mouse lung correlate with time-integrated DNA adduct levels. *Cancer Res* 1995;**55**:1039–44.
19. Roberts DW, Pumford NR, Potter DW, Benson RW, Hinson JA. A sensitive immunochemical assay for acetaminophen-protein adducts. *J Pharmacol Exp Ther* 1987;**241**:527–33.
20. Pumford NR, Hinson JA, Potter DW, *et al.* Immunochemical quantitation of 3-(cystein-*S*-yl)acetaminophen adducts in serum and liver proteins of acetaminophen treated mice. *J Pharmacol Exp Ther* 1989;**248**:190–6.
21. Hinson JA, Roberts DW, Benson RW, *et al.* Mechanism of paracetamol toxicity. *Lancet* 1990;**335**:732.
22. Patel JM. Metabolism and pulmonary toxicity of cyclophosphamide. *Pharmacol Ther* 1990;**47**:137–46.
23. Nelson DR, Koymans L, Kamataki T, *et al.* P450 superfamily: update on new sequences, gene mapping, accession numbers and nomenclature. *Pharmacogenetics* 1996;**6**:1–42.
24. Overby LH, Nishio S, Weir A, *et al.* Distribution of cytochrome P450 1A1 and NADPH-cytochrome P450 reductase in lungs of rabbits treated with 2,3,7,8-tetrachlorodibenzo-*p*-dioxin: ultrastructural immunolocalization and in-situ hybridization

and immunohistochemistry. *Am J Respir Cell Mol Biol* 1994;**11**:386–96.

25. Keith IM, Olson EB, Wilson NM, *et al.* Immunological identification and effects of 3-methylcholanthrene and phenobarbital on rat pulmonary cytochrome P450. *Cancer Res* 1987;**47**:1878–82.

26. Forkert PG, Lord JA, Parkinson A. Alterations in expression of CYP1A1 and NADPH-cytochrome P450 reductase during lung tumor development in SWR/J mice. *Carcinogenesis* 1996;**17**:127–32.

27. Kim JH, Sherman ME, Curriero FC, *et al.* Expression of cytochromes P450 1A1 and 1B1 in human lung from smokers, non-smokers, and ex-smokers. *Toxicol Appl Pharmacol* 2004;**199**:210–19.

28. Saarikoski ST, Husgafvel-Pursiainen K, Hirvonen A, *et al.* Localization of CYP1A1 mRNA in human lung by in situ hybridization: comparison with immunohistochemical findings. *Int J Cancer* 1998;**77**:33–9.

29. Piras E, Franzen A, Fernandez EL, *et al.* Cell-specific expression of CYP2A5 in the mouse respiratory tract: effects of olfactory toxicants. *J Histochem Cytochem* 2003;**51**:1545–55.

30. Putt DA, Ding X, Coon MJ, Hollenberg PF. Metabolism of aflatoxin B1 by rabbit and rat nasal mucosa microsomes and purified cytochrome P450, including isoforms 2A10 and 2A11. *Carcinogenesis* 1995;**16**:1411–17.

31. Ding X, Coon MJ. Immunochemical characterization of multiple forms of cytochrome P-450 in rabbit nasal microsomes and evidence for tissue-specific expression of P-450s NMa and NMb. *Mol Pharmacol* 1990;**37**:489–96.

32. Zhu L-R, Thomas PE, Lu G, *et al.* CYP2A13 in human respiratory tissues and lung cancers: an immunohistochemical study with a new peptide-specific antibody. *Drug Metab Dispos* 2006;**34**:1672–6.

33. Lee M, Dinsdale D. The subcellular distribution of NADPH-cytochrome P450 reductase and isoenzymes of cytochrome P450 in the lungs of rats and mice. *Biochem Pharmacol* 1995;**49**:1387–94.

34. Verschoyle RD, Wolf CR, Dinsdale D. CYP2B1 is responsible for the pulmonary bioactivation of butylated hydroxytoluene, O,O,S-trimethylphosphorothioate and methylcyclopentadienyl manganese tricarbonyl. *J Pharmacol Exp Ther* 1993;**266**:958–63.

35. Porter TD, Khani SC, Coon MJ. Induction and tissue-specific expression of rabbit cytochrome P450IIE1 and IIE2 genes. *Mol Pharmacol* 1989;**36**:61–5.

36. Tindberg N, Ingelman-Sundberg M. Cytochrome P-450 and oxygen toxicity: oxygen-dependent induction of ethanol-inducible cytochrome P-450 (IIE1) in rat liver and lung. *Biochemistry* 1989;**28**:4499–504.

37. Ueng T-H, Tsai J-N, Ju J-M, *et al.* Effects of acetone administration on cytochrome P-450-dependent monooxygenases in hamster liver, kidney, and lung. *Arch Toxicol* 1991;**65**:45–51.

38. Botto F, Seree E, Khyari SE, *et al.* Tissue-specific expression and methylation of the human CYP2E1 gene. *Biochem Pharmacol* 1994;**48**:1095–103.

39. Ding X, Coon MJ. Induction of cytochrome P450 isozyme 3a (P450 IIE1) in rabbit olfactory mucosa by ethanol and acetone. *Drug Metab Dispos* 1990;**18**:742–5.

40. Forkert PG. CYP2E1 is preferentially expressed in Clara cells of murine lung: localization by in-situ hybridization and immunohistochemical methods. *Am J Respir Cell Mol Biol* 1995;**12**:589–96.

41. Forkert PG. 1,1-Dichloroethylene-induced Clara cell damage is associated with in-situ formation of the reactive epoxide: immunohistochemical detection of its glutathione conjugate. *Am J Respir Cell Mol Biol* 1999;**20**:1310–18.

42. Forkert PG, Millen B, Lash LH, *et al.* Pulmonary bronchiolar cytotoxicity and formation of dichloroacetyl lysine protein adducts in mice treated with trichloroethylene. *J Pharmacol Exp Ther* 2006;**316**:520–9.

43. Nhamburo PT, Gonzalez FJ, McBride OW, *et al.* Identification of a new P450 expressed in human lung: complete cDNA sequence, cDNA-directed expression and chromosome mapping. *Biochemistry* 1989;**28**:8060–6.

44. Ritter JK, Owens IS, Negishi M, *et al.* Mouse pulmonary cytochrome P-450 naphthalene hydroxylase: cDNA cloning, sequence and expression in *Saccharomyces cerevisiae*. *Biochemistry* 1991;**30**:11430–7.

45. Wang H, Lanza DL, Yost GS. Cloning and expression of CYP2F3, a cytochrome P450 that bioactivates the selective pneumotoxins 3-methylindole and naphthalene. *Arch Biochem Biophys* 1998; **349**:329–40.

46. Buckpitt AR, Chang AM, Weir A, *et al.* Relationship of cytochrome P450 activity to Clara cell cytotoxicity. IV: Metabolism of naphthalene and naphthalene oxide in microdissected airways from mice, rats and hamsters. *Mol Pharmacol* 1995;**47**:74–81.

47. Kartha JS, Yost GS. Mechanism-based inactivation of lung-selective cytochrome P450 CYP2F enzymes. *Drug Metab Dispos* 2008;**36**:155–62.

48. Simmonds AC, Ghanayem BI, Sharma A, *et al.* Bioactivation of 1,1-dichloroethylene by CYP2E1 and CYP2F2 in murine lung *Drug Metab Dispos* 2004;**32**:1032-9.

49. Saarikoski SI, Rivera SP, Hankinson O, Husgafvel-Pursiainen K. CYP2S1: a short review. *Toxicol Appl Pharmacol* 2005;**207**: S62–9.

50. Saarikoski ST, Wikman HA-L, Smith G, *et al.* Localization of cytochrome P450 CYP2S1 expression in human tissues by in-situ hybridization and immunohistochemistry. *J Histochem Cytochem* 2005;**53**:549–56.

51. Verschoyle RD, Philpot RM, Wolf CR, Dinsdale D. CYP4B1 activates 4-ipomeanol in rat lung. *Toxicol Appl Pharmacol* 1993;**123**:193–8.

52. Wolf CR, Statham CN, McMenamin MG, *et al.* The relationship between the catalytic activities of rabbit pulmonary cytochrome P450 isozymes and the lung specific toxicity of the furan derivative, 4-ipomeanol. *Mol Pharmacol* 1982;**22**:738–44.

53. Buckpitt AR, Statham CN, Boyd MR. In-vivo studies on the target tissue metabolism, glutathione depletion, covalent binding and toxicity of 4-ipomeanol in birds, species deficient in pulmonary enzymes for metabolic activation. *Toxicol Appl Pharmacol* 1982;**65**:38–52.

54. Devereux TR, Jones KG, Bend JR, *et al.* In-vitro metabolic activation of the pulmonary toxin ipomeanol in nonciliated bronchiolar epithelia (Clara) and alveolar type II cells isolated from rabbit lung. *J Pharmacol Exp Ther* 1982;**220**:223–7.

55. Buckpitt AR, Cruikshank MK. Biochemical function of the respiratory tract: metabolism of xenobiotics. In: Roth R (ed.) *Toxicology of the Respiratory System*, pp 159–86. New York: Pergamon Press, 1997.

56. Dahl AR, Hadley WM. Nasal cavity enzymes involved in xenobiotic metabolism: effects on the toxicity of inhalants. *Crit Rev Toxicol* 1991;**21**:345–72.

57. Hukkanen J, Pelkonen O, Hakkola J, Raunio H. Expression and regulation of xenobiotic-metabolizing cytochrome P450 (CYP) enzymes in human lung. *Crit Rev Toxicol* 2002;**32**:391–411.

58. Zhang JY, Wang Y, Prakash C. Xenobiotic-metabolizing enzymes in human lung. *Curr Drug Metab* 2006;**7**:939–48.

59. Forkert PG. Conjugation of glutathione with the reactive metabolites of 1,1-dichloroethylene in murine lung and liver. *Micros Res Tech* 1997;**36**:234–42.

60. Jollow DJ, Mitchell JR, Zampaglione N, Gillette JR. Bromobenzene-induced liver necrosis: protective role of glutathione and evidence for 3,4-bromobenzene oxide as the hepatotoxic metabolite. *Pharmacology* 1974;**11**:151–69.

61. Ketley JN, Habig WH, Jakoby WB. Binding of nonsubstrate ligands to the glutathione S-transferases. *J Biol Chem* 1975;**250**:8670–3.

62. Litwack G, Ketterer B, Arias IM. Ligandin: a hepatic protein which binds steroids, bilirubin, carcinogens and a number of exogenous organic anions. *Nature* 1977;**243**:466–7.

63. Slatter JG, Rashed MS, Pearson PG, Han D-H, Baillie TA. Biotransformation of methyl isocyanate in the rat: evidence for glutathione conjugation as a major pathway of metabolism and implications for isocyanate-mediated toxicities. *Chem Res Toxicol* 1991;**4**:157–61.

64. Dekant W, Vamvakas S, Anders MW. Formation and fate of nephrotoxic and cytotoxic glutathione S-conjugates: cysteine conjugate β-lyase pathway. *Adv Pharmacol* 1994;**27**: 115–62.

65. Bass NM, Kirsch E, Tuff SA, Marks I, Saunders SJ. Ligandin heterogeneity: evidence that two non-identical subunits are the monomers of two distinct proteins. *Biochem Biophys Acta* 1977;**492**:163–75.

66. Scully NC, Mantle TJ. Tissue distribution and subunit structures of the multiple forms of glutathione S-transferase in the rat. *Biochem J* 1981;**193**:367–70.

67. Hayes JD, Clarkson GHD. Purification and characterization of three forms of glutathione S-transferase A: a comparative study of the major YaYa, YbYb and YcYc-containing glutathione S-transferases. *Biochem J* 1982;**207**:459–70.

68. Forkert PG, Geddes BA, Birch DW, Massey TE. Morphologic changes and covalent binding of 1,1-dichloroethylene in Clara and alveolar Type II cells isolated from lungs of mice following in-vivo administration. *Drug Metab Dispos* 1990;**18**:534–9.

69. Devereux TR, Massey TE, Van Scott M, *et al.* Xenobiotic metabolism in human alveolar type II cells isolated by centrifugal elutriation and density gradient centrifugation. *Cancer Res* 1986;**46**:5438–43.

70. Lorenz J, Glatt HR, Fleischmann R, *et al.* Drug metabolism in man and its relationship to that in three rodent species: monooxygenase, epoxide hydrolase, and glutathione S-transferase activities in subcellular fractions of lung and liver. *Biochem Med* 1984;**32**:43–56.

71. Forkert PG, D'Costa D, El-Mestrah M. Expression and inducibility of alpha, pi, and mu glutathione S-transferase protein and mRNA in murine lung. *Am J Respir Cell Mol Biol* 1999;**20**:143–52.

72. Bedetti CD, Singh J, Singh G, *et al.* Ultrastructural localization of rat Clara cell 10 kd secretory protein by the immunogold technique using polyclonal and monoclonal antibodies. *J Histochem Cytochem* 1987;**35**:789–94.

73. Boers JE, Ambergen AW, Thunnissen FB. Number and proliferation of clara cells in normal human airway epithelium. *Am J Respir Crit Care Med* 1999;**159**:1585–91.

74. Evans MJ. Cell death and cell renewal in small airways and alveoli. In: Witschi H, Nettesheim P. (eds) *Mechanisms in Respiratory Toxicology*, pp 189–218. Boca Baton FL: CRC Press, 1982.

75. Reid WD, Ilett KF, Glich JM, Krishna G. Metabolism and binding of aromatic hydrocarbons in the lung: relationship to experimental bronchiolar necrosis. *Am Rev Resp Dis* 1973;**107**:539–51.

76. Griffin KA, Johnson CB, Breger RK, Franklin RB. Effects of inducers and inhibitors of cytochrome P450-linked monooxygenases on the toxicity, in-vitro metabolism and in-vivo irreversible binding of 2-methylnaphthalene in mice. *J Pharmacol Exp Ther* 1982;**221**:517–24.

77. Johnson DE, Riley MG, Cornish HH. Acute target organ toxicity of 1-nitronaphthalene in the rat. *J Appl Toxicol* 1984;**4**:253–7.

78. Lungarella G, Benedetti A, Gardi C, *et al.* Bromotrichloromethane-induced damage to bronchiolar Clara cells. *Res Commun Chem Pathol Pharmacol* 1987;**57**:213–28.

79. Boyd MR, Statham CN, Longo WS. The pulmonary Clara cell is a target for toxic chemicals requiring metabolic activation; studies with carbon tetrachloride. *J Pharmacol Exp Ther* 1980;**212**:109–14.

80. Boyd MR. Evidence for the Clara cell as a site of cytochrome P450-dependent mixed-function oxidase activity in lung. *Nature* (London) 1977;**269**:713.

81. Griffin KA, Johnson CB, Breger RK, Franklin RB. Pulmonary toxicity, hepatic and extrahepatic metabolism of 2-methylnaphthalene in mice. *Toxicol Appl Pharmacol* 1981;**16**:185–96.

82. Adamson IYR, Bowden PH. The pathogenesis of bleomycin-induced pulmonary fibrosis in mice. *Am J Pathol* 1974;**77**: 185–98.

83. Smith LJ. Lung damage induced by butylated hydroxytoluene in mice: biochemical, cellular and morphologic characterization. *Am Rev Respir Dis* 1984;**130**:895–904.

84. Gould VE, Miller J. Sclerosing alveolitis induced by cyclophosphamide. *Am J Pathol* 1975;**81**:513–20.

85. Butler WH. An ultrastructural study of the pulmonary lesion induced by pyrrole derivatives of the pyrrolizidine alkaloids. *J Pathol* 1970;**102**:15–19.

86. Huxtable R, Ciaramitaro D, Eisenstein D. The effect of a pyrrolizidine alkaloid, monocrotaline, and a pyrrole, dehydroretronecine, on the biochemical functions of the pulmonary endothelium. *Mol Pharmacol* 1978;**14**:1189–203.

87. Smith P, Heath D. Paraquat. *Crit Rev Toxicol* 1976;**4**:411–45.

88. Dowsley TF, Forkert PG, Benesch LA, Bolton JL. Reaction of glutathione with electrophilic metabolites of 1,1-dichloroethylene. *Chem-Biol Interact* 1995;**95**:227–44.

89. Dowsley TF, Ulreich JB, Bolton JL, *et al.* Cytochrome-P450 dependent bioactivation of 1,1-dichloroethylene in murine lung: formation of reactive intermediates and glutathione conjugates. *Toxicol Appl Pharmacol* 1996;**139**:42–8.

90. Dowsley TF, Reid K, Petsikas D, *et al.* Cytochrome P450-dependent bioactivation of 1,1-dichloroethylene to a reactive epoxide in human lung and liver microsomes. *J Pharmacol Exp Ther* 1999;**289**:641–8.

91. Forkert PG, Collins KS, Dowsley TF, Ross GM. Immunochemical assay for recognition of 2-*S*-glutathionyl acetate, a glutathione conjugate derived from 1,1-dichloroethylene – epoxide. *J Pharmacol Exp Ther* 1997;**281**:1422–30.

92. Forkert PG. 1,1-Dichloroethylene-induced Clara cell damage is associated with in-situ formation of the reactive epoxide. *Am J Respir Cell Mol Biol* 1999;**20**:1310–18.

93. Battaglia R, Conacher BS, Page BD. Ethyl carbamate (urethane) in alcoholic beverages and foods: a review. *Food Add Contaminants* 1990;**7**:477–96.

94. Nomura T. Urethane (ethyl carbamate) as a co-solvent of drugs commonly used parenterally in humans. *Cancer Res* 1975;**35**:2895–9.

95. Miller JA. The need for epidemiological studies of the medical exposures of Japanese patients to the carcinogen ethyl carbamate (urethane) from 1950 to 1975. *Jpn J Cancer Res* 1991;**82**:1323–4.

96. IARC: International Agency for Research on Cancer. *Urethane.* IARC monographs on the evaluation risk of chemicals to man. Lyon, France: IARC 1974;**13**:851–5.

97. Nettleship A, Henshaw PS, Meyer HL. Induction of pulmonary tumors in mice with ethyl carbamate (urethane). *J Natl Cancer Inst* 1943;**4**:309–19.

98. Dahl GA, Miller EC, Miller JA. Comparative carcinogenicities and mutagenicities of vinyl carbamate, ethyl carbamate, and ethyl *N*-hydroxycarbamate. *Cancer Res* 1980;**40**:1194–203.

99. Thaete LG, Gunning WT, Stoner GD, Malkinson AM. Cellular derivation of lung tumors in sensitive and resistant strains of mice: results at 28 and 56 weeks after urethan treatment. *J Natl Cancer Inst* 1987;**47**:697–701.

100. Park K-K, Liem A, Stewart BC, Miller J. Vinyl carbamate epoxide, a major strong electrophilic, mutagenic and carcinogenic metabolite of vinyl carbamate and ethyl carbamate (urethane). *Carcinogenesis* 1993;**14**:441–50.

101. Lee RP, Forkert PG. Inactivation of cytochrome P-450 (CYP2E1) and carboxylesterase (hydrolase A) enzymes by vinyl carbamate in murine pulmonary microsomes. *Drug Metab Dispos* 1999;**27**:233–9.

102. Forkert PG, Lee RP, Reid K. Involvement of CYP2E1 and carboxylesterase enzymes in vinyl carbamate metabolism in human lung microsomes. *J Pharmacol Exp Ther* 2001;**29**:258–63.

103. Forkert PG, Kaufmann M, Black G, *et al.* Oxidation of vinyl carbamate and formation of 1,*N*⁶-ethenodeoxyadenosine in murine lung. *Drug Metab Dispos* 2007;**35**:713–20.

104. Fernando RC, Nair J, Barbin A, *et al.* Detection of 1,*N*⁶-ethenodeoxyadenosine and 3,*N*⁴-ethenodeoxycytidine by immunoaffinity/³²P-postlabelling in liver and lung DNA of mice treated with ethyl carbamate (urethane) or its metabolites. *Carcinogenesis* 1996;**17**:1711–18.

105. Titis AP, Forkert PG. Strain-related differences in bioactivation of vinyl carbamate and formation of DNA adducts in lungs of A/J, CD-1, and C57BL/6 mice. *Toxicol Sci* 2001;**59**:82–91.

106. Hernandez LG, Forkert PG. In-vivo mutagenicity of vinyl carbamate and ethyl carbamate in lung and small intestine of F1 (Big Blue® × A/J) transgenic mice. *Int J Cancer* 2007;**120**:1464–71.

107. Pandya GA, Moriya M. 1,*N*⁶-ethenodeoxyadenosine, a DNA adduct highly mutagenic in mammalian cells. *Biochemistry* 1996;**35**:11487–92.

108. Moriya M, Zhang W, Johnson F, Grollman AP. Mutagenic potency of exocyclic DNA adducts: marked differences between *Escherichia coli* and simian kidney cells. *Proc Natl Acad Sci USA* 1994;**91**:11899–903.

109. Smith P, Heath D. The ultrastructure and time sequence of the early stages of paraquat lung in rats. *J Path* 1974;**114**:177–84.

110. Dinis-Oliveira RJ, Duarte JA, Sánchez-Navarro A, *et al.* Paraquat poisonings: mechanisms of lung toxicity, clinical features, and treatment. *Crit Rev Toxicol* 2008;**38**:13–71.

111. Sykes BI, Purchase IF, Smith LL. Pulmonary ultrastructure after oral and intravenous dosage of paraquat to rats. *J Path* 1977;**121**:233–41.

112. Gardiner AJ. Pulmonary oedema in paraquat poisoning. *Thorax* 1972;**27**:132–5.

113. Witschi H, Haschek WM, Meyer KR, Ullrich RL, Dalbey WE. A pathogenetic mechanism in lung fibrosis. *Chest* 1980;**78**:395–9.

114. Witschi H, Kacew S, Hirai KL, Cote MG. In-vivo oxidation of reduced nicotinamide–adenine dinucleotide phosphate by paraquat and diquat in rat lung. *Chem-Biol Interact* 1977;**19**:143–60.

115. Bus JS, Aust SD, Gibson JE. Superoxide- and singlet oxygen-catalyzed lipid peroxidation as a possible mechanism for paraquat (methyl viologen) toxicity. *Biochem Biophys Res Commun* 1974;**58**:749–55.

116. Ilett KF, Stripp B, Menard RH, Reid WD, Gillette JR. Studies on the mechanism of the lung toxicity of paraquat: comparison of tissue distribution and some biochemical parameters in rats and rabbits. *Toxicol Appl Pharmacol* 1974;**28**:216–26.

117. Farrington JA, Ebert M, Land EJ, Fletcher K. Bipyridylium quaternary salts and related compounds. V: Pulse radiolysis studies of the reaction of paraquat radical with oxygen: implications for the mode of action of bipyridyl herbicides. *Biochem Biophys Acta* 1973;**314**:372–81.

118. Montgomery MM. Interactions of paraquat with the pulmonary microsomal fatty acid desaturase system. *Toxicol Appl Pharmacol* 1976;**36**:543.

119. Bus JS, Cagen SZ, Olgaard M, Gibson JE. A mechanism of paraquat toxicity in mice and rats. *Toxicol Appl Pharmacol* 1976;**35**:501–13.

120. Tournel G, Cauffiez C, Leclerc J, *et al.* CYP2F1 genetic polymorphism: identification of interethnic variations. *Xenobiotica* 2007;**37**:1433–8.

3

Imaging of drug-induced lung disease

THOMAS E HARTMAN

INTRODUCTION

Chronic and acute lung disease caused by pulmonary drug toxicity is increasingly being diagnosed. Since the review articles in the early 1970s,[1] numerous agents including cytotoxic and non-cytotoxic drugs have been identified as having the potential to cause pulmonary toxicity.[2-7] Early recognition of pulmonary drug toxicity is important because undiagnosed injury can be progressive and fatal whereas cessation of therapy can result in reversal of the lung injury. However, recognition of drug-induced lung disease can be difficult because the clinical manifestations are usually non-specific. The radiological manifestations are also relatively non-specific, but they do reflect the underlying histopathological process.[8-12] Recognition of the imaging appearances of the common histopathological manifestations of drug-induced lung injury and knowledge of the drugs most frequently involved can facilitate diagnosis.

This chapter will review the imaging findings of common drug-induced histopathological injuries including diffuse alveolar damage (DAD), non-specific interstitial pneumonia (NSIP), bronchiolitis obliterans organizing pneumonia (BOOP), eosinophilic pneumonia (EP), diffuse pulmonary haemorrhage (DPH) and pulmonary oedema. Additionally, specific cytotoxic, non-cytotoxic and illicit drugs and the imaging manifestations of their common reactions will be described.

DIFFUSE ALVEOLAR DAMAGE

Diffuse alveolar damage is a common manifestation of drug-induced lung injury that can be divided into an acute exudative phase and a late reparative or proliferative phase.[13] The exudative phase of DAD is most prominent in the first week after lung injury and is characterized by alveolar and interstitial oedema and the formation of hyaline membranes. During this phase, the chest radiographs typically show bilateral confluent opacities which may present initially with a preferential mid and lower lung distribution. These opacities often progress to diffuse opacification of the lungs[10,11] despite corticosteroid therapy. Computed tomography in early DAD typically shows areas of ground-glass attenuation scattered throughout both lungs.[9-11] These will often progress to consolidation and may develop a gravity-dependent posterior distribution (Fig. 3.1).

During the reparative phase of DAD, fibrosis can occur and, depending on the severity of the injury, may progress to honeycomb lung. The fibrosis is not initially evident on the chest radiograph, but CT performed during this phase may show the fibrosis as irregular linear opacities with associated architectural distortion. If there is significant architectural distortion and formation of honeycomb lung this may be visible on both the chest radiograph and CT (Fig. 3.2).[9-11] DAD is typically caused by cytotoxic drugs such as bleomycin, busulphan, and carmustine or other nitrosoureas.

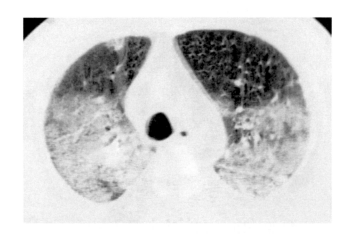

Fig. 3.1 Computed tomography of the chest of a 57-year-old man with DAD secondary to bleomycin. Note the bilateral areas of consolidation in a dependent distribution.

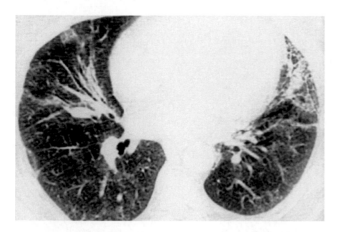

Fig. 3.2 Computed tomography of the chest of a 56-year-old woman with bleomycin toxicity and chronic changes of DAD. Irregular linear opacities in both lower lungs are most marked in the lingula where there is some mild associated traction bronchiectasis and early honeycombing.

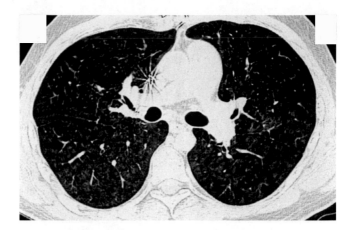

Fig. 3.4 Computed tomography of the chest of a 39-year-old woman with amiodarone toxicity shows diffuse nodular areas of ground-glass attenuation bilaterally. This was due to NSIP presenting in a pattern similar to hypersensitivity pneumonitis.

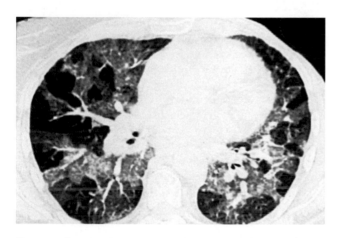

Fig. 3.3 Computed tomography of the chest of a 73-year-old woman shows a cellular NSIP pattern secondary to Cytoxan. Note the patchy areas of ground-glass attenuation in both lungs.

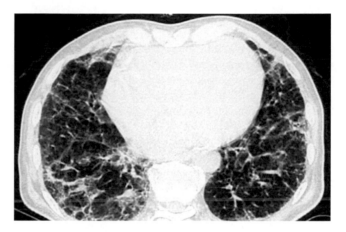

Fig. 3.5 Computed tomography of the lower chest of an 80-year-old man with fibrotic NSIP secondary to amiodarone. There are irregular linear opacities with a basilar and subpleural predominance. There are interspersed areas of scattered ground-glass attenuation in the bases as well.

NON-SPECIFIC INTERSTITIAL PNEUMONIA

Although historically many cases of interstitial pneumonia were thought to be due to usual interstitial pneumonia (UIP), the most commonly encountered form of drug reaction is non-specific interstitial pneumonia.[10] The interstitial inflammation associated with NSIP is typically more homogeneous and cellular, as opposed to the heterogeneous appearance of UIP that shows areas of normal lung, active inflammation and fibrosis.[14]

With the cellular form of NSIP, the chest radiograph usually shows diffuse heterogeneous opacities. CT at this time will show scattered or diffuse areas of ground-glass attenuation (Fig. 3.3).[9–11,15] In some instances the areas of ground-glass attenuation may be nodular and can mimic a hypersensitivity pneumonitis pattern with scattered nodular areas of ground-glass attenuation bilaterally (Fig. 3.4).[16] In the fibrotic form of NSIP, the chest radiograph will often show irregular linear opacities bilaterally with a predominant basal distribution. On CT scans, these are most commonly manifested as irregular linear opacities with associated architectural distortion that

may include traction bronchiectasis.[9–11,17] There may be some associated ground-glass attenuation at this stage as well. There is typically a subpleural and basilar distribution to the findings (Fig. 3.5). Honeycombing is not a predominant finding and, if it is present, usual interstitial pneumonitis should be considered.[18] Amiodarone, methotrexate, rituximab, sirolimus or carmustine are the drugs most commonly associated with an NSIP pattern of pulmonary toxicity.

BRONCHIOLITIS OBLITERANS ORGANIZING PNEUMONIA

BOOP is characterized by the formation of granulation tissue plugs in the respiratory bronchioles, alveolar ducts and adjacent alveolar spaces. Bleomycin, methotrexate and cyclophosphamide are common cytotoxic drugs that can result in a BOOP-like pattern of lung injury.[5–7] Amiodarone, nitrofurantoin and sulfasalazine are some non-cytotoxic causes of drug-induced BOOP.[6]

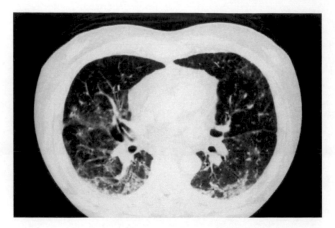

Fig. 3.6   Computed tomography of the chest of a 72-year-old man with a BOOP reaction to doxorubicin. Note the subpleural areas of consolidation posteriorly with scattered areas of ground-glass attenuation bilaterally.

The chest radiograph will demonstrate bilateral areas of consolidation or confluent opacities that are classically seen in a peripheral and subpleural distribution.[19] These may be seen throughout the lung, but have a tendency for a lower lung predominance. CT also classically shows subpleural areas of consolidation (Fig. 3.6) and/or opacities along the bronchovascular bundles, bilaterally. However, sometimes these areas of consolidation will take on a more nodular configuration and can mimic pulmonary nodules.[20] Additionally, centrilobular nodules with associated branching linear opacities ('tree-in-bud' opacities) can be seen.[9] Bronchial dilatation within the involved areas is often an associated finding.

EOSINOPHILIC PNEUMONIA

Eosinophilic pneumonia is characterized histopathologically by the accumulation of eosinophils and macrophages in the alveolar spaces.[21] Typically there is also involvement of the alveolar septa and adjacent interstitium with an infiltrate of eosinophils, lymphocytes and plasma cells. Drugs that can result in an eosinophilic pneumonia pattern include antibiotics, penicillamine, sulfasalazine, nitrofurantoin, L-tryptophan and several non-steroidal anti-inflammatory drugs.

Chest radiographs typically show confluent opacities that may have a peripheral distribution (Fig. 3.7). These are often diffuse, but may have an upper lung predominant distribution.[10] CT will show areas of ground-glass attenuation or consolidation often in a peripheral and subpleural distribution.[9,10,22] Occasionally, the opacities will be more linear than consolidative.[22] Bilateral eosinophilic pleural effusion may be present in severe cases.

DIFFUSE PULMONARY HAEMORRHAGE

Diffuse pulmonary haemorrhage can be a manifestation of drug-induced lung injury with potentially significant morbidity and mortality; however, DPH is an uncommon complication of

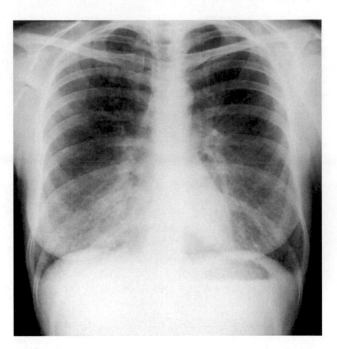

Fig. 3.7   Chest radiograph in a 36-year-old woman with acute eosinophilic pneumonia secondary to L-tryptophan. Note the focal area of consolidation in the right lower lobe medially.

drug therapy. Oral anticoagulants and heparin, abciximab, clopidogrel, amphotericin B, cyclophosphamide, mitomycin and cytarabine (ara-c) are some of the drugs that have been associated with diffuse pulmonary haemorrhage.[5–7]

Pulmonary haemorrhage is characterized by bilateral confluent opacities in the chest radiograph with anaemia and, sometimes, haemoptysis.[23] On CT, DPH is most typically seen as scattered or diffuse areas of ground-glass attenuation.[23] With greater degrees of haemorrhaging, areas of consolidation may also be seen (Fig. 3.8). As the haemorrhaging resolves, a crazy-paving pattern of ground-glass attenuation on a background of reticular opacities may sometimes be seen.[23]

NON-CARDIOGENIC PULMONARY OEDEMA

NCPE is a less common pattern of drug-induced lung involvement. This typically occurs acutely with a close temporal relationship between exposure to the drug and onset of dyspnoea and radiographic changes. The cause is altered capillary permeability and, histopathologically, this is bland pulmonary oedema with proteinaceous fluid filling the alveolar spaces.[24] On occasion, there can be histologic overlap with some of the features of DAD. Aspirin, a number of chemotherapeutic agents,[25] radiographic contrast agents and blood and blood products are associated with the development of NCPE.[6]

The chest radiograph classically shows bilateral confluent perihilar opacities with a normal heart size (Fig. 3.9).[11,26] Less severe cases can show a more interstitial pattern with Kerley's B lines. CT imaging is often not performed in the setting of NCPE, but typically will show scattered bilateral areas of ground-glass attenuation which may have a slight perihilar predominance (Fig. 3.10).[11,27]

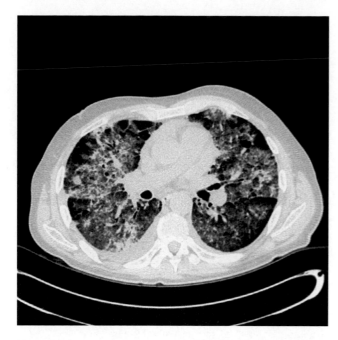

Fig. 3.8 Computed tomography of the chest of a 55-year-old man with diffuse pulmonary haemorrhaging secondary to multiple drugs. There are bilateral areas of ground-glass attenuation with some small areas of consolidation.

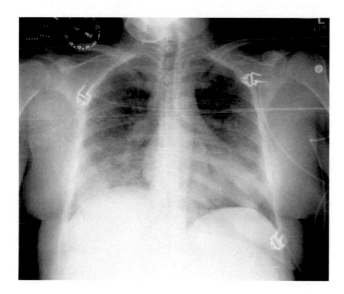

Fig. 3.9 Chest radiograph in a 50-year-old woman with non-cardiogenic pulmonary oedema secondary to transfusion-related acute lung injury. Note the bilateral confluent opacities in the mid and lower lungs.

SPECIFIC DRUGS

Although many drugs can have varying presentations of pulmonary toxicity, there are a few that have more specific patterns of injury associated with their use. There are also some drugs that are recognized for their ability to present with most, if not all, of these previously described patterns of lung injury. In this section, specific drugs will be reviewed as well as some of the complications of specific illicit drugs.

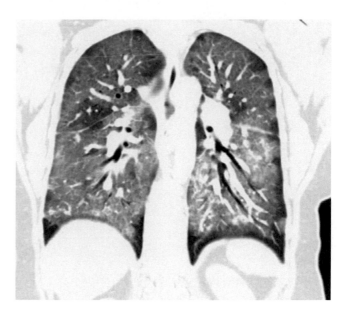

Fig. 3.10 Coronal reconstruction of a CT in a 50-year-old woman with non-cardiogenic pulmonary oedema from transfusion-related acute lung injury shows bilateral ground-glass areas of attenuation with slight sparing of the periphery of the lungs.

Bleomycin

Bleomycin is an antineoplastic antibiotic that results in lung injury in up to 5 per cent of treated patients. There is an increased risk if the total cumulative dose is more than 500 units,[2] but the complication can develop after minidoses have been given. Other risk factors for developing lung injury associated with bleomycin include: oxygen or concomitant chemotherapeutic drug(s); thoracic radiation; elderly patients; and patients in whom bleomycin is re-administered within 6 months of discontinuation.[2]

While DAD is the most common histopathologic manifestation,[28] the fibrotic form of NSIP is probably the most recognized complication of bleomycin therapy radiologically (Fig. 3.11).[10,11] The cellular form that is characterized by subpleural areas of ground-glass attenuation is less commonly seen. A BOOP-like reaction can also occur with bleomycin and this classically presents with subpleural consolidation with a lower lung predominance. However, on occasion, the BOOP changes are nodular and have been reported to mimic metastases.[20]

Methotrexate

Methotrexate is a folic acid analogue used to treat a wide range of benign and malignant conditions. Up to 10 per cent of patients treated with methotrexate will develop lung injury.[9] Methotrexate toxicity is not related to the duration of therapy or total cumulative dose.

An NSIP pattern is the most common histopathologic manifestation of methotrexate-induced lung injury.[11,29] In the cellular cases of methotrexate-induced NSIP, the bilateral areas of ground-glass attenuation are scattered diffusely throughout the lungs as opposed to the more typical lower lung and subpleural distribution (Fig. 3.12). When this occurs,

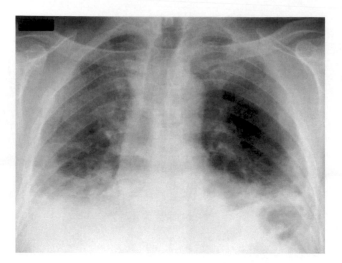

Fig. 3.11 Chest radiograph in a 50-year-old man with bleomycin lung toxicity shows peripheral and basilar fibrotic infiltrates compatible with NSIP.

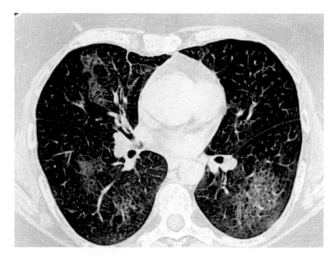

Fig. 3.12 Computed tomography in a 50-year-old woman with methotrexate pulmonary toxicity which was histologically shown to be NSIP. Note the scattered bilateral areas of ground-glass attenuation.

the appearance is more like the CT appearance of a hypersensitivity pneumonitis.[16]

Amiodarone

Amiodarone is a di-iodinated benzofuran derivative used to treat atrial and ventricular tachyarrhythmias. Up to 10 per cent of patients treated with amiodarone will develop lung injury. Typically, amiodarone-induced lung injury occurs more often when the daily dose is greater than 400 mg.[30]

The most common histopathological manifestations of amiodarone-induced lung disease are NSIP and BOOP.[9–11] Because this is an iodinated compound, a unique feature of amiodarone-induced lung disease is the occurrence of high-attenuation infiltrates on CT scans.[31] This is most readily seen in association with the BOOP pattern of reaction and shows focal homogeneous opacities of higher attenuation than

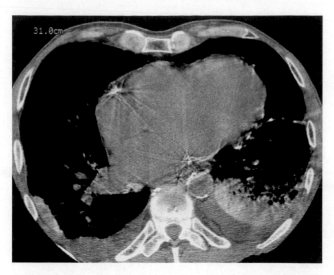

Fig. 3.13 Computed tomography of the chest of a 69-year-old man with amiodarone pulmonary toxicity shows an area of high attenuation consolidation in the left lower lobe posteriorly. There are also small bilateral pleural effusions.

muscle (Fig. 3.13).[31–33] Although non-specific, in the setting of amiodarone treatment, this finding of high-attenuation consolidation is helpful in suggesting the diagnosis of amiodarone toxicity.[33] Iodine from the amiodarone will also accumulate in the reticuloendothelial cells in the liver resulting in high attenuation of the liver parenchyma on CT images as well. Pleural effusion can occur in association with the lung injury owing to pleural inflammation or may be present related to the underlying cardiac disease.

Nitrofurantoin

Nitrofurantoin is an antimicrobial used to treat urinary tract infections. Although nitrofurantoin-induced lung injury is uncommon, this is a frequently encountered cause of pulmonary toxicity because of the frequency with which nitrofurantoin is prescribed.[6] Both acute and chronic drug-induced injury has been reported.

In the acute setting, there is most commonly a cellular NSIP, an eosinophilic pneumonia or BOOP-like pattern of reaction with scattered bilateral areas of ground-glass attenuation or consolidation (Fig. 3.14).[9–11] In the chronic setting, a fibrotic NSIP pattern is the most common histopathological manifestation.[34] In this case, a CT scan shows bilateral subpleural reticular opacities or areas of ground-glass attenuation (Fig. 3.15).[34] There is often associated traction bronchiectasis, indicating the parenchymal fibrosis. In the case of chronic toxicity there may also be a superimposed subacute component with the BOOP-like pattern superimposed on the chronic changes. ANA may stain positive.

Illicit drugs

Use of illicit drugs can cause related lung injury. Cocaine is probably the best recognized illicit drug to cause lung injury.

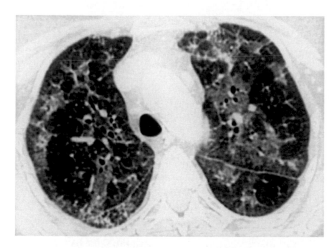

Fig. 3.14   Computed tomography of the chest of a 70-year-old woman with nitrofurantoin pulmonary toxicity shows scattered areas of ground-glass attenuation bilaterally.

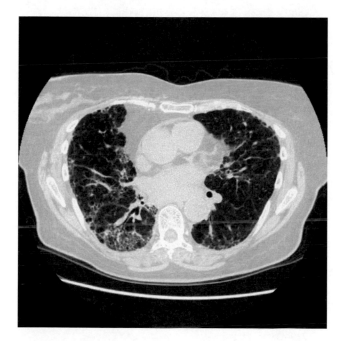

Fig. 3.15   Computed tomography of the chest of a 79-year-old woman with nitrofurantoin pulmonary toxicity showing a chronic NSIP reaction pattern. There are subpleural reticular opacities bilaterally, but greater on the right.

The most common drug injury associated with cocaine use is non-cardiogenic pulmonary oedema, although pulmonary haemorrhage and BOOP can also be seen.[35,36] As with most cases of NCPE, the changes are acutely associated with exposure to the drug and present on imaging as bilateral areas of consolidation or ground-glass attenuation.

Injection of illicit drugs or injection of medication that was supposed to be administered orally can lead to a foreign body granulomatosis that is characterized by centrilobular nodules and 'tree-in-bud' opacities on a chest CT scan (Fig. 3.16).[37] Histologically, there is a perivascular microcrystalline foreign-body deposition with giant cell reaction. Talc, microcrystalline cellulose and cornstarch are common causes of the perivascular granulomas and can cause pulmonary hypertension.

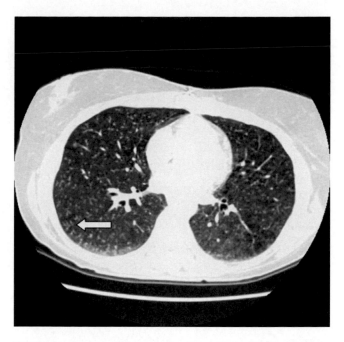

Fig. 3.16   Computed tomography of the chest of a 41-year-old woman with cellulose granulomatosis secondary to talc from intravenous drug injection. Centrilobular nodules and scattered 'tree-in-bud' opacities (arrow) are seen throughout both lungs.

EXTRAPULMONARY DRUG-INDUCED DISEASE

Although this chapter is focused on pulmonary manifestations of drug-induced disease, at least passing mention should be made of extrapulmonary findings related to drug exposures that may be encountered. These include the pleural and mediastinal findings that can be associated with the ergotamines, the 'opacification' caused by exogenous lipid aspiration, and the iatrogenic 'disease' of retained surgical implements.

Ergotamines have long been recognized as causative agents for pleural thickening and effusion as well as for fibrosing mediastinitis. The pleural thickening associated with ergotamines is most typically seen in the lung bases and is bilateral, although it may be asymmetric.[38] The fibrosing mediastinitis associated with ergotamines, most commonly methysergide, is indistinguishable from that caused by histoplasmosis and can lead to narrowing or obstruction of vessels and/or airways in the mediastinum.

Aspiration of exogenous lipids can occur and typically results in areas of consolidation on the chest radiograph. On CT scanning, these areas may present as consolidation or areas of ground-glass attenuation. If there is sufficient accumulation of the lipid material in the lungs, the areas of consolidation will have fat attenuation on CT (Fig. 3.17), allowing a more confident diagnosis of the abnormality.[39]

The most common retained surgical implement seen on imaging is the surgical sponge, also referred to as a gossipyboma. These sponges have a radio-opaque filament within them that allows them to be more easily identified on imaging. The appearance on the chest radiograph is characteristic (Fig. 3.18). Some foreign bodies are radiolucent, and may not be visible on radiographs. In these cases, the diagnosis may be delayed.

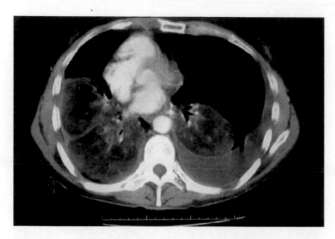

Fig. 3.17 Computed tomography of the chest of a patient with lipoid pneumonia shows large areas of consolidation in both lower lungs. The consolidation is fat attenuation consistent with lipoid pneumonia. There is also a left pleural effusion.

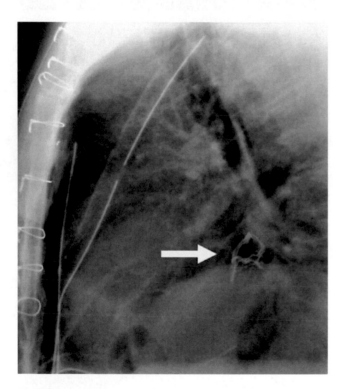

Fig. 3.18 Lateral chest radiograph of an individual after recent sternotomy shows a retained sponge (gossipyboma) in the retrocardiac mediastinum (arrow).

CONCLUSION

Drug-induced pulmonary toxicity is an increasing problem. Imaging in drug-induced pulmonary toxicity is typically non-specific and will show one of the common patterns of drug-induced lung injury. Although many drugs can show various patterns of drug injury, some drugs such as amiodarone with high-attenuation consolidation changes can allow a more confident suggestion of associated toxicity. However, in the vast majority of cases, because of the non-specific findings, the diagnosis hinges on an understanding of the common histopathological and radiological manifestations coupled with knowledge of the drugs usually involved and the resolution upon drug therapy withdrawal. With the introduction of new drugs, the list will constantly be changing and one of the best ways to keep up with the current drugs and their associated pulmonary toxicities is to visit online sites such as Pneumotox® (www.pneumotox.com) which provides updated information on drug-induced respiratory disease.

KEY POINTS

- Drug-induced lung injury can manifest in several different patterns on imaging.
- Multiple drugs can manifest with the same pattern of imaging abnormality.
- Specific drugs can cause different patterns of drug-induced lung injury.
- Some drugs have specific features that make the diagnosis of drug-induced lung injury more likely.
- Knowledge of the typical imaging patterns of drug-induced lung injury and the patterns that specific drugs can cause is key to facilitating the diagnosis of drug-induced lung injury.

REFERENCES

1. Rosenow EC. The spectrum of drug-induced pulmonary disease. *Ann Intern Med* 1972;**77**:977–91.
2. Cooper JAD, White DA, Matthay A. Drug-induced pulmonary disease. I: Cytotoxic drugs. *Am Rev Respir Dis* 1986;**133**:321–40.
3. Cooper JAD, White DA, Matthay A. Drug-induced pulmonary disease. II: Noncytotoxic drugs. *Am Rev Respir Dis* 1986;**133**.488–505.
4. Rosenow EC, Myers JL, Swensen SJ, Pisani RJ. Drug-induced pulmonary disease: an update. *Chest* 1992;**102**: 239–50.
5. Camus P, Costabel U. Drug-induced respiratory disease in patients with hematological diseases. *Semin Respir Crit Care Med* 2005;**26**:458–81.
6. Camus P, Fanton A, Bonniaud P, Camus C, Foucher P. Interstitial lung disease induced by drugs and radiation. *Respiration* 2004;**71**:301–26.
7. Limper AH. Chemotherapy-induced lung disease. *Clin Chest Med* 2004;**25**:53–64.
8. Kuhlman J. The role of chest computed tomography in the diagnosis of drug related reactions. *J Thorac Imaging* 1991;**6**:52–61.
9. Erasmus JJ, McAdams HP, Rossi SE. High-resolution CT of drug-induced lung disease. *Radiol Clin North Am* 2002;**40**:61–72.
10. Rossi S, Erasmus JJ, McAdams HP, *et al*. Pulmonary drug toxicity: radiologic and pathologic manifestations. *Radiographics* 2000;**20**:1245–59.
11. Lindell RM, Hartman TE. Chest imaging in iatrogenic respiratory disease. *Clin Chest Med* 2004;**25**:15–24.

12. Akira M, Ishikawa H, Yamamoto S. Drug-induced pneumonitis: thin-section CT findings in 60 patients. *Radiology* 2002;**224**:852–60.

13. Meyers JL. Pathology of drug-induced lung disease. In: Katzenstein A (ed.) *Surgical Pathology of Non-neoplastic Lung Disease*, pp 81–111. Philadelphia: WB Saunders, 1997.

14. Katzenstein AL, Forelli RF. Nonspecific interstitial pneumonia/fibrosis: histologic features and clinical significance. *Am J Surg Pathol* 1994;**18**:136–47.

15. Park JS, Lee KS, Kim JS, *et al.* Nonspecific interstitial pneumonia with fibrosis: radiographic and CT findings in seven patients. *Radiology* 1995;**195**:645–8.

16. Padley SP, Adler BD, Hansell DM, *et al.* High-resolution computed tomography of drug-induced lung disease. *Clin Radiol* 1992;**46**:232–6.

17. Kim TS, Lee KS, Chung MP, *et al.* Nonspecific interstitial pneumonia with fibrosis: high-resolution CT and pathologic findings. *Am J Roentgenol* 1998;**171**:1645–50.

18. McAdams HP, Rosado-de-Christenson ML, Wehunt WD, *et al.* The alphabet soup revisited: the chronic interstitial pneumonias in the 1990s. *Radiographics* 1996;**15**:1009–33.

19. Muller NL, Staples CA, Miller RR. Bronchiolitis obliterans organizing pneumonia: CT features in 14 patients. *Am J Roentgenol* 1990;**154**:983–7.

20. Glasier CM, Siegel JJ. Multiple pulmonary nodules: unusual manifestation of bleomycin toxicity. *Am J Roentgenol* 1981;**137**:155–6.

21. Pietra GG. Pathologic mechanisms of drug-induced lung disorders. *J Thorac Imaging* 1991;**6**:1–7.

22. Souza CA, Muller NL, Johkoh T, Akira M. Drug-induced eosinophilic pneumonia: high-resolution CT findings in 14 patients. *Am J Roentgenol* 2006;**186**:368–73.

23. Primack SL, Miller RR, Muller NL. Diffuse pulmonary hemorrhage: clinical, pathologic, and imaging features. *Am J Roentgenol* 1995;**164**:295–300.

24. Reed C, Glauser FL. Drug-induced noncardiogenic pulmonary edema. *Chest* 1991;**100**:1120–4.

25. Briasoulis E, Pavlidis N. Noncardiogenic pulmonary edema: an unusual and serious complication of anticancer therapy. *Oncologist* 2001;**6**:153–61.

26. Gluecker T, Capasso P, Schnyder P, *et al.* Clinical and radiologic features of pulmonary edema. *Radiographics* 1999;**19**:1507–31.

27. Storto ML, Kee ST, Golden JA, *et al.* Hydrostatic pulmonary edema: high-resolution CT findings. *Am J Roentgenol* 1995;**165**:817–20.

28. Rosenow EC, Limper AH. Drug-induced pulmonary disease. *Semin Respir Infect* 1995;**10**:86–95.

29. Everts CS, Wescott JL, Bragg DG. Methotrexate therapy and pulmonary disease. *Radiology* 1973;**107**:539–43.

30. Meyers JL, Kennedy JI, Plumb VJ. Amiodarone lung: pathologic findings in clinically toxic patients. *Hum Pathol* 1987;**18**:349–54.

31. Kuhlman JE, Scatarige JC, Fishman EK, Zerhouni EA, Siegelman SS. CT demonstration of high attenuation pleural parenchymal lesions due to amiodarone therapy. *J Comput Assist Tomogr* 1987;**11**:160–2.

32. Kuhlman JE, Tiegen C, Ren H, *et al.* Amiodarone pulmonary toxicity: CT findings in symptomatic patients. *Radiology* 1990;**177**:121–5.

33. Nicholson AA, Hayward C. The value of computed tomography in the diagnosis of amiodarone-induced pulmonary toxicity. *Clin Radiol* 1989;**40**:564–7.

34. Mendez JL, Nadrous HF, Hartman TE, Ryu JH. Chronic nitrofurantoin-induced lung disease. *Mayo Clin Proc* 2005;**80**:1298–302.

35. Ettinger NA, Albin RJ. A review of the respiratory effects of smoking cocaine. *Am J Med* 1989;**87**:664–8.

36. Murray RJ, Albin RJ, Mergner W, Criver GJ. Diffuse alveolar hemorrhage temporally related to cocaine smoking. *Chest* 1988;**93**:427–9.

37. Bendeck SE, Leung AN, Berry GJ, Daniel D, Ruoss SJ. Cellulose granulomatosis presenting as centrilobular nodules. *Am J Roentgenol* 2001;**177**:1151–3.

38. Pfitzenmeyer P, Foucher P, Dennewald G, *et al.* Pleuropulmonary changes induced by ergoline drugs. *Eur Respir J* 1996;**9**:1013–19.

39. Gondouin A, Manzoni P, Ranfaing E, *et al.* Exogenous lipid pneumonia: a retrospective multicentre study of 44 cases in France. *Eur Respir J* 1996;**9**:1463–9.

4

Bronchoalveolar lavage in drug-induced lung disease

FRANCESCO BONELLA, ESRA UZASLAN, JOSUNE GUZMAN, ULRICH COSTABEL

INTRODUCTION

Bronchoalveolar lavage (BAL) was first applied as a research tool to study local immune and inflammatory mechanisms.[1-3] As a diagnostic tool, BAL can serve as a 'window to the lung' to provide complementary information to histopathology from biopsies. BAL is a minimally invasive procedure and has become a powerful investigative tool of pulmonary medicine and a standard diagnostic procedure in patients with interstitial lung disease (ILD).[4-6] In a few of these diseases BAL findings have a specific diagnostic value and can replace lung biopsy.[6,7] The technique has several advantages over biopsy procedures: it is safe, virtually without morbidity, and collects samples from a much larger area of the lungs, therefore giving a more representative view of inflammatory and immunological changes.[8-14] Sometimes even a normal BAL may be useful to exclude some disorders with high probability (e.g. extrinsic allergic alveolitis, eosinophilic pneumonia, alveolar haemorrhage) and to focus attention in other directions.[5]

This chapter is an updated version of a previous publication on this subject.[7] It reviews the changes in BAL cytology and cell differentials in drug-induced lung disease. BAL may non-invasively support a certain clinicopathological pattern of drug-induced lung involvement.

TECHNICAL ASPECTS AND INTERPRETATION OF BAL FINDINGS

Even if the technique of BAL is not completely standardized, guidelines and recommendations for the performance of BAL, the laboratory processing and the analysis of recovered fluid have been published.[4,9,15-17] BAL is generally well tolerated. Side-effects are more or less those of routine fibre-optic bronchoscopy under local anaesthesia. There is practically no mortality, and the complication rate is low, ranging from 0 to 2.3 per cent, compared to 7 per cent with transbronchial biopsy and 13 per cent with surgical lung biopsy.[18,19] There are no absolute contraindications beyond those noted for bronchoscopy.[4,9]

Clinically important is the enumeration of cell differentials after staining with May–Grünwald–Giemsa (MGG) stain. Lymphocyte subpopulations are identified by immunocytochemical methods, immunofluorescence or flow cytometry using monoclonal antibody techniques.[19]

Furthermore a number of soluble components have been measured in BAL fluid. None of them has proven to be useful in clinical settings. Solutes are too non-specific to be of diagnostic value and their prognostic significance is also uncertain.[20,21] In order to obtain a diagnosis, the observation of the morphological appearances of cells and particles is at least as important as the counts of the cell differentials. For example, the morphological features of alveolar proteinosis are almost 100 per cent diagnostic (Table 4.1).[22] BAL should always be interpreted in the context of disease history and clinical, laboratory and radiological findings.

Normal findings in healthy adults without lung disease

Alveolar macrophages are the predominant cell population of the BAL fluid obtained from healthy, non-smoking adults without lung disease. Only small percentages of lymphocytes, neutrophils and other inflammatory cells are present.[4,5,15,22] Cigarette smoking has strong confounding effects on BAL samples.[15,22,23] The total cell yield is 3- to 5-fold higher in smokers, owing to a similar numerical increase in the number of macrophages, leading to a relative decrease of the percentage of lymphocytes. Furthermore, the alveolar macrophages display a characteristic appearance, many of them being larger than those in non-smokers and polymorphic, containing characteristic cytoplasmic inclusion bodies consisting of

Table 4.1   Diagnostic findings with bronchoalveolar lavage (adapted from [22])

BAL finding	Diagnosis
Pneumocystis, fungi, CMV transformed cells	Opportunistic infections
Milky effluent, PAS-positive non-cellular corpuscles, amorphous debris, foamy macrophages	Alveolar proteinosis
Haemosiderin-laden macrophages, intracytoplasmic fragments of red blood cells in macrophages, free red blood cells	Alveolar haemorrhage syndrome
Malignant cells of solid tumours, lymphoma, leukaemia	Malignant infiltrates
Dust particles in macrophages, quantifying asbestos bodies	Dust exposure
Eosinophils > 25%	Eosinophilic lung disease
Positive lymphocyte transformation test to beryllium	Chronic beryllium disease
CD1+ Langerhans cells increased	Langerhans cell histiocytosis
Atypical hyperplastic, bizarre type 2 pneumocytes	Diffuse alveolar damage, drug toxicity
Birefringent particles (talc)	Intravenous talcosis[1]
Oil	Is diagnostic with right stain
Phospholipids	Their absence is diagnostic

CMV, cytomegalovirus; PAS, periodic acid Schiff;[1] secondary to deliberate intravenous injection of drug tablets intended for oral use.

lipids, lipofuscin and other substances (smoker's inclusion bodies).[22]

The normal values of differential cytology following BAL, proposed in the literature, are somewhat variable because only small numbers of normal patients/persons (mostly volunteers) are available for comparison.[15] For practical reasons, the following percentages can be expected as normal in non-smokers (9):

- macrophages > 80 per cent
- lymphocytes < 15 per cent
- neutrophils < 3 per cent
- eosinophils < 0.5 per cent
- mast cells < 0.5 per cent.

The determination of CD4/CD8 ratio is usually performed in the presence of an increase in lymphocytes.[22] The CD4/CD8 ratio has been found to be low in children; this represents the major difference between children and adults. While the normal values for cellular components are only slightly different, the relative proportion of lymphocytes being higher than in adults, and neutrophils being higher in younger children under 8 years of age, explains this mild discrepancy.[24]

BRONCHOALVEOLAR LAVAGE AS A CLINICAL DIAGNOSTIC TOOL

BAL is broadly indicated in every patient with unclear pulmonary shadowing or interstitial lung disease, no matter what cause is suspected. The underlying disorders may be of infectious, non-infectious, immunological or malignant aetiology. BAL may also be indicated in patients with normal chest radiographs when clinical and lung function tests are abnormal and suggestive of diffuse lung disease, or in patients with unexplained pulmonary symptoms in whom a normal BAL finding may allow exclusion of significant, active interstitial lung disease.

BAL in the differential diagnosis

In two recently published international statements on the major interstitial lung diseases, BAL was considered helpful in

achieving the diagnosis in patients with sarcoidosis in the absence of biopsy,[25] and in patients with idiopathic pulmonary fibrosis (IPF) who did not undergo surgical biopsy. In this case BAL and/or transbronchial biopsy are considered requirements for the exclusion of other diseases (one of the four major criteria for making a clinical diagnosis of the disease).[26]

Many disorders in which BAL can directly confirm a particular diagnosis, replacing lung biopsy, are included in the group of the alveolar filling syndromes (see Table 4.1). The abnormal material which accumulates in the alveolar spaces in these syndromes can easily be washed out by lavage and provide a source of diagnosis-specific findings.

In the other diseases in which BAL does not offer a diagnosis, the cellular pattern plays a pivotal role in narrowing the differential diagnosis. The list of diseases with a lymphocytic, a neutrophilic, an eosinophilic, or a mixed cellular pattern is long (Box 4.1). As outlined above, an elevated, normal or decreased CD4/CD8 ratio can further differentiate diseases with an increase in lymphocyte counts (Table 4.2).

Assessment of disease activity

It has still not been determined whether BAL is clinically useful for assessing the activity of disease and to provide prognostic information. In recent years the role of BAL in monitoring interstitial lung involvement and assessing the activity of disease has been particularly addressed in collagen vascular diseases, such as systemic sclerosis or rheumatoid arthritis. Owing to the paucity of good studies, firm conclusions are not possible.[6]

BAL IN DRUG-INDUCED PNEUMONITIS

Many drugs can induce an interstitial lung reaction, either toxic or immune-mediated. In addition to toxic changes of type 2 pneumocytes, any type of alveolitis may be present in BAL (lymphocytic, neutrophilic, eosinophilic or mixed), and also diffuse alveolar haemorrhage (Box 4.2).[4,22,27-29] The changes in BAL cytology so far reported for each individual drug are listed in detail in Table 4.3 (page 36).

Box 4.1   BAL cellular patterns as an adjunct to diagnosis (adapted from [22])

Lymphocytic

- Hypersensitivity pneumonitis
- Berylliosis
- Sarcoidosis
- Tuberculosis
- Connective-tissue disorders
- Drug-induced pneumonitis
- Malignant infiltrates
- Silicosis
- Early asbestosis
- Crohn's disease
- Primary biliary cirrhosis
- HIV infection
- Viral pneumonia
- Radiation-induced pneumonitis

Neutrophilic, with or without eosinophilic

- Idiopathic pulmonary fibrosis
- Desquamative interstitial pneumonia (DIP)
- Acute interstitial pneumonia (AIP)
- Acute respiratory distress syndrome
- Bacterial pneumonia

- Connective-tissue disorders
- Asbestosis
- Wegener's granulomatosis
- Diffuse panbronchiolitis
- Transplant bronchiolitis obliterans
- Idiopathic bronchiolitis obliterans
- Drug-induced reaction

Eosinophilic

- Eosinophilic pneumonia
- Churg–Strauss syndrome
- Hypereosinophilic syndrome
- Allergic bronchopulmonary aspergillosis
- Idiopathic pulmonary fibrosis
- Drug-induced reaction

Mixed cellularity

- Bronchiolitis obliterans organizing pneumonia (BOOP)
- Connective-tissue disorders
- Non-specific interstitial pneumonia (NSIP)
- Drug-induced reaction
- Inorganic dust disease

Table 4.2   CD4/CD8 ratio in diseases with lymphocytic alveolitis

CD4/CD8 increased	CD4/CD8 normal	CD4/CD8 decreased
Sarcoidosis	Tuberculosis	Hypersensitivity pneumonitis
Beryllium disease	Lymphangitic carcinomatosis	Drug-induced pneumonitis
Asbestos-induced alveolitis		Radiation-induced lung injury
Alveolar proteinosis		BOOP
Crohn's disease		Silicosis
Connective-tissue disorders		HIV infection
Drug-induced pneumonitis		Churg–Strauss syndrome

BOOP, bronchiolitis obliterans organizing pneumonia.

The most frequent finding is a lymphocytic alveolitis with the predominance of CD8+ cells, just as in extrinsic allergic alveolitis. In pneumonitis induced by methotrexate and other immunomodulating agents (sirolimus, temsirolimus), CD4+ cells may be preferentially increased.[15,16,30,31,118] Another study reported the opposite about methotrexate.[32] The variation in the CD4/CD8 ratio in methotrexate-induced pneumonitis may reflect the time interval between the last drug administration (high ratio when interval was short, < 10 days) or the use of corticosteroids (high without corticosteroids).[91]

The second frequent pattern is the eosinophilic one. Many new drugs have been reported to induce this BAL pattern (e.g. references 47, 69, 72, 76, 77, 80, 89, 96, 109, 112,

121 and 127). In some cases, the eosinophilic pattern is combined with a lymphocytosis (e.g. references 47, 66, 83, 98, 102, 111, 120 and 121), raising the possibility of a hypersensitivity reaction.[128]

Except for the eosinophilia where a percentage greater than 25 is clinically meaningful and indicative of eosinophilic infiltrates, the magnitude of the increase in lymphocytes or neutrophils is not clinically relevant.

An interesting and new observation is the detection of the alveolar proteinosis pattern in BAL in patients treated with immunomodulating agents.[84] The alveolar proteinosis pattern is characterized by foamy macrophages, cell debris, acellular granules and PAS-positive amorphous material (surfactant

Box 4.2 Overview of BAL patterns in drug-induced interstitial lung disease (for a detailed listing, see Table 4.3)

Lymphocytosis
- Acebutolol
- Amiodarone
- Anagrelide
- Atenolol
- Azathioprine
- Bepridil
- Bleomycin
- Busulfan
- Carbamazepin
- Celiprolol
- Ciprofloxacin
- Cyclophosphamide
- Daptomycin
- Dasatinib
- Diphenylhydantoin
- Doluxetine
- Flecainide
- Gemtuzumab
- Gold
- Imatinib
- Lenalidomide
- Loxoprofen
- Methotrexate
- Minocycline
- Nilutamide
- Nitrofurantoin
- Piperacillin
- Propanolol
- Risedronate
- Simvastatin
- Sulfasalazine
- Temsirolimus
- Thalidomide
- Ticlopidine
- Tolfenamic acid
- Venlafaxin
- Vincristine

Neutrophilia
- Amiodarone
- Bepridil
- Bleomycin
- Busulfan
- Clarithromycin
- Methotrexate
- Minocycline
- Trastuzumab

Eosinophilia
- Ampicillin
- Bepridil
- Carbamazepin
- Clarithromycin
- Cotrimazole
- Daptomycin
- Doluxetine
- Ifenprodil
- Imatinib
- Interferon alpha 2b
- Lenalidomide
- L-tryptophan
- Maloprim
- Meloxicam
- Minocycline
- Nimesulide
- Nitrofurantoin
- Penicillin

- Piperacillin
- Progesterone
- Risedronate
- Sirolimus
- Sulfasalazine
- Temsirolimus
- Tetracycline
- Ticlopidin
- Tolfenamic acid
- Venlafaxin
- Meloxicam

Haemorrhage
- Anticoagulant drugs and platelet GP inhibitors
- Amphotericin B
- Carbimazole
- Cytotoxic drugs
- D-penicillamine
- Gemtuzumab

Cytotoxic reaction
- Bleomycin
- Busulfan
- Cyclophosphamide
- Methotrexate
- Nitrosureas

Alveolar proteinosis
- Busulfan
- Leflunomide
- Sirolimus

proteins and phospholipids).[22] Secondary pulmonary alveolar proteinosis (PAP) as a result of immunosuppressive medications is a rare occurrence and only two reports are available, about leflunomide[84] and busulfan.[129] Epithelial hyperplasia caused by busulfan therapy was probably responsible for the induction of excessive surfactant production, resulting in alveolar proteinosis.[129] The pathogenesis of PAP related to leflunomide immunosuppression is thought to be due to inhibition of macrophage recruitment and phagocytosis.[84] The BAL fluid, in this report, was slightly cloudy in appearance and contained 67 per cent macrophages, 28 per cent neutrophils, 5 per cent lymphocytes and 0 per cent eosinophils and showed the typical aspects of alveolar proteinosis, finally confirmed by lung biopsy.[84]

An emerging topic concerns pulmonary involvement and BAL changes induced by the new class of biological drugs (adalimumab, infliximab, etanercept, gemtuzumab, rituximab and trastuzumab), most of them humanized monoclonal antibodies that selectively bind cytokines or their receptor, and increasingly administrated for collagen vascular diseases and haematological malignancies.[72,123,130–132] The spectrum of lung damage seems to be wide, ranging from alveolar haemorrhage[130] to fibrosis.[131] The only two published reports with BAL findings reveal the presence of neutrophils for trastuzumab[123] and lymphocytes combined with haemorrhage for gemtuzumab.[72] Also for the tyrosine kinase inhibitors (imatinib and dasatinib) some data are available to show, for this new class of antineoplastic drugs, variable BAL findings.[67,77]

Table 4.3 Types of changes in BAL cytology reported for individual drugs (alphabetical order)

Drug name	Ly	Neu	Eos	Mixed	Haem	Cyto	Normal	CD4/CD8	References
Acebutolol	✓								33
Acetaminophen	✓							L	34
Acetylsalicylic acid (aspirin)			✓						35, 36
Amiodarone[a]	✓	✓	✓	✓			✓	L	28, 37–39
Amitriptyline			✓						40
Amphotericin B					✓				33
Ampicillin	✓		✓					H	33, 41
Anagrelide	✓								42
Atenolol	✓			✓					24
Azathioprine	✓				✓				43, 44
Barbiturates			✓						45
BCG therapy (intravesical)	✓							H	33, 46
Bepridil	✓	✓	✓					L	47
Bleomycin	✓	✓	✓						33, 48
Bromocriptine									49
Bucillamine	✓		✓					L	50, 51
Busulphan	✓			✓					52
Carbamazepine	✓	✓	✓			✓		L	53–55
Carbimazole				✓					56
Celiprolol	✓								57
Chlorambucil	✓							L	58
Chloroquine			✓					L	59
Ciprofloxacine	✓							H	60
Clarithromycine		✓	✓						61
Co-trimoxazole				✓				L	62
Cyclophosphamide	✓						✓		63
Cyclothiazide	✓								64
Cyproterone acetate				✓				L	65
Daptomycin	✓		✓						66
Dasatinib	✓	✓		✓				H	67
Disopyramide	✓							H	68
Doluxetine			✓						69
Etoposide	✓								70
Flecainide	✓			✓				L	71
Gemtuzumab	✓			✓					72
Gold (Auranofin)	✓			✓				L, H	73, 74
Heroin			✓						75
Ifenprodil			✓						76
Imatinib	✓		✓						77
Indomethacin			✓						78
Interferon-alpha	✓			✓				L	79
Interferon-alpha 2 beta			✓						80
Interferon-beta	✓			✓				L	81
Isoniazid				✓					82
Lenalidomide	✓							L	83
Leflunomide*				✓					84
Levofloxacin	✓		✓					L	85
Loxoprofen	✓								86, 87

(*Contd*)

Table 4.3 (*Continued*)

Drug name	Ly	Neu	Eos	Mixed	Haem	Cyto	Normal	CD4/CD8	References
Maloprim (pyrimethamine and dapsone)			✓						88
Meloxicam			✓						89
Melphalan	✓								90
Methotrexate	✓		✓	✓				H	30, 31, 91
Minocycline	✓	✓	✓					L	92, 93
Mitomycin C	✓								94
Nalidixic acid			✓						95
Nimesulide			✓						96
Nilutamide	✓	✓						L	97, 98
Nitrofurantoin	✓		✓					H	99
Nitrosureas (Carmustine)	✓							L	100
Oxaliplatin			✓			✓			101, 102
Paclitaxel	✓		✓					L	103
Penicillamine				✓					33
Penicillin			✓						33
Piperacillin	✓							L	103
Phenytoin	✓								104, 105
Pranoprofen			✓						106
Procainamide		✓							107
Progesterone			✓						108
Propranolol	✓							L	109
Rifampicin	✓								110
Risedronate	✓		✓						111
Serrapeptase	✓		✓					L	112, 113
Simvastatin	✓							L	114
Sirolimus	✓				✓			H	115, 116
Sulfasalazine			✓						117
Temsirolimus	✓							H	118
Tetracyclin			✓						33
Thalidomide	✓							N	119
Ticlopidine	✓		✓						120
Tolfenamic acid	✓		✓					H	121
Tosufloxacin									
Tosilate	✓	✓	✓						122
Trastuzumab		✓							123
Trazodone			✓						124
Tryptophan			✓						22, 127
Venlafaxine	✓		✓						125, 126
Vincristine	✓								22

[a] Foamy alveolar macrophages. L, low; H, high; N, normal; Ly, lymphocytosis; Neu, neutrophilia; Eos, eosinophilia; Mixed, mixed cellularity; Haem, alveolar haemorrhage; Cyto, cytotoxic reaction.

The diagnosis of drug-induced lung disease cannot be made by BAL alone and obviously needs additional clinical investigations. BAL is unquestionably useful to exclude other pulmonary problems (including infection) that are commonly associated with chemotherapeutic agents and should be performed in each doubtful case.

DIAGNOSTIC FEATURES OF DRUG–INDUCED LUNG DISEASE

The diagnostic criteria of drug-induced lung disease include:[133]

- a history of drug exposure;

- clinical, imaging, and histopathologic patterns consistent with earlier observations with the same drug;
- exclusion of lung disease other than drug-induced;
- improvement following discontinuation of the suspected drug;
- recurrence of symptoms on rechallenge.

Rechallenge may be done in certain circumstances but is usually considered unethical with drugs causing irreversible pulmonary changes.

BAL may contribute to several of these criteria, such as showing evidence of drug exposure, as is the case with amiodarone. Amiodarone causes characteristic changes in the alveolar macrophage population which shows foamy intracytoplasmic alterations, corresponding to a form of phospolipidosis.[27,28] These changes are not specific, however: foamy macrophages are also seen in extensive allergic alveolitis (EAA) and cryptogenic organizing pneumonitis (COP).

If foamy macrophages are absent, an amiodarone-induced pneumonitis can likely be excluded. On the other hand, a normal BAL cell differential in the presence of foamy macrophages does not exclude amiodarone-induced lung disease.[37–39]

Contribution of BAL to the clinicopathological patterns defining drug-induced lung disease

BAL may often point towards the recognition of specific patterns of drug-induced lung involvement.

CELLULAR NON-SPECIFIC INTERSTITIAL PNEUMONIA (NSIP)

This is a frequent pattern, also called 'hypersensitivity pneumonitis'. BAL shows a high total cell count owing to a marked increase in lymphocytes, usually above 50 per cent, which may be combined with a mild additional increase in neutrophils and other cells. Usually the CD4/CD8 ratio is low.[33,48,98,103,109]

PULMONARY INFILTRATES AND EOSINOPHILIA (CHRONIC OR ACUTE)

This is also known as eosinophilic pneumonia and is caused by many drugs (e.g. references 33, 35, 36, 40, 50, 51, 61, 67, 69, 75, 76, 78, 80, 85, 101, 102, 106, 108, 121, 122, 124 and 125). BAL shows here a significant eosinophilia, usually above 25 per cent.[22,33] A lack of BAL eosinophilia is rather exclusive of pulmonary eosinophilic infiltrates.

ORGANIZING PNEUMONIA (OP)

Drug-related OP is clinically and radiologically similar to OP due to other causes and contexts. BAL usually shows a combination of increased lymphocytes, neutrophils, eosinophils and mast cells.[134] Foamy alveolar macrophages and a few plasma cells can also be noted. The CD4/CD8 ratio is usually low.

CYTOTOXIC REACTION

This associated with cytostatic chemotherapy and resembles diffuse alveolar damage. Histology may show changes such as hyaline membranes, atypia and hyperplasia of type 2 pneumocytes, and diffuse fibroblast proliferation. As in patients with acute respiratory distress syndrome (ARDS) or acute interstitial pneumonia,[135] BAL may show clusters of atypical type 2 pneumocytes, along with a marked increase in neutrophils.

DIFFUSE ALVEOLAR HAEMORRHAGE

This is associated with only a few drugs. BAL is the ideal technique to demonstrate the haemosiderin-laden alveolar macrophages, in addition to the presence of free red blood cells. Importantly, haemosiderin-laden macrophages do not appear earlier than 48 hours after bleeding.[7]

ALVEOLAR PROTEINOSIS

BAL is highly diagnostic of this extremely rare pattern.[22] The constellation of acellular granules, foamy macrophages, cell debris and amorphous material is pathognomonic. However, the idiopathic form of disease or secondary forms associated with haematopoetic malignancies must always be excluded.

Role of BAL in the exclusion of other lung diseases

The major role of BAL is to exclude infections, above all the opportunistic types. This is essential especially in patients with acute febrile interstitial lung disease and in those receiving immunosuppressive therapy. Under administration of corticosteroids, immunosuppressives, disease-modifying antirheumatic drugs (DMARDs) and biological drugs, the development of bacterial, fungal, parasitic (including *Pneumocystis*) and viral infections is common. BAL has a high sensitivity and specificity (>95 per cent) to diagnose *Pneumocystis* pneumonia. Globally, the sensitivity of BAL in the diagnosis of bacterial pneumonia is 60–90 per cent, and in mycobacterial, fungal and viral infections it is 70–80 per cent.[24]

Another important role is played by BAL in the differential diagnosis of interstitial lung disease in patients treated with cardiovascular drugs, an escalating clinical situation. The diagnosis of pulmonary congestion due to left heart failure can be supported by the BAL finding of diffuse alveolar haemorrhaging.[7] Nowadays a few drugs, including anticoagulants, fibrinolytic agents and platelet glycoprotein IIB/IIIA inhibitors (abciximab, clopidogrel and eptifibatide), seem to induce diffuse alveolar haemorrhaging (see Box 4.2, page 35).[56,72,102,130] If BAL cytology is normal, a drug-induced immunological reaction (hypersensitivity pneumonitis, organizing pneumonia, eosinophilic lung disease) can be largely excluded.[7]

BAL in the diagnosis of underlying disorders

BAL can significantly contribute to a diagnosis in diffuse pulmonary malignancies: for metastatic carcinoma and lymphoma the sensitivity reaches 80 and 50 per cent, respectively.[136] BAL findings in interstitial lung disease associated with connective tissue disease or inflammatory bowel disease are non-specific.

Like the histopathological patterns of idiopathic interstitial pneumonias (UIP, NSIP, LIP, DAD, OP and DIP), any type of BAL pattern (lymphocytic, neutrophilic, eosinophilic and mixed) may be present, so that a differentiation between manifestations of systemic immunological disorders and immunological drug-induced disorders is not possible by BAL.[7]

In radiation pneumonitis, BAL fluid may show an increase in lymphocytes and eosinophils.[137] Increased BAL lymphocytes have been found in the contralateral lung following thoracic irradiation for breast cancer.[138] Organizing pneumonia (BOOP) following radiation therapy in breast cancer typically develops a few months after field radiotherapy. BAL shows a mixed and non-specific pattern (lymphocytes, neutrophils, eosinophils and mast cells) with low CD4/CD8 ratio.[139] Nevertheless BAL may be useful to exclude other causes of radiological infiltration, like infections or diffuse spread of malignancy.[7]

KEY POINTS

- BAL is a minimally invasive procedure that has become a powerful investigative tool of pulmonary medicine and a standard diagnostic procedure in patients with interstitial lung disease.

- BAL can contribute to the expected clinicopathological pattern of a given drug-induced lung disease.

- In a few diseases BAL findings have a specific diagnostic value and can replace lung biopsy. However, BAL changes tend to be non-specific for any drug-induced lung disease, and the definitive diagnosis cannot be based on BAL findings alone.

- The technique has several advantages over biopsy procedures. It is safe, virtually without morbidity, and collects samples from a much larger area of the lungs, therefore giving a more representative view of inflammatory and immunological changes.

- BAL is supportive in the differential diagnosis, allowing exclusion of an infective aetiology and detection of lung involvement by underlying diseases (e.g. metastatic cancer or malignant lymphoma).

REFERENCES

1. Reynolds HY, Newball HH. Analysis of proteins and respiratory cells obtained from human lungs by bronchial lavage. *J Lab Clin Med* 1974;**84**:559–73.

2. Yeager HC, Williams MC, Beckman JF, *et al.* Sarcoidosis: analysis of cells obtained by bronchial lavage. *Am Rev Respir Dis* 1976;**113**:96–100.

3. Crystal RG, Roberts WG, Hunninghake GW, *et al.* Pulmonary sarcoidosis: a disease characterized and perpetuated by activated lung T-lymphocytes. *Ann Intern Med* 1981;**94**: 73–94.

4. Klech H, Hutter C (eds). Clinical guidelines and indications for bronchoalveolar lavage (BAL): report of the European Society of Pneumology Task Force on BAL. *Eur Respir J* 1990;**3**:937–74.

5. Costabel U, Guzman J. Bronchoalveolar lavage in interstitial lung disease. *Curr Opin Pulm Med* 2001;**7**:255–61.

6. Costabel U, Guzman J, Bonella F, Ohshimo S. Bronchoalveolar lavage in other interstitial lung diseases. *Semin Respir Crit Care Med* 2007;**28**:514–24.

7. Costabel U, Uzaslan E, Guzman J. Bronchoalveolar lavage in drug-induced lung disease. *Clin Chest Med* 2004;**25**:25–35.

8. Reynolds HY. Use of bronchoalveolar lavage in humans: past necessity and future imperative. *Lung* 2000;**178**:271–93.

9. Klech H, Pohl W (eds). Technical recommendations and guidelines for bronchoalveolar lavage (BAL): report of the ERS Task Group. *Eur Respir J* 1989;**2**:561–85.

10. Hunninghake LGW, Kawanami O, Ferrans VJ, *et al.* Characterisation of the inflammatory and immune effector cells in the lung parenchyma of patients with interstitial lung disease. *Am Rev Respir Dis* 1981;**123**:407–12.

11. Abe S, Munakata M, Nishimura M, *et al.* Gallium-67 scintigraphy, bronchoalveolar lavage, and pathologic changes in patients with pulmonary sarcoidosis. *Chest* 1984;**85**:650–5.

12. Semenzato G, Chilosi M, Ossi E, *et al.* Bronchoalveolar lavage and lung histology: comparative analysis of inflammatory and immunocompetent cells in patients with sarcoidosis and hypersensitivity pneumonitis. *Am Rev Respir Dis* 1985;**132**: 400–4.

13. Campbell DA, Poulter LW, duBois RM. Immunocompetent cells in bronchoalveolar lavage reflect the cell populations in transbronchial biopsies in pulmonary sarcoidosis. *Am Rev Respir Dis* 1985;**132**:1300–6.

14. Paradis IL, Dauber JH, Rabin BS. Lymphocyte phenotypes in bronchoalveolar lavage and lung tissue in sarcoidosis and idiopathic pulmonary fibrosis. *Am Rev Respir Dis* 1986;**133**: 855–60.

15. BAL Cooperative Steering Group. Bronchoalveolar lavage constituents in healthy individuals, idiopathic pulmonary fibrosis, and selected comparison groups. *Am Rev Respir Dis* 1990;**141**:S169–202.

16. Haslam PL, Baughman RP (eds). Report of European Respiratory Society (ERS) Task Force: guidelines for measurement of acellular components and recommendations for standardization of bronchoalveolar lavage (BAL). *Eur Respir Rev* 1999;**9**:25–157.

17. Lam S, Leriche JC, Kijek K, Phillips D. Effect of bronchial lavage volume on cellular and protein recovery. *Chest* 1985;**88**:856–9.

18. Baughman RP. *Bronchoalveolar Lavage.* St Louis: Mosby Year Book, 1992.

19. Costabel U. CD4/CD8 ratios in bronchoalveolar lavage fluid: of value for diagnosing sarcoidosis? *Eur Respir J* 1997;**10**: 2699–700.

20. Costabel U, Bross KJ, Baur R, Rühle H, Matthys Z. Differentialzytologie und Lymphozytensubpopulationen der bronchoalveolären Lavage unter verschiedenen Aufbewahrungsbedingungen. *Prax Klin Pneumol* 1988;**42**:103–5.

21. Haslam PL, Baughman RP. Report of ERS Task Force: guidelines for measurement of acellular components and standardization of BAL. *Eur Respir J* 1999;**14**:245–8.

22. Costabel U. *Atlas of Bronchoalveolar Lavage.* London: Chapman & Hall, 1998.

23. Costabel U, Guzman J. Effect of smoking on bronchoalveolar lavage constituents. *Eur Respir J* 1992;**5**:776–9.

24. ERS Task Force. Bronchioalveolar lavage in children. *Eur Respir J* 2000;**15**:217–31.

25. Hunninghake GW, Costabel U, Ando M, *et al.* ATS/ERS/WASOG Statement on Sarcoidosis. *Sarc Vasc Diffuse Lung Dis* 1999;**16**:149–73.

26. ATS/ERS Statement. Idiopathic pulmonary fibrosis: diagnosis and treatment. *Am J Respir Crit Care Med* 2000;**161**:646–64.

27. Huang MS, Colby T, Goellner JR, Martin WJ. Utility of bronchoalveolar lavage in the diagnosis of drug-induced pulmonary toxicity. *Acta Cytol* 1989;**33**:533–8.

28. Israel-Biet D, Venet A, Caubarrere I, *et al.* Bronchoalveolar lavage in amiodarone pneumonitis: cellular abnormalities and their relevance to pathogenesis. *Chest* 1987;**91**:214–21.

29. White DA, Kris MG, Stover DE. Bronchoalveolar lavage cell populations in bleomycin lung toxicity. *Thorax* 1987;**42**:551–2.

30. Schnabel A, Richter C, Bauerfeind S, Gross WL. Bronchoalveolar lavage cell profile in methotrexate induced pneumonitis. *Thorax* 1997;**52**:377–9.

31. White DA, Rankin JA, Stover DE, Gellene RA, Gupta S. Methotrexate pneumonitis. *Am Rev Respir Dis* 1989;**139**:18–21.

32. Akoun G, Mayaud C, Touboul JL, *et al.* Use of bronchoalveolar lavage in the evaluation of methotrexate lung disease. *Thorax* 1987;**42**:652–5.

33. Israel-Biet D, Danel C, Costabel U, Rossi GA, Wallaert B. The clinical role of BAL in drug induced pneumonitis. *Eur Respir Rev* 1992;**2**:97–9.

34. Kudeken N, Kawakami K, Kakazu T, *et al.* [A case of acetaminophen-induced pneumonitis]. *Nihon Kyobu Shikkan Gakkai Zasshi* 1993;**31**:1585–90.

35. Shimizu K, Shiota S, Nakaya Y, *et al.* [Buffer-induced lung injury manifesting as acute eosinophilic pneumonia]. *Nihon Kyobu Shikkan Gakkai Zasshi* 1997;**35**:1099–103.

36. Uchida K, Sekiguchi S, Matsuda K, Kurihara Y. [Acute eosinophilic pneumonia associated with aspirin]. *Nihon Kyobu Shikkan Gakkai Zasshi* 1997;**35**:873–7.

37. Coudert B, Bailly F, Lombard JN, Andre F, Camus P. Amiodarone pneumonitis: bronchoalveolar lavage findings in 15 patients and review of the literature. *Chest* 1992;**102**:1005–12.

38. Ohar JA, Jackson F, Dettenmeier PA, *et al.* Bronchoalveolar lavage cell count and differential are not reliable indicators of amiodarone-induced pneumonitis. *Chest* 1992;**102**:999–1004.

39. Akoun GM, Cadranel JL, Blanchette G, Milleron BJ, Mayaud CM. Bronchoalveolar lavage cell data in amiodarone-associated pneumonitis: evaluation in 22 patients. *Chest* 1991;**99**:1177–82.

40. Noh H, Lee YK, Kan SW, *et al.* Acute eosinophilic pneumonia associated with amitriptyline in a hemodialysis patient. *Yonsei Med J* 2001;**42**:357–9.

41. Miyashita N, Nakajima M, Kuroki M, *et al.* [Sulbactam/ ampicillin-induced pneumonitis]. *Nihon Kokyuki Gakkai Zasshi* 1998;**36**:684–9.

42. Raghavan M, Mazer MA, Brink DJ. Severe hypersensitivity pneumonitis associated with anagrelide. *Ann Pharmacother* 2003;**37**:1228.

43. Refabert L, Sinnassamy P, Leroy B, *et al.* Azathioprine-induced pulmonary haemorrhage in a child after renal transplantation. *Pediatr Nephrol* 1995;**9**:470–3.

44. Doshi AV, Desai D, Bhaduri A, Udwadia ZF. Lymphocytic interstitial pneumonitis associated with autoimmune hepatitis. *Indian J Gastroenterol* 2001;**20**:76–7.

45. Gali JM, Vilanova JL, Mayos M, *et al.* Febarbamate-induced pulmonary eosinophilia: a case report. *Respiration* 1986;**49**:231–4.

46. Naoki K, Yamaguchi K, Soejima K, *et al.* [A case of interstitial pneumonia induced by intravesical administration of bacillus Calmette–Guerin (BCG)]. *Nihon Kyobu Shikkan Gakkai Zasshi* 1997;**35**:1383–8.

47. Okubo F, Ando M, Ashihara Y, *et al.* A case of drug-induced pneumonitis due to bepridil. *Nihon Kokyuki Gakkai Zasshi* 2006;**44**:17–21.

48. Akoun GM, Cadranel JL, Milleron BJ, D'Ortho MP, Mayaud CM. Bronchoalveolar lavage cell data in 19 patients with drug-associated pneumonitis (except amiodarone). *Chest* 1991;**99**:98–104.

49. Pfitzenmeyer P, Foucher P, Dennewald G, *et al.* Pleuropulmonary changes induced by ergoline drugs. *Eur Respir J* 1996;**9**:1013–19.

50. Kishimoto N, Fujii K. [A case of pulmonary infiltration with eosinophilia (PIE) syndrome induced by bucillamine treatment of rheumatoid arthritis]. *Nihon Kokyuki Gakkai Zasshi* 2002;**40**:321–5.

51. Hara A, Sakamoto O, Matsumoto M, *et al.* [A case of bucillamine-induced interstitial pneumonia]. *Nihon Kyobu Shikkan Gakkai Zasshi* 1992;**30**:1743–8.

52. Vergnon JM, Boucheron S, Riffat J, *et al.* [Interstitial pneumopathies caused by busulfan: histologic, developmental and bronchoalveolar lavage analysis of 3 cases]. *Rev Med Interne* 1988;**9**:377–83.

53. Fernandez Alvarez R, Gullon Blanco JA, Riesgo Alonso C, Molinos Martin L, Martinez Gonzalez-Rio J. [Acute pulmonary toxicity caused by carbamazepine: apropos of a case]. *Arch Bronconeumol* 1994;**30**:471–2.

54. Ben Jemaa M, Kammoun S, Kanoun F, Zghal KM, Ben Hamed S. [Respiratory manifestations of carbamazepine: apropos of a case]. *Rev Pneumol Clin* 1997;**53**:351–4.

55. Wilschut FA, Cobben NA, Thunnissen FB, *et al.* Recurrent respiratory distress associated with carbamazepine overdose. *Eur Respir J* 1997;**10**:2163–5.

56. Calanas-Continente A, Espinosa M, Manzano-Garcia G, *et al.* Necrotizing glomerulonephritis and pulmonary hemorrhage associated with carbimazole therapy. *Thyroid* 2005;**15**:286.

57. Lombard JN, Bonnotte B, Maynadie M, *et al.* Celiprolol pneumonitis. *Eur Respir J* 1993;**6**:588–91.

58. Crestani B, Jaccard A, Israel-Biet D, *et al.* Chlorambucil-associated pneumonitis. *Chest* 1994;**105**:634–6.

59. Mitja K, Izidor K, Music E. [Chloroquine-induced drug hypersensitivity alveolitis]. *Pneumologie* 2000;**54**:395–7.

60. Steiger D, Bubendorf L, Oberholzer M, Tamm M, Leuppi JD. Ciprofloxacin-induced acute interstitial pneumonitis. *Eur Respir J* 2004;**23**:172–5.

61. Terzano C, Petroianni A. Clarithromycin and pulmonary infiltration with eosinophilia. *Br Med J* 2003;**326**:1377–8.

62. Hashizume T, Numata H, Matsushita K. [Drug-induced pneumonitis caused by sulfamethoxazole–trimethoprim]. *Nihon Kokyuki Gakkai Zasshi* 2001;**39**:664–7.

63. Usui Y, Aida H, Kimula Y, *et al.* A case of cyclophosphamide-induced interstitial pneumonitis diagnosed by bronchoalveolar lavage. *Respiration* 1992;**59**:125–8.

64. Kheir A, Chabot F, Lesur O, et al. [Fibrosing pneumopathy induced by cyclothiazide: apropos of a case]. *Rev Mal Respir* 1992;**9**:208-12.

65. Similowski T, Orcel B, Derenne JP. CD8+ lymphocytic pneumonitis in a patient receiving cyproterone acetate. *South Med J* 1997;**90**:1048-9.

66. Hayes D, Anstead MI, Kuhn RJ. Eosinophilic pneumonia induced by daptomycin. *J Infect* 2007;**54**:e211-13.

67. Bergeron A, Réa D, Levy V, et al. Lung abnormalities after dasatinib treatment for chronic myeloid leukemia: a case series. *Am J Respir Crit Care Med* 2007;**176**:814-18.

68. Yamamoto Y, Narasaki F, Futsuki Y, et al. Disopyramide-induced pneumonitis, diagnosed by lymphocyte stimulation test using bronchoalveolar lavage fluid. *Intern Med* 2001;**40**:775-8.

69. Espeleta VJ, Moore WH, Kane PB, Baram D. Eosinophilic pneumonia due to duloxetine. *Chest* 2007;**131**:901-3.

70. Hatakeyama S, Tachibana A, Morita M, Suzuki K, Okano H. [Etoposide-induced pneumonitis]. *Nihon Kyobu Shikkan Gakkai Zasshi* 1997;**35**:210-14.

71. Pesenti S, Lauque D, Daste G, et al. Diffuse infiltrative lung disease associated with flecainide. *Respiration* 2002;**69**: 182-5.

72. Lin TS, Penza SL, Avalos BR, et al. Diffuse alveolar hemorrhage following gemtuzumab ozogamicin. *Bone Marrow Transplant* 2005;**35**:823-4.

73. Tomioka R, King TE. Gold-induced pulmonary disease: clinical features, outcome, and differentiation from rheumatoid lung disease. *Am J Respir Crit Care Med* 1997;**155**:1011-20.

74. Okuda Y, Takasugi K, Imai A, et al. [A case of rheumatoid arthritis complicated with auranofin-induced acute interstitial pneumonitis]. *Ryumachi* 1991;**31**:54-61.

75. Brander PE, Tukiainen P. Acute eosinophilic pneumonia in a heroin smoker. *Eur Respir J* 1993;**6**:750-2.

76. Hashizume T, Numata H. A case of eosinophilic pneumonia possibly due to ifenprodil. *Nihon Kokyuki Gakkai Zasshi* 2001;**39**:868-70.

77. Bergeron A, Bergot E, Vilela G, et al. Hypersensitivity pneumonitis related to imatinib mesylate. *J Clin Oncol* 2002;**20**:4271-2.

78. Oishi Y, Sando Y, Tajima S, et al. Indomethacin induced bulky lymphadenopathy and eosinophilic pneumonia. *Respirology* 2001;**6**:57-60.

79. Ogata K, Koga T, Yagawa K. Interferon-related bronchiolitis obliterans organizing pneumonia. *Chest* 1994;**106**:612-13.

80. Hoffman SD, Hammadeh R, Shah N. Eosinophilic pneumonitis secondary to pegylated interferon alpha-2b and/or ribavirin. *Am J Gastroenterol* 2003;**98**:S152.

81. Kourakata H, Tanabe Y, Mikami O, Sato K, Suzuki E. [A case of interferon beta-induced pneumonia]. *Nihon Kokyuki Gakkai Zasshi* 2000;**38**:687-91.

82. Hatakeyama S, Tatibana A, Suzuki K, Okano H, Oka T. [Isoniazid-induced pneumonitis]. *Nihon Kokyuki Gakkai Zasshi* 1998;**36**:448-52.

83. Thornburg A, Abonour R, Smith P, Knox K, Twigg HL. Hypersensitivity pneumonitis-like syndrome associated with the use of lenalidomide. *Chest* 2007;**131**:1572-4.

84. Wardwell NR, Miller R, Ware LB. Pulmonary alveolar proteinosis associated with a disease-modifying antirheumatoid arthritis drug. *Respirology* 2006;**11**:663-5.

85. Fujimori K, Shimatsu Y, Suzuki E, Arakawa M, Gejyo F. [Levofloxacin-induced eosinophilic pneumonia complicated by bronchial asthma]. *Nihon Kokyuki Gakkai Zasshi* 2000;**38**:385-90.

86. Fujita K, Sakamoto O, Matsumoto M, Kohrogi H, Suga M. A case of loxoprofen sodium-induced bronchiolitis obliterans organizing pneumonia. *Nihon Kokyuki Gakkai Zasshi* 2003;**41**:878-83.

87. Ii T, Doutsu Y, Ashitani J, et al. [A case of loxoprofen-induced pulmonary eosinophilia]. *Nihon Kyobu Shikkan Gakkai Zasshi* 1992;**30**:926-9.

88. Begbie S, Burgess KR. Maloprim-induced pulmonary eosinophilia. *Chest* 1993;**103**:305-6.

89. Karakatsani A, Chroneou A, Koulouris NG, Orphanidou D, Jordanoglou J. Meloxicam-induced pulmonary infiltrates with eosinophilia: a case report. *Rheumatology* (Oxford) 2003;**42**:1112-13.

90. Liote H, Gauthier JF, Prier A, et al. [Acute, reversible, interstitial pneumopathy induced by melphalan]. *Rev Mal Respir* 1989;**6**:461-4.

91. Fuhrman C, Parrot A, Wislez M, et al. Spectrum of CD4 to CD8 T-cell ratios in lymphocytic alveolitis associated with methotrexate-induced pneumonitis. *Am J Respir Crit Care Med* 2001;**164**:1186-91.

92. Bando T, Fujimura M, Noda Y, et al. Minocycline-induced pneumonitis with bilateral hilar lymphadenopathy and pleural effusion. *Intern Med* 1994;**33**:177-9.

93. Sitbon O, Bidel N, Dussopt C, et al. Minocycline pneumonitis and eosinophilia: a report on eight patients. *Arch Intern Med* 1994;**154**:1633-40.

94. Lenci G, Muller-Quernheim J, Lorenz J, Schweden F, Ferlinz R. [Toxic lung damage caused by mitomycin C]. *Pneumologie* 1994;**48**:197-201.

95. Dan M, Aderka D, Topilsky M, Livni E, Levo Y. Hypersensitivity pneumonitis induced by nalidixic acid. *Arch Intern Med* 1986;**146**:1423-4.

96. Perng DW, Su HT, Tseng CW, Lee YC. Pulmonary infiltrates with eosinophilia induced by nimesulide in an asthmatic patient. *Respiration* 2005;**72**:651-3.

97. Pfitzenmeyer P, Foucher P, Piard F, et al. Nilutamide pneumonitis: a report on eight patients. *Thorax* 1992;**47**:622-7.

98. Akoun GM, Liote HA, Liote F, Gauthier-Rahman S, Kuntz D. Provocation test coupled with bronchoalveolar lavage in diagnosis of drug (nilutamide)-induced hypersensitivity pneumonitis. *Chest* 1990;**97**:495-8.

99. Brutinel WM, Martin WJ. Chronic nitrofurantoin reaction associated with T-lymphocyte alveolitis. *Chest* 1986;**89**: 150-2.

100. Lena H, Desrues B, Le Coz A, Quinquenel ML, Delaval P. Severe diffuse interstitial pneumonitis induced by carmustine. *Chest* 1994;**105**:1602-3.

101. Gagnadoux F, Roiron C, Carrie E, Monnier-Cholley L, Lebeau B. Eosinophilic lung disease under chemotherapy with oxaliplatin for colorectal cancer. *Am J Clin Oncol* 2002;**25**:388-90.

102. Trisolini R, Lazzari Agli L, Tassinari D, et al. Acute lung injury associated with 5-fluorouracil and oxaliplatinum combined chemotherapy. *Eur Respir J* 2001;**18**:243-5.

103. Fujimori K, Yokoyama A, Kurita Y, Uno K, Saijo N. Paclitaxel-induced cell-mediated hypersensitivity pneumonitis: diagnosis using leukocyte migration test, bronchoalveolar

lavage and transbronchial lung biopsy. *Oncology* 1998;**55**:340–4.

104. Chamberlain DW, Hyland RH, Ross DJ. Diphenylhydantoin-induced lymphocytic interstitial pneumonia. *Chest* 1986;**90**:458–60.

105. Munn NJ, Baughman RP, Ploysongsang Y, Wirman JA, Bullock WE. Bronchoalveolar lavage in acute drug-hypersensitivity pneumonitis probably caused by phenytoin. *South Med J* 1984;**77**:1594–6.

106. Fujimori K, Shimatsu Y, Suzuki E, Gejyo F, Arakawa M. [Pranoprofen-induced lung injury manifesting as acute eosinophilic pneumonia]. *Nihon Kokyuki Gakkai Zasshi* 1999;**37**:401–5.

107. Goldberg SK, Lipschutz JB, Ricketts RM, Fein AM. Procainamide-induced lupus lung disease characterized by neutrophil alveolitis. *Am J Med* 1984;**76**:146–50.

108. Bouckaert Y, Robert F, Englert Y, *et al.* Acute eosinophilic pneumonia associated with intramuscular administration of progesterone as luteal phase support after IVF: case report. *Human Reproduct* 2004;**19**:1806–10.

109. Akoun GM, Milleron BJ, Mayaud CM, Tholoniat D. Provocation test coupled with bronchoalveolar lavage in diagnosis of propranolol-induced hypersensitivity pneumonitis. *Am Rev Respir Dis* 1989;**139**:247–9.

110. Kunichika N, Miyahara N, Kotani K, *et al.* Pneumonitis induced by rifampicin. *Thorax* 2002;**57**:1000–1.

111. Arai T, Inoue Y, Hayashi S, Yamamoto S, Sakatani M. Risedronate induced BOOP complicated with sarcoidosis. *Thorax* 2005;**60**:613–14.

112. Sasaki S, Kawanami R, Motizuki Y, *et al.* [Serrapeptase-induced lung injury manifesting as acute eosiniphilic pneumonia]. *Nihon Kokyuki Gakkai Zasshi* 2000;**38**:540–4.

113. Hirahara K, Saitoh T, Terada I, *et al.* [A case of pneumonitis due to serrapeptase]. *Nihon Kyobu Shikkan Gakkai Zasshi* 1989;**27**:1231–6.

114. Lantuejoul S, Brambilla E, Brambilla C, Devouassoux G. Statin-induced fibrotic nonspecific interstitial pneumonia. *Eur Respir J* 2002;**19**:577–80.

115. Morelon E, Stern M, Israel-Biet D, *et al.* Characteristics of sirolimus-associated interstitial pneumonitis in renal transplant patients. *Transplantation* 2001;**72**:787–90.

116. Bauer C, Lidove O, Lamotte C, *et al.* Sirolimus-associated interstitial pneumonitis in a renal transplant patient. *Rev Medecine Interne* 2006;**27**:248–52.

117. Valcke Y, Pauwels R, Van der Straeten M. Bronchoalveolar lavage in acute hypersensitivity pneumonitis caused by sulfasalazine. *Chest* 1987;**92**:572–3.

118. Duran I, Siu LL, Oza AM, *et al.* Characterisation of the lung toxicity of the cell cycle inhibitor temsirolimus. *Eur J Cancer* 2006;**42**:1875–80.

119. Onozawa M, Hashino S, Sogabe S, *et al.* Thalidomide-induced interstitial pneumonitis. *J Clin Oncol* 2005;**23**:2425–6.

120. Persoz CF, Cornella F, Kaeser P, Rochat T. Ticlopidine-induced interstitial pulmonary disease: a case report. *Chest* 2001;**119**:1963–5.

121. Nakatsumi Y, Nomura M, Yasui M, *et al.* [A case of eosinophilic pneumonia due to tolfenamic acid]. *Nihon Kyobu Shikkan Gakkai Zasshi* 1993;**31**:1322–6.

122. Kimura N, Miyazaki E, Matsuno O, Abe Y, Tsuda T. [Drug-induced pneumonitis with eosinophilic infiltration due to tosufloxacin tosilate]. *Nihon Kokyuki Gakkai Zasshi* 1998;**36**:618–22.

123. Vahid B, Mehrotra A. Trastuzumab (Herceptin)-associated lung injury. *Respirology* 2006;**11**:655–8.

124. Salerno SM, Strong JS, Roth BJ, Sakata V. Eosinophilic pneumonia and respiratory failure associated with a trazodone overdose. *Am J Respir Crit Care Med* 1995;**152**:2170–2.

125. Fleisch MC, Blauer F, Gubler JG, Kuhn M, Scherer TA. Eosinophilic pneumonia and respiratory failure associated with venlafaxine treatment. *Eur Respir J* 2000;**15**:205–8.

126. Drent M, Singh S, Gorgels AP, *et al.* Drug-induced pneumonitis and heart failure simultaneously associated with venlafaxine. *Am J Respir Crit Care Med* 2003;**167**:958–61.

127. Bogaerts Y, Van Renterghem D, Vanvuchelen J, *et al.* Interstitial pneumonitis and pulmonary vasculitis in a patient taking an L-tryptophan preparation. *Eur Respir J* 1991;**4**:1033–6.

128. Read WL, Mortimer JE, Picus J. Severe interstitial pneumonitis associated with docetaxel administration. *Cancer* 2002;**94**:847–53.

129. Watanabe K, Sueishi K, Tanaka K, *et al.* Pulmonary alveolar proteinosis and disseminated atypical mycobacteriosis in a patient with busulfan lung. *Acta Pathol Japon* 1990;**40**:63–6.

130. Alexandrescu DT, Dutcher JP, O'Boyle K, *et al.* Fatal intra-hemorrhage after rituximab in a patient with non-Hodgkin lymphoma. *Lymphoma* 2004;**45**:2321–5.

131. Schoc A, Laan-Baalbergen NE, Huizniga TWJ, Breedveld FC, Laar JM. Pulmonary fibrosis in a patient with rheumatoid arthritis treated with adalimumab. *Arthritis Rheum* 2006;**55**:157–9.

132. Phillips K, Weinblatt M. Granulomatous lung disease occurring during etanercept treatment. *Arthritis Rheum* 2005;**53**:618–20.

133. Camus P. Drug induced infiltrative lung diseases. In: Schwarz MI, King TE (eds) *Interstitial Lung Disease*, 4th edn, pp 485–534. London: BC Decker, 2003.

134. Costabel U, Teschler H, Guzman J. Bronchiolitis obliterans organizing pneumonia: the cytological and immunocytological profile of bronchoalveolar lavage. *Eur Respir J* 1992;**5**:791–7.

135. Bonaccorsi A, Cancellieri A, Chilosi M, *et al.* Acute interstitial pneumonia: report of a series. *Eur Respir J* 2003;**21**:187–91.

136. Semenzato G, Poletti V. Bronchoalveolar lavage in lung cancer. *Respiration* 1992;**59**(Suppl 1):44–6.

137. Gibson PG, Bryant DH, Morgan GW, *et al.* Radiation-induced pneumonitis: a hypersensitivity pneumonitis? *Ann Intern Med* 1988;**109**:288–91.

138. Martin C, Romero S, Sanchez-Paya J, *et al.* Bilateral lymphocytic alveolitis: a common reaction after unilateral thoracic irradiation. *Eur Respir J* 1999;**13**:727–32.

139. Crestani B, Valeyre D, Roden S, *et al.* Bronchiolitis obliterans organising pneumonia syndrome primed by radiation therapy to the breast. *Am J Respir Crit Care Med* 1998;**158**:1929–35.

Pathology of drug-induced respiratory disease

WILLIAM D TRAVIS, DOUGLAS B FLIEDER

INTRODUCTION

Adverse effects of medications on the lungs are an increasing problem, given the rapidly growing number of drugs used for therapeutic and illicit purposes.[1–5] In addition to a wide variety of clinical manifestations (Box 5.1),[1] there is a broad spectrum of histologic patterns of pulmonary reactions to drugs (Box 5.2).[2,6,7] The diagnosis of pulmonary drug toxicity is complicated because it requires multidisciplinary correlation between clinical, radiological, laboratory and – when available – pathology material.[2,6,7] In the majority of pulmonary drug reactions the diagnosis is made using only clinical, laboratory and radiological features without a lung biopsy. In fact, fewer than 10 per cent of cases of drug-induced interstitial lung disease include histologic data.[1]

Most lung biopsies performed in the setting of potential drug toxicity are transbronchial rather than surgical biopsies, so it may be difficult to assess the complete extent or spectrum of the histological reaction. Another problem is that few drugs cause a single specific histological reaction. Further, in many

---

Box 5.1 Clinical syndromes/diseases associated with pulmonary drug reactions (modified from [2] and [7])

Airway dysfunction or respiratory insufficiency

- Bronchospasm or asthma
- Cough
- Irreversible airflow obstruction
- Emphysema
- Alveolar hypoventilation
- Respiratory muscle dysfunction

Alveolar processes

- Non-cardiogenic pulmonary oedema
- Alveolar haemorrhage

Diffuse lung processes

- Acute respiratory distress syndrome
- Interstitial lung disease, acute, subacute or chronic
- Hypersensitivity pneumonitis
- Eosinophilic syndromes

- Systemic lupus erythematosus-like syndromes
- Organizing pneumonia
- Metastatic calcification
- Foreign body granulomatosis

Localized or nodular infiltrates

Vascular disease

- Pulmonary arterial hypertension
- Vasculitis
- Thromboemboli
- Veno-occlusive disease
- Thrombotic microangiopathy

Pleural disease

- Pleuritis or pleural effusion
- Haemothorax
- Pneumothorax

Box 5.2   Histology patterns associated with pulmonary drug reactions (modified from [2] and [7])

Interstitial

- Diffuse alveolar damage
- Organizing pneumonia
- Usual interstitial pneumonia-like pattern
- Diffuse cellular interstitial infiltrates +/− granulomas
- Non-specific interstitial pneumonia
- Giant cell interstitial pneumonia
- Granulomatous interstitial pneumonia
- Lymphocytic interstitial pneumonia
- Metastatic calcification
- Bronchiolitis
- Bronchiectasis

Alveolar

- Pulmonary oedema
- Alveolar haemorrhage
- Alveolar proteinosis-like reaction
- Eosinophilic pneumonia, acute or chronic
- Panacinar emphysema and bullous lung disease

Vascular

- Foreign body giant cell reaction (intravenous drug abuse)
- Small vessel angiitis
- Pulmonary arterial hypertension
- Pulmonary veno-occlusive disease

Pleural

- Pleuritis
- Pleural effusion

Box 5.3   Criteria for diagnosis of drug reactions (reproduced from [8])

- Correct identification of the drug in question: Was the patient taking the drug? What dose? What duration?
- Exclusion of other primary or secondary lung diseases
- Temporal eligibility: appropriate latent period (exposure – toxicity)
- Remission of symptoms with removal of challenge; recurrence with re-challenge
- Singularity of drug: what other drugs was the patient taking?
- Characteristic pattern of reaction to specific drug – previous documentation?
- Quantification of drug levels that confirm abnormal levels (especially for overdoses)
- Degree of certainty of drug reaction:
  - causative
  - probable
  - possible

This chapter reviews the spectrum of pathological manifestations of pulmonary drug toxicities, and the role of lung biopsy interpretation in the context of multidisciplinary correlation.

DIAGNOSTIC CRITERIA FOR DRUG REACTIONS

In 1976, Irey proposed a stringent set of diagnostic criteria for drug reactions (Box 5.3).[8] Documenting that the patient was actually taking the drug under consideration is essential. The treatment regimen including the dose and duration of therapy is important, because some drugs are known to cause toxicity only above a specific threshold dose or cumulative dose. Timing of onset of pulmonary symptoms relative to taking the drug is valuable information because certain drugs have well recognized latency periods. Resolution of symptoms after cessation of the medication supports drug toxicity. One of the strongest pieces of evidence confirming drug toxicity is development of the lung toxicity after a re-challenge of the patient with the same drug following resolution of symptoms after initial withdrawal. However, seldom do the clinical circumstances permit this to occur.

One of the greatest challenges in determining drug toxicity is that most patients are taking multiple medications. If the patient is on a single drug, that allows for the most definitive documentation of drug toxicity. In some cases, one of the multiple drugs has such a characteristic pattern of lung injury, in contrast to the other medications, it is possible to implicate a single drug in the toxicity. Specific patterns of lung injury tend to be associated with certain medications, so observing one of these patterns in a patient on a specific drug can be very helpful in determining drug toxicity and occurrence of an uncharacteristic pattern of lung injury is against it. Serum or urine drug levels, particularly for drug overdoses, may be helpful.

A common problem in the lung is that underlying disease such as collagen vascular disease can cause the same spectrum

reports in the literature where pathology was available the histological descriptions are limited or not reported according to current concepts of classification. In addition, few patients are treated with a single drug, so it is often problematic to sort out which drug may be responsible for the pulmonary toxicity. Issues of drug dosage, timing of therapy related to onset of pulmonary symptoms and latency period are details that are in some cases essential to determine drug toxicity, but frequently they are elusive or impossible to track down. A proper drug history is often not obtained or provided to the pathologist.

If a clinician withdraws drug therapy but also gives steroids, improvement may not necessarily be linked to the drug withdrawal, but could be due to the steroid treatment. Finally, underlying disease or alternative exposures such as inhalation of antigens from birds may cause similar reactions to those caused by drug toxicity, further complicating the decision whether a drug is the cause of pulmonary symptoms. Therefore if a patient with a specific lung biopsy finding is taking a drug known to cause that specific reaction, that does not necessarily mean that the drug is the cause of the lung disease.

of patterns of lung injury that can be caused by the drugs the patients are taking for their underlying disease.

After all the available information regarding each of these issues is analysed, the potential for drug toxicity is categorized into (1) causative, (2) probable, or (3) possible.[8]

PATHOGENESIS OF DRUG-INDUCED PULMONARY DISEASE

Drugs can cause lung injury through multiple mechanisms, including:

- the oxidant/antioxidant system (i.e. nitrofurantoin);
- the immune system (i.e. methotrexate, bleomycin);[9,10]
- matrix repair (i.e. penicillamine, gold salts);
- proteases/antiproteases (i.e. colchicine);
- interfering with lipid metabolism (i.e. amiodarone);
- the central nervous system (i.e. opiates, tranquillizers).[1,11,12]

Not every drug affects the lung with a single mechanism of action. Methotrexate, for example, can present with multiple pulmonary manifestations including diffuse alveolar damage, hypersensitivity pneumonitis, eosinophilic pneumonia, and cellular interstitial pneumonia.[13,14] Drug toxicity can result from direct or indirect injury to the lung. Direct toxicity can be through either toxic or idiosyncratic mechanisms. Indirect effects may result from suppressive effects on the central nervous system, immune system or haematopoiesis. Pulmonary complications related to these systems include secondary effects such as aspiration pneumonia, infection, haemorrhage due to thrombocytopenia, or thrombosis due to a hypercoagulable state.

METHODS OF DIAGNOSIS

Pathological assessment of the lungs for the diagnosis of drug toxicity can be approached by surgical lung biopsy, bronchoscopic biopsy or bronchoalveolar lavage (BAL). In fatal cases, autopsy material may also be informative. It is easier to determine the histological pattern of the lung reaction in larger samples of lung tissue such as surgical lung biopsies. However, regardless of the size of the specimen, obtaining a diagnostic lesion is the key and this can happen in small specimens such as bronchoscopic biopsies.

BAL plays more of a role in excluding infection or malignancy than in diagnosing a specific pattern of lung injury. However, it can be useful in documenting eosinophilia, lymphocytosis, haemorrhage, and cytologically atypical pneumocytes, each of which could suggest a reaction compatible with drug toxicity.[6] Lipid-rich macrophages may suggest lipoid pneumonia in a patient taking oil-based laxatives or nose drops. Gold may be retained in macrophages, and foamy macrophages may indicate the effects of amiodarone, but these are markers of treatment and not necessarily toxicity.[15,16]

Transbronchial lung biopsy (TBB) can be helpful in diagnosing a limited number of conditions including malignancy, infection and certain diffuse lung diseases. Therefore one of the more common roles of TBB is to exclude malignancy or infection. It is helpful if diagnostic lesions are sampled of

entities such as diffuse alveolar damage, eosinophilic pneumonia, cellular and/or granulomatous interstitial pneumonia, organizing pneumonia, lipoid pneumonia and sarcoidosis. Surgical lung biopsies are most helpful in diagnosing drug toxicity because they give the greatest opportunity to evaluate the pathological pattern of lung injury. It is most helpful for biopsies to be obtained from multiple lobes of the lung and for the thoracic surgeon to choose the site of the biopsy based on high-resolution CT findings. With small specimens, it is particularly important to correlate the pathology findings with the clinical and high-resolution CT findings in order to be sure that all the results fit together and that there is not a sampling problem; however this is also true with surgical lung biopsies.

CLINICAL, RADIOLOGICAL AND PATHOLOGICAL CORRELATION NEEDED TO ESTABLISH DRUG TOXICITY

Pathologists are in the best position to help the clinician if sufficient clinical and radiological information is provided to aid in interpreting lung biopsies. The final diagnosis of pulmonary drug toxicity is based not on pathology alone, but on correlation of histological findings with clinical, laboratory and radiological data. Accurate classification of the appropriate histological pattern is critical. Clinical data may be important – such as peripheral eosinophilia for the diagnosis of eosinophilic pneumonia and acute respiratory failure with requirement of mechanical ventilation for the diagnosis of diffuse alveolar damage. Haemoptysis would be consistent with alveolar haemorrhage. Rapid onset of symptoms would fit with an acute reaction such as diffuse alveolar damage, pulmonary oedema or bronchospasm. Gradual onset would fit more with reactions such as chronic interstitial pneumonias or organizing pneumonia. With some drugs such as BCNU, the pulmonary symptoms may occur years after discontinuation of the drug.[17,18]

Correlation with a high-resolution CT scan (HRCT) is also important in determining the overall diagnosis. This is essential because sometimes the pathology findings can be trumped by CT findings. For example, if a biopsy shows only a cellular interstitial pneumonia pattern but a CT scan shows consolidation with minimal or no ground-glass opacities, it is likely that there is a sampling problem and the final diagnosis is more likely to represent organizing pneumonia. Another example would be a biopsy that shows a non-specific interstitial pneumonia (NSIP) pattern while the CT shows bilateral, basilar honeycombing more consistent with usual interstitial pneumonia (UIP). In such a case the clinical–radiological–pathological diagnosis is more likely to be UIP with a sampling problem rather than NSIP. Sometimes localized findings on CT may raise questions about histological findings that are more suggestive of diffuse parenchymal disease. Clinical, radiological and pathological correlation may also be helpful in excluding an opportunistic infection using a combination of culture, serological, radiological and histological data.

In lung biopsy material, the presence of necrosis, necrotizing granulomas, acute pneumonia or infarction may raise concern for infection. In such cases, organisms can be pursued using a variety of histochemical or immunohistochemical special stains. These approaches are discussed in detail elsewhere.[19]

Once the overall clinical–radiological–pathological diagnosis is determined, one must examine the factors related to the drug(s) in question – such as the dose, timing and duration of therapy relative to onset of pulmonary symptoms, history of underlying disease or prior therapy, and whether the lung reaction is appropriate for the drug in question.[7,8,11,12,20,21]

HISTOLOGICAL PATTERNS OF DRUG TOXICITY

Drugs can have a wide variety of patterns of toxicity in the lung (see Box 5.2). Some drugs may cause a single pattern, but sometimes the patterns are mixed or not exactly the same in appearance, as seen in an idiopathic setting. As a result pathologists often find it challenging to interpret lung biopsies in the setting of drug toxicity. Determination of the correct histological pattern(s) of lung injury is the primary role of the pathologist.

After the lung reaction pattern has been classified accurately, one can begin to sort out the differential diagnosis of possible causes. This is problematic, because few reactions are specific and can be associated with a variety of causes. Examples include the histological pattern of UIP which could be seen in collagen vascular disease, pneumoconiosis, chronic hypersensitivity pneumonitis, as a familial condition, as an idiopathic condition (known as idiopathic pulmonary fibrosis), as well as drug toxicity.[19,22] In some settings the absence of an histology finding such as foamy macrophages in a patient receiving amiodarone may influence a decision whether lung injury is due to drug toxicity.

Part of the clinical correlation process includes identification of the potential offending drug and knowledge of the established patterns of lung toxicity. This requires careful evaluation of published literature on the subject. Comprehending the literature presents its own set of difficulties since for many drugs detailed pathology descriptions are not available or they may be published using old terminology or concepts. The pulmonary drug toxicity website Pneumotox® (www.pneumotox.com) may be a helpful reference. While some forms of drug toxicity are well recognized, one should be alert for new emerging patterns of drug toxicity.

The remainder of this chapter will focus on the major histological patterns of drug toxicity that are likely to be seen in surgical lung biopsies. These patterns can be divided into interstitial, alveolar, vascular and pleural categories. One could debate the distinction between interstitial and alveolar, since there is overlap between these patterns. This is one reason why the American Thoracic Society/European Respiratory Society Classification of Idiopathic Interstitial Pneumonias uses the term 'diffuse parenchymal lung disease' as a general term to describe these disorders.[22]

INTERSTITIAL LESIONS

Diffuse alveolar damage

The histological hallmark of diffuse alveolar damage (DAD) is hyaline membranes in the acute phase (Fig. 5.1).[6,7] As the process progresses into the organizing phase, there is proliferation

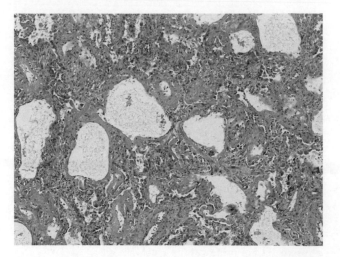

Fig. 5.1   Diffuse alveolar damage associated with taxol therapy. The lung shows diffuse interstitial thickening with hyaline membranes lining the alveolar walls. The alveolar wall thickening is caused by loose connective tissue proliferation and mild chronic inflammation.

of Type II pneumocytes and loose connective tissue in the alveolar septa. DAD can be associated with a wide variety of conditions including infection, collagen vascular disease, trauma, shock and uraemia, and it may be idiopathic. For this reason, careful clinical correlation needs to be made to exclude these causes before suggesting a diagnosis of drug toxicity. The pathologist can look for histological clues to infection such as foci of necrosis or acute pneumonia, viral cytopathic effects or organisms using special stains. Infection can also be excluded using stains for fungi (such as Gomori methenamine silver) or mycobacteria (such as Ziehl–Neelsen). If there are accumulations of neutrophils suggesting acute pneumonia or microabscesses, Gram stains (Brown & Brenn or Brown & Hopps) may be needed to exclude bacterial infection.

Approximately 10 per cent of cases of DAD may be caused by drug toxicity.[23] The list of drugs that cause DAD is very long and it includes various chemotherapeutic agents such as BCNU, bleomycin, methotrexate and cyclophosamide, in addition to amiodarone, amitriptyline, gold salts, nitrofurantoin, paraquat and propofol.[1,6,7,24,25] Recently gefitinib has been recognized to cause DAD.[26]

Organizing pneumonia

This pattern consists of patchy areas of consolidation consisting of intraluminal plugs of loose connective tissue within distal airways including alveoli, alveolar ducts and bronchioles (Fig. 5.2). The architecture of the lung is preserved, interstitial chronic inflammation is usually mild, and dense scarring fibrosis is absent. This pattern can be seen in a wide variety of settings including infection, collagen vascular disease, eosinophilic pneumonia, vasculitis, as a secondary reaction adjacent to other mass lesions, and as an idiopathic condition known as cryptogenic organizing pneumonia.[19,22] The latter condition is also known as idiopathic bronchiolitis obliterans organizing pneumonia (BOOP). The term 'COP' would not be used in the setting of a known drug reaction; in this situation the term BOOP

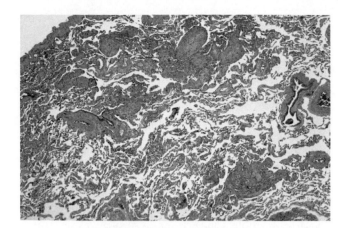

Fig. 5.2 Organizing pneumonia pattern associated with bleomycin therapy. The lung shows patchy nodular foci of loose connective tissue proliferating within alveolar spaces and alveolar ducts.

could be used clinically with mention of the specific drug association (e.g. 'BOOP due to amiodarone toxicity'). Various drugs are known to cause this reaction, including amiodarone, bleomycin, interferons, nitrofurantoin, radiation and statins.[1,11,12,20] With bleomycin, the organizing pneumonia can present radiologically with discrete nodules.

Chronic interstitial pneumonia

This category encompasses a broad spectrum of histological findings including UIP, NSIP, or otherwise unclassified interstitial pneumonias with a cellular and fibrosing pattern that may be difficult to classify. Sometimes there may be a mixed pattern of lung injury with additional lesions such as organizing pneumonia (Fig. 5.3).

These are the terms currently utilized for classification of idiopathic interstitial pneumonias; but as they have been formally recognized only recently,[19,22] much of the older published literature uses terminology that is less precise and it is often difficult to be certain of the pathology pattern described.

The UIP pattern consists of patchy dense scarring often showing honeycombing and fibroblastic foci at the edges of the scars. The lower lobes and periphery of the lung are more severely affected. NSIP shows more uniform lung involvement by a spectrum of interstitial cellular inflammation and/or fibrosis. In this condition honeycombing is characteristically absent, but small amounts of organizing pneumonia may be present.

There are long lists of drugs that can cause these patterns, but the most common are chemotherapeutic agents (BCNU, cyclophosphamide, procarbazine), antiarrhythmics (amiodarone), antihypertensives (quinidine), anti-inflammatories (sulfasalazine), anticonvulsants (phenytoin), and ergots. Prolonged inflammatory drug reactions may ultimately progress to fibrosing patterns.

Granulomatous inflammation

Granulomatous interstitial lesions may take a variety of forms, including non-caseating granulomas in a lymphatic

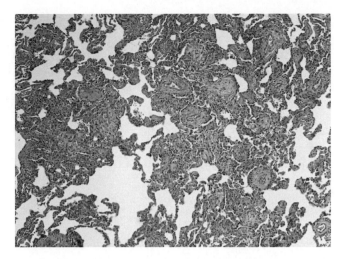

Fig. 5.3 Cellular interstitial and organizing pneumonia with rituximab therapy. The alveolar walls are thickened by a mild to moderate chronic interstitial inflammatory infiltrate. In addition there are small scattered foci of organizing pneumonia within alveolar spaces.

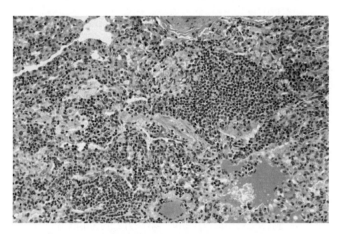

Fig. 5.4 Cellular interstitial pneumonia with poorly formed granulomas in a patient with rheumatoid arthritis who was on methotrexate therapy. The lung shows a cellular interstitial pneumonia with numerous lymphocytes and plasma cells. In addition several small non-caseating granulomas are present (top centre and right).

distribution with a sarcoid-like pattern, nodular confluent non-caseating granulomas, or scattered poorly formed granulomas within a cellular and/or fibrosing interstitial pneumonia (Fig. 5.4). The presence of granulomas may raise the differential diagnosis of infection (usually mycobacterial or fungal) or hypersensitivity pneumonitis. The presence of necrosis usually favours infection, but granulomas with or without necrosis can be seen with topical instillation in the bladder with BCG.

Drugs known to cause granulomatous reactions in the lung that may or may not be associated with interstitial inflammation and a bronchiolocentric distribution include acebutolol, cocaine, etanercept, fluoxetine hydrochloride, methotrexate, nitrofurantoin, procarbazine, sirolimus and mineral oil.[6,7,11,12,20]

Lymphocytic interstitial pneumonia

LIP consists of a severe interstitial lymphocytic inflammatory infiltrate within alveolar septa. Rarely it has been reported in the setting of drug toxicity associated with dilantin therapy.[27]

ALVEOLAR LESIONS

Pulmonary eosinophilia

Eosinophilic pneumonia is a common form of drug toxicity, occurring in association with a wide variety of drugs including acetaminophen, amiodarone, ampicillin, bleomycin, captopril, carbamazepine, clindamycin, clozapine, dapsone, erythromycin, hydrochlorothiazide, ibuprofen, isoniazid, mesalamine, methotrexate, minocycline, naproxen, nitrofurantoin, nonsteroidal anti-inflammatory agents, paclitaxel, penicillin, phenytoin, procarbazine, propranolol, propylthiouracil, rifampin, simvastin, streptomycin, sulfasalazine, tetracycline, venlafaxine and zafirlukast.[20]

Eosinophilic pneumonia has several distinct clinical presentations, including simple, acute, chronic and secondary. If the biopsy findings are not definitive but raise the consideration of eosinophilic pneumonia, documentation of peripheral eosinophilia may be helpful in supporting the diagnosis, although in some situations (e.g. acute eosinophilic pneumonia) peripheral eosinophilia is typically absent. Since tissue eosinophils are very sensitive to steroids and may vanish from lung tissues rapidly after initiation of therapy, if the patient has received steroids the pathologist should be informed as they are interpreting the lung biopsy findings.

Histologically, eosinophilic pneumonia consists of accumulation of eosinophils within alveolar spaces (Fig. 5.5). Prominent alveolar macrophages and/or organizing pneumonia may develop as the eosinophilic pneumonia resolves. Interstitial chronic inflammation and/or thickening, allergic granulomas and secondary vasculitis may also occur. Acute cases may also show hyaline membranes with the pattern of diffuse alveolar damage.

Pulmonary oedema

Non-cardiogenic pulmonary oedema is usually diagnosed based on clinical criteria rather than by pathology as these patients seldom come to lung biopsy. The onset of pulmonary oedema after administration of a new drug may suggest this diagnosis. It can be difficult to diagnose pulmonary oedema on biopsy because it requires exclusion of other histological lesions and it raises the consideration of acute fibrinous pneumonia or very early DAD (Fig. 5.6). Recognized drugs responsible for NCPE include aspirin, adrenaline, amphotericin B, carbamazepine, colchicine, cytosine arabinoside, docetaxel, etoposide, fluoresceine, heroin, hydrochlorothiazide, ibuprofen, interleukin-2, lidocaine, methotrexate, naloxone, paraquat, plasma (fresh frozen), radiocontrast material, sulfonamides, terbutaline and vinblastine.[20]

Pulmonary haemorrhage

Diffuse pulmonary haemorrhage is a potential complication of drug toxicity. The drugs known to cause this include abciximab, aspirin, amiodarone, azathioprine, carbamazepine, cocaine, ciclosporin, docetaxel, fibrinolytics, heparin, hydralazine, mitomycin C, nitrofurantoin, oral anticoagulants, paraquat, penicillamine, phenytoin, platelet receptor inhibitors such as abciximab and clopidogrel, retinoic acid, sirolimus, streptokinase and urokinase.[20,28–30]

The histology of acute and/or chronic pulmonary haemorrhage consists of intra-alveolar accumulation of red blood cells and/or hemosiderin-laden macrophages. Since fresh haemorrhaging is a common artifact of a biopsy procedure, one has to be sure the haemorrhage is pathological. Clues to this include clinical findings compatible with haemorrhage, such as haemoptysis and diffuse alveolar infiltrates on a CT scan. The presence of hemosiderin-laden macrophages also supports the existence of an ongoing haemorrhage process.

Other histological findings that are compatible with a haemorrhage process include neutrophilic capillaritis or other acute

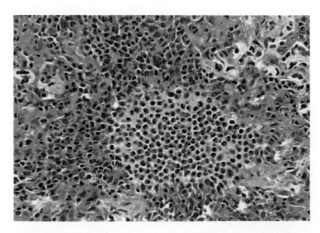

Fig. 5.5 Eosinophilic pneumonia in a patient with colitis treated with mesalamine. The interstitium and alveolar space are filled with numerous eosinophils.

Fig. 5.6 Alveolar haemorrhage and pulmonary oedema in a patient one day after exposure to paraquat. The alveolar spaces are filled with fresh red blood from acute haemorrhage and fluid from pulmonary oedema.

reactive changes such as prominent pneumocyte hyperplasia with focal disruption of alveolar walls.

To conclude that a pulmonary haemorrhage problem is caused by drug toxicity requires exclusion of other causes of pulmonary haemorrhage syndromes such as vasculitis (especially Wegener's granulomatosis or microscopic polyangiitis) and collagen vascular disease (especially systemic lupus erythematosus).

Alveolar proteinosis

Alveolar proteinosis consists of intra-alveolar accumulation of proteinaceous exudates. These are typically accompanied by dense eosinophilic clumps, cholesterol clefts and holes that have sharp round borders. This reaction has been described best in the setting of leukaemia or myelodysplastic syndromes in association with busulfan and imatinib.[31,32] It has also recently been described with sirolimus.[33]

Aspiration

Aspiration may be an indirect complication of drug toxicity if patients are impaired and thus predisposed to aspiration. Exogenous lipoid pneumonia is one of the most common forms of aspiration, resulting in either nodular masses or diffuse infiltrates that histologically are characterized by varying sized vacuoles associated with a spectrum of inflammation, histiocytes, giant cells and fibrosis. BAL can be useful in the diagnosis of lipoid pneumonia by identification of lipid-laden macrophages or identification of specific oils such as squalene using gas chromatography with mass spectrometry.[23,34,35] However, more rarely substances such as barium or kayexalate (sodium polystyrene sulfonate) may be aspirated. Kayexalate is an enterally administered cation-exchange resin used for treatment of hyperkalaemia. The latter appears as basophilic, irregular, sharply angulated, fragmented particles with little tissue reaction.[19,36]

VASCULAR LESIONS

Vasculitis

Vasculitis is a rare complication of drug toxicity.[37–39] Medium-sized vessel vasculitis seen in Wegener's granulomatosis and Churg–Strauss syndrome is not usually associated with drug reactions, although one recent case points to a WG-like syndrome developing in an ulcerative colitis patient who was taking mesalazine.[40] Churg–Strauss syndrome has developed also in patients treated with leukotriene antagonists.[41,42]

Pulmonary hypertension

A few drugs are known to cause pulmonary hypertension, including mitomycin, aminorex, fenfluramine (Fig. 5.7) combined with phenteramine (fenphen) and methamphetamine.[43–45]

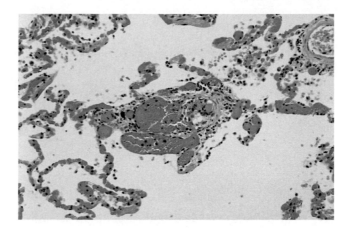

Fig. 5.7 Pulmonary hypertension in a patient who died as a complication of fenfluramine toxicity. The parent artery shows a dilatation lesion with a cluster of very thin-walled, dilate branches.

Histological changes consist of medial thickening, intimal proliferation and, in severe cases, plexiform lesions.

Pulmonary veno-occlusive disease

Rare cases of PVOD are reported in the setting of drug toxicity.[46,47] On biopsy the veins show narrowing and occlusion by vascular medial fibrous thickening. The drugs reported to cause PVOD include BCNU, bleomycin, cyclophosphamide, etoposide, gemcitabine and mitomycin.

Vascular thrombosis

Vascular thrombosis associated with drug toxicity can have several manifestations. Conventional pulmonary thromboembolism can occur in association with oral contraceptives or antipsychotic drugs.[48] In addition, a thrombotic microangiopathic syndrome resembling haemolytic uraemic syndrome with renal insufficiency occurs most commonly with mitomycin C.[49]

Embolism of foreign material

A variety of foreign substances can reach the lungs through intravascular spread, through iatrogenic mechanisms, intravenous catheters or drug abuse. These include crospovidone, talc, acrylate and silicone.[50–52]

Pleural disease

Drug-induced pleural disease can cause pleural effusion, acute pleuritis, chronic fibrous pleuritis, eosinophilic pleuritis and pneumothorax.[53,54] Drugs known to cause pleural disease include amiodarone, minoxidil, methysergide, bromocriptine, bleomycin, mitomycin, procarbazine, methotrexate and cyclophosphamide. Pleural thickening is also described in association with ergoline drugs. Pleuritis can also occur in the setting of drug-induced lupus.[55] More than 80 drugs have been documented to be associated with a drug-induced lupus-like syndrome or with exacerbation of underlying systemic lupus erythematosus (SLE). The most common drugs to cause this are procainamide, hydralazine, chlorpromazine and quinidine.[55] Pneumothorax may also result from drug toxicity; it has been reported in a few patients following chemotherapy with bleomycin, BCNU as well as radiation and amiodarone use.[56–58]

KEY POINTS

- Pathologists play an important role in the diagnosis of pleuropulmonary drug toxicity.

- Because biopsies of the lung or pleura are performed only in the most difficult cases, accurate histological diagnosis is essential. Sampling errors are common because the tissue obtained may not represent the diagnostic pathology seen on imaging.

- For pathologists to make the best diagnosis, it is important to provide them with sufficient clinical information, including drug(s) administered, with dose, duration of therapy and timing of onset of symptoms relative to taking the drug.

- Most histological reactions to drugs are non-specific and can be caused by underlying conditions, so exclusion of other causes is needed along with knowledge of prior patterns of drug therapy to implicate a specific drug as a cause of a toxic reaction.

- With many new and novel therapeutic agents, pathologists and clinicians alike need to be alert for possible new emerging forms of drug toxicity.

REFERENCES

1. Camus P, Bonniaud P, Fanton A, et al. Drug-induced and iatrogenic infiltrative lung disease. Clin Chest Med 2004;25:479–519.
2. Flieder DB, Travis WD. Pathologic characteristics of drug-induced lung disease. Clin Chest Med 2004;25:37–45.
3. Muller NL, White DA, Jiang H, Gemma A. Diagnosis and management of drug-associated interstitial lung disease. Br J Cancer 2004;91(Suppl 2):S24–30.
4. Cleverley JR, Screaton NJ, Hiorns MP, et al. Drug-induced lung disease: high-resolution CT and histological findings. Clin Radiol 2002;57:292–9.
5. Erasmus JJ, McAdams HP, Rossi SE. Drug-induced lung injury. Semin Roentgenol 2002;37:72–81.
6. Myers JL, El-Zammar O. Pathology of drug-induced lung disease. In: Katzenstein A (ed.). Surgical Pathology of Non-neoplastic Lung Disease, 4th edn, pp 85–125. China: Saunders Elsevier, 2006.
7. Travis WD, Colby TV, Koss MN, et al. Drug and radiation reactions. In: King DW (ed.) Nonneoplastic Disorders of the Lower Respiratory Tract, pp 321–50. Washington, DC: American Registry of Pathology, 2002.
8. Irey NS. Teaching Monograph: Tissue reactions to drugs. Am J Pathol 1976;82:613–47.
9. Holoye PY, Luna MA, Mackay B, Bedrossian CW. Bleomycin hypersensitivity pneumonitis. Ann Intern Med 1978;88:47–9.
10. Sostman HD, Matthay RA, Putman CE, Smith GJ. Methotrexate-induced pneumonitis. Medicine (Baltimore) 1976;55:371–88.
11. Cooper JAD, White DA, Matthay RA. Drug-induced pulmonary disease. 1: Cytotoxic drugs. Am Rev Respir Dis 1986;133:321–40.
12. Cooper JAD, White DA, Matthay RA. Drug-induced pulmonary disease. 2: Noncytotoxic drugs. Am Rev Respir Dis 1986;133:488–505.
13. Lateef O, Shakoor N, Balk RA. Methotrexate pulmonary toxicity. Expert Opin Drug Safety 2005;4:723–30.
14. Imokawa S, Colby TV, Leslie KO, Helmers RA. Methotrexate pneumonitis: review of the literature and histopathological findings in nine patients. Eur Respir J 2000;15:373–81.
15. Akoun GM, Cadranel JL, Blanchette G, et al. Bronchoalveolar lavage cell data in amiodarone-associated pneumonitis: evaluation in 22 patients. Chest 1991;99:1177–82.
16. Garcia JG, Munim A, Nugent KM, et al. Alveolar macrophage gold retention in rheumatoid arthritis. J Rheumatol 1987;14:435–8.
17. Brandes AA, Tosoni A, Amista P, et al. How effective is BCNU in recurrent glioblastoma in the modern era? A phase II trial. Neurology 2004;63:1281–4.
18. O'Driscoll BR, Hasleton PS, Taylor PM, et al. Active lung fibrosis up to 17 years after chemotherapy with carmustine (BCNU) in childhood. New Engl J Med 1990;323:378–82.
19. Travis WD, Colby TV, Koss MN, et al. Non-neoplastic Disorders of the Lower Respiratory Tract. Washington, DC: American Registry of Pathology, 2002.
20. Camus P, Fanton A, Bonniaud P, et al. Interstitial lung disease induced by drugs and radiation. Respiration 2004;71:301–26.
21. Camus PH, Foucher P, Bonniaud PH, Ask K. Drug-induced infiltrative lung disease. Eur Respir J Suppl 2001;32:93s–100s.
22. Travis WD, King TE, Bateman ED, et al. ATS/ERS International Multidisciplinary Consensus Classification of Idiopathic Interstitial Pneumonia. Am J Respir Crit Care Med 2002;165:277–304.
23. Parambil JG, Myers JL, Aubry MC, Ryu JH. Causes and prognosis of diffuse alveolar damage diagnosed on surgical lung biopsy. Chest 2007;132:50–7.
24. Chondrogiannis KD, Siontis GC, Koulouras VP, et al. Acute lung injury probably associated with infusion of propofol emulsion. Anaesthesia 2007;62:835–7.
25. Kam PC, Cardone D. Propofol infusion syndrome. Anaesthesia 2007;62:690–701.
26. Camus P, Kudoh S, Ebina M. Interstitial lung disease associated with drug therapy. Br J Cancer 2004;91(Suppl 2):S18–23.
27. Chamberlain DW, Hyland RH, Ross DJ. Diphenylhydantoin-induced lymphocytic interstitial pneumonia. Chest 1986;90:458–60.
28. Schwarz MI, Fontenot AP. Drug-induced diffuse alveolar hemorrhage syndromes and vasculitis. Clin Chest Med 2004;25:133–40.

29. Conley M, Patino G, Romick B, *et al.* Abciximab-induced alveolar hemorrhage after percutaneous coronary intervention. *Can J Cardiol* 2008;**24**:149-51.

30. Kilaru PK, Schweiger MJ, Kozman HA, Weil TR. Diffuse alveolar hemorrhage after clopidogrel use. *J Invasive Cardiol* 2001;**13**:535-7.

31. Wagner U, Staats P, Moll R, *et al.* Imatinib-associated pulmonary alveolar proteinosis. *Am J Med* 2003;**115**:674.

32. Aymard JP, Gyger M, Lavallee R, *et al.* A case of pulmonary alveolar proteinosis complicating chronic myelogenous leukemia: a peculiar pathologic aspect of busulfan lung? *Cancer* 1984;**53**:954-6.

33. Pedroso SL, Martins LS, Sousa S, *et al.* Pulmonary alveolar proteinosis: a rare pulmonary toxicity of sirolimus. *Transpl Int* 2007;**20**:291-6.

34. Kanaji N, Bandoh S, Takano K, *et al.* Positron emission tomography: positive squalene-induced lipoid pneumonia confirmed by gas chromatography–mass spectrometry of bronchoalveolar lavage fluid. *Am J Med Sci* 2008;**335**: 310-14.

35. Spatafora M, Bellia V, Ferrara G, Genova G. Diagnosis of a case of lipoid pneumonia by bronchoalveolar lavage. *Respiration* 1987;**52**:154-6.

36. Fenton JJ, Johnson FB, Przygodzk RM, *et al.* Sodium polystyrene sulfonate (kayexalate) aspiration: histologic appearance and infrared microspectrophotometric analysis of two cases. *Arch Pathol Lab Med* 1996;**120**:967-9.

37. Imai H, Nakamoto Y, Hirokawa M, *et al.* Carbamazepine-induced granulomatous necrotizing angiitis with acute renal failure. *Nephron* 1989;**51**:405-8.

38. Somogyi A, Muzes G, Molnar J, Tulassay Z. Drug-related Churg–Strauss syndrome? *Adverse Drug React Toxicol Rev* 1998;**17**:63-74.

39. Voorburg AM, van Beek FT, Slee PH, *et al.* Vasculitis due to gemcitabine. *Lung Cancer* 2002;**36**:203-5.

40. Yano S, Kobayashi K, Kato K, Nishimura K. A limited form of Wegener's granulomatosis with bronchiolitis obliterans organizing pneumonitis-like variant in an ulcerative colitis patient. *Intern Med* 2002;**41**:1013-15.

41. Keogh KA. Leukotriene receptor antagonists and Churg–Strauss syndrome: cause, trigger or merely an association? *Drug Safety* 2007;**30**:837-43.

42. Shimbo J, Onodera O, Tanaka K, Tsuji S. Churg–Strauss syndrome and the leukotriene receptor antagonist pranlukast. *Clin Rheumatol* 2005;**24**:661-2.

43. Silvani P, Camporesi A. Drug-induced pulmonary hypertension in newborns: a review. *Curr Vasc Pharmacol* 2007;**5**:129-33.

44. Chin KM, Channick RN, Rubin LJ. Is methamphetamine use associated with idiopathic pulmonary arterial hypertension? *Chest* 2006;**130**:1657-63.

45. Walker AM, Langleben D, Korelitz JJ, *et al.* Temporal trends and drug exposures in pulmonary hypertension: an American experience. *Am Heart J* 2006;**152**:521-6.

46. Gagnadoux F, Capron F, Lebeau B. Pulmonary veno-occlusive disease after neoadjuvant mitomycin chemotherapy and surgery for lung carcinoma. *Lung Cancer* 2002;**36**:213-15.

47. Vansteenkiste JF, Bomans P, Verbeken EK, *et al.* Fatal pulmonary veno-occlusive disease possibly related to gemcitabine. *Lung Cancer* 2001;**31**:83-5.

48. Hamanaka S, Kamijo Y, Nagai T, *et al.* Massive pulmonary thromboembolism demonstrated at necropsy in Japanese psychiatric patients treated with neuroleptics including atypical antipsychotics. *Circ J* 2004;**68**:850-2.

49. Gordon LI, Kwaan HC. Thrombotic microangiopathy manifesting as thrombotic thrombocytopenic purpura/hemolytic uremic syndrome in the cancer patient. *Semin Thromb Hemost* 1999;**25**:217-21.

50. Ganesan S, Felo J, Saldana M, *et al.* Embolized crospovidone (poly[*N*-vinyl-2-pyrrolidone]) in the lungs of intravenous drug users. *Mod Pathol* 2003;**16**:286-92.

51. Pelz DM, Lownie SP, Fox AJ, Hutton LC. Symptomatic pulmonary complications from liquid acrylate embolization of brain arteriovenous malformations. *Am J Neuroradiol* 1995;**16**:19-26.

52. Schmid A, Tzur A, Leshko L, Krieger BP. Silicone embolism syndrome: a case report, review of the literature, and comparison with fat embolism syndrome. *Chest* 2005;**127**: 2276-81.

53. Huggins JT, Sahn SA. Drug-induced pleural disease. *Clin Chest Med* 2004;**25**:141-53.

54. Lee CH, Park KU, Nah DY, Won KS. Bilateral spontaneous pneumothorax during cytotoxic chemotherapy for angiosarcoma of the scalp: a case report. *J Korean Med Sci* 2003;**18**:277-80.

55. Huggins JT, Sahn SA. Causes and management of pleural fibrosis. *Respirology* 2004;**9**:441-7.

56. Doll DC. Fatal pneumothorax associated with bleomycin-induced pulmonary fibrosis. *Cancer Chemother Pharmacol* 1986;**17**:294-5.

57. Durant JR, Norgard MJ, Murad TM, *et al.* Pulmonary toxicity associated with bischloroethylnitrosourea (BCNU). *Ann Intern Med* 1979;**90**:191-4.

58. Hsu JR, Chang SC, Perng RP. Pneumothorax following cytotoxic chemotherapy in malignant lymphoma. *Chest* 1990; **98**:1512-13.

6

Drug allergy in lung disease

AMITAVA GANGULI, MUNIR PIRMOHAMED

INTRODUCTION

Drugs are used for their benefits in treating or curing disease. Although there are inherent risks with using all drugs,[1] in most cases the harm that ensues from drug therapy is mild and brief. An area of concern with many drugs is their propensity to cause allergic reactions, which manifest in many different ways, and can vary in severity. Fortunately, for reasons that are not clear, these allergic reactions affect only a minority of patients, so their perceived harm/benefit balance is still considered to be positive. In this chapter, we provide a general overview of allergic drug reactions, focusing where necessary on those affecting the lung.

DEFINITIONS

An *adverse reaction* has been defined as any noxious or unintended reaction to a drug that is administered in a standard dose, by the proper route, for the purpose of prophylaxis, diagnosis or treatment.[2] Most adverse reactions – between 80 and 90 per cent – are of this 'type A' category; that is, they are an augmentation of the known pharmacology of the drug, show a good dose–response relationship, and can be alleviated by dose reduction or drug withdrawal. *Idiosyncratic* adverse reactions, which also include the immune-mediated or allergic reactions, fall into the so-called 'type B' category. These reactions are bizarre, cannot be readily explained on the basis of the known pharmacology of the drug, and do not show a simple dose–response relationship. They cannot be easily reproduced in animal models, unlike the type A reactions.[3]

A *drug allergy* is an immunologically mediated reaction that exhibits specificity and recurrence on re-exposure to the offending drug.[4] In most cases the diagnosis of drug allergy is clinically based since we do not have readily available sensitive and specific laboratory tests that can pinpoint whether the immune system (or what part of it) is involved.[5] Furthermore, the clinical diagnosis that a drug is responsible may be difficult because there may be different clinical manifestations with the same drug in different patients, or many different drugs can cause the same manifestation. These manifestations can mimic disease that is not drug-induced, which clearly needs to be excluded through specific investigations. The problem of making a clinical diagnosis becomes particularly difficult when patients are on multiple drugs, which is becoming commonplace.[6] Some simple rules need to be followed and these are covered later in the chapter. However, this also emphasizes that there is a need to understand the mechanisms of these allergic reactions so that laboratory tests to ascribe causality more precisely to individual therapeutic agents can be developed.

EPIDEMIOLOGY OF ADVERSE DRUG REACTIONS

Adverse drug reactions account for 3–7 per cent of all hospital admissions and occur in 10–15 per cent of hospitalized patients.[7-9] They are the most common cause of iatrogenic illness, complicating 5–15 per cent of therapeutic drug courses. In the United States, more than 100 000 deaths annually and 2.2 million hospitalizations have been attributed to serious adverse drug reactions.[9] The majority of these fall into the type A category.

Drug allergy is relatively uncommon, although accurate figures of the incidence are difficult to find.[10] It is thought to account for less than 10 per cent of all adverse drug reactions. Estimates suggest that drug allergy occurs in 1–2 per cent of all admissions and 3–5 per cent of hospitalized patients.[11] While it can be difficult to make a correct diagnosis of drug allergy, the converse is also true in that many patients, especially children, are misdiagnosed as being 'allergic' to various medications – particularly antibiotics – and end up carrying

Box 6.1 Drugs implicated in causing pulmonary allergic reactions

Anti-inflammatory agents

- Anti-TNF therapies
- Azathioprine
- Gold
- Leflunomide
- Methotrexate
- Penicillamine

Cardiovascular drugs

- Amiodarone
- Flecainide
- Hydralazine
- Procainamide

Anti-infective agents

- Abacavir
- Cephalosporin
- Ethambutol
- Isoniazid
- Minocycline
- Nitrofurantoin
- Penicillins
- Sulfonamides
- Tetracycline

CNS acting drugs

- Carbamazepine
- Fluoxetine

Phenothiazines

- Phenytoin
- Trazodone
- Tricyclic antidepressants

Miscellaneous drugs

- Appetite suppressants (dexfenfluramine, fenfluramine)
- Bromocriptine
- Cocaine
- Diphenhydramine
- Gefitinib
- Methysergide
- Paclitaxel
- Phenindione
- Propylthiouracil

this label into adulthood.[12] These patients are frequently treated with alternative medications that may be more toxic, less effective and more expensive. This in turn may result in increased morbidity and mortality, and in the context of antibiotics it can lead to resistance.

The incidence of drug allergy in the context of lung disease is not clearly known, but there are data on particular drugs, for example nitrofurantoin, an antimicrobial prescribed for urinary tract infections. Nitrofurantoin-induced pneumonitis, first described in 1965, occurs in fewer than 1 in 1000 patients, with acute toxicity nearly 10 times as prevalent as chronic.[13] Toxicity can occur at low doses and does not show a simple dose relationship. Of the 921 cases of adverse reaction to nitrofurantoin reported to the Swedish Adverse Drug Reaction Committee from 1966 to 1976, the two reactions reported most frequently – acute pulmonary reactions (43 per cent) and allergic reactions (42 per cent) – are both characteristic of acute hypersensitivity.[14] Mixed interstitial and alveolar infiltrates on chest X-ray are seen in 16 per cent of patients with pneumonitis and are associated with a mortality of 0.5 per cent. Approximately 70 per cent of patients with chronic interstitial lung disease fail to improve or show significant residual pulmonary abnormalities, and mortality has been reported at 8–10 per cent.[15]

Amiodarone, an antiarrhythmic, can cause lung toxicity, which varies in manifestations from lung opacities to acute respiratory distress syndrome (which may be rapidly fatal). The range of manifestations suggests that different mechanisms may be involved, at least some of which may be immunologic.[16] Lung effects have been witnessed in up to 6 per cent of patients taking at least 400 mg amiodarone per day; and although lower doses are stated to be safer, acute lung involvement is seen also at 200 mg/day.[17]

Methotrexate, an immunosuppressant used in the treatment of cancer, rheumatoid arthritis and psoriatic arthritis, can lead to an immune-mediated pneumonitis, the prevalence

of which ranged from 0.8 to 6.9 per cent in reported studies.[18] Most episodes occur within the first year of use, and the prevalence may be higher in patients with rheumatoid arthritis than with psoriatic arthritis, while the highest incidence has been reported in patients with primary biliary cirrhosis.[18]

Apart from the three drugs mentioned above, many others can produce immunological reactions affecting the lung (Box 6.1).[16] It is virtually impossible to predict who will develop allergic lung disease from a drug, or its severity.

CLINICAL PATTERNS OF PULMONARY DRUG ALLERGY

The pulmonary manifestations of allergic drug reactions vary between drugs and patients. They can include anaphylaxis, lupus-like reactions, organizing pneumonia, (BOOP), bronchospasm and laryngospasm, and alveolar or interstitial pneumonitis.[16] These adverse reactions may occur in isolation or may be observed as part of a generalized drug hypersensitivity syndrome that has been reported with many compounds, including for example anticonvulsants.[19]

Drug anaphylaxis is an IgE-mediated reaction that can result in severe and life-threatening respiratory symptoms.[20,21] The reaction involves the release of mast cell mediators following multivalent binding of IgE molecules, and results in a multitude of physiological effects that include both upper and lower airway obstruction. True anaphyalxis has to be distinguished from anaphylactoid reactions that do not involve IgE, but rather result from direct actions of the drugs on the mast cells. Similarly, aspirin – which is known to cause asthma – cannot be classified as being due to an immune reaction, but is probably a manifestation of its pharmacological actions in some susceptible patients.

Patients with drug-induced acute pneumonitis may present with acute respiratory distress.[16] The classical features of fever, rash and eosinophilia may not always be present. Drugs that are commonly associated with this form of disease include non-steroidal anti-inflammatories (NSAIDs), nitrofurantoin and sulphasalazine. With persistent use or poor elimination of these drugs, the pulmonary lesions may progress to a more chronic form with interstitial fibrosis. Acute pulmonary reactions caused by fibrogenic agents such as methotrexate also appear to be mediated by hypersensitivity mechanisms.[18]

Drugs may exacerbate underlying lupus or indeed directly cause it.[22] Patients may suffer a wide range of symptoms, with pulmonary disease occurring in 50–75 per cent of patients. Clinical manifestations include pleural effusions, diffuse interstitial pneumonitis and alveolar infiltrates.

The drug hypersensitivity syndrome is a severe, multisystem reaction that typically develops within 2 months of starting a drug.[23] This is a potentially life-threatening syndrome with a complex of symptoms including fever, sore throat, rash, lymphadenopathy, hepatitis, nephritis, and leucocytosis with eosinophilia. Pulmonary involvement is not uncommon and patients may present with acute respiratory distress due to florid pneumonitis. Bronchiolitis obliterans with organizing pneumonia (BOOP, also known as cryptogenic organizing pneumonitis or COP) has also been reported. Drugs associated with this form of acute lung injury include the aromatic anticonvulsants (phenytoin and carbamazepine), quinolone antibiotics and phenindione.

RISK FACTORS FOR DRUG ALLERGY

The risk factors can be broadly divided into those that are related to either the drug or the patient (such as age, sex, concurrent illness and previous reactions to related drugs). However, the situation is quite complex and it is important to remember that it is not one factor that increases the risk of drug hypersensitivity but the interplay of many.

Nature of the drug

An important risk factor for drug hypersensitivity is the chemistry and molecular weight of the drug.[4] Larger drugs with greater structural complexity (e.g. non-human proteins) are more likely to be immunogenic. Most drugs have a small molecular weight (less than 1 kDa), but may still become immunogenic by coupling with carrier proteins – such as albumin – to form simple chemical-carrier complexes (hapten).[24] Not all highly allergenic drugs are intrinsically chemically reactive, but may be metabolized to reactive species that are protein reactive and are thought to be responsible for the ultimate clinical sensitization.[25] Various functional groups on the drug molecule increase the risk for formation of chemically reactive intermediates.[26]

Route of administration

Topical, intramuscular and intravenous administration are more likely to cause hypersensitivity reactions,[20] and recent evidence suggests that drugs eluting from an intracoronary stent may cause hypersensitivity. These effects are caused by the efficiency of antigen presentation and the high concentrations of circulating drug antigen rapidly achieved with intravenous therapy. Oral administration is less likely to result in drug hypersensitivity. Although this is partly related to the lower C_{max} and T_{max} when compared to intravenous administration, tolerance mechanisms in the gut may also play a role.[20]

Degree of exposure (dose, duration, frequency)

There is some evidence that sensitization is more likely with higher drug doses and prolonged administration, but we do not have robust clinical evidence for most drugs with respect to dose dependency. An exception seems to be amiodarone where higher doses lead to a higher prevalence of lung toxicity.[20] Also of importance in predisposing to allergic reactions are intermittent courses of moderate drug doses that predispose to sensitization, while prolonged treatment without free intervals may be less likely to cause similar allergic reactions. For example, rifampicin causes a hypersensitivity reaction with intermittent therapy but is less likely to do so with chronic therapy.[20]

Age and gender

Host-related factors can predispose to drug allergy probably by acting on the way the drug is handled in the body or by unknown pharmacodynamic mechanisms. Most studies have shown that women are more often affected by allergies than are men (65–70 compared to 30–35 per cent).[11] Differences can, however, depend on the age group considered and on the type of reaction (cutaneous reaction rates were 35 per cent higher in females than males) and on the culprit drug. Increasing age is believed to be a factor in the development of drug-induced lupus syndromes.[22] Decreased drug clearance and increased medication usage in these individuals may be responsible for higher rates of drug-induced lupus reactions.

Concomitant factors

Epidemiological studies have shown that concomitant factors can increase the risk of drug hypersensitivity. For example, infection with human immunodeficiency virus (HIV) increases the risk of sulfamethoxazole hypersensitivity, and EBV infection increases the risk of penicillin allergy.[27] Whether this also operates in relation to lung involvement is unclear. Other factors associated with an increased risk of hypersensitivity drug reaction include systemic lupus erythematosus and an atopic diathesis.[20] Concomitant lung disease increases the risk of pneumonitis associated with methotrexate.[18]

Genetic susceptibility

The fact that only a minority of individuals taking a drug are affected by an allergic reaction suggests that individual genetic

Table 6.1 Host factors known to predispose to immune-mediated adverse drug reactions (adapted from [28])

Factor	At-risk group	Drug(s)
Age	Older	Many
Sex	Female	Many
HLA	HLA-B7,D22,D23	Insulin
	HLA-B*5701	Abacavir
	HLA-B*1502	Carbamazepine
	HLA-B*5802	Allopurinol
	HLA-DR9	Penicillin
Infection	HIV	Cotrimoxazole
	EBV	Ampicillin

HLA, human leucocyte antigen.

and/or environmental predisposing factors are important.[28] Given the immune basis of these reactions, an important area for studying genetic predisposition is the major histocompatibility complex on chromosome 6. Particular attention has focused on the HLA genes that encode a group of proteins on the surface of cells that are involved in presenting foreign materials or antigens to the immune system. Thus, individuals who possess certain HLA antigens have a higher risk of allergic or hypersensitivity reactions with particular drugs (Table 6.1). However, it is important to note that each HLA gene usually contributes to the risk of developing the hypersensitivity reaction, and usually is neither necessary nor sufficient by itself to predispose to the reaction.

The importance of HLA genes in predisposing to allergic drug reactions is further emphasized by the differences in incidence of these reactions between ethnic groups. For example, severe bullous cutaneous adverse reactions such as toxic epidermal necrolysis seem to be more common in Chinese than in Caucasians. Intuitively, one would expect that allergic reactions with pulmonary involvement also show (at least partly) a predisposition with certain HLA alleles, but this has not been tested, largely because such studies require the availability of samples from a well phenotyped, adequately powered cohort of affected patients.

DIAGNOSIS OF PULMONARY DRUG ALLERGY

The diagnosis of drug allergy is one of exclusion.[16] It is based on a detailed history of the onset of symptoms/signs combined with ascertainment of the temporal relationship between the appearance of those symptoms and drug use/discontinuation. Thorough documentation of current and previous drug use and prior reactions is also important. Knowledge of the patient's background, comorbidities and family history may highlight specific host factors. Deciding which drug of the multiple drugs a patient may be taking is responsible can be difficult, although following a few simple rules (remembered by the mnemonic TREND) may help in the identification of the culprit drug (Box 6.2).[6]

A drug reaction can involve almost any system in the body, so a comprehensive clinical examination should be performed.

Dermatological reactions are the most common and skin rashes should therefore be described carefully. Macroscopic features (e.g. macular, papular or vasculitic) and distribution of the rash (e.g. mucous membranes) are important and need to be documented. Pulmonary manifestations can be wide-ranging with physical signs dependent on the region of the respiratory tract affected.[16] In acute life-threatening anaphylaxis, upper airway signs may include stridor, and auscultation may reveal wheeze indicating constriction of the smaller airways. Pleuropneumopathic abnormalities such as pneumonitis and pleural effusions may take longer to present but can be diagnosed on auscultation. Other organ systems, particularly renal, liver and cardiovascular, should be assessed too.

Specific diagnostic criteria for pulmonary reactions induced by amiodarone and methotrexate have been published and are summarized in Box 6.3.[29,30]

Given that there are no specific tests routinely available to implicate drug involvement, when drug-induced pulmonary involvement is suspected, general investigations may be helpful in identifying the aetiology. A full blood count may show the presence of eosinophilia which is commonly associated with drug allergy. Assessment of arterial blood gases may reveal the degree of hypoxia being suffered by the patient, and provide a proxy for the degree of lung involvement. A chest X-ray may reveal a wide array of abnormalities ranging from benign-looking pulmonary infiltrates (even in asymptomatic patients) to more significant lung pathology similar to that observed in acute respiratory distress syndrome. Where necessary, a chest film should be followed by high-resolution CT scanning. This may need to be repeated after drug withdrawal to ensure resolution of lung involvement particularly in patients with interstitial disease. Spirometry is a useful tool but, for more detail of the extent of lung involvement, gas transfer studies should be performed too. If cardiovascular and hepatic manifestations are suspected, an electrocardiogram and liver function tests need to be performed.

There is a need to develop tests that can implicate certain drugs in causing the allergic reactions. Such tests would need to reliably identify the drug responsible for the allergic reaction: they would need to be reproducible (implicating the same culprit every time the test is done); transportable between laboratories; feasible, timely and cost-effective; and show good diagnostic characteristics (with minimal false-negative and false-positive results). Various tests are being used as research tools but none to date has fulfilled all the characteristics of an ideal test to allow it to be used routinely in different clinical settings. Some of these laboratory tests are listed in Table 6.2.

MECHANISMS OF ALLERGIC DRUG REACTIONS

Immunological reactions caused by drugs broadly fall into the classification proposed by Coombs and Gell,[31] recently extended by Pichler[23] (Table 6.3). However, it is important to note that the same drug may cause different types of reaction in different patients, while one patient may have more than one type of ongoing immunological process accounting for his or her clinical manifestations.

Box 6.2   Steps to determine whether a drug is responsible for an adverse reaction (reproduced with permission from [6])

TREND =

Temporal relationship

What is the timing between the start of drug therapy and the reaction?
Most reactions occur soon after commencing drug therapy; anaphylactic reactions can occur within hours, while hypersensitivity reactions typically take 2–6 weeks. Other reactions such as bone density changes may be delayed for years,

Rechallenge

What happens when the patient is rechallenged with the drug?
Recurrence on rechallenge provides good evidence that the drug is responsible for the adverse effect. This is, however, rarely possible, particularly for serious reactions, because of the danger to the patient. More rapid occurrence following re-exposure to the drug than on secondary exposure indicates an immune-mediated pathogenesis.

Exclusion

Have concomitant drugs and other non-drug causes been excluded?
An adverse drug reaction is a diagnosis of exclusion since there are no specific laboratory tests available. It is important to exclude non-drug causes clinically as well as by performing relevant investigations.

Novelty

Has the reaction been reported before?
If the reaction is well recognized, it may be mentioned in the literature from the manufacturer, or may have been reported in the literature. An opportunity should always be taken to systemically search reference databases such as MEDLINE, which can provide valuable insight into the appropriate management of patients with what may be a relatively rare reaction. Drug Information Centres can also provide useful information.

Dechallenge

Does the reaction improve when the drug is withdrawn or dose reduced?
Most, but not all, reactions improve on drug withdrawal although the recovery phase can be prolonged. In rare instances, an autoimmune phenomenon may be set up, and thus the reaction will not improve on drug withdrawal.

Type I IgE–mediated reactions

IgE-mediated, type I, immediate or anaphylactic reactions may involve the skin (urticaria and eczema), eyes (conjunctivitis), nasopharynx (rhinorrhea, rhinitis), bronchopulmonary tissues (asthma) and the gastrointestinal tract (gastroenteritis).[21] The reaction may cause a range of symptoms ranging from minor inconvenience to a serious reaction that rapidly results in death. The reactions usually occur 15–30 minutes from the time of exposure to the antigen, but they may on occasions have a delayed onset (10–12 hours). Immediate hypersensitivity is mediated by IgE antibodies to the drug allergen (Fig. 6.1). The primary cellular component in the hypersensitivity reaction is the mast cell or basophil, with the accessory cells (platelets, neutrophils and eosinophils) amplifying the clinical effects.[32] In sensitized patients (and sensitization may be covert), multivalent IgE antibodies – which have very high affinity for their receptors on mast cells and basophils – cross-link, triggering the release of various pharmacologically active substances. Mast cell degranulation is preceded by increased calcium ion influx, which is a crucial process; ionophores which increase cytoplasmic calcium also promote degranulation, whereas agents that deplete cytoplasmic calcium suppress degranulation. Furthermore, cyclic nucleotides appear to play a significant role in the modulation of immediate hypersensitivity reactions, although their exact function is not well understood. Substances that alter cAMP and cGMP levels significantly alter the allergic symptoms. An increase in intracellular cAMP seems to relieve allergic symptoms, particularly bronchopulmonary symptoms, and is the basis of therapy, while agents that decrease cAMP or stimulate cGMP aggravate the allergic reaction. The agents that are released from mast cells and their effects are listed in Table 6.4. Beta-lactam antibiotics, such as penicillins and cephalosporins, are the most likely to induce a type I reaction. Penicillin-induced anaphylaxis has been reported to occur in about 1 in 5000 to 10 000 courses, with a fatality rate of 10 per cent.[33]

Type II (cytotoxic) hypersensitivity

This form of hypersensitivity reaction is perhaps least important with respect to lung involvement in allergic drug reactions. The drug/metabolite binds to a cell surface, and is recognized as being foreign by an IgM antibody or certain subclasses of IgG antibodies (IgG1 and IgG3) which specialize in complement fixation.[23] Complement fixation activates the complement cascade which leads to cell lysis. The onset of action

> **Box 6.3**   Diagnostic criteria for acute pneumonitis caused by amiodarone and methotrexate (modified from [29] and [30])
>
> *Amiodarone-induced pneumonitis*
>
> - Initial presentation as worsening dyspnoea
> - Chest X-ray shows diffuse interstitial or alveolar infiltrates
> - Pulmonary function testing shows a decline > 15% in the diffusion capacity of the lung for carbon monoxide (DLCO) or in total lung capacity
> - Left heart failure is excluded
> - Bronchoalveolar lavage fluid reveals 'foamy' cytoplasm in alveolar macrophages (phospholipidosis) and CD8+ lymphocytosis
> - Lung biopsy findings include pulmonary interstitium inflammation, fibrosis, pneumocyte hyperplasia, and hyaline membranes
> - Withdrawal of amiodarone with or without steroid therapy reverses some or all of the abnormalities.
>
> *Methotrexate-induced pneumonitis*
>
> - Acute onset dyspnoea
> - Fever > 38.0°C
> - Tachypnoea at least 28/min and dry cough
> - Radiological evidence of pulmonary interstitial or alveolar infiltrates
> - White blood cell count up to 15.0×10^9 with or without eosinophilia
> - Negative blood and sputum cultures (mandatory)
> - Restrictive defect and decreased DLCO on pulmonary function tests
> - $PO_2 < 7.5$ kPa on air
> - Histopathology consistent with bronchiolitis or interstitial pneumonitis with giant cells granulomas and without evidence of infection
>
> Definite: 6 criteria present
> Probable: 5 of 9 criteria present
> Possible: 4 of 9 criteria present

varies with the drug and cell type, but it usually takes longer than 24 hours before clinical effects become noticeable.

Type III immune complex mediated hypersensitivity

Type III (immune complex) reactions develop when a drug combines with antibodies to form immune complexes, the deposition of which on basement membranes causes tissue damage through activation of complement. The drug usually binds directly with IgG or IgM antibodies. Once the complement cascade is initiated, neutrophil and macrophage chemotaxis occurs resulting in the release of large amounts of lysosomal enzymes which cause destruction of the surrounding basement epithelial cell membranes. Drug-induced lupus syndromes fall into the category of type III reactions (although other mechanisms may also be operating in conjunction).[22] Renal involvement in drug-induced lupus is rare, as is the presence of anti-double-stranded DNA antibodies, but other autoantibodies are common; these include antihistone antibodies from procainamide, hydralazine or phenytoin; perinuclear antineutrophil cytoplasmic antibody (p-ANCA) from minocycline; and SS-A and SS-B from thiazides. In the lungs, desposition of these immune complexes may cause diffuse interstitial pneumonitis. Procainamide, an antiarrhythmic agent, can be regarded as a paradigm drug that is commonly associated with drug-induced lupus. Between 50 and 90 per cent of patients taking procainamide for longer than 2 months develop serum antinuclear antibodies (ANAs); 10–20 per cent of these ANA-positive patients develop symptomatic drug-induced SLE, with 40–80 per cent of these patients having pulmonary manifestations which (in up to 40 per cent) are accompanied by bibasilar pulmonary infiltrates.

Type IV (delayed–type) hypersensitivity

T-cell-mediated reactions are probably the most prevalent type of reaction to drugs, and have been implicated in many different types of allergic reactions, including those involving the lungs.[23] The fundamental mechanism here involves an interaction between the HLA (on antigen presenting cells) and the T-cell receptor which, together with the interaction with co-stimulatory molecules, leads to the formation of an immunological synapse, that in turn leads to an immune response characterized by T-cell clonal proliferation, the secretion of various cytokines (typically Th1 type cytokines) and chemokines which may mediate the involvement of different organ systems.[34] What is unclear is how the drug

Table 6.2   Investigation of drug-induced lung allergy

1. General tests

Non-invasive tests

Full blood count	Eosinophilia may be present
C-reactive protein	Raised due to inflammation
Liver function tests	Abnormalities may be present specifically in drug hypersensitivity syndrome
Autoantibodies	Antinuclear antibody, antihistone antibody, perinuclear antibody, anti SS-A and anti SS-B
Arterial blood gases	Hypoxia with interstitial lung involvement
Pulmonary function tests	Restrictive and/or obstructive pattern and decreased transfer factor
Chest X-ray	Reticulonodular shadowing
High-resolution CT scanning	Pulmonary infiltrates, interstitial fibrosis

Invasive tests

Bronchoalveolar lavage	Eosinophils and/or lymphocytes
Lung biopsy	Alveolitis, alveolar fibrosis

2. Specific tests

Immune reaction	In vitro	In vivo
IgE-mediated (type I)	RAST, serum tryptase	Skin tests
Cytotoxic (type II)	Direct or indirect Coombs' test	
Immune complex (type III)	Complement studies, autoantibodies	Tissue biopsy for immunofluorescence
T-cell-mediated (type IV)	Lymphocyte transformation test	Patch testing

RAST, radioallergosorbent test.

Table 6.3   Mediators of IgE-mediated type I hypersensitivity reactions

Mediators	Effects
Primary mediators	
Histamine	Bronchoconstriction, mucus secretion, vasodilatation, vascular permeability
Serotonin	Vascular permeability, small muscle contraction
ECF-A	Eosinophil chemotaxis
NCF-A	Neutrophil chemotaxis
Proteases	Mucus secretion, connective tissue degradation
Secondary mediators	
Leukotriene B_4	Basophil attractant
Leukotriene C_4, D_4	Same as histamine but 1000 times more potent
Prostaglandins D_2	Oedema and pain
PAF	Platelet aggregation and heparin release, microthrombi formation

PAF, platelet-activating factor; ECF-A, eosinophil chemotactic factor; NCF-A, neutrophil chemotactic factor.

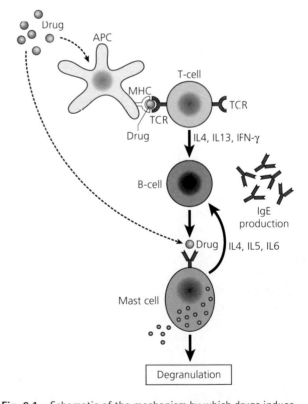

Fig. 6.1   Schematic of the mechanism by which drugs induce anaphylaxis. A drug allergen leads to IgE antibody production, which can then bind in a multivalent fashion on mast cell surfaces leading to degranulation and release of vasoactive mediators (see Table 6.3).

Table 6.4 Classification of drug hypersensitivity reactions (adapted from [23] and [31])

Immune reaction	Mechanism	Clinical manifestations	Timing of reactions
Type I (IgE-mediated)	Drug–IgE complex binding to mast cells with release of histamine, mediators	Urticaria, angioedema, bronchospasm, anaphylaxis	Minutes to hours after drug exposure
Type II (cytotoxic)	Cell destruction mediated by an interaction between IgG or IgM antibodies, complement, and a drug antigen associated with cell membranes	Haemolytic anaemia, neutropenia, thrombocytopenia	Variable
Type III (immune complex)	Tissue deposition of drug–antibody complexes with complement activation and inflammation	Serum sickness, fever, rash, arthralgias, lymphadenopathy, pneumonitis, glomerulonephritis, vasculitis	1–3 weeks after drug exposure
Type IV (delayed, cell-mediated)			
Type IVa	Th1 reactions with high IFN-γ/TNF secretion and involves monocyte/macrophage activation	Tuberculin reaction, contact dermatitis	2–7 days after drug exposure
Type IVb	Eosinophilic inflammation, Th2 reactions with high IL4/IL5/IL13 secretion; associated with IgE-mediated type I reaction	Chronic asthma, chronic allergic rhinitis	
Type IVc	Cytotoxic T-cell-driven reactions	Contact dermatitis	
Type IVd	T-cell-dependent neutrophilic inflammation	Acute generalized exanthematous pustulosis	

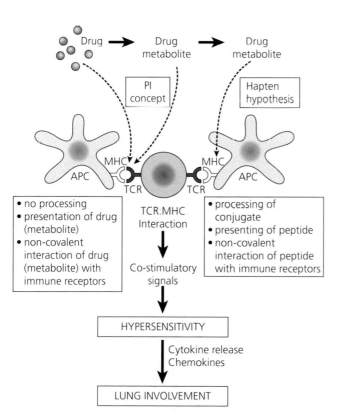

Fig. 6.2 Schematic of the mechanisms by which drugs induce T-cell-mediated adverse reactions. Drugs can act as antigens either via the classical hapten pathway and/or by interacting with the T-cell receptor (the PI hypothesis). This leads to clonal proliferation with tissue damage occurring as a result of various mechanisms, including cytokine and chemokine release. MHC, major histocompatibility complex; TCR, T-cell receptor; APC, antigen presenting cell.

or its metabolites interact with the immune system to form the antigen.

Two complementary hypotheses, the hapten hypothesis and the PI hypothesis (Fig. 6.2) both of which can be supported by experimental evidence, have been put forward to explain the mechanism by which drugs can induce a T-cell reaction.

It is widely believed that drugs, in line with other low-molecular-weight compounds, are not immunogenic *per se*, and therefore must form stable adducts with endogenous proteins to initiate an immune response.[24] This is the basis of the hapten hypothesis of drug hypersensitivity, in which the critical step is the formation of adducts (drug–protein conjugates) between the drug (metabolite) and an endogenous protein.[4,35] The immunochemical literature shows that compounds with a molecular weight of less than 1 kDa must be covalently bound to a high-molecular-weight protein(s) to act as an effective immunogen. Classical studies by Landsteiner and Jacobs[36,37] showed that chemicals that can bind covalently to protein are potent sensitizing agents; for example, the model hapten dinitrofluorobenzene reacts spontaneously with lysine groups in autologous proteins, and is a potent contact sensitizer.[24] Protein adducts may be formed by two mechanisms: either by direct chemical reaction (e.g. penicillin is able to do this because of an inherently unstable chemical structure) or by generation of electrophilic metabolites through metabolism in the body, which then react with nucleophilic groups on proteins.[4] The majority of drugs belong to the latter group, and hence it has been postulated that metabolism to chemically reactive metabolites (a process termed 'bioactivation') is a crucial step in the pathogenesis of allergic drug reactions.

The fundamental concept that protein conjugation is an obligatory step in the process of immune recognition of drugs has been challenged by the observation that T-cells from patients hypersensitive to a number of drugs undergo proliferation in an antigen-processing independent (but MHC-restricted) manner.[38,39] This involves labile, reversible binding of a drug to the MHC complexes on antigen-presenting cells. This has been termed the 'pharmacological interaction with immune receptors' (PI or p-i) concept.[23] The relevance of these in-vitro findings to T-cell activation in vivo still needs to be defined. Nevertheless, these studies show unequivocally a T-cell response to drug-derived antigen and alert us to the possibility of novel mechanisms of antigen presentation in cells from hypersensitive patients, which do not operate in animal models of chemical immunogenicity.[40]

The mechanism of a type IV hypersensitivity reaction involving the lung has recently been investigated with the anticoagulant phenindione.[41] Phenindione causes hypersensitivity in 1.5–3 per cent of patients. A 46-year-old female, 30 days after starting phenindone, developed a generalized maculopapular erythematous rash associated with a temperature of 39°C, generalized lymphadenopathy, mild jaundice, and bilateral patchy consolidation on the chest X-ray. Her eosinophil count was elevated at 2.5×10^9/L (normal range 0–0.4), whereas alanine aminotransferase level peaked at over 30 times the upper limit of normal. T-cells taken from the patient proliferated in the presence of phenindione, but not from four controls. T-cell clones generated from the patient were characterized as being MHC class Ii restricted, and CD4+ bearing the $\alpha\beta$ T-cell receptor. The clones secreted IFN-γ (consistent with this being a Th1 response) and IL5 (consistent with the eosinophilia). In terms of the mechanism of antigen presentation, two types of clones were detected, those that bound the drug directly on the MHC and those that bound processed peptide. The reasons why the lung was involved were not investigated – it would be interesting to determine whether CCR6 and CXCR3, which regulate antigen-induced T-cell homing to human asthmatic airways,[42] are also implicated in drug-induced pneumonitis. This investigation of phenindione hypersensitivity leading to pneumonitis represents the most comprehensive investigation that has been performed to date with respect to drug-induced lung allergy.

TREATMENT AND PREVENTION

In order to ensure appropriate treatment of drug-induced allergic lung disease, it is first important to suspect the drug as the aetiological agent, make the diagnosis and then promptly withdraw the drug. A detailed description of the treatments available for each different type of allergic manifestation is beyond the scope of this chapter; suffice it to say that treatment is essentially supportive and symptomatic. While steroids are often used for the treatment of acute drug-induced lung disease, the evidence for their effectiveness is largely anecdotal.

There are essentially no good preventive strategies, although the use of the lowest effective dose of drugs implicated in causing lung allergy for the shortest period of time should be usual practice. In some cases, exclusion of patients with prior lung disease may be justified; although there is evidence that prior lung disease increases the risk of pneumonitis with methotrexate,[18] even with other drugs where no

such evidence exists, these patients have lower respiratory reserve, and would run a more severe disease course if an allergic reaction did occur. A possible preventive strategy is to monitor the patients closely; however, the timing and nature of the monitoring undertaken is often arbitrary, and not universally practised.[43] Indeed it could be argued that such monitoring would not be effective unless the necessary tests were performed regularly, in which case they may be considered to be cost-ineffective and impractical.

KEY POINTS

- The diagnosis of drug allergy is usually clinically based because there are no readily available sensitive and specific laboratory tests.

- The clinical diagnosis that a particular drug is responsible may be difficult because there are different clinical manifestations with the same drug in different patients, and several drugs can cause the same manifestation.

- The lack of specific treatment, monitoring and diagnostic strategies for allergic drug reactions involving the lung highlights the lack of research in this area. There is a need to undertake fundamental research to define mechanisms and individual susceptibility factors.

- Given the relative rarity of these reactions and the lack of an animal model, there is a need to undertake multicentre collaborative studies that allow the identification, clinical and laboratory characterization of patients together with biobanking of samples.

REFERENCES

1. Aronson JK. Adverse drug reactions: no farewell to harms. Br J Clin Pharmacol 2007;**63**(2):131–5.
2. Pirmohamed M, Breckenridge AM, Kitteringham NR, Park BK. Adverse drug reactions. Br Med J 1998;**316**:1295–8.
3. Pirmohamed M, Park BK. The adverse effects of drugs. Hosp Med 1999;**60**:348–52.
4. Park BE, Pirmohamed M, Kitteringham NR. Role of drug disposition in drug hypersensitivity: a chemical, molecular, and clinical perspective. Chem Res Toxicol 1998;**11**:969–88.
5. Pirmohamed M, Naisbitt DJ, Gordon F, Park BK. The danger hypothesis: potential role in idiosyncratic drug reactions. Toxicology 2002;**181/182**:55–63.
6. Pirmohamed M. Anticipating, investigating and managing the adverse effects of drugs. Clin Med 2005;**5**:23–6.
7. Pirmohamed M, James S, Meakin S, et al. Adverse drug reactions as cause of admission to hospital: prospective analysis of 18820 patients. Br Med J 2004;**329**:15–19.
8. Howard RL, Avery AJ, Slavenburg S, et al. Which drugs cause preventable admissions to hospital? A systematic review. Br J Clin Pharmacol 2007;**63**(2):136–47.
9. Lazarou J, Pomeranz BH, Corey PN. Incidence of adverse drug reactions in hospitalized patients: a meta-analysis of prospective studies. J Am Med Assoc 1998;**279**:1200–5.

10. Vervloet D, Durham S. Adverse reactions to drugs. *Br Med J* 1998;**316**:1511–14.

11. Gomes ER, Demoly P. Epidemiology of hypersensitivity drug reactions. *Curr Opin Allergy Clin Immunol* 2005;**5**:309–16.

12. Gruchalla RS, Pirmohamed M. Clinical Practice. Antibiotic allergy. *New Engl J Med* 2006;**354**:601–9.

13. D'Arcy PF. Nitrofurantoin. *Drug Intell Clin Pharm* 1985;**19**:540–7.

14. Holmberg L, Boman G, Bottiger LE, *et al.* Adverse reactions to nitrofurantoin: analysis of 921 reports. *Am J Med* 1980;**69**:733–8.

15. Jick SS, Jick H, Walker AM, Hunter JR. Hospitalizations for pulmonary reactions following nitrofurantoin use. *Chest* 1989;**96**:512–15.

16. Ozkan M, Dweik RA, Ahmad M. Drug-induced lung disease. *Cleve Clin J Med* 2001;**68**:782–5,789–95.

17. Chang SN, Hwang JJ, Hsu KL, *et al.* Amiodarone-related pneumonitis. *J Formos Med Assoc* 2007;**106**:411–17.

18. Saravanan V, Kelly CA. Reducing the risk of methotrexate pneumonitis in rheumatoid arthritis. *Rheumatology* (Oxford) 2004;**43**(2):143–7.

19. Michael JR, Rudin ML. Acute pulmonary disease caused by phenytoin. *Ann Intern Med* 1981;**95**:452–4.

20. Adkinson NF, Wheeler B. Risk factors for IgE dependent reactions to penicillin. In: Ganderton JW (ed.) *XI International Congress of Allergology and Clinical Immunology*, pp 515–40. London: Macmillan, 1983.

21. Weiss ME, Adkinson MF. Immediate hypersensitivity reactions to penicillin and related antibiotics. *Clin Allergy* 1988;**18**:515–40.

22. Antonov D, Kazandjieva J, Etugov D, Gospodinov D, Tsankov N. Drug-induced lupus erythematosus. *Clin Dermatol* 2004;**22**:157–66.

23. Pichler WJ. Delayed drug hypersensitivity reactions. *Ann Intern Med* 2003;**139**:683–93.

24. Park BK, Coleman JW, Kitteringham NR. Drug disposition and drug hypersensitivity. *Biochem Pharmacol* 1987;**36**:581–90.

25. Pirmohamed M, Madden S, Park BK. Idiosyncratic drug reactions: metabolic bioactivation as a pathogenic mechanism. *Clin Pharmacokin* 1996;**31**:215–30.

26. Park BK, Naisbitt DJ, Gordon SF, Kitteringham NR, Pirmohamed M. Metabolic activation in drug allergies. *Toxicology* 2001;**158**(1/2):11–23.

27. Vilar FJ, Naisbitt DJ, Park BK, Pirmohamed M. Mechanisms of drug hypersensitivity in HIV-infected patients: the role of the immune system. *J HIV Ther* 2003;**8**(2):42–7.

28. Pirmohamed M. Genetic factors in the predisposition to drug-induced hypersensitivity reactions. *Aaps J* 2006;**8**(1):E20–6.

29. Oren S, Turkot S, Golzman B, *et al.* Amiodarone-induced bronchiolitis obliterans organizing pneumonia (BOOP). *Respir Med* 1996;**90**(3):167–9.

30. Searles G, McKendry RJ. Methotrexate pneumonitis in rheumatoid arthritis: potential risk factors. Four case reports and a review of the literature. *J Rheumatol* 1987;**14**:1164–71.

31. Coombs RRA, Gell PGH. Classification of allergic reactions responsible for clinical hypersensitivity and disease. In: Gell PGH (ed.) *Clinical Aspects of Immunology*, pp 575–96. Oxford: Oxford University Press, 1968.

32. Naisbitt DJ, Gordon SF, Pirmohamed M, Park BK. Immunological principles of adverse drug reactions: the initiation and propagation of immune responses elicited by drug treatment. *Drug Safety* 2000;**23**:483–507.

33. Rudolph AH, Price EV. Penicillin reactions among patients in venereal disease clinics: a national survey. *J Am Med Assoc* 1973;**223**:499–501.

34. Naisbitt DJ, Pirmohamed M, Park BK. Immunological principles of T-cell-mediated adverse drug reactions in skin. *Expert Opin Drug Safety* 2007;**6**(2):109–24.

35. Uetrecht JP. New concepts in immunology relevant to idiosyncratic drug reactions: the 'danger hypothesis' and innate immune system. *Chem Res Toxicol* 1999;**12**:387–95.

36. Landsteiner K, Jacobs J. Studies on the sensitisation of animals with simple chemical compounds. *J Exp Med* 1935;**61**:643–56.

37. Landsteiner K, Jacobs J. Studies on the sensitisation of animals with simple chemical compounds II. *J Exp Med* 1936;**64**:625–31.

38. Schnyder B, Mauri-Hellweg D, Zanni M, *et al.* Direct, MHC-dependent presentation of the drug sulfamethoxazole to human alpha/beta T-cell clones. *J Clin Invest* 1997;**100**:136–41.

39. Schnyder B, Burkhart C, Schnyder-Frutig K, *et al.* Recognition of sulfamethoxazole and its reactive metabolites by drug-specific CD4+ T-cells from allergic individuals. *J Immunol* 2000;**164**:6647–54.

40. Farrell J, Naisbitt DJ, Drummond NS, *et al.* Characterization of sulfamethoxazole and sulfamethoxazole metabolite-specific T-cell responses in animals and humans. *J Pharmacol Exp Ther* 2003;**306**:229–37.

41. Naisbitt DJ, Farrell J, Chamberlain PJ, *et al.* Characterization of the T-cell response in a patient with phenindione hypersensitivity. *J Pharmacol Exp Ther* 2005;**313**:1058–65.

42. Thomas SY, Banerji A, Medoff BD, Lilly CM, Luster AD. Multiple chemokine receptors, including CCR6 and CXCR3, regulate antigen-induced T-cell homing to the human asthmatic airway. *J Immunol* 2007;**179**:1901–12.

43. Pirmohamed M, Ferner RE. Monitoring drug treatment: criteria used for screening tests should apply to monitoring. *Br Med J* 2003;**327**:1179–81.

DRUG-INDUCED RESPIRATORY EMERGENCIES

7

Drug-induced acute upper airway obstruction

MICHAEL LIPPMANN, GANESAN MURALI

INTRODUCTION

Upper airway obstruction is usually seen in the setting of an infectious process such as epiglotitis, inflammation, trauma, haemorrhage or a neoplasm. However, in the last few decades it has been recognized that various drugs and anaesthetics can be associated with severe airway compromise (Box 7.1). The groups of drugs known as angiotensin-converting enzyme inhibitors (ACEIs), agents that have been shown to reduce mortality from heart failure, are by far the most common class of drugs associated with this complication.[1] In this chapter we will review the incidence, clinical manifestations, treatment and outcome of patients with drug-induced upper airway obstruction.

ANGIOTENSIN-CONVERTING ENZYME INHIBITORS AND AIRWAYS DISEASE

ACEI-induced cough

The most commonly reported adverse effects of this class of drug are chronic nagging cough and angioedema. The incidence of cough has been variously reported at 5–20 per cent, is more than twice as common in women, and usually starts 1 week to 6 months after initiation of therapy.[2] Sebastin and colleagues, in a cross-sectional outpatient Veterans Administration (VA) population comparing ACEI and hydrochlorthiazide (HCTZ), described an incidence of cough of 19 per cent in ACEI users versus 9 per cent in the HCTZ group. The implicated drugs were captopril and enalapril. The study suffers from using a self-reporting questionnaire and addressed only a VA male population.[3] Coulter and Edwards described 33 cases from New Zealand, finding the incidence of cough higher in those on enalapril (2.8 vs. 1.1 per cent) with a higher incidence in women. There was no increase in the incidence of cough in those with congestive heart failure or asthma. The time to the development of cough from initiation of the ACEI ranged

Box 7.1 Drugs reported to cause upper airway obstruction

- Antibiotics: penicillin, cephalosporins, co-trimoxazole, ciprofloxacin, tetracyclines, vancomycin
- Anaesthetic agents
- Neuromuscular blocking drugs
- Angiotensin-converting enzyme inhibitors (ACEIs)
- Anticoagulants
- Intravenous radio-contrast media
- 'Crack' cocaine
- Others: latex rubber, psyllium, isocyanate, eucalyptus

between 4 weeks and 15 months with no relationship to dose. All resolved their cough within 3–5 days after drug cessation. In all nine patients who were rechallenged with the drug the cough promptly returned.[4] In another retrospective review of 209 patients taking enalapril, the incidence of cough was 10.5 per cent, again being more common in women. Cough occurred within 5 months of initiation of therapy, was unrelated to dose, underlying disease, smoking or concomitant medications, and resolved within 2 weeks of cessation in all.[5] Simon and co-workers looked at the incidence of cough in a retrospective, consecutive review of 500 patients and found different incidences of cough: 13 per cent in a university practice and 25 per cent in a private practice.[6] The reasons for this were not entirely apparent and may be a feature of self-reporting.

The mechanism of ACEI cough is poorly understood. Various hypotheses have included: excess kinins; increased sensitivity to stimuli such as capascin and methacholine; an increase in substance P; an increase in prostaglandins and leukotrienes; and genetic factors. In a retrospective series by Kaufman et al. 8 of 9 patients with ACEI-induced cough were demonstrated to have increased bronchial hyperreactivity with methacholine.[7] Caution was advised in the use of this class of drugs in patients with reactive airways. Subsequent prospective studies in asthmatics failed to identify asthma as a risk factor for ACEI-induced cough.[8] Boulet et al. confirmed this by finding normal bronchial hyperresponsiveness to methacholine in 15 patients on captopril, 4 of whom had cough.[9] However, Bucca et al. demonstrated increased responsiveness of the extrathoracic airways to inhaled histamine in 9 patients with captopril-induced cough, which reversed upon drug cessation.[10]

Recently published guidelines for the management of ACEI-induced cough by the American College of Chest Physicians emphasizes cessation of the drug as the only effective method of treatment, with resolution of cough occurring 1–4 weeks after cessation with a delay of up to 3 months in some patients. In those in whom cessation is not an option, pharmacological treatment with sodium cromoglycate, theophylline, sulindac, indomethacin, amlodopine, nifedipine, ferrous sulfate or picotamide may be viable options.[11] It is anticipated that the introduction of the new class of angiotensin-converting enzyme receptor blockers will be associated with a much lower incidence of cough.[12]

ACEI-induced angioedema

The most serious adverse effect of this class of drug is angioedema leading to airway compromise. The incidence has been reported as 0.1–0.2 per cent in most series.[2,13–15] With an estimated 35 million patients on ACEI, theoretically 100 000 new cases can be expected yearly.[16] It is apparent that genetic and racial factors make African Americans (or possibly all black individuals) at higher risk for this complication. In a review of the Tennessee Medicaid population, Brown et al. described 82 cases of angioedema over a 6-year period; 53 African American (AA), 29 Caucasian.[17] The relative risk (RR) of this complication was 4.5 for AA. Of the 8 patients requiring intensive care management, 7 were AA and 4 of these needed intubation. There were no deaths. Other risk factors were occurrence of angioedema within the first 30 days of ACEI administration (RR = 4.6), and a 2.2 RR for either enalapril or lisinopril as compared to captopril. There was no relationship to dose, sex, age or concomitant medications. In a study from a British teaching hospital, Gibbs et al. described 20 cases over a 6-year period, 65 per cent of whom were black.[18] Seventy per cent occurred within 4 weeks of drug initiation, but three had taken the medication for 2–4 years. Eight required hospital admission, three to the intensive care unit, resulting in one death from laryngeal obstruction. What was astonishing is that the ACEI was continued in 50 per cent of patients after discharge. In an effort to determine the reason for the increased incidence in blacks, Gainer et al. assessed the sensitivity to bradykinins, the leading theory as to the mechanism of ACEI angioedema, in an AA population.[19] Intradermal injections of bradykinin at two doses were administered in a double-blind fashion. It was hypothesized that urinary kallikrein levels were lower in AA, hence their endogenous bradykinin levels may be lower thus making them more sensitive to the effects of bradykinin. There was a 70 per cent larger wheal size in AA to either dose of bradykinin, with half of the hypertensive AA group having larger wheal sizes. Therefore there appears to be increased sensitivity to bradykinin in normotensive and hypertensive African Americans, suggesting that there is up-regulation and increased sensitivity to exogenous bradykinin.

Another group apparently at high risk from ACEI angioedema are transplant patients. In a retrospective review of cardiac and renal transplants, Abbosh et al. demonstrated the prevalence of angioedema in cardiac transplant patients to be 24 times and renal transplant patients 5 times that of the general population.[20] Once again, African Americans made up the majority of patients (80 per cent) with occurrences from 3 days to 6 years (mean 19 months) after starting therapy. In our comprehensive review of 45 patients with ACEI-induced angioedema, 91 per cent of the patients were African Americans, a possible reflection of the demographics of our urban neighborhood.[13]

Some evidence exists that patients with sleep apnoea, who already have a decreased upper airway diameter, may be at increased risk.[21,22] In one study, nitric oxide (a marker of airway inflammation) was used to demonstrate that ACEIs might contribute to obstructive sleep apnoea by inducing upper airway inflammation.[22] Thus caution is advised in the use of ACEIs in people with significant sleep apnoea. Past facial or head and neck surgery or external trauma may also be risk factors for the development of angioedema.[21,23]

CLINICAL PRESENTATION OF ANGIOEDEMA

The vast majority of patients develop angioedema within one week and often within hours of drug initiation. However, there are numerous reports of patients taking the drug for years prior to their first episode of angioedema.[24–29] A summary of the typical patient characteristics is given in Table 7.1.

Enalapril appears to be the most common ACEI associated with this complication, but virtually every marketed ACEI has been reported.[30] There appears to be no sex predominance in the reviewed studies. What is surprising is that the reaction may be poorly identified as drug-related, so the drug is continued with the patients suffering repeated, and potentially life-threatening episodes, of angioedema and tongue oedema.[15,31] The rate of recurrent angioedema in those with continued use of ACEI was 10 times that in those in whom the drug was stopped. It was apparent that physicians attributed the angioedema to causes other than this class of drug.

Angioedema can be classified as type 1 (oedema limited to the face), type 2 (oedema to the floor of the mouth, base of tongue and uvula), and type 3 (oropharyngeal, glottic and supraglottic oedema). In one study these occurred in, respectively, 57, 26 and 17 per cent of patients.[32] Facial swelling, tongue oedema, stridor and dyspnoea were the most prominent clinical presentations. The median time from initial onset of symptoms to hospital presentation was 10 hours. Fifty per cent of patients with type 3 where admitted to the intensive care unit, while only 32 per cent with type 2 required ICU admission. Of those requiring airway interventions, six had type 2 and six had type 3. All intubated patients were African Americans. It is generally recommended that patients with types 2 and 3 angioedema be monitored in an intensive care setting with equipment available to perform fibre-optic intubation as anatomic landmarks are distorted. The degree of orophayngeal oedema was quantified in one other study.[33] In their four cases, three had type 3 and one type 2 oedema, all responding to medical therapy. In the series by Agah et al. odonophagia and tongue oedema were good predictors of those who required hospitalization.[34]

Table 7.1 Characteristics of patients with angioedema due to angiotensin-converting enzyme inhibitor (ACEI) intake

Reference	Patients	Age	Sex	Race	Interval or range[a]	ACEI	Hosp[b]	Intub/trach[c]
34	15	42	NA	NA	50% 1 wk	E/L	NA	NA
32	74	60	75% F	96% AA	NA	NA	34	11
56	64	63	38M/26F	NA	12 mo (0–156)	E	6	1
28	9	52	4M/5F	NA	Months	E	2	2
21	5	64	2M/3F	NA	2 d to 10 mo	E	2	2
30	86	NA	35M/51F	23% AA	6 mo	E	2	NA
23	6	65	NA	NA	2 d to months	E	6	NA
25	3	64	1M/2F	Cau	12–33 mo	E	NA	NA
26	6	64	2M/4F	NA	12 h	E	6	2
14	38	NA	NA	NA	1 wk to 11 mo	E	17	4
27	36	61	20M/16F	75% AA	Hours to years	E	36	5
33	4	70	3M/1F	NA	5 mo to 10 yr	E	4	NA
18	20	60	9M/11F	65% AA	70% 4 wk	E	8	2
15	82	NA	NA	65% AA	25% 1 mo; 75% > 1 mo	E/L	25	4
29	19	56	12M/7F	NA	6 h to 13 d	E/C	NA	NA
13	45	62	16M/29F	91% AA	1 d to 5 yr	L/E/Q	45	5

[a] Mean interval or range from drug initiation; [b] Number hospitalized; [c] Number intubated or trached.
NA, not available; M/F, male/female; AA, African American; Cau, Caucasian; E, enalapril; C, captopril; L, lisinopril; Q, quinipril.

Dean *et al.* reporting from the Columbus, Ohio, coroner's office, noted the presence of massive tongue swelling in seven autopsied cases of ACEI-induced angioedema over a 2-year period. Notable was that all subjects were African Americans being treated for hypertension. They also had lip, pharyngeal and laryngeal swelling with negative toxicology for ethanol or illicit drugs.[35]

An isolated case of uvular oedema was attributed to the reinstitution of lisinopril in a 40-year-old African American.[36] Medical management was successful as there was no airway compromise. Thus it is crucial that a careful ENT exam, preferably with nasopharyngoscopy, be performed to both quantify the degree of obstruction and determine whether the patient requires a monitored unit.

MECHANISM OF ANGIOEDEMA

The hypothesized mechanism of ACEI angioedema is the inhibition of bradykinin degradation by the drug, which acts as a kininase II inhibitor, as kininase II is identical to angiotensin-converting enzyme. The resulting increase in circulating bradykinin leads to a vasodilatory state, altering capillary permeability and leading to oedemagenesis, increased post-capillary venule pressure with the release of other vasoactive peptides.[2,21] In addition there may be genetic factors such as a deficiency of other bradykinin-metabolizing enzymes such as carboxypeptidase-N and aminopeptidase-P.[13] Nussberger *et al.* measured bradykinin levels in patients with hereditary angioedema, acquired angioedema, and one patient with angioedema due to captopril.[37] Levels all rose during an acute attack of angioedema in all conditions, with the level rising 15-fold in the ACEI-induced cases, falling back to normal after drug withdrawal. In a subsequent paper the same authors described four additional patients with ACEI-induced angioedema, three of whom had elevated bradykinin levels during remission.[16,38] Anderson and deShazo studied the

cutaneous wheal and flare response to bradykinin, histamine and codeine in 10 normal subjects following the ingestion of the ACEI captopril.[39] Five of the 10 patients developed flushing after the bradykinin skin test. They concluded that inhibition of bradykinin metabolism by ACEI is the cause of angioedema and not substance P where an expected increase in wheal and flare to histamine and codeine would have occurred. All these studies support the proposed mechanism of ACEI-induced angioedema outlined in Fig. 7.1.

TREATMENT OF ANGIOEDEMA

Fibre-optic nasopharyngoscopy is vital in the evaluation and management of any patient with suspected angioneurotic oedema presenting with dyspnoea, stridor, and lip and tongue swelling. In the study by Bentsianov *et al.* of 70 cases of angioneurotic oedema, 24 cases (34 per cent) due to ACEI, the finding of pharyngeal and laryngeal oedema mandated close observation.[40] Fourteen patients required airway intervention: six tracheotomies, four nasal intubations and four oral intubations. The finding of isolated tongue and facial oedema indicated a benign course. Nasopharyngoscopy may have to be repeated at intervals to monitor airway obstruction (the oedema may progress so that the airway lumen becomes impossible to reliably identify) and allow for the removal of artificial airways. As indicated in Table 7.1, a minority of patients require intubation or urgent tracheostomy.

A careful ENT exam will identify those who are at highest risk for airway compromise and mandate their admission to a monitored unit. It is wise to place a tracheostomy tray at the bedside of any patient with airway compromise not yet intubated. As the course of a patient with angioedema is difficult to predict, close observation in an intensive care unit is advised. It is probably best to consider elective intubation of all patients whose speech or control of secretions is impaired even in those with normal oxygenation. A laryngeal mask

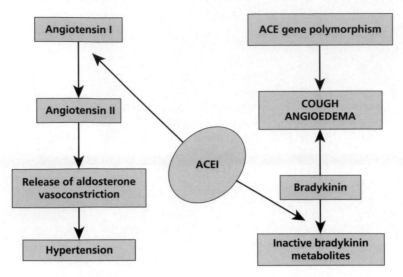

Fig. 7.1   Mechanism of induction of cough and angioedema by an angiotensin-converting enzyme inhibitor (ACEI).

airway (LMA) is an effective temporary method if the airway cannot be immediately visualized while awaiting definitive intubation or, rarely, tracheostomy.

The ventilator management of such patients is fairly straightforward. Most patients do not have intrinsic lung disease and can be weaned easily to a spontaneous mode of ventilation (pressure support or T-piece) once the airway oedema has resolved. The cuff leak test may be useful in assessing the integrity of the upper airway and may be used in the weaning protocol.[41,42] However, caution should be advised in its interpretation, as it may not be adequately sensitive or specific.[43] No studies have utilized the cuff leak test to assess resolution of laryngeal oedema induced by ACEI. Patients should be monitored after extubation, because in a few rebound airway obstruction may develop.

There are no controlled or evidence-based studies evaluating any of the supportive medical therapies for angioedema. Almost all authors recommend intravenous corticosteroids and antihistamines (H1 blockers), while some add adrenaline (epinephrine) and H2 blockers. Most mild cases of angioedema principally limited to the tongue can be managed with an antihistamine alone.[44] A summary of the preferences of several authors is outlined in Table 7.2.

There is an isolated report of improvement in a patient with life-threatening angioedema after the administration of fresh frozen plasma (FFP). The patient had failed therapy with antihistamines, steroids and adrenaline (epinephrine) and required intubation. Within 2 hours of FFP administration there was dramatic improvement in the tongue oedema, allowing extubation. It is postulated that FFP may replace deficient kininase II that is required to break down excess bradykinin.[45]

It is crucial that patients with ACEI- or ARB-induced angioedema never receive these classes of drug again.

ANGIOTENSIN RECEPTOR BLOCKERS

If bradykinin is actually the primary initiator of ACEI angioedema, the angiotensin receptor blockers (ARBs) – which do

Table 7.2   Treatment recommendations for angioedema induced by angiotensin-converting enzyme inhibitors, with percentages of patients receiving intervention, where available

Reference	Treatment recommendations
34	Steroids 70%, H1 blockers 80%, epinephrine 30%
28	Steroids and antihistamines 100%
30	Steroids 23%, H1 blockers 41%, epinephrine 1%
27	Intravenous diphenhydramine and decadron, nebulized racemic epinephrine
17	Steroids 74%, H1 blockers 100%, epinephrine 31%, H2 blockers 12%, bronchodilators 5%
13	Steroids and H1 blockers 100%, H2 blockers 90%, epinephrine 30%
32	Steroids, H1 blockers, high-humidity face tents

not affect the bradykinin pathway – ought to be relatively safe. The losartan intervention trial[46] identified only six cases of angioedema in a total of 4600 treated patients, resulting in an incidence of 0.1 per cent – a small number but similar to the reported incidence of ACEI angioedema. Nevertheless, since the release and widespread use of ARBs there have been numerous reports of associated angioedema.

The first reports were in patients who had experienced cough due to captopril and angioedema due to accupril or enalapril and then switched to losartan. The angioedema developed minutes to 2 months after the institution of the drug, with all patients responding to medical therapy.[47-49] Chiu *et al.*, in a retrospective review, described three patients with recurrent losartan-induced angioedema.[50] One required a tracheostomy, all received steroids and antihistamines, and all survived. C-1 esterase levels were normal in the one subject in which it was measured. Thirteen cases were reported from the Netherlands occurring 24 hours to 16 months after losartan initiation; 3/13 (23 per cent) had previous angioedema due to ACEI.[51-53] Caution was advised in instituting ARBs in patients with a history of ACEI-related, hereditary or idiopathic angioedema. A single case report appears to link neurological symptoms of a transient ischaemic attack to losartan.[54]

In a multicentre, double-blind randomized study, 135 patients with a history of ACEI-induced cough were randomly assigned to receive 50 mg of losartan, 20 mg of lisinopril or 25 mg of HCTZ.[55] The incidences of cough were 29, 71 and 34 per cent, respectively. Cough occurred 3–4 weeks after initiation of the drug in all groups. Although the incidence of cough was less with losartan it still occurred in almost one-third of the patients.

In a retrospective analysis of 54 patients with ACEI angioedema, 26 of whom were switched to an ARB, Cicardi *et al.* suggested the relative safety of ARBs, with only two patients with persistent oedema.[56] It has been suggested that this category of drugs may also increase bradykinin – although the mechanism is unclear.

Thus, although the incidence is small, caution seems to be best when adjusting therapy in those with a history of ACEI angioedema. Patients who are switched to an ARB after an episode of ACEI angioedema should have a strong justifiable reason for the switch, such as intractable congestive heart failure, uncontrolled hypertension, considerable proteinuria, or sensitivities to other medications. Proper education of the patient is warranted.

OTHER AGENTS ASSOCIATED WITH UPPER AIRWAY OEDEMA

There are isolated case reports of other substances associated with upper airway oedema. Isocyanate, a compound used in polyurethane resins, was implicated in epiglottic dysfunction requiring surgical intervention.[57] Pathologic evaluation of the surgically removed epiglottis revealed lymphoid hyperplasia and a follicular reaction consistent with exposure to isocyanate.

Psyllium, a bulk laxative, has been associated with severe laryngeal oedema and death after ingestion. Healthcare workers appear to be at high risk because of repeated exposure and sensitization. In a review of the literature, 63 patients were identified with allergic reactions to psyllium, 30 of whom had anaphylaxis.[58] Anaphylaxis was more likely to occur after ingestion, while inhalation was associated with rhinitis and bronchospasm. Treatment is wholly supportive.

DRUGS PRODUCING IMMUNE-MEDIATED UPPER AIRWAY COMPROMISE

Examples of drugs causing immune-mediated upper airway compromise include penicillin, aspirin, methotrexate, non-steroidal anti-inflammatory drugs (NSAIDs), radio-contrast media, anaesthetic agents, and blood products (including immunoglobulin products). *Anaphylaxis* occurs in patients who have been previously sensitized to an allergen and is independent of dose, whereas anaphylactoid reactions are dose-dependent and do not require previous exposure or sensitization to an allergen. The incidence of anaphylaxis of all types is reportedly rising and may be related to the increased allergens in various settings.[59]

PENICILLIN

Epidemiology of penicillin–related immune reactions

A penicillin-induced immune-mediated upper airway problem is part of the anaphylactic reaction, and the drug is the most common class of antibiotic associated with this reaction. Anaphylaxis due to penicillin is a dreaded adverse effect that has been reported in 0.2 per cent (1.2/10 000 injections), with a fatality incidence of 0.05 per cent (0.31/10 000 injections).[60] Penicillin is the most frequent cause of anaphylaxis in humans and has been estimated to be responsible for 75 per cent of anaphylactic deaths in the United States.[61] Estimates of non-fatal drug-induced anaphylaxis to penicillin range from 0.7 to 10 per cent among the general population. Anaphylaxis to penicillin occurs most commonly in adults aged 20–49 years, but it has been observed in all age groups.

The penicillin class of drugs differ at their side-chains, so that cross-sensitivity to all the drugs may occur. There is perhaps a 5–10 per cent cross-sensitivity between penicillin derivatives, cephalosporins and carbapenems.[62]

Mechanisms of immune reactions due to penicillin

Box 7.2 outlines the various drug-related upper airway obstructions. Penicillin can give rise to anaphylaxis via the formation of (1) specific IgE antibodies (type I) and mast-cell degranulation on re-exposure in a sensitized individual, (2) complement-mediated haemolytic anaemia (type II), (3) interstitial nephritis, delayed onset of a vasculitis-like skin rash, arthritis and respiratory distress give rise to immune complexes which are deposited

---

Box 7.2   Mechanisms of drug-related upper airway obstruction

Immune mechanisms causing angioedema and/or anaphylaxis

- Type I–IV reactions
- Complement activation
- Direct action on mast cells, basophils and eosinophils
- Binding of antibodies
- Act as haptens against which an immune response develops

Non-immune-mediated mechanisms causing upper airway obstruction

- Angiotensin-converting enzyme inhibitors (bradykinin pathway)
- Anticoagulants (local haematoma formation)
- Airway injury (thermal–'crack' cocaine)
- Anaesthetic agents and neuromuscular blockers (immune, anatomic/physiological factors)

in blood vessel walls and in tissues (type III); and (4) (type IV) delayed contact dermatitis. The penicillin molecule can form several haptens *in vivo* and *in vitro*. The penicilloyl hapten is responsible for the majority of the reactions. IgE-mediated hypersensitivity is manifested in only a small proportion.

Immediate allergic reaction to penicillin may occur within 1 hour after exposure. The pertinent clinical features are cardiovascular collapse with acute hypotension, rhinitis, respiratory distress with wheezing, laryngeal oedema, abdominal pain, urticaria and angioedema. Serum tryptase concentrations (peak 1 hour after anaphylactic reactions), serum and urinary histamine metabolites may be useful in the diagnosis.

Management of anaphylaxis

Practice guidelines for the management of drug-related anaphylaxis have been recommended.[61] Airway protection, oxygen and cardiopulmonary resuscitation are the primary steps of management. The other immediate pharmacological treatments include:

- intramuscular or subcutaneous adrenaline (epinephrine: adults, 0.2–0.5 mL of a 1:1000 dilution every 10–15 minutes up to a maximum of 1.0 mL per dose);
- parenteral diphenhydramine (25–50 mg);
- intravenous hydrocortisone, primarily for a late response;
- intravenous fluids and vasopressors for hypotension.

In a study of 124 fatal anaphylactic reactions in the UK by Pumphrey, iatrogenic causes, foods and venoms made up the bulk of aetiologies.[63] Of the 55 iatrogenic causes, anaesthetic agents and antibiotics made up the majority, with the median time of cardiac or respiratory arrest occurring within 5 minutes of ingestion. Although 28 per cent were initially resuscitated, all died within 30 days from hypoxic brain damage. Adrenaline was administered in 62 per cent of patients, but before the arrest in only 14 per cent. The ready availability of adrenaline-containing allergy kits as well as prompt distinction between an asthma attack and an anaphylactic reaction are instrumental in saving lives.

In the event of penicillin being absolutely necessary, skin testing with the allergenic components may be cautiously undertaken. If hypersensitivity is confirmed, a rapid desensitization programme can be undertaken, with gradually increasing intradermal and subsequently oral doses of the drug being given over a number of hours.

RADIO-CONTRAST MEDIA-INDUCED ANAPHYLAXIS

More than 10 million procedures are performed in the United States each year using radio-contrast media (RCM). Anaphylactic reactions due to RCM are speculated to be independent of IgE antibody-mediated mast-cell degranulation. The reactions may be due to direct complement activation or basophil degranulation. A history of seafood allergy is not a definite predictor of risk, and conflicting reports have led to premedication with corticosteroids prior to contrast administration.[64] The frequency of anaphylactoid reactions is reported to be between

0.22 and 1 per cent of patients who receive RCM. Fewer anaphylactoid reactions and mortality have been reported with the newer, lower-osmolarity RCMs.[65]

Anaphylactoid deaths caused by lower-osmolar RCMs are as low as 1 in 168 000 administrations, compared to 1 in 75 000 with conventional RCMs.[64] In a Japanese study involving 337 647 cases, the prevalence of severe adverse reaction after use of a high-osmolar RCM was 0.22 per cent, compared to 0.04 per cent with the use of a lower osmolar RCM.[65] The frequency of reactions is equally distributed between males and females and occurs most often between the ages of 20 and 50 years. Reactions are less frequent in children and the elderly but are generally then more severe.

An increased risk and severity of anaphylactoid reactions to RCMs has been linked to the concomitant use of beta-blocker drugs. One study found that risk of moderate to severe anaphylactoid reaction was 2.6 times higher in patients taking beta-blocking agents than in those who were not taking these medications.[66] Clinical features of RCM-related anaphylactoid reaction include vasomotor reactions such as nausea, vomiting, flushing or warmth, and a combination of pulmonary, cardiovascular, gastrointestinal and cutaneous symptoms.

Prevention of RCM-induced reactions

Proper hydration and the use of a lower osmolar RCM, especially in high-risk patients (with asthma and cardiovascular disease, on beta-blockers), are of prime importance in preventing serious adverse effects. A combination of H1 antihistamines and corticosteroids has been shown to be effective in reducing reactions to radio-contrast media. Greenberger and Patterson reported the pretreatment use of prednisone 50 mg orally 13 hours, 7 hours and 1 hour before the procedure, and diphenhydramine 50 mg and ephedrine 25 mg 1 hour before the procedure.[67] Significant clinical improvement has been reported with the use of glucagon in anaphylactoid shock potentiated by beta-blockers during coronary angiography.[68,69]

ANAESTHETIC AGENTS AND UPPER AIRWAY OBSTRUCTION

Adverse outcomes associated with respiratory events in the American Society of Anesthesiologists' Closed Claims Project were evaluated. Respiratory events accounted for 762 of 2046 cases (37 per cent). Airway obstruction occurred in 56 claims (3 per cent of the database) and 39 cases occurred in the upper airway.[70] In another review of 104 claims associated with induction of anaesthesia, 37 per cent of the patients were obese, and failed awake intubation occurred in 12 cases owing to upper airway obstruction.[71]

A variety of anaesthetic drugs has the potential to cause adverse reactions by a number of mechanisms, such as immune-mediated or anatomic/physiological factors related to the oropharynx. The incidence of allergic reactions during general anaesthesia has been reported to be between 1 in 4000 and 1 in 25 000, with a 3.4 per cent mortality rate.[61]

There have been several case reports describing spontaneous ignition and fire in the operating room after the use of sevoflurane due to a chemical reaction with carbon dioxide absorbers containing barium hydroxide. Considerable melting of the plastic components of the anaesthesia machine was noted (no fatalities occurred). This is a potential problem with all halogenated anaesthetics and attention should be given to the avoidance of strong alkaline absorbents when using these agents.[72,73]

Immune-mediated anaesthetic drug reactions

Anaesthetic agents can cause anaphylactoid reactions and upper airway oedema due to mast-cell degranulation, and IgE- or non-IgE-mediated mechanisms (Box 7.3). In a 6-year Norwegian study, 83 cases of anaphylaxis during anaesthesia were reported. Neuromuscular blockers were by far the most frequently associated agents (93 per cent). Suxamethonium was the most common agent, followed by rocuronium and vecuronium. Latex had accounted for 3.6 per cent.[74] Injectable agents such as succinyl choline, d-tubocurarine, pancuronium, thiopental, protamine, preoperative antibiotics, blood products and topical agents such as povidone iodine and latex have all been reported to cause IgE- and non-IgE-mediated anaphylaxis. True IgE-mediated allergy can be caused by suxamethonium. Bivalence (ability to cross-link receptors) is a mechanism by which succinyl choline acts as a hapten to trigger the allergic response. Local anaesthetic agents, particularly amino amides such as lidocaine, can cause rash and other allergic reactions including anaphylaxis. Some amino amides contain methylparaben, a compound structurally similar to para-aminobenzoic acid (PABA), which may account for the resultant stimulatory immune effect.

The first problem for the anaesthetist is to recognize an adverse event as being due to the drugs administered, particularly if anaphylaxis occurs intraoperatively. The investigators must have access to a full list of administered drugs including fluids and blood products, as the latter group is associated with direct mast-cell degranulation and complement activation. Because of the potential complexity of anaesthetic reactions, it is advisable that a clinical immunologist be involved early after such a reaction to provide guidance.

Anatomic/physiological factors causing upper airway obstruction during anaesthesia

Oropharyngeal factors such as underlying obstructive sleep apnoea, local trauma or pre-existing inflammation may become clinically evident with administration of anaesthetic agents (Box 7.4).

Obstructive sleep apnoea (OSA) is an anatomical problem. A review of the anatomy of the oropharyngeal muscles assists in the understanding of the mechanism of upper airway obstruction during sleep. Movement of the tongue occurs with the help of four strong muscles: genioglossus for protrusion; hypoglossus for depression; styloglossus for pulling the tongue upwards and backwards; and palatoglossus for pulling the posterior part of the tongue upwards. The hypoglossal nerve innervates all the tongue muscles except the palatopharyngeus, which is supplied by the vagus-pharyngeal plexus.

Box 7.3 Anaesthetic agents that may cause anaphylaxis/upper airway obstruction

- Narcotics: codeine, morphine, meperidine (Demerol)
- Neuromuscular blocking agents: succinyl choline, D-tubocurarine, pancuronium
- Sympathomimetics: ephedrine
- Barbiturates: thiopental
- Others: protamine, latex, ethylene oxide, lidocaine, propofol, chloral hydrate

---

Box 7.4 Oropharyngeal factors during anaesthesia causing upper airway obstruction

- Pre-existing swelling of supraglottic structures: large tongue, tonsils and uvula
- Primary disease, previous radiation treatment
- Loss of upper airway muscle tone during sleep
- Collapse of upper airway structures during deep inspiratory efforts
- Laryngospasm caused by irritation of lidocaine on the cords
- Functional stridor and paradoxical vocal cord movement
- Panic-related spasm

Patients with OSA have a structurally smaller pharyngeal airway than non-apnoeic patients. The genioglossus is of great importance in the pathophysiology of sleep apnoea as it contracts during inspiration and increased phasic activity of the genioglossus is seen during the awake state in patients with OSA. The phasic activity of genioglossus and tensor veli palatini decreases with sleep onset. Sleep deprivation and use of alcohol, sedatives and anaesthetic agents such as propofol also reduce the activity of the genioglossus.[75–79] The loss of genioglossus muscle activity is associated with a decrease in airway patency and collapse of the pharyngeal airway. Hypoglossal nerve stimulation contracts the genioglossus and has been studied for maintaining upper airway patency in OSA.[80] Continuous positive airway pressure (CPAP) acts as a pneumatic splint of the upper airway and is used for the management of OSA.

Total airway obstruction during local anaesthesia has been described. Local anaesthesia of the larynx decreases upper airway muscle contractions during inspiration and a decrease in upper airway calibre.[81,82] Lidocaine depresses the laryngeal muscles and normal laryngeal function. Lidocaine also inhibits laryngeal receptors, resulting in the upper airway mucosa collapsing during inspiration. Upper airway obstruction during recovery may be attributed to the residual effects of neuromuscular blockers and anaesthetic agents and loss of muscle support of upper airway structures.[83]

Upper airway obstruction (UAO) during anaesthesia may be predicted preoperatively by the following: evaluation of static structural abnormalities by preoperative plain X-ray; computed tomography; magnetic resonance imaging; ultrasonography;

dynamic airflow obstruction by spirometry with flow–volume loops; dynamic negative airway pressure to assess sedative-induced UAO. However, the practical utility of any of these investigations as part of airway evaluation is not established to predict severe airway obstruction after anaesthesia.

Management of anaesthesia-related upper airway obstruction

Airway management in a patient with significant oral and head and neck pathology should be a joint decision between anaesthetist and surgeon. Use of non-invasive ventilation such as CPAP or BIPAP may relieve UAO induced by conscious sedation. Oral or nasotracheal intubation is to be considered in patients with significant airway compromise and hypoxaemia. Well-trained personnel and equipment for an emergency surgical airway should always be readily available in the operating room and post-op recovery room.[84]

ANTICOAGULANT–INDUCED HAEMATOMA CAUSING UPPER AIRWAY OBSTRUCTION

Spontaneous haemorrhaging into the parapharyngeal and submandibular spaces is a rare complication of anticoagulant therapy.[85–90] Massive haemorrhaging causing airway obstruction may occur. In most of the cases when diagnosed early and managed promptly the prognosis is good, with or without surgical intervention, and the haematomas resolve completely. The most common presentation is sore throat, hoarseness and swelling in the oral cavity or neck following an episode of straining (sneeze, cough or vomiting), dysphagia, and drooling or respiratory distress. An examination may confirm the parapharyngeal swelling or a bluish sublingual mass.

Lepore *et al.* explained how airway obstruction might result from submandibular bleeding (pseudo-Ludwig phenomenon) due to warfarin.[88] Swelling of the floor of the mouth displaces the tongue upward and backward leading to respiratory distress. Failure of visualization of the fauces and bulging of the posterior pharyngeal wall are signs of impending airway compromise.

In a case series of 12 patients presenting with UAO secondary to warfarin-induced sublingual haematoma, endotracheal intubation was needed in two patients, emergency cricothyrotomy in two, and tracheotomy in five. Fresh frozen plasma and vitamin K were given to reverse anticoagulation.[90]

Management of anticoagulant-induced haematoma

Rapid reversal of elevated prothrombin time and international normalization ratio (INR) with fresh frozen plasma and vitamin K is necessary and ICU monitoring is essential. As oral intubation may be difficult owing to the obstructive effects of the sublingual haematoma, the nasal route is preferred. Direct visualization by naso-endoscope of the upper airway and intubation performed by an experienced operator is ideal to avoid

an invasive surgical airway. Surgical drainage of the haematoma is not indicated, as it usually resolves over the next few days. There are insufficient data supporting use of corticosteroids and antibiotics.

COCAINE AND THE UPPER AIRWAY

Illicit cocaine and 'crack' cocaine inhalation can affect the upper airway and cause chronic cough, chronic rhino-sinusitis, epistaxis, nasal septal perforation, acute glossitis, epiglottitis, oropharyngeal ulcers, pseudo-vasculitis and uvular angioedema and thermal injury (Quincke's disease).[91–94] Cocaine has local anaesthetic properties and may lead to loss of the protective airway reflex. Paradoxical vocal cord movement due to this action of cocaine and laryngeal burns after inhalation of hot particles of 'crack' have been reported. Cocaine has been shown to impair alveolar macrophage function, cytokine production and activation of polymorphonuclear leucocytes that result in lung injury.[95,96] Inhalation of reheated cocaine freebase residue leads to persistent cough productive of black sputum.[97] Bronchoalveolar lavage yields carbonaceous material with alveolar macrophages.[95]

Cocaine is extracted from the leaves of *Erythroxolon coca*. Freebase cocaine is prepared by mixing cocaine with baking soda and boiling with water. Once cooled, a solvent such as ether or alcohol is added. The separated solvent is then evaporated while smoking, which may lead to mucosal injuries of the upper aerodigestive tract. In addition, inhalation of hot particulate debris from the smoking pipe and the wire mesh screen may contribute to the upper airway injury.

In a retrospective chart review, seven patients with upper airway burns due to smoking 'crack' or freebase cocaine were reported.[98] Symptoms of laryngeal burn due to cocaine include cough, dysphonia, stridor, globus sensation or foreign body aspiration. Mucosal oedema of the aryepiglottic folds, arytenoids, epiglottis and posterior pharyngeal wall, pooling of secretions in the pyriform sinuses, glottic narrowing and bullous mucosal lesions on the epiglottis were noted on nasopharyngoscopy. Two of the seven patients needed an emergent tracheostomy due to stridor. All patients received intravenous corticosteroids, antibiotics, humidified oxygen and aerosolized adrenaline (epinephrine). Topical viscous lidocaine 2 per cent has also been used in thermal injury of the upper airway. In the presence of odynophagia, upper gastrointestinal endoscopy may be required to evaluate for thermal injury and foreign bodies.[99]

Pulmonary complications of illicit drug use are discussed further in Chapter 12.

> **KEY POINTS**
>
> - Angiotensin-converting enzyme inhibitors (ACEIs) are by far the most common class of drugs associated with airway obstruction. It is crucial that patients with ACEI-induced angioedema never receive this class of drug again. Patients who are switched to an ARB after an episode of ACEI angioedema should have a strong justifiable reason for the switch, since relapse may occur.

- Anaphylaxis occurs in patients who have been previously sensitized to an allergen and is independent of dose, whereas anaphylactoid reactions are dose-dependent and do not require previous exposure or sensitization to an allergen.
- An increased risk and severity of anaphylactoid reactions to radio-contrast media has been linked to the concomitant use of beta-blocker drugs.
- A variety of anaesthetic drugs has the potential to cause adverse upper airway reactions by a number of mechanisms.

REFERENCES

1. Garg R, Yusuf S. Overview of randomized trials of angiotensin-converting enzyme inhibitors on mortality and morbidity in patients with heart failure. *J Am Med Assoc* 1995;**273**:1450–6.
2. Israeli ZH, Hall WD. Cough and angioneurotic edema associated with angiotensin-converting enzyme inhibitor therapy: a review of the literature and pathophysiology. *Ann Int Med* 1992;**117**:234–42.
3. Sebastin J, Mckinney WP, Kaufman J, Young MJ. Angiotensin-converting enzyme inhibitors and cough. *Chest* 1991;**99**:36–9.
4. Coulter DM, Edwards IR. Cough associated with captopril and enalapril. *Br Med J* 1987;**294**:1521–3.
5. Gibson GR. Enalapril-induced cough. *Arch Intern Med* 1989;**149**:2701–3.
6. Simon S, Black H, Moser M, Berland WE. Cough and ACE inhibitors. *Arch Intern Med* 1992;**152**:1698–700.
7. Kaufman J, Cassanova J, Riendl P, Schlueter D. Bronchial hyperreactivity and cough due to ACE inhibitors. *Chest* 1989;**95**:544–8.
8. Kaufman J, Schmitt S, Barnard J, Busse W. Angiotensin-converting enzyme inhibitors in patients with bronchial responsiveness and asthma. *Chest* 1992;**101**:922–5.
9. Boulet LP, Milot J, Lampron N, Lacourciere Y. Pulmonary function and airway responsiveness during long-term therapy with captopril. *J Am Med Assoc* 1989;**261**:413–16.
10. Bucca C, Rolla G, Pinna G, Oliva A, Bugiani M. Hyperresponsiveness of the extrathoracic airway in patients with captopril-induced cough. *Chest* 1990;**98**:1133–7.
11. Dicpinigaitis PV. Angiotensin-converting enzyme inhibitor-induced cough: ACCP evidence-based clinical practice guidelines. *Chest* 2006;**129**:169S–73S.
12. Lacourciere Y, Lefebvre J, Nakhle G, *et al.* Association between cough and angiotensin converting enzyme inhibitors versus angiotensin II antagonists: the design of a prospective controlled study. *J Hypertens* 1994;**12**(Suppl 2):S49–53.
13. Sondhi D, Lippmann M, Murali G. Airway compromise due to angiotensin-converting enzyme inhibitor-induced angioedema. *Chest* 2004;**126**:400–4.
14. Slater E, Merrill DD, Guess HA, *et al.* Clinical profile of angioedema associated with angiotensin converting enzyme inhibition. *J Am Med Assoc* 1988;**260**:967–70.
15. Brown NJ, Snowden M, Griffin MR. Recurrent angiotensin-converting enzyme inhibitor-associated angioedema. *J Am Med Assoc* 1997;**278**:232–3.
16. Nussberger J, Cugno M, Cicardi M. Bradykinin-mediated angioedema. *New Engl J Med* 2002;**347**:621–2.
17. Brown NJ, Ray WA, Snowden M, Griffin MR. Black Americans have an increased rate of angiotensin converting enzyme inhibitor-associated angioedema. *Clin Pharmacol Ther* 1996;**60**:8–13.
18. Gibbs CR, Lip GYH, Beevers DG. Angioedema due to ACE inhibitors: increased risk in patients of African origin. *Br J Clin Pharmacol* 1999;**48**:861–5.
19. Gainer JV, Nadeau JH, Ryder D, Brown N. Increased sensitivity to bradykinin among African Americans. *J Allergy Clin Immunol* 1996;**98**:283–7.
20. Abbosh J, Anderson J, Levine A, Kupin W. Angiotensin converting enzyme inhibitor-induced angioedema more prevalent in transplant patients. *Ann Allergy Asthma Immunol* 1999;**82**:473–6.
21. Jain M, Armstrong L, Hall J. Predisposition to and late onset of upper airway obstruction following angiotensin-converting enzyme inhibitor therapy. *Chest* 1992;**102**:872–4.
22. Cicolin A, Mangiardi L, Mutani R, Bucca C. Angiotensin-converting enzyme inhibitors and obstructive sleep apnea. *Mayo Clin Proc* 2006;**81**:53–5.
23. Megerian C, Arnold JE, Berger M. Angioedema: 5 years' experience, with a review of the disorder's presentation and treatment. *Laryngoscope* 1992;**102**:256–60.
24. Mchaourab A, Sarantopoulos C, Stowe DF. Airway obstruction due to late-onset angioneurotic edema from angiotensin-converting enzyme inhibition. *Can J Anesth* 1999;**46**:975–8.
25. Schiller PI, Langauer SM, Haefeli S, *et al.* Angiotensin-converting enzyme inhibitor-induced angioedema: late onset, irregular course, and potential role of triggers. *Allergy* 1997;**52**:432–5.
26. Seidman MD, Lewandowski CA, *et al.* Angioedema related to angiotensin-converting enzyme inhibitors. *Otolaryngol Head Neck Surg* 1990;**102**:727–31.
27. Thompson T, Frable MAS. Drug-induced, life-threatening angioedema revisited. *Laryngoscope* 1993;**103**:10–12.
28. Gunkel AR, Thurner KH, *et al.* Angioneurotic edema as a reaction to angiotensin-converting enzyme inhibitors. *Am J Otolaryngol* 1996;**17**:87–91.
29. Wood S, Mann R, Rawlins M. Angioedema and urticaria associated with angiotensin converting enzyme inhibitors. *Br Med J* 1987;**294**:91–2.
30. Kostis JB, Kim HJ, Rusnak J. Incidence and characteristics of angioedema associated with enalapril. *Arch Intern Med* 2005;**165**:1637–42.
31. Golden WE, Cleves M, Heard JK, *et al.* Frequency and recognition of angiotensin-converting enzyme inhibitor-associated angioneurotic edema. *Clin Perform Quality Health Care* 1993;**1**:205–7.
32. Chiu AG, Newkirk KA, Davidson B, *et al.* Angiotensin-converting enzyme inhibitor-induced angioedema: a multicenter review and algorithm for airway management. *Ann Otol Rhinol Laryngol* 2001;**111**:834–40.
33. Zanoletti E, Bertino G, Malvezzi L, *et al.* Angioneurotic edema of the upper airways and antihypertensive therapy. *Acta Otolaryngol* 2003;**123**:960–4.
34. Agah R, Bandi C, Guntupalli KK. Angioedema: the role of ACE inhibitors and factors associated with poor clinical outcome. *Intens Care Med* 1997;**23**:793–6.

35. Dean DE, Schultz DL, Powers RH. Asphyxia due to angiotensin converting enzyme (ACE) inhibitor mediated angioedema of the tongue during the treatment of hypertensive heart disease. *J Forens Sci* 2001;**46**:1239–43.

36. Kuo D, Barish RA. Isolated uvular angioedema associated with ACE inhibitor use. *J Emerg Med* 1995;**13**:327–30.

37. Nussberger J, Cugno M, Amstutz C, *et al.* Plasma bradykinin in angio-edema. *Lancet* 1998;**351**:1693–7.

38. Cugno M, Nussberger J, Cicardia M, Agostoni A. Bradykinin and the pathophysiology of angioedema. *Int Immunopharm* 2003;**3**:311–17.

39. Anderson MW, deShazo RD. Studies of the mechanism of angiotensin-converting enzyme (ACE) inhibitor-associated angioedema: the effect of an ACE inhibitor on cutaneous responses to bradykinin, codeine, and histamine. *J Allergy Clin Immunol* 1990;**85**:856–8.

40. Bentsianov BL, Parhiscar A, Azer M, Har El G. The role of fiberoptic nasopharyngoscopy in the management of the acute airway in angioneurotic edema. *Laryngoscope* 2000;**110**:2016–19.

41. Miller RL, Cole R. Association between reduced cuff leak volume and postextubation stridor. *Chest* 1996;**110**:1035–40.

42. Chung YH, Chao TY, Chiu CT, Lin MC. The cuff-leak test is a simple tool to verify severe laryngeal edema in patients undergoing long-term mechanical ventilation. *Crit Care Med* 2006;**34**:409–14.

43. Wani A, Delgado J, Becker C, Adams S, Lippmann ML. The cuff-leak test as a predictor of successful extubation in mechanically ventilated patients. *AJR CCM* 2005;**2**:A158 (abstract issue).

44. Sica D, Black H. ACE inhibitor-related angioedema: can angiotensin-receptor blockers be safely used? *J Clin Hypertension* 2002;**4**:375–80.

45. Karim MY, Masood A. Fresh-frozen plasma as a treatment for life-threatening ACE-inhibitor angioedema. *J Allergy Clin Immunol* 2002;**109**:370–1.

46. Dahlof B, Devereux RB, *et al.* Cardiovascular morbidity and mortality in the Losartan Intervention For Endpoint reduction in hypertension study (LIFE): a randomized trial against atenolol. *Lancet* 2002;**359**:995–1003.

47. Acker CG, Greenberg A. Angioedema induced by the angiotensin II blocker losartan. *New Engl J Med* 1995;**333**:1572.

48. Boxer M. Accupril and cozaar-induced angioedema in the same patient. *J Clin Immunol* 1996;**98**:471.

49. Abdi R, Deng V, Lee C, Ntoso K. Angiotensin II receptor blocker-associated angioedema: on the heels of ACE inhibitor angioedema. *Pharacotherapy* 2002;**22**:1173–5.

50. Chiu A, Krowiak E, Deeb Z. Angioedema associated with angiotensin II receptor antagonists: challenging our knowledge of angioedema and its etiology. *Laryngoscope* 2001;**111**:1729–31.

51. Van Rijnsoever EW, Kwee Zuiderwijk JM, Feenstra J. Angioneurotic edema attributed to the use of losartan. *Arch Intern Med* 1998;**158**:2063–5.

52. Warner KK, Visconti JA, Tschampel MM. Angiotensin II receptor blockers in patients with ACE inhibitor-induced angioedema. *Ann Pharmacol* 2000;**34**:526–8.

53. Fuchs SA, Koopmans RP, *et al.* Are angiotensin II receptor antagonists safe in patients with previous angiotensin-converting enzyme inhibitor-induced angioedema? *Hypertension* 2001;**37**:1.

54. Rupprecht R, Vente C. Angioedema due to losartan. *Allergy* 1999;**54**:81–2.

55. Lacourciere Y, Brunner H, Irwin R, *et al.* Effects of modulators of the renin–angiotensin–aldosterone system on cough. *J Hypertension* 1994;**12**:1387–93.

56. Cicardi M, Zingale L, *et al.* Angioedema associated with angiotensin-converting enzyme inhibitor use. *Arch Intern Med* 2004;**164**:910–13.

57. Sales JH, Kennedy K. Epiglottic dysfunction after isocyanate inhalation exposure. *Arch Otolaryngol Head Neck Surg* 1990;**116**:725–7.

58. Khalili B, Bardana EJ, Yunginger JW. Psyllium-associated anaphylaxis and death: a case report and review of the literature. *Ann Allergy Asthma Immunol* 2003;**91**:79–84.

59. Neugut AI, Ghatak AT, Miller RL. Anaphylaxis in the United States. *Arch Intern Med* 2001;**161**:15–21.

60. Kelkar PS, Li JT. Cephalosporin allergy. *New Engl J Med* 2001;**345**:804–9.

61. JACI. The diagnosis and management of anaphylaxis: an updated practice parameter. *J Allergy Clin Immunol* 2005;**115**:S483–523.

62. Markowitz M, Kaplan E, Cuttica R, *et al.* Allergic reactions to long-term benzathine penicillin prophylaxis for rheumatic fever. *Lancet* 1991;**337**:1308–10.

63. Pumphrey RSH. Lessons for management of anaphylaxis from a study of fatal reactions. *Clin Exper Allergy* 2000;**30**:1144–50.

64. Leder R. How well does a history of seafood allergy predict the likelihood of an adverse reaction to IV contrast material? *Am J Roentgenol* 1997;**169**:906–7.

65. Katayama H, Yamaguchi K, Kozuka T, *et al.* Adverse reactions to ionic and nonionic contrast media: a report from the Japanese Committee on the Safety of Contrast Media. *Radiology* 1990;**175**:621–8.

66. Lang DM, Alpern MB, Visintainer PF, Smith ST. Increased risk for anaphylactoid reaction from contrast media in patients on beta-adrenergic blockers or with asthma. *Ann Intern Med* 1991;**115**.270–6.

67. Greenberger PA, Patterson R. The prevention of immediate generalized reactions to radio contrast media in high-risk patients. *J Allergy Clin Immunol* 1991;**87**:867–72.

68. Zaloga GP, Delacey W, Holmboe E, *et al.* Glucagon reversal of hypotension in a case of anaphylactoid shock. *Ann Intern Med* 1986;**105**:65–6.

69. Javeed N, Javeed H, Javeed S, *et al.* Refractory anaphylactoid shock potentiated by beta-blockers. *Cathet Cardiovasc Diagn* 1996;**39**:383–4.

70. Cheney FW, Posner KL, Caplan RA. Adverse respiratory events infrequently leading to malpractice suits: a closed claims analysis. *Anesthesiology* 1991;**75**:932–9.

71. Peterson GN, Domino KB, Caplan RA, *et al.* Management of the difficult airway: a closed claims analysis. *Anesthesiology* 2005;**103**:33–9.

72. Wu J, Previte J, Adler E, *et al.* Spontaneous ignition, explosion, and fire with sevoflurane and barium hydroxide lime. *Anesthesiology* 2004;**101**:534–7.

73. Castro BA, Freedman LA, Craig WL, Lynch C. Explosion within an anesthesia machine: baralyme, high fresh gas flows and sevoflurane concentration. *Anesthesiology* 2004;**101**:557–9.

74. Harboe T, Guttormsen AB, Irgens A, *et al.* Anaphylaxis during anesthesia in Norway: a 6-year single-center follow-up study. *Anesthesiology* 2005;**102**:897–903.
75. Crawford MW, Rohan D, Macgowan CK, *et al.* Effect of propofol anesthesia and continuous positive airway pressure on upper airway size and configuration in infants. *Anesthesiology* 2006;**105**:45–50.
76. Mathru M, Esch O, Lang J, *et al.* Magnetic resonance imaging of the upper airway: effects of propofol anesthesia and nasal continuous positive airway pressure in humans. *Anesthesiology* 1996;**84**:253–5.
77. Eastwood PR, Platt PR, Shepherd K, *et al.* Collapsibility of the upper airway at different levels of propofol anesthesia. *Anesthesiology* 2005;**103**:470–7.
78. Leiter JC, Knuth SL, Krol RC, Bartlett D. The effect of diazepam on genioglossal muscle activity in normal human subjects. *Am Rev Respir Dis* 1985;**132**:216–19.
79. Kuna ST, Sant'Ambrogio G. Pathophysiology of upper airway closure during sleep. *J Am Med Assoc* 1991;**266**:1384–9.
80. Goding GS, Eisele DW, Testerman R, *et al.* Relief of upper airway obstruction with hypoglossal nerve stimulation in the canine. *Laryngoscope* 1998;**108**:162–9.
81. Liistro G, Stanescu DC, Veriter C, *et al.* Upper airway anesthesia induces airflow limitation in awake humans. *Am Rev Respir Dis* 1992;**46**:581–5.
82. Ho AM, Chung DC, To EW, Karmakar MK. Total airway obstruction during local anesthesia in a non-sedated patient with a compromised airway. *Can J Anaesth* 2004;**51**:838–41.
83. D'Honneur G, Lofaso F, Drummond GB, *et al.* Susceptibility to upper airway obstruction during partial neuromuscular block. *Anesthesiology* 1998;**88**:371–8.
84. Nicklas RA, Bernstein IL, Li JT, *et al.* The diagnosis and management of anaphylaxis. XIX: Anaphylaxis during general anesthesia, the intraoperative period, and the postoperative period. *J Allergy Clin Immunol* 1998;**101**:S465–528.
85. Duong TC, Burtch G, Shatney CH. Upper airway obstruction as a complication of oral anticoagulation therapy. *Crit Care Med* 1986;**14**:830–1.
86. Bachman P, Gaussorgues P, Pignat J. Pulmonary edema secondary to warfarin-induced sublingual and laryngeal hematoma. *Crit Care Med* 1987;**15**:1074–5.
87. Basavaraj S, Sharp JF. Spontaneous parapharyngeal haemorrhage as a complication of anticoagulation therapy. *Internet J Otorhinolaryngol* 2002;**1**(2).
88. Lepore ML. Upper airway obstruction induced by warfarin sodium. *Arch Otolaryngol* 1976;**102**:505–6.
89. Rosenbaum L, Thurman P, Krantz SB. Upper airway obstruction as a complication of oral anticoagulation therapy. *Arch Intern Med* 1979;**139**:1151–3.
90. Cohen AF, Warman SP. Upper airway obstruction secondary to warfarin-induced sublingual hematoma. *Arch Otolaryngol Head Neck Surg* 1989;**115**:718–20.
91. Cregler LL, Mark H. Medical complications of cocaine abuse. *New Engl J Med* 1986;**315**:1495–500.
92. Friedman D, Wolfsthal SD. Cocaine-induced pseudovasculitis. *Mayo Clin Proc* 2005;**80**:671–3.
93. Shanti CM, Lucas CE. Cocaine and the critical care challenge. *Crit Care Med* 2003;**31**:1851–9.
94. Kestler A, Keyes L. Uvular angioedema. *New Engl J Med* 2003;**349**:867.
95. Baldwin GC, Tashkin DP, Buckley DM, *et al.* Marijuana and cocaine impair alveolar macrophage function and cytokine production. *Am J Respir Crit Care Med* 1997;**156**:1606–13.
96. Baldwin GC, Buckley DM, Roth MD, *et al.* Acute activation of circulating polymorphonuclear neutrophils following in-vivo administration of cocaine. *Chest* 1997;**111**:698–705.
97. Klinger JR, Bensadoun E, Corrao WM. Pulmonary complications from alveolar accumulation of carbonaceous material in a cocaine smoker. *Chest* 1992;**101**:1171–3.
98. Meleca RJ, Burgio DL, Carr RM, Lolachi CM. Mucosal injuries of the upper aero digestive tract after smoking crack or freebase cocaine. *Laryngoscope* 1997;**107**:620–5.
99. Ginsberg GC, Lipman TO. Endoscopic diagnosis of thermal injury to the laryngopharynx after crack cocaine ingestion. *Gastrointest Endosc* 1993;**39**:838–9.

8

Drug-induced bronchospasm

K SURESH BABU, JAYMIN MORJARIA

INTRODUCTION

Pulmonary manifestations of drugs occur in various forms, and one of the most significant effects that may need urgent medical attention is bronchospasm. Drug-induced bronchospasm is a common symptom that is triggered by various drugs. It ranges in severity from mild to severe, and even fatal from post-anoxic brain damage. Bronchospasm can present as an isolated event or as a representation of drug-induced anaphylaxis. The list of culprit drugs is extensive and some of them are listed in Box 8.1. For an understanding of the pharmacology of drug-induced bronchospasm, it is essential to understand how bronchomotor tone is controlled in the airways. Alteration in the neural control of airways and the airway smooth muscle tone are responsible for bronchospasm.

NEURAL REGULATION OF AIRWAYS

The human airways consist of the trachea and dichotomously dividing bronchi which undergo branching for 20–23 generations. The airways consist of the airway epithelium, a submucosal layer where mucous glands are present, the airway smooth muscles, and the outer cartilaginous layer with connective tissue. The smooth muscle layer extends from the major bronchi to the respiratory bronchioles. Contraction of the airway smooth muscle is responsible for narrowing of the bronchial lumen and bronchospasm. The airway diameter is controlled by the autonomic nervous system and the afferent receptors with their afferent fibres travelling via the vagus nerves.

The autonomic nervous system influences many aspects of the airway function, including airway smooth muscle (ASM) tone, airway secretions, blood flow, epithelial cell function, mediator release and migration of inflammatory cells. Changes in bronchomotor tone are regulated by a balance between bronchoconstrictor and bronchodilatory pathways. Any events or processes that alter the balance can lead to decrease in the airway lumen and consequent airway obstruction. The autonomous nervous system comprises efferent pathways mediated through cholinergic, adrenergic and non-adrenergic/non-cholinergic (NANC) mechanisms. While the parasympathetic system is the key regulator of airway tone by innervation of the ASM, the sympathetic nervous system is important in the control of airway blood flow and secretions.

Reflex-mediated alteration of the airway smooth muscle tone is evoked by stimulation of the airway afferent nervous system.[1] The receptors in the airways belong to two classes. The myelinated vagal afferents arise from the slowly adapting pulmonary stretch receptors (SARs) and the rapidly adapting pulmonary stretch receptors (RARs). C-fibres are unmyelinated nerve endings in the airways and respond to mechanical distortion, H^+ ions and mediators like bradykinin.

The SAR sensory end-organ lies among ASM cells and their activation reduces the airway cholinergic tone. SARs are essentially mechanoreceptors and are affected by changes in airway wall tension. RARs are myelinated nerve endings located in the epithelium and they are predominantly located in the proximal airways and in the distal large airways. Exogenously administered bronchoconstrictors activate the RARs which respond to both mechanical and chemical stimuli. RARs are also known as 'irritant receptors'. Activation of the RARs in the trachea and the large bronchi is believed to cause cough, bronchoconstriction and increased mucus secretion. Various mediators like histamine and prostaglandins stimulate RARs, so RARs may have a function in the airway response to pathological states. C-fibre endings are the predominant sensory fibres emerging from the lung. These are again stimulated by both chemical and mechanical stimuli. These include exogenous agents like capsaicin and hypertonic saline and endogenous mediators like histamine, bradykinin, serotonin and prostaglandins.

The cholinergic nervous system is the predominant bronchoconstrictor pathway providing the resting bronchomotor

Box 8.1   Drugs reported to cause bronchospasm

Acetaminophen	Cisapride	Heparin	– Atracurium	Ranitidine
Acetylcysteine	Clarithromycin	Heroin (particularly	– D-tubocurarine	Rifampicin
ACE inhibitors	Cocaine	when insufflated)	– Mivacurium	Risperidone
Adalimumab	Curares	Hydrocortisone	– Rapacuronium	Sirolimus
Adenosine	Cyclophosphamide	Iodine radio-contrasts	– Suxamethonium	Streptokinase
Aminoglycoside	Cytokines	Inhaled pentamidine	Nitrofurantoin	Sulphamides/
antibiotics	D-pencillamine	Insulin (non-recombinant)	NSAIDs	sulphonamides
Amiodarone	Desensitization	Interferon-alpha	Oxytocin	Taxanes
Amphotericin B	extracts	Interferon-beta	Paclitaxel	– Paclitaxel
Antidepressants	Desipramine	Isoflurane	Pamidronate	Tetracycline
– Phenelzine	Dipyridamole	Isotretinoin	Penicillamine	Timolol
– Tranylcypromine	Epipodophyllotoxins	Latanaprost	Penicillins	ophthalmic
– Isocarboxacid	– Teniposide	Lidocaine	Pentamidine	Urokinase
– L-asparaginase	– Etoposide	Losartan	Platinum	Venlafaxine
Aspirin	Ergots	Melphalan	chemotherapeutic	Verapamil
Beta-blockers	Erythromycin	Mesalamine	agents	Verteporfin
– Propranalol	Ethanol injections	Methimazole	– Carboplatin	Vinblastine
– Labetalol	Etoposide	Methotrexate	– Cisplatin	Vinca alkaloids
– Atenolol	Excipients	Methylprednisolone	Propofol	Vinorebline
– Metoprolol	– Cremophor EL	Metoclopramide	Propylthiouracile	Vindesine
– Timolol	– EDTA	Midazolam	Prostaglandin-$F_{2\alpha}$	Zanamivir
Betahistine	– PABA	Naftidrofuryl	Protamine	
Carbamazepine	Gemcitabine	Neuromuscular blocking	Polyethylene glycol	
Cephalosporins	Halothane	agents	Radio-contrast media	

tone. Efferents run from vagal nuclei in the brain-stem through the vagus nerves towards small ganglions in the bronchial wall. The neurotransmitter is acetylcholine (Ach) which exerts its effects on excitatory muscarinic (M) receptors. M1 and M2 receptors are localized to nerves and parasympathetic ganglia in the airways,[2] and the M3 subtypes are localized to the ASM, submucosal glands and airway epithelium mediating bronchoconstriction, vasodilation and increased mucus secretion.[3] Cholinergic bronchoconstriction may also be mediated by inflammatory mediators by facilitating Ach release from parasympathetic ganglia. Furthermore, inhibitory muscarinic receptors (autoreceptors) have been demonstrated in airways. These pre-junctional receptors inhibit Ach release and may serve to limit vagal bronchoconstriction. In humans, these belong to the M2 subtype. It is believed that, in asthma, the pre-junctional auto-inhibitory M2 cholinergic receptor function is impaired, leading to airway obstruction through uncontrolled release of Ach. This defect in muscarinic autoreceptor could then lead to exaggerated cholinergic reflex in asthma as the normal feedback inhibition of Ach release may be lost. This probably explains the severe bronchoconstriction that occurs with beta-blockers in asthma. It is unclear why in some patients certain drugs induce mild symptoms while in others severe symptoms of bronchoconstriction.

MECHANISMS OF DRUG-INDUCED BRONCHOSPASM

Mechanisms that underlie drug-induced bronchospasm are diverse and our understanding is limited in view of the complex neuro-humoral factors regulating the airway tone, bronchial reactivity and the role of different cytokines on the airway smooth muscle. While some of these mechanisms have been defined, some of the reactions and the mechanisms underlying them are still unclear. Possible mechanisms include:

- IgE-mediated anaphylaxis;
- non-IgE-mediated anaphylactoid reactions;
- aspirin-induced asthma (AIA) (including NSAIDs);
- other mechanisms (beta-blockers, inhaled bronchodilators, vinca alkaloids).

RISK FACTORS

Drug-induced bronchospasm is common in certain groups and depends on patient factors and drug-related factors (Box 8.2). Drug-induced bronchospasm most often occurs in patients known to have asthma.[4] Atopic status and increased

Box 8.2   Risk factors for drug-induced bronchospasm

Patient-related

- Age
- Sex
- Genetics
- Atopy
- Acquired immune deficiency syndrome (AIDS)
- Smoking
- Systemic lupus erythematosus (SLE)
- Previous hypersensitivity to drug
- Liver disease
- Renal insufficiency
- Respiratory infection

Drug-related

- Macromolecular size
- Bivalency
- Haptens
- Route of administration
- Dose of drug
- Duration of treatment

Aggravating factors

- Asthma
- Pregnancy
- Confounding drugs
- GERD

bronchial reactivity are also known to be predisposing factors for drug-induced bronchospasm. Adverse drug reactions occur mainly in young and middle-aged adults and are twice as common in women. Genetic factors and familial predisposition to antimicrobial drugs have also been reported.[5] A Swiss study reported that analgesics and NSAIDs were the major offenders, accounting for nearly 24 per cent of the reactions, of which 64.5 per cent were classified as serious.[6] Of note, serious drug-induced bronchospasm was also frequently implicated with anti-infective agents with reported rates of 0.7–10 per cent,[7] cardiovascular drugs and excipients. Sensitization may be dependent on the route of administration, and repeated administration may play a role as well,[5] with the intravenous route giving rise to more severe reactions.

IgE-MEDIATED BRONCHOSPASM

IgE-mediated drug reactions have been studied extensively and the pathophysiological processes that trigger these reactions have been well characterized. IgE and IgG antibodies are generated by B-lymphocytes in response to the exposure to a specific antigen. IgE molecules bind to high-affinity FcεRI receptors on the surface of tissue mast cells and circulating basophils,[8] and antigen cross-linking of these molecules results in degranulation of mast cells, releasing both preformed

(histamine, heparin, tryptase, chymase and tumour necrosis factor) and newly synthesized mediators over minutes (platelet-activating factor, prostaglandin D2, sulphidoleukotrienes, leukotriene (LT)-B4, LT-C3, LT-D4 and LT-E4) and over hours (interleukin (IL)-4, IL-5, IL-13 and granulocyte macrophage colony stimulating factor).[9] This results in fluid influx (leading to upper airway obstruction), vasodilation (in combination with fluid extravastion from the vascular space causes distributive hypovolaemic shock), and smooth muscle contraction akin to asthma – causing lower airway obstruction due to bronchospasm, mucosal oedema and mucous plugging. Treatment should be targeted according to pathophysiology.

The hapten hypothesis

Drugs such as peptide hormones (insulin), owing to their macromolecular form, are intrinsically immunogenic. However, most drugs have a molecular mass of less than 1 kDa and are incapable alone of inducing an immune response in their innate state. For these agents to become effective immunogens, they must bind covalently to high-molecular-weight proteins.[10] Drug–protein conjugates or protein adducts are formed either by direct chemical interaction or by generation of electrophilic metabolites that react with nucleophilic groups on proteins.[11] This forms the basis of the hapten hypothesis (Fig. 8.1).[12] Some drugs, such as penicillin, can be directly chemically reactive as a result of the instability of their molecular structure. Others, such as anaesthetic agents, must be metabolized, or bioactivated, to a reactive form before an immune response can be initiated.[13] Bioactivation is usually followed by a bioinactivation. In some cases, however, genetic or environmental factors may disturb the equilibrium between these two processes, leading to increased synthesis or decreased elimination of reactive drug metabolites. Once formed, these reactive species may bind occasionally to larger macromolecular targets, to form an immunogenic complex, and induce an immune response.

The pharmacological interaction with immune receptors (PI) concept

The hapten hypothesis circumvents the blindness of the immune system to low-molecular-weight substances by coupling with a macromolecular carrier. However, some drugs appear to mount an immune response at the first encounter before an immune response had time to evolve. In the PI (or p-i) concept, certain drugs are considered to be able to activate T-cells in a direct way by drug binding to T-cell receptors and subsequent cell activation.[14] Pichler proposed that many drugs can bind reversibly to the MHC T-cell receptor, much like a super-antigen, and this can stimulate an immune response leading to a drug reaction.[15]

Immune reactions to antimicrobial agents

PENICILLIN AND OTHER β-LACTAM DRUGS

Allergy to β-lactam drugs is frequently reported, and in particular penicillin allergy. Severe anaphylactic reactions with

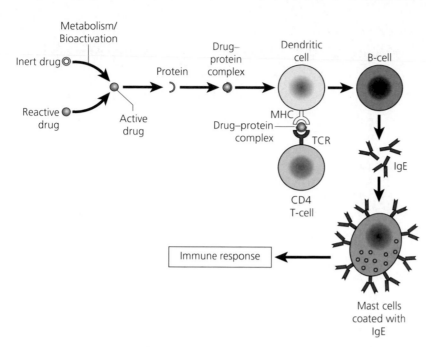

Fig. 8.1   Hapten hypothesis. The inert drug may be metabolized to a reactive metabolite or the drug may itself be a reactive. These reactive molecules bind to protein making the drug–protein complex 'foreign'. The foreign peptide is taken up by dendritic cells and then presented to helper T-cells (CD4) via MHC-II for processing. Following the processing, B-cells are activated to synthesis IgE molecules which bind to mast cells leading to the immune response as a result of mast-cell degranulation. IgE, immunoglobulin E; MHC, major histocompatibilty complex; TCR, T-cell receptor.

β-lactam agents can and do occur but are rare. It was reported that after 32 430 injections of benzathine penicillin during 2736 patient-years of observation, 3.2 per cent had an allergic reaction. Of these reactions, four were anaphylactic, an incidence of 0.2 per cent.[16] Despite this rarity, penicillin-induced anaphylaxis continues to be the most common cause of anaphylaxis in humans, accounting for approximately 75 per cent of fatal anaphylactic cases in the United States each year.[7,17]

Penicillins contain both a β-lactam ring and a thiazolidine ring. The former makes penicillin intrinsically reactive and unstable. This ring structure readily opens, allowing the carbonyl group to form amide linkages with amino groups of lysine residues on nearby proteins.[18] Since 95 per cent of penicillin molecules bind to proteins in this manner, the antigenic determinant formed, benzyl penicilloyl, has been termed the major penicillin determinant. Coupled to a weakly immunogenic carrier, penicilloyl determinants, penicilloyl polylysine (PPL) too have demonstrated IgE-mediated responses in humans. It is believed that patients with a positive history but negative skin test with PPL and minor determinants mixture are unlikely to have IgE-mediated reactions on penicillin readministration.[19] The side-chain group that distinguishes the different penicillins may also elicit the production of IgE antibodies that are clinically significant.[20] Different penicillins may be cross-reactive, not only by virtue of their shared β-lactam and thiazolidine rings, but also by the shared or similar side-chain determinants.

In contrast to the penicillins, our knowledge of the immunochemistry of cephalosporins is even more limited. A literature review by Lin of more than 15 000 subjects treated with four cephalosporins reported that 8.1 per cent of those with a history of penicillin allergy had reactions, versus 1.9 per cent who had no such history.[21] A study to evaluate the risk of administering a cephalosporin to penicillin-allergic subjects reported that, of the 135 subjects with positive skin-test results, 6 had reactions (reaction rate 4.4 per cent), whereas only 2 of 351 subjects with negative skin-test results reacted (reaction rate 0.6 per cent).[22] These data indicate that subjects who have penicillin-specific IgE antibodies may be at increased

risk for a reaction to cephalosporins; however, other studies have not reported similar findings.[23,24]

Management

Patients with known or presumed IgE antibodies to a β-lactam drug may undergo desensitization if that drug is required for treatment.[25] Acute drug desensitization involves the administration of incremental doses of a drug over a period of hours to days and is a process whereby a subject is converted from a drug-sensitive state to a state of drug tolerance. Not only is the desensitized state antigen-specific, it is also antigen-dependent, requiring the continuous presence of antigen. While desensitization is being performed, vital signs and peak flow values are monitored. Although most of our experience with drug desensitization has been derived from penicillin, this principle has been successfully applied to numerous other drugs, including non-IgE-dependent drugs such as aspirin and NSAIDs.[26,27] If a subject who has a history of a cephalosporin allergy requires another cephalosporin, one of two approaches may be considered:

1. Perform a graded challenge with a cephalosporin that does not share side-chain determinants with the original cephalosporin.
2. Perform cephalosporin skin-testing, although such testing is not standardized and the negative predictive value is unknown.[28] Both oral and intravenous protocols are available for penicillin (Tables 8.1 and 8.2).

SULPHONAMIDES

Sulphonamides are metabolized in the liver by *N*-acetylation, yielding non-toxic metabolites, and by cytochrome P450 catalysed *N*-oxidation, yielding reactive hydroxylamines[29] that then oxidize to nitroso species.[30] These reactive nitroso metabolites are reduced by glutathione and then excreted. When the capacity for glutathione conjugation is exceeded, however, these metabolites may be directly cytotoxic; or through the

Table 8.1 Oral penicillin desensitization protocol (interval between doses is 15 minutes)

Step	Penicillin (mg/mL)	Amount (mL)	Dose (mg)	Cumulative dose (mg)
1	0.5	0.1	0.05	0.05
2	0.5	0.2	0.1	0.15
3	0.5	0.4	0.2	0.35
4	0.5	0.8	0.4	0.75
5	0.5	1.6	0.8	1.55
6	0.5	3.2	1.6	3.15
7	0.5	6.4	3.2	6.35
8	5.0	1.2	6.0	12.35
9	5.0	2.4	12.0	24.35
10	5.0	5.0	25.0	49.35
11	50.0	1.0	50.0	100.0
12	50.0	2.0	100.0	200.0
13	50.0	4.0	200.0	400.0
14	50.0	8.0	400.0	800.0

Table 8.2 Intravenous penicillin desensitization protocol using a continuous infusion pump (interval between doses is 15 minutes)

Step	Penicillin (mg/mL)	Flow rate (mL/h)	Dose (mg)	Cumulative dose (mg)
1	0.01	6	0.015	0.015
2	0.01	12	0.03	0.045
3	0.01	24	0.06	0.105
4	0.1	5	0.125	0.23
5	0.1	10	0.25	0.48
6	0.1	20	0.5	1.0
7	0.1	40	1.0	2.0
8	0.1	80	2.0	4.0
9	0.1	160	4.0	8.0
10	10.0	3	7.5	15.0
11	10.0	6	15.0	30.0
12	10.0	12	30.0	60.0
13	10.0	25	62.5	123.0
14	10.0	50	125.0	250.0
15	10.0	100	250.0	500.0
16	10.0	200	500.0	1000.0

formation of an N^4-sulfonamidoyl hapten they may haptenate protein carriers, forming immunogenic complexes.[29,31] The fact that sulphonamide-induced reactions usually occur after several days of treatment suggests that an immune mechanism may be involved. Direct evidence of this involvement has come from studies that have reported the presence of both specific T-lymphocytes[32] and pro-inflammatory cytokines[33] in the involved tissues of patients with sulphonamide-induced reactions. In addition, IgE antibodies have been detected in patients reporting immediate-type hypersensitivity reactions to sulfamethoxazole.

Management

Currently there are minimal diagnostic tools for assessing subjects with sulphonamide-induced reactions. Despite the fact that little is known about the mechanisms responsible for sulphonamide reactions in patients with HIV/AIDS, several 'desensitization' protocols have been empirically developed and used quite successfully.[34,35] Of note, the desensitization protocols that have been developed are more typical of a cautious graded challenge than of rapid desensitization methods that are typically used in other antibiotic desensitization.

LOCAL ANAESTHETICS AND HALOTHANE

The allergic trigger for local anaesthetics was found to be para-aminobenzoic acid (PABA), generated as an intermediate metabolite of ester hydrolysis. Sensitivity to PABA may occur through exposure to ester local anaesthetics, or to cosmetics and foods that contain preservatives that are antigenically similar. In addition, sulphonamide-containing substances (antibiotics, diuretics) structurally resemble PABA, so cross-reactivity may occur when sulphonamide-sensitive patients are then exposed to ester local anaesthetics.[36]

The development of amide local anaesthetics in the 1940s reduced reports of allergic reactions. Allergy to amides is extremely rare, with estimates of less than 1 per cent of reported allergic reactions to amide local anaesthetics representing true immune system-mediated responses.[37] Local anaesthetics have molecules too small ($< 300\,Da$) to be intrinsically antigenic, but can bind to plasma or tissue proteins as a hapten that possesses antigenic properties. A similar process of haptenation is thought to occur in halothane intermediates following hepatic metabolism.[38]

HEPARIN

Heparin, a mucopolysaccharide with a molecular weight of 6–20 kDa, can elicit several types of immunologically mediated reactions, such as asthma and anaphylaxis.[39] Skin-testing does not appear to be useful in elucidating immediate-type hypersensitivity responses to heparin, and cross-reactivity has been demonstrated between heparin, the low-molecular-weight heparins and donaporoid sodium. It is therefore recommended that the new direct thrombin inhibitors (argatroban and lep-irudin) be used in patients allergic to heparin.[40]

In 2007 and 2008, severe allergic reactions to heparin were reported in the United States. The serious adverse events included allergic or hypersensitivity-type reactions, with symptoms of oral swelling, nausea, vomiting, sweating, shortness of breath, and cases of severe hypotension requiring treatment. Most events developed within minutes of heparin initiation. Following investigation, the FDA found a heparin mimic (as much as 5–20 per cent of the product's active ingredient) blended with the real thing. This was identified as oversulphated chondroitin sulphate and was responsible for the adverse reactions and death with the use of heparin.

INSULIN AND PROTAMINE

Reactions to insulin therapy have occurred since the launch of animal insulin in 1922. With the initiation of recombinant human insulin, however, the incidence of insulin-induced allergic reactions has decreased. Anaphylactic reactions have been reported, though are rare, and IgE antibodies have been demonstrated,[41] sometimes causing acute pulmonary hypertension. Reactions to protamine-containing insulins may be caused by the protamine component in the insulin preparation and not the insulin itself. Protamine sulphate is a low-molecular-weight polycationic protein that is used to reverse the anticoagulant properties of heparin, and it is also complexed to insulin to delay absorption. Dykewicz et al. reported on two patients with diabetes who had anaphylaxis in response to neutral protamine Hagedorn insulin and to protamine, indicating that protamine-specific IgE antibodies were responsible for the anaphylactic reactions.[42] Inhaled insulin (Exubera), initially approved for the management of diabetes and then withdrawn from the market, had respiratory adverse events reported in approximately 11.6 per cent.[43] It is not entirely certain, but the reactions might be due to the increased immunogenicity of inhaled insulin arising from the structural changes in insulin during manufacture of the powder, or due to differences in immune responses related to different routes of drug delivery, storage conditions, or pre-sensitization with subcutaneous insulin used previously; and pulmonary insulin may have an increased antibody response. This suggests that site of delivery and susceptibility of the recipient may significantly affect immune responses.[44,45]

Anaesthetic neuromuscular blocking agents

Muscle relaxants are responsible for 59 per cent of perioperative anaphylactic reactions.[46] The incidence of suspected anaphylaxis in France is estimated to be 1 in 13 000 anaesthetics, whereas true anaphylaxis to neuromuscular blocking agents (NBAs) is approximately 1 in 6500. There is an increasing awareness of adverse reactions, particularly those relating to anaesthesia and surgery where a wide variety of substances may be administered intravenously.[47] While cardiovascular collapse and skin changes are most commonly reported, bronchospasm which may be transient is the most challenging, being responsible for 75 per cent of the deaths. Although rare, anaphylactic reactions in anaesthesia may be fatal if not recognized or treated promptly. NBAs are responsible for 60–70 per cent

of reactions under anaesthesia. The antigen is a quaternary ammonium group usually found in foods, drugs, cosmetics and hair products, hence requiring no previous exposure to the anaesthetic agent. Of the NBAs, suxamethonium[48] and atracuronium appear to have an incidence of reactions greater than that predicted by their market share, while pancuronium and vecuronium appear to be safer both on incidence of reactions and on positive skin-tests in reactors.[49]

In a recent case report, cisatracurium given during anaesthesia precipitated severe persistent bronchospasm and central cyanosis needing continuous intravenous infusion of adrenaline (epinephrine) to break the severe refractory bronchial hyperresponsiveness.[50] Elevated serum tryptase, histamine and cisatracurium-specific IgE were demonstrated.

Plasma expanders (discussed later) as well as thiopentone (1 in 14 000) have also been reported as triggers of anaphylactic reactions.

During anaesthesia, anaphylaxis may occur at any time, but is more common (90 per cent) at or shortly after induction. Clinical manifestations are not consistent, making diagnosis difficult. Cardiovascular collapse is the most common presentation (88 per cent) followed by erythema (45 per cent), bronchospasm (36 per cent), angioedema (24 per cent), rash (13 per cent) and urticaria (8.5 per cent).

Mast-cell tryptase if elevated is very suggestive of mast-cell degranulation but does not suggest an IgE-mediated reaction. However, mast-cell tryptase can be useful to distinguish between a drug reaction secondary to mast-cell degranulation and a reaction due to a direct pharmacological action of the drug on the muscle. Muscle relaxants causing anaphylaxis can be identified by skin prick tests (SPTs) which are standardized or by specific IgE tests which are not standardized.

Management

The sensitivity of skin tests to muscle relaxants is greater than 95 per cent.[51] In the absence of an established predictive value of tests for the occurrence or perioperative anaphylactic reaction, there is no demonstrated evidence of systemic preoperative screening in the general population. Furthermore, there is no evidence for the need to investigate sensitization against anaesthetic agents in patients who are atopic or sensitized to a substance that they would not be exposed to during anaesthesia. Certain groups of patients, who present with documented allergy to muscle relaxants, who have had an unexplained reaction during a previous general anaesthesia, who have a history of allergy to local anaesthetics and who belong to a high-risk group for sensitization, need an allergy work-up. The latter includes children subjected to multiple surgeries (e.g. spina bifida), and patients allergic to avocado, kiwi, banana, sweet pepper, figs, chestnut, melon or papaya.[52]

Chemotherapeutic agents

Platinum compounds (cisplatin, carboplatin and oxaliplatin) commonly induce hypersensitivity reactions. This normally occurs over courses of drug administration before an anaphylactic reaction is produced, thus making the reaction immunogenic. Skin-testing has been used to identify patients at risk of development of reactions, and a negative test result has been shown to have an extremely high negative predictive value (96 per cent).[53] Desensitization protocols have been developed for patients with positive skin test results, but not with much success.[54,55]

Excipient Cremophor EL

Cremophor EL is a non-ionic surfactant derived from castor oil that has been shown to lead to histamine release and hypotension in dogs.[56] Cremophor EL is an excipient in some drugs like paclitaxel and intravenous ciclosporin. In a case report, it has been demonstrated that the excipient Cremophor EL produced severe bronchospasm, while subsequent oral administration that did not contain Cremophor EL precipitated no reaction.[57] Other drugs with this excipient and other such excipients may produce similar anaphylactic reactions. Intradermal testing with Cremophor EL and basophil tests support the presence of IgE antibodies to Cremophor EL.

Management of IgE-mediated drug-induced bronchospasm

CLINICAL HISTORY

Evaluation of drug allergy must begin with a precise and detailed history, including clinical symptoms and their timing and duration in relation to drug exposure.[5] However, patient history plays a limited role in predicting the outcome of the penicillin skin-test. It was shown that only 17–46 per cent of patients with a history of anaphylaxis to penicillin had a positive penicillin skin-test.[58,59] Hence, on clinical grounds the diagnosis of a patient's illness as a reaction that was drug-induced is often challenging. Physicians preferred the 'better safe than sorry' strategy over the possible clinical approaches in a small survey.[60] Ideally, all patients who have experienced anaphylaxis should be referred to a specialist clinical immunologist/allergist; however, the practical reality is that this is difficult in many areas.[9]

CLINICAL FEATURES

The hallmark of the anaphylactic reaction is the onset of manifestation within seconds to minutes after introduction of the antigen. There may be upper or lower airway obstruction or both. Symptoms include swelling of the tongue, laryngeal oedema and stridor suggestive of upper airway obstruction, while bronchospasm is associated with a feeling of tightness in the chest and/or audible wheezing. Patients with bronchial asthma are predisposed to severe involvement of the lower airways. Intensely pruritic urticarial eruptions and angioedema may also be present. Symptoms can be severe enough to lead to vascular collapse and death if not managed immediately.

DIAGNOSTIC TESTING

Some diagnostic tests such as skin-testing are available, especially with suspected penicillin-allergic patients. Thus, when reagents are available, it is an important clinical option to consider, rather than turning immediately to broad-spectrum antibiotics.[10] This is not only relevant because of the increasing prevalence of antibiotic resistance,[60] but also because the antibiotic costs for patients allergic to penicillin are 63 per cent higher than for those who are not.[61,62]

Skin-testing

The presence of drug-specific IgE antibodies indicates that the patient is at risk of an IgE-mediated reaction if re-exposed to that agent. Skin-testing is especially useful for polypeptides that are multivalent and of a large molecular weight, such as anti-lymphocyte globulin, toxoids, insulin and streptokinase. Skin-testing is unreliable for small-molecular-weight agents like antibiotics, except for penicillin, because the relevant immunogenic determinants for most antibiotics and drugs have not been identified. Despite this lack of knowledge, beneficial information may be obtained if skin-tests are done with non-irritative antibiotic concentrations. A positive result suggests the presence of drug-specific IgE antibodies. In contrast, a negative result could mean either that drug-specific IgE antibodies are absent or that they are present but not detectable because an inappropriate immunogen was used as the testing reagent.

In-vitro tests have shown moderate cross-reactivity between cephalosporin and penicillins,[23,61] which may be due to the shared β-lactam ring and similar side-chains. Of note, up to 10 per cent of patients with penicillin allergy can have an adverse reaction to cephalosporins; and no validated skin-test or anti-cephalosporin IgE antibody assays are available.[22] Pumphrey and Davis, in a study of fatal anaphylaxis in the UK from 1992 to 1997, revealed that 6 of 12 deaths were from the first course of a cephalosporin.[63] Thus, the Joint Task Force on Practice Parameters recommends that, if a substitute with a non-β-lactam microbial agent is unavailable for patients who need a cephalosporin, patients with a history of an immediate-type hypersensitivity reaction to penicillin should undergo penicillin skin-testing.[64] If penicillin skin tests are negative, a cephalosporin can be administered with a less than 1 per cent risk of immediate adverse reaction. However, if the penicillin skin-test is positive, then the patient should either avoid β-lactam antimicrobials or be considered for cephalosporin desensitization.

Radioimmunoassay

Radioimmunoassays may detect serum IgE antibodies to certain drugs (penicillin and succinyl choline). The tests have the same limitations as *in-vitro* skin-tests, and they are less sensitive.[65] Thus, clinicians must be cautioned about using these assays in the diagnostic assessment of an antibiotic-allergic individual.

Mast-cell tryptase

Tryptase is a valuable marker of systemic mast-cell degranulation.[66] Tryptase exists in the α and β forms, of which the β form is a measure of mast-cell activation, and concentrations are raised in mast-cell-dependent anaphylactic (IgE-mediated) and anaphylactoid (non-IgE-mediated) reactions.[65] The half-life of tryptase in plasma is 2 hours and the enzyme is not prone to rapid degradation. For these reasons, measurement of tryptase concentrations is favoured over measurement of serum histamine. Serum samples between 30 minutes and 5 hours after the event, when compared with baseline concentrations taken weeks after, may confirm or exclude the diagnosis.[5] However, despite the usefulness of this test in assessing both immune and non-immune drug reactions, it is important to note that β tryptase concentrations can be normal if the reaction does not involve haemodynamic changes.[67]

In the acute setting of an anaphylaxis and anaphylactoid reaction, small doses of intramuscular adrenaline is the most important therapeutic agent.[68] The usual doses for intramuscular injection are from 50 to 250 μg for children up to 12 years and 500 μg for adolescents and adults – given as solution diluted as 1 in 1000 (1 mg/mL). Intravenous administration is potentially hazardous and runs the risk of provoking a hypertensive crisis, tachyarrhythmias, myocardial infarction, or pulmonary oedema.[69] Supplementary treatment includes supine posture, intravenous fluids, oxygen, parenteral antihistamines, and corticosteroids. Bronchospasm usually is reversible with adrenaline, and β$_2$-agonists are beneficial.

NON-IgE-MEDIATED DRUG-INDUCED BRONCHOSPASM

Non-IgE-mediated anaphylactoid or pseudo-allergic reactions resemble anaphylaxis, but various mechanisms without the generation of IgE antibodies mediate these reactions. Possible mechanisms of non-immune anaphylactic (pseudo-allergic) reactions are listed in Box 8.3.

General anaesthesia induction drugs

Bronchospasm has been reported after the administration of neuromuscular blocking drugs. In 1999, Laxenaire published the fourth French survey of anaphylaxis during general anaesthesia between July 1994 and December 1996.[70] From the 1648 patients identified and skin-tested, 692 were accounted as IgE-mediated (clinical symptoms and positive skin-test), and another 611 were judged to be anaphylactoid in nature (clinical symptoms and negative skin-test).

Some non-depolarizing muscle relaxants (rapacuronium and mivacurium) have been implicated in severe bronchospasm in patients.[71,72] Bronchospasm due to neuromuscular blocking drugs has been attributed to the tendency of these drugs to release histamine as determined by skin-test reactivity and *in-vitro* histamine release from basophils.[73] The potential mechanism involved is by blockade of M2 muscuranic

---

Box 8.3 Possible mechanisms of non-immune (pseudo-allergic) reactions

- Direct release of mediators (e.g. histamine)
- Direct activation of the complement system
- Activation of the coagulation system
- Interaction with the kallikrein–kinin system
- Shift in eicosanoid metabolism towards leukotriene formation
- Platelet activation
- Psychoneurogenic reactions

---

receptors on pre-junctional parasympathetic nerve, leading to increased release of acetylcholine and thereby resulting in M3 muscuranic receptor-mediated airway smooth muscle bronchoconstriction.[74] Mivacurium is a neuromuscular blocking agent with a short half-life of around 30 minutes due to a fast hydrolysis by pseudocholinestrases. This can cause massive amounts of histamine release, inducing severe refractory bronchospasm.[75] Pancuronium, considered the agent of choice in asthmatics, has also been reported to cause bronchospasm.[76] Skin-test reactivity may not correlate well with the likelihood of a patient developing bronchospasm, suggesting that other mechanisms other than histamine release are involved.[77]

Chemotherapy agents

TAXANES

In studies using taxanes, as many as 42 per cent of the patients who received paclitaxel had so-called 'hypersensitivity' reactions, and 2 per cent of these were severe.[78] Most of the reactions occurred during first exposure to the drug. Although the symptoms were consistent with an IgE-mediated mechanism, there is no period of sensitization; hence the likely mechanism is direct mast-cell degranulation. Docetaxel, a sister agent, is also associated with a high incidence of reactions and does not contain Cremophor EL,[56] found in paclitaxel, which is known to have immunogenic properties. Thus, it is likely that the taxane moiety may be the aetiologic agent responsible for these reactions.

Prophylaxis using corticosteroid therapy and antihistamines is routinely administered prior to taxane use to reduce the incidence and severity of 'hypersensitivity' reaction.[78] Essentially, patients who have had reactions to taxane therapy are able to undergo successful re-treatment, with pre-treatment of antihistamines and corticosteroids.[55] Furthermore, for patients having recurrent reactions despite pre-medication, a desensitization protocol is available.[78]

ASPARAGINASE

This is a bacterial polypeptide protease that depletes tumour cells of aspargine. It is used in the treatment of acute lymphoblastic leukaemia. Approximately 25–35 per cent of patients receiving this drug develop reactions of anaphylactoid symptoms.[79] Most patients require repeated exposure before a reaction develops, and antiasparginase antibodies have been reported in a few cases.[80] Hence, intradermal testing is generally performed prior to treatment and if more than a week of treatment has elapsed since the previous dose. Other newer agents have been developed to treat patients who have had asparginase-induced reaction. A rapid desensitization protocol for asparaginase-induced reactions has been developed,[81] but is not routinely used.

EPIPODOPHYLLOTOXINS

Etoposide and teniposide are antimitotic agents used in the treatment of ovarian germ-cell tumours, small-cell lung carcinoma, testicular tumours, and many other tumours.[82] Reactions range from 6 to 41 per cent, and there is a 0.7–14 per cent incidence of anaphylaxis with epipodophyllotoxins.[83,84] As these reactions occurred with the first dose, it is unlikely to be immunologically mediated. Probably, like taxanes, they may cause direct mast-cell mediator release. There are no standard prophylaxis regimens, and fewer than 50 per cent who have had a drug-induced reaction are able to tolerate the drug on re-administration.[81]

Iodinated contrast media

It is estimated that approximately 50 million procedures involving iodinated contrast media (CM) administration are made per year worldwide.[85] However, although rare, anaphylactoid reactions to CM are a principal worry of radiologists and cardiologists. Minor reactions occur in 5–10 per cent of the patients.[86] Severe reactions occur in 1 in 14 000 investigations and significant bronchospasm occurs in approximately 12 per cent of these severe reactions.[87]

The majority of patients receiving CM develop a subclinical decrease in airflow as measured by their peak expiratory flows or FEV_1. This decrease occurs in patients with or without an allergic history and tends to be higher in those with a history of allergy.[88] Factors that contribute to an acute reaction to CM include a history of asthma, allergy, or atopy and medications, like β-blockers, recombinant IL-2, aspirin and NSAIDs.[89,90]

If CM were to follow the phenomenon of IgE-mediated reaction, then it would have to involve haptenization due to their small size (<1 kDa). This theory was proposed involving up to several hundred molecules aggregating together and then invalidated. Activation of the complement system with activation of mast cells and basophils via anaphylatoxins C3a and C5a receptors has been proposed and studied since the 1970s.[91,92] Alternative hypotheses include direct secretory effects on mast cells or basophils related to local changes in osmolarity and ion concentration.[93] It is not clear whether the use of lower osmolarity agents can reduce the incidence of bronchospasm. Dawson and colleagues compared a low osmolarity agent iopamidol with that of sodium iothalamate, an agent with high osmolarity, and found no change in FEV_1 with iopamidol compared with a 7.5 per cent decrease with sodium iothalamate.[94]

The notable feature of CM reactions making them anaphylactoid rather than anaphylactic reactions is that they occur in patients not previously exposed to contrast media.[95] Furthermore, CM molecules are highly hydrophilic, making them unable to form covalent bonds with proteins. They are also not metabolized and thus cannot be transformed into compounds which behave like haptens.[91] Another mechanism proposed involves a non-covalent interaction of drugs to major histocompatibility peptide complex and fitting into T-cell receptors.[96]

Numerous clinical studies have evaluated the prophylactic value of drugs, mainly antihistamines and corticosteroids, but the results are unclear and highly variable.[97] Any prevention depends on the mechanism involved, but the mechanism of CM-induced allergy-like reaction remains unclear. Administration of β₂-adrenergic agonists can abolish the bronchospasm.

Opiates

Morphine, meperidine, codeine and narcotic analogues are non-specific histamine liberators by binding to specific mast-cell receptors. Occasional wheezing has been reported. The effect is postulated to be dependent on the dose and the potency of the opiate. Pulmonary mast-cell degranulation may be a possible mechanism for opiate-induced bronchospasm.[98] Intradermal and intravenous morphine and synthetic opioid administration stimulate histamine release from mast cells through a direct pharmacologic mechanism, independent of an opiate receptor.[99,100] Histamine release occurs in one-fifth of patients receiving postoperative IV analgesia with morphine or heroin. Another mechanism may relate to opioid inhibition of cholinesterases, which has been demonstrated in animal species;[101] the clinical significance of this in humans is unknown. In patients with a positive history of reaction to an opiate, a non-narcotic analgesic should be selected. However, if pain is not controlled, reactions can be attenuated by pre-administration of antihistamines or a graded challenge, which most patients tolerate in small doses.

Other agents

Anaphylactoid reactions have been reported after administration of many colloid volume expanders (dextran, gelatin, hydroxyethyl, starch, and human serum albumin).[102] Life-threatening anaphylactoid reactions to mannitol have been suggested, probably due to hyperosmolar-dependent histamine release. An effective graded challenge protocol may be used to prevent severe anaphylactoid reactions to dextran contained in iron–dextran complexes.[103] This may be life-saving in patients who require parenteral iron.

General management of non–IgE–mediated bronchospasm

The general principles of management of anaphylactoid reactions are similar to those of anaphylactic reactions, with the exception that no IgE-mediated immune response is present and hence tests such as skin and radioimmunoassay testing are negative. There is great similarity in the symptomatology between the two.

The tools available for assessment of anaphylactoid reactions are similarly limited. In addition to limited knowledge of relevant antigenic determinants and pathogenic mechanisms, the clinical relevance of available tests must be determined.[65] Often there is no correlation between the presence of IgG or IgM antibodies and pathogenesis. Similarly, a lack of correlation is noted with lymphocyte-transformation testing. A marked proliferative response induced in a drug-allergic patient's lymphocytes when cultured in the presence of the suspected agent indicates specific T-cell sensitization, but whether this observation is clinically pertinent is questionable.[10]

ASPIRIN–INDUCED BRONCHOSPASM

Aspirin or acetyl salicylic acid (ASA) was synthesized in 1897 by Felix Hoffmann and since its introduction several allergic and pseudo-allergic reactions have been reported. Aspirin along with NSAIDs can induce asthmatic attacks, urticaria and anaphylactic reactions in individuals with or without a background of asthma or urticaria. The acute bronchospasm by aspirin was first reported by Cooke (1919), and in 1920 van der Veer described the first death from asphyxia due to aspirin. In 1922, Widal and colleagues described the clinical symptoms of aspirin sensitivity, asthma and nasal polyps and performed the first desensitization. Later, Samter and Beers characterized the disease associated with aspirin sensitivity and thus 'Samter's triad' came into use.

EPIDEMIOLOGY AND PREVALENCE

The prevalence of aspirin sensitivity varies from 3 to 21 per cent. A questionnaire survey in Western Australia has shown the prevalence of aspirin sensitivity to be 10–11 per cent.[104] In a meta-analysis the prevalence of aspirin-induced asthma (AIA) was higher when determined by oral provocation testing (21 per cent in adults and 5 per cent in children) than by verbal history (3 per cent in adults and 2 per cent in children). There was also cross-sensitivity to other NSAIDs (ibuprofen, 98 per cent; diclofenac sodium, 93 per cent; naproxen, 100 per cent).[105] Aspirin-induced bronchospasm is more common in women and is uncommon in children. The syndrome usually begins in adulthood at around age 30 years. Once developed, aspirin sensitivity remains throughout life.

TYPES OF REACTIONS TO ASPIRIN AND NSAIDS

Allergic and pseudo-allergic reactions to aspirin/NSAIDs have been classified into six different types (Table 8.3).[106] In addition to these six reactions, NSAID-induced aseptic meningitis and NSAID-induced hypersensitivity pneumonitis have been described. This classification is based on the presence or absence of underlying disease and the existence of cross-reaction to other cyclooxygenase (COX) inhibitors. Aspirin/NSAID-induced rhinitis/bronchospasm and urticaria/angioedema are linked to the effects of inhibition of COX enzymes,[107] while the remaining seems to be independent of COX inhibition.

PATHOPHYSIOLOGY

The basis of bronchospasm precipitated by aspirin and NSAIDs is based on the pharmacological inhibition of cyclooxygenase (COX) in arachidonic acid (AA) metabolism. Metabolism of arachidonic acid via the cyclooxygenase pathway leads to the generation of prostanoids, which are important signalling molecules produced during physiological as well as in inflammatory conditions. The airway inflammatory cells produce pro-inflammatory prostanoids PGD_2 and $PGF_{2\alpha}$ and anti-inflammatory prostaglandins like PGE_2.[108,109]

Table 8.3 Types of aspirin-induced reactions

Type of reaction	Underlying disease	Cross-reactions	Description
NSAID-induced rhinitis/asthma	Rhinitis, nasal polyps, sinusitis, asthma	Aspirin/NSAIDs	Affects upper and lower airways. The reaction is dose-dependent, with low dose causing lesser symptoms while a higher dose leads to severe reactions. The syndrome develops slowly over several decades and individuals are usually not allergic to other allergens including house dustmite, grass pollen or animal dander. Ingestion of NSAIDs can lead to upper airway symptoms to lower airway bronchospasm.
NSAID-induced urticaria/angioedema	Chronic idiopathic urticaria	Aspirin/NSAIDs	In patients with ongoing urticaria, symptoms flare up within a few hours of ingestion of aspirin/NSAIDs. Symptoms improve when symptoms of urticaria improve. These reactions are dependent on the activity of urticaria and on the dose of the drug. Patients are not usually allergic to aspirin/NSAIDs, but these drugs increase the irritability of the skin.
Single-drug-induced anaphylaxis	None	None	These reactions are IgE-mediated and occur after exposure to a specific NSAID/aspirin in otherwise normal individuals. The reactions occur in patients with a history of urticaria and are due to the patient getting sensitized to a specific drug hapten to form specific IgE antibodies.
Multiple-drug-induced anaphylactoid reactions	None	Aspirin/NSAIDs	These are due to histamine release from mast cells in skin/gut or vascular spaces. This can lead to flushing, rhinitis and bronchospasm.
Aspirin/NSAID-induced urticaria/angioedema	None	Aspirin/NSAIDs	These patients do not have urticaria. They develop urticaria/angioedema with multiple NSAIDs similar to chronic idiopathic urticaria, but the symptoms develop only when they ingest aspirin/NSAIDs.
Single-drug- or NSAID-induced blended reaction	Asthma, rhinitis, urticaria, none	–	The patients who do not fit into any of the above categories are classified into this group. Reactions include urticaria/angioedema, rhinitis and bronchospasm.

It is recognized that there are at least two COX enzymes: COX-1 and COX-2. COX 1 and 2 are coded by two different genes. Recently COX-3 and two smaller forms of COX-1 – derived from COX-1 by alternative splicing of COX-1 mRNA – have been identified.[110] COX-1 is the housekeeping enzyme (i.e. expressed in cells at baseline) while COX-2 is induced during inflammation and mainly enhances synthesis of inflammatory prostanoids. Bronchial biopsy studies show no difference in expression of COX-1 or COX-2 between subjects with AIA and non-AIA.[111] In patients with AIA, the COX enzymes are very sensitive to ASA/NSAIDs, and studies have shown that inhibition of COX-1 but not COX-2 is responsible for precipitating bronchospasm.[112,113] PGE_2 is a key mediator in aspirin-induced asthma. PGE_2 is pro-inflammatory in sepsis and arthritis but blunts and protects against bronchoconstriction following aspirin challenge. The levels of PGE_2 and thromboxane B_2 were decreased in patients with AIA and in the aspirin-tolerant asthma (ATA) groups.[114]

Cysteinyl leukotrienes (Cys LT) are derived from arachidonic acid via the 5-lipoxygenase (5-LO) pathway. The cellular biosynthesis of leukotrienes (LTs) involves a protein named FLAP (five lipoxygenase activating protein) which transports arachidonic acid into the cytosol to be acted on by the enzyme 5-LO. Sequential catalytic action of 5-LO on arachidonic acid yields LTA_4, which is further hydroxylated to LTB_4 or converted into the first of the cysteinyl LTs, LTC_4, by LTC_4 synthase. LTC_4 is exported to the extracellular space where it forms LTD_4, which in turn is cleaved to form the 6-cysteinyl analogue of LTC_4 known as LTE_4. The cysteinyl LTs exert their biological action by binding to two types of G-protein coupled 7-transmembrane receptors, Cys LT1 and Cys LT2.

Aspirin-induced bronchoconstriction is thought to be caused by the shunting of arachidonic acid metabolism away from the cyclooxygenase pathway towards the lipoxygenase pathway. This results in the increased production of leukotrienes with the resultant bronchoconstriction. Consistent with this finding, bronchoconstriction in patients with AIA can be inhibited by leukotriene receptor antagonists (LTAs).[115] Furthermore, in AIA subjects there is up-regulation of the 5-LO pathway resulting in increased Cys LT production. LTC_4 synthase is the rate-limiting enzyme for the synthesis of Cys LTs, and bronchial biopsy studies have revealed an overexpression of LTC_4 synthase in patients with AIA as compared to ATA patients.[116] In subjects with AIA and rhinosinusitis there was an increased expression of Cys LT1 receptor on leucocytes.[117] Therefore, in AIA the over-production of Cys LT is accompanied by up-regulation of Cys LT1 receptors on inflammatory cells which enhances their ability to become activated. Lipoxins, mediators derived from AA metabolism by combined action of different lipoxygenase/cyclooxygenase enzymes, are involved in resolution of inflammation. There is evidence to suggest that patients with AIA have a reduced capacity for biosynthesis of lipoxins.[118]

The initializing event in AIA appears to be the interruption of the synthesis of PGE_2. PGE_2 has profound regulatory effects on other inflammatory systems. It reduces LT synthesis by inhibiting 5-LO, inhibits cholinergic transmission, prevents mediator release from mast cells, and prevents ASA-precipitated bronchoconstriction.[119] Hence, it is possible that PGE_2 may act as a brake for the inflammatory responses. In AIA the

pharmacological action of COX-1 inhibition is to accelerate depletion of the protective PGE_2 which is more vulnerable because of a functional deficiency of COX-2.[120] This putative pathway could generate 5-LO products and account for the effects of aspirin-induced asthma. The removal of the PGE_2 brake by ASA/NSAIDs leads to exaggerated cysteinyl LT synthesis. This along with an over-expression of Cys LT1 receptors in inflammatory cells in subjects with AIA and reduced capacity for the biosynthesis of lipoxins could lead to precipitation of bronchospasm.

CLINICAL FEATURES

AIA is common in females and symptoms include persistent rhinitis followed by asthma, ASA sensitivity and nasal polyposis initially described by Samter in 1968.[121] Asthma and ASA sensitivity manifest an average of 1–5 years after the onset of rhinitis. After ingestion of ASA or NSAIDs, an acute asthma exacerbation occurs after 3 hours accompanied by profuse rhinorrhoea, conjunctival injection, peri-orbital oedema and sometimes a scarlet flushing of the face and neck. Fifty per cent of the patients with AIA have chronic, severe, corticosteroid-dependent asthma, 30 per cent have moderate asthma controlled with inhaled steroids, and the remaining 20 per cent are the mild and intermittent asthmatics. Bronchoconstriction may be severe and life-threatening requiring hospital admissions and at times may require mechanical ventilation. Up to 25 per cent of hospital admissions for acute asthma requiring mechanical ventilation may be due to NSAID ingestion.[122]

MANAGEMENT

In most cases the clear history should enable the physician to make a diagnosis. Most patients have moderate or severe persistent asthma. In doubtful cases, carefully controlled challenge testing with aspirin or other NSAIDs is justified, but this should be done in a hospital with full facilities for resuscitation. Lysine aspirin by inhalation could be used for challenge testing although it is of value in diagnosis of analgesic sensitivity rather than in assessing sensitivity to specific NSAIDs. Nasal provocation tests with lysine aspirin can be used as it has been found to be safe, simple and specific in the diagnosis of AIA.[123] In the longer term, patients should be advised to avoid aspirin and aspirin-containing products by any route. NSAIDs that cross-react with aspirin should also be avoided (Box 8.4).

Anti-leukotriene drugs are currently used in the treatment of AIA.[124] There are two classes of anti-LT drugs, the 5-LO inhibitors (i.e. Zileuton) and the specific cysteinyl LT receptor antagonists (i.e. zafirlukast, montelukast and pranlukast). LT-modifying drugs have been found to attenuate the aspirin-induced bronchial reactions in AIA patients. Anti-LTs also induce bronchodilation in patients with AIA. Salmeterol, a long acting β_2-agonist, has also been found to be effective in the management of AIA and also attenuated the bronchial hyperresponsiveness to lysine-ASA.[125]

Aspirin desensitization can be used in patients with aspirin/NSAID sensitivity.[126] Small incremental doses of aspirin are ingested over the course of 2–3 days until 400–650 mg of aspirin is tolerated. Aspirin can then be administered daily, with doses of 100–300 mg used for desensitization. After each

Box 8.4 Non-steroidal anti-inflammatory drugs (NSAIDs) that cross-react with aspirin

Predominant COX-1 and COX-2 inhibitors

- Piroxicam
- Indomethacin
- Sulindac
- Tolmetin
- Diclofenac
- Naproxen
- Naproxen sodium
- Ibuprofen
- Fenoprofen
- Ketoprofen
- Flubriprofen
- Mefenamic acid
- Meclofenamate
- Ketorolac
- Etodolac
- Diflusinal
- Oxyphenbutazone
- Phenylbutazone

Poor COX-1 and COX-2 inhibitors

- Acetaminophen
- Salsalate

Relative inhibitors of COX-2

- Nimesulide
- Meloxicam
- Nabumetone

Selective COX-2 inhibitors

- Celecoxib
- Rofecoxib
- Meloxicam
- Etodolac

dose of aspirin, there is a refractory period of 2–5 days during which aspirin and other COX inhibitors can be taken with impunity. It is possible to maintain the tolerance state for a long time by the administration of aspirin at proper intervals.[127] Although the precise mechanism of aspirin desensitization is still unclear, studies have shown a substantial decline in the peripheral monocyte synthesis of LTB_4 in AIA individuals after aspirin desensitization.[128] In addition, the cysteinyl LT receptors were down-regulated, thereby reducing the effectiveness of the same load of leukotrienes. It was observed that in aspirin-induced urticaria, mast-cell degranulation did not occur after aspirin desensitization.[129] The increase in urinary LTE_4 during aspirin challenge and its reduction after desensitization is inversely proportional to thromboxane B_2 and suggests a 'shunting' of the arachidonic acid metabolites of the target cells.[130] These findings suggest a change in the balance between the inflammatory mediators after desensitization leading to aspirin tolerance in the AIA population.

OTHER DRUGS CAUSING BRONCHOSPASM

In this section a few groups of drugs will be discussed individually with their respective management strategies.

Bronchospasm due to vinca alkaloids

Vinca alkaloids (vinorelbine, vinblastine) are used as chemotherapeutic agents. Bronchospasm is an acute reaction that has been reported in patients taking vinca alkaloids with an incidence of up to 5 per cent.[131] Although acute breathlessness and severe bronchospasm are seldom reported with vinca alkaloids, in combination with mitomycin the respiratory symptoms increase dramatically. Vinca alkaloid-induced bronchospasm responds well with bronchodilator agents.

Bronchospasm caused by inhalational agents

Occasionally non-specific bronchospasm has been reported with inhalational agents used to treat chronic obstructive pulmonary disease (COPD) and asthma. This is generally due to the excipient and substances in the preparation rather than the bronchodilator agent itself.[132] The carrier agent, the mode of delivery, pH, osmolarity, or the temperature may play a role too. Inhaled albuterol, salmeterol, sodium chromoglycate, and corticosteroid can induce bronchospasm. Also nebulized bronchodilators (β-adrenergic and anticholinergic) have been reported to induce bronchospasm, possibly owing to the tonicity of the solution, or added preservatives or stabilizers.[133]

Edentate disodium (EDTA), a stabilizing agent, and benzoalkonium chloride (BAC), an antibacterial agent, are often present as preservative or stabilizing agents in inhalers and nebulizer solutions used in the treatment of asthma and COPD.[133,134] Furthermore, sulphites and chlorobutanol have also been used as additives to non-sterile products to prevent bacterial growth in nebulizer solutions; however the former has also been reported to induce bronchospasm.[132] Steroid-dependent asthmatics seem to be at a higher risk. Also of note, it has been reported that in asthmatic patients the use of inhaled corticosteroids (ICS) containing ethanol (an excipient) can trigger bronchoconstriction, probably due to the direct contact of ethanol and the bronchial mucosa.[135] Thus, the use of ethanol-free formulations of ICS prevents this side-effect.

Like the bronchodilator inhalational agents inducing paradoxical bronchospasm, nebulized antibiotics, pentaminidine and *N*-acetyl cysteine also may induce bronchospasm.[136] The precise mechanism for pentamidine-induced bronchospasm is not clear. Theories include non-specific irritation from the inhaled particle, histamine release, and inhibition of cholinesterases by pentamidine which has been demonstrated *in vitro*.[137]

It has been shown that the use of Respimat Soft Mist Inhaler (SMI), an innovative device designed to improve lung deposition and an environmentally friendly alternative to conventional chlorofluorocarbon containing metered-dose inhaler (CFC-MDI), delivers extremely low absolute amounts of BAC and EDTA to the lungs.[134,138] Thus, common Respimat SMI may be a safer option than MDIs with regard to induction of paradoxical bronchoconstriction in patients with asthma or COPD. The use of preservative-free bronchodilator nebulizer solutions does not result in clinically significant bacterial contamination if they are dispersed in sterile unit-dose vials, in volumes and concentrations that do not require modification by their users. Thus, it may be best practice to avoid using nebulizer solutions that contain preservatives.[133]

Cough and bronchospasm due to inhaled pentamidine can be eliminated by pre-treatment with inhaled salbutamol. When patients do not respond to inhaled treatment or give a history of increased wheezing after inhaled treatment, a change in the specific agents administered may be appropriate.

Insufflated heroin has become an important aetiological agent of patients who present for acute bronchospasm in the emergency departments mainly in large metropolitan areas in the United States.[98] Pulmonary mast-cell degranulation by a direct pharmacological effect and a possible allergic mechanism may be involved in its pathophysiology.

Bronchospasm caused by β–adrenergic antagonists

Oral, intravenous and ophthalmic preparations that block β-adrenergic receptors may induce bronchospasm among patients with asthma or subclinical bronchial hyperreactivity.[4] In normal individuals, beta-blockers neither induce bronchospasm nor increase bronchial reactivity to inhaled methacholine.[139] Beta-blockers may induce bronchoconstriction by blocking the β_2-adrenergic receptors in bronchial smooth muscle. Although bronchospasms have been reported in patients without a prior history of asthma, most of these patients had a family history of atopy or asthma or had a positive reaction to methacholine challenge, suggesting that beta-blockers unmasked underlying airway hyperreactivity.[88] The bronchoconstriction induced may be mild, severe, near-fatal or fatal, especially in patients with pre-existing asthma. Unlike asthmatics, normal subjects, even at high doses, rarely have bronchospasm induced by beta-blockers.[88]

The exact mechanisms are unknown as human bronchial smooth muscle has no direct sympathetic innervation. As β-adrenergic agonists are used for the treatment of bronchospasm, it follows that antagonizing β-adrenergic receptors may precipitate bronchial hyperreactivity. It has been suggested that the circulating catecholamines have a tonic smooth muscle relaxing effect and that this is abolished by β-adrenergic antagonists.[140] An alternative idea is that propranolol may work on bronchial smooth muscle via the parasympathetic nervous system by antagonizing inhibitory pre-synaptic β_2-adrenergic receptors on cholinergic nerves.[141,142] Beta-blockers may also have an effect on mast-cell degranulation as pre-treatment with sodium cromoglycate blocks propranolol-induced bronchoconstriction. Furthermore, beta-blockers have been implicated in depressing central responses to carbon dioxide and promotion of mast-cell degranulation.[142]

Beta-blockers have a significant impact on the prognosis of patients with cardiovascular disease, especially those with hypertension, coronary artery disease and chronic heart failure.[143] A number of studies have shown that cardioselective (β_1) beta-blockers are not only safe but are beneficial in patients with coexisting airways and coronary disease.[144] Although theoretically cardioselective, beta-blockers should abolish the incidence of bronchospasm, but there are reports of alteration in pulmonary function with β_1-selective blockers.[145]

A meta-analysis by Salpeter *et al.* evaluated data from 29 clinical trials to observe the pulmonary effects of a single dose and chronic use of oral β_1-selective blockers in patients who had reversible airways disease.[146] Although β_1-selective blockers have low affinity to β_2-receptors, the selectivity is incomplete. Long-term dosing did not change the forced expiratory volume in 1 second (FEV_1) or symptoms, or rescue medication used from placebo in patients who had mild-to-moderate airways disease. However, a single administration of a β_1-selective blocker decreased FEV_1 by just under 8 per cent more than placebo. There was a greater than 13 per cent increase in FEV_1 with β-agonist response.

The treatment of choice of beta-blocker-induced bronchospasm is inhaled anticholinergic agents as β-agonists are ineffective owing to the β-receptor blockade. Because the likelihood of an adverse reaction depends on the selectivity of the drug and its plasma concentration, the bronchoconstrictor effects may be minimized by the use of a slow-release drug delivery system that provides a stable concentration. Such drugs are best avoided in asthma and in COPD; however, if administered, then great caution is needed.[145]

There are numerous drugs that have been implicated in inducing bronchospasm in case reports, but it is often difficult to postulate the underlying mechanism of bronchospasm. These drugs include ethanol injections,[147] angiotensin-converting enzyme inhibitors[4] and oxytocin.[148]

KEY POINTS

- Drug-induced bronchospasm is a common symptom that is triggered by various drugs. It can be severe and even fatal.

- It is common in certain groups and depends on patient factors and drug-related factors. It most often occurs in patients known to have asthma.

- Possible mechanisms include IgE-mediated anaphylaxis, non-IgE-mediated anaphylactoid reactions, aspirin-induced asthma and other mechanisms (beta-blockers, inhaled bronchodilators, vinca alkaloids etc).

- Atopic status and increased bronchial reactivity are also known to be predisposing factors for drug-induced bronchospasm.

- Adverse drug reactions occur mainly in young and middle-aged adults and are twice as common in women. Genetic factors and familial predisposition to antimicrobial drugs have also been reported.

REFERENCES

1. Canning BJ, Fischer A. Neural regulation of airway smooth muscle tone. *Respir Physiol* 2001;**125**(1/2):113–27.
2. van Koppen CJ, Blankesteijn WM, Klaassen AB, *et al.* Autoradiographic visualization of muscarinic receptors in human bronchi. *J Pharmacol Exp Ther* 1988;**244**:760–4.
3. Barnes PJ. Muscarinic receptor subtypes in airways. *Life Sci* 1993;**52**:521–7.
4. Takafuji S, Nakagawa T. Drug-induced pulmonary disorders. *Intern Med* 2004;**43**:169–70.

5. Vervloet D, Durham S. Adverse reactions to drugs. *Br Med J* 1998;**316**:1511–14.

6. Leuppi JD, Schnyder P, Hartmann K, Reinhart WH, Kuhn M. Drug-induced bronchospasm: analysis of 187 spontaneously reported cases. *Respiration* 2001;**68**:345–51.

7. Neugut AI, Ghatak AT, Miller RL. Anaphylaxis in the United States: an investigation into its epidemiology. *Arch Intern Med* 2001;**161**:15–21.

8. Payne V, Kam PC. Mast cell tryptase: a review of its physiology and clinical significance. *Anaesthesia* 2004;**59**:695–703.

9. Brown SG. Anaphylaxis: clinical concepts and research priorities. *Emerg Med Australas* 2006;**18**(2):155–69.

10. Gruchalla R. Understanding drug allergies. *J Allergy Clin Immunol* 2000;**105**(6 Pt 2):S637–44.

11. Park BK, Coleman JW, Kitteringham NR. Drug disposition and drug hypersensitivity. *Biochem Pharmacol* 1987;**36**:581–90.

12. Njoku DB, Talor MV, Fairweather D, et al. A novel model of drug hapten-induced hepatitis with increased mast cells in the BALB/c mouse. *Exp Mol Pathol* 2005;**78**:87–100.

13. Riley RJ, Leeder JS. In-vitro analysis of metabolic predisposition to drug hypersensitivity reactions. *Clin Exp Immunol* 1995;**99**:1–6.

14. Pichler WJ, Beeler A, Keller M, et al. Pharmacological interaction of drugs with immune receptors: the p-i concept. *Allergol Int* 2006;**55**(1):17–25.

15. Pichler WJ. Pharmacological interaction of drugs with antigen-specific immune receptors: the p-i concept. *Curr Opin Allergy Clin Immunol* 2002;**2**:301–5.

16. International Rheumatic Fever Study Group. Allergic reactions to long-term benzathine penicillin prophylaxis for rheumatic fever. *Lancet* 1991;**337**:1308–10.

17. Joint Task Force on Practice Parameters, American Academy of Allergy, Asthma and Immunology, American College of Allergy, Asthma and Immunology, and the Joint Council of Allergy, Asthma and Immunology. The diagnosis and management of anaphylaxis. *J Allergy Clin Immunol* 1998;**101**(6 Pt 2): S465–528.

18. Levine BB. Immunochemical mechanisms of drug allergy. *Annu Rev Med* 1966;**17**:23–38.

19. Sogn DD, Evans R, Shepherd GM, et al. Results of the National Institute of Allergy and Infectious Diseases Collaborative Clinical Trial to test the predictive value of skin testing with major and minor penicillin derivatives in hospitalized adults. *Arch Intern Med* 1992;**152**:1025–32.

20. Baldo BA. Penicillins and cephalosporins as allergens: structural aspects of recognition and cross-reactions. *Clin Exp Allergy* 1999;**29**:744–9.

21. Lin RY. A perspective on penicillin allergy. *Arch Intern Med* 1992;**152**:930–7.

22. Kelkar PS, Li JT. Cephalosporin allergy. *New Engl J Med* 2001;**345**:804–9.

23. Anne S, Reisman RE. Risk of administering cephalosporin antibiotics to patients with histories of penicillin allergy. *Ann Allergy Asthma Immunol* 1995;**74**:167–70.

24. Novalbos A, Sastre J, Cuesta J, et al. Lack of allergic cross-reactivity to cephalosporins among patients allergic to penicillins. *Clin Exp Allergy* 2001;**31**:438–43.

25. Solensky R. Drug desensitization. *Immunol Allergy Clin North Am* 2004;**24**:425–43,vi.

26. Patterson R, Deswarte R, Greenberger P, et al. *Drug Allergy and Protocols for Management of Drug Allergies.* Providence, RI: Oceanside Publications, 1995.

27. Sullivan TJ. Drug allergy. In Adkinson NF, et al. (eds) *Middleton's Allergy: Principles and Practice.* 4th edn. Mosby, 1993.

28. Bernstein I, et al., Joint Task Force on Practice Parameters, the American Academy of Allergy, Asthma and Immunology, and the Joint Council of Allergy, Asthma and Immunology, and the Joint Council of Allergy and Immunology. Disease management of drug hypersensitivity: a practice parameter. *Ann Allergy Asthma Immunol* 1999;**83**:665–700.

29. Cribb AE, Spielberg SP. Hepatic microsomal metabolism of sulfamethoxazole to the hydroxylamine. *Drug Metab Dispos* 1990;**18**:784–7.

30. Naisbitt DJ, Hough SJ, Gill HJ, et al. Cellular disposition of sulphamethoxazole and its metabolites: implications for hypersensitivity. *Br J Pharmacol* 1999;**126**:1393–407.

31. Naisbitt DJ, Gordon SF, Pirmohamed M, et al. Antigenicity and immunogenicity of sulphamethoxazole: demonstration of metabolism-dependent haptenation and T-cell proliferation *in vivo*. *Br J Pharmacol* 2001;**133**:295–305.

32. Pichler WJ, Schnyder B, Zanni MP, Hari Y, von Greyerz S. Role of T cells in drug allergies. *Allergy* 1998;**53**:225–32.

33. Yawalkar N, Shrikhande M, Hari Y, et al. Evidence for a role for IL-5 and eotaxin in activating and recruiting eosinophils in drug-induced cutaneous eruptions. *J Allergy Clin Immunol* 2000;**106**:1171–6.

34. Absar N, Daneshvar H, Beall G. Desensitization to trimethoprim/sulfamethoxazole in HIV-infected patients. *J Allergy Clin Immunol* 1994;**93**:1001–5.

35. Yoshizawa S, Yasuoka A, Kikuchi Y, et al. A 5-day course of oral desensitization to trimethoprim/sulfamethoxazole (T/S) in patients with human immunodeficiency virus type-1 infection who were previously intolerant to T/S. *Ann Allergy Asthma Immunol* 2000;**85**:241–4.

36. McLure HA, Rubin AP. Review of local anaesthetic agents. *Minerva Anestesiol* 2005;**71**(3):59–74.

37. Finucane BT. Allergies to local anesthetics: the real truth. *Can J Anaesth* 2003;**50**:869–74.

38. Martin JL, Kenna JG, Martin BM, et al. Halothane hepatitis patients have serum antibodies that react with protein disulfide isomerase. *Hepatology* 1993;**18**:858–63.

39. MacLean JA, Moscicki R, Bloch KJ. Adverse reactions to heparin. *Ann Allergy* 1990;**65**:254–9.

40. Aijaz A, Nelson J, Naseer N. Management of heparin allergy in pregnancy. *Am J Hematol* 2001;**67**:268–9.

41. Wessbecher R, Kiehn M, Stoffel E, Moll I. Management of insulin allergy. *Allergy* 2001;**56**:919–20.

42. Dykewicz MS, Kim HW, Orfan N, Yoo TJ, Lieberman P. Immunologic analysis of anaphylaxis to protamine component in neutral protamine Hagedorn human insulin. *J Allergy Clin Immunol* 1994;**93**(1 Pt 1):117–25.

43. Siekmeier R, Scheuch G. Inhaled insulin: does it become reality? *J Physiol Pharmacol* 2008;**59**(Suppl 6):81–113.

44. Tlaskalova-Hogenova H, Tuckova L, et al. Mucosal immunity: its role in defense and allergy. *Int Arch Allergy Immunol* 2002;**128**:77–89.

45. Fineberg SE, Kawabata T, Finco-Kent D, Liu C, Krasner A. Antibody response to inhaled insulin in patients with type 1 or type 2 diabetes: an analysis of initial phase II and III

inhaled insulin (Exubera) trials and a two-year extension trial. *J Clin Endocrinol Metab* 2005;**90**:3287–94.

46. Porri F, Lemiere C, Birnbaum J, *et al.* Prevalence of muscle relaxant sensitivity in a general population: implications for a preoperative screening. *Clin Exp Allergy* 1999;**29**:72–5.

47. Watkins J. The allergic reaction to intravenous induction agents. *Br J Hosp Med* 1986;**36**:45–8.

48. Watkins J. Adverse reaction to neuromuscular blockers: frequency, investigation, and epidemiology. *Acta Anaesthesiol Scand Suppl* 1994;**102**:6–10.

49. Fisher MM, Baldo BA. The diagnosis of fatal anaphylactic reactions during anaesthesia: employment of immunoassays for mast cell tryptase and drug-reactive IgE antibodies. *Anaesth Intens Care* 1993;**21**:353–7.

50. Briassoulis G, Hatzis T, Mammi P, Alikatora A. Persistent anaphylactic reaction after induction with thiopentone and cisatracurium. *Paediatr Anaesth* 2000;**10**:429–34.

51. Moneret-Vautrin DA, Gueant JL, Kamel L, *et al.* Anaphylaxis to muscle relaxants: cross-sensitivity studied by radioimmunoassays compared to intradermal tests in 34 cases. *J Allergy Clin Immunol* 1988;**82**(5 Pt 1):745–52.

52. Mertes PM, Laxenaire MC. Allergic reactions occurring during anaesthesia. *Eur J Anaesthesiol* 2002;**19**:240–62.

53. Zanotti KM, Rybicki LA, Kennedy AW, *et al.* Carboplatin skin testing: a skin-testing protocol for predicting hypersensitivity to carboplatin chemotherapy. *J Clin Oncol* 2001;**19**:3126–9.

54. Broome CB, Schiff RI, Friedman HS. Successful desensitization to carboplatin in patients with systemic hypersensitivity reactions. *Med Pediatr Oncol* 1996;**26**:105–10.

55. Markman M, Kennedy A, Webster K, *et al.* Paclitaxel-associated hypersensitivity reactions: experience of the gynecologic oncology program of the Cleveland Clinic Cancer Center. *J Clin Oncol* 2000;**18**:102–5.

56. Lorenz W, Reimann HJ, Schmal A, *et al.* Histamine release in dogs by Cremophor E1 and its derivatives: oxethylated oleic acid is the most effective constituent. *Agents Actions* 1977;**7**(1):63–7.

57. Ebo DG, Piel GC, Conraads V, Stevens WJ. IgE-mediated anaphylaxis after first intravenous infusion of cyclosporine. *Ann Allergy Asthma Immunol* 2001;**87**:243–5.

58. Gadde J, Spence M, Wheeler B, Adkinson NF. Clinical experience with penicillin skin testing in a large inner-city STD clinic. *J Am Med Assoc* 1993;**270**:2456–63.

59. Green GR, Rosenblum AH, Sweet LC. Evaluation of penicillin hypersensitivity: value of clinical history and skin testing with penicilloyl-polylysine and penicillin G: a cooperative prospective study of the penicillin study group of the American Academy of Allergy. *J Allergy Clin Immunol* 1977;**60**:339–45.

60. Solensky R, Earl H, Gruchalla R. Clinical approach to penicillin allergic patients: a survey [abstract]. *J Allergy Clin Immunol* 1999;**103**:S33.

61. Project Team of The Resuscitation Council (UK). Emergency medical treatment of anaphylactic reactions. *Resuscitation* 1999;**41**:93–9.

62. Park MA, Li JT. Diagnosis and management of penicillin allergy. *Mayo Clin Proc* 2005;**80**:405–10.

63. Pumphrey RS, Davis S. Under-reporting of antibiotic anaphylaxis may put patients at risk. *Lancet* 1999;**353**:1157–8.

64. Joint Task Force on Practice Parameters, the American Academy of Allergy, Asthma and Immunology, and the Joint Council of Allergy, Asthma and Immunology. Executive summary of disease management of drug hypersensitivity: a practice parameter. *Ann Allergy Asthma Immunol* 1999; **83**(6 Pt 3):665–700.

65. Gruchalla RS. Clinical assessment of drug-induced disease. *Lancet* 2000;**356**:1505–11.

66. Schwartz LB, Sakai K, Bradford TR, *et al.* The alpha form of human tryptase is the predominant type present in blood at baseline in normal subjects and is elevated in those with systemic mastocytosis. *J Clin Invest* 1995;**96**:2702–10.

67. Schwartz LB, Metcalfe DD, Miller JS, Earl H, Sullivan T. Tryptase levels as an indicator of mast-cell activation in systemic anaphylaxis and mastocytosis. *New Engl J Med* 1987;**316**:1622–6.

68. Chamberlain D, Project Team of the Resuscitation Council (UK). Emergency medical treatment of anaphylactic reactions. *J Accid Emerg Med* 1999;**16**:243–7.

69. Pumphrey RS. Lessons for management of anaphylaxis from a study of fatal reactions. *Clin Exp Allergy* 2000;**30**:1144–50.

70. Laxenaire MC. [Epidemiology of anesthetic anaphylactoid reactions. Fourth multicenter survey (July 1994 to December 1996).] *Ann Fr Anesth Reanim* 1999;**18**:796–809.

71. Meakin GH, Pronske EH, Lerman J, *et al.* Bronchospasm after rapacuronium in infants and children. *Anesthesiology* 2001;**94**:926–7.

72. Naguib M. How serious is the bronchospasm induced by rapacuronium? *Anesthesiology* 2001;**94**:924–5.

73. Buckland RW, Avery AF. Histamine release following pancuronium: a case report. *Br J Anaesth* 1973;**45**:518–21.

74. Levy JH, Pitts M, Thanopoulos A, *et al.* The effects of rapacuronium on histamine release and hemodynamics in adult patients undergoing general anesthesia. *Anesth Analg* 1999;**89**:290–5.

75. Bishop MJ, O'Donnell JT, Salemi JR. Mivacurium and bronchospasm. *Anesth Analg* 2003;**97**:484–5, table of contents.

76. Heath ML. Bronchospasm in an asthmatic patient following pancuronium. *Anaesthesia* 1973;**28**:437–40.

77. O'Callaghan AC, Scadding G, Watkins J. Bronchospasm following the use of vecuronium. *Anaesthesia* 1986;**41**:940–2.

78. Eisenhauer EA, ten Bokkel Huinink WW, Swenerton KD, *et al.* European–Canadian randomized trial of paclitaxel in relapsed ovarian cancer: high-dose versus low-dose and long versus short infusion. *J Clin Oncol* 1994;**12**:2654–66.

79. Billett AL, Carls A, Gelber RD, Sallan SE. Allergic reactions to Erwinia asparaginase in children with acute lymphoblastic leukemia who had previous allergic reactions to *Escherichia coli* asparaginase. *Cancer* 1992;**70**:201–6.

80. Woo MH, Hak LJ, Storm MC, *et al.* Anti-asparaginase antibodies following *E. coli* asparaginase therapy in pediatric acute lymphoblastic leukemia. *Leukemia* 1998; **12**:1527–33.

81. Zanotti KM, Markman M. Prevention and management of antineoplastic-induced hypersensitivity reactions. *Drug Safety* 2001;**24**:767–79.

82. Shepherd GM. Hypersensitivity reactions to chemotherapeutic drugs. *Clin Rev Allergy Immunol* 2003;**24**:253–62.

83. Kellie SJ, Crist WM, Pui CH, *et al.* Hypersensitivity reactions to epipodophyllotoxins in children with acute lymphoblastic leukemia. *Cancer* 1991;**67**:1070–5.

84. O'Dwyer PJ, King SA, Fortner CL, Leyland-Jones B. Hypersensitivity reactions to teniposide (VM-26): an analysis. *J Clin Oncol* 1986;**4**:1262–9.

85. Christiansen C. Late-onset allergy-like reactions to X-ray contrast media. *Curr Opin Allergy Clin Immunol* 2002;**2**:333–9.

86. Bertrand P, Rouleau P, Alison D, Chastin I. Use of peak expiratory flow rate to identify patients with increased risk of contrast medium reaction: results of preliminary study. *Invest Radiol* 1988;**23**(Suppl 1):S203–5.

87. Ansell G. Adverse reactions to contrast agents: scope of problem. *Invest Radiol* 1970;**5**:374–91.

88. Meeker DP, Wiedemann HP. Drug-induced bronchospasm. *Clin Chest Med* 1990;**11**:163–75.

89. Morcos SK. Review Article. Effects of radiographic contrast media on the lung. *Br J Radiol* 2003;**76**:290–5.

90. Morcos SK, Thomsen HS. Adverse reactions to iodinated contrast media. *Eur Radiol* 2001;**11**:1267–75.

91. Eloy R, Corot C, Belleville J. Contrast media for angiography: physicochemical properties, pharmacokinetics and biocompatibility. *Clin Mater* 1991;**7**:89–197.

92. Szebeni J. Hypersensitivity reactions to radiocontrast media: the role of complement activation. *Curr Allergy Asthma Rep* 2004;**4**:25–30.

93. Speck U, Bohle F, Krause W, *et al.* Delayed hypersensitivity to X-ray CM: possible mechanisms and models. *Acad Radiol* 1998;**5**(Suppl 1):S162–5; discussion S6.

94. Dawson P, Pitfield J, Britton J. Contrast media and bronchospasm: a study with iopamidol. *Clin Radiol* 1983;**34**:227–30.

95. Sandstrom C. Secondary reactions from contrast media and the allergy concept. *Acta Radiol* 1955;**44**:233–42.

96. Pichler WJ. Modes of presentation of chemical neoantigens to the immune system. *Toxicology* 2002;**181/182**: 49–54.

97. Idee JM, Pines E, Prigent P, Corot C. Allergy-like reactions to iodinated contrast agents: a critical analysis. *Fund Clin Pharmacol* 2005;**19**:263–81.

98. Krantz AJ, Hershow RC, Prachand N, *et al.* Heroin insufflation as a trigger for patients with life-threatening asthma. *Chest* 2003;**123**:510–17.

99. Barke KE, Hough LB. Opiates, mast cells and histamine release. *Life Sci* 1993;**53**:1391–9.

100. Doenicke A, Moss J, Lorenz W, Hoernecke R. Intravenous morphine and nalbuphine increase histamine and catecholamine release without accompanying hemodynamic changes. *Clin Pharmacol Ther* 1995;**58**:81–9.

101. Galli A, Ranaudo E, Giannini L, Costagli C. Reversible inhibition of cholinesterases by opioids: possible pharmacological consequences. *J Pharm Pharmacol* 1996;**48**:1164–8.

102. Ring J, Messmer K. Incidence and severity of anaphylactoid reactions to colloid volume substitutes. *Lancet* 1977;**1**:466–9.

103. Monaghan MS, Glasco G, St John G, Bradsher RW, Olsen KM. Safe administration of iron dextran to a patient who reacted to the test dose. *South Med J* 1994;**87**:1010–12.

104. Vally H, Taylor ML, Thompson PJ. The prevalence of aspirin intolerant asthma (AIA) in Australian asthmatic patients. *Thorax* 2002;**57**:569–74.

105. Jenkins C, Costello J, Hodge L. Systematic review of prevalence of aspirin induced asthma and its implications for clinical practice. *Br Med J* 2004;**328**:434.

106. Proceedings for the International Conference on Environmental Allergic Disease, 2 November 2000. *Ann Allergy Asthma Immunol* 2001;**87**(6 Suppl 3):1–63.

107. Vane JR. The mode of action of aspirin and similar compounds. *J Allergy Clin Immunol* 1976;**58**:691–712.

108. Churchill L, Chilton FH, Resau JH, *et al.* Cyclooxygenase metabolism of endogenous arachidonic acid by cultured human tracheal epithelial cells. *Am Rev Respir Dis* 1989;**140**:449–59.

109. Holgate ST, Burns GB, Robinson C, Church MK. Anaphylactic- and calcium-dependent generation of prostaglandin D_2 (PGD_2), thromboxane B_2, and other cyclooxygenase products of arachidonic acid by dispersed human lung cells and relationship to histamine release. *J Immunol* 1984;**133**: 2138–44.

110. Chandrasekharan NV, Dai H, Roos KL, Evanson NK, *et al.* COX-3, a cyclooxygenase-1 variant inhibited by acetaminophen and other analgesic/antipyretic drugs: cloning, structure, and expression. *Proc Natl Acad Sci USA* 2002;**99**:13926–31.

111. Sousa A, Pfister R, Christie PE, *et al.* Enhanced expression of cyclo-oxygenase isoenzyme 2 (COX-2) in asthmatic airways and its cellular distribution in aspirin-sensitive asthma. *Thorax* 1997;**52**:940–5.

112. Dahlen B, *et al.*, Celecoxib in Aspirin-Intolerant Asthma Study Group. Celecoxib in patients with asthma and aspirin intolerance. *New Engl J Med* 2001;**344**:142.

113. Yoshida S, Ishizaki Y, Onuma K, *et al.* Selective cyclo-oxygenase 2 inhibitor in patients with aspirin-induced asthma. *J Allergy Clin Immunol* 2000;**106**:1201–2.

114. Szczeklik A, Mastalerz L, Nizankowska E, Cmiel A. Protective and bronchodilator effects of prostaglandin E and salbutamol in aspirin-induced asthma. *Am J Respir Crit Care Med* 1996;**153**:567–71.

115. Christie PE, Smith CM, Lee TH. The potent and selective sulfidopeptide leukotriene antagonist, SK&F 104353, inhibits aspirin-induced asthma. *Am Rev Respir Dis* 1991;**144**:957–8.

116. Cowburn AS, Sladek K, Soja J, *et al.* Overexpression of leukotriene C4 synthase in bronchial biopsies from patients with aspirin-intolerant asthma. *J Clin Invest* 1998;**101**:834–46.

117. Sousa AR, Parikh A, Scadding G, Corrigan CJ, Lee TH. Leukotriene-receptor expression on nasal mucosal inflammatory cells in aspirin-sensitive rhinosinusitis. *New Engl J Med* 2002;**347**:1493–9.

118. Sanak M, Levy BD, Clish CB, *et al.* Aspirin-tolerant asthmatics generate more lipoxins than aspirin-intolerant asthmatics. *Eur Respir J* 2000;**16**:44–9.

119. Szczeklik A, Stevenson DD. Aspirin-induced asthma: advances in pathogenesis and management. *J Allergy Clin Immunol* 1999;**104**:5–13.

120. Szczeklik A, Sanak M. The broken balance in aspirin hypersensitivity. *Eur J Pharmacol* 2006;**533**(1/3):145–55.

121. Samter M, Beers RFJ. Intolerance to aspirin: clinical studies and consideration of its pathogenesis. *Ann Intern Med* 1968;**68**:975–83.

122. Marquette CH, Saulnier F, Leroy O, *et al.* Long-term prognosis of near-fatal asthma: a 6-year follow-up study of 145 asthmatic patients who underwent mechanical ventilation for a near-fatal attack of asthma. *Am Rev Respir Dis* 1992;**146**:76–81.

123. Milewski M, Mastalerz L, Nizankowska E, Szczeklik A. Nasal provocation test with lysine-aspirin for diagnosis of aspirin-sensitive asthma. *J Allergy Clin Immunol* 1998;**101**:581–6.

124. Drazen JM, Israel E, O'Byrne PM. Treatment of asthma with drugs modifying the leukotriene pathway. *New Engl J Med* 1999;**340**:197–206; erratum **340**:663.

125. Szczeklik A, Dworski R, Mastalerz L, *et al.* Salmeterol prevents aspirin-induced attacks of asthma and interferes with eicosanoid metabolism. *Am J Respir Crit Care Med* 1998;**158**:1168–72.

126. Berges-Gimeno MP, Stevenson DD. Nonsteroidal anti-inflammatory drug-induced reactions and desensitization. *J Asthma* 2004;**41**:375–84.

127. Szmidt M, Grzelewska-Rzymowska I, Rozniecki J. Tolerance to aspirin in aspirin-sensitive asthmatics: methods of inducing the tolerance state and its influence on the course of asthma and rhinosinusitis. *J Investig Allergol Clin Immunol* 1993;**3**(3):156–9.

128. Juergens UR, Christiansen SC, Stevenson DD, Zuraw BL. Inhibition of monocyte leukotriene B4 production after aspirin desensitization. *J Allergy Clin Immunol* 1995;**96**:148–56.

129. Grzelewska-Rzymowska I, Szmidt M, Rozniecki J, Grzegorczyk J. Aspirin-induced neutrophil chemotactic activity (NCA) in patients with aspirin-sensitive urticaria after desensitization to the drug. *J Investig Allergol Clin Immunol* 1994;**4**:28–31.

130. Salazar Villa RM, Zambrano Villa S. [Asthma induced by aspirin and arachidonic acid metabolites. 3: Role of leukotrienes.] *Rev Alerg Mex* 1994;**41**(2):51–7.

131. Hohneker JA. A summary of vinorelbine (Navelbine) safety data from North American clinical trials. *Semin Oncol* 1994;**21**(5 Suppl 10):42–6; discussion 6–7.

132. Asmus MJ, Sherman J, Hendeles L. Bronchoconstrictor additives in bronchodilator solutions. *J Allergy Clin Immunol* 1999;**104**(2 Pt 2):S53–60.

133. Beasley R, Hendeles L. Preservatives in nebulizer solutions: risks without benefit, a further comment. *Pharmacotherapy* 1999;**19**:473–4.

134. Hodder R, Pavia D, Dewberry H, *et al.* Low incidence of paradoxical bronchoconstriction in asthma and COPD patients during chronic use of Respimat soft mist inhaler. *Respir Med* 2005;**99**:1087–95.

135. Antonicelli L, Micucci C, Bonifazi F. Bronchospasm induced by inhalant corticosteroids: the role of ethanol. *Allergy* 2006;**61**:146–7.

136. Erjavec Z, Woolthuis GM, de Vries-Hospers HG, *et al.* Tolerance and efficacy of amphotericin B inhalations for prevention of invasive pulmonary aspergillosis in haematological patients. *Eur J Clin Microbiol Infect Dis* 1997;**16**:364–8.

137. Alston TA. Inhibition of cholinesterases by pentamidine. *Lancet* 1988;**2**(8625):1423.

138. Koehler D, Pavia D, Dewberry H, Hodder R. Low incidence of paradoxical bronchoconstriction with bronchodilator drugs administered by Respimat Soft Mist inhaler: results of phase II single-dose crossover studies. *Respiration* 2004;**71**:469–76.

139. Grieco MH, Pierson RN. Mechanism of bronchoconstriction due to beta adrenergic blockade: studies with practolol, propranolol, and atropine. *J Allergy Clin Immunol* 1971;**48**:143–52.

140. Babu KS, Marshall BG. Drug-induced airway diseases. *Clin Chest Med* 2004;**25**:113–22.

141. Myers JD, Higham MA, Shakur BH, Wickremasinghe M, Ind PW. Attenuation of propranolol-induced bronchoconstriction by frusemide. *Thorax* 1997;**52**:861–5.

142. Trembath PW, Taylor EA, Varley J, Turner P. Effect of propranolol on the ventilatory response to hypercapnia in man. *Clin Sci* (London) 1979;**57**:465–8.

143. Ashrafian H, Violaris AG. Beta-blocker therapy of cardiovascular diseases in patients with bronchial asthma or COPD: the pro viewpoint. *Prim Care Respir J* 2005;**14**:236–41.

144. Everly MJ, Heaton PC, Cluxton RJ. Beta-blocker underuse in secondary prevention of myocardial infarction. *Ann Pharmacother* 2004;**38**:286–93.

145. Miki A, Tanaka Y, Ohtani H, Sawada Y. Betaxolol-induced deterioration of asthma and a pharmacodynamic analysis based on beta-receptor occupancy. *Int J Clin Pharmacol Ther* 2003;**41**:358–64.

146. Salpeter S, Ormiston T, Salpeter E. Cardioselective beta-blocker use in patients with reversible airway disease. *Cochrane Database Syst Rev* 2001(2):CD002992.

147. Stefanutto TB, Halbach V. Bronchospasm precipitated by ethanol injection in arteriovenous malformation. *Am J Neuroradiol* 2003;**24**:2050–1.

148. Cabestrero D, Perez-Paredes C, Fernandez-Cid R, Arribas MA. Bronchospasm and laryngeal stridor as an adverse effect of oxytocin treatment. *Crit Care* 2003;**7**:392.

9

Drug-induced pulmonary oedema and acute respiratory distress syndrome

TEOFILO LEE-CHIONG JR, RICHARD A MATTHAY

INTRODUCTION

Non-cardiogenic pulmonary oedema and acute respiratory distress syndrome (ARDS) are common pulmonary complications related to medication use. Unfortunately, aside from a few specific medications, there are no consistently reliable investigations that can help distinguish the two clinical presentations or identify the causative agent.

INCIDENCE AND EPIDEMIOLOGY

Numerous medications have been reported to cause pulmonary oedema or ARDS. A large proportion of published articles on drug-related pulmonary toxicities contains only single case reports (Tables 9.1 and 9.2).[1] More comprehensive discussions on drug-related pulmonary toxicities may be found in the drug information leaflet enclosed with each medication, in pharmaceutical textbooks, and via electronic resources, and interested readers are encouraged to refer to these publications for additional information.

DRUGS IMPLICATED IN PULMONARY OEDEMA

ASPIRIN

Pulmonary oedema has been noted to occur in about 20–30 per cent of salicylate-intoxicated patients who are over 30 years old.[2,3] The likelihood of developing pulmonary oedema was greater in older patients, in those with a history of chronic salicylate ingestion or history of smoking, and in patients with neurologic abnormalities, proteinuria, and serum salicylate levels greater than 40 mg/dL.[3] Risk is also associated with the ingested dosage of the drug. In the paediatric age group, pulmonary oedema secondary to salicylate intoxication is correlated with a higher initial anion gap, greater serum salicylate levels, younger age, hypocarbia, and lower serum potassium.[4]

BETA-2 ADRENOCEPTOR AGONISTS

In a review of 1407 patients treated with parenteral isoxsuprine over a 7-year period, the incidence of pulmonary oedema was noted to be about 0.5 per cent.[5] The incidence of pulmonary oedema appears to be less following therapy with isoxsuprine than terbutaline.[6] The likelihood of developing a pulmonary complication following intravenous tocolysis is greater with excessive co-administration of tocolytic agent and intravenous fluids, infusion longer than 24 hours, twin or multiple gestations, presence of anaemia, sustained tachycardia, and undiagnosed cardiopulmonary disease.[5,7]

CONTRAST AGENTS

Pulmonary oedema can develop following intravenous administration of radio-contrast media. The incidence of pulmonary toxicity varies depending on the agent, dose and injection rate.[8] Similarly, the state of hydration at the time of contrast media infusion may also alter the risk of adverse effects, with the incidence of pulmonary oedema possibly lower in well-hydrated compared to dehydrated subjects.[9] Greater physiological and pathological changes occur with contrast materials possessing higher osmolarity, viscosity and iodine content. There are no apparent differences in risk of pulmonary oedema between non-anionic and anionic agents with similar viscosities and iodine content.[10]

OTHER AGENTS

Pulmonary oedema occurred in 15 per cent of patients with severe chronic heart failure (ejection fractions of less than 30 per cent) who were given amlodipine.[11] The incidence of

Table 9.1 Agents that can cause pulmonary oedema

Medication	Characteristics
Acetazolamide	Carbonic anhydrase inhibitor used as a diuretic, as an agent to reduce intraocular pressure in glaucoma, as prophylaxis of acute mountain sickness, and as adjunctive therapy in seizure disorders[27]
Amiloride	Potassium-sparing diuretic used for treating hypertension and congestive heart failure
Amlodipine	Dihydropyridine calcium-channel blocking agent indicated for the therapy of hypertension and stable angina pectoris[11]
Aspirin	Non-steroidal anti-inflammatory agent[2–4]
Beta-2 adrenoceptor agonists	Agents used as bronchodilators or as tocolytic agents in the management of preterm labour[5–7,33–35]
Boric acid	Astringent used in a variety of powder, ointment and ophthalmic preparations
Calusterone	Synthetic androgen indicated primarily for treating inoperable or metastatic breast cancer in postmenopausal women
Chlorothiazide/ hydrochlorothiazide	Thiazide diuretic
Chlordiazepoxide	Benzodiazepine agent that facilitates GABA-mediated CNS neurotransmission and that has been used to treat anxiety disorders and alcohol withdrawal symptoms
Cocaine	Agent with adrenergic agonist, central nervous system stimulant, vasoconstrictive and anaesthetic properties
Codeine	Opioid commonly used as an analgesic, antidiarrhoeal or antitussive agent
Cytarabine arabinoside	Antineoplastic agent
Desipramine	Tricyclic antidepressant
Dextran	Formulation consisting of branched glucose residues that possesses colloidal osmotic, antithrombotic and antiplatelet activity
Dextrose	Simple sugar used as a fluid and caloric replacement
Dipyridamole	Phosphodiesterase inhibitor that is capable of inhibiting platelet aggregation and is indicated in the prevention of thromboembolism; it also possesses vasodilatory properties and is used in myocardial thallium imaging
Donepezil	Centrally active acetylcholinesterase inhibitor indicated for patients with Alzheimer's disease
Epinephrine/adrenaline	Sympathomimetic agent with vasopressor, bronchodilator and cardiac stimulant properties
Epoprostenol	Prostaglandin with vasodilator properties that is administered by continuous infusion in patients with pulmonary hypertension
Ergonovine	Ergot alkaloid with uterine smooth muscle contractile properties used for the prevention and treatment of postpartum haemorrhage
Ethchlorvynol	Hypnotic agent[36–41]
Fluorescein	Diagnostic indicator dye agent
Gemcitabine	Antineoplastic agent used for a number of solid tumours including pancreatic cancer, non-small-cell lung cancer, advanced breast cancer and ovarian cancer[16,17]
Heparin	Parenteral anticoagulant and antithrombotic agent
Heroin	Opioid analgesic
Hydrochlorothiazide	Diuretic[52–54]
Immune globulin	Agent indicated as replacement therapy in immunoglobulin G (IgG) deficient patients or as an immunomodulator for autoimmune disorders
Irofulven	Antineoplastic agent given for solid tumours and acute leukaemias
Ketamine	Non-barbiturate agent with anaesthetic, analgesic and sedative properties
Metaraminol	Sympathomimetic amine vasopressor agent
Methadone	Long-acting narcotic analgesic used for the treatment of acute and chronic pain, and for maintenance therapy for narcotic addiction

Table 9.1   *(Continued)*

Medication	Characteristics
Methotrexate	Immunosuppressive and antineoplastic agent[22]
Methylene blue	Agent used for drug-induced methemoglobinaemia, cystitis or urethritis
Minocycline	Tetracycline-derived antimicrobial agent[23]
Mitomycin	Antineoplastic antibiotic
Monoctanoin	Formulation used to dissolve cholesterol gallstones[13,14]
Morphine	Opioid analgesic[24]
Muromonab-CD3	Immunosuppressive agent that prevents T-lymphocyte activation and used primarily to reverse episodes of transplant rejection
Nalmephene	Opioid antagonist indicated for the treatment of opioid overdose and for reversal of opioid-related respiratory depression
Naloxone	Opioid antagonist used to reverse opioid overdose
Naproxen	Non-steroidal anti-inflammatory agent
Nifedipine	Calcium-channel blocker used for the treatment of hypertension, angina, migraine headaches and Raynaud's disease
Nisoldipine	Dihydropyridine calcium channel-blocking agent used for managing hypertension and stable angina[59]
Nitric oxide	Pulmonary vasodilator given for pulmonary hypertension
Nitrofurantoin	Antimicrobial agent
Nitrogen mustard	Alkylating antineoplastic agent used to treat Hodgkin's disease
Oxycodone	Oral narcotic analgesic given for the relief of acute and chronic pain
Oxytocin	Agent with uterine smooth muscle contractile properties used for the induction of labour or prevention of post-partum bleeding
Paclitaxel	Antineoplastic agent used to treat patients with non-small-cell lung, breast and ovarian cancers
Paraldehyde	Sedative-hypnotic agent
Pentazocine	Narcotic agonist and antagonist
Pentostatin	Purine analogue antineoplastic agent indicated for the treatment of hairy-cell leukaemia
Phenothiazines	Antipsychotic agents[26]
Phenylbutazone	Non-steroidal anti-inflammatory agent
Phenylephrine	Sympathomimetic agent and potent vasoconstrictor
Polyethylene glycol	Osmotic laxative agent
Potassium iodide	Agent that is used as expectorant, an antithyroid medication and a dermatological preparation
Propoxyphene	Opioid narcotic agent
Propranolol	Beta-adrenergic blocker indicated for the management of hypertension, post-myocardial infarction, angina and arrhythmias
Protamine	Basic protein that binds and inactivates the effects of heparin
Ribavirin	Antiviral agent
Rofecoxib	Non-steroidal anti-inflammatory agent that selectively inhibits cyclooxygenase-2
Sodium chloride	Saline formulation
Streptokinase	Fibrinolytic agent
Tegafur	Pro-drug of 5-fluorouracil with activity against head and neck, breast and gastrointestinal cancers
Tretinoin	All-*trans*-retinoic acid (ATRA) is used for therapy of acute promyelocytic leukaemia
Troglitazone	Oral antidiabetes agent that decreases hepatic glucose output and peripheral insulin resistance
Verapamil	Calcium-channel blocking antihypertensive agent that is also used for the management of atrial tachyarrhythmias and angina
Zinc compounds	Mild astringent or dietary supplement for zinc deficiency

Table 9.2   Agents that can cause acute respiratory distress syndrome

Medication	Characteristics
Amiodarone	Antiarrhythmic agent
Amitryptyline	Tricyclic antidepressant
Ampicillin	Semi-synthetic penicillin antibiotic agent
Aprotinin	Haemostatic agent with antifibrinolytic and antiplasmin activities used prophylactically in patients undergoing coronary artery bypass graft surgery to reduce excessive perioperative bleeding[45]
Atenolol	Beta-adrenergic blocking agent used in the therapy of hypertension, arrhythmias, angina pectoris and myocardial infarction[42]
Chlorine gas	Gaseous element
Colchicine	Anti-inflammatory alkaloid agent used primarily for gouty arthritis
Ciclosporin	Cyclic polypeptide immunosuppressant used for the prevention and treatment of transplant rejection[48]
Cytarabine	Antineoplastic agent used for the treatment of leukaemias
Deferoxamine	Iron-chelating agent used for both acute and chronic iron overload[46]
Desipramine	Tricyclic antidepressant
Fluvastatin	Agent indicated as an adjunct to dietary fat restriction for the treatment of hyperlipidaemia
Heroin	Opioid agent
Imipramine	Tricyclic antidepressant
Infliximab	Antibody to tumour necrosis factor-alpha used for treating Crohn's disease or severe rheumatoid arthritis[43]
Lidocaine	Local anaesthetic and antiarrhythmic agent
Lymphocyte immune globulin	Immunosuppressant agent used for the management of transplant rejection
Nitrofurantoin	Antimicrobial agent[50]
Propylthiouracil	Antithyroid agent indicated for the treatment of patients with hyperthyroidism
Recombinant granulocyte–macrophage colony stimulating factor	Agent that preferentially stimulates granulocyte–macrophage progenitor cells
Vinca alkaloids	Antineoplastic agents[18-20]

pulmonary oedema following immunosupression for orthotopic liver transplantation with monoclonal antibody (OKT3) administration was estimated at 41 per cent.[12] Pulmonary oedema was noted in approximately 1 per 1000 patients during intrabiliary infusion of monooctanoin.[13,14] In one study, 30 per cent of patients with massive tricyclic antidepressant ingestion developed abnormal chest radiographic findings including pulmonary oedema.[15] Finally, non-cardiogenic pulmonary oedema has been described following administration of anticancer chemotherapy.[16,17]

DRUGS IMPLICATED IN ACUTE RESPIRATORY DISTRESS SYNDROME

ARDS has been reported following administration of vinca alkaloids. Combination therapy with vinblastine and mitomycin may increase the risk of pulmonary toxicity compared with vinblastine therapy alone, with the incidence of pulmonary toxicity following mitomycin and vinca alkaloid combination

estimated at 3–6 per cent.[18,19] Risk may be higher among patients with ovarian cancer than for those with other tumours.[20]

AETIOLOGIES

Most cases of drug-induced pulmonary oedema and ARDS appear to be idiosyncratic reactions rather than common adverse reactions. Except for a few agents, notable of which are the drugs of abuse, pulmonary toxicity is independent of drug dosage or duration of therapy.

Aetiology of pulmonary oedema

For most agents, the pathogenetic mechanisms responsible for pulmonary oedema remain incompletely understood. Postulated explanations include increased pulmonary vascular permeability (aspirin and cytosine arabinoside),[21] hypersensitivity reaction (methotrexate),[22] lymphocytic over-activation and massive cytokine release (minocycline),[23] immunoglobulin and complement deposition in the lung or brain-stem neurogenic reflex (opioids),[24] vasoactive myocardiotoxic agents released by damaged T-cells (OKT3),[12] capillary leak syndrome and cardiac toxicity (interleukin-2),[25] neurogenic processes, possibly from a disturbance of hypothalamic function (phenothiazines),[26] cross-reactivity to an associated compound (acetazolamide with an allergy to sulphonamide),[27] intrapulmonary sequestration of granulocytes and IgG deposition in alveolar membranes (hydrochlorothiazide, GM-CSF toxicity, and all-trans-retinoic acid),[28] vasoconstrictor-induced systemic hypertension (alpha agonists),[29] hypotension (tricyclic antidepressants),[15] and inadvertent aspiration (contrast agents).[30]

CONTRAST AGENTS

The pathogenesis of pulmonary oedema secondary to contrast agents appears to be due to chemical irritation of the pulmonary endothelium and enhanced formation of endothelium-derived prostacyclin.[31] It seems unrelated to classical anaphylaxis, contrast media overdose, sodium and fluid overload, or acute myocardial infarction.[32]

BETA-2 ADRENOCEPTOR AGONISTS

The mechanism responsible for the development of pulmonary oedema during tocolytic therapy using β_2 adrenoceptor agonists is most likely non-cardiogenic. A key factor is plasma volume expansion secondary to fluid and sodium retention.[33] Beta-2 sympathomimetics possess antidiuretic effects and administration can reduce urine output and elevate central venous pressure.[34] Other possible mechanisms for the development of pulmonary oedema secondary to tocolytic therapy include decreased oncotic pressure, capillary leakage, and unmasking of an underlying asymptomatic peripartum cardiomyopathy.[35]

ETHCHLORVYNOL

Pulmonary oedema during administration of ethchlorvynol is believed to be due to an increase in extravascular lung tissue water secondary to greater alveolar membrane permeability or increase in lymph flow.[36,37] It has been postulated that the ethchlorvynol-induced increase in vascular permeability is due to the creation of irregular gaps between endothelial cells, as well as the formation of subendothelial blebs.[38] Ethchlorvynol-induced acute lung injury does not appear to be related to prostaglandin, cyclooxygenase or histamine activity since oedema formation following ethchlorvynol administration is not attenuated by pre-treatment with the cyclooxygenase inhibitors, indomethacin, ibuprofen, histamine receptor blockers or prostaglandin synthetase inhibitors.[39,40] Complement activation and leucocyte migration also play insignificant pathogenetic roles.[41]

Aetiology of ARDS

Postulated pathophysiologic mechanisms for drug-related ARDS include capillary leak syndrome, hypervolaemia, anaphylaxis,

acute drug hypersensitivity (atenolol),[42] delayed hypersensitivity reaction (infliximab),[43] direct lung cytotoxicity, complement-mediated leucocyte and platelet destruction (antilymphocyte globulin),[44] microthrombosis of the small pulmonary arterioles (aprotinin),[45] free-radical generation (desferrioxamine),[46] reaction to associated compounds (ciclosporin with toxicity to cremaphor, the solubilizing agent in the intravenous formulation),[47] increased adhesiveness of neutrophils to the pulmonary endothelium induced by expression of the glycoproteins CD11B and CD18 on the neutrophil surface, enhanced release of superoxide anions, and increased platelet-activating factor production by neutrophils (GM-CSF).[48] Underlying cardiovascular diseases may either increase the risk of developing pulmonary oedema or simply confound the diagnosis.

CLINICAL PRESENTATIONS

Apart from a history of exposure (usually recent) to drugs, clinical features are generally indistinguishable from other causes of pulmonary oedema and ARDS.

Presentation of pulmonary oedema

Patients may present with dyspnoea, cough, chest discomfort or pain, or respiratory failure requiring mechanical ventilatory support. Fever has been noted in some patients. The all-trans-retinoic acid (ATRA) syndrome is a life-threatening complication of ATRA therapy of acute promyelocytic leukaemia. Toxicity, including fever, respiratory distress, pulmonary infiltrates, weight gain and oedema, commonly occurs 1 to 11 days after therapy is started.[49] Prevention by pre-treatment with dexamethasone has been described.

Physical examination findings include tachypnoea, cyanosis, hypotension or hypertension, tachycardia or bradycardia, bilateral crepitations, rales, and oedema. Unlike cardiogenic pulmonary oedema, jugular venous distention or S3 gallop is absent. Pink frothy fluid can be present during endotracheal intubation.

Presentation of ARDS

Typical manifestations include dyspnoea, chest discomfort, tachypnoea, cough and hypoxaemia. Hypotension, fever, chills and respiratory failure can develop. Rales, respiratory distress, cyanosis, tachypnoea, tachycardia and rashes may be appreciated on physical examination.[50]

INVESTIGATIONS

Diagnosis rests on exclusion of other disorders that may present in a similar fashion (e.g. congestive heart failure, acute respiratory distress syndrome, diffuse pulmonary haemorrhage or respiratory infection). Laboratory evaluation and respiratory function test results are generally non-specific. Rarely is lung biopsy indicated.

Investigations for pulmonary oedema

Arterial blood gas (ABG) analysis commonly reveals the presence of arterial hypoxaemia, hypocapnia or hypercapnia, metabolic acidosis, and an increased alveolar–arterial oxygen gradient (P(A-a)O_2). Pulmonary function testing may disclose mildly decreased diffusing capacity.

Results from laboratory evaluation are typically non-specific. Beta-2 adrenegergic therapy, when provided with large amounts of isotonic solution, can give rise to decreased haematocrit, haemoglobin and total protein levels.[51] Immunological studies after exposure to hydrochlorothiazide may demonstrate decreased serum IgG_1 and IgG_4, increased serum IgM and IgE,[28] negative antinuclear antibodies and lymphocyte transformation test, and normal blood lymphocyte subpopulations and complement levels.[52]

Chest radiographic findings range from mild interstitial oedema to frank pulmonary oedema. These include bilateral interstitial and alveolar filling infiltrates, increased pulmonary blood volume, septal lines, peribronchial cuffing, ground-glass opacification, consolidation, nodules and bronchograms. Pleural effusions are infrequent. Cardiomegaly and pulmonary vascular redistribution are generally absent (Fig. 9.1).

Haemodynamic studies confirm the presence of non-cardiogenic oedema, with low cardiac output,[53] and a normal[54] or mildly elevated[55] pulmonary capillary wedge pressure. In patients with pulmonary oedema related to tocolytic therapy with β_2 adrenoceptor agonists, haemodynamic features include an increase in cardiac output, left ventricular ejection fraction and mean pulmonary artery pressure, and a reduction in total resistance of the pulmonary vessels, diastolic blood pressure and end-systolic volume.[56] Pulmonary capillary wedge pressure can be normal,[57] but may also rise progressively and exceed the threshold for the development of pulmonary congestion.[58] Administration of α-agonists can produce systemic vasoconstriction, increased left ventricular afterload, decreased left ventricular compliance, and decreased cardiac output.[29] Pulmonary oedema related to administration of calcium-channel blockers may be associated with a decrease in both systemic vascular resistance and mean arterial pressure.[59]

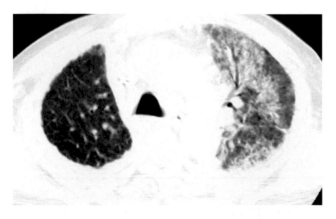

Fig. 9.1 Unilateral left-sided gemcitabine toxicity in a man being treated for a right-sided mesothelioma. Photograph provided by David Lynch MD, National Jewish Medical and Research Center, Denver, Colorado.

Investigations for ARDS

Hypoxaemia and hypocapnia (or hypercapnia) are common. Leucocytosis and eosinophilia can be present as well. Pulmonary function testing can demonstrate decreased lung compliance. Restrictive or obstructive ventilatory defect, and diminished diffusing capacity, have been described. Chest radiographs may reveal bilateral interstitial and alveolar filling infiltrates.[50] Cardiomegaly and pulmonary vascular redistribution are generally absent.

Haemodynamically, precapillary pulmonary arterial hypertension may be noted. Lung injury, including interstitial and alveolar pulmonary oedema, haemorrhage, alveolitis, consolidation, eosinophilic pneumonia, and hyaline membrane formation can be evident on open lung biopsy or autopsy.[43]

TREATMENT/MANAGEMENT

Clinical recovery is often rapid, with few long-term sequelae once drug administration is discontinued.

Pulmonary oedema

Oxygen therapy, morphine and diuretic therapy are often prescribed for pulmonary oedema. Mechanical ventilatory support along with positive end-expiratory pressure is occasionally required. Some cases had been treated with corticosteroids but their therapeutic role is uncertain. Corticosteroids have also been used for prophylaxis against contrast media-induced pulmonary oedema and ATRA syndrome.[60] A case of pulmonary oedema and bradycardia from inadvertent overdosage of donepezil was successfully reversed with atropine administration.[61]

Pulmonary oedema secondary to aspirin toxicity resolves with declining levels of serum salicylate.[3]

Management of tocolytic-related pulmonary oedema secondary to administration of β_2 adrenoceptor agonists consists of withdrawal of the offending agent, oxygen supplementation, restriction of fluid intake, diuretic therapy and normalization of acid–base status. If feasible, labour should be allowed to proceed while fluid status, vital signs, oxygenation and electrolytes are closely monitored. Mechanical ventilation with positive end-expiratory pressure may be required for patients with severe respiratory failure. The risk of tocolytic-induced pulmonary oedema can be minimized by using the lowest possible tocolytic perfusion rate and limiting infusion duration to less than 48 hours.[62]

Acute respiratory distress syndrome

Oxygen supplementation, intubation and mechanical ventilatory support may be required for respiratory failure. Corticosteroids, antihistamines, diuretics, β-adrenergic agonists, vasopressors or vasodilators, and bronchodilators have been given in some cases but their therapeutic role is uncertain.

COMPLICATIONS

Clinical course is generally benign and self-limiting, but severe reactions including respiratory failure, cardiac arrhythmias and hypotension can develop. Symptoms can recur following re-challenge with the same medication.

> **KEY POINTS**
>
> - Many medications may cause pulmonary oedema or ARDS. Most of the published articles on drug-related pulmonary toxicities consist of case reports.
> - Most cases appear to be idiosyncratic reactions and are independent of drug dosage or duration of therapy.
> - Clinical features are generally indistinguishable from other causes of pulmonary oedema and ARDS.
> - Diagnosis requires exclusion of other disorders that may present in a similar fashion, such as congestive heart failure, ARDS, diffuse pulmonary haemorrhage or respiratory infection.
> - Laboratory evaluation, pulmonary function testing and chest radiographs are generally non-specific. Rarely is lung biopsy indicated.
> - Clinical course is typically benign and self-limiting, but symptoms can recur following re-challenge with the same medication.

REFERENCES

1. Lee-Chiong TL, Matthay RA. Medications causing pulmonary edema and ARDS. *Clin Chest Med* 2004;**25**:95–104.
2. Walters JS, Woodring JH, Stelling CB, Rosenbaum HD. Salicylate-induced pulmonary edema. *Radiology* 1983;**146**:289–93.
3. Heffner JE, Sahn SA. Salicylate-induced pulmonary edema: clinical features and prognosis. *Ann Intern Med* 1981;**95**:405–9.
4. Fisher CJ, Albertson TE, Foulke GE. Salicylate-induced pulmonary edema: clinical characteristics in children. *Am J Emerg Med* 1985;**3**:33–7.
5. Nimrod C, Rambihar V, Fallen E, Effer S, Cairns J. Pulmonary edema associated with isoxsuprine therapy. *Am J Obstet Gynecol* 1984;**148**:625–9.
6. Robertson PA, Herron M, Katz M, Creasy RK. Maternal morbidity associated with isoxsuprine and terbutaline tocolysis. *Eur J Obstet Gynecol Reprod Biol* 1981;**11**:371–8.
7. Gabriel R, Harika G, Saniez D, Durot S, Quereux C, Wahl P. Prolonged intravenous ritodrine therapy: a comparison between multiple and singleton pregnancies. *Eur J Obstet Gynecol Reprod Biol* 1994;**57**:65–71.
8. Bock JC, Barker BC, Federle MP, Lewis FR. Iodinated contrast media effects on extravascular lung water, central blood volume, and cardiac output in humans. *Invest Radiol* 1990;**25**:938–41.
9. Zack A, Mare K, Violante M. The aggravating effect of dehydration on pulmonary edema induced by ionic and nonionic contrast media. *Invest Radiol* 1985;**20**:212–16.
10. Moore DE, Carroll FE, Dutt PL, Reed GW, Holburn GE. Comparison of nonionic and ionic contrast agents in the rabbit lung. *Invest Radiol* 1991;**26**:134–42.
11. Packer M, et al., Prospective Randomized Amlodipine Survival Evaluation Study Group. Effect of amlodipine on morbidity and mortality in severe chronic heart failure. *New Engl J Med* 1996;**335**:1107–14.
12. Golfieri R, Giampalma E, Lalli A, et al. Pulmonary complications from monoclonal antibody (OKT₃) immunosuppression in patients who have undergone an orthotopic liver transplant. *Radiol Med* (Torino) 1994;**87**(1/2):58–64.
13. Hine LK, Arrowsmith JB, Gallo-Torres HE. Monooctanoin-associated pulmonary edema. *Am J Gastroenterol* 1988;**83**:1128–31.
14. Shustack A, Noseworthy TW, Johnston RG, et al. Noncardiogenic pulmonary edema during intrabiliary infusion of mono-octanoin. *Crit Care Med* 1986;**14**:659–60.
15. Shannon M, Lovejoy FH. Pulmonary consequences of severe tricyclic antidepressant ingestion. *J Toxicol Clin Toxicol* 1987;**25**:443–61.
16. Briasoulis E, Pavlidis N. Noncardiogenic pulmonary edema: an unusual and serious complication of anticancer therapy. *Oncologist* 2001;**6**:153–61.
17. Briasoulis E, Froudarakis M, Milionis HJ, et al. Chemotherapy-induced noncardiogenic pulmonary edema related to gemcitabine plus docetaxel combination with granulocyte colony-stimulating factor support. *Respiration* 2000;**67**:680–3.
18. Rao SX, Ramaswamy G, Levin M, McCravey JW. Fatal acute respiratory failure after vinblastine–mitomycin therapy in lung carcinoma. *Arch Intern Med* 1985;**145**:1905–7.
19. Hoelzer KL, Harrison BR, Luedke SW, Luedke DW. Vinblastine-associated pulmonary toxicity in patients receiving combination therapy with mitomycin and cisplatin. *Drug Intell Clin Pharm* 1986;**20**:287–9.
20. Ozols RF, Hogan WM, Ostchega Y, Young RC. MVP (mitomycin, vinblastine, and progesterone): a second-line regimen in ovarian cancer with a high incidence of pulmonary toxicity. *Cancer Treat Rep* 1983;**67**:721–2.
21. Haupt HM, Hutchins GM, Moore GW. Ara-C lung: noncardiogenic pulmonary edema complicating cytosine arabinoside therapy of leukemia. *Am J Med* 1981;**70**:256–61.
22. Lascari AD, Strano AJ, Johnson WW, Collins JG. Methotrexate-induced sudden fatal pulmonary reaction. *Cancer* 1977;**40**:1393–7.
23. MacNeil M, Haase DA, Tremaine R, Marrie TJ. Fever, lymphadenopathy, eosinophilia, lymphocytosis, hepatitis, and dermatitis: a severe adverse reaction to minocycline. *J Am Acad Dermatol* 1997;**36**(2 Pt 2):347–50.
24. Bruera E, Miller MJ. Non-cardiogenic pulmonary edema after narcotic treatment for cancer pain. *Pain* 1989;**39**:297–300.
25. Conant EF, Fox KR, Miller WT. Pulmonary edema as a complication of interleukin-2 therapy. *Am J Roentgenol* 1989;**152**:749–52.
26. Li C, Gefter WB. Acute pulmonary edema induced by overdosage of phenothiazines. *Chest* 1992;**101**:102–4.
27. Gerhards LJ, van Arnhem AC, Holman ND, Nossent GD. Fatal anaphylactic reaction after oral acetazolamide (Diamox) for glaucoma. *Neder Tijdschr Geneeskunde* 2000;**144**:1228–30.

28. Bernal C, Patarca R. Hydrochlorothiazide-induced pulmonary edema and associated immunologic changes. *Ann Pharmacother* 1999;**33**:172–4.

29. Kalyanaraman M, Carpenter RL, McGlew MJ, Guertin SR. Cardiopulmonary compromise after use of topical and submucosal alpha-agonists: possible added complication by the use of beta-blocker therapy. *Otolaryngol Head Neck Surg* 1997;**117**:56–61.

30. Wells HD, Hyrnchak MA, Burbridge BE. Direct effects of contrast media on rat lungs. *Can Assoc Radiol J* 1991;**42**:261–4.

31. Paajanen H, Uotila P. Effect of contrast media on the formation of prostacyclin in isolated rat lungs. *Acta Radiol Diagn* (Stockholm) 1985;**26**:777–83.

32. Greganti MA, Flowers WM. Acute pulmonary edema after the intravenous administration of contrast media. *Radiology* 1979;**132**:583–5.

33. Armson BA, Samuels P, Miller F, Verbalis J, Main EK. Evaluation of maternal fluid dynamics during tocolytic therapy with ritodrine hydrochloride and magnesium sulfate. *Am J Obstet Gynecol* 1992;**167**:758–65.

34. Grospietsch G, Fenske M, Dietrich B, *et al.* Effects of the tocolytic agent fenoterol on body weight, urine excretion, blood hematocrit, hemoglobin, serum protein, and electrolyte levels in nonpregnant rabbits. *Am J Obstet Gynecol* 1982;**143**:667–72.

35. Hes R, Stolk B, Twaalfhoven FC, Meinders AE. A patient with pulmonary edema following use of beta-sympathomimetics (tocolytic agents). *Neder Tijdschr Geneeskunde* 1991;**135**:668–71.

36. Fischer P, Glauser FL, Millen JE, Lewis J, Egan P. The effects of ethchlorvynol on pulmonary alveolar membrane permeability. *Am Rev Respir Dis* 1977;**116**:901–6.

37. Garnett R, Fairman RP, Glauser FL. Ethchlorvynol-induced pulmonary edema: a chronically instrumented, awake sheep model mimicking human disease. *Am J Med Sci* 1987;**294**:317–23.

38. Wysolmerski R, Lagunoff D. The effect of ethchlorvynol on cultured endothelial cells: a model for the study of the mechanism of increased vascular permeability. *Am J Pathol* 1985;**119**:505–12.

39. Sprague RS, Stephenson AH, Dahms TE, Lonigro AJ. Effect of cyclooxygenase inhibition on ethchlorvynol-induced acute lung injury in dogs. *J Appl Physiol* 1986;**61**:1058–64.

40. Fairman RP, Falls R, Millen JE, Glauser FL. Histamine receptor and prostaglandin synthetase blockers in ethchlorvynol-induced pulmonary edema in canines. *Clin Toxicol* 1980;**16**:25–31.

41. Fairman RP, Glauser FL, Falls R. Increases in lung lymph and albumin clearance with ethchlorvynol. *J Appl Physiol* 1981;**50**:1151–5.

42. Guilaume B, Moualla M, Bugnon D, Clementi E, Berruchon J. Acute respiratory distress syndrome in a patient treated with atenolol. *Rev Mal Respir* 1998;**15**(4):541–3.

43. Riegert-Johnson DL, Godfrey JA, Myers JL, Hubmayr RD, Sandborn WJ, Loftus EV Jr. Delayed hypersensitivity reaction and acute respiratory distress syndrome following infliximab infusion. *Inflamm Bowel Dis* 2002;**8**(3):186–91.

44. Dean NC, Amend WC, Matthay MA. Adult respiratory distress syndrome related to antilymphocyte globulin therapy. *Chest* 1987;**91**:619–20.

45. Vucicevic Z, Suskovic T. Acute respiratory distress syndrome after aprotinin infusion. *Ann Pharmacother* 1997;**31**:429–32.

46. Tenenbein M, Kowalski S, Sienko A, Bowden DH, Adamson IY. Pulmonary toxic effects of continuous desferrioxamine administration in acute iron poisoning. *Lancet* 1992;**339**:699–701.

47. Mackie FE, Umetsu D, Salvatierra O, Sarwal MM. Pulmonary capillary leak syndrome with intravenous cyclosporin A in pediatric renal transplantation. *Pediatr Transplant* 2000;**4**(1):35–8.

48. Verhoef G, Boogaerts M. Treatment with granulocyte-macrophage colony stimulating factor and the adult respiratory distress syndrome. *Am J Hematol* 1991;**36**:285–7.

49. Martin Del Pozo M, Cisneros De La Fuente E, Solano F, Martin ML, De La Serna J. The retinoic acid syndrome, a complication of acute promyelocytic leukemia therapy. *Ann Med Intern* 2001;**18**:195–200.

50. Chudnofsky CR, Otten EJ. Acute pulmonary toxicity to nitrofurantoin. *J Emerg Med* 1989;**7**:15–19.

51. Fenske M, Grospietsch G, Dietrich B, *et al.* Fenoterol-induced changes of urine excretion, body weight, blood hematocrit, hemoglobin, total protein values and serum electrolyte levels in pregnant rabbits. *Gynecol Obstet Invest* 1982;**14**:273–82.

52. Shieh CM, Chen CH, Tao CW, Tsai JJ, Perng RP. Hydrochlorothiazide-induced pulmonary edema: a case report and literature review. *Zhonghua Yi Xue Za Zhi* (Taipei) 1992;**50**:495–9.

53. Frierson JH, Marvel SL, Thomas GM. Hydrochlorothiazide-induced pulmonary edema with severe acute myocardial dysfunction. *Clin Cardiol* 1995;**18**:112–14.

54. Levay ID. Hydrochlorothiazide-induced pulmonary edema. *Drug Intell Clin Pharm* 1984;**18**:238–9.

55. Leser C, Bolliger CT, Winnewisser J, Burkart F, Perruchoud AP. Pulmonary oedema and hypotension induced by hydrochlorothiazide. *Monaldi Arch Chest Dis* 1994;**49**:308–10.

56. Wolff F, Carstens V, Fischer JH, Behrenbeck D, Bolte A. Cardiopulmonary effects of betamimetic tocolytic and glucocorticoid therapy in pregnant women. *Arch Gynecol* 1986;**239**:49–58.

57. Russi EW, Spaetling L, Gmur J, Schneider H. High permeability pulmonary edema (ARDS) during tocolytic therapy: a case report. *J Perinat Med* 1988;**16**(1):45–9.

58. Hadi HA, Abdulla AM, Fadel HE, Stefadouros MA, Metheny WP. Cardiovascular effects of ritodrine tocolysis: a new noninvasive method to measure pulmonary capillary pressure during pregnancy. *Obstet Gynecol* 1987;**70**:608–12.

59. Barjon JN, Rouleau JL, Bichet D, Juneau C, De Champlain J. Chronic renal and neurohumoral effects of the calcium entry blocker nisoldipine in patients with congestive heart failure. *J Am Coll Cardiol* 1987;**9**:622–30.

60. Mare K, Violante M, Zack A. Pulmonary edema following high intravenous doses of diatrizoate in the rat: effects of corticosteroid pretreatment. *Acta Radiol Diagn* (Stockholm) 1985;**26**:477–82.

61. Shepherd G, Klein-Schwartz W, Edwards R. Donepezil overdose: a tenfold dosing error. *Ann Pharmacother* 1999;**33**:812–15.

62. Bader AM, Boudier E, Martinez C, *et al.* Etiology and prevention of pulmonary complications following beta-mimetic mediated tocolysis. *Eur J Obst Gynecol Reprod Biol* 1998;**80**(2):133–7.

10

Pulmonary complications of blood transfusion

PATRICIA M KOPKO, MARK A POPOVSKY

INTRODUCTION

The two most common pulmonary complications of blood transfusion are volume overload and transfusion-related acute lung injury (TRALI). The volume overload that can accompany transfusion is a result of infusion of blood components more rapidly than the circulatory system can accommodate. The symptoms, diagnosis and treatment are identical to volume overload from infusion of other fluids. Therefore, this chapter will focus on TRALI, which is a significant pulmonary complication that can follow transfusion of any blood component.

Transfusion-related acute lung injury was first described as a distinct clinical entity in 1985.[1] However, earlier medical literature contains rare case reports of acute pulmonary oedema in the absence of myocardial injury or circulatory overload after transfusion. These cases probably represent TRALI. The first of these cases was published in 1951.[2] These reactions were given a variety of descriptive terms including 'non-cardiogenic pulmonary oedema',[3] 'allergic pulmonary oedema',[4] 'leukoagglutinin transfusion reaction'[5] and 'hypersensitivity reaction'.[6] Owing to the infrequent nature of these reports, there was a perception that this clinical disorder was extremely uncommon. It is only in the last 20 years that TRALI has been recognized as a significant cause of transfusion-related morbidity and mortality.

DEFINITIONS

ACUTE LUNG INJURY

TRALI is a form of acute lung injury (ALI). A definition of ALI was published in 1994 as a result of the North American–European Consensus Conference.[7] ALI was defined as hypoxaemia with PaO_2/FIO_2 ratio of $\leq 300\,mmHg$, bilateral pulmonary oedema on frontal chest radiograph, and pulmonary artery occlusion pressure of $\geq 18\,mmHg$, with no clinical evidence of left atrial hypertension. Several risk factors for ALI have been defined and include sepsis, pneumonia, aspiration of gastric contents, burns, disseminated intravascular coagulation (DIC), near drowning, massive transfusion and fracture of long bones.

TRANSFUSION-RELATED ACUTE LUNG INJURY

Two definitions of TRALI have recently been published.[8,9] Both definitions rely on the North American–European Consensus definition of ALI. They define TRALI as the presence of ALI within 6 hours of transfusion in the absence of other risk factors for ALI. The two definitions treat ALI, after transfusion, in the presence of other risk factors for ALI differently. The National Heart Lung and Blood Institute (NHLBI) working group definition allows for ruling out other risk factors as the mechanistic cause of the lung injury and defines these cases as TRALI. Conversely, cases in which the mechanistic cause of the lung injury is thought to be unrelated to transfusion are considered ALI, not TRALI. The Canadian Consensus Conference does not make this distinction. It uses the term 'possible TRALI' for ALI after transfusion in the presence of other risk factors for ALI.

INCIDENCE

The true incidence of TRALI is not known. Most of the published estimates of TRALI are based on observational studies performed at a single institution. These estimates range from 1 in 1300 to 1 in 5000 transfusions.[1,10] The wide range of published data is most likely explained by differences in the studies.[9] Incidence studies have not all used the same definition of

TRALI. Some of the studies relied on the results of testing for human leucocyte antibodies or human neutrophil antibodies to make the diagnosis,[1] while some studies are based on a clinical diagnosis.[11] In general, studies with an active surveillance mechanism report a higher incidence than studies with a passive surveillance mechanism. Additionally, the denominator varies between the studies. While some studies determine the exact number of components transfused, other studies estimate this number.

Until recently TRALI has been the third leading cause of transfusion-related mortality reported to the United States Food and Drug Administration.[12] The leading causes of transfusion-related mortality were haemolytic and septic transfusion reactions. However, starting in 2001, TRALI has become the most common cause of transfusion-related mortality reported in the United States.[13] TRALI is now responsible for close to 20 per cent of transfusion-related fatalities reported in the US. Given this, it may appear that the incidence of TRALI is increasing. However, the increased reporting of TRALI is most likely related to increased awareness of the disorder. Kopko *et al.* performed a look-back of recipients of fresh frozen plasma from a female donor with a strong antibody against human neutrophil antigen-3a(5b).[14] Of the 50 recipients who were investigated, 14 were too ill to evaluate for a reaction to the transfusion. The remaining 36 patients were evaluated for evidence of a transfusion reaction. Seven mild to moderate reactions were identified in 6 patients, and 8 severe reactions were identified in 8 patients. A severe reaction was defined as need for mechanical ventilation or new onset of pulmonary oedema. Of the 15 reactions identified in the look-back, 7 were reported to the transfusion service, while only 2 were reported to the facility that collected the implicated blood components. This study supports the hypothesis that the increased reporting of TRALI is not due to an increase in the frequency of the disorder, but to increased clinical awareness of TRALI as a complication of transfusion.

CLINICAL PRESENTATION

Although TRALI can occur up to 6 hours after transfusion, it typically presents during transfusion or shortly after (within 1 or 2 hours). Signs and symptoms include dyspnoea with acute onset of bilateral pulmonary oedema and severe hypoxia.[15,16] Fever, tachycardia, hypotension and hypertension followed by hypotension can also be present during the reaction. The pulmonary oedema can be visualized on roentgenograms within minutes to hours of the reaction, as the classic 'white-out' pattern of infiltration. In some patients, early roentgenograms display a patchy infiltrate that rapidly progresses to involve the entirety of both lung fields. This pattern of progression can also be seen in the dependent lung regions.

The hypotension that accompanies TRALI can be profound and is often unresponsive to fluid administration. Unlike ALI secondary to other causes, about 80 per cent of TRALI patients show marked clinical improvement within 96 hours of onset of symptoms.[1,15] The remaining 20 per cent of patients either have a protracted clinical course or succumb to the disorder. The mortality rate of TRALI has been estimated to be between 5 and 15 per cent.[1]

In one of the largest series of TRALI cases published, 100 per cent of the patients required oxygen support, with 72 per cent requiring mechanical ventilation.[1] Once clinical symptoms resolve, no permanent sequelae have been associated with TRALI.

IMPLICATED BLOOD COMPONENTS

TRALI has been associated with the transfusion of all plasma-containing blood components including whole blood, packed red blood cells, granulocytes collected by apheresis, platelet concentrates, apheresis platelets, fresh frozen plasma and cryoprecipitate.[1,17-21] There are a few case reports of TRALI following the administration of intravenous immune globulin.[22] The risk of TRALI from a type of blood component appears to be related to the volume of plasma in the component.[23] Components that contain large amounts of plasma such as whole blood, plasma and apheresis platelets have been disproportionately implicated in TRALI cases.[24]

MECHANISM OF TRALI

The precise mechanism is unknown. However, there is growing evidence that TRALI is an immune-mediated event. Two immunological triggers – the infusion of antibodies directed against cognate antigen in the recipient, and the infusion of biological response modifiers – have been implicated in the pathogenesis of TRALI.

Antibodies

There are numerous case reports and series of case reports of antibodies directed against human leucocyte antigens (HLA), both class I and class II, and human neutrophil antigens (HNA)[1,17,25,26] in TRALI. A number of these cases have been investigated sufficiently to determine the specificity of the antibody and the presence of the cognate antigen in the transfusion recipient. When antibody is present, it is identified in the transfused component in 85–95 per cent of cases, and antibody is identified in the transfusion recipient in about 10 per cent of cases.[1,6,17,25] In these cases, antibody in the patient corresponds to antigen(s) present on white blood cells in the transfused blood component. Rare cases of inter-donor TRALI have also been reported.[20,27,28] In these cases antibody in one blood component interacts with antigen(s) on the white blood cells of another transfused blood component.

Antibodies to white blood cells can be formed after exposure to foreign antigens. In blood donors the most common cause of antibody formation is pregnancy. Antibodies can also be formed as a result of exposure to foreign antigens through transfusion or transplantation. Since blood donors represent a healthy population, the majority of blood donors with antibodies to white cell antigens are females who developed antibody during pregnancy.

Early case reports of TRALI often included granulocyte and/or lymphocyte cross-matches as part of the investigation. The interpretation of these case reports can be misleading.

Since the surface of granulocytes routinely expresses both granulocyte and HLA class I antigens, a positive granulocyte cross-match does not automatically mean that a granulocyte antibody is present. The converse is also true. Many so-called granulocyte antigens are expressed on lymphocytes. Thus the presence of a positive lymphocyte cross-match does not always mean that an HLA antibody is present. Additionally, most of the lymphocyte cross-matches in case reports are unfractionated (i.e. the T- and B-cells have not been separated). It is extremely difficult to detect HLA class II antibodies with an unfractionated cross-match. This is because HLA class II antigens are expressed on B-lymphocytes and are not routinely expressed on unactivated T-lymphocytes. Since T-cells represent the overwhelming majority of lymphocytes, an antibody to HLA class II is typically not detected in an unfractionated lymphocyte cross-match.

Recent advances in the field of histocompatibility testing have greatly increased the ability to detect and determine the specificity of HLA class I and class II antibodies. Instead of detecting antibodies using cells with a full complement of known HLA antigens, newer methodologies are based on isolation of single HLA antigens. The single antigens are attached to a solid surface or latex bead.

In the case of solid surface assays, a reaction well is coated with HLA antigen(s) and antibody is detected by enzyme-linked immunosorbent assay (ELISA). Latex beads can also be coated with HLA antigen(s). Antibody adheres to the latex bead and is detected by flow cytometry or Luminex.

In early case reports of TRALI a high percentage of cases was attributed to granulocyte antibodies. In the last decade, a much larger percentage of cases has been attributed to HLA class I and II antibodies. The reasons for this difference may be:

- the association of HLA class II antibodies in TRALI was first published in 2001;[25]
- there is an increased ability to accurately detect and identify HLA antibodies;
- the detection and identification of HNA antibodies remains difficult and is performed in only a few laboratories.

Although HNA antibodies are now reported in a smaller percentage of reported cases, it should be noted that TRALI associated with the infusion of antibodies directed against HNA-3a has been reported in a number of fatal TRALI cases.[14,26]

Biological response modifiers

In the most general sense, biological response modifiers (BRMs) are generated by white blood cells and accumulate during storage of cellular blood components. Cellular blood components that have demonstrated increased levels of BRMs include red blood cell and platelet components. BRMs are lipophilic, chloroform soluble and consist of a mixture of lyso-phosphatidyl cholines (lyso-PCs).[29,30] These compounds enhance or prime the neutrophil (PMN) oxidative burst.

A series of 90 TRALI reactions that occurred at a single hospital has been published by Silliman et al.[31] These investigators measured the levels of BRMs in pre- and post-transfusion samples from the last 51 patients who experienced a TRALI reaction. Significantly more PMN priming activity ($p < 0.05$)

was identified in the post-transfusion samples than the pre-transfusion samples and controls. BRMs have also been identified after the infusion of autologous blood.[32] In this case report the patient had donated two units of autologous blood in preparation for radical prostatectomy. After transfusion of 60 mL of the first unit the patient experienced marked hypotension. The unit was discontinued and the second unit was started. After transfusion of 40 mL of the second unit, the patient experienced hypotension and oxygen desaturation. The symptoms ceased after each of the units was discontinued. The plasma of the second unit contained significantly more PMN priming ability than control plasma or the patient's convalescent plasma.

Khan et al. have recently associated soluble CD40 ligand with TRALI.[33] Soluble CD40 ligand is a pro-inflammatory mediator that accumulates in stored cellular blood components, particularly platelets. In a series of TRALI cases, platelets implicated in reactions contained significantly more soluble CD40 ligand than platelets not involved in TRALI reactions. Soluble CD40 ligand resulted in priming of the neutrophil PMN oxidase, and promoted cytotoxicity of endothelial cells.

Two-event model of TRALI

There is a growing appreciation that the pathophysiology of TRALI is more complex than the infusion of leucocyte antibody into a patient with the corresponding antigen or the infusion of BRMs. While one study in blood donors demonstrated that approximately 20 per cent of blood components contain HLA class I and/or class II antibodies,[34] the prevalence of such antibodies depends, in part, on the proportion of multiparous donors in the donor pool. However, only about 1 in 5000 transfusions results in TRALI. Thus, a two-event model of TRALI has been proposed.

In this model, two events are needed for TRALI to occur in a patient. In the first event, the generation of biologically active compounds related to stress is thought to alter the pulmonary microvascular environment.[31,35] These events are thought to cause neutrophil sequestration and priming in the pulmonary microvasculature. In the second event, the infusion of antibodies or BRMs into this altered pulmonary microvascular environment results in activation of neutrophils, endothelial cell damage and disruption of the underlying basement membrane. The result of this damage is that fluid containing large amounts of protein leaks into the pulmonary alveoli, accounting for the pulmonary oedema.

The two-event model of TRALI is supported by a number of studies. Silliman et al. performed a retrospective study of 10 patients who had experienced a TRALI reaction and 10 control patients.[36] The control patients had experienced febrile or urticarial reactions. All patients were assessed for the presence of a predisposing factor that would put them at risk of developing TRALI. Stress factors that were hypothesized to represent a first event included infection, cytokine administration, recent surgery and massive transfusion. All the patients who developed TRALI and only two of the control group patients had one of these clinical factors.

The two-event mechanism of TRALI has been reproduced in an ex-vivo rat lung model.[31] The lungs were perfused with one of the following solutions: buffered salt solution, 5 per cent human plasma in saline, 5 per cent supernatant from day 0 or day 5 platelets or lipid extract from day 0 or day 5 platelets. Pulmonary oedema resulted after infusion of plasma and lipid extract from day 5 platelets only if the lungs were pre-treated with lipopolysaccharide (LPS) to simulate sepsis. Infusion of buffered salt solution, human plasma, supernatant from day 0 platelets and lipid extract from day 0 platelets did not result in pulmonary oedema even when the lungs were pre-treated with LPS.

The two-event model of TRALI has been reproduced in an *in-vitro* system using human microvascular endothelial cell (HMVECs) and HNA-3a antibody.[37] HMVECs were activated with endotoxin or buffer. HNA-3a positive or negative PMNs were added to the test system. The culture was then incubated with plasma with or without HNA-3a antibody. The fMLP respiratory burst was activated and HMVECs were damaged only when the HMVECs were activated with endotoxin, HNA-3a positive PMNs were present and HNA-3a antibody was present. If any of the three components was missing from the test system the fMLP respiratory burst was not activated and the HMVECs were not damaged.

Additionally, evidence from look-back studies supports a two-event model of TRALI. Kopko *et al.* performed a look-back study of prior donations from a frequent apheresis plasma donor with a strong antibody to HNA-3a (granulocyte 5b antigen).[14] Only 36 per cent of recipients developed reactions to infusion of blood components from this donor. However, HNA-3a is present in greater than 90 per cent of the population.[38] Other look-back studies, performed in recent years, have also demonstrated previously unrecognized cases of TRALI,[39,40] while some studies have not found any previously unrecognized cases of TRALI in recipients of blood components from a donor with a leucocyte antibody, even if the recipient possessed the cognate leucocyte antigen.[41,42] A common observation in these studies is that the frequency of observed reactions is less than would be predicted based on the frequency of the antigen. These studies support the hypothesis that infusion of antibodies into a patient with the cognate antigen is not sufficient to cause TRALI. An additional factor in the development of TRALI may be the density of cognate antigen in the recipient. This hypothesis is supported by a recent study performed by Sachs *et al.*[43] These investigators utilized an *ex-vivo* rat lung model to reproduce TRALI. TRALI developed when neutrophils expressing 70 per cent HNA-2a were present in the system. When neutrophils expressing 30 per cent cognate antigen were used, TRALI did not develop. The studies of Sachs *et al.* also suggest that the factors necessary for TRALI to develop may be different for each individual. If antigen density is a factor in the development of TRALI, then an individual with higher expression of antigen on their leucocytes may be susceptible to TRALI after a less significant 'first hit' than an individual with a lower antigen density on their leucocytes.

DIAGNOSIS

There are no diagnostic tests or pathognomonic signs to diagnose TRALI. Therefore, TRALI is a diagnosis of exclusion. The most essential part of diagnosing TRALI is to recognize non-cardiogenic pulmonary oedema in the setting of transfusion.

Other causes of respiratory distress that are secondary to volume overload and cardiogenic causes need to be ruled out in order to make the diagnosis. In the setting of transfusion, circulatory overload from infusion of the blood components is the primary differential. Normal to decreased central venous and pulmonary pressures are consistent with a diagnosis of TRALI. Cardiogenic causes of pulmonary oedema, such as myocardial infarction, also need to be excluded.

Transient leucopenia,[44] pulmonary oedema fluid/plasma protein ratio > 0.75,[45] and post-transfusion/pre-transfusion β-natriuretic peptide ratio of < 1.5 are laboratory values suggestive of TRALI.[46] Other laboratory findings that can be consistent with TRALI include demonstration of HLA class I or II antibodies or neutrophil specific antibodies in either the donor or the recipient. A positive lymphocytotoxic or reverse lymphocytotoxic cross-match also supports the diagnosis. Although laboratory studies can support the diagnosis of TRALI, the diagnosis should be based on clinical findings. Additionally, laboratory studies, particularly histocompatibility and neutrophil testing, are often not available for extended periods of time. The diagnosis and treatment of TRALI cannot wait for the results of these studies to be available and must be based on the patient's clinical situation.

Non-cardiogenic pulmonary oedema in the setting of transfusion should be investigated as a potential case of TRALI. If alternate causes of the pulmonary oedema are not identified, the case should be reported to the hospital transfusion service as soon as possible. In turn, the hospital transfusion service should report the case, including the identification numbers of all blood components transfused in the 6 hours prior to first symptoms, to the blood collection agency. This will allow the blood collection agency to investigate donors and appropriately defer donors with HLA antibodies.

TREATMENT

Once the diagnosis of TRALI is entertained, therapeutic measures should be started as quickly as possible. The two key elements in the treatment of TRALI are respiratory support and fluid administration. The respiratory support should be appropriate for the clinical symptoms. In almost all cases, oxygen administration is required.[1] Intubation and mechanical ventilation are usually needed in severe cases.

Fluid administration and pressor support should also be administered as appropriate for the clinical situation. Administration of diuretics, which are often required in other types of pulmonary oedema, should probably be avoided. Although no clinical studies are available to support this recommendation, the aetiology of pulmonary oedema in TRALI is related to microvascular injury, not fluid overload. Case reports of fatal outcome after diuretic administration in TRALI have been published.[47]

Because TRALI occurs infrequently, there are no published clinical trials to guide treatment. Case reports that advocate the use of corticosteroids and other immunosuppressive agents in TRALI have been published.[18,48] The administration of steroids in these cases did not prevent the need for mechanical ventilation. Clinical trials have demonstrated that there is no clinical benefit in the use of corticosteroids in the early acute phase of ARDS[49] and persistent ARDS.[50] The routine use of corticosteroids in ARDS cannot be recommended and their use has been demonstrated to be deleterious in the persistent phase of ARDS.[51]

There is a single case report of the administration of albumin infusion resulting in immediate and drastic clinical improvement in a patient with TRALI.[52] The authors of this case report hypothesize that albumin reduces capillary permeability, resulting in amelioration of ongoing pulmonary oedema. No recommendations regarding the use of albumin in TRALI can be made until further studies are performed.

MANAGEMENT OF FUTURE TRANSFUSIONS

Since the majority of cases of TRALI are related, at least in part, to the infusion of leucocyte antibodies, no special measures are indicated to reduce the risk of recurrence in most cases. However, if antibodies are demonstrated in the transfusion recipient, leucocyte-reduced blood components should be used for transfusion of cellular components to reduce the risk of recurrence. This recommendation is based on our understanding of the mechanism of TRALI and the possibility that removing leucocytes from transfused cellular components, for patients known to possess leucocyte antibodies, will reduce the risk of TRALI. Since TRALI occurs relatively infrequently, there are no data to support this recommendation.

PREVENTION

All strategies aimed at reducing the incidence of TRALI are complicated by lack of a profile to identify recipients at greatest risk, lack of a diagnostic test, and the need to balance the effect of any deferral strategy on the availability of the blood supply.

Several recommendations to reduce the risk of TRALI have been made. These recommendations include the following:[15]

1. Donors implicated in TRALI should be permanently deferred or subsequent donations should be limited to plasma-poor components, including frozen-deglycerolized or washed red blood cells. These steps would prevent the use of plasma from donors implicated in TRALI reactions.
2. Multiparous donors should be screened for HLA class I and II and granulocyte antibodies prior to donation or their donations should be limited to plasma-poor blood components. This approach has been implemented at one centre.[53] No cases of TRALI were reported at this centre in the 15 months following implementation.
3. Plasma from female donors should not be used for transfusion. This approach has been implemented at the United Kingdom's National Blood Transfusion Service.[54]

It was possible to implement strategy number 3 because the majority of plasma from whole blood is not used for transfusion. Instead the plasma is used to manufacture albumin, IVIg and clotting factors. This strategy would not be possible to reduce the risk of TRALI from red blood cells and platelets because there is not an excess of these components for transfusion.

In the United States, the American Association of Blood Banks has issued a bulletin aimed at reducing the risk of TRALI from plasma-rich blood components.[55] The bulletin recommends reducing the risk of TRALI by either deferring donors at increased risk of having antibody or screening these donors for antibody. Although compliance with these recommendations is voluntary, measures that affect plasma components, and those that affect apheresis platelets, were due to be implemented in the near future.

As the significance of TRALI as a major cause of transfusion-related morbidity and mortality is increasingly appreciated, there is growing interest in implementing effective TRALI prevention strategies. A number of factors should be kept in mind during this evaluation process. First, donor histories are not necessarily accurate, in that women who have had alloimmunizing exposures to fetal leucocytes through either ectopic pregnancy or abortion may neglect to report this part of their medical history. Second, donors can be immunized by transfusion. Deferring females or multiparous females will not prevent this population from donating, unless the donor's blood was tested for HLA and granulocyte-specific antibodies. Third, red blood cells, random-donor platelet concentrates, and cryoprecipitate may contain less than 50 mL of plasma, but such components have been infrequently associated with TRALI. Deferring multiparous donors (estimated to be 5–50 per cent of donor bases) or donors with an HLA or HNA antibody from all donations would result in a significant loss of blood donors. Tests for HLA and granulocyte antibodies are time-consuming and are not routinely available in many blood centres or hospital blood banks. Finally, deferring donors based on gender, parity, transfusion history or presence of antibody will undoubtedly reduce the risk of TRALI, but will not eliminate it. None of these strategies will reduce the risk of TRALI from biological response modifiers. Clearly, more sensitive and specific measures are needed for a more effective strategy to prevent TRALI.

KEY POINTS

- The two most common pulmonary complications of blood transfusion are volume overload and transfusion-related acute lung injury. These may be difficult to differentiate. It has only been in the last 20 years that TRALI has been recognized as a significant cause of transfusion-related morbidity and mortality.
- Acute lung injury has been defined as hypoxaemia with PaO_2/FIO_2 ratio of ≤ 300 mmHg, bilateral pulmonary oedema on frontal chest radiograph, and pulmonary artery occlusion pressure of ≤ 18 mmHg, with no clinical evidence of left atrial hypertension.
- TRALI has been associated with the transfusion of all plasma-containing blood components including whole blood, packed red blood cells, granulocytes collected by apheresis, platelet concentrates, apheresis platelets, fresh frozen plasma and cryoprecipitate.

(Contd)

- TRALI typically presents during transfusion or shortly after (within 1–6 hours). Signs and symptoms include dyspnoea with acute onset of bilateral pulmonary oedema and severe hypoxia. Fever, tachycardia, hypotension and hypertension followed by hypotension can also be present during the reaction.
- High-volume plasma from female multiparous donors exposed particularly to the risk of developing trali.

REFERENCES

1. Popovsky MA, Moore SB. Diagnostic and pathogenetic considerations in transfusion-related acute lung injury. *Transfusion* 1985;**25**:573–7.
2. Barnard RD. Indiscriminate transfusion: a critique of case reports illustrating hypersensitivity reactions. *NY State J Med* 1951;**51**:2399–401.
3. Carilli AD, Ramanamurty MV, Chang YS, *et al.* Noncardiogenic pulmonary edema following blood transfusion. *Chest* 1978;**74**:310–12.
4. Kernoff PBA, Durrant IJ, Rizza CR, Wright FW. Severe allergic pulmonary oedema after plasma transfusion. *Br J Haematol* 1972;**23**:777–81.
5. Ward HN. Pulmonary infiltrates associated with leukoagglutinin transfusion reactions. *Ann Intern Med* 1970;**73**:689–94.
6. Wolf CFW, Canale VC. Fatal pulmonary hypersensitivity reaction to HL-A incompatible blood transfusion: report of a case and review of the literature. *Transfusion* 1976;**16**:135–40.
7. Bernard GR, Artigas A, Righam KL, *et al.* The American–European consensus conference on ARDS: definitions, mechanisms, relevant outcomes, and clinical trial coordination. *Am J Respir Crit Care Med* 1994;**149**:818–24.
8. Toy P, Popovsky MA, Abraham E, *et al.* Transfusion-related acute lung injury: definition and review. *Crit Care Med* 2005;**33**:721–6.
9. Kleinman S, Caulfield T, Chan P, *et al.* Toward an understanding of transfusion-related acute lung injury: statement of a consensus panel. *Transfusion* 2004;**44**:1774–89.
10. Silliman CC, Boshkov LK, Mehdizadehkashi Z, *et al.* Transfusion-related acute lung injury: epidemiology and a prospective analysis of etiologic factors. *Blood* 2003;**101**:454–62.
11. Wallis JP, Lubenko A, Wells AW, Chapman CE. Single hospital experience of TRALI. *Transfusion* 2003;**43**:1053–9.
12. Sazama K. Reports of 355 transfusion-associated deaths: 1976 through 1985. *Transfusion* 1990;**30**:583–90.
13. Goldman M, Webert KE, Arnold DM, *et al.* Proceedings of a consensus conference: towards an understanding of TRALI. *Transfus Med Rev* 2005;**19**:2–31.
14. Kopko PM, Marshall CS, MacKenzie MR, *et al.* Transfusion-related acute lung injury: report of a clinical look-back investigation. *J Am Med Assoc* 2002;**287**:1968–71.
15. Popovsky MA, Chaplin HC, Moore SB. Transfusion-related acute lung injury: a neglected, serious complication of hemotherapy. *Transfusion* 1992;**32**:589–92.
16. Popovsky MA, Haley NR. Further characterization of transfusion-related acute lung injury: demographics, clinical and laboratory features and morbidity. *Immunohematology* 2000:**16**:157–9.
17. Kopko PM, Paglierioni TG, Popovsky MA, *et al.* TRALI: correlation of antigen-antibody and monocytes activation in donor–recipient pairs. *Transfusion* 2003;**43**:177–84.
18. Ou SM, Wan HL, Kuo HY, *et al.* Transfusion-related acute lung injury: a report of two cases and the possible role of steroid. *Acta Anaesthesiol Taiwanica* 2004;**42**:159–63.
19. Ramanathan RK, Triulzi DF, Logan TF. Transfusion-related acute lung injury following random donor platelet transfusion: a report of two cases. *Vox Sang* 1997;**73**:43–5.
20. Virchis AE, Patel RK, Contreras M, *et al.* Acute non-cardiogenic lung oedema after platelet transfusion. *Br Med J* 1997;**314**:880–2.
21. Reese EP, McCullough JJ, Craddock PR. An adverse pulmonary reaction to cryoprecipitate in a hemophiliac. *Transfusion* 1975;**15**:583–8.
22. Rizk A, Gorson K, Kenney L, Weinstein R. Transfusion-related acute lung injury after the infusion of IVIG. *Transfusion* 2001;**41**:264–8.
23. Kopko PM, MacKenzie M, Holland PV, Popovsky M. Transfusion-related acute lung injury: disproportionate involvement of plasma components. *Transfusion* 2002;**42**:29S(abstract).
24. SHOT 2003 Annual Report available at www.shotuk.org.
25. Kopko PM, Popovsky MA, MacKenzie MR, *et al.* Human leukocyte antigen class II antibodies in transfusion-related acute lung injury. *Transfusion* 2001;**41**:1244–8.
26. Davoren A, Curtis BR, Shulman IA, *et al.* TRALI due to granulocyte-agglutinating human neutrophil antigen-3a (5b) alloantibodies in donor plasma: a report of two fatalities. *Transfusion* 2003;**43**:641–5.
27. Eastlund DT, McGrath PC, Burkart P. Platelet transfusion reaction associated with interdonor HLA incompatibility. *Vox Sang* 1988;**55**:157–60.
28. O'Connor JC, Strauss RG, Goeken NE, Knox LB. A near-fatal reaction during granulocyte transfusion of a neonate. *Transfusion* 1988;**28**:173–6.
29. Silliman CC, Clay LK, Thurman GW, *et al.* Partial characterization of lipids that develop during the routine storage of blood and prime the neutrophil NADPH oxidase. *J Lab Clin Med* 1994;**124**:684–94.
30. Silliman CC, Dickey WO, Paterson AJ, *et al.* Analysis of the priming activity of lipids generated during routine storage of platelet concentrates. *Transfusion* 1996;**36**:133–9.
31. Silliman CC, Bjornsen AJ, Wyman TH, *et al.* Plasma and lipids from stored platelets cause acute lung injury in an animal model. *Transfusion* 2003;**43**:633–40.
32. Covin RB, Ambruso DR, England KM, *et al.* Hypotension and acute pulmonary insufficiency following transfusion of autologous red blood cells during surgery: a case report and review of the literature. *Transfus Med* 2004;**14**:375–83.
33. Khan SY, Kelher MR, Heal JM, *et al.* Soluble CD40 ligand accumulates in stored blood components, prime neutrophils through CD40, and is a potential cofactor in the development of transfusion-related acute lung injury. *Blood* 2006;**108**:2455–62.
34. Bray RA, Harris SB, Josephson CD, *et al.* Unappreciated risk factors for transplant patients: HLA antibodies in blood components. *Hum Immunol* 2004;**65**:240–4.
35. Silliman CC, Ambruso DR, Boshkov LK. Transfusion-related acute lung injury. *Blood* 2005;**105**:2266–73.

36. Silliman CC, Paterson AJ, Dickey DF, *et al.* The association of biologically active lipids with the development of transfusion-related acute lung injury: a retrospective study. *Transfusion* 1997;**37**:719–26.

37. Silliman CC, Curtis BR, Kopko PM, *et al.* Donor antibodies to HNA-3a implicated in TRALI reactions prime neutrophils and cause PMN-mediated damage to human pulmonary microvascular endothelial cells in a two-event, in-vitro model. *Blood* 2007;**109**:1752–5.

38. Van Leeuwen A, Eernise JG, Van Rood JJ. A new leukocyte group with two alleles: leukocyte group five. *Vox Sang* 1964;**9**:431–7.

39. Cooling L. Transfusion-related acute lung injury. *J Am Med Assoc* 2002;**288**:315–16.

40. Nicolle AL, Chapman CE, Carter V, Wallis P. Transfusion-related acute lung injury caused by two donors with anti-human leucocyte antigen class II antibodies: a look-back investigation. *Transfus Med* 2004;**64**:225–30.

41. Win N, Ranasinghe E, Lucas G. Transfusion-related acute lung injury: a 5-year look-back study. *Transfus Med* 2002;**12**:387–9.

42. Toy P, Hollis-Perry M, Jun J, Nakagawa M. Recipients of blood from a donor with multiple HLA antibodies: a look back study of transfusion-related acute lung injury. *Transfusion* 2004;**44**:1683–8.

43. Sachs UJH, Hattar K, Weissmann N, *et al.* Antibody-induced neutrophil activation as a trigger for transfusion-related acute lung injury in an ex-vivo rat lung model. *Blood* 2006;**107**:1217–19.

44. Nakagawa M, Toy P. Acute and transient decrease in neutrophil count in transfusion-related acute lung injury: cases at one hospital. *Transfusion* 2004;**44**:1689–94.

45. Yost CS, Matthay MA, Gropper MA. Etiology of acute pulmonary edema during liver transplantation. *Chest* 2001;**119**:219–23

46. Zhou L, Giacherio D, Cooling L, Davenport RD. Use of B-natriuretic peptide as a diagnostic marker in the differential diagnosis of transfusion-associated circulatory overload. *Transfusion* 2005;**45**:1056–63.

47. Levy GJ, Shabot MM, Hart ME, *et al.* Transfusion-associated non-cardiogenic pulmonary edema: report of a case and a warning regarding treatment. *Transfusion* 1986;**26**:278–81.

48. Ganguly S, Carrum G, Nizzi F, *et al.* Transfusion-related acute lung injury following allogeneic stem cell transplant for acute myeloid leukemia. *Am J Hematol* 2004;**75**:48–51.

49. Bernard GR, Luce JM, Spring CL, *et al.* High-dose corticosteroids in patients with the adult respiratory distress syndrome. *New Engl J Med* 1987;**317**:1565–70.

50. National Heart, Lung and Blood Institute acute respiratory distress syndrome (ARDS) clinical trials network. Efficacy and safety of corticosteroids for persistent acute respiratory distress syndrome. *New Engl J Med* 2006;**354**:1671–84.

51. Suter PM. Lung inflammation in ARDS: friend or foe? *New Engl J Med* 2006;**354**:1739–42.

52. Djalali AG, Moore KA, Kelly E. Report of a patient with severe transfusion-related acute lung injury after multiple transfusions, resuscitated with albumin. *Resuscitation* 2005;**66**:225–30.

53. Insunza A, Romon I, Gonzalez-Ponte ML, *et al.* Implementation of a strategy to prevent TRALI in a regional blood center. *Transf Med* 2004;**14**:157–64.

54. SHOT. Annual report for 2004 available at www.shotuk.org/SHOTREPORT2004.pdf.

55. Association Bulletin 06-07. *Transfusion-related Acute Lung Injury.* AABB, 2006. Available at www.aabb.org/Conent/Members_Area/Association_Bulletins/ab06-07.htm (accessed January 2007).

11

Drug-induced alveolar haemorrhage

ABIGAIL R LARA, MARVIN I SCHWARZ

INTRODUCTION

Diffuse alveolar haemorrhage (DAH) is a syndrome defined by specific clinical and pathological parameters. It is characterized by accumulation of red blood cells in the alveolar spaces. Most commonly, accumulation occurs due to damage of the alveolar capillaries and less frequently from the precapillary or postcapillary venules.[1]

The clinical spectrum of DAH includes cough, progressive dyspnoea and varying degrees of haemoptysis, which may lead to acute respiratory failure. Characteristic laboratory features include a falling haematocrit and sequential bronchial alveolar lavage (BAL) fluid with increasing red blood cell counts.

There are many causes of DAH. It is most commonly seen in systemic vasculitides and connective tissue disorders, but the focus of this chapter will be on DAH as a complication of medical therapy.

DEFINITIONS

Diffuse alveolar haemorrhage is the clinicopathologic syndrome of the accumulation of red blood cells in the alveolar spaces. From a histopathological standpoint, there are several different causes which lead to DAH, including pulmonary capillaritis, bland pulmonary haemorrhage and diffuse alveolar damage (Fig. 11.1). Other miscellaneous histologies that are associated with DAH exist, but they are beyond the scope of this discussion.

Pulmonary capillaritis was first described in 1957 by Spencer[2] and is the most frequent histological pattern associated with DAH. It is characterized by damage of the interstitium including the alveolar capillaries that allows red blood cells and fibrin to leak into the alveolar spaces. The interstitium expands owing to the presence of oedema, fibrinoid necrosis, infiltrating inflammatory cells, and free red blood cells.[3] Another

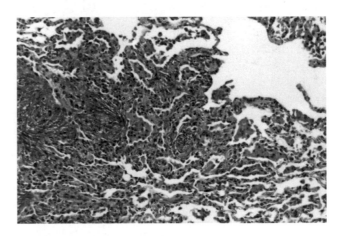

Fig. 11.1 Lung biopsy showing capillaritis, pulmonary haemorrhage and diffuse alveolar damage.

characteristic finding is that of neutrophilic interstitial infiltration with many of the neutrophils appearing apoptotic or fragmented (leukocytoclasis) and pyknotic. Other histological features include capillary and rare arteriolar thrombosis, organizing pneumonia, and Type II cell hyperplasia. In severe cases of pulmonary capillaritis, there is complete destruction of the alveolar walls and replacement with necrotic tissue.

In bland pulmonary haemorrhage, the alveolar interstitium is not inflamed, oedematous or necrotic. The most common histopathological finding is that of Type II epithelial cell hyperplasia and interstitial fibrosis, which develops after repeated episodes.[4] Bronchoalveolar lavage shows many haemosiderin-laden macrophages and haemorrhaging.[5]

Diffuse alveolar damage (DAD) is diagnosed by the histological presence of alveolar–septal and intra-alveolar oedema, capillary congestion and microthrombi with intra-alveolar hyaline membrane formation.[6] It refers to a response of the lung to an acute injury from a variety of sources.[7–12] Diffuse alveolar haemorrhage results from widespread injury of both the epithelial lining and the alveolar basement membrane.

INCIDENCE AND EPIDEMIOLOGY

The overall frequency with which drug-induced DAH occurs is essentially unknown given the lack of lung tissue available for histological confirmation. However, the syndrome is increasingly recognized with antithyroid drugs and platelet glycoprotein inhibitors.

AETIOLOGIES

Pulmonary capillaritis

Numerous drugs have been associated with pulmonary capillaritis. Many have been reported as single case reports in the literature. In this chapter, discussion will focus on those drugs that have the strongest evidence for causing pulmonary capillaritis. Propylthiouracil (PTU), a common antithyroid drug, has been shown to cause a systemic small vessel vasculitis with DAH due to pulmonary capillaritis, leucocytoclastic vasculitis of the skin, and a focal segmental necrotizing glomerulonephritis. PTU commonly induces antineutrophil cytoplasmic antibodies (ANCA), ranging from 4 to 46 per cent of all patients treated with PTU.[13] The development of vasculitis may occur in a fraction of such patients, at any time during treatment, and presents with a viral-like prodrome. The exact pathogenesis for the development of PTU-induced ANCA vasculitis is unknown; however, several potential mechanisms suggest interactions between PTU and myoperoxidase (MPO) and anti-MPO antibodies. Levels of PTU-induced anti-MPO ANCA have been shown to vary during disease onset and relapse.[14,15]

All-*trans*-retinoic acid (ATRA) is used to induce remission in patients with acute promyelocytic leukaemia. A complication of treatment with ATRA is known as 'retinoic acid syndrome' which is characterized by fever, respiratory distress, capillary leak, weight gain and pleural effusions, peripheral oedema, thromboembolic events, intermittent hypotension and the clinical features of DAH.[16] Histology reveals capillary infiltration of leukaemic or maturing myeloid cells with *in-situ* arterial thrombosis and occasionally capillaritis.

Leukotriene antagonists are often used in the treatment and management of asthma. Several reports describe Churg–Strauss syndrome developing after the introduction of leukotriene antagonists, and one case describes a relapse of the syndrome.[17–19]

Rarely, diphenylhydantoin has been associated with pulmonary capillaritis.[20]

Bland pulmonary haemorrhage

Excess anticoagulation during warfarin or heparin therapy, or poisoning with superwarfarin rotenticides, may cause DAH in which the only clinical findings are dyspnoea, unexplained anaemia and evolving pulmonary infiltrates.[21] Drug-induced thrombocytopenia has also been implicated as a causative factor in the development of DAH.[22,23]

The use of platelet inhibitors following percutaneous coronary interventions with stents has been associated with DAH. Abciximab, a platelet glycoprotein IIb/IIIa inhibitor, and clopidogrel, a platelet aggregation inhibitor, have been extensively associated with the development of DAH.[24,25] Increased risk of bleeding occurs in patients with the following risk factors and comorbidities: female, elderly, underweight, prolonged or complicated procedures, COPD, pulmonary hypertension and a high pulmonary capillary wedge pressure.[26,27] Platelet counts may remain normal.

Primary pulmonary hypertension is a rare disease, which is characterized by progressive elevation in pulmonary artery pressures and vascular resistance and is associated with a poor prognosis. The standards of care are aimed at decreased pulmonary vascular resistance. Epoprostenol is a prostacyclin which acts as a potent pulmonary artery dilator and an inhibitor of platelet aggregation.[28] Owing to the increased risk of thromboembolic events, anticoagulation is recommended as standard therapy.[29] When used in combination, particularly high doses of epoprostenol, the risk of DAH may exceed 20 per cent; this can be fatal.[30]

Several other medications have been associated with the development of DAH. Amiodarone-induced DAH typically occurs following treatment of 400 mg daily for 2 months[31,32] and may also occur after intravenous treatment.[33] Histology may show foamy macrophages with lamellar inclusions along with a non-specific cellular interstitial pneumonitis, organizing pneumonia and bland haemorrhage.

Penicillamine is a rare cause of DAH that may present after 12 months of therapy, but may be delayed up to 20 years. Owing to the associated glomerulonephritis that is often found in this syndrome, it is also known as 'penicillamine-induced Goodpasture's' or 'pneumo-renal' syndrome. In comparison to Goodpasture's syndrome, penicillamine-induced Goodpasture's syndrome lacks deposition of tissue immunoglobulin.[34–37]

The pulmonary toxicity most commonly associated with nitrofurantoin and methotrexate therapy is pulmonary fibrosis, but DAH may be a rare complication of both.[38–42] Finally, sirolimus, a macrocyclic triene immunosuppressive agent, has been associated with lymphocytic alveolitis as well as a bland pulmonary haemorrhage and has slow resolution after cessation of drug therapy.[43,44] A few cases have been described in patients receiving the CD20-bearing lymphocyte depletor rituximab.[45,46]

Diffuse alveolar damage

Diffuse alveolar damage represents the lungs' response to acute injury and has multiple causes (see Box 11.1). The most common causes of DAD with associated DAH are the cytotoxic drugs used to treat solid tumours and acute leukaemia as well as the preconditioning cytotoxic therapy used prior to bone marrow transplantation.[46,47] DAH may occur in 10–20 per cent of all autologous bone marrow transplant recipients and in the context of stem-cell transplantation and is a poor prognostic indicator, with mortality reported as high as 50–100 per cent.[48]

Cytotoxic drugs most commonly injure the lung via a DAD process; however, if the injury is severe enough DAH may

Box 11.1 Causes and underlying histologies associated with diffuse alveolar haemorrhage

Pulmonary capillaritis

- Diphenylhydantoin
- Retinoic acid toxicity
- Propylthiouracil
- Hydralazine

Bland pulmonary haemorrhage

- Excess anticoagulation
- Penicillamine
- Amiodarone
- Nitrofurantoin
- High-dose epoprostenol

Diffuse alveolar damage

- Bone marrow transplantation
- 'Crack' cocaine inhalation
- Cytotoxic drug therapy
- Radiation therapy
- Paraquat
- Acetyl salicylic acid (aspirin)
- Heroin

occur, particularly in the setting of respiratory failure.[50–52] In DAH following preconditioning therapy for bone marrow transplantation or treatment for acute leukaemia, thrombocytopenia may be a contributing factor.

Smoking or freebasing 'crack' cocaine is a common cause of a spectrum of alveolar injury with DAD and varying degrees of DAH, although there is one reported case of capillaritis.[53] The injury can be a direct result from cocaine itself or from the solvents that it is mixed with. Injury occurs within hours of the event and is typically short-lived and reversible if the exposure is limited.[7,54,55] Rarely, anti-GBM disease has been detected following cocaine poisoning.[56]

CLINICAL PRESENTATION AND INVESTIGATIONS

The clinical presentation of DAH, irrespective of the underlying aetiology, runs the spectrum of acute to insidious to fulminate respiratory failure. An abnormal chest radiograph with diffuse pulmonary infiltrates, cough and dyspnoea along with a falling haematocrit should alert the physician to the presence of DAH. The cardinal sign of DAH, haemoptysis, may be absent in up to one-third of all patients with DAH even with extensive intra-alveolar bleeding.[57] In these cases, DAH can be diagnosed only by performing sequential BAL revealing a haemorrhagic return and should be performed in all cases of suspected drug-induced DAH unless otherwise contraindicated.

Lung tissue is rarely required to make the diagnosis of drug-induced DAH and should be considered only if withdrawal of the offending agent is ineffective and systemic corticosteroids have not shown improvement.

Owing to both the rarity of drug-induced DAH and the number of drugs that may cause DAH, there currently do not exist any large-scale clinic or scientific investigations.

TREATMENT/MANAGEMENT

Central to the treatment and management of drug-induced DAH is removal of the offending agent and institution of systemic corticosteroids to decrease the inflammatory nature of the acute injury if improvement is not seen. There is no standard treatment algorithm with regard to steroids for drug-induced DAH. Various doses have been used with variable success, ranging from moderate to high-dose corticosteroids.[58,59]

In drug-induced coagulation disorders, treatment should be primarily aimed at restoration of normal bleeding times with vitamin K and fresh frozen plasma as well as transfusion with platelets if thrombocytopenia is thought to be a contributing factor.

In drug-induced capillaritis with life-threatening DAH, rapid institution of plasmapheresis may be life-saving.[60] Recombinant-activated human factor VII (rFVIIa) has been shown to be successful in life-threatening DAH associated with small vessel vasculitis and bone marrow transplantation in a dose range of 80–120 ug/kg of body weight.[61–63] Restoration of haemostasis with rFVIIa is driven by the enhancement of thrombin generation on the surface of activated platelets, which bypasses the traditional FVIIa/tissue factor complex.[64]

In cases of ATRA-induced retinoic acid syndrome, discontinuation of the ATRA is not recommended, rather institution of high-dose corticosteroids is the mainstay of treatment. Mitomycin therapy, which is used for the treatment of lung and bladder cancer, is associated with a haemolytic uraemic syndrome with DAD. Mitomycin should be immediately discontinued with institution of both corticosteroids and plasmapheresis; blood transfusion should be avoided if possible as it may worsen the syndrome.[65]

COMPLICATIONS

Most cases of drug-induced DAH represent single episodes, as avoidance of the offending agent should be followed; thus complications of the syndrome are often associated with the acute event. Complications include respiratory failure and anaemia. In rare instances of repeated episodes of DAH due to vasculitis or idiopathic pulmonary haemosiderosis, chronic interstitial fibrosis and progressive obstructive lung disease develop.[66,67]

FUTURE DIRECTIONS FOR THERAPY

Each case of drug-induced DAH should be evaluated and treated on an individual basis as the mechanisms of DAH vary between drug classes.

REFERENCES

1. Fontenot AP, Schwarz MI. Diffuse alveolar hemorrhage. In: Schwarz MI (ed.) *Interstitial Lung Disease*, 4th edn. Hamilton: Decker, 2003.
2. Spencer H. Pulmonary lesions in polyarteritis nodosa. *Br J Tuberc Dis Chest* 1957;**51**:123–30.
3. Travis WD, Colby TV, Lombard C, Carpenter HA. A clinicopathologic study of 34 cases of diffuse pulmonary hemorrhage with lung biopsy confirmation. *Am J Surg Pathol* 1990;**14**:1112–25.
4. Lombard CM, Colby TV, Elliott CG. Surgical pathology of the lung in anti-basement membrane antibody-associated Goodpasture's syndrome. *Hum Pathol* 1989;**20**:445–51.
5. Finley TN, Aronow A, Cosentino AM, Golde DW. Occult pulmonary hemorrhage in anticoagulated patients. *Am Rev Respir Dis* 1975;**112**:23–9.
6. Haselton P (ed.). *Spencer's Pathology of the Lung*. New York: McGraw-Hill, 1996.
7. Forrester JM, Steele AW, Waldron JA, Parsons PE. Crack lung: an acute pulmonary syndrome with a spectrum of clinical and histopathologic findings. *Am Rev Respir Dis* 1990;**142**:462–7.
8. Haim DY, Lippmann ML, Goldberg SK, Walkenstein MD. The pulmonary complications of crack cocaine: a comprehensive review. *Chest* 1995;**107**:233–40.
9. Gross NJ. Pulmonary effects of radiation therapy. *Ann Intern Med* 1977;**86**:81–92.
10. Pratt PC. Pathology of pulmonary oxygen toxicity. *Am Rev Respir Dis* 1974;**110**(6 Pt 2):51–7.
11. Sisson JH, Thompson AB, Anderson JR, *et al*. Airway inflammation predicts diffuse alveolar hemorrhage during bone marrow transplantation in patients with Hodgkin disease. *Am Rev Respir Dis* 1992;**146**:439–43.
12. Israel-Biet D, Labrune S, Huchon GJ. Drug-induced lung disease: 1990 review. *Eur Respir J* 1991;**4**:465–78.
13. Slot MC, Links TP, Stegeman CA, Tervaert JW. Occurrence of antineutrophil cytoplasmic antibodies and associated vasculitis in patients with hyperthyroidism treated with antithyroid drugs: a long-term followup study. *Arthritis Rheum* 2005;**53**:108–13.
14. Zhao MH, Chen M, Gao Y, Wang HY. Propylthiouracil-induced anti-neutrophil cytoplasmic antibody-associated vasculitis. *Kidney Int* 2006;**69**:1477–81.
15. Ye H, Gao Y, Guo XH, Zhao MH. Titre and affinity of propylthiouracil-induced anti-myeloperoxidase antibodies are closely associated with the development of clinical vasculitis. *Clin Exp Immunol* 2005;**142**:116–19.
16. Nicolls MR, Terada LS, Tuder RM, Prindiville SA, Schwarz MI. Diffuse alveolar hemorrhage with underlying pulmonary capillaritis in the retinoic acid syndrome. *Am J Respir Crit Care Med* 1998;**158**:1302–5.
17. Green RL, Vayonis AG. Churg-Strauss syndrome after zafirlukast in two patients not receiving systemic steroid treatment. *Lancet* 1999;**353**:725–6.
18. Solans R, Bosch JA, Selva A, Orriols R, Vilardell M. Montelukast and Churg–Strauss syndrome. *Thorax* 2002;**57**:183–5.
19. Stirling RG, Chung KF. Leukotriene antagonists and Churg–Strauss syndrome: the smoking gun. *Thorax* 1999;**54**:865–6.
20. Green RJ, Ruoss SJ, Kraft SA, *et al*. Pulmonary capillaritis and alveolar hemorrhage: update on diagnosis and management. *Chest* 1996;**110**:1305–16.
21. Barnett VT, Bergmann F, Humphrey H, Chediak J. Diffuse alveolar hemorrhage secondary to superwarfarin ingestion. *Chest* 1992;**102**:1301–2.
22. Fireman Z, Yust I, Abramov AL. Lethal occult pulmonary hemorrhage in drug-induced thrombocytopenia. *Chest* 1981;**79**:358–9.
23. Awadh N, Ronco JJ, Bernstein V, Gilks B, Wilcox P. Spontaneous pulmonary hemorrhage after thrombolytic therapy for acute myocardial infarction. *Chest* 1994;**106**:1622–4.
24. Usman MH, Shah MA, ul-Islam T, *et al*. Abciximab and fatal pulmonary hemorrhage. *Heart Lung* 2006;**35**:423–6.
25. Kilaru PK, Schweiger MJ, Kozman HA, Weil TR. Diffuse alveolar hemorrhage after clopidogrel use. *J Invasive Cardiol* 2001;**13**:535–7.
26. Aguirre FV, *et al*. EPIC Investigators. Bleeding complications with the chimeric antibody to platelet glycoprotein IIb/IIIa integrin in patients undergoing percutaneous coronary intervention. *Circulation* 1995;**91**:2882–90.
27. Khanlou H, Tsiodras S, Eiger G, *et al*. Fatal alveolar hemorrhage and Abciximab (ReoPro) therapy for acute myocardial infarction. *Cathet Cardiovasc Diagn* 1998;**44**:313–16.
28. Whittle BJ, Moncada S, Vane JR. Comparison of the effects of prostacyclin (PGI2), prostaglandin E1 and D2 on platelet aggregation in different species. *Prostaglandins* 1978;**16**:373–88.
29. Fuster V, Steele PM, Edwards WD, *et al*. Primary pulmonary hypertension: natural history and the importance of thrombosis. *Circulation* 1984;**70**:580–7.
30. Ogawa A, Matsubara H, Fujio H, *et al*. Risk of alveolar hemorrhage in patients with primary pulmonary hypertension: anticoagulation and epoprostenol therapy. *Circ J* 2005;**69**:216–20.
31. Vizioli LD, Cho S. Amiodarone-associated hemoptysis. *Chest* 1994;**105**:305–6.
32. Iskandar SB, Abi-Saleh B, Keith RL, Byrd RP, Roy TM. Amiodarone-induced alveolar hemorrhage. *South Med J* 2006;**99**:383–7.
33. Iskander S, Raible DG, Brozena SC, *et al*. Acute alveolar hemorrhage and orthodeoxia induced by intravenous amiodarone. *Catheter Cardiovasc Interv* 1999;**47**:61–3.

34. Gavaghan TE, McNaught PJ, Ralston M, Hayes JM. Penicillamine-induced 'Goodpasture's syndrome': successful treatment of a fulminant case. *Aust NZ J Med* 1981; **11**:261–5.

35. Sternlieb I, Bennett B, Scheinberg IH. D-penicillamine induced Goodpasture's syndrome in Wilson's disease. *Ann Intern Med* 1975;**82**:673–6.

36. Matloff DS, Kaplan MM. D-penicillamine-induced Goodpasture's-like syndrome in primary biliary cirrhosis: successful treatment with plasmapheresis and immunosuppressives. *Gastroenterology* 1980;**78**(5 Pt 1): 1046–9.

37. Gibson T, Burry HC, Ogg C. Letter: Goodpasture syndrome and D-penicillamine. *Ann Intern Med* 1976;**84**:100.

38. Bucknall CE, Adamson MR, Banham SW. Non fatal pulmonary haemorrhage associated with nitrofurantoin. *Thorax* 1987;**42**:475–6.

39. Averbuch SD, Yungbluth P. Fatal pulmonary hemorrhage due to nitrofurantoin. *Arch Intern Med* 1980;**140**:271–3.

40. Lisbona A, Schwartz J, Lachance C, Frank H, Palayew MJ. Methotrexate-induced pulmonary disease. *J Can Assoc Radiol* 1973;**24**:215–20.

41. Boggess KA, Benedetti TJ, Raghu G. Nitrofurantoin-induced pulmonary toxicity during pregnancy: a report of a case and review of the literature. *Obstet Gynecol Surv* 1996;**51**:367–70.

42. Meyer MM, Meyer RJ. Nitrofurantoin-induced pulmonary hemorrhage in a renal transplant recipient receiving immunosuppressive therapy: case report and review of the literature. *J Urol* 1994;**152**:938–40.

43. Champion L, Stern M, Israel-Biet D, *et al.* Brief communication: sirolimus-associated pneumonitis: 24 cases in renal transplant recipients. *Ann Intern Med* 2006;**144**:505–9.

44. Vlahakis NE, Rickman OB, Morgenthaler T. Sirolimus-associated diffuse alveolar hemorrhage. *Mayo Clin Proc* 2004;**79**:541–5.

45. Alexandrescu DT, Dutcher JP, O'Boyle K, *et al.* Fatal intra-alveolar haemorrhage after rituximab in a patient with non-Hodgkin lymphoma. *Leukemia Lymphoma* 2004;**45**:2321–5.

46. Heresi GA, Farver CF, Stoller JK. Interstitial pneumonitis and alveolar haemorrhage complicating use of rituximab: case report and review of the literature. *Respiration* 2008;**76**:449–53.

47. Pietra GG. Pathologic mechanisms of drug-induced lung disorders. *J Thorac Imaging* 1991;**6**(1):1–7.

48. Robbins RA, Linder J, Stahl MG, *et al.* Diffuse alveolar hemorrhage in autologous bone marrow transplant recipients. *Am J Med* 1989;**87**:511–18.

49. Jules-Elysee K, Stover DE, Yahalom J, White DA, Gulati SC. Pulmonary complications in lymphoma patients treated with high-dose therapy autologous bone marrow transplantation. *Am Rev Respir Dis* 1992;**146**:485–91.

50. Agusti C, Ramirez J, Picado C, *et al.* Diffuse alveolar hemorrhage in allogeneic bone marrow transplantation: a postmortem study. *Am J Respir Crit Care Med* 1995;**151**:1006–10.

51. Kahn FW, Jones JM, England DM. Diagnosis of pulmonary hemorrhage in the immunocompromised host. *Am Rev Respir Dis* 1987;**136**:155–60.

52. Chao NJ, Duncan SR, Long GD, Horning SJ, Blume KG. Corticosteroid therapy for diffuse alveolar hemorrhage in autologous bone marrow transplant recipients. *Ann Intern Med* 1991;**114**:145–6.

53. Garcia-Rostan y Perez GM, Garcia Bragado F, Puras Gil AM. Pulmonary hemorrhage and antiglomerular basement membrane antibody-mediated glomerulonephritis after exposure to smoked cocaine (crack): a case report and review of the literature. *Pathol Int* 1997;**47**:692–7.

54. Baldwin GC, Choi R, Roth MD, *et al.* Evidence of chronic damage to the pulmonary microcirculation in habitual users of alkaloidal ('crack') cocaine. *Chest* 2002;**121**:1231–8.

55. Murray RJ, Albin RJ, Mergner W, Criner GJ. Diffuse alveolar hemorrhage temporally related to cocaine smoking. *Chest* 1988;**93**:427–9.

56. Lazor R, Bigay-Game L, Cottin V, *et al.* Alveolar hemorrhage in anti-basement membrane antibody disease: a series of 28 cases. *Medicine (Baltimore)* 2007;**86**:181–93.

57. Peces R, Navascues RA, Baltar J, *et al.* Antiglomerular basement membrane antibody-mediated glomerulonephritis after intranasal cocaine use. *Nephron* 1999;**81**:434–8.

58. Zamora MR, Warner ML, Tuder R, Schwarz MI. Diffuse alveolar hemorrhage and systemic lupus erythematosus: clinical presentation, histology, survival, and outcome. *Medicine (Baltimore)* 1997;**76**:192–202.

59. Nakamori Y, Tominaga T, Inoue Y, Shinohara K. Propylthiouracil (PTU)-induced vasculitis associated with antineutrophil antibody against myeloperoxidase (MPO-ANCA). *Intern Med* 2003;**42**:529–33.

60. Frankel SR, Eardley A, Lauwers G, Weiss M, Warrell RP. The 'retinoic acid syndrome' in acute promyelocytic leukemia. *Ann Intern Med* 1992;**117**:292–6.

61. Klemmer PJ, Chalermskulrat W, Reif MS, *et al.* Plasmapheresis therapy for diffuse alveolar hemorrhage in patients with small-vessel vasculitis. *Am J Kidney Dis* 2003;**42**:1149–53.

62. Betensley AD, Yankaskas JR. Factor VIIa for alveolar hemorrhage in microscopic polyangiitis. *Am J Respir Crit Care Med* 2002;**166**:1291–2.

63. Henke D, Falk RJ, Gabriel DA. Successful treatment of diffuse alveolar hemorrhage with activated factor VII. *Ann Intern Med* 2004;**140**:493–4.

64. Pastores SM, Papadopoulos E, Voigt L, Halpern NA. Diffuse alveolar hemorrhage after allogeneic hematopoietic stem-cell transplantation: treatment with recombinant factor VIIa. *Chest* 2003;**124**:2400–3.

65. Monroe DM, Hoffman M, Oliver JA, Roberts HR. Platelet activity of high-dose factor VIIa is independent of tissue factor. *Br J Haematol* 1997;**99**:542–7.

66. Torra R, Poch E, Torras A, Bombi JA, Revert L. Pulmonary hemorrhage as a clinical manifestation of hemolytic-uremic syndrome associated with mitomycin C therapy. *Chemotherapy* 1993;**39**:453–6.

67. Schwarz MI, Mortenson RL, Colby TV, *et al.* Pulmonary capillaritis: the association with progressive irreversible airflow limitation and hyperinflation. *Am Rev Respir Dis* 1993;**148**:507–11.

68. Brugiere O, Raffy O, Sleiman C, *et al.* Progressive obstructive lung disease associated with microscopic polyangiitis. *Am J Respir Crit Care Med* 1997;**155**:739–42.

12

Pulmonary complications of illicit drug use

DEEPA LAZARUS, ANNE O'DONNELL

INTRODUCTION

Illicit drugs are a truly global phenomenon. As a result of this and the illicit nature of the trade, reliable statistics are hard to obtain. An estimated 3 per cent of the global population (185 million people) consumes drugs annually.[1] By far the most widely abused substance is cannabis, followed by amphetamine-like stimulants.[1] The lungs are unique in their exposure to the environment and the circulation; they are susceptible to damage from illicit drugs that are used by intravenous injection and by inhalation.[2] Thoracic complications are the most common complications associated with illicit drug use and consist of infective and non-infective complications affecting the pleura pulmonary parenchyma, soft tissues, heart and mediastinum.

This chapter provides an update on the pulmonary and thoracic effects of illicit drug use at the exclusion of systemic bacterial infection,[3] abused drug-induced HIV or HVC infection, and the risks associated with illicit manufacture of abused substances,[4] or risks to forensic personnel involved in post-mortem examination fatalities from abused substances.[5,6]

INFECTIVE COMPLICATIONS OF ILLICIT DRUG USE

Acute non-cardiac pulmonary oedema

The most common street drug associated with non-cardiogenic pulmonary oedema (NCPE) is heroin. It is a fairly infrequent complication of heroin overdose: the incidence was reported to be 2.1 per cent in a recent case series by Sporer and Dorn.[7] NCPE related to heroin overdose usually presents as a combination of persistent hypoxia after resolution of opiate respiratory depression along with frothy pink-tinged pulmonary secretions and a characteristic radiographic pattern of fluffy diffuse pulmonary infiltrates.[7] The mechanism of heroin-induced pulmonary oedema (HIPE) has been debated. Haemodynamic studies performed in patients with HIPE show elevated pulmonary artery pressures and normal pulmonary capillary wedge pressures.[8] It was suggested that hypoxia-related pulmonary hypertension resulted in oedema formation. Bronchoalveolar lavage (BAL) studies performed by Katz and colleagues showed that HIPE fluid had protein levels that were nearly similar to serum levels, considerably higher than levels found in cardiogenic pulmonary oedema. Therefore, increased capillary permeability was suggested as a possible mechanism for this.[9] However, later studies by Dettmeyer and colleagues found no difference in the number of IgE positive cells or in alveolar staining for collagen or lamin in lung tissue for patients who died of heroin intoxication compared to patients who had sudden cardiovascular death.[10] Other theories suggested include histamine mediated capillary leak.[11,12] The central nervous system depression that accompanies opiate use suggests that neurogenic pulmonary oedema may also contribute to opiate-induced acute lung injury. Acute lung injury manifests as bilateral, perihilar areas of increased opacity or attenuation, usually without pleural effusion or cardiomegaly. High-resolution CT may show multifocal ground-glass attenuation associated with septal thickening.[13]

The treatment of heroin-induced pulmonary oedema includes supportive care, and administration of the specific narcotic antagonist naloxone which is administered intramuscularly. HIPE carries a very good prognosis. In the largest case series published by Sporer et al., 33 per cent of patients required mechanical ventilation and all but one patient was extubated within 24 hours.[7] Though an extended observation period of 12–24 hours is often quoted,[14,15] a 1–2 hours observation period may be sufficient as HIPE is an infrequent complication of heroin overdose and tends to occur early.[7,16,17]

Though heroin is the most common drug associated with the development of non-cardiogenic pulmonary oedema, it has been reported with overdoses of ecstasy (MDMA), methadone,[18] propoxyphene,[19] codeine,[20] buprenorphine,[21,22] and nalbuphine.[23] Evaluation of heroin levels in blood is required in cases of acute pulmonary oedema seen in the emergency department.

Pulmonary granulomatosis

Foreign body granulomas are an infrequent complication of intravenous drug use and occur mainly with injection of substances intended for oral use. They are predominantly vascular and perivascular in location though interstitial granulomas are also found. They form as a result of the reaction to the embolization of insoluble materials used as fillers and binders in the tablets, including talc and crospovidone.[24,25] Talc is the most frequent filler used and has been reported in a patient who developed fatal pulmonary hypertension secondary to use of injected crushed methylphenidate tablets which resulted in large refractile foreign body granulomas predominantly in the perivascular and intravascular location resulting in pulmonary hypertension and cor pulmonale.[26] Talc-containing foreign body granulomas have also been described in patients who injected alpha-sympathomimetic agents that were obtained from nasal inhalers.[27] The intravenous talc particles trapped within the pulmonary arterioles and capillaries induce pulmonary endothelial injury,[24] which results in occlusion of the pulmonary vessels by thrombosis or fibrosis and in severe cases of hypertrophy and angiomatoid malformation of the pulmonary arteries[27-30] this results in pulmonary hypertension. Transvascular migration of the talc particles results in the formation of perivascular and interstitial granulomas (Figs 12.1 and 12.2), and may result in interstitial fibrosis in association.[24]

Patients present with progressive dyspnoea on exertion and an unproductive cough. Precocious emphysema has been well documented in these patients. Pulmonary function tests show reduced DLCO. Right-heart catheterization may show pulmonary hypertension depending on the stage of the disease. Radiographically, injection talcosis may appear as irregular nodular areas of increased opacity in the middle and upper areas of the lungs that may coalesce to form conglomerate masses.[31] Diffuse small nodules may be the first manifestation of talc-induced lung disease, with coalescence and enlargement of the nodules being observed as the disease progresses.[32] On occasion, lymphadenopathy is present.[31] In the later stages of talc-induced pulmonary disease, upper-lobe conglomerate masses that resemble progressive massive fibrosis as seen in silicosis may develop.[32] These areas of increased opacity are often superimposed on a background of numerous small nodules and lower lobe-predominant emphysema (see below).

Computed tomography findings show diffuse fine nodular pattern, nodules combined with lower-lobe emphysema, pure emphysema or ground-glass infiltrates. The disease will progress even after cessation of intravenous drug use. In a 10-year follow-up study of six intravenous drug users with pulmonary talcosis by Paré and co-workers, all six developed severe respiratory dysfunction and three died of progressive pulmonary disease.[33]

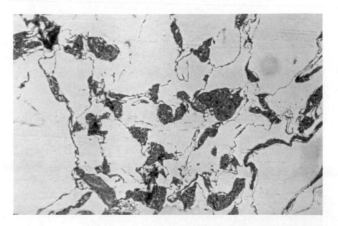

Fig. 12.1 Photomicrograph of haematoxylin- and eosin-stained lung biopsy specimen showing emphysematous changes and talc granulomas.

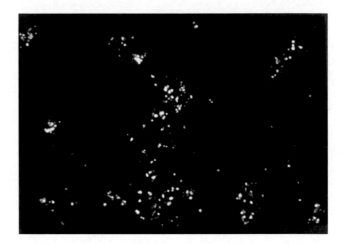

Fig. 12.2 Polarized light microscopy of talc granulomas.

Alveolar haemorrhage

Cocaine inhalation has been linked to alveolar haemorrhage independently and through anti-basement membrane antibody syndrome. Other agents that have been implicated in alveolar haemorrhage are marijuana, firesmoke and various forms of hydrocarbons. Patients present with flu-like symptoms associated with cough dyspnoea and haemoptysis. Chest imaging reveals infiltrates predominantly in the lower lobes. Bronchoalveolar lavage has the highest yield in diagnosis and shows macroscopically pink or red lavage return (Fig. 12.3).[34]

There has been a recent practice of adulterating cannabis with glass beads or sand to increase the weight and boost profits by the dealers. A case report by Just and colleagues describes the potential risks of this procedure. A patient who smoked adulterated cannabis joints presented with dyspnoea and diffuse bilateral upper-lobe predominant infiltrates which resolved completely after cessation of cannabis smoking.[35] The symptoms and infiltrates were temporally related to the smoking of cannabis adulterated with sand grit.

Other reported pulmonary parenchymal complications of drug abuse include eosinophilic lung disease due to cocaine[36] and bronchiolitis obliterans organizing pneumonia (BOOP).[37]

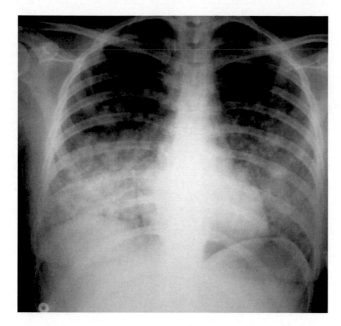

Fig. 12.3 Acute diffuse alveolar haemorrhage. Diagnosis was established by a bloody BAL return. The patient eventually admitted to inhaled cocaine abuse. A tox screen was supportive.

Precocious emphysema and bullous pulmonary damage

Many users of illicit substances are tobacco smokers. However, recently many cases of bullous lung disease and precocious emphysema have been reported which cannot be attributed to tobacco smoking alone. Emphysema associated with tobacco smoking typically presents at an older age (the exception would be α_1 antitrypsin deficiency). Almost all the patients reported are young and have a lower pack-year history of smoking. There is a distinct difference in the radiological appearance of these bullous changes in users of predominantly smoked substances and other intravenous agents compared to individuals who inject substances intended for oral use.

In a case series reported by Johnson and colleagues, four patients who smoked marijuana had large bullous lesions in a paraseptal distribution in the upper lung zones. The average smoking history of these patients was 26-pack years.[38] Goldstein and colleagues compared two groups of patients with bullous pulmonary damage and intravenous drug abuse; both groups were smokers. The drug users had large bullous lesions confined exclusively to the upper lobes while the non-users had bullae that varied in size, were generally diffuse and were rarely limited to the upper lobes.[37] Both groups of patients had reduced diffusing capacity for carbon monoxide (DLCO), mild hypoxia and moderately severe to severe obstruction on pulmonary function tests;[39] in contrast, patients who inject methylphenidate tablets have bibasilar disease similar to that seen in α_1 antitrypsin deficiency.[40] Though experimental and clinical evidence is lacking, recurrent foreign body emboli leading to the formation of thin-walled cavities which coalesce to form larger bullous lesions causing airflow obstruction are thought to be the cause of these lesions.[39]

Airway complications

Marijuana is the most commonly smoked illicit substance. When smoked, marijuana causes a significant bronchodilator response in healthy adults as well as stable asthmatics. It can also reverse the bronchoconstrictor effects of inhaled methacholine and exercise-induced bronchospasm in asthmatic subjects.[41] However, tolerance to the bronchodilator effects of marijuana has been demonstrated after several weeks of use.[42] Habitual marijuana smokers have an increased prevalence of chronic cough, sputum production and wheeze and increased frequency of acute bronchitic episodes.[43–45] Interestingly, recent studies have disproved the notion that chronic marijuana use can cause chronic airflow obstruction. A study by Tashkin and co-workers comparing respiratory symptoms and lung function in habitual heavy marijuana smokers (mean of three joints/day > 15 years) alone, marijuana and tobacco smokers, smokers of tobacco alone, and non-smokers could not demonstrate abnormalities in even the most sensitive measures of early functional impairment, including flow rates at low lung volumes and indices derived from single-breath nitrogen washout. Moreover, regular use of marijuana was not associated with any abnormality in single-breath diffusing capacity for carbon monoxide, a sensitive physiological indicator for emphysema.[43]

The same authors also did a long-term follow-up study to evaluate whether habitual marijuana smoking leads to progressive decline in lung function over a period, as happens in COPD. No effect of even heavy marijuana smoking (three joints/day) on FEV_1 decline was observed, nor an additive effect of marijuana on tobacco.[46] The above studies contradict the findings of various older studies in which mild chronic airflow obstruction was noted with marijuana use.[44,45,47] Cocaine freebasing, on the other hand, causes acute bronchospasm; but, as with marijuana, studies have failed to demonstrate any adverse long-term effects on ventilatory mechanics as indicated by normal spirometric findings even in heavy habitual smokers.[48,49] Certain studies, however, have demonstrated a decrease in DLCO, a sensitive marker of the integrity of the alveolar capillary membrane.[48–50]

There have been various case reports in the literature of heroin-induced status asthmaticus.[51] It is still unclear whether asthma is a prerequisite for heroin-induced bronchospasm. These patients presenting with sudden onset of asthma symptoms tend to have a protracted course and are generally unresponsive to β-adrenergic agents.[51]

Inhalational injury to the airway

Many cases of airway injury and burns secondary to inhalation of illicit substances have been reported. Taylor and Bernard reported a case of severe thermal injury to the conducting airways due to ether ignition.[52] The thermal injuries resulted in long-term tracheal narrowing that required multiple surgeries. Mayo-Smith and Spinale reported three cases of thermal epiglottitis secondary to inhalation of metal pieces of 'crack' pipes.[53] Oh and colleagues reported a series of 48 patients who had explosive burns related to inhalation of butane gas. Of these, 12 demonstrated inhalation injury; 3 developed pulmonary complications and died.[54] 'Huffing' describes the practice of inhaling gasoline fumes. Sheridan reported four patients who had burns and inhalation injury related to this practice.[55] Bronchiectasis secondary to heroin and cocaine has also been described.[56]

Pneumothorax and pneumomediastinum

The pleural complications are mainly pneumothorax and pneumomediastinum. It may be related to smoking practices or to injection by a friend rather than a particular drug.[2] Pneumomediastinum results from the deep inhalation of a drug, followed by coughing, which causes increased alveolar pressure and subsequent dissection of air into the pulmonary interstitium. Air may then enter the mediastinum or, rarely, the pericardial space.[57]

'Shotgunning' refers to the practice of inhaling smoke then exhaling it with positive pressure into another person's mouth. 'Pocket shot' refers to the attempted intravenous injection into the supraclavicular fossa which may cause a pneumothorax.[57]

A recent study by Tashkin and co-workers found no association between marijuana smoking and lung cancer (oral presentation, ATS, 2006).

INFECTIVE COMPLICATIONS OF ILLICIT DRUG USE

Pneumonia

Respiratory infections are among the most frequent sequele of drug use and patients may present with atypical clinical and radiographic findings.[58–60] Drug users have a 10-fold increase in the risk of community-acquired pneumonia.[61] The range of organisms is similar to that of community-acquired pneumonia in the general population. Aspiration pneumonia as well as pneumonia caused by *Staphylococcus aureus*, *S. pneumoniae*, *Haemophilus influenzae* and *Klebsiella pneumoniae* are among the most common reasons for hospitalization.[62]

Inhaling cocaine and marijuana also predisposes drug users to upper respiratory tract infections, including sinusitis and rarely invasive pulmonary aspergillosis.

Septic pulmonary emboli

Septic pulmonary emboli result from recurrent emboli of infected material which almost invariably contains *Staphylococcus aureus* (80 per cent), *S. albus* (19 per cent) or occasionally *Candida* species or Gram-negative organisms.[61] The source of the infection is either subcutaneous infection at the site of injection or tricuspid endocarditis or both.[63,64] Patients with infective endocarditis who abuse intravenous drugs may have underlying structural valvular lesions that may be secondary to the injection of particulate matter along with the illicit drugs, which can produce transient or permanent damage to the tricuspid valve.[65] Patients present with fever, dyspnoea and haemoptysis; radiographically, there are peripheral fluffy infiltrates with a tendency to cavitate (Fig. 12.4).

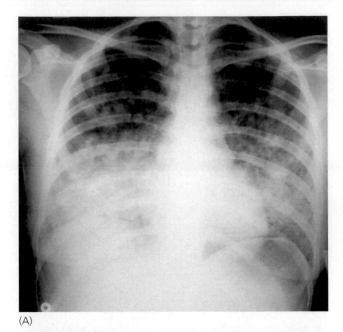

(A)

(B)

Fig. 12.4   Chest X-ray and computed tomography demonstrating multiple rounded densities in a peripheral distribution consistent with septic emboli to the lungs.

Pulmonary tuberculosis

Tuberculosis is more common in intravenous drug users than in the general population owing to the risk factors associated with their lifestyle (alcoholism, overcrowding and depressed immunity).[66] Tuberculosis may also be transmitted through a practice known as 'shotgunning' (smoking and inhaling a drug and then expelling the smoke into another person's mouth), a common practice among smokers of 'crack' cocaine. Drug users are two to six times at higher risk of contracting tuberculosis. It may present atypically, especially if the patient is infected with HIV. Often mediastinal adenopathy may be the only radiological finding. Directly observed therapy for tuberculosis is recommended for this patient population.

The increased prevalence of HIV seropositivity in this patient population also places them at high risk for developing opportunistic infections.

Miscellaneous conditions

Clandestine subcutaneous injection of silicone in varied parts of the body is illegally practised to alter body shape. Injection may be delivered by unqualified personnel, and results in silicone embolism to the pulmonary circulation. The disease is characterized by sudden or delayed development of disseminated bilateral alveolar pulmonary opacities and deteriorated gas exchange. Subcutaneous silicone may be present in the breast area on imaging as well. Diffuse alveolar haemorrhage may be present. Bronchoalveolar lavage and lung biopsy may show characteristic lipid droplets. The disease shares features with the fat embolism syndrome, including neurological involvement. Fatalities may occur, especially in those patients with neurological involvement.

Inhaled nitrites pose the risk of developing significant life-threatening methemoglobinaemia. Methemoglobin is a ferric (Fe^{3+}) oxidized instead of the normal ferrous (Fe^{2+}) state haemoglobin, a poor oxygen carrier. Emergent recognition is warranted. The condition may follow exposure to inhaled nitrites in so-called 'poppers', as named from the brisk noise of opening the rubber stopper of nitrite mini-vial. The condition is suspected when slate-grey cyanosis and low pulse O_2 saturation (SpO_2) develop, along with a normal dissolved PaO_2 and calculated SaO_2. Multi-wavelength examination of an arterial blood sample is required to quantitate methemoglobin and confirm the diagnosis. Management includes high-flow oxygen, cessation of exposure to the agent, intravenous methylene blue, and in severe cases hyperbaric oxygen therapy.

KEY POINTS

- Thoracic complications are the most common, consisting of infective and non-infective complications affecting the pleura pulmonary parenchyma, soft tissues, heart and mediastinum.

- Infective complications include acute non-cardiac pulmonary oedema, pulmonary granulomatosis, alveolar haemorrhage, precocious emphysema, bullous pulmonary damage, airway complications, pneumothorax and pneumomediastinum.

- Infective complications include pneumonia, septic pulmonary emboli and pulmonary tuberculosis.

- The most widely abused substance is cannabis, followed by amphetamine-like stimulants. Other common but illicit drugs include heroin, cocaine, marijuana and hydrocarbons.

- Foreign body granulomas are an infrequent complication and occur mainly with injection of substances intended for oral use.

REFERENCES

1. United Nations Office on Drugs and Crime. *World Drug Report 2004*. Available at www.unodc.org/unodc/data-and-analysis/WDR.html
2. Wolff AJ, O'Donnell AE. Pulmonary effects of illicit drug use. *Clin Chest Med* 2004;**25**:203–16.
3. Finn SP, Leen E, English L, O'Briain DS. Autopsy findings in an outbreak of severe systemic illness in heroin users following injection site inflammation: an effect of *Clostridium novyi* exotoxin? *Arch Pathol Lab Med* 2003;**127**:1465–70.
4. Lineberry TW, Bostwick JM. Methamphetamine abuse: a perfect storm of complications. *Mayo Clin Proc* 2006;**81**:77–84.
5. Willers-Russo LJ. Three fatalities involving phosphine gas, produced as a result of methamphetamine manufacturing. *J Forens Sci* 1999;**44**:647–52.
6. Burgess JL. Phosphine exposure from a methamphetamine laboratory investigation. *J Toxicol Clin Toxicol* 2001;**39**:165–8.
7. Sporer KA, Dorn E. Heroin related noncardiogenic pulmonary edema: a case series. *Chest* 2001;**120**:1628–32.
8. Gopinathan K, Saroja J, Spears R, *et al.* Hemodynamic studies in heroin induced acute pulmonary edema. *Circulation* 1970;**61**(Suppl 3):111–44.
9. Katz S, Aberman A, Frand UI, *et al.* Heroin pulmonary edema: evidence for increased pulmonary capillary permeability. *Am Rev Respir Dis* 1972;**106**:472–4.
10. Dettmeyer R, Schmidt P, Musshoff F, Dreisvogt C, Madea B. Pulmonary edema in fatal heroin overdose. Immunohistological investigations with IgE, collagen IV and laminin: no increase of defects of alveolar–capillary membranes. *Forens Sci Int* 2000;**110**:87–96.
11. Pietra GG, Szidon JP, Leventhal MM, *et al.* Histamine and interstitial pulmonary edema in the dog. *Circ Res* 1971;**29**:323–37.
12. Brigham KL, Owen PJ. Increased sheep lung vascular permeability caused by histamine. *Circ Res* 1975;**37**:647–57.
13. Gotway MB, Marder SR, Hanks DK, *et al.* Thoracic complications of illicit drug use: an organ system approach. *Radiographics* 2002;**22**:S119–35.
14. Steinberg AD, Karliner JS. The clinical spectrum of heroin pulmonary edema. *Arch Intern Med* 1968;**122**:122–7.
15. Goldfrank L, Weisman R, Flomenbaum NE, *et al.* Opioids. In: *Goldfrank's Toxicologic Emergencies*, pp 769–783. Norwalk, CT: Appleton & Lange, 1994.
16. Osterwalder JJ. Patients intoxicated with heroin or heroin mixtures: how long should they be monitored? *Eur J Emerg Med* 1995;**2**:97–101.
17. Christenson J, Etherington J, Grafstein E, *et al.* Early discharge of patients with presumed opioid overdose: development of a clinical prediction rule. *Acad Emerg Med* 2000;**7**:1110–18.
18. Kjeldgaard JM, Hahn GW, Heckenlively JR, *et al.* Methadone-induced pulmonary edema. *J Am Med Assoc* 1971;**218**:882–3.
19. Bogartz, LJ, Miller, WC. Pulmonary edema associated with propoxyphene intoxication. *J Am Med Assoc* 1971;**215**:259–62.
20. Sklar J, Timms RM. Codeine-induced pulmonary edema. *Chest* 1977;**72**:230–1.
21. Gould DB. Letter: Buprenorphine causes pulmonary edema just like all other mu-opioid narcotics. *Chest* 1995;**107**:1478–9.
22. Thammakumpee G, Sumpatanukule P. Noncardiogenic pulmonary edema induced by sublingual buprenorphine. *Chest* 1994;**106**:306–8.
23. Stadnyk A, Grossman RF. Nalbuphine-induced pulmonary edema. *Chest* 1986;**90**:773–4.

24. Hind CRK. Pulmonary complications of intravenous drug misuse. *Thorax* 1990;**45**:891–8.

25. Ganesan S, Felo J, Saldana M, *et al.* Embolized crospovidone (poly[*N*-vinyl-2-pyrrolidone]) in the lungs of intravenous drug users. *Modern Pathol* 2003;**16**:286–92.

26. Lewman LV. Fatal pulmonary hypertension from intravenous injection of methylphenidate (Ritalin) tablets. *Hum Pathol* 1972;**1**:67–70.

27. Robertson CH, Reynolds RC, Wilson JE. Pulmonary hypertension and foreign body granulomas in intravenous drug abusers: documentation by cardiac catheterization and lung biopsy. *Am J Med* 1976;**61**:657–64.

28. Arnett EN, Attle WE, Russo JV, Roberts WC. Intravenous injection of talc-containing drugs intended for oral use: a cause of pulmonary granulamatosis and pulmonary hypertension. *Am J Med* 1976;**60**:718–19.

29. Wendt VE, Puro HE, Shapiro J, *et al.* Angiothrombotic pulmonary hypertension in adults 'blue velvet' addiction. *J Am Med Assoc* 1964;**188**:755–7.

30. Tomashefski JF, Hirsch CS. The pulmonary vascular lesions of intravenous drug abuse. *Hum Pathol* 1980;**11**:133–45.

31. Feigin DS. Talc: understanding its manifestations in the chest. *Am J Roentgenol* 1986;**146**:295–301.

32. Fraser RA, Muller NL, Colman N, Paré PD. *Diagnosis of Diseases of the Chest*, pp 2566–9. Philadelphia, PA: Saunders, 1999.

33. Paré JP, Cote G, Fraser RS. Long-term follow-up of drug abusers with intravenous talcosis. *Am Rev Respir Dis* 1989;**139**:233–41.

34. Lazor R, Bigay-Game L, Cottin V. Alveolar hemorrhage in anti-basement membrane antibody disease: a series of 28 cases. *Medicine* (Baltimore) 2007;**86**:181–93.

35. Just N, Delourme J, Delattre C. An unusual case of patchy ground-glass opacity. *Thorax* 2009;**64**:74.

36. Pi OH, Balter MS. Cocaine induced eosinophilic lung disease. *Thorax* 1992;**47**:478–9.

37. Bishay A, Amchentsev A, Saleh A. A hitherto unreported pulmonary complication in an IV heroin user. *Chest* 2008;**133**:549–51.

38. Johnson MK, Smith RP, Morrison D, *et al.* Large lung bullae in marijuana smokers. *Thorax* 2000;**55**:340–2.

39. Goldstein DS, Karpel JP, Appel D, Williams MH. Bullous pulmonary damage in users of intravenous drugs. *Chest* 1986;**89**:266–9.

40. Stern EJ, Frank MS, Schmutz JF, *et al.* Panlobular emphysema caused by IV injection of methylphenidate (Ritalin): findings on chest radiographs and CT scans. *Am J Roentgenol* 1994;**162**:555–60.

41. Tashkin DP, Shapiro BJ, Lee EY. Effects of smoked marijuana in experimentally induced asthma. *Am Rev Respir Dis* 1975;**112**:377–86.

42. Tashkin DP, Shapiro BJ, Lee EY, *et al.* Sub acute effects of heavy marihuana smoking on pulmonary function in healthy men. *New Engl J Med* 1976;**294**:125–9.

43. Tashkin DP, Coulson AH, Clark VA, *et al.* Respiratory symptoms and lung function in habitual, heavy smokers of marijuana alone, smokers of marijuana and tobacco, smokers of tobacco alone, and nonsmokers. *Am Rev Respir Dis* 1987;**135**:209–16.

44. Bloom JW, Kaltenborn WT, Paoletti P, *et al.* Respiratory effects of non-tobacco cigarettes. *Br Med J* 1987;**295**:1516–18.

45. Taylor DR, Poulton R, Moffitt TE, *et al.* The respiratory effects of cannabis dependence in young adults. *Addiction* 2000;**95**:1669–77.

46. Tashkin DP, Simmons MS, Sherrill D, *et al.* Heavy habitual marijuana smoking does not cause an accelerated decline in FEV_1 with age: a longitudinal study. *Am J Respir Crit Care Med* 1997;**155**:141–8.

47. Sherrill DL, Krzyzanowski M, Bloom JW, *et al.* Respiratory effects of non-tobacco cigarettes: a longitudinal study in general population. *Int J Epidemiol* 1991;**20**:132–7.

48. Tashkin DP, Khalsa ME, Gorelick D, *et al.* Pulmonary status of habitual cocaine smokers. *Am Rev Respir Dis* 1992;**145**:92–100.

49. Itkonen J, Schnoll S, Glassroth J. Pulmonary dysfunction in freebase cocaine users. *Arch Intern Med* 1984;**144**:2195–7.

50. Weiss RD, Goldenheim PD, Mirin SM, *et al.* Pulmonary dysfunction in cocaine smokers. *Am J Psychiatry* 1981;**138**:1110–12.

51. Cygan J, Trunsky M, Corbridge T. Inhaled heroin-induced status asthmaticus: five cases and a review of the literature. *Chest* 2000;**117**:272–5.

52. Taylor RF, Bernard GR. Airway complications from free-basing cocaine. *Chest* 1989;**95**:47–67.

53. Mayo-Smith MF, Spinale J. Thermal epiglottitis in adults: a new complication of illicit drug use. *J Emerg Med* 1997;**15**:483–5.

54. Oh SJ, Lee SE, Burm JS, *et al.* Explosive burns during abusive inhalation of butane gas. *Burns* 1999;**25**:341–4.

55. Sheridan RL. Burns with inhalation injury and petrol aspiration in adolescents seeking euphoria through hydrocarbon inhalation. *Burns* 1996;**22**:566–7.

56. Schachter EN, Basta W. Bronchiectasis following heroin overdose: a report of two cases. *Chest* 1973;**63**:363–6.

57. McCarroll KA, Roszler MH. Lung disorders due to drug abuse. *J Thorac Imaging* 1991;**6**:30–5.

58. Cheidegger C, Zimmerli W. Infectious complications in drug addicts: seven-year review of 269 hospitalized narcotics abusers in Switzerland. *Rev Infect Dis* 1989;**11**:486–93.

59. Louria DB, Hensle T, Rose J. The major medical complications of heroin addiction. *Ann Intern Med* 1967;**67**:1–22.

60. Tumbarello M, Tacconelli E, de Gaetano K, *et al.* Bacterial pneumonia in HIV-infected patients: analysis of risk factors and prognostic indicators. *J Acquir Immune Defic Syndr Hum Retrovirol* 1998;**18**:39–45.

61. Hind CR. Pulmonary complications of intravenous drug misuse. 2: Infective and HIV related complications. *Thorax* 1990;**45**:957–61.

62. Boschini A, Smacchia C, Di Fine M, *et al.* Community-acquired pneumonia in a cohort of former injection drug users with and without human immunodeficiency virus infection: incidence, etiologies, and clinical aspects. *Clin Infect Dis* 1996;**23**:107–13.

63. Chambers HF, Korzeniowski OM, Sande MA. *Staphylococcus aureus* endocarditis: clinical manifestations in addicts and non-addicts. *Medicine* 1983;**62**;170–7.

64. Lynch JP, Ravikrishnan KP. Infectious cardiopulmonary complications of drug abuse. *Med Intern* 1984;**5**:76–101.

65. Sande MA, Lee BL, Mill J, Chambers HF. Endocarditis in intravenous drug users. In: Kaye D (ed.) *Infective Endocarditis*, p 345. New York: Raven Press, 1992.

66. Brown M, Stimmel B, Taub RN. Immunological dysfunction in heroin addicts. *Arch Intern Med* 1974;**134**:1001–3.

13

Iatrogenic tracheobronchial and chest injury

MARIOS FROUDARAKIS, DEMOSTHENES MAKRIS, DEMOSTHENES BOUROS

INTRODUCTION

Tracheobronchial and chest injuries due to iatrogenic causes are not common complications. They concern multiple chest organs from various iatrogenic interventions. Their incidence varies according to the injured organ and the intervention.

Tracheobronchial ruptures (TBRs) are life-threatening lesions that may be a rare complication of tracheal intubation. Tracheal intubation may also be the cause of tracheal stenosis. TBRs also occur after chest radiation therapy with or without chemotherapy in patients with bronchogenic carcinoma, sometimes associated with tracheoesophageal fistula. Surgical procedures may also cause ruptures of different intrathoracic organs or vessels, necessitating immediate re-intervention. Insertion of drainage tubes or biopsy devices for both parenchymal or pleural lesions may cause pulmonary, vascular or diaphragmatic injuries. Also, insertion of vascular catheters such as the subclavicular type, and pulmonary artery catheterization, may lead to life-threatening vascular perforation.

This chapter reviews the tracheobronchial, pulmonary and pleural injuries caused during interventions, their incidence, symptoms and management.

TRACHEOBRONCHIAL INJURIES

Ruptures

Iatrogenic tracheobronchial lacerations are rare complications. Their incidence varies from 0.005 per cent for orotracheal intubations[1] to 0.19 per cent for double-lumen intubations,[2] and to 0.23 per cent for tracheostomies. Repeated attempts in a difficult intubation and/or the presence of tracheal abnormalities may be predisposing factors. Cuff-related tracheal ruptures may also occur in mechanically ventilated patients related to an overinflation of the cuff.[2] Tracheal ruptures also occur after chest radiation therapy with or without chemotherapy in patients with bronchogenic carcinoma, sometimes associated with tracheoesophageal fistula. Considering that there are a great number of intubations, diagnostic and therapeutic procedures, these complication rates are extremely low. However, their true incidence is probably underestimated, as we do not know the exact number of interventional procedures and because many complications go unregistered.[3] It seems that female patients undergoing endotracheal/endobronchial interventional procedures are more likely to present iatrogenic ruptures.[2]

The mechanism of iatrogenic tracheobronchial rupture is different from injuries due to traumas, leading to different appearance and therapeutic options. Traumatic rupture is usually the result of blunt chest trauma.[4] It appears horizontal or irregular involving carina and often extended into the main stem bronchi. Iatrogenic rupture usually presents as longitudinal lacerations of the posterior tracheal wall, centrally or laterally located, which is disrupted from the cartilaginous armature, and rarely involves the bronchial tree.[4] In such cases, the mechanism seems to be a direct laceration from an endotracheal tube tip caught in a fold of a flaccid posterior tracheal membrane, while advancing the tube.[5] In less than 10 per cent of the cases tracheal lacerations are associated with laryngeal/vocal cord iatrogenic lesions.

The diagnosis is often delayed, since tracheobronchial rupture is not always evident. In mechanically ventilated patients who have undergone an interventional procedure and do not comply with the ventilator, a tracheobronchial laceration may be suspected, as well as a soft-tissue emphysema, pneumothorax and/or pneumomediastinum. In awake patients with a recent intervention, persistent cough, dyspnoea and haemoptysis (usually massive) are symptoms occurring after tracheal ruptures. Bronchoscopy will reveal the site and the extent of the tracheobronchial rupture.

Early surgical repair is the classical therapeutic option for tracheobronchial ruptures.[1,2] Mechanically ventilated patients,

Table 13.1 Outcome of patients with tracheobronchial rupture in published series, according to the mode of treatment

Reference	Patients	Surgery: patients treated/cured	Conservative: patients treated/cured
5	30	2/0 (0%)	28/24 (85.7%)
8	5	0	5/3 (60%)
9	18	1/0 (0%)	17/14 (82.3%)
10	13	10/10 (100%)	3/3 (100%)
6	8	1/1 (100%)	7/2 (28.7%)
11	12	4/4 (100%)	8/7 (87.5%)
12	9	8/8 (100%)	1/1 (100%)
7	9	6/5 (83.3%)	3/2 (75%)
2	9	8/8 (100%)	1/1 (100%)
13	6	5/3 (60%)	1/1 (100%)
1	10	6/6 (100%)	4/4 (100%)
Totals	129	51/45 (88.2%)	73/62 (85%)

especially those in whom the rupture results from intubation, are suitable for surgical repair. Surgery in these patients is of high risk, with a reported mortality of approximately 70 per cent.[6] Such a high mortality in critically ill patients receiving mechanical ventilation demands that alternatives to high-risk surgery be considered.[5]

In selected cases, iatrogenic tracheal lacerations may be treated in a conservative way.[4] Criteria for patients' operability are debatable. Criteria in favour of conservative treatment are the presence of oesophageal injury, mediastinitis, progressive pneumomediastinum, respiratory and/or cardiac and/or haemodynamic instability, or active sepsis.[4] Some authors consider the size and the depth of the defect as important criteria, advocating surgery whenever full-thickness or lengthy lacerations are present.[1,7]

Survival after iatrogenic tracheobronchial rupture is related to the extent of the rupture, the neighbouring organs involved, the complications that occurred before and after treatment, as well as the underlying disease. The mortality rate is between 20 and 71.4 per cent with a conservative approach,[5,6] and between 10 and 55 per cent for surgical repair (Table 13.1). However, data on the conservative approach concern mainly critically ill patients, with severe underlying disease, being intubated in an emergency situation.[5]

Stenosis

Tracheal stenosis is another iatrogenic situation related to multiple attempts at emergency intubation, prolonged endotracheal intubation, tracheostomy and/or excessive inflation of the cuff in mechanically ventilated patients. Its incidence is reduced by the use of low-pressure endotracheal tubes. Other rare causes are sleeve resection and pulmonary transplantation.[14] The incidence of tracheal stenosis is higher than that of tracheobronchial ruptures. In prospective studies it has been shown that stenosis occurs in 10–19 per cent of intubated patients. Tracheal strictures usually occur on the first 2–3 cm of trachea, under the vocal cords. The pathogenesis is related

to the chronic inflammation that the tube and/or cuff produce on the trachea leading to wall damage (malacia). Histological findings are wall alterations of cicatricial tracheal stenosis with subacute mucosa and submucosal inflammation, chondritis and degeneration of the cartilaginous support.[15]

Strictures may be associated with granuloma formation on the tracheal wall.[16,17] Granulations may also be the result of the endotracheal/endobronchial treatment of either iatrogenic stenosis or tracheomalacia occurring after stent placement, but also after surgical resection. Granulation tissue is composed of various inflammatory cell types including plasma cells, lymphocytes, eosinophils, histiocytes, foamy macrophages and neutrophils, embedded in oedematous, mucinous material with a great number of capillaries. Mitoses are absent.

Symptoms in tracheal stenosis include dyspnoea in 80 per cent of cases, cough in 45 per cent, wheezing, and stridor. Haemoptysis may be associated to cough in less than 10 per cent of the cases.[18] Diagnosis may be confused with asthma. The history of recent endotracheal and/or endobronchial intervention together with the characteristic pattern of flow/volume curve will support the diagnosis, which is confirmed by fibre-optic bronchoscopy. Fibre-optic bronchoscopy will report the length of the stenosis, its distance from the vocal cords and carina, and its form.[19] This information is important for accurate treatment.

Two types of stenosis can be observed. A 'web-like stenosis' is a short (<1 cm) membranous stenosis without damage to the cartilages. A complex tracheal stenosis is longer, with circumferential hourglass-like contraction, scarring, or malacia without recognition of the typical tracheal structure with the rings. Surgical sleeve resection is considered the standard treatment.[19]

Both flexible and rigid bronchoscopy are adequate to treat tracheobronchial iatrogenic stenosis.[20] Flexible bronchoscopy is usually performed under local anaesthesia, but where there is a therapeutic indication the patient should be sedated to avoid cough and discomfort that may affect the outcome of the procedure. Rigid bronchoscopy is performed under general anaesthesia, with either spontaneous ventilation or jet ventilation. The main goal of rigid bronchoscopy is interventional bronchoscopy. In therapeutic bronchoscopy, the large size of the operative channel of the rigid bronchoscope allows an immediate aspiration and the introduction of rigid instruments such as forceps, rigid needles, rigid cryoprobes or stents.[20] Owing to the rigid structure, the bronchoscope can easily control a bronchial bleeding with tamponade and open the bronchial stenoses with gentle dilatation.

The membranous web-like short stenosis, without cartilage damage, is the indication for endobronchial therapy. In tracheal stenotic lesions, the laser-assisted resection can be curative.[21] High-frequency electrocoagulation is performed with rigid or flexible monopolar probes. The thermal action of these probes, introduced in the tissue, induces retraction and dehydration. Results and immediate efficacy are comparable to laser action. Iatrogenic tracheal stenosis, with or without granulomas, are also an excellent indication of cryotherapy.[20]

Endotracheal stents may also be placed in the trachea, after local treatment of the stenosis by electrocautery or laser or cryotherapy and successive dilatations.[21] For iatrogenic tracheal stenosis we mostly use silicone-type Endoxane stents, which we introduce through the rigid bronchoscope after stenosis dilatation. In cases of inoperable benign tracheal stenosis associated with trachea wall lesions, the indication of stent placement is required.[22] The stent should be removed 1–2 years later, resulting in definitive resolution of the stenosis in up to 50 per cent of cases. The main complications are a high migration rate (in about 18 per cent) and development of granulomas (in 15 per cent).[23] However, newer silicone stents have a low rate of granuloma formation (7 per cent), and no migration has been reported.[24] Metallic stents have a low migration rate, but the same rate of granuloma formation,[18] and their extraction after a long period is difficult.[25] However, some reports show positive results in extracting metallic covered stents with the newer nitinol stents. Therefore, careful choice of the stent is important for accurate treatment.

The indication for surgical resection of iatrogenic tracheal stenosis is classically the complex stenosis with circumferential hourglass-like contraction, scarring, or malacia affecting tracheal cartilage.[26] Surgical sleeve resection of the tracheal rings, with primary end-to-end anastomosis, is considered the standard curative treatment.[27,28] However, in some patients (10–15 per cent), even after radical surgical sleeve resection, tracheal stenosis will recur.[29,30] In patients with web-like stenosis, when repeated accurate endotracheal treatment is conducted with laser dissection, dilatation and stent placement, failure has been noted in 5 of 15 patients (30 per cent); two had definitive tracheostomy, two had sleeve resection, and one died from disease unrelated to the procedure (carcinoma).[26] In patients with complex stenosis, when repeated endotracheal treatment was performed with dilatations and stent placement, 12 out of 18 patients were finally cured (6 of them were unsuitable for operation and had a definitive stent placement), and 4 patients had to undergo a surgical sleeve resection as salvage treatment.[26]

While this less aggressive, low-risk approach is more acceptable for elderly patients, definitive surgical treatment is probably more appropriate for younger patients who are less likely to tolerate prolonged limitations in their lifestyle. However, patients with severe inflammatory changes must be excluded from surgery and treated only when the stenosis appears stabilized after a suitable period of observation. In fact, restoration of a healthy mucosa is mandatory to obtain good results. In this setting, endoscopic management may be particularly valuable to provide time to reduce the damage due to inflammation and oedema, while awaiting definitive surgical treatment (Tables 13.2 and 13.3).

PARENCHYMAL LUNG INVOLVEMENT

Iatrogenic parenchymal injuries

Iatrogenic injuries of the lung parenchyma most commonly follow diagnostic needle punctures of the lung, such as percutaneous transthoracic biopsies, or therapeutic procedures, such as tube thoracostomy and surgical interventions.[35] Transbronchial biopsies produce relatively small parenchymal trauma that can be complicated rarely by bleeding and pneumothorax.[36–38]

Table 13.2 Outcome of patients with tracheobronchial stenosis treated conservatively

Reference	Patients	Cured
26	33	22 (70%)
21	81	56 (69.1%)
31	14	13 (92.8%)
24	13	9 (69.2%)
23	11	8 (72.7%)
34	19	19 (100%)
Totals	179	156 (87.1%)

Table 13.3 Outcome of patients with tracheobronchial stenosis undergoing surgery

Reference	Patients	Cured
27	200	175 (87.5%)
28	120	117 (97.5%)
32	11	8 (72.7%)
33	18	18 (100%)
30	28	25 (89%)
29	503	471 (93.7%)
Totals	880	814 (92.5%)

Iatrogenic pulmonary parenchymal injuries are rare complications with an incidence that varies widely according to the procedure. Transthoracic needle biopsy of lung lesions can be complicated by pneumothorax and/or pulmonary haemorrhage with rates that reach over 30 per cent, although pneumothorax that needs drainage occurs in 5–12 per cent and haemorrhage/haemoptysis in 2 per cent; morbidity is very low (0.1 per cent).[36–38] Iatrogenic injuries of the parenchyma of the lung produced by percutaneous tube thoracostomy (PTT) may be less frequent in the clinical practice but can be more serious than transthoracic needle biopsies.[35] Nevertheless, PTT remains the most widely performed procedure to manage blunt and penetrating chest traumatisms and complicated pleural effusions. The reported incidence of lung injury due to PTT in large retrospective series ranges from 0.2 to 4 per cent, while the mortality attributable to complications related to PTT in this population has been reported at 0–0.7 per cent.[39]

However, the incidence of lung injury related to PTT – or secondary to surgical interventions – may be underestimated because lung injury was not evaluated systematically by computed tomography in published series. Many lung injuries may remain unrecognized, corresponding to cases with unexplained sustained air leak or recurrent pneumothoraces. Indeed, previous investigations and reports suggest that up to 24 per cent of cases with these two complications might have been secondary to chest tube malpositioning and erosion of the lung by the tube. Regarding transbronchial biopsies, lung laceration and serious lung injury are practically impossible. The incidence of serious moderate self-limited bleeding has been reported as 15 per cent,[36] while the incidences of brisk haemorrhage or pneumothorax requiring drainage are very low (1 and 2–3 per cent, respectively).[36–38]

Iatrogenic injury of the lung parenchyma results from unintentional application of exogenous force through rigid, sharp instruments (trocar, biopsy needles, surgical instruments) on lung tissue, during surgical procedures, tube thoracostomy, and lung biopsies. The result is either contusion of lung parenchyma and/or pulmonary haemorrhage when a vein or artery is damaged during the procedure. Consequently air leak, pulmonary bleeding, parenchymal contusion and rarely air embolism may occur. Given concerns about damage to the underlying lung when using a rigid sharp trocar or instruments, both surgical and medical techniques have been advanced. Regarding percutaneous trocar insertion for pleural drainage, many physicians prefer the blunt dissection technique of the chest wall and digital exploration of the pleural space, or the use of small indwelling catheters through a guide-wire (Seldinger technique) to diminish the rate of adverse outcomes.[39,40]

However, experience and expertise are still required to know when such treatment is necessary and how to place a tube adequately with minimum or no complication. Imaging helps in determining the need for, and the optimal positioning of, the thoracostomy tube.[40] Nevertheless, we should acknowledge that the decision to place a thoracostomy tube in emergency cases (ie. tension pneumothorax, chest injuries) may need to be made before an imaging procedure can be obtained and the risk of lung perforation may be increased. These cases are not rare; in previous published series they represented up to 20 per cent of the cases which needed to be treated with a thoracostomy.[35]

Air leak, pulmonary vascular injuries and parenchymal contusion are the most common causes of the clinical manifestation of iatrogenic injury depending on the causal mechanism and the underlying condition of the lung. Patients usually experience respiratory or general symptoms such as dyspnoea, cough, pleuritic pain, haemoptysis, haemodynamic instability, hypoxaemia or problems in ventilation if they are supported by mechanical ventilation.[35] The severity of the clinical manifestations of parenchymal injury is affected by the volume of contusion and vascular injury. Alveolar haemorrhage and parenchymal destruction are maximal during the first 24 hours after injury and then usually resolve within 7 days.

Respiratory distress is common after lung trauma, with hypoxaemia and hypercapnia greatest after about 72 hours.[41] Patients may rarely experience severe illness or a more globular or diffuse pattern of injury (ALI or ARDS) which may require acute resuscitation and advanced life support measures.[42] However, physicians should be alert to the potential risk and significant mortality that accompany pulmonary contusions.[43] Pulmonary vascular injuries may result in bleeding and rarely in development of pseudo-aneurysm, and these incidents can lead to significant blood loss. In some cases simultaneous hydro/haemo-pneumothorax, and injuries to intercostals or intrathoracic vessels, diaphragm and sympathetic chain, may accompany parenchymal injuries. Secondary complications such as residual or recurrent pneumothoraces or nosocomial infections may follow lung parenchymal injury depending on the causal mechanism, the severity of trauma and the underlying condition/disease; their rate ranges between 9 and 36 per cent.

In all these cases the complication is immediately apparent during or after the procedure. The sudden development of extensive extra-alveolar air concomitant with tube placement indicates an air leak that may be secondary to laceration of lung tissue. However, it should be emphasized that complications may initially be clinically unapparent,[35] or physicians may not recognize them in the context of pre-existing pulmonary and pleural disease.[44] Haematoma around the chest tube or the site of percutaneous transthoracic needle biopsy does not always occur in iatrogenic injuries due to drainage tube thoracotomy, so clinical manifestations may be subtle. Patients may even experience marked clinical and radiographic improvement immediately after the procedure, followed later by clinical and/or radiographic deterioration.[35,44]

A plausible explanation for this discrepancy may be that the tube initially evacuates air or pleural fluid as it passes through the pleural space on the path to or from the parenchyma. Radiography and particularly CT may be helpful for diagnosis of lung injury. Lung laceration can be recognized easily by an abrupt or gradual increase in pulmonary or pleural opacity after a diagnostic or therapeutic procedure.[44] The same is true if pneumothorax and/or chest wall emphysema develop or dramatically increase following the procedure.

However, injuries due to malpositioning of a chest tube may be difficult to spot, particularly when only anteroposterior portable chest radiographs are available. Such films have indicated malpositioning only in a minority of patients with confirmed malpositioning by CT scan. Posteroanterior and lateral examinations frequently provide considerably more information than do anteroposterior films and CT may be helpful in more complicated cases.[40] However, diagnostic difficulties may exist even when CT is available. Fissures may be difficult or impossible to delineate on CT, especially if pulmonary consolidation is present, and further uncertainty may exist as to whether the tube penetrates or impresses a peripheral pleural surface.

The overall condition of the patient and the type and severity of the pulmonary injury produced by lung penetration will dictate further care. Most patients eventually respond to measures that have already been planned, such as scheduled drainage after operation on the lungs, and then further surgical therapies are not required. In patients with iatrogenic injuries of the lung after tube thoracostomy, removal of the tube alone can be successful and additional drainage or surgery may be avoided.[39,45] In cases with prominent lung injury it may be of value to quantify and note changes in the extent of the pulmonary contusions and PaO_2/FIO_2 ratio during the first 24 hours in order to determine the need for ventilatory assistance and to predict outcome.[43] In more complicated cases or when secondary complications occur, such as retained haemothorax or empyema, subsequent drainage, management of infection or even surgical management including lobectomy may be necessary.[44] Thus, 4–16 per cent of patients who had an injury secondary to placement of a thoracostomy tube required further drainage, while 1–3 per cent required an operative intervention to repair lung lacerations in recent series.[35]

Lung injury secondary to positive–pressure ventilation

It is well known that ventilatory strategies, used to improve oxygenation and gas exchange in critically ill patients, are also critical for patients' outcomes. Large volumes and/or high positive pressures generated by a ventilator have the potential to lead to shear injury, atelectasis, trauma, collapse, reopening and over-distention of lung units which all are critical determinants in generating lung injury.[46]

Mechanical ventilation also causes subtle injury (biochemical injury, otherwise called barotrauma) – the release of mediators from the lung.[47,48] This applies to all mechanically ventilated patients and has important clinical implications when excessive volumes or pressures are used. Inflammatory mediators such as TNF-α produced by the epithelial lining cells have been found to be increased in the bronchoalveolar lavage (BAL) fluid in subjects treated with high tidal volumes (15 mL/kg body weight) compared to those ventilated with lower volumes (6 mL/kg). The release of these mediators occurs within a few hours of mechanical ventilation. These mediators can attract neutrophils and other inflammatory cells that worsen the lung injury.

The clinical importance has been well demonstrated in ARDS. The ARDS Net Study I demonstrated a 22 per cent relative decrease in mortality in the group treated with a tidal volume of 6 mL/kg predicted body weight, versus 12 mL/kg.[46] The reason for the decreased mortality in the smaller tidal-volume group is not known, but it was *not* due to reduced barotrauma as the incidence of that was essentially identical in both groups. Moreover, it was *not* due to differences in oxygenation; in fact, the lower tidal-volume group, which had the better survival, had lower PaO_2/FIO_2 ratios for the first couple of days than the higher-tidal volume group. The ARDS Net investigators suggested that this difference in mortality may be due to differences in mediator release; levels of IL-6 decreased significantly more quickly over time in the lower tidal-volume group.

Finally, changes in intra-thoracic pressure during mechanical ventilation may have a number of effects, including an increase in alveolar–capillary permeability, a decrease in cardiac output, and a decrease in organ perfusion. These can all lead to end-organ dysfunction as well.

Miscellaneous causes of parenchymal injury

Laparoscopic adjustable gastric banding has become the most common bariatric operation in recent years owing to its low perioperative comorbidity, short hospitalization and low complication rate. Published evidence showed that most early and late complications were band-related complications (such as band slippage, pouch dilation, band erosion and wound-related). However, a recently published 10-year follow-up study on patients after the LAP-BAND operation demonstrated that 54 per cent of the patients experienced postoperative complications requiring hospital treatment. The most important late complications were oesophagitis (30 per cent) and obstruction due to pouch dilatation (21 per cent). However, chronic cough and community-acquired lung abscess has been reported rarely in previously well immunocompetent patients, and its occurrence was suggested to be related to oesophageal reflux disease and aspiration of gastric contents.[49]

Percutaneous vertebroplasty with cement is a technique that was developed to treat pain in patients with metastases to the spine.[50,51] The operation may halt the progression of vertebral collapse by strengthening the vertebral body, and it can provide within the first 48 hours an analgesic effect that persists for at least 6 months. It achieves spine stabilization by preventing the micromovements responsible for vertebral pain. Although it is considered as a safe technique, adverse events secondary to this method have been reported. Local complications may be due to foraminal venous leakage, causing radicular irritation, owing to polymerization-generated heat or ischaemia. Systemic complications may arise due to cement migration through the vena cava. Although these complications are rare, percutaneous vertebroplasty has been associated with clinically full-blown embolism and at least one fatal case has been reported in the literature.

PLEURAL INJURY

Iatrogenic injury of the chest may result in accumulation of air or fluid or both in the pleural cavity without significant concomitant injury of the lung parenchyma or the chest wall. Iatrogenic pneumothorax or haemothorax may be the result of various diagnostic or therapeutic procedures.

Iatrogenic pneumothorax

Pneumothorax is defined as the presence of air in the pleural cavity. An iatrogenic pneumothorax occurs secondary to a medical or surgical procedure. Most common causes are transthoracic needle aspiration or biopsy, percutaneous tube thoracostomy, transbronchial biopsy, central venous cannulation and central venous port placement in oncology patients, thoracocentesis, pacemaker placement, colonoscopy and laparoscopy, mechanical positive-pressure ventilation and non-invasive positive-pressure ventilation, especially in cystic fibrosis patients, cardiopulmonary resuscitations, and percutaneous radio-frequency ablation of lung neoplasms. Less common causes are tracheostomy, gastric tube placement, colonoscopy, pericardiocentesis and acupuncture.[52]

The incidence and causes of iatrogenic pneumothorax vary considerably according to the causal procedure. The rate of pneumothorax after transbronchial biopsies is 2–7.5 per cent.[36] Percutaneous radio-frequency ablation of lung neoplasms is associated with rates that reach 30 per cent and a minority of them require chest tube placement as therapy.[53] The incidence of iatrogenic pneumothorax in the intensive care setting following diagnostic procedures (or rarely endotracheal intubation) was reported as 3–15 per cent.[54] In ARDS patients due to barotrauma, the rates may be as high as 87 per cent and the development of a pneumothorax in this group of patients appears to be an independent predictor for mortality.

Pneumothorax in the setting of non-invasive ventilation (NIV) is not frequent and it mainly affects patients with cystic fibrosis. Cystic fibrosis is characterized by progressive pulmonary infection and apical subpleural bronchiectatic cysts that often rupture. However, the incidence of iatrogenic pneumothorax in patients using NIV is less than 5 per cent, which is similar to pneumothorax rates with cystic fibrosis patients who do not use non-invasive ventilation. In the future the incidence of NIV-associated pneumothorax may increase as NIV is used more widely in patients with cystic fibrosis.[55] Pneumothorax after endotracheal tube suctioning has been reported, especially in infants.[56] Central venous catheter placement for total parenteral nutrition, chemotherapy, or vasopressors, and central venous pressure monitoring may be complicated by pneumothorax in 1–6 per cent of patients.[57]

Post-acupuncture pneumothorax has been reported and in some cases the outcome was fatal. Thus, this complication may be serious although under-reported. The choice of acupoints and the technique of acupuncture are the realm of this specialty. However, certain acupoints are located in potentially dangerous areas of the body. 'Jianjing', for instance, is situated right over the apex of the lung and is associated with the risk of pneumothorax if the needling technique is faulty. Therefore these complications may be avoided when therapists have sufficient basic medical knowledge.[58]

The most plausible mechanism for iatrogenic pneumothorax is that the rupture of pulmonary parenchyma and pleura produces air leak into lung interstitium and pleural space.[59] High alveolar pressures – as in mechanical ventilation – may induce further air leak and pneumothorax development. There are no large studies investigating the factors that are associated with higher incidence of iatrogenic pneumothorax. It seems reasonable that the presence of obstructive airways disease and emphysema,[60,61] intractable coughing, increased duration of the procedure, cavitary or small lesions and positive-pressure air ventilation are parameters that have to be taken into account to minimize the risk for iatrogenic pneumothoraces. No increased risk of pneumothorax has been noted secondary to parenchymal biopsy with the lack of experience by the operator, and multiple numbers of needle passes. In contrast, for central venous catheterization, the experience of the operator, difficult placement with multiple numbers of needle passes, and placement on the left side seem to be factors with higher risk of iatrogenic pneumothorax.[62]

The natural history of iatrogenic pneumothorax may be different from that of other types. The needle puncture site usually heals within 24 hours so the outcome is expected to be better in comparison with primary or secondary pneumothorax. Patients with iatrogenic pneumothorax can be asymptomatic or they may experience dyspnoea and chest pain, depending on the amount of air in the pleural space and the presence or not of underlying disease such as emphysema or simultaneous parenchymal injury.[59] It should be underlined that, although iatrogenic pneumothorax appears usually within the first hour of the intervention, late-onset pneumothorax has been reported in many cases, especially in central venous catheterization.

Delayed development of pneumothorax following central vein catheterization is a less commonly recognized yet well-documented complication that can develop days after placement and occurs in fewer than 0.5 per cent of all central venous access attempts. The onset of delayed hydrothorax has been typically described as a result of initial catheter misplacement or delayed perforation of the vessel wall by the catheter tip rather than extrusion of the proximal catheter port. Pneumothoraces in the mechanically ventilated patient tend to enlarge rapidly and frequently develop tension (60–96 per cent). The risk of pneumothorax increases with time on the ventilator and with the presence of underlying lung disease.[63]

Diagnosis of iatrogenic pneumothorax is established when there is evidence of accumulation of air in the pleural cavity on clinical examination or in chest radiography following a medical or surgical procedure.[59] Many authors have suggested that postprocedural radiography is not indicated following catheter replacement over a guide-wire or following a clinically uneventful placement of a catheter by an experienced operator with limited (< 3) needle passes. However, it should be underlined that catheter malpositioning cannot be accurately identified clinically. In the intensive care setting where even a small pneumothorax is significant, detection on the portable ICU radiograph is difficult. In addition, if there is a densely consolidated lung or severe chronic lung disease, the radiographic appearance and distribution of a pneumothorax will be altered. In these patients, the lung will often collapse in a non-uniform manner, and areas of aerated lung may cause the appearance of lung markings lateral to the pleural margin. In patients with ARDS, the radiological signs of a tension pneumothorax may not be obvious. Thus, up to 30 per cent of pneumothoraces are missed on the supine or semi-erect radiograph, with about half progressing to tension.[64]

Chest computed tomography (CCT) is more sensitive than simple radiography for the diagnosis of small pneumothoraces and may be helpful when there are comorbidities. Although there is no consensus for the systematic application of CCT as a diagnostic tool in these cases, any new hyperlucency seen on the chest radiograph of a rapidly deteriorating mechanically ventilated patient should be considered a pneumothorax until proven otherwise by CCT.[64]

Treatment of a patient with iatrogenic pneumothorax is variable. Most patients require treatment for 4–7 days; Sassoon et al. reported prolonged hospitalization in 8 per cent of patients.[60] Development of treatment protocols for pneumo- and haemothorax is complicated by insufficient tracking and reporting of these iatrogenic events. It has been recommended that observation and oxygen supplementation may be effective in pneumothorax management (Box 13.1).[65] However, the average rate of absorption of air from the pleural cavity is very slow. The lung has a mean rate of re-expansion of about 2 per cent per day. Therefore, observation is a conservative approach that can be considered for patients with first episodes of iatrogenic pneumothorax who are asymptomatic

---

Box 13.1   Clinical criteria for observation of pneumothorax irrespective of cause (modified from [52])

Clinical stability
- Stable vital signs
- Ability to speak in full sentences
- Normal oxygenation

Lack of progression
- Unchanging pneumothorax signs over 4 hours

Access to emergency care
- Proximity to emergency care
- Sufficient resources to return for worsening cases

and have a small pneumothorax, which corresponds to either:

- less than 20 per cent as estimated by Light's formula (size as a percentage $= [1 - (\text{lung diameter})^3/\text{hemithorax}^3] \times 100$); or
- a small rim of air (distance of cupula to lung apex $< 3\,\text{cm}$).

If a patient has minimal symptoms or a larger pneumothorax, air aspiration is recommended.[65]

In primary spontaneous pneumothorax, simple manual aspiration has proved safe and effective in up to 60 per cent of patients, allowing half of the population to be managed in the emergency room. The use of a small-calibre indwelling catheter attached to a one-way valve or to a water-seal device allows air evacuation to continue if the initial manual aspiration attempt fails. Especially for patients with underlying diseases such as emphysema and sustaining iatrogenic pneumothorax, chest tube placement is appropriate. Small-bore tubes have proved successful in mixed pneumothorax populations, and it was shown recently in a prospective study that a high 1-week success rate (85 per cent), with no or short hospitalization, and a reasonable 1-year recurrence rate (24 per cent) can be achieved. In this respect, a simplified stepwise therapeutic approach that includes initial observation or simple aspiration followed by chest drainage (or chest drainage from the beginning) followed later by further interventions seems a reasonable approach to management that may minimize morbidity, limit the cost of treatment and provide a rationale for future guidelines.[65]

Iatrogenic pleural effusion

In some cases, diagnostic and therapeutic procedures are responsible for abnormal accumulation of fluid in the pleura.[66,67] The incidence, the characteristics and the outcome of iatrogenic pleural effusions vary widely according to the cause of the effusion (Box 13.2). Pleural effusion may complicate the percutaneous insertion of central venous catheters – usually subclavian – owing to vascular erosion or to late migration of the catheter, or to aortocoronary bypass. Half of the hydrothorax cases associated with placement of central lines occur in the first 2 days of placement and half later. The incidence is estimated as 0.4–1 per cent. It is usually associated with iatrogenic pneumothorax. Symptoms and signs are similar to those of pleuritis. Dyspnoea and chest pain are the most common symptoms, while 10 per cent of patients may experience no symptoms. Tension hydrothorax has been described as well. The onset of the complication may be acute or indolent and symptoms may be delayed up to 60 days. This is a major cause of delays in diagnosis (0–11 days) or misdiagnosis.[67]

Diagnosis is mainly based on radiography and occasionally on the characteristics of the accumulated fluid (milky fluid, high concentration of glucose according to the type of parenteral fluids used). The pleural fluid has been reported as clear, milky, serous, sanguineous or haemorrhagic. Demonstration of extravascular extravasation of radio contrast after injection of contrast material is another way to diagnose directly the occurrence of such a complication.[67]

Treatment of pleural effusion secondary to central venous catheters is based on immediate removal of the catheter and

aspiration of the fluid. The insertion of a chest tube for pleural drainage should be recommended in patients with respiratory compromise or haemothorax.[67]

The institution of nasogastric tubes for enteral feeding has the potential for complications, especially when metallic tips and stiffening introducing stylets are used.[68] Perforation of the oesophagus by the feeding tube has been sporadically reported but malpositioning of the tube into the tracheobronchial tree is less rare. In the latter case, a malpositioned tube could perforate into the pleural space, with potential development of pneumonitis and empyema. Therefore, institution of the tubes and confirmation of desired position, especially in a critically ill patient, are mandatory before feeding is initiated.

The incidence of pleural effusions immediately after surgery in patients undergoing coronary artery bypass grafting (CABG) has been reported to be 42–89 per cent. This wide variability in the incidence of post-CABG effusions is probably related to the technique used to diagnose the pleural effusion because higher rates have been reported when ultrasound or computed tomography is used. The incidence of pleuropulmonary complications is higher with use of inferior mesenteric artery (IMA) grafts compared with saphenous vein grafts.[69]

Most effusions are diagnosed within the first 2–3 weeks after surgery. Patients with late effusions have presented 40–90 days after surgery. Most post-CABG effusions are small, are left-sided, and regress spontaneously. Occasionally a patient may develop a moderate to large symptomatic pleural effusion after CABG. The prevalence of large effusions after CABG is not definitely known but has been reported to be 1–4 per cent. Early post-CABG effusions are more bloody (median haematocrit of 0.20), with high numbers of both red and white blood

---

Box 13.2 Most common iatrogenic causes of pleural effusion and haemothorax (modified from [67])

Causes of pleural effusion

- Vascular erosion by central venous catheters
- Perforation of pleura with a nasogastric tube
- Translumbar aortography
- Mammoplasty
- Radiation therapy
- Endoscopic oesophageal sclerotherapy
- Ovarian hyperstimulation syndrome
- Fluid overload
- Coronary artery bypass surgery
- Abdominal surgery

Causes of haemothorax

- Vascular erosion by central venous catheters
- Chest tube placement
- Cardiac or thoracic surgery
- Transthoracic biopsy of pleural or lung lesions
- Thoracocentesis
- Transbronchial biopsies

cells, mostly neutrophils and eosinophils. Late effusions show a lymphocyte predominance.[70,71]

The pathogenesis of the post-CABG effusion remains obscure. The pathogenesis of early effusions is probably related to trauma during surgery that results in bleeding into the pleural space. Late pleural effusions probably have a different pathogenesis. Because of the late presentation, it is unlikely to be related to blood in the pleural space at the time of surgery. The high lymphocyte count suggests an immunologic aetiology similar to postcardiotomy syndrome and Dressler syndrome. It has been suggested that late lymphocytic pleural effusions after CABG may represent a subset of patients with a variant of, or a limited type of, postcardiotomy syndrome involving only the pleura. However, several other factors may also contribute to the formation of effusions in patients undergoing CABG. These may include congestive heart failure, pulmonary embolism, atelectasis and the use of drugs such as amiodarone, procainamide hydrochloride, and beta-adrenergic blocking agents. It is therefore important to include these in the differential diagnosis of effusions occurring in patients who have undergone CABG. Patients should be managed initially with therapeutic thoracocentesis and an anti-inflammatory agent such as indomethacin which is particularly useful in late pleural effusions, where an immunologic basis for the effusion is speculated. However, if the effusion recurs, a second therapeutic thoracocentesis with the possible addition of oral prednisone therapy for a brief period is recommended. When the effusion continues to recur, appropriate investigations to exclude other causes of effusion and then more aggressive measures such as tube thoracostomy with pleurodesis should be considered.[70,71]

Radiation exposure can cause pleural effusion, either secondary to radiation pleuritis and systemic venous hypertension or by lymphatic obstruction from mediastinal fibrosis. It occurs in about 10 per cent of irradiated patients.[72] It occurs simultaneously with radiation pneumonitis and the pleural fluid has no special characteristics. The pleural effusion develops usually within 6 months of completing radiation (4000–6000 rad) but it may also occur 1–2 years following intensive mediastinal radiation.[73] Pleural effusions are usually small and rarely reach half of the hemithorax. The pleural effusion is usually resolved spontaneously, gradually and slowly. It may take up to 40 months for the fluid to disappear.

Haemothorax

Iatrogenic haemothorax is defined as accumulation of pleural fluid with an haematocrit at least 50 per cent that of peripheral blood, secondary to a medical or surgical procedure. The incidence of iatrogenic haemothorax is lower than that of iatrogenic pneumothorax but it is difficult to estimate precisely since these events are not registered systematically. Most common causes are thoracic surgery, placement of chest tubes or central venous catheters, thoracocentesis, and transthoracic or transbronchial biopsies.[67] Sources of bleeding include the thoracic cage, lungs, the diaphragm, blood vessels and mediastinum. Significant bleeding occurs when the intercostal arteries or the internal mammary artery are damaged. Although rarely, anticoagulation (heparin, aspirin, ticlopidine) should be also

considered as a potential cause of haemothorax including those massive in size, usually occurring within the first week of therapy for thromboembolism.[67] Anticoagulation can be associated with life-threatening haemothorax which complicates chest tube placement.

Symptoms and signs are similar to those of pleuritis and of bleeding. General symptoms depend on the amount of blood in the pleural space and range from minimum to life-threatening shock. Loss of more than 250 mL gives symptoms that help in diagnosis, whereas with massive haemothorax pallor, hypotension and cardiovascular collapse may be present. Haemothorax has the potential to be complicated by clotted or retained haemothorax, empyema, fibrothorax or trapped lung depending on the underlying condition and management.

The radiographic findings of haemothorax correlate with the amount of fluid accumulated. However, this pointer may not be sensitive in small pneumothoraces or when another lung disease is also present. In such cases CT of the chest can provide further information.[74] In cases where central line placement is complicated by haemothorax, radiographic findings may appear contralateral to the site of the line. Thus, contralateral radiographic findings should not dissuade consideration of a procedure-associated pneumothorax.

In the setting of haemothorax the main therapeutic options are chest tube placement, thoracoscopy or video-assisted thoracic surgery and early or late thoracostomy. Pleural drainage allows removal of the blood with monitoring of blood loss and can be helpful to decide whether further surgical interventions are necessary. Rapid loss of 1–2 L of blood or ongoing blood loss of exceeding 200 mL/h through a chest tube is an indication for thoracostomy. In cases of severe blood loss, enough blood is matched and general measures for resuscitation are required to correct hypovolaemia and to support an unstable patient.

CARDIOVASCULAR INJURIES

Cardiac catheterization involves some risks, as with every invasive procedure. The most common complications resulting from cardiac catheterization are vascular-related, including external bleeding at the arterial puncture site, haematomas, and pseudoaneurysms. Less frequent complications include cardiac arrhythmias, pericardial tamponade and renal failure. The most serious complications are stroke, myocardial infarction and death resulting from clotting or rupture in one of the coronary or cerebral vessels.[57] The risk of complications from cardiac catheterization is higher in patients over the age of 60, in those who have severe heart failure, and in those with advanced valvular disease.

Pulmonary artery catheterization is a known cause of iatrogenic lung injury. Pulmonary catheter misplacement has been reported to occur in 19 per cent of internal jugular vein and 16 per cent of subclavian vein cannulations.[75] Following internal jugular cannulation, the most common complication is inadvertent puncture of the internal carotid artery,[76] occurring in 10 per cent of procedures. Moreover, distal placement of the catheter may lead to vascular trauma and thromboembolic events. Rupture of the chordae of the tricuspid valve and rupture of the pulmonary artery are uncommon. The latter may be associated with high mortality. A broken Swan–Ganz catheter fragment may also end up in the right pulmonary artery, representing a foreign body. If the fragment catheter remains in place then the patient is at risk for infection, sepsis and ultimately death.

A large survey of pulmonary artery catheterization in 6245 patients revealed a low incidence of morbidity associated with the procedure.[77] The incidence rate of intrapulmonary haemorrhage was 0.064 per cent, of minor pulmonary infarcts 0.064 per cent, perforation of right ventricle in one patient 0.016 per cent, and death from uncontrollable pulmonary haemorrhage in another patient (0.016 per cent). Patients who undergo cardiac surgery or patients with pulmonary hypertension are at increased risk to present these complications. Anticoagulation may compound the risk.

Asymptomatic ruptures can also occur and they may present with sudden appearance of localized air-space opacification around the catheter tip, representing focal haemorrhage. However, this finding may not be obvious since it may be obscured by extensive pulmonary oedema. Another complication is the pulmonary artery pseudo-aneurysm which is a late complication of pulmonary artery rupture, and most of them are diagnosed by 2 weeks. The pseudo-aneurysm is identified as a well-defined nodule in an area of focal lung consolidation.[78] The diagnosis will be confirmed by contrast-enhanced CT that reveals a characteristic halo of ground-glass opacification which represents surrounding haemorrhaging. The rate of re-bleeding is high (of the order of 30–40 per cent) and the mortality rate approaches 70 per cent. The pseudo-aneurysm can be obliterated effectively with embolotherapy.

Thromboembolic complications have a reported incidence between 1 and 11 per cent. Thrombus formation may occur from persistent wedging of the catheter in small arteries, or it may originate from catheter contact with the vessel wall. This complication may lead to subsequent pulmonary embolism and pulmonary infarction. Catheter insertion or manipulation is rarely associated with cardiac perforation, which unfortunately is uniformly fatal unless promptly diagnosed. Haemopericardium rapidly develops and should be considered in any patient who presents with clinical deterioration and radiographic findings suggestive of cardiac tamponade.

Intra-aortic balloon pump placement for temporary circulatory support and management of cardiac ischaemia may be complicated with limb ischaemia in 11–33 per cent of cases.[79,80] Females and patients with diabetes or peripheral vascular disease are at greater risk. One-fifth of patients who present symptoms will develop a chronic limb, and one-third of the patients who develop limb ischaemia will require surgical intervention. Embolic events are rare. Infectious complications are associated with prolonged intra-aortic balloon pump support. On the other hand, bleeding is not infrequent and thrombocytopenia due to platelet consumption may rarely require balloon removal.

Cardiac laceration, pericarditis, haemopericardium and pericardial tamponade have been reported as rare iatrogenic complications.[81] Acute pericarditis following diagnostic or therapeutic procedures have been rarely reported shortly after endoscopic variceal sclerotherapy. In the case of sclerotherapy, the possible pathogenesis is the involvement of the pericardium in an inflammatory reaction that develops in the oesophageal wall and surrounding tissues. Symptoms presented by the patients vary. In a case of cardiac laceration the clinical presentation may be dramatic. In cases where pericarditis is the main issue, symptoms may be mild and this may result in underestimation of the clinical condition. However, prompt diagnosis, follow-up and treatment should be applied in cases with cardiac tamponade or constrictive pericarditis development. Late complications may be fatal and are consequent upon inflammation, sepsis or thromboembolism.

ESOPHAGEAL INVOLVEMENT AND MEDIASTINITIS

Iatrogenic oesophageal injury, perforation, esophageal fistulas and mediastinitis are rare complications that have been reported to be associated with diagnostic procedures like endoscopy or mediastinoscopy or endoscopic ultrasound-guided fine-needle aspiration, photodynamic therapy, endotracheal intubation, resuscitation and percutaneous dilatational tracheostomy. Mediastinoscopy is often performed to permit sampling of mediastinal lymph nodes or masses and surgical complications may occur in up to 9 per cent of patients.[82]

Although uncommon and rarely reported, pharyngoesophageal perforation following endotracheal intubation may result in severe airway complications that include mediastinitis, retropharyngeal abscess and pneumothorax and pneumonia. Laryngeal masks and Combitubes seem to carry higher risk compared to orotracheal tubes. The use of an intubating laryngeal mask has been reported to result in oesophageal perforation while an endotracheal tube was passed through the mask.[83] A retrospective analysis underlined that its use in the pre-hospital setting is associated with a notable incidence of serious complications (20.7 per cent) and complications directly related to trauma from the insertion of the Combitube were estimated as 4.3 per cent.

Fracture of the sternum during resuscitation may result in iatrogenic haematoma. Fracture of the thorax wall is not infrequent, but in most cases it has no adverse effects. Abscesses of the mediastinum, and empyema of both pleural cavities, have been rarely reported to follow resuscitation attempts.[82]

There are distinct clinical patterns of presentation, which depend on the level at which the oesophagus is perforated. Proximal thoracic perforations led to signs in the left chest region (effusion/pneumothorax) while with distal perforation the signs were on the right side. Depending on the defect size and the overall situation, symptoms and signs can be initially mild or severe. Pain, pyrexia and tachycardia are typical signs of mediastinal inflammation. Attention should be paid especially after difficult intubations because perforating traumas of the pharynx, larynx or oesophagus following intubation may temporarily remain undetected. Consecutive development of an abscess or mediastinitis can cause a potential vital threat to the patient who may have a poor prognosis. Although mediastinitis is a consistent finding, one should consider that pneumonitis, pneumothorax and pneumomediastinum may coexist.[82]

The majority of iatrogenic injuries to the oesophagus can be managed successfully by conservative measures and pleural drainage, with surgical procedures reserved for large disruptions of the esophagus, intra-abdominal perforations, and cases that do not respond to appropriate conservative measures. Patients with caustic injury to the oesophagus have a risk for the development of penetrating injury, and this may be one indicator of the severity of scarring. Dilatation disruption of a localized stricture has a good long-term prognosis for the oesophagus and may even cure the stricture.[84] Despite its rare occurrence, oesophageal perforation during intubation is associated with the poorest outcome, especially when the diagnosis and treatment are delayed.

DIAPHRAGMATIC INJURY

Iatrogenic diaphragmatic injury, drug-related ventilatory depression due to peripheral neuromuscular blockage or due to iatrogenic alterations of the central ventilatory command (central rhythmogenesis, central drive to breathe, transmission from the brain-stem to spinal motorneurons) may lead to diaphragmatic dysfunction. Anaesthetics, acute or chronic administration of neurotropic, non-anaesthetic agents, procedures that lead to ventilatory depression, hiccups, dyskinesia, iatrogenic alterations of the transmission of the ventilatory command, drug-related myopathy (i.e. corticosteroid myopathy), iatrogenic trauma of roots and nerves, iatrogenic diaphragm fatigue and respiratory muscle injury due to mechanical ventilation are all potential causes of diaphragmatic dysfunction.[85] Injuries include diaphragmatic hernia following laparoscopic gastric banding, laparoscopic procedures, phrenic nerve injury after coronary artery grafting, unilateral diaphragmatic paralysis following bronchial artery embolization for haemoptysis, or supraclavicular brachial plexus block.

When there are respiratory symptoms that are associated with alveolar hypoventilation or a restrictive ventilatory defect, and in the absence of parenchymal or pleural abnormalities on the chest radiograph, iatrogenic causes must be evoked, exactly as they are in the presence of interstitial lung disease. When diaphragmatic dysfunction is traumatic, it may not be intrinsically lethal and the incapacity incurred by unilateral paralysis is mild compared to the clinical spectrum of bilateral diaphragm paralysis.[86] Phrenic nerve injury may have a highly morbid effect in patients undergoing coronary artery bypass grafting, especially when there is another underlying pathology such as chronic obstructive pulmonary disease (COPD).

KEY POINTS

- The incidence of iatrogenic complications is probably underestimated in surveys because many go unregistered.
- Tracheobronchial ruptures are life-threatening lesions that may be a rare complication of tracheal intubation.
- They occur also after chest radiation therapy with or without chemotherapy in patients with bronchogenic carcinoma, sometimes associated with tracheoesophageal fistula.
- Tracheal intubation may be the cause of stenosis of the trachea.

- Surgical procedures may cause ruptures of intrathoracic organs or vessels, necessitating immediate re-intervention.
- Insertion of drainage tubes or biopsy devices for both parenchymal or pleural lesions may cause pulmonary, vascular or diaphragmatic injuries.
- Insertion of vascular catheters and pulmonary artery catheterization may lead to life-threatening vascular perforation.

REFERENCES

1. Borasio P, Ardissone F, Chiampo G. Post-intubation tracheal rupture: a report on ten cases. Eur J Cardiothorac Surg 1997;12:98–100.
2. Massard G, Rouge C, Dabbagh A, et al. Tracheobronchial lacerations after intubation and tracheostomy. Ann Thorac Surg 1996;61:1483–7.
3. Schneider T, Storz K, Dienemann H, Hoffmann H. Management of iatrogenic tracheobronchial injuries: a retrospective analysis of 29 cases. Ann Thorac Surg 2007;83:1960–4.
4. Lampl L. Tracheobronchial injuries: conservative treatment. Interact Cardiovasc Thorac Surg 2004;3:401–5.
5. Conti M, Pougeoise M, Wurtz A, et al. Management of postintubation tracheobronchial ruptures. Chest 2006;130:412–18.
6. Meyer M. Iatrogenic tracheobronchial lesions: a report on 13 cases. Thorac Cardiovasc Surg 2001;49:115–19.
7. Kaloud H, Smolle-Juettner FM, Prause G, List WF. Iatrogenic ruptures of the tracheobronchial tree. Chest 1997;112:774–8.
8. Beiderlinden M, Adamzik M, Peters J. Conservative treatment of tracheal injuries. Anesth Analg 2005;100:210–14.
9. Gomez-Caro Andres A, Moradiellos Diez FJ, Ausin Herrero P, et al. Successful conservative management in iatrogenic tracheobronchial injury. Ann Thorac Surg 2005;79:1872–8.
10. Carbognani P, Bobbio A, Cattelani L, et al. Management of postintubation membranous tracheal rupture. Ann Thorac Surg 2004;77:406–9.
11. Jougon J, Ballester M, Choukroun E, et al. Conservative treatment for postintubation tracheobronchial rupture. Ann Thorac Surg 2000;69:216–20.
12. Mussi A, Ambrogi MC, Ribechini A, et al. Acute major airway injuries: clinical features and management. Eur J Cardiothorac Surg 2001;20:46–52.
13. Marty-Ane CH, Picard E, Jonquet O, Mary H. Membranous tracheal rupture after endotracheal intubation. Ann Thorac Surg 1995;60:1367–71.
14. Sonett JR, Keenan RJ, Ferson PF, Griffith BP, Landreneau RJ. Endobronchial management of benign, malignant, and lung transplantation airway stenoses. Ann Thorac Surg 1995;59:1417–22.
15. Marquette CH, Mensier E, Copin MC, et al. Experimental models of tracheobronchial stenoses: a useful tool for evaluating airway stents. Ann Thorac Surg 1995;60:651–6.
16. Bromberg D, Turtz MG. Tracheal granulation: a complication of pediatric tracheotomy. Ann Otol Rhinol Laryngol 1977;86:639–43.

17. Bassett HF. Etiology of tracheal stenosis following cuffed intubation. *Proc R Soc Med* 1971;**64**:890–2.

18. Saad CP, Murthy S, Krizmanich G, Mehta AC. Self-expandable metallic airway stents and flexible bronchoscopy: long-term outcomes analysis. *Chest* 2003;**124**:1993–9.

19. Mehta AC, Harris RJ, De Boer GE. Endoscopic management of benign airway stenosis. *Clin Chest Med* 1995;**16**:401–13.

20. Bolliger CT, Sutedja TG, Strausz J, Freitag L. Therapeutic bronchoscopy with immediate effect: laser, electrocautery, argon plasma coagulation and stents. *Eur Respir J* 2006;**27**:1258–71.

21. Cavaliere S, Foccoli P, Farina PL. Nd:YAG laser bronchoscopy: a five-year experience with 1396 applications in 1000 patients. *Chest* 1988;**94**:15–21.

22. Baugnee P-E, Marquette C-H, Ramon P, Darras J, Wurtz A. Traitement endoscopique des sténoses trachéales post-intubation: a propos de 58 cas. *Rev Mal Resp* 1995;**12**:585–92.

23. Noppen M, Meysman M, Claes I, D'Haese J, Vincken W. Screw-thread vs. Dumon endoprosthesis in the management of tracheal stenosis. *Chest* 1999;**115**:532–5.

24. Vergnon JM, Costes F, Polio JC. Efficacy and tolerance of a new silicone stent for the treatment of benign tracheal stenosis: preliminary results. *Chest* 2000;**118**:422–6.

25. Madden BP, Loke TK, Sheth AC. Do expandable metallic airway stents have a role in the management of patients with benign tracheobronchial disease? *Ann Thorac Surg* 2006;**82**:274–8.

26. Brichet A, Verkindre C, Dupont J, *et al.* Multidisciplinary approach to management of postintubation tracheal stenoses. *Eur Respir J* 1999;**13**:888–93.

27. Bisson A, Bonnette P, el Kadi NB, *et al.* Tracheal sleeve resection for iatrogenic stenoses (subglottic laryngeal and tracheal). *J Thorac Cardiovasc Surg* 1992;**104**:882–7.

28. Couraud L, Jougon JB, Velly JF. Surgical treatment of nontumoral stenoses of the upper airway. *Ann Thorac Surg* 1995;**60**:250–9.

29. Grillo HC, Donahue DM, Mathisen DJ, Wain JC, Wright CD. Postintubation tracheal stenosis: treatment and results. *J Thorac Cardiovasc Surg* 1995;**109**:486–92.

30. Wynn R, Har-El G, Lim JW. Tracheal resection with end-to-end anastomosis for benign tracheal stenosis. *Ann Otol Rhinol Laryngol* 2004;**113**:613–17.

31. Sheski FD, Mathur PN. Long-term results of fiberoptic bronchoscopic balloon dilation in the management of benign tracheobronchial stenosis. *Chest* 1998;**114**:796–800.

32. van den Boogert J, Hans Hoeve LJ, Struijs A, Hagenouw RR, Bogers AJ. Single-stage surgical repair of benign laryngotracheal stenosis in adults. *Head Neck* 2004;**26**:111–17.

33. Mansour KA, Lee RB, Miller JI. Tracheal resections: lessons learned. *Ann Thorac Surg* 1994;**57**:1120–5.

34. Mayse ML, Greenheck J, Friedman M, Kovitz KL. Successful bronchoscopic balloon dilation of nonmalignant tracheobronchial obstruction without fluoroscopy. *Chest* 2004;**126**:634–7.

35. Landay M, Oliver Q, Estrera A, *et al.* Lung penetration by thoracostomy tubes: imaging findings on CT. *J Thorac Imaging* 2006;**21**:197–204.

36. Chechani V. Bronchoscopic diagnosis of solitary pulmonary nodules and lung masses in the absence of endobronchial abnormality. *Chest* 1996;**109**:620–5.

37. Zavala DC. Diagnostic fiberoptic bronchoscopy: techniques and results of biopsy in 600 patients. *Chest* 1975;**68**:12–19.

38. Credle JWF, Smiddy JR, Elliott RC. Complications of fiberoptic bronchoscopy. *Am Rev Respir Dis* 1974;**109**:67–72.

39. Deneuville M. Morbidity of percutaneous tube thoracostomy in trauma patients. *Eur J Cardiothorac Surg* 2002;**22**:673–8.

40. Mirvis SE, Tobin KD, Kostrubiak I, Belzberg H. Thoracic CT in detecting occult disease in critically ill patients. *Am J Roentgenol* 1987;**148**:685–9.

41. Cohn SM. Pulmonary contusion: review of the clinical entity. *J Trauma* 1997;**42**:973–9.

42. Miller PR, Croce MA, Bee TK, *et al.* ARDS after pulmonary contusion: accurate measurement of contusion volume identifies high-risk patients. *J Trauma* 2001;**51**:223–8.

43. Tyburski JG, Collinge JD, Wilson RF, Eachempati SR. Pulmonary contusions: quantifying the lesions on chest X-ray films and the factors affecting prognosis. *J Trauma* 1999;**46**:833–8.

44. Baldt MM, Bankier AA, Germann PS, *et al.* Complications after emergency tube thoracostomy: assessment with CT. *Radiology* 1995;**195**:539–43.

45. Daly RC, Mucha P, Pairolero PC, Farnell MB. The risk of percutaneous chest tube thoracostomy for blunt thoracic trauma. *Ann Emerg Med* 1985;**14**:865–70.

46. The Acute Respiratory Distress Syndrome Network. Ventilation with lower tidal volumes as compared with traditional tidal volumes for acute lung injury and the acute respiratory distress syndrome. *New Engl J Med* 2000;**342**:1301–8.

47. Tremblay L, Valenza F, Ribeiro SP, Li J, Slutsky AS. Injurious ventilatory strategies increase cytokines and c-fos m-RNA expression in an isolated rat lung model. *J Clin Invest* 1997;**99**:944–52.

48. Tremblay LN, Miatto D, Hamid Q, Govindarajan A, Slutsky AS. Injurious ventilation induces widespread pulmonary epithelial expression of tumor necrosis factor-alpha and interleukin-6 messenger RNA. *Crit Care Med* 2002;**30**:1693–700.

49. Gentil B, Etienne-Mastroianni B, Cordier JF. [Chronic cough after laparoscopic adjustable gastric banding]. *Rev Mal Respir* 2003;**20**:451–4.

50. Cortet B, Cotten A, Boutry N, *et al.* Percutaneous vertebroplasty in patients with osteolytic metastases or multiple myeloma. *Rev Rhum* (Engl Ed) 1997;**64**:177–83.

51. Cotten A, Dewatre F, Cortet B, *et al.* Percutaneous vertebroplasty for osteolytic metastases and myeloma: effects of the percentage of lesion filling and the leakage of methyl methacrylate at clinical follow-up. *Radiology* 1996;**200**:525–30.

52. Strange C, Jantz M. Pneumothorax. In: Bouros D (ed.) *Pleural Diseases,* pp 661–76. New York: Marcel-Dekker, 2004.

53. Yamagami T, Kato T, Hirota T, *et al.* Pneumothorax as a complication of percutaneous radiofrequency ablation for lung neoplasms. *J Vasc Interv Radiol* 2006;**17**:1625–9.

54. de Lassence A, Timsit JF, Tafflet M, *et al.* Pneumothorax in the intensive care unit: incidence, risk factors, and outcome. *Anesthesiology* 2006;**104**:5–13.

55. Haworth CS, Dodd ME, Atkins M, Woodcock AA, Webb AK. Pneumothorax in adults with cystic fibrosis dependent on

nasal intermittent positive pressure ventilation (NIPPV): a management dilemma. *Thorax* 2000;**55**:620–2.

56. Alpan G, Glick B, Peleg O, Amit Y, Eyal F. Pneumothorax due to endotracheal tube suction. *Am J Perinatol* 1984;**1**:345–8.

57. Hagley MT, Martin B, Gast P, Traeger SM. Infectious and mechanical complications of central venous catheters placed by percutaneous venipuncture and over guidewires. *Crit Care Med* 1992;**20**:1426–30.

58. Peuker ET, White A, Ernst E, Pera F, Filler TJ. Traumatic complications of acupuncture: therapists need to know human anatomy. *Arch Fam Med* 1999;**8**:553–8.

59. Noppen M, Scrammel F. Pneumothorax. *Eur Respir Monograph* 2002;**7**(22):279–96.

60. Sassoon CS, Light RW, O'Hara VS, Moritz TE. Iatrogenic pneumothorax: etiology and morbidity. Results of a Department of Veterans Affairs Cooperative Study. *Respiration* 1992;**59**:215–20.

61. Cox JE, Chiles C, McManus CM, Aquino SL, Choplin RH. Transthoracic needle aspiration biopsy: variables that affect risk of pneumothorax. *Radiology* 1999;**212**:165–8.

62. Bo-Linn GW, Anderson DJ, Anderson KC, McGoon MD. Percutaneous central venous catheterization performed by medical house officers: a prospective study. *Cathet Cardiovasc Diagn* 1982;**8**:23–9.

63. de Latorre FJ, Tomasa A, Klamburg J, *et al.* Incidence of pneumothorax and pneumomediastinum in patients with aspiration pneumonia requiring ventilatory support. *Chest* 1977;**72**:141–4.

64. Kollef MH. Risk factors for the misdiagnosis of pneumothorax in the intensive care unit. *Crit Care Med* 1991;**19**:906–10.

65. Tschopp JM, Rami-Porta R, Noppen M, Astoul P. Management of spontaneous pneumothorax: state of the art. *Eur Respir J* 2006;**28**:637–50.

66. Froudarakis ME. Diagnostic work-up of pleural effusions. *Respiration* 2008;**75**:4–13.

67. Plataki M, Bouros D. Iatrogenic and rare pleural effusions. In: Bouros D (ed.) *Pleural Diseases*, pp 897–913. New York: Marcel-Dekker, 2004.

68. Bankier AA, Wiesmayr MN, Henk C, *et al.* Radiographic detection of intrabronchial malpositions of nasogastric tubes and subsequent complications in intensive care unit patients. *Intens Care Med* 1997;**23**:406–10.

69. Daganou M, Dimopoulou I, Michalopoulos N, *et al.* Respiratory complications after coronary artery bypass surgery with unilateral or bilateral internal mammary artery grafting. *Chest* 1998;**113**:1285–9.

70. Light RW, Rogers JT, Cheng D, Rodriguez RM. Large pleural effusions occurring after coronary artery bypass grafting. *Ann Intern Med* 1999;**130**:891–6.

71. Aarnio P, Kettunen S, Harjula A. Pleural and pulmonary complications after bilateral internal mammary artery grafting. *Scand J Thorac Cardiovasc Surg* 1991;**25**:175–8.

72. Hietala SO, Hahn P. Pulmonary radiation reaction in the treatment of carcinoma of the breast. *Radiography* 1976;**42**:225–30.

73. Whitcomb ME, Schwarz MI. Pleural effusion complicating intensive mediastinal radiation therapy. *Am Rev Respir Dis* 1971;**103**:100–7.

74. Trupka A, Waydhas C, Hallfeldt KK, *et al.* Value of thoracic computed tomography in the first assessment of severely injured patients with blunt chest trauma: results of a prospective study. *J Trauma* 1997;**43**:405–12.

75. Kaiser CW, Koornick AR, Smith N, Soroff HS. Choice of route for central venous cannulation: subclavian or internal jugular vein? A prospective randomized study. *J Surg Oncol* 1981;**17**:345–54.

76. Eckhardt WF, Iaconetti J, Kwon JS, Brown E, Troianos CA. Inadvertent carotid artery cannulation during pulmonary artery catheter insertion. *J Cardiothorac Vasc Anesth* 1996;**10**:283–90.

77. Shah KB, Rao TL, Laughlin S, El-Etr AA. A review of pulmonary artery catheterization in 6245 patients. *Anesthesiology* 1984;**61**:271–5.

78. Ferretti GR, Thony F, Link KM, *et al.* False aneurysm of the pulmonary artery induced by a Swan–Ganz catheter: clinical presentation and radiologic management. *Am J Roentgenol* 1996;**167**:941–5.

79. Kantrowitz A, Wasfie T, Freed PS, *et al.* Intraaortic balloon pumping 1967 through 1982: analysis of complications in 733 patients. *Am J Cardiol* 1986;**57**:976–83.

80. Gottlieb SO, Brinker JA, Borkon AM, *et al.* Identification of patients at high risk for complications of intraaortic balloon counterpulsation: a multivariate risk factor analysis. *Am J Cardiol* 1984;**53**:1135–9.

81. Noffsinger AE, Blisard KS, Balko MG. Cardiac laceration and pericardial tamponade due to cardiopulmonary resuscitation after myocardial infarction. *J Forens Sci* 1991;**36**: 1760–4.

82. Kaminski Z, Lukasik M. [Mediastinal abscesses and empyema of the pleural cavity after iatrogenic fracture of the sternum]. *Pol Merkur Lekarski* 1997;**3**:279–80.

83. Vezina D, Lessard MR, Bussieres J, Topping C, Trepanier CA. Complications associated with the use of the Esophageal-Tracheal Combitube. *Can J Anaesth* 1998;**45**:76–80.

84. Panieri E, Millar AJ, Rode H, Brown RA, Cywes S. Iatrogenic esophageal perforation in children: patterns of injury, presentation, management, and outcome. *J Pediatr Surg* 1996;**31**:890–5.

85. Similowski T, Straus C. Iatrogenic-induced dysfunction of the neuromuscular respiratory system. *Clin Chest Med* 2004;**25**:155–66.

86. Eleftheriades J, Singh M, Tang P, *et al.* Unilateral diaphragm paralysis: etiology, impact, and natural history. *J Cardiovasc Surg* (Torino) 2008;**49**:289–95.

ONCOLOGY AND ALLIED CONDITIONS

14

Pulmonary reactions to chemotherapeutic agents: the 'chemotherapy lung'

FABIEN MALDONADO, ANDREW H LIMPER

INTRODUCTION

Since the discovery in the 1940s that nitrogen mustard, a chemical warfare agent, could be effective in the treatment of non-Hodgkin's lymphoma, chemotherapeutic agents have been increasingly used for the treatment of a variety of neoplastic and inflammatory conditions. These compounds encompass a heterogeneous group of medications with various mechanisms of action, which are frequently responsible for significant adverse reactions, occasionally requiring discontinuation of therapy. It is estimated that as many as 10–20 per cent of patients undergoing some type of chemotherapy will develop respiratory symptoms during the course of treatment.[1,2] These adverse effects may represent a real diagnostic challenge for the clinician as the clinical expression of these reactions can vary considerably between agents and between individual patients. Although the overall mortality rate of chemotherapy-associated lung toxicity is difficult to determine, it has been estimated that respiratory complications, including opportunistic infections secondary to immunosuppression, may be fatal in as many as 3 per cent of patients receiving chemotherapeutic agents.[3,4]

PRESENTATIONS AND DIAGNOSIS

The diagnosis of chemotherapy-induced lung toxicity virtually always remains one of exclusion.[5] The various presentations are rarely specific enough to be readily linked to a particular aetiological agent.[1,6] Furthermore, patients undergoing intensive therapy for malignancy are frequently substantially immunosuppressed, either from the chemotherapy itself or as a consequence of their underlying condition. These patients represent easy targets for a host of opportunistic infections. Atypical presentations of infections caused by more common pathogens are also fairly common in these patients. In addition, the use of combination chemotherapy generally further complicates the identification of the responsible medication. In many clinical settings it may be exceedingly difficult to implicate one agent over another. Finally, the timeframe that characterizes chemotherapy-induced lung toxicity is highly variable and its onset may be delayed considerably. Most often, respiratory complications from chemotherapy typically occur in a subacute fashion, with symptom onset several months after the initiation of chemotherapy. On the other hand, cases of drug-induced hypersensitivity reactions in the lungs may present earlier, frequently within days to weeks after the initiation of therapy. These factors often lead to significant clinical confusion, especially since adverse drug reactions may also coexist with infection or neoplastic involvement of the lungs. Therefore, the diagnosis of chemotherapy-associated lung disease must generally remain a diagnosis of exclusion. It can usually only be established after all alternative diagnoses have been carefully excluded. Nonetheless, carefully establishing a definite or presumptive diagnosis of chemotherapy-associated lung disease is critical, since it often leads to profound changes in patient management, with discontinuation of the drug and, sometimes, administration of prednisone or other anti-inflammatory agents.

The diagnosis of chemotherapy-induced lung disease is often suggested by the onset of non-specific respiratory symptoms, consistent radiographic findings, and compatible pulmonary function studies. Such a diagnosis may be supported by characteristic pathological findings on microscopic examination, though lung biopsy is not required in most cases. This constellation of findings should prompt the clinician to consider the diagnosis of drug adverse reaction and to act accordingly, with discontinuation of the suspected agent(s).

Patients will often present with the insidious onset of dyspnoea on exertion, dry cough and often fever. Chills, however, are not usually reported. Constitutional symptoms such as weight loss may be prominent in subacute presentations.

Dry crackles suggestive of pulmonary fibrosis are the typical physical finding. Wheezing is rare and may result from significant bronchospasm, suggesting a hypersensitivity type of reaction. Clubbing is not typically reported in cases of chemotherapy-associated lung disease, most likely related to the subacute nature of most of these reactions.

Radiographic findings rarely implicate a specific agent, except perhaps in the case of methotrexate-induced pneumonitis, in which typical hilar lymphadenopathy suggests reaction to this agent.[7,8] However, in most cases of chemotherapy-associated lung disease the radiographic pattern is one of interstitial pneumonitis, which is impossible to distinguish from interstitial infiltrates related to common inflammatory or infectious aetiologies.[9] Ground-glass and alveolar infiltrates are occasionally seen on the chest computed tomography (CT) scan. Pleural effusions and nodular tumour-like infiltrates are much less common, but have been classically described with certain agents (discussed below). A restrictive pattern is present on the pulmonary function studies of most patients with chemotherapy-induced lung disease.[10] A decrease in diffusing capacity for carbon monoxide (DLCO), and in some situations in forced vital capacity and pulmonary capillary blood volume, may precede clinically overt interstitial lung disease. These parameters may prove to be valuable assets in screening for respiratory complications during the course of chemotherapy, although the evidence for supporting this practice remains scarce. As a result, screening pulmonary functions studies are not routinely indicated. Gallium-uptake scans are of limited value and rarely employed in recent years, except perhaps in the case of methotrexate-induced lung toxicity.[11]

Bronchoscopy with bronchoalveolar lavage (BAL) and biopsies is not required for establishing the diagnosis, but this procedure is employed in most patients because of its great value in ruling out differential diagnoses, in particular metastatic disease or infectious processes. When the diagnosis is unsettled, lung tissue biopsies, whether transbronchial or surgical, are occasionally needed. The presence of atypical 'bizarre' type 2 pneumocytes is characteristic, although not pathognomonic, of cytotoxic chemotherapy-induced lung toxicity.[12] These cellular atypia do not systematically imply the presence of clinically significant lung toxicity as they are frequently observed in the absence of any respiratory symptoms in patients treated with cytotoxic agents. For obvious reasons, it is additionally important to differentiate these cellular atypia from the neoplastic changes observed in metastatic lung disease. Numerous histopathological patterns have been reported in chemotherapy-associated lung disease. These include diffuse alveolar damage (DAD), bronchiolitis obliterans with organizing pneumonitis (BOOP), granuloma formation and alveolar proteinosis, among others.[13] Fibrin and collagen deposition leading to diffuse fibrosis is often the final common pathway for many chemotherapeutic agents. Once again, these patterns, although highly suggestive in the right context, remain fairly non-specific, and the absence of an obvious alternative diagnosis (particularly infection or malignancy) is often the strongest argument for an adverse drug reaction.

Clinicians directly or indirectly involved in the care of patients undergoing chemotherapy must constantly remain aware of the possibility of drug-induced pulmonary toxicities, as a growing list of agents is implicated (Box 14.1). Cytotoxic and non-cytotoxic chemotherapeutic agents may injure the lungs in many different ways, and the early recognition of these adverse reactions requires a high level of suspicion. Limitation of the total cumulative chemotherapy dose, avoidance of high fraction of

---

Box 14.1 Selected chemotherapeutic agents with associated pulmonary toxicities

Antibiotics

- Bleomycin
- Mitomycin C

Alkylating agents

- Busulfan
- Cyclophosphamide
- Chlorambucil
- Melphalan

Antimetabolites

- Methotrexate
- 6-Mercaptopurine
- Azathioprine
- Cytosine arabinoside
- Gemcitabine
- Fludarabine

Nitrosoamines

- Bischloroethyl nitrosourea (BCNU)
- Chloroethyl cyclohexyl nitrosourea (CCNU)
- Methyl-CCNU

Podophyllotixins

- Etoposide
- Paclitaxel
- Docetaxel

Novel antitumour agents

- All-*trans*-retinoic acid (ATRA)
- Erlotinib
- Gefitinib
- Imatinib mesylate
- Irinotecan

Immune modulatory agents used in malignancy

- Interferons
- Interleukin-2
- Tumour necrosis factor alpha

Other miscellaneous chemotherapy agents

- Procarbazine
- Zinostatin
- Vinblastine

inspired oxygen or concomitant radiotherapy also are important priorities when contemplating the use of these agents. Discontinuation of the chemotherapeutic agent with or without administration of corticosteroids represents the single most important intervention when the diagnosis is considered probable.

ANTIBIOTIC CHEMOTHERAPEUTIC AGENTS

Bleomycin

Bleomycin was first isolated from the microbe *Streptomyces verticillus* in 1966. It has been the cornerstone of numerous chemotherapeutic regimens for the treatment of a variety of neoplasms including germ-cell tumours, Hodgkin and non-Hodgkin's lymphomas, genitourinary tract malignancies, malignant pleural effusions, and squamous cell carcinoma of the head, neck and oesophagus. The pulmonary toxicity of bleomycin was discovered early in its use and has since been the main factor limiting its use.

Although the pathogenesis of bleomycin-induced lung toxicity remains incompletely understood, it is nonetheless the most common and also the most widely studied of chemotherapeutic agents responsible for the development of a 'chemotherapy lung'. It is estimated that up to 46 per cent of patients treated with this agent will demonstrate evidence of drug-induced pneumonitis.[14] There is frequent progression to pulmonary fibrosis, pathologically indistinguishable from idiopathic usual interstitial pneumonitis (UIP). The mortality rate is relatively high – around 3 per cent of all patients treated in one large series.[4] It is up to 10 per cent in those who develop pulmonary toxicity.

Bleomycin appears to exert its main antineoplastic effect by direct cellular cytotoxicity,[14] resulting microscopically in cytological atypia with the characteristic presence of atypical type 2 pneumocytes and a relative paucity of type 1 pneumocytes. Prevention of tumour angiogenesis is another mechanism of action that has been attributed to bleomycin. Bleomycin complexes ferrous ions and oxygen molecules, and induces the generation of free radicals responsible for cellular death, through directly fragmenting the DNA of tumour cells in a cell-cycle dependent fashion.[15] Unfortunately this is also one mechanism by which bleomycin exerts its untoward side-effects. This mechanism further explains why supplemental oxygen has been documented to worsen bleomycin-induced lung toxicity in certain clinical scenarios. Bleomycin also directly stimulates production of various inflammatory cytokines, which in addition to the free radical generation promotes endothelial injury with secondary infiltration of the lung by inflammatory cells. Subsequently, reparative fibroblast activation is initiated and collagen deposition ensues.

Bleomycin is excreted predominantly by the kidneys. Accordingly, the presence of renal dysfunction has been associated with a higher incidence of adverse pulmonary toxicity. Bleomycin is further inactivated by an enzyme known as bleomycin hydrolase, which is found mainly in the liver and spleen. This enzyme is relatively absent in the lung and the skin, which may explain both the pulmonary toxicity and the cutaneous scleroderma-like changes frequently described with bleomycin.[14]

Of all various risk factors identified for toxicity, the cumulative dose received appears to be the most reliable predictor of bleomycin-induced lung injury. Because of the cell cycle-dependent mechanism of toxicity, repeated bleomycin dosing preferentially targets type 2 pneumocytes, which proliferate to replace injured type 1 pneumocytes, among other functions. A cumulative dose greater than 450 units is associated with a significantly higher incidence of lung toxicity, with a mortality rate in excess of 10 per cent in those patients who receive greater than 550 units.[16] However, even cumulative doses of less than 50 units have, in some instances, resulted in significant pulmonary toxicity. Rapid rates of intravenous infusion may also contribute to the development of inflammation and fibrosis. Although intramuscular injections of bleomycin theoretically have a lower rate of systemic delivery, recent studies have not confirmed a reduced incidence of chemotherapy-associated pulmonary toxicity.

The age of the patient receiving bleomycin also clearly plays a role in predicting toxicity. There is convincing evidence that older patients, in particular those older than 70 years, develop bleomycin-induced pneumonitis with greater consistency and at lower doses than younger patients. In addition, higher rates of bleomycin-induced pneumonitis have been described in paediatric populations, with development of lung toxicity in up to 70 per cent of children treated with bleomycin for rhabdomyosarcomas.[17] Although no clear scientific explanation exists for these age-related differences, it seems plausible to invoke immature or altered mechanisms of inactivation of free radicals, impaired kidney function, and a higher prevalence of pre-existing lung conditions in patients at the extremes of age.

Patients undergoing chemotherapy with bleomycin often require other therapeutic agents, some of which have shown to promote lung toxicity in a synergistic manner. In particular, supplemental oxygen, frequently required in the perioperative period or during episodes of respiratory distress, can dramatically worsen bleomycin-induced lung toxicity (Fig. 14.1).[18,19] Synergistic bleomycin toxicity with oxygen can occur whether used concomitantly or in a delayed manner. In particular, such oxygen sensitivity may persist for months and maybe years after chemotherapy with bleomycin. The presumed mechanism is the enhanced production of free radicals in the presence of supplemental oxygen. Furthermore, radiotherapy also exerts its therapeutic and adverse effects by generating free radicals. Expectedly, concomitant radiotherapy, in particular when the chest is irradiated, has proven to be a dangerous treatment when combined with bleomycin, with a significantly increased incidence of pulmonary toxicity and fibrosis. Bleomycin may also reactivate a prior radiation-induced pulmonary toxicity, a phenomenon known as 'radiation recall'. Interestingly, this classic interaction has been debated in recent literature. Earlier reports had suggested a potentially deleterious effect of granulocyte-stimulating factor when added to bleomycin, presumably by stimulating the production of various cytokines, but this has not been confirmed by more recent data. Finally, a number of other cytotoxic agents appear to also exert a synergistic toxic effect when combined with bleomycin, the most frequent being cyclophosphamide and vincristine.

By far, interstitial pneumonitis is the most common clinical presentation of bleomycin-induced lung toxicity. It can

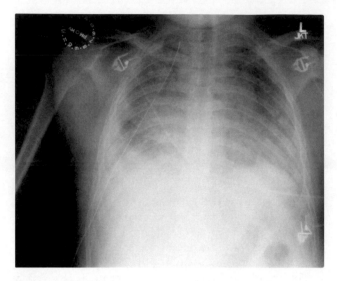

Fig. 14.1 Bleomycin toxicity precipitated by oxygen administration. The chest radiograph was obtained from a 28-year-old male with a mediastinal germ-cell tumour treated with bleomycin, etoposide and cisplatin. He developed respiratory distress and alveolar and interstitial infiltration 24 h after a thoracotomy for removal of the residual tumour.

ultimately progress to fibrosis. This interstitial pneumonitis is usually of subacute onset and may progress over months after the administration of bleomycin. The symptoms are non-specific and usually consist of a dry cough with the insidious onset of dyspnoea on exertion. Dry 'Velcro'-type crackles may be observed in the later fibrosing phase of the disease. Likewise, the radiographic studies are usually fairly non-specific, demonstrating an interstitial pattern, with a predominantly peripheral and subpleural location on the chest CT scan. The pulmonary function studies typically reveal hypoxaemia, a restrictive pattern and a decreased DLCO. Bronchoscopy with BAL, with or without biopsies, is generally important in ruling out infectious or malignant aetiologies, as bleomycin-induced pneumonitis remains a diagnosis of exclusion. Histologically, the presence of atypical 'bizarre' type 2 pneumocytes in conjunction with a relative paucity of type 1 pneumocytes is characteristic, although not diagnostic. The role of serial pulmonary function studies with DLCO measurements in monitoring the onset of subclinical pulmonary toxicity is of unclear value, but is still frequently recommended by many experts. Vital capacity and pulmonary capillary blood volume may be better predictors of clinically overt pulmonary toxicity.[20] Serial chest X-rays or CT scans are not recommended for screening purposes at this time. Chest CT scans were abnormal in 38 per cent of patients treated with bleomycin compared to 15 per cent when evaluated with chest X-rays only in one series.[21]

Bleomycin may also induce a hypersensitivity pneumonitis with acute onset of fever and peripheral or BAL eosinophilia.[16,22] Pneumothorax and pneumomediastinum with the acute onset of chest pain, and occasionally the confusing onset of peripheral nodular infiltrates, have also been described. The latter usually requires a pathological diagnosis in order to rule out metastatic disease with confidence. Histologically, these nodules exhibit features consistent with BOOP.

The treatment of bleomycin-induced lung injury is largely empirical. Discontinuation of the offending agent remains the mainstay of therapy. Limiting the cumulative dose of bleomycin, especially in older patients, and avoiding high fraction of inspired oxygen and concomitant radiotherapy agents are clearly indicated as preventive measures.

Alternative agents may need to be considered, especially in the face of kidney dysfunction. Numerous agents have been shown to prevent or limit bleomycin-induced toxicity in animal models, but none of them has definitely been shown to be of benefit in humans. High doses of corticosteroids may prevent or limit pulmonary fibrosis and are currently recommended if there is progression to fibrosis, although the evidence for this therapy remains largely anecdotal.[14]

Mitomycin C

Mitomycin C is another antibiotic chemotherapeutic agent associated with significant lung toxicity. This agent was isolated from a strain of *Streptomyces caespitosis*. It is a very commonly used alkylating agent, which has demonstrated benefit in a variety of neoplasms including bladder tumours, prostate, lung, breast and cervical cancer. The main limiting factor in its use is the development of pulmonary toxicity, which shares several common features with that induced by bleomycin, along with other features more specific to mitomycin C. Adverse pulmonary reactions have been reported in 8–39 per cent of treated patients.[23,24] Bone marrow suppression and various gastrointestinal manifestations are other commonly reported side-effects of this agent.

The probability of mitomycin C-induced lung toxicity directly parallels the cumulative dose received, and pulmonary fibrosis is highly unlikely at doses less than 30 mg/m^2.[23] High fractions of inspired oxygen and radiotherapy, in a manner parallel to that described for bleomycin, seem to worsen lung toxicity in a synergistic fashion.[25] Concomitant administration of other chemotherapeutic agents, namely bleomycin, doxorubicine, cyclophosphamide and vinca alkaloids, may also exert synergistic toxicity, although it may become difficult to blame one drug over another in many clinical situations.

The most common pulmonary adverse effect reported with mitomycin C is interstitial pneumonitis, usually within a year of administration, with secondary progression to pulmonary fibrosis. The subacute onset of dry cough and dyspnoea on exertion suggests the diagnosis, which is supported by the presence of interstitial infiltrates on the imaging studies and a restrictive pattern and hypoxaemia on pulmonary function studies. The DLCO tends to fall early during the subclinical stage of the disease, but does not necessarily herald overt pneumonitis, making it a rather unreliable predictive tool clinically.[26] Treatment with high doses of intravenous corticosteroids before the onset of fibrosis has occasionally been associated with dramatic responses and is currently the recommended treatment of choice, along with discontinuation of the drug.[27]

Mitomycin C has been associated with a number of other adverse reactions in the lung. Specifically, mitomycin C can precipitate a type of thrombotic microangiopathy similar to the haemolytic uraemic syndrome. This reaction is particularly prevalent when 5-fluorouracil and blood transfusions are administered before mitomycin C chemotherapy.[28] This unique reaction is typically associated with the presence of a non-cardiogenic pulmonary oedema in approximately 50 per cent of cases. This presentation, although relatively rare, is associated with mortality rates around 50 per cent and can be as high as 95 per cent if complicated by acute respiratory distress syndrome (ARDS). The presence of haemolytic anaemia, thrombocytopenia, acute renal failure and diffuse pulmonary infiltrates strongly suggests this diagnosis. Pathologically, examination of the lungs and kidneys demonstrates evidence of microangiopathic changes with intracapillary thrombi and intimal hyperplasia of the arterioles. Treatment remains essentially supportive. Plasmapheresis may be of benefit, in addition to high doses of steroids and initiation of dialysis if necessary.[29]

Other less common pulmonary respiratory complications of chemotherapy with mitomycin C have been described. These include transient episodes of bronchospasm and exudative pleural effusions. These manifestations remain extremely rare.

Other antibiotic chemotherapeutic agents

Other agents have occasionally been associated with the development of interstitial pneumonitis with secondary pulmonary fibrosis. Because of the nature of chemotherapeutic regimens, it is sometimes difficult to specifically identify the medication directly responsible for an individual case of pulmonary toxicity. Doxorubicine is an anthracycline agent with well-described cumulative cardiac toxicity, which may sometimes lead to the development of cardiogenic pulmonary oedema. However, in addition, there are rare reports of doxorubicine-induced interstitial pneumonitis. Furthermore, doxorubicine may exacerbate mitomycin C-induced pneumonitis. Similarly, mitoxantrone has also been associated with a variant of subacute interstitial pneumonitis.[30] In addition, treatment with actinomycin D can specifically reactivate prior radiotherapy-induced lung injury, which can lead to the progression of pulmonary fibrosis ('radiation recall').[17]

ALKYLATING AGENTS

Busulfan

Busulfan is an alkylating chemotherapeutic agent essentially used in the treatment of myeloproliferative disorders. This agent was discovered in 1961 and was the first chemotherapeutic agent found to be responsible for significant lung toxicity, as initially reported by Oliner et al. that same year.[31] The incidence of busulfan-induced pulmonary toxicity is estimated at about 6 per cent overall, with a wide range of 2.5–43 per cent.[32] The reported mortality is extremely high, approximating 80 per cent in some series.[17] There is no known effective treatment for busulfan lung toxicity.

The adverse effects of busulfan on the lungs are unique in many ways. The exact pathogenesis remains unclear, as there are no experimental models of busulfan-induced lung damage. Interestingly, no risk factor has yet been clearly identified. Of note, the cumulative dose does not seem to directly correlate

with the incidence of lung injury, neither does the age of the patient. Concomitantly administered cytotoxic agents or radiotherapy may, however, worsen the lung toxicity.

Clinically, busulfan-induced pulmonary toxicity results in the subacute onset of cough, fever, weight loss and dyspnoea on exertion. These symptoms typically occur more than four years after initiation of chemotherapy. Occasionally, these symptoms may occur within a few months, and some cases have been reported as late as 10 years after initiation of treatment. The chest X-ray generally shows characteristic alveolar and interstitial infiltrates which should suggest the diagnosis and prompt the clinician to discontinue the drug. Pulmonary function studies are frequently abnormal with evidence of a restrictive pattern and a decreased DLCO. Bronchoscopic examination may be of importance in ruling out alternative diagnoses, in particular infectious or neoplastic processes. Histologically, atypical type 2 pneumocytes are present with a relatively decreased number of type 1 pneumocytes. In addition, lymphocytic and plasma cell infiltration can be observed in lavage. Discontinuation of the offending agent is mandatory. Corticosteroids may also be used, although their efficacy remains unclear at best. Unfortunately, the majority of patients progress to respiratory failure and death.

The alveolar infiltrates observed on imaging studies presumably result from intra-alveolar deposition of cellular debris. This process may occasionally be so prominent as to lead to the development of an alveolar proteinosis pattern. This alveolar proteinosis is poorly responsive to whole lung lavage, and the role for corticosteroids in this situation remains largely undefined.[33,34]

Cyclophosphamide

Cyclophosphamide is another chemotherapeutic alkylating agent, which has been shown to adversely affect the lungs. This agent is commonly used in a variety of neoplasms, in particular haematological malignancies, usually in combination with other drugs. It is also increasingly used in a number of autoimmune and inflammatory diseases. The incidence of cyclophosphamide-induced lung toxicity is low, affecting less than 1 per cent of all patients treated, although this may be an underestimation.[16] One of the largest series published from a large tertiary referral medical centre identified only six patients over a period of 20 years for whom exposure to cyclophosphamide was the only identifiable reason for their pulmonary symptoms.[35]

Supplemental oxygen and radiotherapy may increase the likelihood of developing pulmonary toxicity, presumably via additional oxidative stress. Concomitant administration of other cytotoxic agents, in particular bleomycin, has also proven to enhance the risk of pulmonary toxicity from cyclophosphamide. The combination of cyclophosphamide and carmustine (BCNU) in preparation for bone marrow transplant seems to offer particular risk. Therefore, it may be important to limit the cumulative dose of cyclophosphamide.

Two main patterns of lung injury from cyclophosphamide have been identified.[35] In the first form, the acute onset of fever, cough and shortness of breath may occur early in the course of treatment, usually within 6 months after initiating

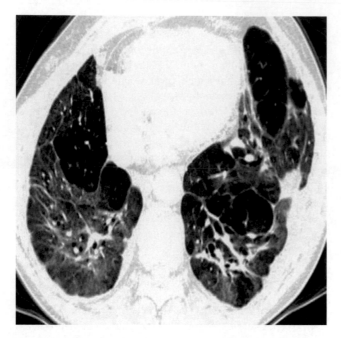

Fig. 14.2   Chronic lung fibrosis related to cyclophosphamide therapy. The computed tomography scan demonstrates pulmonary parenchymal and pleural fibrotic reaction to cyclophosphamide, which was administered to this 52-year-old woman for lymphoplasmacytic lymphoma. This fibrotic reaction was appreciated approximately 1 year after administration of the agent.

therapy. The chest X-ray in this form shows diffuse reticular and sometimes nodular infiltrates, whereas the chest CT scan typically shows peripheral ground-glass infiltrates. This form of cyclophosphamide-induced lung injury usually responds promptly to discontinuation of therapy and administration of corticosteroids. In contrast, a late-onset pneumonitis can also occur and is characterized by a more insidious progression, months or sometimes years after initiation of therapy. This later form usually results in an irreversible and diffuse fibrosis with pleural thickening, and may occur months to years after discontinuation of the drug (Fig. 14.2). Steroids and discontinuation of the drug only alter the course of the disease in a limited number of cases. Interestingly, there have been well-documented cases of re-challenge with cyclophosphamide that did not result in pulmonary toxicity, although for obvious reasons this is not usually recommended. Finally, rare cases of pulmonary oedema attributed to cyclophosphamide and pleural effusions have been described.

Chlorambucil

Chlorambucil is an alkylating agent used in the treatment of chronic lymphocytic leukaemia, Hodgkin's disease and low-grade non-Hodgkin's lymphoma. It has occasionally been associated with the development of lung toxicity, usually within 6 months to a year of initiation of therapy. Most features of chlorambucil-induced lung toxicity are common to other alkylating agents, and usually occur after doses in excess of 2 g.[1]

The reported mortality is high. Discontinuation of the offending agent and corticosteroids may be of benefit.

Melphalan

Higher doses of melphalan may induce pulmonary toxicity.[36] This alkylating agent has been used extensively in the treatment of multiple myeloma and ovarian cancer. There are relatively few cases reported in the literature. These cases usually consist of interstitial pneumonitis, sometimes reversible after discontinuation of the drug and administration of corticosteroids. This adverse effect is unlikely to be underestimated given the frequent long-term use of melphalan in the treatment of multiple myeloma.

Ifosfamide

Ifosfamide is another alkylating agent structurally related to cyclophosphamide. It has been used as part of therapeutic regimens for a variety of neoplasms, including lung, testicular, breast, cervical and ovarian cancers. A few case reports of subacute interstitial pneumonitis are described in the literature. One case of methemoglobinaemia induced by ifosfamide has also been described.[37]

Procarbazine

Procarbazine is an additional alkylating agent used in the treatment of Hodgkin's lymphoma and glioblastoma mutiforme. This agent has occasionally been associated with pulmonary toxicity.[38] Although relatively rare, the most common manifestation is the development of an interstitial pneumonitis, sometimes associated with eosinophilia, suggesting a hypersensitivity mechanism. Progression of procarbazine lung toxicity to pulmonary fibrosis is uncommon, and reversibility after discontinuation of the drug is the rule. Pleural effusions have also been occasionally reported with this drug.

Oxaliplatine

Oxaliplatine is an alkylating agent used in advanced colon and rectal carcinomas. It has been associated with laryngeal dysaesthesia, and in association with 5-fluorouracil has precipitated at least one case of diffuse alveolar damage.

ANTIMETABOLITES

Methotrexate

Methotrexate interferes with the metabolism of folic acid, thereby directly affecting cellular DNA synthesis functions. Methotrexate has been used extensively in a number of neoplastic and inflammatory conditions including rheumatoid arthritis, psoriasis and sarcoidosis. Its mechanism of action is also the basis for its numerous adverse effects that specifically target replicating DNA producing cells. Bone marrow suppression, mucositis, alopecia and gastrointestinal side-effects frequently lead to discontinuation of treatment. Pulmonary toxicity has also been extensively reported, although less commonly than those other side-effects. Pulmonary toxicity occurs in less than 10 per cent of patients.[39,40] Fortunately, the associated mortality rate is relatively low.

Many cases of methotrexate-induced lung toxicity have been well documented given its widespread use. Methotrexate is distinguished from many other chemotherapy-induced lung diseases because it usually leads to an acute type of pneumonitis, occurring typically within days or weeks of initiation of therapy. It has been shown to present after various doses of methotrexate, including occasionally minimal doses of the drug.[41] Occasionally, methotrexate lung toxicity may also present as a more traditional subacute type of pneumonitis with rare progression to pulmonary fibrosis, occurring months to years after starting the chemotherapy.[42] Methotrexate has also been reported in association with non-cardiogenic pulmonary oedema and may predispose patients to a variety of opportunistic lung infections, including *Pneumocytis jiroveci*, *Nocardia*, atypical mycobacterial and viral pneumonias.[43] Intrathecal injections of methotrexate have also reportedly caused acute adverse pulmonary reactions.

The acute onset of non-productive cough, fever and dyspnoea should suggest the possibility of methotrexate toxicity. Peripheral eosinophilia is present in more than 50 per cent of cases, supporting a hypersensitivity mechanism rather than direct cytotoxicity.[16] Radiographic studies may show mixed alveolar and interstitial infiltrates, sometimes associated with hilar lymphadenopathy, which is highly typical of methotrexate-induced lung toxicity. Pleural effusions may be observed in 10 per cent of patients. Pulmonary function studies are frequently normal, but may sometimes reveal a restrictive pattern with decreased DLCO. The decrease in DLCO does not precede the clinical symptoms, and screening pulmonary function studies are not routinely recommended.[44]

Bronchoscopy with BAL and biopsies is also often helpful in establishing the presence of methotrexate lung toxicity. First, BAL is valuable in ruling out alternative diagnoses, such as infection. Second, BAL may provide additional information supporting the diagnosis of methotrexate-induced lung toxicity. Bronchoscopy may reveal a predominance of lymphocytes, while transbronchoscopic biopsies often demonstrate poorly defined granulomas without the cellular atypia seen with cytotoxic agents, once again supporting a hypersensitivity-type reaction.[12,13] While these features are highly supportive of methotrexate toxicity, they are not pathognomonic.

As a rule, discontinuation of the offending drug is mandatory and results in a usually rapid resolution of symptoms. Rare cases may progress to secondary pulmonary fibrosis, and a course of corticosteroids may then be indicated. Spontaneous remissions without discontinuation of the drug have also been reported, and numerous patients have been re-challenged with methotrexate without recurrence of symptoms. However, reintroduction of this drug is not recommended in most cases. Prompt initiation of appropriate antibiotic therapy is important when infection is suspected, as both entities may coexist and overlap.

Cytosine arabinoside

Cytosine arabinoside is used in the treatment of acute myeloid leukaemia and in preparation for bone marrow transplantation. Rare cases of pulmonary toxicity have been reported with this drug, the pathogenesis of which remains unclear. Non-cardiogenic pulmonary oedema occurred in 13–28 per cent of patients treated with cytosine arabinoside in two series.[45,46] More than 40 per cent developed respiratory symptoms within a month of completing therapy. Pathological examination of the lung parenchyma reveals the presence of intra-alveolar proteinaceous material without evidence of cellular atypia. The treatment is essentially supportive and consists of discontinuation of the drug with a possible role for diuretics and corticosteroids. The reported mortality rate is significantly higher than for most other cytotoxic agents.

Gemcitabine

Gemcitabine is structurally related to cytosine arabinoside. It has recently been included in the treatment of non-small-cell lung cancers, pancreatic and breast cancers. Overall, gemcitabine has been relatively well tolerated, although significant pulmonary adverse reactions have been reported.[47,48] Dyspnoea heralds the onset of pulmonary toxicity which develops in approximately 1 per cent of patients. Non-cardiogenic pulmonary oedema occurs in 0.1–7 per cent of the cases (Fig. 14.3).[49] Cases of alveolar haemorrhages, pleural effusions, interstitial fibrosis and rarely veno-occlusive disease have been described. The pathogenesis of these pulmonary manifestations is not clear, and the treatment is once again largely supportive. Most patients improve after discontinuation of the drug and administration of corticosteroids. Occasionally rapid progression to

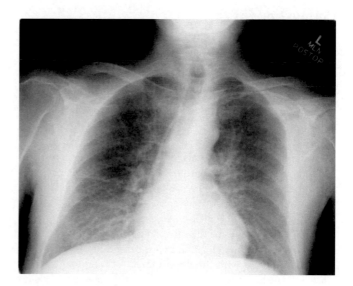

Fig. 14.3 Gemcitabine-associated lung toxicity. This 50-year-old female developed bilateral alveolar and interstitial infiltrates after receiving gemcitabine for metastatic cholangiocarcinoma. As the infiltrates resolved she subsequently developed bilateral pleural effusions, which are also described with gemcitabine toxicity.

respiratory failure and death may occur. Thrombotic micro-angiopathy has also been described with this agent.

Fludarabine

Fludarabine is a nucleoside analogue used in the treatment of chronic lymphocytic leukaemia and indolent non-Hodgkin's lymphomas. It has been reportedly associated with several types of lung toxicities. One series of 105 patients with lymphoproliferative disorders treated with fludarabine reported an incidence of pulmonary complications of 8.6 per cent.[50] The onset of dyspnoea and dry cough suggests the diagnosis. Imaging studies are fairly non-specific with the presence of alveolar and interstitial infiltrates. Nodular infiltrates have been described.[51] Pathologically, non-specific inflammatory changes with fibrosis may be seen. Fludarabine may also predispose to opportunistic lung infections, and invasive sampling of lung tissue, such as with bronchoscopy, may be required to establish the diagnosis. Most patients seem to respond to discontinuation of the drug and administration of corticosteroids, although fatal cases have been reported.

Azathioprine and 6-mercaptopurine

Azathioprine and its metabolite, 6-mercaptopurine, have been used extensively in a variety of cancers and inflammatory conditions. They have rarely been associated with the development of interstitial pneumonitis, although their direct responsibility has not clearly been established.[52] It is nonetheless important for the clinician to remain aware of the possibility of drug-induced pulmonary toxicity when faced with unexplained respiratory manifestations in patients receiving these agents.

Carmustine (BCNU)

Carmustine is an alkylating agent related to nitrogen mustard. This drug is used in the treatment of various haematological malignancies and solid tumours, including lymphomas, breast cancers and melanomas. Its ability to cross the blood–brain barrier makes this agent particularly valuable in the treatment of brain tumours. Unfortunately, it has been associated with several types of pulmonary toxicity, with the very characteristic potential to cause delayed-onset pulmonary fibrosis. Carmustine-related lung toxicity and fibrosis typically involves the upper lobes.[53,54] This process may occur many years after discontinuation of treatment.[55]

The likelihood of developing carmustine-induced pulmonary toxicity seems to be directly related to the cumulative dose received. In particular, doses in excess of 1400 mg/m² have been clearly associated with the development of lung toxicity in at least half of the patients, although lower doses may be just as harmful.[56,57] The incidence of lung toxicity during and after treatment with carmustine is estimated at between 1.5 and 20 per cent, although studies suggest that most patients will develop some degree of pulmonary fibrosis if given enough time.

There are two typical patterns of lung involvement reported. The first pattern of toxicity is an acute interstitial pneumonitis

with secondary progression to pulmonary fibrosis, which seems to be directly related to the dose of carmustine administered. The second type appears as delayed onset of pulmonary fibrosis, which may present more insidiously usually within 6 months to 3 years after chemotherapy. This delay may sometimes exceed several decades. Cyclophosphamide and radiotherapy may affect the lungs in a synergistic fashion. Pre-existing lung conditions may also worsen the pulmonary toxicity.

Clinically, the patient may complain of the onset of dry cough and shortness of breath. The presence of fever is rare and an alternative diagnosis should be considered. The sudden onset of chest pain after treatment with carmustine should suggest the presence of a pneumothorax, which can sometimes be related to carmustine-induced fibrobullous changes in the apices of the lungs.[58] The radiographic features of carmustine-induced lung toxicity are non-specific, except for the typical involvement of the upper lobes when the onset is delayed. Pulmonary function studies are similarly non-specific, with a restrictive pattern and a decrease in DLCO. A decrease in DLCO may suggest subclinical lung involvement and usually warrants discontinuation of the drug. Corticosteroids may be of benefit if administered early, but do not seem to alter the progression of the upper lobe fibrosis once it is established.

Other nitrosoureas

Cases of pulmonary fibrosis have been associated, although to a lesser degree, with other nitrosoureas. These agents include lomustine (CCNU), semustine (methyl-CCNU), fotemustine (CENU) and chlorozotocin (DCNU). Pulmonary veno-occlusive disease has been described with these agents. Pneumothoraces have rarely been reported.

TUBULIN-ACTING AGENTS

Vinblastine and vinorelbine

These agents exert their chemotherapeutic effect by binding tubulin and preventing the formation of microtubules, therefore interfering with an essential step in the process of cell replication. Vinblastine is used in various chemotherapeutic regimens to treat Hodgkin's and non-Hodgkin's lymphoma, germ-cell tumours, breast and renal tumours, and is also used to treat Langerhans cell histiocytosis. This agent has been rarely associated with lung toxicity, although frequently when co-administered with other agents potentially toxic to the lungs, such as mitomycin C.[59,60] Interstitial pneumonitis, bronchospasm, lung nodules and non-cardiogenic pulmonary oedema have been reported. There are several case reports of pulmonary toxicity induced by vinorelbine, although the evidence remains scarce.

Podophyllotoxins and taxanes

ETOPOSIDE (VP-16)

Etoposide (VP-16) is a chemotherapeutic agent commonly used in the treatment of small-cell and non-small-cell lung cancers. This drug has rarely been associated with the development of pulmonary toxicity, usually after prolonged treatment, although cases have occurred after a single cycle of chemotherapy.[61] Pathologic examination of lung tissues shows diffuse alveolar damage and the presence of atypical type 2 pneumocytes. Discontinuation of the drug and administration of corticosteroids is usually recommended.

PACLITAXEL AND DOCETAXEL

Paclitaxel and docetaxel are chemotherapeutic agents used in the treatment of breast, ovarian and lung cancers. There are well-documented cases of pulmonary toxicity induced by paclitaxel in the literature, although the exact incidence remains elusive. Patients typically experience the onset of dry cough, dyspnoea and chest tightness within a few minutes of administration of the drug, suggesting a type 1 hypersensitivity reaction.[62–64] IgE antibodies directed against paclitaxel or maybe its vehicle, Cremophor EL, are thought to be responsible. This reaction may occur in as many as 30 per cent of patients, and may be somewhat preventable by premedication with steroids. Lung auscultation may reveal significant wheezing, suggesting bronchospasm. Radiographic findings are non-specific, typically consisting of transient interstitial infiltrates, suggesting a hypersensitivity mechanism. Cases of noncardiogenic pulmonary oedema have been described. Rare cases of interstitial pneumonitis presumably induced by docetaxel have also been reported. The combination of docetaxel and gemcitabine may be particularly harmful.

ZINOSTATIN

Zinostatin has been implicated in the development of drug-induced interstitial pneumonitis after prolonged courses of treatment. It has also been shown to induce pulmonary endothelial hypertrophy.[65]

OTHER CHEMOTHERAPEUTIC AGENTS

Irinotecan and topotecan

Irinotecan is a semisynthetic camptothecin, which has been approved for the treatment of metastatic colorectal cancer. Earlier reports from Japan suggested an incidence of lung toxicity of approximately 1.8 per cent, characterized by the onset of shortness of breath and fever.[66,67] Imaging studies are usually non-specific, and may show reticulonodular infiltrates or occasionally pleural effusions. This contrasts with a reported incidence of cough and dyspnoea in excess of 20 per cent in US studies with this agent, although most of these patients had previously been exposed to 5-fluorouracil and many of them had also been previously diagnosed with thoracic malignancies.[68] Other studies have suggested a much lower incidence of drug-induced pulmonary toxicity, perhaps in fewer than 1 per cent of the patients receiving this medication. Radiotherapy and pre-existing lung morbidities may increase the likelihood of pulmonary toxicity.

Topotecan is a similar chemotherapeutic agent, which has rarely been associated with pulmonary toxicity. Rare reports

of obliterative bronchiolitis presumably induced by topotecan can be found in the literature.[69]

All-*trans*-retinoic acid

ATRA has been used with excellent results for the treatment of promyelocytic leukaemia, or acute myeloblastic leukaemia M3 according to the French/American/British classification. The basis for this treatment is the presence of a mutation in the retinoic acid receptor-alpha gene in patients with these disorders, which to a certain extent can be corrected by high doses of ATRA. ATRA thus allows for the maturation of leukaemic cells and can therefore induce disease remission. It is usually used in association with other agents, namely daunorubicin and cytosine arabinoside. ATRA may also prevent the onset of disseminated intravascular coagulation which frequently complicates this type of leukaemia.

The main complication of the treatment with retinoic acid is the development of the so-called 'retinoic acid syndrome' in approximately 25 per cent of patients receiving this medication.[70–72] This capillary leak syndrome is heralded by the onset of shortness of breath, fever and diffuse pulmonary infiltrates resulting from non-cardiogenic pulmonary oedema. Pericardial and pleural effusions are also common, and may further impair the patient's respiratory status. Hypotension and renal failure may be present as well. This syndrome usually occurs between 2 and 21 days after initiation of treatment. Its pathogenesis remains poorly understood, but it may involve a sudden massive release of cytokines by the newly differentiated myeloid cells and adhesion of matured granulocytes to the pulmonary endothelium. High leucocyte counts have been associated with an increased incidence of the retinoic acid syndrome, although inconsistently. No other obvious risk factors have been clearly identified. Microscopic examinations of affected lungs typically show diffuse infiltration by mature myeloid cells. An association with Sweet syndrome and diffuse alveolar haemorrhage has also been described.[73] The mortality rate with ATRA-associated pulmonary toxicity may be as high as 9 per cent.

Discontinuation of the drug is the priority. Administration of high doses of intravenous steroids is often helpful in the majority of cases. Interestingly, ATRA can usually be reintroduced after resolution of the retinoic aid syndrome. Preventive treatment with corticosteroids before administration of ATRA may also be beneficial for patients receiving this medication.[74]

Tyrosine kinase inhibitors

Gefitinib is a selective tyrosine kinase inhibitor that has been primarily used in the treatment of non-small-cell lung carcinomas that are not responsive to conventional chemotherapy. It is now also used in the treatment of breast, ovarian and colon cancers. Numerous cases of gefitinib-induced respiratory failure have been reported and the overall incidence may be of the order of 1 per cent with a mortality rate approaching 30 per cent.[75] Radiotherapy, prior history of lung fibrosis, or lung infection may predispose to this pulmonary complication.

Patients present with gefitinib-induced pulmonary toxicity exhibiting sudden onset of fever, cough and rapidly progressive shortness of breath. Progression to fibrosis is possible but rare. Cases of diffuse alveolar haemorrhage have been reported. CT scan of the chest typically shows diffuse ground-glass infiltrates. Biopsies of lung tissue may reveal a diffuse alveolar damage pattern. The pathogenesis of gefitinib-induced pulmonary toxicity remains poorly understood, although it is likely that the drug may interfere with the epidermal growth factor receptor, thereby impairing the alveolar repair process.[48]

Discontinuation of this offending agent and administration of corticosteroids results in significant improvement in the majority of cases. Unfortunately, a minority of patients progress to respiratory failure despite these interventions.

Imatinib mesylate is another tyrosine kinase inhibitor, which was specifically designed to treat chronic myelogenous leukaemia. This myeloproliferative disorder is characterized by the bcr-abl rearrangement responsible for the production of an aberrant tyrosine kinase targeted by imatinib mesylate. Rare cases of pulmonary complications presumably secondary to imatinib mesylate have been described.[76,77] Interstitial pneumonitis may present early in the course of the disease, but late-onset fibrosis has also been described. This agent has also been associated with the development of cardiac insufficiency with pleural effusions and pulmonary oedema on that basis. Discontinuation of the drug should be considered in the presence of unexplained pulmonary infiltrates.

Interferons

Interferons are immune modulatory proteins secreted by various cells including B- and T-cell lymphocytes, natural killer lymphocytes, and fibroblasts. There are three types of interferons: alpha, beta and gamma. They are involved in a host of reactions and are characterized by their antiviral and anti-oncogenic properties. Expectedly, they have been used in a broad variety of neoplastic, infectious, inflammatory and immunologic diseases.

Several different types of pulmonary reaction have been described with the use of interferons. A granulomatous inflammatory process has been described, clinically and pathologically indistinguishable from sarcoidosis.[78] Pathology can confirm the presence of non-caseating granulomas in lymph nodes, spleen, skin and liver. Second, a subacute onset of interstitial lung disease has been reported, within weeks to months of initiation of treatment. Both presentations have been shown to respond to discontinuation of treatment and administration of corticosteroids. In the subacute form, biopsies typically show a lymphocytic infiltrate with predominance of CD8 lymphocytes, although a bronchiolitis obliterans with organizing pneumonitis pattern has also been described.[79,80] In addition, four patients reportedly developed acute respiratory failure during treatment with interferon-γ for idiopathic pulmonary fibrosis in one series.[81] Finally, severe exacerbation of bronchospasm has been reported in asthmatic patients receiving interferons.[82]

Tumour necrosis factor

TNF-α is a cytokine produced chiefly by macrophages. It has been tried in advanced malignancies as an investigational agent, though its use has greatly diminished in recent years.

In one study, all 27 patients treated demonstrated decreased DLCO on screening pulmonary function studies. These abnormalities occurred several weeks after initiation of therapy, and severe and life-threatening diffuse lung injury reactions have also occurred in some patients receiving this agent.[83,84]

Interleukin-2

The main pulmonary toxicity induced by interleukin-2 is the development of a capillary leak syndrome. Interleukin-2 has been used in the past as an investigational agent for a variety of advanced malignancies and has been associated with pulmonary symptoms in half the patients treated. This toxicity has greatly limited its use recently. Both cardiogenic and non-cardiogenic pulmonary oedema have been reported.[85–87] They usually resolve promptly on discontinuation of therapy. Cases of hypersensitivity pneumonitis and exacerbation of bronchospasm have also been described.

Granulocyte–macrophage stimulating factor (GM–CSF) and granulocyte stimulating factor (G–CSF)

These agents are frequently used in the post-chemotherapy period to stimulate bone marrow production. They potentiate chemotherapy-induced lung toxicity, perhaps via a recall phenomenon. G–CSF has also been associated with the development of non-cardiogenic pulmonary oedema.[88]

KEY POINTS

- It is essential to maintain a high index of suspicion that unexplained pulmonary disease may be caused by solo chemotherapy medication, or chemotherapy regimen.
- Many of the drug-induced pulmonary reactions can be reversible if the drug is stopped in time. If necessary, judicious use of corticosteroids may also be beneficial.
- There are almost no diagnostic studies available that will definitively confirm the presence of drug-induced lung disease, and infectious and neoplastic processes must be excluded in these patients.
- If there is a question of whether or not a medication can be the cause of a particular pulmonary abnormality, one option is to call the medical director of the drug manufacturer listed in the *Physicians' Desk Reference* (PDR), or to contact the FDA. In addition, a valuable website (Pneumotox®) with a wealth of information on drug-induced pulmonary reactions can be assessed at www.pneumotox.com.

REFERENCES

1. Rosenow EC, Limper AH. Drug-induced pulmonary disease. *Semin Respir Infect* 1995:**10**;86–95.
2. Snyder LS, Hertz MI. Cytotoxic drug-induced lung injury. *Semin Respir Infect* 1988:**3**;217–28.

3. Sleijfer S. Bleomycin-induced pneumonitis. *Chest* 2001;**120**:617–24.

4. Simpson AB, Paul J, Graham J, Kaye SB. Fatal bleomycin pulmonary toxicity in the west of Scotland 1991–1995: a review of patients with germ cell tumours. *Br J Cancer* 1998;**78**:1061–6.

5. Limper AH, Rosenow EC. Drug-induced interstitial lung disease. *Curr Opin Pumon Med* 1996;**2**:396–404.

6. Camus PH, Foucher P, Bonniaud PH, Ask K. Drug-induced infiltrative lung disease. *Eur Respir J Suppl* 2001;**32**:93s–100s.

7. Carson CW, Cannon GW, Egger MJ. Pulmonary disease during the treatment of rheumatoid arthritis with low-dose pulse methotrexate. *Semin Arthritis Rheum* 1987;**16**:186–95.

8. Gispen JG, Alarcaon GS, Johnson JJ. Toxicity of methotrexate in rheumatoid arthritis. *J Rheumatol* 1987;**14**:73–9.

9. Cleverley JR, Screaton NJ, Hiorns MP, Flint JD, Muller NL. Drug-induced lung disease: high resolution CT and histological findings. *Clin Radiol* 2002;**57**:292–9.

10. Limper AH, Rosenow EC. Drug-induced pulmonary disease. In: Murray JF, Nadel JA (eds) *Textbook of Respiratory Medicine*, 3rd edn. New York: Saunders, 2000.

11. Lin WY, Kao CH, Wang SJ, Yeh SH. Lung toxicity in chemotherapeutic agents detected by TC-99m DTPA radioaerosol inhalation lung scintigraphy. *Neoplasma* 1995;**42**:133–5.

12. Myers JL. Pathology of drug-induced lung disease. In: Katzenstein AL, Askin F (eds) *Surgical Pathology of Non-neoplastic Lung Disease*, 2nd edn. Philadelphia, PA: Saunders, 1990.

13. Camus P, Fanton A, Bonniaud P, Camus C, Foucher P. Interstitial lung disease induced by drug and radiation. *Respiration* 2004;**71**:301–26.

14. Sleijfer S. Bleomycin-induced pneumonitis. *Chest* 2001;**120**:617–24.

15. Chandler DB. Possible mechanisms of bleomycin-induced fibrosis. *Clin Chest Med* 1990;**11**:21.

16. Limper AH. Chemotherapy-induced lung disease. *Clin Chest Med* 2004;**25**:53–64.

17. Abid SH, Malhotra V, Perry MC. Radiation-induced and chemotherapy-induced pulmonary injury. *Curr Opin Oncol* 2001;**13**:242–8.

18. Ingrassia TS, Ryu JH, Trastek VF, et al. Oxygen-exacerbated bleomycin pulmonary toxicity. *Mayo Clin Proc* 1991;**66**:173–8.

19. Goldiner PL, Carlon GC, Cvitkovic E, et al. Factors influencing postoperative mortality and mortality in patients treated with bleomycin. *Br Med J* 1978;**1**:1664–7.

20. Sleijfer S, van der Mark TW, Schraffordt Koops H, Mulder NH. Decrease in pulmonary function during bleomycin-containing combination chemotherapy for testicular cancer: not only a bleomycin effect. *Br J Cancer* 1995;**71**:120–3.

21. Bellamy EA, Husband JE, Blaquiere RM. Bleomycin-related lung damage: CT evidence. *Radiology* 1985;**156**:155–8.

22. Jules-Elyse K, White DA. Bleomycin induced pulmonary toxicity. *Clin Chest Med* 1990;**11**:1–20.

23. Verweij J, van Zenten T, Souren T, et al. Prospective study on the dose relationship of mitomycin C-induced interstitial pneumonitis. *Cancer* 1987;**60**:756–61.

24. Okuno SH, Frytak S. Mitomycin lung toxicity: acute and chronic phases. *Am J Clin Oncol* 1997;**20**:282–4.

25. Klein DS, Wilds PR. Pulmonary toxicity of antineoplastic agents: anaesthetic and postoperative implications. *Can Anaes Soc J* 1983;**30**:399.

26. Castro M, Veeder MH, Mailliard JA, Tazelaar HD, Jett JR. A prospective study of pulmonary function in patients receiving mitomycin. *Chest* 1996;**109**:939–44.

27. Chang AY, Kuebler JP, Pandya KJ, et al. Pulmonary toxicity induced by mitomycin C is highly responsive to glucocorticoids. *Cancer* 1986;**57**:2285–90.

28. Shelton R, Slaughter D. A syndrome of microangiopathic hemolytic anemia, renal impairment, and pulmonary edema in chemotherapy-treated patients with adenocarcinoma. *Cancer* 1986;**58**:1428–36.

29. Lyman NW, Michaelson R, Viscuso R L, et al. Mitomycin-induced hemolytic-uremic syndrome: successful treatment with corticosteroids and intense plasma exchange. *Arch Intern Med* 1983;**143**:1617–18.

30. Hinson JM, McKibben AW. Chemotherapy-associated lung injury. In: Perry MC (ed.) *The Chemotherapy Source Book*, 3rd edn. New York: Lippincott Williams & Wilkins, 2001.

31. Oliner H, Schwartz R, Rubio F, Dameshek W. Interstitial pulmonary fibrosis following busulfan therapy. *Am J Med* 1961;**31**:134.

32. Fernandez HF, Tran HT, Albrecht F, et al. Evaluation of safety and pharmacokinetics of administering intravenous busulfan in a twice-daily or daily schedule to patients with advanced hematologic malignant disease undergoing stem cell transplantation. *Biol Blood Marrow Transplant* 2002;**8**:486–92.

33. Aymard JP, Gyger M, Lavallee R, et al. A case of pulmonary alveolar proteinosis complicating chronic myelogenous leukemia. *Cancer* 1984;**53**:954–6.

34. Watanabe K, Sueishi K, Tanake K, et al. Pulmonary alveolar proteinosis and disseminated atypical mycobacteriosis in a patient with busulfan lung. *Acta Pathol Jpn* 1990;**40**:63–6.

35. Malik SW, Myers JL, DeRemee RA, Specks U. Lung toxicity associated with cyclophosphamide use: two distinct patterns. *Am J Respir Crit Care Med* 1996;**154**:1851–6.

36. Akasheh MS, Freytes CO, Vesole DH. Melphalan-associated pulmonary toxicity following high-dose therapy with autologous hematopoietic stem cell transplantation. *Bone Marrow Transplant* 2000;**26**:1107–9.

37. Hadjiliadis D, Govert J A. Methemoglobinemia after infusion of ifosfamide chemotherapy. *Chest* 2000;**118**:1208–10.

38. Mahmood T, Mudad R. Pulmonary toxicity secondary to procarbazine. *Am J Clin Oncol* 2002;**25**:187–8.

39. Zisman DA, McCune WJ, Tino J, Lynch JP. Drug-induced pneumonitis: the role of methotrexate. *Sarcoidosis Vasculitis Diffuse Lung Dis* 2001;**18**:243–52.

40. Rosenow EC, Myers JL, Swensen SJ, Pisani RJ. Drug-induced pulmonary disease. *Chest* 1992;**102**:239.

41. St Clair EW, Rice JR, Snyderman R. Pneumonitis complicating low-dose methotrexate therapy in rheumatoid arthritis. *Arch Intern Med* 1985;**145**:2035–8.

42. Lateef O, Shakoor N, Balk RA. Methotrexate pulmonary toxicity. *Expert Opin Drug Safety* 2005;**4**:723–30.

43. Hilliquin P, Renoux M, Perrot S, et al. Occurrence of pulmonary complications during methotrexate therapy in rheumatoid arthritis. *Br J Rheumatol* 1996;**35**:441.

44. Cottin V, Tebib J, Massonet B, *et al.* Pulmonary function in patients receiving long-term low-dose methotrexate. *Chest* 1996;**109**:933.

45. Andersson BS, Luna MA, Yee C, *et al.* Fatal pulmonary failure complicating high-dose cytosine arabinoside therapy in acute leukemia. *Cancer* 1990;**65**:1079–84.

46. Jehn J, Goldel N, Rienmauller R, *et al.* Noncardiogenic pulmonary edema complicating intermediate and high-dose Ara C treatment for relapsed acute leukemia. *Med Oncol Tumor Pharmacother* 1988;**5**:41–7.

47. Gupta N, Ahmed I, Steinberg H, *et al.* Gemcitabine-induced pulmonary toxicity: case report and review of the literature. *Am J Clin Oncol* 2002;**25**:96–100.

48. Dimopoulou I, Bamias A, Lyberopoulos P, Dimopoulos MA. Pulmonary toxicity from novel antineoplastic agents. *Ann Oncol* 2006;**17**:372–9.

49. Sauer-Heilborn A, Kath R, Schneider CP, Hoffken K. Severe non-haematological toxicity after treatment with gemcitabine. *J Cancer Res Clin Oncol* 1999;**125**:637–40.

50. Helman DL, Byrd JC, Ales NC, Shorr AF. Fludarabine-related pulmonary toxicity: a distinct clinical entity in chronic lymphoproliferative syndromes. *Chest* 2002;**122**:785–90.

51. Garg S, Garg MS, Basmaji N. Multiple pulmonary nodules, an unusual presentation of fludarabine pulmonary toxicity: case report and review of the literature. *Am J Hematol* 2002;**70**:241–5.

52. Bedrossian CWM, Sussman J, Conklin RH, Kahan B. Azathioprine-associated interstitial pneumonitis. *Am J Clin Pathol* 1984;**82**:148–54.

53. Weiss RB, Poster DS, Penta JS. The nitrosureas and pulmonary toxicity. *Cancer Treat Rev* 1981;**8**:111.

54. Durant JR, Norgard MJ, Murad TM, *et al.* Pulmonary toxicity associated with bischloroethylnitrosurea (BCNU). *Ann Intern Med* 1979;**90**:191.

55. O'Driscoll BR, Hasleton PS, Taylor PM, *et al.* Active lung fibrosis up to 17 years after chemotherapy with carmustine (BCNU) in childhood. *New Engl J Med* 1990;**323**:378.

56. Weinstein AS, Diener-West M, Nelson DF, Pakuris E. Pulmonary toxicity of carmustine in patients treated for malignant glioma. *Cancer Treat Rep* 1986;**70**:943.

57. Kalaycioglu M, Kavuru M, Tuason L, Bolwell B. Empiric prednisone therapy for pulmonary toxic reaction after high-dose chemotherapy containing carmustine (BCNU). *Chest* 1995;**107**:482.

58. Wilson KS, Brigden ML, Alexander S, Worth A. Fatal pneumothorax in 'BCNU lung'. *Med Pediatr Oncol* 1982;**10**:195–9.

59. Rao SX, Ramaswamy G, Levin M, *et al.* Fatal acute respiratory failure after vinblastine–mitomycin therapy in lung carcinoma. *Arch Intern Med* 1985;**145**:1905–7.

60. Hoelzer KL, Harrison BR, Luedke SW, *et al.* Vinblastine-associated pulmonary toxicity in patients receiving combination therapy with mitomycin and cisplatin. *Drug Intel Clin Pharm* 1986;**20**:287–9.

61. Gurjal A, An T, Valdivieso M, Kalemkerian GP. Etoposide-induced pulmonary toxicity. *Lung Cancer* 1999;**26**:109–12.

62. Rowinsky EK, Donehower RC. Paclitaxel (Taxol). *New Engl J Med* 1995;**332**:1004–14.

63. Shannon VR, Price KJ. Pulmonary complications of cancer therapy. *Anesth Clin N Amer* 1998;**16**:563–86.

64. Fujimori K, Yokoyama A, Kurita Y, Uno K, Saijo N. Paclitaxel-induced cell-mediated hypersensitivity pneumonitis: diagnosis using leukocyte migration test, bronchoalveolar lavage and transbronchial biopsy. *Oncology* 1998;**55**:340–4.

65. Calvo DB, Legha SS, McKelvey EM, Bodey GP, Dail DH. Zinostatin-related pulmonary toxicity. *Cancer Treat Rep* 1981;**65**:165–7.

66. Fukuoka M, Niitnai H, Suzuki A, *et al.* A phase II study of CPT-11, a derivative of camptothecin, for previously untreated non small cell lung cancer. *J Clin Oncol* 1992;**10**:16–20.

67. Masuda N, Fukuoka M, Kusunoki Y, *et al.* CPT-11: a new derivative of camptothecin for the treatment of refractory or relapsed small cell lung cancer. *J Clin Oncol* 1992;**10**:1225–9.

68. Madarnas Y, Webster P, Shorter AM, *et al.* Irinotecan-associated pulmonary toxicity. *Anti-Cancer Drugs* 2000;**11**:709–13.

69. Edgerton CC, Gilman M, Roth BJ. Topotecan-induced bronchiolitis. *South Med J* 2004;**97**:699–701.

70. Frankel SR, Eardley A, Lauwers G, *et al.* The 'retinoic acid syndrome' in acute promyelocytic leukemia. *Ann Intern Med* 1992;**117**:292.

71. Vahdat L, Maslak P, Miller WH, *et al.* Early mortality and the retinoic acid syndrome in acute promyelocytic leukemia: impact of leukocytosis, low-dose chemotherapy, PMN/RAR-alpha isoform, and CD13 expression in patients treated with all-*trans*-retinoic acid. *Blood* 1994;**84**:3843.

72. De Botton S, *et al.*, European APL Group. Incidence, clinical features, and outcome of all trans-retinoic acid syndrome in 413 cases of newly diagnosed acute promyelocytic leukemia. *Blood* 1998;**92**:2712.

73. Nicolls MR, Terada LS, Tuder RM, Prindiville SA, Schwarz MI. Diffuse alveolar hemorrhage with underlying pulmonary capillaritis in the retinoic acid syndrome. *Am J Respir Crit Care Med* 1998;**158**:1302–5.

74. Wiley JS, Firkin FC. Reduction of pulmonary toxicity by prednisolone prophylaxis during all-*trans*-retinoic acid treatment of acute promyelocytic leukemia. *Leukemia* 1995;**9**:774–8.

75. Cohen MH, Williams GA, Sridhara R, *et al.* FDA drug approval summary: gefitinib (ZD 1839) (Iressa) tablets. *Oncologist* 2003;**8**:303–6.

76. Ma CX, Hobday TJ, Jett JR. Imatinib mesylate-induced interstitial pneumonitis. *Mayo Clin Proc* 2003;**78**:1578–9.

77. Goldsby R, Pulsipher M, Adams R, *et al.* Unexpected pleural effusions in 3 pediatric patients treated with STI-571. *J Pediatr Hematol Oncol* 2002;**24**:694–5.

78. Rubinowitz AN, Naidich DP, Alinsonorin C. Interferon-induced sarcoidosis. *J Comp Assist Tomog* 2003;**27**:279–83.

79. Kumar KS, Russo MW, Borczuk AC, *et al.* Significant pulmonary toxicity associated with interferon and ribavirin therapy for hepatitis C. *Am J Gastro* 2002;**97**:2432–40.

80. Patel M, Ezzat W, Pauw KL, Lowsky R. Bronchiolitis obliterans organizing pneumonia in patient with chronic myelogenous leukemia developing after initiation of interferon and cytosine arabinoside. *Eur J Haem* 2001;**67**:318–21.

81. Honores I, Nunes H, Groussard O, *et al.* Acute respiratory failure after interferon-gamma therapy of end-stage pulmonary fibrosis. *Am J Respir Crit Care Med* 2003;**167**:953–7.

82. Bini EJ, Weinshel EH. Severe exacerbation of asthma: a new side effect of interferon-alpha in patients with asthma and chronic hepatitis C. *Mayo Clin Proc* 1999;**74**:367–70.

83. Hocking DC, Philips PG, Ferro TJ, *et al.* Mechanisms of pulmonary edema induced by tumor necrosis factor-alpha. *Circ Res* 1990;**67**:68–77.

84. Kuei JH, Tashkin DP, Figlin RA. Pulmonary toxicity of recombinant human tumor necrosis factor. *Chest* 1989;**96**:334–8.

85. Lazarus DS, Kurnick JT, Kradin RL. Alterations in pulmonary function in cancer patients receiving adoptive immunotherapy with tumor-infiltrating lymphocytes and interleukin-2. *Am Rev Respir Dis* 1990;**141**:193–8.

86. Saxon RR, Klein JS, Bar MH, *et al.* Pathogenesis of pulmonary edema during interleukin-2 therapy: correlation of chest radiographic and clinical findings in 54 patients. *Am J Roentgenol* 1991;**156**:281–5.

87. Conant EF, Fox KR, Miller WT. Pulmonary edema as a complication of interleukin-2 therapy. *Am J Roentgenol* 1989;**152**:749–52.

88. Vial T, Descoates J. Clinical toxicity of cytokines used as hematopoetic growth factors. *Drug Safety* 1995;**13**:371–406.

15

Pulmonary reactions to novel chemotherapeutic agents and biomolecules

ALBERT J POLITO

INTRODUCTION

Advances in our understanding of disease mechanisms and in the ability to synthesize novel pharmacological agents have revolutionized the treatment of a wide array of disorders, including infections, autoimmune diseases and cancer. 'Biomolecules' is the broad term used to describe both synthetically manufactured versions of endogenous cytokines and monoclonal antibodies directed at specific molecular targets. In the field of cancer chemotherapy, additional novel compounds have been developed that inhibit a variety of different targets, including tyrosine kinases, serine-threonine kinase and the proteasomes.

Along with these new agents has come a host of newly described drug-induced lung syndromes, including infection to specific microorganisms, ranging from isolated case reports to much larger case series. These syndromes are frequently a direct result of the biological targeting and the attendant changes in the molecular milieu. This chapter discusses the individual agents and their clinical pulmonary manifestations and management.

INTERFERONS

Interferons (IFNs) are a family of related glycoproteins with antiviral and immunomodulatory properties. Their biological effects include regulation of T-cell-mediated immunity, enhancement of natural killer cell activity, and activation of macrophages in response to viral infection.[1] They have been used to treat a variety of conditions, including collagen vascular diseases, malignancies, neurological diseases and infections, notably viral hepatitis C infection. The vast majority of patients experience some constitutional symptoms during IFN therapy, and a smaller number develop immune-related disorders resulting from the profound influence of these drugs on immune regulation. Not surprisingly, the unique immunobiology of the lungs is fertile ground for such adverse effects.

Interferon-α

Through its stimulation of T-helper-1 (Th1) responses and macrophage activation, IFN-α can induce the formation of granulomas.[2] In the clinical setting these effects have translated into sarcoidosis induced by IFN-α, either as a de-novo condition or a relapse of previously diagnosed sarcoidosis. A recent review identified 60 published reports of this entity in the English literature.[3] The epidemiological profile includes a mean age of 49.8 years for affected patients. There is a slight female predominance (1.3:1), and the most common site of involvement is the lungs, with a frequency (70 per cent) slightly less than with naturally occurring sarcoidosis (90 per cent).[4] A variety of radiographic findings may be seen, including mediastinal and hilar adenopathy, interstitial infiltrates and diffuse nodularity. Only a few cases in the literature have reported pulmonary function test results; obstruction, restriction and a gas transfer defect, either alone or in combination, have been observed. As with naturally occurring sarcoidosis, the second major organ affected is the skin, and hypercalcaemia and involvement of the cardiac, neurologic, hepatic and articular systems have all been reported.[3,4]

Since IFN-α is the mainstay therapy for chronic hepatitis C infection, it is not surprising that the majority of reported cases (52/60) have been in the population with this viral infection.[3] While sarcoidosis may occur when IFN-α is used as monotherapy, most affected patients have received a combination of IFN-α and ribavirin. This synthetic guanosine analogue may potentiate the sarcoid-inducing effects of IFN-α by enhancing Th1 responses and inhibiting Th2 cytokine production.[5] Sarcoidosis occurs with both the standard and pegylated forms of IFN-α. The mean time to onset of symptoms after initiation of therapy is 11.4 months (range 1–60), and the diagnosis is established most often by skin or lung biopsies.[3]

Approximately half of affected patients achieve spontaneous remission of the disease after diminution or discontinuation of therapy.[4] Corticosteroids have been used successfully in some patients, and there are occasional reports of spontaneous resolution despite continuation of IFN-α therapy.[6,7] A small percentage of patients appear to develop a persistent, relapsing course of sarcoidosis.

Because of the non-specific and insidious presentation of sarcoidosis, the overall incidence of this disease in the setting of IFN-α therapy is likely underestimated, as the symptoms may be interpreted as constitutional adverse effects of the drug. A baseline chest X-ray and a complete set of pulmonary function tests should be performed in all patients on the initiation of IFN-α therapy. The later development of dyspnoea or skin lesions during therapy warrants a directed evaluation for sarcoidosis, as does the development of unexpectedly severe or prolonged IFN-α-related constitutional symptoms. Once sarcoidosis is diagnosed, the decision whether to discontinue therapy, decrease the dosage, or initiate corticosteroid treatment must be individualized based on the benefits of IFN-α and the severity of the symptoms, including the specific organs involved. The potential for corticosteroids to increase the viral load of hepatitis C also must be taken into account in making this decision.

A spectrum of non-sarcoid lung toxicity from IFN-α also may be seen, albeit infrequently. Bronchiolitis obliterans organizing pneumonia (BOOP) usually resolves with discontinuation of IFN-α and administration of corticosteroids,[8,9] although progressive, fatal BOOP from respiratory failure has been described.[10] Interstitial pneumonitis also may occur, and cessation of drug treatment with or without corticosteroids may result in improvement.[11] Prolonged need for corticosteroids has been observed,[9] and there are isolated reports of progression to death from acute respiratory distress syndrome (ARDS), with open lung biopsy showing diffuse alveolar damage from acute interstitial pneumonitis.[12] Although inhaled IFN-α has been shown to induce bronchospastic reactions in asthmatic patients,[13] the data on subcutaneously administered IFN-α in the setting of asthma are conflicting. There is a single report of two patients with mild asthma in whom treatment with IFN-α resulted in severe exacerbations,[14] but improvements in asthma control also have been observed with IFN-α therapy in corticosteroid-dependent asthmatic patients.[15]

Interferon-β

IFN-β has similar biological actions to IFN-α, but pulmonary adverse effects have been reported much less frequently. There has been no consistent association with an individual disease state. Only three case reports of sarcoidosis occurring with IFN-β therapy have been published – in the settings of renal cell carcinoma,[16] multiple myeloma[17] and multiple sclerosis.[18] BOOP has been reported in two patients treated with IFN-β, one with multiple sclerosis[19] and the other with malignant melanoma.[20] Interstitial pneumonitis, respiratory failure and asthma effects have not been described.

Interferon-γ

IFN-γ has a different spectrum of pulmonary adverse effects from IFN-α and IFN-β. It has few United States FDA-approved indications (osteopetrosis and chronic granulomatous disease), but investigations as a potential therapy for malignancy and for idiopathic pulmonary fibrosis (IPF) have received wide attention. Several reports have suggested that IFN-γ potentiates the deleterious effects of radiation in the treatment of lung cancer.[21,22] In one of these, a pilot study of IFN-γ combined with radiotherapy for non-small-cell lung cancer, 44 per cent of treated patients developed severe or fatal radiation pneumonitis, compared to a 9.5 per cent incidence with radiotherapy alone in a prior study.

An early phase I trial investigated the use of IFN-γ in combination with interleukin-2 (IL-2) for patients with metastatic cancer.[23] The major dose-limiting toxicity was pulmonary and manifested as rales and dyspnoea. This toxicity was related to the dose of both IFN-γ and IL-2 (the adverse lung effects of the latter are discussed below).

IPF has been an area of intense investigation for the potential therapeutic use of IFN-γ. The drug is proven to be of no benefit in this disease, but new insights into the possible lung toxicity of IFN-γ were gained during the several years when it was prescribed widely in an off-label capacity. Honoré and colleagues described four patients with IPF in whom ARDS developed in a fulminant manner soon after initiation of IFN-γ therapy.[24] The clinical presentations were similar, with increasing dyspnoea, fever, new areas of ground-glass infiltrates, rapidly worsening hypoxaemia, and the need for mechanical ventilation. All four patients died of respiratory failure. In two patients, lung pathology was available and revealed diffuse alveolar damage superimposed on pre-existing usual interstitial pneumonia. Suggested predisposing factors for these effects of IFN-γ included severe pre-treatment pulmonary function impairment and a history of familial lung fibrosis. Later studies comparing the incidence rate of acute exacerbations in IFN-γ-treated IPF patients versus controls will help to determine whether these events are drug-related.

INTERLEUKINS

Interleukin-2

Interleukin-2 (IL-2) is an endogenous cytokine that is produced by activated T-cells and natural killer cells and has direct activating effects on several subsets of immune cells.[25] In its recombinant form, it is used as an immunoactivating cancer treatment for metastatic melanoma and renal cell carcinoma, but its clinical utility has been limited by the development of a vascular leak syndrome.[26] Increased microvascular permeability results in weight gain, peripheral oedema, oliguria and non-cardiogenic pulmonary oedema. This presents within a few days after IL-2 administration as a subacute clinical syndrome characterized by dyspnoea, cough and hypoxaemia. Radiographic imaging demonstrates pleuropulmonary abnormalities in approximately 75 per cent of patients, with the most common findings being bilateral perihilar alveolar or interstitial infiltrates and pleural effusions.[27]

Non-cardiogenic pulmonary oedema is a dose-related phenomenon that usually resolves within a few days of treatment discontinuation, but it can be very severe and lead to the need for intubation and mechanical ventilation. Monitoring of weight gain may help to identify affected patients at an earlier stage. Greater experience administering IL-2 since its first clinical trials in 1984 has led to a progressive reduction in morbidity and mortality from the drug, with the incidence of intubation dropping to 3 per cent.[28] Subcutaneous administration has a lower incidence of pulmonary toxicity than intravenous use,[29] and some investigators have suggested that locoregional administration of nebulized IL-2 for treatment of pulmonary or mediastinal metastases carries very little toxicity.[30] Corticosteroids are helpful in reducing the flu-like side-effects of IL-2 therapy, but conflicting data exist on whether they can reduce the incidence of the vascular leak syndrome.[31,32]

HAEMATOPOIETIC GROWTH FACTORS

Granulocyte colony-stimulating factor

Granulocyte colony-stimulating factor (G-CSF) is a haematopoietic growth factor that is used to stimulate neutrophil recovery following chemotherapy-induced myelosuppression. Although G-CSF is generally safe and well tolerated, there have been a number of reports of acute lung injury (ALI) or ARDS when it is used in combination with chemotherapeutic agents with known lung toxicity, including bleomycin, busulfan, cyclophosphamide and methotrexate.[33,34] Susceptibility to this type of pulmonary adverse effect may be linked to human leucocyte antigens (HLA) B51 and B52.[34] Most patients have received at least three courses of chemotherapy, suggesting that repeated endothelial damage by chemotherapy agents enhances G-CSF-induced alveolar inflammation.[33,35] The treatment approach is discontinuation of G-CSF with or without addition of corticosteroids, and the mortality rate is 25 per cent. Interstitial pneumonitis also has been described,[33,36] as have isolated exudative pleural effusions.[37]

G-CSF is sometimes used in healthy volunteers donating granulocytes to neutropenic relatives undergoing bone marrow transplantation, as well as for mobilization of peripheral blood stem cells for stem-cell transplantation. In these settings, monotherapy with G-CSF rarely has been reported to cause non-cardiogenic pulmonary oedema,[38] ALI[39] or pulmonary haemorrhage.[40] A case of fatal ARDS also has been reported in an elderly man inappropriately administered G-CSF for anaemia.[41] The mechanism for the lung toxicity in these cases is unknown, but G-CSF-induced neutrophil activation has been postulated. Transient, asymptomatic gas exchange disturbances, with increased arterial carbon dioxide tension and increased alveolar–arterial oxygen gradient, also have been described during peripheral stem-cell mobilization by G-CSF in healthy donors.[42]

Granulocyte–macrophage colony-stimulating factor

Experience with granulocyte–macrophage colony-stimulating factor (GM-CSF) is not as extensive as with G-CSF, as its range

of indications is much narrower. As with G-CSF, it has been used for reversal of neutropenia and for progenitor mobilization, and direct therapeutic efficacy for treatment of cancer has also been evaluated. In an early phase I trial of GM-CSF in patients with lung cancer, the dose-limiting toxicity was pulmonary in all patients.[43] A vascular leak syndrome with non-cardiogenic pulmonary oedema and pleural effusions has been described.[44] When administered to a series of 16 patients with acute leukaemia as a 'cytokine priming' manoeuvre, there were three deaths from acute respiratory failure characterized by pulmonary infiltrates and oedema.[45] Addition of GM-CSF to chemotherapy and radiation therapy for limited stage small-cell lung cancer resulted in significantly more 'toxic' deaths than for the combination modality alone in a randomized, multicentre trial, with most of the deaths from pulmonary complications (pneumonitis, pulmonary oedema, pulmonary embolism, aspiration pneumonia and haemoptysis).[46] Alternative means of delivery hold promise, as administration of nebulized GM-CSF as a lung-specific biological therapy in patients with lung metastases showed positive antitumour effects without documented pulmonary toxicity.[47]

Macrophage colony-stimulating factor

Clinical experience with macrophage colony-stimulating factor (M-CSF, also called CSF-1) is fairly limited, as its specificity for cells of the monocyte lineage has prevented broad clinical utility.[48] Few studies have commented on potential pulmonary side-effects. Zamkoff and colleagues administered a range of M-CSF dosages to 20 patients with advanced malignancies and reported acute dyspnoea in three subjects.[49] The investigators felt that the dyspnoea was related to the rate of M-CSF infusion. Another phase I trial of 36 patients with cancer documented one patient with chronic obstructive pulmonary disease who experienced dyspnoea and oxygen desaturation at the highest dose of M-CSF in combination with IFN-γ.[50] Most other studies on M-CSF have shown no pulmonary toxicity.

TUMOUR NECROSIS FACTOR-α INHIBITORS

Tumour necrosis factor-α (TNF-α) is a central cytokine in the inflammatory cascade responsible for the pathogenesis of rheumatoid arthritis and other autoimmune disorders. It is found in high concentrations in the rheumatoid joint, and TNF-α inhibitors suppress arthritis in an experimental animal model.[51] These observations formed the basis for human studies that led to the approval of TNF-α inhibitors for clinical use. This class of drugs has revolutionized the therapeutic approach to rheumatoid arthritis and has had a major impact on the treatment of Crohn's disease, psoriatic arthritis and ankylosing spondylitis.

Initial pre-approval clinical trials for the TNF-α inhibitors reported significant efficacy with minimal adverse effects, but post-marketing surveillance revealed substantial numbers of infectious complications,[52] particularly bacterial infections, tuberculosis endemic mycoses and Pneumocystis pneumonia. The lung is a major site of infection (of note, the rate of infection is increased in rheumatoid arthritis compared to control subjects), and non-infectious pulmonary complications also have been described.

Infliximab, adalimumab and etanercept

Three TNF-α inhibitors have been approved by the Food and Drug Administration (FDA).[52]

- Infliximab is a chimeric (human/murine) IgG1 monoclonal antibody to TNF-α.
- Adalimumab is a recombinant fully human IgG1 monoclonal antibody to TNF-α.
- Etanercept works by a different mechanism, as it is a soluble dimeric fusion protein that combines two extracellular binding domains of the TNF-α with the Fc portion of human IgG1.

Much attention has been focused on the infectious complications of anti-TNF-α therapy. It is important to recognize that, even in the absence of such therapy, patients with rheumatoid arthritis have approximately twice the background risk of serious infection as individuals without the disease.[53] Nonetheless, the risk of certain infections, most notably tuberculosis, in the setting of anti-TNF-α therapy definitely is increased (Table 15.1). Most of the cases of tuberculosis have occurred with infliximab.

The pre-approval clinical trials for infliximab did not suggest an increased incidence of tuberculosis. In one phase III study of 428 patients comparing placebo against four different infliximab regimens, there was a single case of disseminated fatal tuberculosis in one of the active drug treatment arms.[54] Initial post-marketing surveillance by the FDA Adverse Event Reporting System (AERS), however, revealed 70 cases of tuberculosis in approximately 147 000 infliximab-treated patients worldwide.[55] Since then, multiple additional cases have appeared in the medical literature, and reports to the FDA AERS totalled 335 as of September 2002.[56] Tuberculosis in this setting usually occurs within the first several months of infliximab therapy, with a median onset of 12 weeks after the start of treatment, and most cases are due to reactivation of latent infection.[55] Extrapulmonary and disseminated disease is common.

The incidence of tuberculosis in infliximab-treated patients likely depends on the background incidence of tuberculosis in the country considered and practice regarding BCG vaccination.

Etanercept was approved by the FDA for clinical use within several months of the approval of infliximab, but subsequent cases of tuberculosis were much fewer. The AERS data to

Table 15.1   Infections associated with TNF-α inhibitor therapy

Definite associations (strong data)	Possible associations (weaker data)
Mycobacterium tuberculosis	Aspergillus species
Histoplasma capsulatum	Candida species
Coccidioides immitis	Cryptococcus neoformans
Pneumocystis jiroveci (carinii)	Legionella pneumophila
Listeria monocytogenes	
Viral upper respiratory tract infections	

September 2002 reported 39 cases of tuberculosis associated with etanercept.[56] As with infliximab, extrapulmonary presentations may be seen. Several reasons have been suggested for the difference in tuberculosis rates of these two agents. First, they utilize different mechanisms of action for neutralization of TNF-α. In addition, the recipient populations for the two drugs differ, as a large proportion of the patients treated with infliximab have Crohn's disease. There may be inherent variability in the susceptibility to tuberculosis of individuals with differing target diseases. The delivery modes of the two drugs also may contribute to the difference in tuberculosis risk; intravenous administration of infliximab results in higher blood levels than subcutaneous administration of etanercept. Finally, infliximab has been used more than etanercept in Europe, where rates of latent tuberculosis infection are higher than in the United States.[57]

The third TNF-α inhibitor, adalimumab, arrived on the market after the association of this class of drugs with tuberculosis was established. As a result, screening for latent tuberculosis had become the standard of practice before its approval, and accordingly, fewer reports of adalimumab-associated tuberculosis have appeared. Post-marketing surveillance in the United States identified 17 cases of tuberculosis to June 2005.[58]

Discussion of adverse effects

The development of tuberculosis in the setting of TNF-α inhibitor use is not surprising. TNF-α is essential in the control of intracellular pathogens and in the formation and maintenance of granulomas. It also directly activates macrophages which then engulf and kill mycobacteria.[57] While tuberculosis is the primary mycobacterial risk for TNF-α inhibitors, occasional cases of non-tuberculous mycobacterial infection have been reported.[56]

In 2005, the US Centers for Disease Control and Prevention (CDC) published recommendations for the prevention of tuberculosis in the setting of TNF-α inhibitor use.[59] Patients should be screened carefully for risk factors, including: birth or residence in an area where tuberculosis is prevalent; history of residence in a prison, homeless shelter or long-term care facility; substance abuse; and healthcare employment involved in the care of tuberculosis patients. Tuberculin skin-testing is mandatory unless there is a known history of positive reactivity. Published guidelines should be used to interpret the skin-test, though the standard of practice now suggests 5 mm of induration to be a positive result when infliximab is the planned therapy. Chest radiographs should be obtained in all patients to look for evidence of latent infection. The preferred treatment for latent tuberculosis infection is 9 months of daily isoniazid, but there is no consensus of opinion on when TNF-α inhibitor therapy can be started in such patients. While completion of the entire course of antituberculous therapy is preferable, it also has been suggested that anti-TNF-α therapy can be initiated 1–2 months after the start of prophylaxis.[57,60] Development of tuberculosis, however, has been reported despite adequate prophylaxis.[61] In all patients receiving these agents, a high degree of vigilance and suspicion for tuberculosis must be maintained, and febrile or respiratory illnesses should be worked up thoroughly for this possibility.

In recent years, a special blood test, the interferon-γ release assay (IGRA), has been used increasingly to test for latent tuberculosis infection. While specific recommendations from the CDC for the use of the IGRA are evolving, there are few circumstances when it can be helpful in screening. IGRA testing consistently shows better specificity than tuberculin skin-testing in BCG-vaccinated populations. Because the IGRA only requires a single visit rather than the two visits for tuberculin skin-testing, it also may be the best option for historically mobile populations such as the homeless.

Histoplasmosis, the most prevalent endemic mycosis in the United States, also may develop in the setting of TNF-α inhibitor treatment. Host defence to *Histoplasma capsulatum* shares similarities to the immune response to *Mycobacterium tuberculosis*, in that cell-mediated immunity, including TNF-α secretion, plays a central role.[62] Lee and colleagues provided the first review of the FDA AERS for this infection to July 2001 and identified 10 cases of histoplasmosis, nine of whom received infliximab and one who received etanercept.[63] All of the patients resided in states known to be endemic for histoplasmosis, and all had received at least one additional immunosuppressive medication, such as prednisone, methotrexate or azathioprine. Pulmonary symptoms and lung infiltrates were present in almost all patients, but disseminated disease also was observed. The severity of the infection was pronounced, with one fatality and all but one individual requiring intensive care.

In a recent review, 84 cases of histoplasmosis occurring in the setting of anti-TNF-α therapy were identified in the medical literature.[64] The breakdown of cases for the individual drugs was 72 (86 per cent) with infliximab, 8 (10 per cent) with etanercept, and 4 (5 per cent) with adalimumab. No major differences in clinical presentation were seen in comparison with the index report of 10 patients. Almost all affected individuals were from an endemic area, and all had received concomitant treatment with an immunosuppressive medication. Amphotericin B is the drug of choice for histoplasmosis in this setting, and chronic suppression with itraconazole has been suggested if ongoing anti-TNF-α therapy is necessary.[65]

Screening for exposure to histoplasmosis with skin-test reactivity or serology is not recommended for patients about to start TNF-α inhibitor therapy. Because of the apparent excess risk with infliximab, however, careful weighing of the risks and benefits of the drug is essential, especially if the drug is being considered as potential treatment in a patient with a history of residence in an area endemic for *H. capsulatum*.

Other pulmonary fungal infections have been reported. An association between the TNF-α inhibitors and coccidioidomycosis has been described in patients residing in the southwestern US, an area endemic for *Coccidioides immitis*.[66] Aspergillosis and candidiasis also may occur,[56] though the reported frequencies may be skewed by the study populations; most cases have been in bone marrow transplant recipients receiving TNF-α inhibitors as treatment for acute graft-versus-host disease.[67] A recent review identified 28 published cases of cryptococcosis in TNF-α-treated individuals, with pneumonia being the predominant clinical manifestation.[64]

Pneumocystis jiroveci (carinii), now considered a fungus, was identified through the FDA AERS as the cause of pneumonia in 84 patients treated with infliximab to December 2003.[68] The mortality rate for these patients was 27 per cent (23/84). The authors recommended consideration of prophylaxis against this infection, but this has not been adopted universally, given the low overall incidence of *Pneumocystis* pneumonia in this setting. An increased risk for *Pneumocystis* in the setting of infliximab treatment is plausible, as reductions in both the peripheral blood CD4 count[69] and tissue-based (gut mucosa) CD4 counts[70] have been reported in patients with Crohn's disease treated with this drug.

Among bacterial pathogens, *Listeria monocytogenes* represents the most commonly reported infection affecting recipients of TNF-α inhibitors, but it typically causes meningitis or sepsis rather than pneumonia.[71] *Legionella pneumophila* pneumonia has been reported infrequently, but some investigators have suggested that its incidence may be increased by treatment with these agents.[72,73] In a French series of 10 affected patients, the relative risk of *L. pneumophila* pneumonia in the setting of anti-TNF-α therapy was estimated to be between 16.5 and 21.0 compared with the relative risk in France overall.[73] An unexpected finding in that series was that adalimumab accounted for six of the cases, while infliximab and etanercept each were implicated in two.

Viral pneumonias do not appear to have increased frequency, though viral upper respiratory tract infections do.[54] There is a single report of a parasitic adverse event, in which addition of adalimumab to low-dose prednisone and methotrexate in a Filipino man with rheumatoid arthritis resulted in life-threatening hyperinfection with *Strongyloides stercoralis*.[74]

A summary of the recommendations for prevention of infection in the setting of TNF-α inhibitor therapy is given in Box 15.1.

Anti-TNF-α medications rarely may cause non-infectious pulmonary complications. A cause-and-effect relationship is not always demonstrated. Accelerated formation of rheumatoid lung nodules has been described with etanercept,[75] as has

---

Box 15.1 Recommendations for the prevention of infection in the setting of TNF-α inhibitor therapy

1. Screen thoroughly for risk factors for exposure to *Mycobacterium tuberculosis*.

2. Obtain baseline chest radiographs to look for evidence of prior exposure.

3. Perform tuberculin skin-testing in all patients unless there is a history of known positivity.

4. Follow published guidelines for interpretation of tuberculin skin test. Five millimetres of induration should be considered as a positive result in all patients scheduled for potential infliximab therapy.

5. Treat latent tuberculosis infection with 9 months of isoniazid.

6. Advise patients to avoid activities with potential *Histoplasma capsulatum* exposure (e.g. cleaning chicken coops, spelunking).

7. Advise patients to avoid soft cheeses and other unpasteurized milk products (potential sources of *Listeria monocytogenes* exposure).

the development of a pneumonitis characterized by non-necrotizing granulomas, lymphohistiocytic infiltrates, and negative cultures and special stains for microorganisms.[76] Similarly, 'bronchiocentric allergic granulomatosis of the lung' was reported in a patient treated with infliximab for ankylosing spondylitis.[77] Acute progression of underlying rheumatoid interstitial lung disease after addition of either etanercept or infliximab has been described,[78,79] and a potentiating effect on the pulmonary toxicity of methotrexate has been postulated in patients concurrently on this drug.[80,81] ARDS may occur, and in one such case associated with infliximab, an open lung biopsy revealed eosinophilic pneumonia; the patient fully recovered with intravenous corticosteroids.[82]

MONOCLONAL ANTIBODIES USED IN THE TREATMENT OF CANCER

Monoclonal antibody technology has led to major advances in the treatment of cancer. Targeting a specific antigen on malignant cells ideally should result in a more specific therapeutic outcome and fewer adverse effects.[83] The first monoclonal antibody approved by the FDA for the treatment of cancer was rituximab in 1997, and many others have followed, each with its own mode of action and therapeutic indication. Pulmonary toxicities, while uncommon except for rituximab, have been observed for most of these agents (Table 15.2).

Rituximab

Rituximab is a chimeric (human/murine) anti-CD20 antibody that is now considered part of the standard regimen for the treatment of relapsed or refractory, low-grade or follicular, CD20+ B-cell non-Hodgkin's lymphomas.[83] Its role has expanded beyond the realm of oncology, as it has been approved for use in rheumatoid arthritis, and the potential for activity in other autoimmune diseases, including Wegener's granulomatosis and systemic lupus erythematosus, is being investigated.[84]

In early clinical trials of rituximab, 38 per cent of patients developed pulmonary adverse events, but most of these were of low severity and included cough, rhinitis, bronchospasm, dyspnoea and sinusitis. Only 4 per cent of patients had respiratory symptoms classified as grade 3 or 4.[85] Case reports of severe interstitial pneumonitis have appeared intermittently in the medical literature. A recent review of rituximab-induced interstitial lung disease by Wagner and colleagues identified 16 published reports.[86] The current number of cases is 45. The average age of affected patients was 65, and 75 per cent had a diagnosis of non-Hodgkin's lymphoma. Interstitial lung disease occurred with rituximab monotherapy as well as in combination with cytotoxic chemotherapy, and it developed after an average of four cycles of the drug. Dyspnoea, fever and cough were common complaints. Objective data revealed ground-glass opacities, alveolitis and diffuse infiltrates on high-resolution chest CT scan, and a combined restrictive defect and gas transfer defect on pulmonary function testing. All patients were treated with corticosteroids, which led to complete recovery in 10 (63 per cent), but the other six individuals (37 per cent) died. The investigators postulated that rituximab-induced TNF-α secretion might be the central mediator in the pathogenesis of the interstitial lung disease, and they suggested that TNF-α inhibitor therapy therefore might have a role in severe cases, including patients who worsen despite corticosteroids. This therapeutic approach has never actually been carried out in any published case. ARDS following rituximab infusion has been described in two patients,[87,88] though it is worth noting that some of the interstitial pneumonitis cases also fit the consensus criteria for ARDS.[89] Moreover, the presenting features in several of the reported cases suggest the possibility of tumour lysis pneumopathy rather than a direct drug-related effect.

Rarely, BOOP has been described, both with rituximab-based combination chemotherapy[90] and after single-agent rituximab.[91] All cases showed complete resolution with corticosteroid treatment. Two reports of alveolar haemorrhage occurring with the use of rituximab each noted peripheral blood eosinophilia and loosely formed, non-necrotizing granulomas on lung biopsy, supporting the notion of a hypersensitivity-like drug reaction.[92,93] One of the episodes was fatal, despite the use of corticosteroids, the role of which is uncertain.

Trastuzumab

Trastuzumab is a humanized monoclonal antibody that blocks stimulation of neoplastic cell proliferation through selective binding of the human epidermal growth factor receptor 2 protein (HER2). It is an important part of the approach to breast cancer treatment, both with metastatic disease and in the adjuvant setting.[94] While the major adverse effect appears to be trastuzumab-induced cardiomyopathy, incidents of pulmonary toxicity also have appeared in the literature.

Table 15.2 Pulmonary adverse events caused by monoclonal antibodies used in the treatment of cancer

Monoclonal antibody	Pulmonary adverse event(s)
Rituximab	Interstitial pneumonitis BOOP ARDS Diffuse alveolar haemorrhage Hypersensitivity pneumonitis Bronchospasm
Trastuzumab	Interstitial pneumonitis BOOP ARDS Bronchospasm/anaphylaxis
Bevacizumab	Pulmonary haemorrhaging from site of tumour
Gemtuzumab ozogamicin	Pulmonary oedema ARDS
Alemtuzumab	Bronchospasm Diffuse alveolar haemorrhage Pulmonary infections (tuberculosis, aspergillosis)

ARDS, acute respiratory distress syndrome; BOOP, organizing pneumonitis.

In an early retrospective analysis of the side-effects of trastuzumab in approximately 25 000 patients, 74 (0.3 per cent) experienced a serious infusion-related reaction.[95] The majority of these reactions occurred during or shortly after the first infusion and were characterized primarily by respiratory symptoms, including bronchospasm, hypoxaemia, anaphylaxis and ARDS. Of those patients, 9 eventually died, but 33 were able to continue trastuzumab therapy with no recurrence of infusion reactions.

A seminal report of two trials that confirmed the benefit of trastuzumab in the adjuvant setting described rare cases of interstitial pneumonitis.[96] In one of the trials, four patients in the trastuzumab group developed interstitial pneumonitis, one of whom died. Five patients in the other trial had grade 3+ pneumonitis or pulmonary infiltrates, with one death. No further characterization of the affected patients was provided. In a cohort of 24 patients administered trastuzumab for its potential role in non-small-cell lung carcinoma, two patients developed grade 3 hypoxaemia and one had grade 1 pneumonitis.[97] One additional patient developed fatal pulmonary toxicity with rapidly progressive infiltrates and respiratory failure, consistent with an acute drug-induced pneumonitis. The affected individual had previously received radiation therapy, and the authors stated that it was unclear whether such radiation exposure predisposed to the development of lung toxicity with trastuzumab. BOOP occurring 6 weeks after the start of trastuzumab therapy also has been reported.[98]

Bevacizumab

Since angiogenesis is one of the hallmarks of cancer, inhibition of this process has great appeal as a therapeutic intervention. Bevacizumab is a humanized monoclonal antibody that selectively binds vascular endothelial growth factor (VEGF), the major regulator of angiogenesis. It is approved for use in advanced cases of colorectal cancer and non-small-cell, non-squamous-cell lung carcinoma, and its activity against other solid tumours is being investigated.[99]

The most unique pulmonary side-effect with bevacizumab is its potential for local complications at the site of the tumour. A randomized phase II trial of 99 patients with advanced non-small-cell lung cancer compared paclitaxel and carboplatin alone with paclitaxel and carboplatin plus bevacizumab.[100] Two distinct patterns of bleeding were observed in the 66 patients receiving the monoclonal antibody therapy. Minor mucocutaneous bleeding, characterized primarily by epistaxis, was fairly common and did not require discontinuation of the drug. Six patients experienced life-threatening pulmonary haemorrhage, including four fatal events. Squamous histology, central tumour location close to major blood vessels, and tumour necrosis or cavitation at baseline or during therapy were all associated with an increased risk of serious bleeding.

These observations prompted a phase III trial of paclitaxel and carboplatin alone or with bevacizumab.[101] Patients with advanced non-small-cell lung cancer were eligible, but those with predominant squamous cell carcinoma histology were excluded. Eight of the 427 patients in the bevacizumab group developed life-threatening pulmonary haemorrhage, with five deaths. Thus, the exclusion of squamous-cell carcinoma histology resulted in a drop in the incidence of this complication from 9.1 per cent (in the phase II trial) to 1.9 per cent. Pulmonary haemorrhage has not been reported in patients receiving bevacizumab in colorectal, breast, prostate or renal-cell cancer trials.[102]

Gemtuzumab ozogamicin

Gemtuzumab ozogamicin is a humanized anti-CD33 monoclonal antibody linked to a semisynthetic derivative of calicheamicin, a cytotoxic antibiotic. Its primary indication is for patients with CD33 positive acute myeloid leukaemia (AML) in first relapse who are 60 years of age or older and who are not candidates for other cytotoxic chemotherapy. Pre-approval testing identified dyspnoea and transient hypoxaemia as pulmonary adverse reactions, but more serious respiratory events were observed in post-marketing safety data.[103] During the first 6 months after approval, there were eight patients who experienced pulmonary events immediately after or within 24 hours of administration of the drug. Of these events, five were ARDS, and three were reported as 'pulmonary oedema'. Six of the patients had pre-treatment leucocyte counts greater than 60 000/μL. Although pulmonary leucostasis can have a similar clinical presentation in patients with very high blast counts, the temporal association of the events with drug administration suggested a drug reaction. To reduce the risk of severe pulmonary toxicity, the authors of the report recommended reduction in peripheral blast counts to below 30 000/μL with hydroxyurea or leukapheresis prior to gemtuzumab ozogamicin administration.[103]

Alemtuzumab

The CD52 antigen is expressed on normal and neoplastic B- and T-lymphocytes, monocytes and natural killer cells. Alemtuzumab, a humanized monoclonal antibody directed against CD52, is approved for the treatment of fludarabine refractory chronic lymphocytic leukaemia (CLL) and is effective in treating a number of lymphoid malignancies.[104] Bronchospasm is the primary lung toxicity that has been observed with this agent. Severe grade 3/4 bronchospasm can occur during or shortly after infusion,[105,106] and co-administration of a tapering regimen of corticosteroids appears to reduce the frequency and severity of such infusion-associated side-effects.[106] A single case of diffuse alveolar haemorrhage has been reported as a complication of alemtuzumab therapy following renal transplantation.[107] Because there was no evidence of vasculitis, the investigators postulated that it occurred as a result of thrombocytopenia and direct epithelial lung toxicity, both caused by the drug.

Profound lymphopenia typically develops within several weeks after administration of alemtuzumab. As a result, a high incidence of infectious complications has been observed. Pulmonary tuberculosis and pulmonary aspergillosis each have been reported.[108] Cytomegalovirus reactivation is a frequent development, but most commonly this has been identified through viral surveillance, and clinically significant pneumonitis is exceedingly rare.[109]

MONOCLONAL ANTIBODIES USED IN OTHER DISEASES

Omalizumab

Omalizumab is a humanized monoclonal antibody directed against immunoglobulin E (IgE) that not only reduces unbound IgE concentrations but also promotes down-regulation of IgE receptors.[110] In patients with moderate-to-severe asthma with documented allergic sensitivities, it may be considered as additional treatment if inhaled corticosteroids do not provide adequate control.

An analysis of 12 omalizumab clinical trials enrolling 5328 subjects with moderate-to-severe allergic asthma showed an excellent safety profile for the drug, with no statistically significant differences from placebo.[111] Anaphylaxis, however, occurred in three patients in clinical trials, and it also was observed in the post-marketing setting. As a result, the FDA issued a black-box warning on the product label regarding the potential for anaphylaxis, with a range of symptoms including bronchospasm, angioedema of the tongue and throat, and hypotension. It was estimated that at least 0.2 per cent of the 57 300 patients who received omalizumab between June 2003 and December 2006 experienced anaphylaxis. This is a comparable rate to anaphylaxis caused by penicillins, aspirin or non-steroidal anti-inflammatory drugs.[110,112]

In 40 per cent of cases, anaphylaxis and anaphylactoid reactions occurred after the first dose of omalizumab, but these events also were observed after up to 2 years of therapy. Most episodes occurred within the first 2 hours after the injection, but 30 per cent were delayed reactions and occurred up to 24 hours later.[110] Neither respiratory failure nor death has been reported.

The current recommendation from a task force on omalizumab is for administration of the drug only in medical facilities equipped with appropriate staffing and the capability to treat anaphylaxis. Additional recommendations are for direct observation for 2 hours after the first three treatments and for 30 minutes after all subsequent injections.[112]

Recent evidence suggests that omalizumab may be associated with the development of Churg–Strauss syndrome.[113]

Abciximab

Platelet activation and aggregation are central to the development of acute coronary syndromes, and the development of several inhibitors targeting the platelet glycoprotein IIb/IIIa receptor has been a major advance in cardiac care. Abciximab, a chimeric (human/murine) monoclonal antibody Fab fragment, was the first of these inhibitors to undergo clinical evaluation.[114]

A rare but potentially under-recognized complication of abciximab therapy is diffuse alveolar haemorrhage.[115] Presenting radiographic signs are non-specific and may include new pulmonary infiltrates in a pattern similar to cardiogenic pulmonary oedema. Haemoptysis is not a consistent finding, but a sudden fall in haemoglobin in association with haemoptysis should alert the physician to the possibility of pulmonary haemorrhage.

The incidence of alveolar haemorrhage has been reported as high as 0.7 per cent, with some cases resulting in death.[116] Risk factors for its development include the presence of chronic lung disease, congestive heart failure, and elevated pulmonary capillary wedge pressures during left heart catheterization. Increased hydrostatic pressure, with resultant rupture of capillaries, is a plausible mechanism; vasculitis has not been seen on pathologic examination. Rigorous attention to heparin dosing is essential in patients administered abciximab, as higher doses may predispose to this pulmonary complication. Treatment is aimed at supportive measures and should include discontinuation of all antiplatelet and anticoagulant agents.[115] The more traditional, non-biological glycoprotein IIb/IIIa inhibitors, such as clopidogrel and tirofiban, can produce a similar syndrome of diffuse alveolar haemorrhage.

Basiliximab

A few cases of non-cardiogenic pulmonary oedema have been described following induction therapy with the chimeric anti-IL-2 receptor monoclonal antibody basiliximab in paediatric recipients who received a renal transplant.[117] Anaphylaxis has rarely been reported with this agent.[118]

Anti-CD28 monoclonal antibody TGN1412

A recent report serves as a sobering example of the potential for harm induced by biomolecules.[119] TGN1412, an anti-CD28 monoclonal antibody, is a 'superagonist' that directly stimulates and expands T-cell populations. In a phase I trial, six healthy young male volunteers received TGN1412, and all became critically ill within 12–16 hours after infusion. The systemic inflammatory response syndrome (SIRS), hypotension and renal failure requiring dialysis characterized the course of all six patients, and two required mechanical ventilation for acute respiratory failure. All subjects ultimately survived. 'Cytokine storm' (rapid release of proinflammatory cytokines stored in designated normal T-cell clones) induced by the monoclonal antibody was implicated as the mechanism for the severe clinical manifestations.

TYROSINE KINASE INHIBITORS

Gefitinib

Gefitinib is an oral epidermal growth factor receptor (EGFR) tyrosine kinase inhibitor with activity against non-small-cell lung cancer.[120] Soon after its approval for treatment of advanced non-small-cell lung cancer in Japan in July 2002, reports of treatment-associated interstitial lung disease appeared.[121] Expanded global use of gefitinib revealed a disparity in the incidence of interstitial lung disease, with a significantly higher reporting rate in Japan compared with the rest of the world. A 2003 overview of the available data on gefitinib-associated interstitial lung disease estimated the incidence at 1.86 per cent in approximately 39 600 treated patients in Japan, a markedly higher rate than the estimated 0.34 per cent incidence in approximately 53 150 treated patients outside of Japan.[122] In a more recent prospective epidemiologic Japanese cohort, the incidence rate for gefitinib-associated interstitial lung disease over 12 weeks of therapy was 4.0 per cent, and the mortality rate from this lung toxicity was 31.6 per cent.[123]

The onset of interstitial lung disease is usually within the first month after initiation of gefitinib treatment.[123,124] Risk factors for this adverse event have been identified (Box 15.2),[123,125] and a range of radiographic patterns on chest CT scan has been observed (Box 15.3).[126] Diffuse alveolar damage is the typical finding on pathologic examination of lung tissue.[122] Corticosteroids and discontinuation of gefitinib are the mainstays of management.[124]

The mechanism of gefitinib-associated interstitial lung disease has not been elucidated. Even more unclear is the reason for the disparity observed between Japan and the rest of the world. An increased genetic susceptibility in the Japanese population is one possible explanation; some increase in the reporting rates also may be due simply to a greater awareness of the disease in Japan.[122,125]

Erlotinib

Erlotinib is another oral EGFR tyrosine kinase inhibitor, and its chemical structure is similar to that of gefitinib. It also shares a similar spectrum of activity against non-small-cell lung cancer.[120] Erlotinib-associated interstitial lung disease has been described, but the incidence is less well defined than with gefitinib. In a phase III trial of paclitaxel and carboplatin with or without concurrent erlotinib in patients with advanced non-small-cell lung cancer, there were five interstitial lung disease-like events in the 526 patients on erlotinib (1.0 per cent) and one in the 533 patients on placebo (0.2 per cent).[127] All of these events were fatal. A subsequent independent review of these data suggested that a possible/probable link between erlotinib use and fatal interstitial lung disease occurred in only 3/526 patients (0.6 per cent).[128] This percentage

was higher than the previously reported rate of gefitinib-associated interstitial lung disease in the non-Japanese population (0.34 per cent).[122] While this might suggest a higher overall risk of lung toxicity with erlotinib, the medical literature argues otherwise, as there have been only occasional related case reports,[129,130] in comparison with the preponderance of documented gefitinib-associated cases. To date, there has not been any suggestion of an increased incidence of erlotinib-associated interstitial lung disease in Japan.

Ground-glass infiltrates have been described most commonly, but alveolar consolidations may also occur.[128–130] Pathological examinations typically have been done after death. Diffuse alveolar damage is the classic finding, but there is a single report of BOOP found on a surgical lung biopsy.[128] Despite discontinuation of erlotinib and institution of corticosteroid therapy, the mortality rate is exceedingly high.

Imatinib

Imatinib is a potent inhibitor of the BCR-ABL (breakpoint cluster region-Abelson) family of tyrosine kinases. It has broad clinical applicability, including treatment of Philadelphia chromosome-positive acute lymphoblastic leukaemia (ALL) and chronic myelogenous leukaemia (CML), gastrointestinal stromal tumours, and systemic mastocytosis.[131] In a large trial of 1106 patients randomized to receive imatinib versus interferon-α plus low-dose cytarabine for chronic-phase chronic myelogenous leukaemia, cough and dyspnoea were reported in 14 per cent and 7 per cent, respectively, of the imatinib-treated individuals.[132] In most of the medical literature on imatinib, these symptoms typically have been ascribed to a fluid retention syndrome, with features including pleural effusions, pulmonary oedema and congestive heart failure.[133,134]

Parenchymal lung disease also can be caused by imatinib. In a report of 27 cases of imatinib-associated interstitial lung disease, six radiographic patterns were identified (Box 15.4).[135] The majority of patients in this series responded well to corticosteroid treatment, and most additional published cases have demonstrated reversibility in the interstitial changes.[136–138] There is a paucity of literature on the pathology of imatinib-associated lung disease, but separate reports have noted eosinophilic infiltration[137] or lymphocytic infiltration[138] in areas of interstitial fibrosis. Other disease associations have been suggested, including the possible development of pulmonary alveolar proteinosis[139] or lymphomatoid granulomatosis,[140] but these links to imatinib exposure remain unproven.

Box 15.2   Risk factors for gefitinib-associated interstitial lung disease (data from [120] and [122])

- Japanese ethnicity
- Male sex
- Older age
- History of smoking
- Pre-existing interstitial lung disease
- Poor performance status

Box 15.3   Radiographic patterns on CT imaging of gefitinib-associated interstitial lung disease (data from [123])

- Non-specific faint ground-glass infiltrates
- Multifocal airspace consolidations (similar to bronchiolitis obliterans organizing pneumonia)
- Patchy ground-glass infiltrates with interlobular septal thickening
- Extensive ground-glass, airspace consolidation, and traction bronchiectasis (similar to acute interstitial pneumonia)

Box 15.4   Radiographic patterns on high-resolution CT imaging of imatinib-associated interstitial lung disease (data from [132])

- Hypersensitivity pneumonitis
- Interstitial pneumonia
- Bronchiolitis obliterans organizing pneumonia
- Nodular infiltrates
- Peribronchovascular bundle infiltrates
- Diffuse alveolar damage

Dasatinib

Another BCR-ABL tyrosine kinase inhibitor, dasatinib, is two log more potent than imatinib and has activity against imatinib-resistant Philadelphia chromosome-positive ALL or CML.[141] Pleural effusions are the most common pulmonary adverse effect of dasatinib therapy. In a series of 138 patients with CML treated with dasatinib, 48 (35 per cent) developed a pleural effusion, of which 78 per cent were exudative.[142] Hypertension, a history of cardiac disease, and use of a twice-daily schedule (rather than once daily) were identified on multivariate analysis as risk factors associated with the development of pleural effusions.

The finding of exudative chemistries in most pleural effusions argued against fluid retention as the mechanism underlying the pulmonary toxicity of dasatinib. An immune-mediated mechanism was supported by a separate series of 40 patients treated with dasatinib for CML.[143] Nine of the 40 (22.5 per cent) developed dyspnoea, cough and chest pain, and six of those had pleural effusions. All of the effusions were exudative, and there was a lymphocyte predominance in all but one who had a neutrophil predominance; pleural biopsy also showed a lymphocytic infiltrate. Parenchymal abnormalities also were observed, with seven of the nine showing ground-glass infiltrates, septal thickening and/or alveolar opacities. In the patients who had a bronchoalveolar lavage (BAL), a lymphocytic alveolitis most commonly was found. The pleural and parenchymal abnormalities resolved in all patients with interruption of dasatinib administration (one also received corticosteroids), and they did not recur in three of four patients in whom the drug was reintroduced at a lower dose, even though the lower dose was still on a twice-daily schedule.

RAPAMYCIN ANALOGUES

The mammalian target of rapamycin (mTOR) is a serine-threonine kinase involved in regulation of cell growth, proliferation and apoptosis.[144] Sirolimus, a rapamycin analogue, has potent immunosuppressive properties through inhibition of the kinase activity, and it has a central role in the prevention of allograft rejection in transplant patients. While sirolimus also has potential anticancer effects, its poor aqueous solubility and instability have been limiting factors in this area of clinical development.[145] Derivatives of sirolimus, most notably temsirolimus, have better pharmacological profiles and are being investigated more extensively as cancer treatments.

Temsirolimus

Temsirolimus, an inhibitor of mTOR, is active against a variety of tumours, including renal-cell carcinoma, breast carcinoma, endometrial carcinoma, glioblastoma multiforme and gastrointestinal neuroendocrine tumours. Duran and colleagues retrospectively analysed patients treated with temsirolimus for advanced cancers of either endometrial or neuroendocrine origin, and eight (36 per cent) out of 22 patients developed pulmonary abnormalities consistent with drug-induced pneumonitis.[145] Two different radiographic patterns,

ground-glass opacities and lung parenchymal consolidation, were seen. Half of the patients were asymptomatic, and in those with symptoms, dyspnoea and dry cough were most common. The management approach was variable, as some subjects required discontinuation of the drug (with subsequent improvement) and others did fine remaining on temsirolimus in spite of the radiographic abnormalities.

Based on the data from a phase II trial investigating temsirolimus as a treatment for advanced renal-cell carcinoma, the pulmonary toxicity does not appear to be dose-dependent.[146] In that study, the incidence of non-specific pneumonitis was 5 per cent (6 out of 111 patients). Adverse pulmonary toxicity events also have been reported at lower incidence rates in studies of temsirolimus for breast cancer (1 per cent),[147] glioblastoma multiforme (3 per cent),[148] and mantle-cell lymphoma (3 per cent).[149] One possible explanation for the discrepancy in the incidence rate between these studies and the Duran study is that the latter included routine chest CT scans in the follow-up of most patients.[145] The mechanism of action of temsirolimus-induced lung disease has not been established. Patients improve on corticosteroids.

PROTEASOME INHIBITORS

Cellular protein homeostasis, including control of those proteins involved in cell cycle progression and apoptosis, is achieved through the proteasome. This multi-enzyme complex now is recognized as an important potential target in therapies directed against cancer.[150]

Bortezomib

The FDA recently approved the first proteasome inhibitor, bortezomib, for the treatment of multiple myeloma and mantle-cell lymphoma. Early studies did not suggest any lung toxicity, but a report in 2006 from Miyakoshi and colleagues described severe pulmonary complications in 4 of 13 Japanese patients treated for multiple myeloma.[151] The clinical presentation was characterized by asthma-like symptoms and fever, followed by respiratory failure with pulmonary infiltrates. In three of the patients, respiratory failure developed after repeated administration of bortezomib, and corticosteroids resulted in improvement; one of those three still ultimately died of respiratory failure, and autopsy showed diffuse alveolar damage. The fourth patient developed pulmonary complications immediately after the first dose of bortezomib, showed no improvement after corticosteroids, and died of respiratory failure.

A subsequent 2006 article reported the results of a questionnaire sent to Japanese physicians regarding their experience with bortezomib, which, at the time, was not yet approved for use in Japan.[152] Seven (15 per cent) of 46 patients were identified as having bortezomib-induced pulmonary complications, including three who died from respiratory failure. Multivariate analysis revealed that a prior stem-cell transplant might increase the risk of lung injury from bortezomib and that concomitant use of corticosteroids might reduce the risk. The possibility of a higher incidence in Japanese patients has been raised, as very few reports have come from outside of Japan.[153]

SUMMARY

As new biomolecules and novel chemotherapeutic agents have been developed, new presentations of drug-induced lung disease have been observed. In some cases, such as with interferon-α-induced sarcoidosis and TNF-α-inhibitor-associated tuberculosis, the adverse effects are not surprising, given the nature of the biomolecules and their specific roles in host immunity. Many of the effects, however, are more random and variable. The potential for a genetic predisposition to drug-induced lung disease is an area of intense interest, and the differing experience with gefitinib in the Japanese population compared with the rest of the world highlights this issue. Awareness of the pulmonary toxicities of these newer molecules is crucial as their use becomes more routine, and vigilance for as yet unobserved adverse effects on the lungs is equally important.

KEY POINTS

- Interferon-α therapy can induce the clinical syndrome of sarcoidosis. Vigilance for the development of this problem is crucial, as the symptoms can be easily mistaken for side-effects of interferon-α.

- All patients considered for TNF-α inhibitor therapy should be screened appropriately for exposure to tuberculosis prior to initiation of treatment using novel tests. During therapy, patients must be monitored closely for the development of tuberculosis and other infections.

- Rituximab-induced interstitial lung disease, while uncommon, is potentially fatal and should be treated with corticosteroids.

- Bevacizumab should be avoided in the setting of non-small-cell lung cancer with squamous histology, central tumour location near major blood vessels, or tumour necrosis/cavitation, as it can cause potentially life-threatening pulmonary haemorrhage.

- Omalizumab can cause anaphylaxis and should be administered in settings where there is the capability to treat this adverse event. The incidence of anaphylaxis with omalizumab is comparable to the incidence with penicillins, aspirin or NSAIDs.

- Diffuse alveolar haemorrhage is a rare but potentially under-recognized complication of abciximab therapy. Radiographic imaging may show a pattern similar to cardiogenic pulmonary oedema.

- Gefitinib-associated interstitial lung disease is significantly more common in Japanese individuals than in patients of other ethnicities.

- Interstitial lung disease can develop as a complication of treatment with tyrosine kinase inhibitors, including gefitinib, erlotinib, imatinib, dasatinib and several mTOR inhibitors.

REFERENCES

1. Baron S, Tyring SK, Fleischmann WR, et al. The interferons: mechanisms of action and clinical applications. J Am Med Assoc 1991;**266**:1375–83.
2. Agostini C, Basso U, Semenzato G. Cells and molecules involved in the development of sarcoid granuloma. J Clin Immunol 1998;**18**:184–92.
3. Goldberg HJ, Fiedler D, Webb A, et al. Sarcoidosis after treatment with interferon-α: a case series and review of the literature. Respir Med 2006;**100**:2063–8.
4. Alazemi S, Campos MA. Interferon-induced sarcoidosis. Int J Clin Pract 2006;**60**:201–11.
5. Ning Q, Brown D, Parodo J, et al. Ribavirin inhibits viral-induced macrophage production of TNF, IL-1, the procoagulant fgl2 prothrombinase and preserves Th1 cytokine production but inhibits Th2 cytokine response. J Immunol 1998;**160**:3487–93.
6. Wendling J, Descamps V, Grossin M, et al. Sarcoidosis during combined interferon alfa and ribavirin therapy in 2 patients with chronic hepatitis C. Arch Dermatol 2002;**138**:546–7.
7. Leclerc S, Myers RP, Moussalli J, et al. Sarcoidosis and interferon therapy: report of five cases and review of the literature. Eur J Intern Med 2003;**14**:237–43.
8. Ogata K, Koga T, Yagawa K. Interferon-related bronchiolitis obliterans organizing pneumonia. Chest 1994;**106**:612–13.
9. Kumar KS, Russo MW, Borczuk AC, et al. Significant pulmonary toxicity associated with interferon and ribavirin therapy for hepatitis C. Am J Gastroenterol 2002;**97**:2432–40.
10. Kalambokis G, Stefanou D, Arkoumani E, et al. Fulminant bronchiolitis obliterans organizing pneumonia following 2 days of treatment with hydroxyurea, interferon-α and oral cytarabine ocfosfate for chronic myelogenous leukaemia. Eur J Haematol 2004;**73**:67–70.
11. Chin K, Tabata C, Sataka N, et al. Pneumonitis associated with natural and recombinant interferon alfa therapy for chronic hepatitis C. Chest 1994;**105**:939–41.
12. Abi-Nassif S, Mark EJ, Fogel RB, Hallisey RK. Pegylated interferon and ribavirin-induced interstitial pneumonitis with ARDS. Chest 2003;**124**:406–10.
13. Krasnowska M, Malolepszy J, Liebhart E, Inglot AD. Inhaled human interferon alpha induces bronchospastic reactions in asthmatics. Arch Immunol Ther Exp (Warsz) 1992;**40**:75–8.
14. Bini EJ, Weinshel EH. Severe exacerbation of asthma: a new side effect of interferon-alpha in patients with asthma and chronic hepatitis C. Mayo Clin Proc 1999;**74**:367–70.
15. Simon HU, Seelbach H, Ehmann R, Schmitz M. Clinical and immunological effects of low-dose IFN-α treatment in patients with corticosteroid-resistant asthma. Allergy 2003;**58**:1250–5.
16. Abdi EA, Nguyen GK, Ludwig RN, Dickout WJ. Pulmonary sarcoidosis following interferon therapy for advanced renal cell carcinoma. Cancer 1987;**59**:896–900.
17. Bobbio-Pallavicini E, Valsecchi C, Tacconi F, et al. Sarcoidosis following beta-interferon therapy for multiple myeloma. Sarcoidosis 1995;**12**:140–2.
18. Mehta CL, Tyler RJ, Cripps DJ. Granulomatous dermatitis with focal sarcoidal features associated with recombinant interferon beta-1b injections. J Am Acad Dermatol 1998;**39**:1024–8.

19. Ferriby D, Stojkovic T. Clinical picture: bronchiolitis obliterans with organizing pneumonia during interferon β-1a treatment. *Lancet* 2001;**357**:751.

20. Kimura Y, Okuyama R, Watanabe H, *et al.* Development of BOOP with interferon-β treatment for malignant melanoma. *J Eur Acad Dermatol Venereol* 2006;**20**:999–1000.

21. Shaw EG, Deming RL, Creagan ET, *et al.* Pilot study of human recombinant interferon gamma and accelerated hyperfractionated thoracic radiation therapy in patients with unresectable stage IIIA/B nonsmall cell lung cancer. *Int J Radiat Oncol Biol Phys* 1995;**31**:827–31.

22. van Zandwijk N, Groen HJM, Postmus PE, *et al.* Role of recombinant interferon-gamma maintenance in responding patients with small cell lung cancer: a randomised phase III study of the EORTC Lung Cancer Cooperative Group. *Eur J Cancer* 1997;**33**:1759–66.

23. Redman BG, Flaherty L, Chou TH, *et al.* A phase I trial of recombinant interleukin-2 combined with recombinant interferon-gamma in patients with cancer. *J Clin Oncol* 1990;**8**:1269–76.

24. Honoré I, Nunes H, Groussard O, *et al.* Acute respiratory failure after interferon-γ therapy of end-stage pulmonary fibrosis. *Am J Respir Crit Care Med* 2003;**167**:953–7.

25. Smith KA. Interleukin-2: inception, impact, and implications. *Science* 1988;**240**:1169–76.

26. Baluna R, Vitetta ES. Vascular leak syndrome: a side effect of immunotherapy. *Immunopharmacology* 1997;**37**:117–32.

27. Vogelzang PJ, Bloom SM, Mier JW, Atkins MB. Chest roentgenographic abnormalities in IL-2 recipients: incidence and correlation with clinical parameters. *Chest* 1992;**101**:746–52.

28. Kammula US, White DE, Rosenberg SA. Trends in the safety of high dose bolus interleukin-2 administration in patients with metastatic cancer. *Cancer* 1998;**83**:797–805.

29. Guida M, Abbate I, Casamassima A, *et al.* Long-term subcutaneous recombinant interleukin-2 as maintenance therapy: biological effects and clinical implications. *Cancer Biother* 1995;**10**:195–203.

30. Huland E, Heinzer H, Mir TS, Huland H. Inhaled interleukin-2 therapy in pulmonary metastatic renal cell carcinoma: six years of experience. *Cancer J Sci Am* 1997;**3**(Suppl 1):S98–105.

31. Vetto JT, Papa MZ, Lotze MT, *et al.* Reduction of toxicity of interleukin-2 and lymphokine-activated killer cells in humans by the administration of corticosteroids. *J Clin Oncol* 1987;**5**:496–503.

32. Mier JW, Vachino G, Klempner MS, *et al.* Inhibition of interleukin-2-induced tumor necrosis factor release by dexamethasone: prevention of an acquired neutrophil chemotaxis defect and differential suppression of interleukin-2-associated side effects. *Blood* 1990;**76**:1933–40.

33. Azoulay E, Attalah H, Harf A, *et al.* Granulocyte colony-stimulating factor or neutrophil-induced pulmonary toxicity: myth or reality? Systematic review of clinical case reports and experimental data. *Chest* 2001;**120**:1695–701.

34. Takatsuka H, Takemoto Y, Mori A, *et al.* Common features in the onset of ARDS after administration of granulocyte colony-stimulating factor. *Chest* 2002;**121**:1716–20.

35. Karlin L, Darmon M, Thiéry G, *et al.* Respiratory status deterioration during G-CSF-induced neutropenia recovery. *Bone Marrow Transplant* 2005;**36**:245–50.

36. Niitsu N, Iki S, Muroi K, *et al.* Interstitial pneumonia in patients receiving granulocyte colony-stimulating factor during chemotherapy: survey in Japan 1991–96. *Br J Cancer* 1997;**76**:1661–6.

37. Busmanis IA, Beaty AE, Basser RL. Isolated pleural effusion with hematopoietic cells of mixed lineage in a patient receiving granulocyte-colony-stimulating factor after high-dose chemotherapy. *Diagn Cytopathol* 1998;**18**:204–7.

38. Kitamura S, Kinouchi K, Fukumitsu K, *et al.* A risk of pulmonary edema associated with G-CSF pretreatment. *Masui* 1997;**46**:946–50.

39. Arimura K, Inoue H, Kukita T, *et al.* Acute lung injury in a healthy donor during mobilization of peripheral blood stem cells using granulocyte-colony stimulating factor alone. *Haematologica* 2005;**90**:e27–9.

40. Kopp HG, Horger M, Faul C, *et al.* Granulocyte colony-stimulating factor-induced pulmonary hemorrhage in a healthy stem cell donor. *J Clin Oncol* 2007;**25**:3174–8.

41. Ruiz-Argüelles GJ, Arizpe-Bravo D, Sánchez-Sosa S, *et al.* Fatal G-CSF-induced pulmonary toxicity. *Am J Hematol* 1999;**60**:82–3.

42. Yoshida I, Matsuo K, Teshima T, *et al.* Transient respiratory disturbance by granulocyte-colony-stimulating factor administration in healthy donors of allogeneic peripheral blood progenitor cell transplantation. *Transfusion* 2006;**46**:186–92.

43. Bukowski RM, Murthy S, McLain D, *et al.* Phase I trial of recombinant granulocyte-macrophage colony-stimulating factor in patients with lung cancer: clinical and immunologic effects. *J Immunother Emphasis Tumor Immunol* 1993;**13**:267–74.

44. Emminger W, Emminger-Schmidmeier W, Peters C, *et al.* Capillary leak syndrome during low dose granulocyte-macrophage colony-stimulating factor (rh GM-CSF) treatment of a patient in a continuous febrile state. *Blut* 1990;**61**:219–21.

45. Wiley JS, Jamieson GP, Cebon JS, *et al.* Cytokine priming of acute myeloid leukemia may produce a pulmonary syndrome when associated with a rapid increase in peripheral blood myeloblasts. *Blood* 1993;**82**:3511–12.

46. Bunn PA, Crowley J, Kelly K, *et al.* Chemoradiotherapy with or without granulocyte–macrophage colony-stimulating factor in the treatment of limited-stage small-cell lung cancer: a prospective phase III randomized study of the Southwest Oncology Group. *J Clin Oncol* 1995;**13**:1632–41.

47. Anderson PM, Markovic SN, Sloan JA, *et al.* Aerosol granulocyte macrophage-colony stimulating factor: a low toxicity, lung-specific biological therapy in patients with lung metastases. *Clin Cancer Res* 1999;**5**:2316–23.

48. Hübel K, Dale DC, Liles WC. Therapeutic use of cytokines to modulate phagocyte function for the treatment of infectious diseases: current status of granulocyte colony-stimulating factor, granulocyte–macrophage colony-stimulating factor, macrophage colony-stimulating factor, and interferon-γ. *J Infect Dis* 2002;**185**:1490–501.

49. Zamkoff KW, Hudson J, Groves ES, *et al.* A phase I trial of recombinant human macrophage colony-stimulating factor by rapid intravenous infusion in patients with refractory malignancy. *J Immunother* 1992;**11**:103–10.

50. Weiner LM, Li W, Holmes M, *et al.* Phase I trial of recombinant macrophage colony-stimulating factor and recombinant

γ-interferon: toxicity, monocytosis, and clinical effects. *Cancer Res* 1994;**54**:4084–90.

51. Williams RO, Feldmann M, Maini RN. Anti-tumor necrosis factor ameliorates joint disease in murine collagen-induced arthritis. *Proc Natl Acad Sci USA* 1992;**89**:9784–8.

52. Olsen NJ, Stein CM. New drugs for rheumatoid arthritis. *New Engl J Med* 2004;**350**:2167–79.

53. Doran MF, Crowson CS, Pond GR, *et al.* Frequency of infection in patients with rheumatoid arthritis compared with controls: a population-based study. *Arthritis Rheum* 2002;**46**:2287–93.

54. Maini R, St Clair EW, Breedveld F, *et al.* Infliximab (chimeric anti-tumour necrosis factor α monoclonal antibody) versus placebo in rheumatoid arthritis patients receiving concomitant methotrexate: a randomized phase III trial. *Lancet* 1999;**354**:1932–9.

55. Keane J, Gershon S, Wise RP, *et al.* Tuberculosis associated with infliximab, a tumor necrosis factor α-neutralizing agent. *New Engl J Med* 2001;**345**:1098–104.

56. Wallis RS, Broder MS, Wong JY, *et al.* Granulomatous infectious diseases associated with tumor necrosis factor antagonists. *Clin Infect Dis* 2004;**38**:1261–5.

57. Gardam MA, Keystone EC, Menzies R, *et al.* Anti-tumour necrosis factor agents and tuberculosis risk: mechanisms of action and clinical management. *Lancet Infect Dis* 2003;**3**:148–55.

58. Schiff MH, Burmester GR, Kent JD, *et al.* Safety analyses of adalimumab (HUMIRA) in global clinical trials and US postmarketing surveillance of patients with rheumatoid arthritis. *Ann Rheum Dis* 2006;**65**:889–94.

59. Winthrop KL, Siegel JN, Jereb J, *et al.* Tuberculosis associated with therapy against tumor necrosis factor α. *Arthritis Rheum* 2005;**52**:2968–74.

60. Crum NF, Lederman ER, Wallace MR. Infections associated with tumor necrosis factor-α antagonists. *Medicine* 2005;**84**:291–302.

61. Raychaudhuri S, Shmerling R, Ermann J, Helfgott S. Development of active tuberculosis following initiation of infliximab despite appropriate prophylaxis. *Rheumatology* (Oxford) 2007;**46**:887–8.

62. Newman SL. Cell-mediated immunity to *Histoplasma capsulatum*. *Semin Respir Infect* 2001;**16**:102–8.

63. Lee JH, Slifman NR, Gershon SK, *et al.* Life-threatening histoplasmosis complicating immunotherapy with tumor necrosis factor α antagonists infliximab and etanercept. *Arthritis Rheum* 2002;**46**:2565–70.

64. Tsiodras S, Samonis G, Boumpas DT, Kontoyiannis DP. Fungal infections complicating tumor necrosis factor α blockade therapy. *Mayo Clin Proc* 2008;**83**:181–94.

65. Wood KL, Hage CA, Knox KS, *et al.* Histoplasmosis after treatment with anti-tumor necrosis factor-α therapy. *Am J Respir Crit Care Med* 2003;**167**:1279–82.

66. Bergstrom L, Yocum DE, Ampel NM, *et al.* Increased risk of coccidioidomycosis in patients treated with tumor necrosis factor α antagonists. *Arthritis Rheum* 2004;**50**:1959–66.

67. Couriel D, Saliba R, Hicks K, *et al.* Tumor necrosis factor-α blockade for the treatment of acute GVHD. *Blood* 2004;**104**:649–54.

68. Kaur N, Mahl TC. *Pneumocystis jiroveci (carinii)* pneumonia after infliximab therapy: a review of 84 cases. *Dig Dis Sci* 2007;**52**:1481–4.

69. Kaur N, Mahl TC. *Pneumocystis carinii* pneumonia with oral candidiasis after infliximab therapy for Crohn's disease. *Dig Dis Sci* 2004;**49**:1458–60.

70. Baert FJ, D'Haens GR, Peeters M, *et al.* Tumor necrosis factor α antibody (infliximab) therapy profoundly down-regulates the inflammation in Crohn's ileocolitis. *Gastroenterology* 1999;**116**:22–8.

71. Slifman NR, Gershon SK, Lee JH, *et al.* Listeria monocytogenes infection as a complication of treatment with tumor necrosis factor α-neutralizing agents. *Arthritis Rheum* 2003;**48**:319–24.

72. Wondergem MJ, Voskuyl AE, van Agtmael MA. A case of legionellosis during treatment with a TNF-α antagonist. *Scand J Infect Dis* 2004;**36**:310–11.

73. Tubach F, Ravaud P, Salmon-Céron D, *et al.* Emergence of *Legionella pneumophila* pneumonia in patients receiving tumor necrosis factor-α antagonists. *Clin Infect Dis* 2006;**43**:e95–100.

74. Krishnamurthy R, Dincer HE, Whittemore D. *Strongyloides stercoralis* hyperinfection in a patient with rheumatoid arthritis after anti-TNF-α therapy. *J Clin Rheumatol* 2007;**13**:150–2.

75. Cunnane G, Warnock M, Fye KH, Daikh DI. Accelerated nodulosis and vasculitis following etanercept therapy for rheumatoid arthritis. *Arthritis Rheum* 2002;**47**:445–9.

76. Yousem SA, Dacic S. Pulmonary lymphohistiocytic reactions temporally related to etanercept therapy. *Mod Pathol* 2005;**18**:651–5.

77. Braun J, Brandt J, Listing J, *et al.* Treatment of active ankylosing spondylitis with infliximab: a randomised controlled multicentre trial. *Lancet* 2002;**359**:1187–93.

78. Ostör AJ, Chilvers ER, Somerville MF, *et al.* Pulmonary complications of infliximab therapy in patients with rheumatoid arthritis. *J Rheumatol* 2006;**33**:622–8.

79. Hagiwara K, Sato T, Takagi-Kobayashi S, *et al.* Acute exacerbation of preexisting interstitial lung disease after administration of etanercept for rheumatoid arthritis. *J Rheumatol* 2007;**34**:1151–4.

80. Kramer N, Chuzhin Y, Kaufman LD, *et al.* Methotrexate pneumonitis after initiation of infliximab therapy for rheumatoid arthritis. *Arthritis Rheum* 2002;**47**:670–1.

81. Lindsay K, Melsom R, Jacob BK, Mestry N. Acute progression of interstitial lung disease: a complication of etanercept particularly in the presence of rheumatoid lung and methotrexate treatment. *Rheumatology* (Oxford) 2006;**45**:1048–9.

82. Riegert-Johnson DL, Godfrey JA, Myers JL, *et al.* Delayed hypersensitivity reaction and acute respiratory distress syndrome following infliximab infusion. *Inflamm Bowel Dis* 2002;**8**:186–91.

83. Cersosimo RJ. Monoclonal antibodies in the treatment of cancer, part 1. *Am J Health Syst Pharm* 2003;**60**:1531–48.

84. Rastetter W, Molina A, White CA. Rituximab: expanding role in therapy for lymphomas and autoimmune diseases. *Annu Rev Med* 2004;**55**:477–503.

85. Rituxan (rituximab) package insert. South San Francisco, CA: Biogen Idec Inc. and Genentech Inc., 2008. Accessed on 19 August 2008 at www.gene.com/gene/products/information/pdf/rituxan-prescribing.pdf.

86. Wagner SA, Mehta AC, Laber DA. Rituximab-induced interstitial lung disease. *Am J Hematol* 2007;**82**:916–19.

87. Saito B, Nakamaki T, Adachi D, *et al.* Acute respiratory distress syndrome during the third infusion of rituximab in a patient with follicular lymphoma. *Int J Hematol* 2004;**80**:164–7.

88. Montero AJ, McCarthy JJ, Chen G, Rice L. Acute respiratory distress syndrome after rituximab infusion. *Int J Hematol* 2005;**82**:324–6.

89. Ware LB, Matthay MA. The acute respiratory distress syndrome. *New Engl J Med* 2000;**342**:1334–49.

90. Macartney C, Burke E, Elborn S, *et al.* Bronchiolitis obliterans organizing pneumonia in a patient with non-Hodgkin's lymphoma following R-CHOP and pegylated filgrastim. *Leuk Lymphoma* 2005;**46**:1523–6.

91. Biehn SE, Kirk D, Rivera MP, *et al.* Bronchiolitis obliterans with organizing pneumonia after rituximab therapy for non-Hodgkin's lymphoma. *Hematol Oncol* 2006;**24**:234–7.

92. Alexandrescu DT, Dutcher JP, O'Boyle K, *et al.* Fatal intra-alveolar hemorrhage after rituximab in a patient with non-Hodgkin lymphoma. *Leuk Lymphoma* 2004;**45**:2321–5.

93. Heresi GA, Farver CF, Stoller JK. Interstitial pneumonitis and alveolar hemorrhage complicating use of rituximab: case report and review of the literature. *Respiration* 2008;**76**:449–53.

94. Hudis CA. Trastuzumab: mechanism of action and use in clinical practice. *New Engl J Med* 2007;**357**:39–51.

95. Cook-Bruns N. Retrospective analysis of the safety of Herceptin immunotherapy in metastatic breast cancer. *Oncology* 2001;**61**:58–66.

96. Romond EH, Perez EA, Bryant J, *et al.* Trastuzumab plus adjuvant chemotherapy for operable HER2-positive breast cancer. *New Engl J Med* 2005;**353**:1673–84.

97. Clamon G, Herndon J, Kern J, *et al.* Lack of trastuzumab activity in nonsmall cell lung carcinoma with overexpression of erb-B2. 39810: a phase II trial of Cancer and Leukemia Group B. *Cancer* 2005;**103**:1670–5.

98. Radzikowska E, Szczepulska E, Chabowski M, Bestry I. Organising pneumonia caused by trastuzumab (Herceptin) therapy for breast cancer. *Eur Respir J* 2003;**21**:552–5.

99. Kamba T, McDonald DM. Mechanisms of adverse effects of anti-VEGF therapy for cancer. *Br J Cancer* 2007;**96**:1788–95.

100. Johnson DH, Fehrenbacher L, Novotny WF, *et al.* Randomized phase II trial comparing bevacizumab plus carboplatin and paclitaxel with carboplatin and paclitaxel alone in previously untreated locally advanced or metastatic non-small-cell lung cancer. *J Clin Oncol* 2004;**22**:2184–91.

101. Sandler A, Gray R, Perry MC, *et al.* Paclitaxel–carboplatin alone or with bevacizumab for non-small-cell lung cancer. *New Engl J Med* 2006;**355**:2542–50.

102. Herbst RS, Sandler AB. Non-small cell lung cancer and antiangiogenic therapy: what can be expected of bevacizumab? *Oncologist* 2004;**9**:19–26.

103. Bross PF, Beitz J, Chen G, *et al.* Approval summary: gemtuzumab ozogamicin in relapsed acute myeloid leukemia. *Clin Cancer Res* 2001;**7**:1490–6.

104. Ravandi F, O'Brien S. Alemtuzumab in CLL and other lymphoid neoplasms. *Cancer Invest* 2006;**24**:718–25.

105. Uppenkamp M, Engert A, Diehl V, *et al.* Monoclonal antibody therapy with CAMPATH-1H in patients with relapsed high- and low-grade non-Hodgkin's lymphomas: a multicenter phase I/II study. *Ann Hematol* 2002;**81**:26–32.

106. Rieger K, Von Grünhagen U, Fietz T, *et al.* Efficacy and tolerability of alemtuzumab (CAMPATH-1H) in the salvage treatment of B-cell chronic lymphocytic leukemia: change of regimen needed? *Leuk Lymphoma* 2004;**45**:345–9.

107. Sachdeva A, Matuschak GM. Diffuse alveolar hemorrhage following alemtuzumab. *Chest* 2008;**133**:1476–8.

108. Wendtner CM, Ritgen M, Schweighofer CD, *et al.* Consolidation with alemtuzumab in patients with chronic lymphocytic leukemia (CLL) in first remission: experience on safety and efficacy within a randomized multicenter phase III trial of the German CLL Study Group (GCLLSG). *Leukemia* 2004;**18**:1093–101.

109. Laurenti L, Piccioni P, Cattani P, *et al.* Cytomegalovirus reactivation during alemtuzumab therapy for chronic lymphocytic leukemia: incidence and treatment with oral ganciclovir. *Haematologica* 2004;**89**:1248–52.

110. Hendeles L, Sorkness CA. Anti-immunoglobulin E therapy with omalizumab for asthma. *Ann Pharmacother* 2007;**41**:1397–410.

111. Deniz YM, Gupta N. Safety and tolerability of omalizumab (Xolair), a recombinant humanized monoclonal anti-IgE antibody. *Clin Rev Allergy Immunol* 2005;**29**:31–48.

112. Cox L, Platts-Mills TAE, Finegold I, *et al.* American Academy of Allergy, Asthma & Immunology/American College of Allergy, Asthma and Immunology Joint Task Force Report on omalizumab-associated anaphylaxis. *J Allergy Clin Immunol* 2007;**120**:1373–7.

113. Wechsler ME, Wong DA, Miller MK, Lawrence-Miyasaki L. Churg–Strauss syndrome in patients treated with omalizumab. *Chest* 2009;**136**:507–18.

114. Mazzaferri EL, Young JJ. Abciximab: a review and update for clinicians. *Expert Rev Cardiovasc Ther* 2008;**6**:609–18.

115. Iskandar SB, Kasasbeh ES, Mechleb BK, *et al.* Alveolar hemorrhage: an underdiagnosed complication of treatment with glycoprotein IIb/IIIa inhibitors. *J Interven Cardiol* 2006;**19**:356–63.

116. Ali A, Hashem M, Rosman HS, *et al.* Use of platelet glycoprotein IIb/IIIa inhibitors and spontaneous pulmonary hemorrhage. *J Invas Cardiol* 2003;**15**:186–8.

117. Dolan N, Waldron M, O'Connell M, *et al.* Basiliximab induced non-cardiogenic pulmonary edema in two pediatric renal transplant recipients. *Pediatr Nephrol* 2009;**24**:2261–5.

118. Baudouin V, Crusiaux A, Haddad E, *et al.* Anaphylactic shock caused by immunoglobulin E sensitization after retreatment with the chimeric anti-interleukin-2 receptor monoclonal antibody basiliximab. *Transplantation* 2003;**76**:459–63.

119. Suntharalingam G, Perry MR, Ward S, *et al.* Cytokine storm in a phase 1 trial of the anti-CD28 monoclonal antibody TGN1412. *New Engl J Med* 2006;**355**:1018–28.

120. Metro G, Finocchiaro G, Toschi L, *et al.* Epidermal growth factor receptor (EGFR) targeted therapies in non-small cell lung cancer (NSCLC). *Rev Recent Clin Trials* 2006;**1**:1–13.

121. Inoue A, Saijo Y, Maemondo M, *et al.* Severe acute interstitial pneumonia and gefitinib. *Lancet* 2003;**361**:137–9.

122. Camus P, Kudoh S, Ebina M. Interstitial lung disease associated with drug therapy. *Br J Cancer* 2004;**91**:S18–23.

123. Kudoh S, Kato H, Nishiwaki Y, *et al.* Interstitial lung disease in Japanese patients with lung cancer: a cohort and nested case–control study. *Am J Respir Crit Care Med* 2008;**177**: 1348–57.

124. Kataoka K, Taniguchi H, Hasegawa Y, *et al.* Interstitial lung disease associated with gefitinib. *Respir Med* 2006;**100**: 698–704.

125. Ando M, Okamoto I, Yamamoto N, *et al.* Predictive factors for interstitial lung disease, antitumor response, and survival in non-small-cell lung cancer patients treated with gefitinib. *J Clin Oncol* 2006;**24**:2549–56.

126. Endo M, Johkoh T, Kimura K, Yamamoto N. Imaging of gefitinib-related interstitial lung disease: multi-institutional analysis by the West Japan Thoracic Oncology Group. *Lung Cancer* 2006;**52**:135–40.

127. Herbst RS, Prager D, Hermann R, *et al.* TRIBUTE: a phase III trial of erlotinib hydrochloride (OSI-774) combined with carboplatin and paclitaxel chemotherapy in advanced non-small-cell lung cancer. *J Clin Oncol* 2005;**23**:5892–9.

128. Yoneda KY, Shelton DK, Beckett LA, Gandara DR. Independent review of interstitial lung disease associated with death in TRIBUTE (paclitaxel and carboplatin with or without concurrent erlotinib) in advanced non-small cell lung cancer. *J Thorac Oncol* 2007;**2**:537–43.

129. Vahid B, Esmaili A. Erlotinib-associated acute pneumonitis: report of two cases. *Can Respir J* 2007;**14**:167–70.

130. Makris D, Scherpereel A, Copin MC, *et al.* Fatal interstitial lung disease associated with oral erlotinib therapy for lung cancer. *BMC Cancer* 2007;**7**:150.

131. Piccaluga PP, Rondoni M, Paolini S, *et al.* Imatinib mesylate in the treatment of hematologic malignancies. *Expert Opin Biol Ther* 2007;**7**:1597–611.

132. O'Brien SG, Guilhot F, Larson RA, *et al.* Imatinib compared with interferon and low-dose cytarabine for newly diagnosed chronic-phase chronic myeloid leukemia. *New Engl J Med* 2003;**348**:994–1004.

133. Park YH, Park HJ, Kim BS, *et al.* BNP as a marker of the heart failure in the treatment of imatinib mesylate. *Cancer Lett* 2006;**243**:16–22.

134. Ishii Y, Shoji N, Kimura Y, Ohyashiki K. Prominent pleural effusion possibly due to imatinib mesylate in adult Philadelphia chromosome-positive acute lymphoblastic leukemia. *Intern Med* 2006;**45**:339–40.

135. Ohnishi K, Sakai F, Kudoh S, Ohno R. Twenty-seven cases of drug-induced interstitial lung disease associated with imatinib mesylate. *Leukemia* 2006;**20**:1162–4.

136. Bergeron A, Bergot E, Vilela G, *et al.* Hypersensitivity pneumonitis related to imatinib mesylate. *J Clin Oncol* 2002;**20**:4271–2.

137. Yokoyama T, Miyazawa K, Kurakawa E, *et al.* Interstitial pneumonia induced by imatinib mesylate: pathologic study demonstrates alveolar destruction and fibrosis with eosinophilic infiltration. *Leukemia* 2004;**18**:645–6.

138. Lin JT, Yeh KT, Fang HY, Chang CS. Fulminant, but reversible interstitial pneumonitis associated with imatinib mesylate. *Leuk Lymphoma* 2006;**47**:1693–5.

139. Wagner U, Staats P, Moll R, *et al.* Imatinib-associated pulmonary alveolar proteinosis. *Am J Med* 2003;**115**:674.

140. Salmons N, Gregg RJ, Pallalau A, *et al.* Lymphomatoid granulomatosis in a patient previously diagnosed with a gastrointestinal stromal tumour and treated with imatinib. *J Clin Pathol* 2007;**60**:199–201.

141. Talpaz M, Shah NP, Kantarjian H, *et al.* Dasatinib in imatinib-resistant Philadelphia chromosome-positive leukemias. *New Engl J Med* 2006;**354**:2531–41.

142. Quintás-Cardama A, Kantarjian H, O'Brien S, *et al.* Pleural effusion in patients with chronic myelogenous leukemia treated with dasatinib after imatinib failure. *J Clin Oncol* 2007;**25**:3908–14.

143. Bergeron A, Réa D, Levy V, *et al.* Lung abnormalities after dasatinib treatment for chronic myeloid leukemia: a case series. *Am J Respir Crit Care Med* 2007;**176**:814–18.

144. Vignot S, Faivre S, Aguirre D, Raymond E. mTOR-targeted therapy of cancer with rapamycin derivatives. *Ann Oncol* 2005;**16**:525–37.

145. Duran I, Siu LL, Oza AM, *et al.* Characterisation of the lung toxicity of the cell cycle inhibitor temsirolimus. *Eur J Cancer* 2006;**42**:1875–80.

146. Atkins MB, Hidalgo M, Stadler WM, *et al.* Randomized phase II study of multiple dose levels of CCI-779, a novel mammalian target of rapamycin kinase inhibitor, in patients with advanced refractory renal cell carcinoma. *J Clin Oncol* 2004;**22**:909–18.

147. Chan S, Scheulen ME, Johnston S, *et al.* Phase II study of temsirolimus (CCI-779), a novel inhibitor of mTOR, in heavily pretreated patients with locally advanced or metastatic breast cancer. *J Clin Oncol* 2005;**23**:5314–22.

148. Galanis E, Buckner JC, Maurer MJ, *et al.* Phase II trial of temsirolimus (CCI-779) in recurrent glioblastoma multiforme: a North Central Cancer Treatment Group Study. *J Clin Oncol* 2005;**23**:5294–304.

149. Witzig TE, Geyer SM, Ghobrial I, *et al.* Phase II trial of single-agent temsirolimus (CCI-779) for relapsed mantle cell lymphoma. *J Clin Oncol* 2005;**23**:5347–56.

150. Adams J. The proteasome: a suitable antineoplastic target. *Natl Rev Cancer* 2004;**4**:349–60.

151. Miyakoshi S, Kami M, Yuji K, *et al.* Severe pulmonary complications in Japanese patients after bortezomib treatment for refractory multiple myeloma. *Blood* 2006;**107**:3492–4.

152. Gotoh A, Ohyashiki K, Oshimi K, *et al.* Lung injury associated with bortezomib therapy in relapsed/refractory multiple myeloma in Japan: a questionnaire-based report from the Lung Injury by Bortezomib Joint Committee of the Japanese Society of Hematology and the Japanese Society of Clinical Hematology. *Int J Hematol* 2006;**84**: 406–12.

153. Ohri A, Arena FP. Severe pulmonary complications in African-American patient after bortezomib therapy. *Am J Ther* 2006;**13**:553–5.

16

Radiation-induced lung disease

MAX M WEDER, M PATRICIA RIVERA

INTRODUCTION

Radiation therapy of the chest has become part of standard regimens to treat limited-stage small-cell and locally advanced (stage III) non-small-cell lung cancer (NSCLC), and it is also used in other types of malignancies such as breast cancer and lymphoma. Recent advances in radiation techniques were aimed at minimizing radiation exposure to normal tissue. With the use of three-dimensional conformal radiation, it has been possible to target chest tumours much more precisely and to considerably exceed the traditional cut-off dose of 60 Gy (1 Gy equals 100 rad). These advances have resulted in improved local tumour control and prolonged survival in patients with locally advanced NSCLC. However, even with the most advanced planning techniques, normal tissue structures will inevitably be exposed to radiation, which may result in morbidity and even mortality. Radiation toxicity may develop acutely within a short time span after radiation exposure or occur as late as several years after therapy has been completed.

EPIDEMIOLOGY

It is difficult to estimate how frequently radiation injury to the lungs truly occurs. The reported incidence is 5–20 per cent in patients with lung cancer and 5–15 per cent in patients with breast cancer and lymphoma.[1] These differences are not surprising since there are substantial differences in treatment protocols and chosen clinical end-points for radiation injury. Among other factors, the frequency of radiation injury depends on the radiation dose, the volume of irradiated lung, the radiation technique, and whether neoadjuvant, adjuvant, concurrent or consolidation chemotherapy is used. Non-specific symptoms of radiation pneumonitis contribute to misdiagnosis and under-reporting of this disease. In addition, there is a discrepancy between patients who have only radiographic

evidence of radiation injury and those who have clinical signs of radiation pneumonitis. In patients with breast cancer, symptomatic radiation pneumonitis occurs in 0–10 per cent, while radiographic abnormalities can be detected in 27–40 per cent. In lung cancer, clinical pneumonitis occurs in 5–15 per cent, whereas radiographic changes consistent with radiation injury can be detected in up to two-thirds of all patients. It is unclear to what extent these latter changes may actually be related to radiation or to the tumour itself.

PATHOPHYSIOLOGY

Even though processes of radiation injury may start immediately after radiation exposure, the consequences may be clinically unapparent and histopathological changes may take weeks, months, sometimes even years to develop. Radiation injury to the lung is a combination of the direct cytotoxic effects of radiation and the activation of pro-inflammatory cytokine cascades that promote the development of pulmonary fibrosis.

Radiation is directly cytotoxic to normal lung tissue. It causes localized release of energy that is sufficient to damage DNA structures and cause clonogenic death of pulmonary epithelial cells. The release of free oxygen radicals promotes cellular injury. Over-expression of manganese superoxide dismutase, an enzyme capable of neutralizing free oxygen radicals, has been shown to protect against radiation injury in animal models.[2]

Radiation exposure causes up-regulation of cytokines that play an important role in the development of pulmonary fibrosis.[3] The induction of TGF-β has been studied most extensively. The level of TGF-β after radiation exposure has been used to predict the development of radiation-induced pulmonary fibrosis.[4] Pro-inflammatory cytokines like TNF-α, IL-1 and fibrogenic IL-6 are up-regulated following radiation exposure. Platelet-derived growth factor (PDGF) and basic fibroblast growth factor (bFGF) are potent stimulants of fibroblast proliferation and are also up-regulated following radiation exposure.[5]

PATHOLOGY

Pathological changes that occur after radiation exposure can be divided into several phases. Immediately following radiation, the bronchial mucosa is hyperaemic, congested and infiltrated with leucocytes. Bronchial hypersecretion may occur. Increased capillary permeability results in pulmonary oedema which is followed by an exudative alveolitis that is associated with degeneration of the alveolar epithelium and endothelium. Type I pneumocytes are sloughed and surfactant levels are initially increased. This immediate phase is generally asymptomatic and may not produce radiographic abnormalities. It is followed by hypersecretion of thick mucus which is due to proliferation of goblet cells and ciliary dysfunction.

The clinical equivalent of the next phase which occurs 2–3 months after radiation exposure is radiation pneumonitis. Obliteration of pulmonary capillaries results in microthrombosis. Detachment of epithelial and endothelial cells with decrease in surfactant production promotes leakage of a fibrinous exudate into the alveoli that transform into hyaline membranes. Giant-cell formation along the endothelial lining may be seen along with hyperplasia and atypia of type 2 pneumocytes.

Subsequently, the process of radiation injury may resolve with dissolution of alveolar oedema and hyaline membranes, or it may progress to fibrosis with fibroblastic collagen deposition and interstitial thickening.

The final phase of fibrosis may become apparent as early as 6 months after the exposure, but it is possible that progression occurs slowly over several years and remains clinically silent. There is prominence of myofibroblasts in the interstitium and alveolar space with increase in collagen deposition. The progressive narrowing and obliteration of alveolar spaces corresponds with decrease in lung volume.

CLINICAL PRESENTATION

Radiation pneumonitis

Symptoms of radiation pneumonitis usually occur 1–6 months after completion of radiation therapy. They are fairly non-specific and usually have an insidious onset. Typical complaints include a cough which is usually non-productive, exertional dyspnoea and fever which is usually low-grade. A pleural reaction may present with pleuritic-type chest pain. Patients with radiation pneumonitis can feel quite ill and may demonstrate generalized malaise and weight loss, which can be mistaken for tumour progression.

Radiation fibrosis

If radiation pneumonitis progresses to fibrosis, symptoms mainly include those of chronic dyspnoea, although a dry cough may persist. Patients complain about worsening dyspnoea that

may occur at rest in severe cases. In advanced stages, there may be signs of pulmonary hypertension and increased right heart strain.

Radiation-induced oesophagitis

Radiation-induced oesophagitis remains the main dose-limiting acute toxicity in radiotherapeutic management of thoracic neoplasms. The oesophageal mucosa is subject to constant regeneration and high cell turnover, which renders it very susceptible to radiation injury. The basal epithelial layer is usually affected in acute oesophageal radiation injury, resulting in mucosal thinning that may progress to denudation. Symptoms usually occur within 2–3 weeks after initiation of radiotherapy and consist of dysphagia, odynophagia and substernal chest discomfort. In severe cases, decreased food and fluid intake may result in dehydration and weight loss and warrant discontinuation or interruption of radiation therapy.

Endoscopically, radiation-induced oesophagitis usually presents with mucosal inflammation and ulceration, but perforation is rarely observed. After discontinuation of radiation, the mucosa usually regenerates and symptoms resolve after a period of approximately 3 weeks.

Late effects of oesophageal radiation injury occur with a median time onset of 6 months after completion of radiation therapy and are related to fibrosis and scarring following the initial insult.[6] The most common clinical manifestation is dysphagia related to oesophageal stricture or dysmotility, which may be the result of fibrosis and muscular or nerve damage. Less common presentations include chronic oesophageal ulcerations and rare formation of fistulas. Modified barium swallow is useful to identify oesophageal strictures and dysmotility, classically demonstrating repetitive non-peristaltic waves above and below the area that has been directly exposed to radiation.

The risk of developing oesophageal radiation injury depends on the radiation dose, radiation technique and the use of concurrent chemotherapy. In patients receiving conventional radiotherapy for oesophageal cancer, a total radiation dose of 60 Gy was estimated to result in a 5 per cent incidence of oesophageal complications at 5 years when one-third of the oesophagus was exposed to radiation.[7] With three-dimensional conformal radiation, the risk of late complications was found to be related to the percentage of oesophageal volume and surface receiving radiation doses exceeding 50 Gy. A maximum point dose of 69 Gy in patients receiving radiotherapy alone and 58 Gy with concurrent use of chemotherapy have also been found to predict the development of severe oesophageal toxicity.[8] Patients with pre-existing oesophageal mucosal injury related to gastroesophageal reflux disease or tumour invasion appear to be at increased risk. Higher incidences of severe toxicity have also been observed with hyperfractionated dose regimens and concurrent as opposed to sequential chemoradiation. It has also been observed that HIV-positive patients are more susceptible to develop oesophageal toxicities of radiation treatment.[9]

Therapy of acute oesophageal radiation toxicity is mainly directed towards alleviation of symptoms. Topical anaesthetics, acid-lowering agents, sucralfate, promotility agents and analgesics can be used individually or combined to provide symptomatic relief. Patients should also be screened for co-existing *Candida* mucositis and receive appropriate antifungal therapy when found to be infected.

The ability of amifostine (an agent capable of neutralizing free oxygen radicals that has also been tried in radiation pneumonitis) to prevent radiation-induced oesophageal injury has not been confirmed consistently across trials.

Patients with oesophageal strictures are usually subjected to endoscopic balloon dilatation. Although serial procedures may be required, this approach usually provides good symptomatic relief. The tissue damage caused by radiation-induced oesophageal injury subjects the patient to a higher risk of procedure-related complications such as oesophageal perforation.

RARE FORMS OF RADIATION INJURY

Sporadic radiation pneumonitis

Sporadic radiation pneumonitis refers to a disease entity that is different from classic radiation pneumonitis both clinically and pathogenically. In contrast to classic radiation pneumonitis, sporadic radiation pneumonitis involves areas of the lung that are outside the radiation port. Its occurrence is rare and unpredictable. The severity of symptoms is out of proportion to the volume of irradiated lung. However, the syndrome usually resolves without sequelae of radiation fibrosis. Laboratory findings suggest that radiation may induce a T-cell-mediated inflammatory response that is quite similar to hypersensitivity pneumonitis.[10] Bronchoalveolar lavage demonstrates marked CD4-predominant lymphocytic alveolitis. In cases of unilateral chest radiation, BAL findings do not differ significantly between the radiated and the non-radiated lung.[6]

Radiation recall

Radiation recall refers to an inflammatory reaction in previously irradiated tissue following the administration of certain pharmacological agents. This phenomenon has been most commonly described in the skin.[11] Recall toxicities involving other organs including the lungs have been reported. There is no specific time window during which radiation recall can occur, and this reaction may develop years after completion of radiation exposure. Also, the occurrence of radiation recall is independent of previous manifestations of radiation injury. Radiation recall has been most commonly associated with the administration of various chemotherapeutic agents, but it can also occur with other classes of drugs, including antituberculous therapy, antibiotics, tamoxifen and statins.

Recall pneumonitis resembles regular radiation pneumonitis both clinically and radiographically. The pathophysiological mechanism responsible for recall pneumonitis is unknown. Drug hypersensitivity reactions, as well as changes in epithelial stem cells and altered expression of various proteins, have been proposed.[12] Radiation recall pneumonitis responds favourably to withdrawal of the offending drug and to anti-inflammatory therapy, usually in the form of corticosteroids.

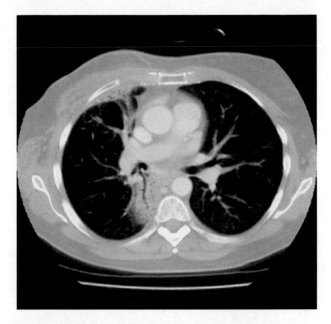

Fig. 16.1 Computed tomography of chest demonstrating bronchiolitis obliterans organizing pneumonia after radiation for breast cancer.

Bronchiolitis obliterans organizing pneumonia

BOOP is a distinct pathologic entity that is characterized by diffuse migratory, patchy peripheral air-space opacities. Like sporadic radiation pneumonitis, it usually resolves without clinical sequelae. Pathologically, BOOP is characterized by excessive formation of granulation tissue in small airways with associated chronic inflammation of adjacent alveoli, which causes organizing pneumonia. Radiation-induced BOOP has almost exclusively been described as a consequence of radiation therapy for breast cancer (Fig. 16.1).[13] This may be related to the tangential-beam radiation technique that is commonly used, which limits radiation exposure primarily to subpleural regions of the lung. In most affected individuals, BOOP lesions arise in irradiated areas and subsequently spread to non-irradiated tissue. Sporadic case reports mention radiation-induced BOOP following radiation therapy for lymphoma and lung cancer.[14] Usually, radiation-induced BOOP occurs 6–10 weeks after completion of radiation therapy,[15] although much longer intervals have been reported.[16] Clinically, radiation-induced BOOP presents with low-grade fever, non-productive cough and dyspnoea. Radiographically, it is characterized by migrating patchy, diffuse ground-glass opacities that frequently have a subpleural or peribronchial distribution. Infiltrates tend to spread from irradiated to non-irradiated lung areas. Less frequently, it may manifest with a single area of consolidation or diffuse reticular infiltrates. Radiation-induced BOOP usually responds favourably to corticosteroid treatment, but has a tendency to relapse when therapy is withdrawn.[17]

Pneumothorax

Spontaneous pneumothorax following radiation therapy has occasionally been reported, mainly as a consequence of mantle field radiation for Hodgkin's and non-Hodgkin's lymphoma.[12] Typically, the radiation dose exceeds 30 Gy in patients treated with mantle field radiation, and most of these patients demonstrate evidence of radiation fibrosis. Pneumothoraces may be recurrent, rarely occur bilaterally and have a mean time to onset of 16 months following completion of radiation therapy. The volume is usually small, and most pneumothoraces re-expand without further intervention.[18]

Bronchial stenosis

Bronchial stenosis as a consequence of radiation therapy has mostly been described in patients receiving palliative

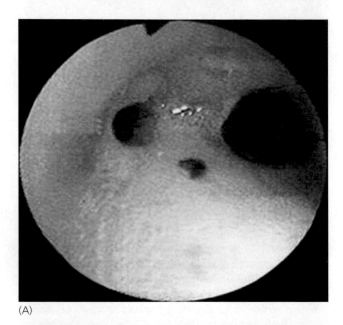

(A)

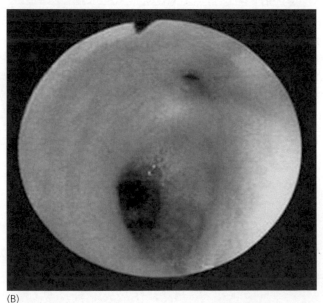

(B)

Fig. 16.2 (a) Bronchostenosis of the left mainstem bronchus following combined chemotherapy and high-dose radiation. (b) Bronchostenosis of the right upper lobe following combined therapy.

endoluminal brachytherapy.[19] Reports on bronchial stenosis following external beam radiation are scarce, but recent evidence suggests that this complication may occur more frequently with high-dose three-dimensional conformal radiation (Fig. 16.2).[20] With this technique, a relatively high radiation dose is applied to a target area that may be considerably smaller than with conventional radiation techniques. This could explain the occurrence of normal tissue complications that were considered rare with conventional radiation.

Clinically, bronchial stenosis may present with nonproductive cough and dyspnoea due to atelectatic lung areas. Patients with bronchial stenosis are at risk to develop postobstructive pneumonia. If infection is ruled out, a trial of corticosteroids to reduce mucosal oedema and improve aeration of dependent lung areas is reasonable. In refractory cases where patients remain symptomatic, endobronchial dilatation and/or bronchoscopic stenting of the obstructed airway may be considered.

Pericardial disease

Radiation-induced pericardial disease is well described in the literature, commonly in the setting of mediastinal radiation for Hodgkin's lymphoma.[21] Manifestations may vary in time onset and severity and depend on the amount of pericardial radiation exposure. Post-radiation pericarditis rarely occurs with radiation doses below 40 Gy.[22] Most patients present with asymptomatic pericardial effusions, which can occur in 20–40 per cent of patients receiving radiation to the chest and usually occurs 12–18 months after completion of therapy. The development of pericardial tamponade is rare and the progression to constrictive–effusive pericarditis is even more uncommon. Effusive–constrictive pericarditis is characterized by markedly increased intrapericardial pressure that fails to normalize after the effusion has been drained. In a large retrospective analysis looking at 1184 patients with pericarditis and various underlying diseases, only 15 patients with effusive–constrictive pericarditis were identified, two of whom had previous radiation exposure.[23]

Cranial nerve palsy

Cranial nerve palsy is a rare complication of radiation therapy for head and neck cancer. The latency period with which cranial nerve palsy occurs after completion of radiation treatment varies considerably and can be as long as 10 years.[24] The hypoglossal, vagal and recurrent laryngeal nerves seem to be affected most frequently, resulting in hoarseness and dyspnoea due to vocal cord paralysis, dysphagia and dysarthria. Involvement of the other cranial nerves with either single or multiple cranial nerve involvement has also been described. Higher individual fraction size of the radiation dose may increase the risk of developing cranial nerve palsy. Since cranial nerve palsy is a common presentation of tumour recurrence and may also occur as a result of infectious and non-infectious disease processes, a thorough work-up needs to be conducted in these patients and radiation-induced cranial nerve palsy should be considered a diagnosis of exclusion.

The exact aetiology of radiation-induced cranial nerve palsy is unknown. Fibrosis of the surrounding tissue leading to mechanical compression of the affected nerve is one possible mechanism.[25] This might explain why cranial nerves that pass through the anterior portion of the neck (hypoglossal, vagal and recurrent laryngeal nerve), which receive high doses of radiation during treatment of head and neck cancer, are more commonly affected by radiation-induced cranial nerve palsy.

Cranial nerve palsy may also contribute to the development of dysphagia, but destruction of the normal head and neck anatomy by the cancer itself, surgical resection or radiation probably plays a more significant pathophysiological role.

Patients with long-term dysphagia carry a significant risk for aspiration and subsequent lung damage. The type of treatment (radiation alone, surgery alone, chemoradiation, or postoperative radiation) and advanced cancer stage have not been associated with an increased risk of severe dysphagia and aspiration.[26] The cough reflex in these patients may often be suppressed, and aspiration therefore may be clinically silent.[27] Thus, further diagnostic work-up to assess for aspiration should be considered in all patients with persistent dysphagia.

Chylothorax

Chylothorax refers to the presence of lymphatic fluid in the pleural space, caused by a leak of the thoracic duct or one of its divisions. This condition usually occurs as a consequence of chest trauma, as an iatrogenic complication of chest or head and neck surgery, oesophageal sclerotherapy, central venous cannulation, or as a result of tumour invasion in the setting of metastatic cancer, lymphoma or chronic lymphocytic leukaemia. Chylothorax is a rare complication of radiation therapy and has been reported in patients with lymphoma, non-small-cell lung cancer and oesophageal cancer. The latency period may be considerable and chylothorax has been observed to occur as late as 23 years after completion of radiation therapy.[28] The exact pathophysiologic mechanism of radiation-induced chylothorax is unknown. Mechanical lymphatic obstruction due to radiation-induced mediastinal fibrosis is one possible aetiology.

The diagnosis of chylothorax is established with pleural fluid analysis demonstrating a high level of triglycerides ($>110\,mg/dL$ or $1.24\,mmol/L$) and a low level of cholesterol ($<250\,mg/dL$ or $6.45\,mmol/L$). Lipoprotein analysis to confirm the presence of chylomicrons can be considered when the diagnosis is doubtful. The pleural fluid usually has a milky appearance.

Conservative treatment options are limited and mainly consist of serial thoracenteses and a diet that is rich in medium-chain triglycerides to reduce the flow of chyle. Surgical options in refractory cases include pleurodesis, ligation of the thoracic duct (usually performed as a thorascopic procedure), and pleuro-peritoneal shunting.

Fibrosing mediastinitis

Fibrosing mediastinitis refers to the presence of excessive fibrotic tissue in the mediastinum that has a tendency to invade and obstruct normal mediatinal structures. It usually occurs as a result of an autoimmune process or as a result of granulomatous infections, most commonly histoplasmosis. Fibrosing mediastinitis can also occur as a rare late complication of radiation therapy. Symptoms are usually mild and non-specific until obstruction of mediatinal structures such as airways, oesophagus, pulmonary arteries and veins, and the superior vena cava occurs after a considerable latency period.[29] Depending on the structure involved, symptoms may be quite variable and include cough, dyspnoea, chest pain, haemoptysis, recurrent pulmonary infections and the superior vena cava syndrome. No medical therapy has proven to be effective. Endoluminal stenting of the affected structures has been employed successfully in patients suffering from mechanical obstruction due to fibrosing mediastinitis.

DIAGNOSIS

Clinical clues

The diagnosis of radiation pneumonitis may not always be obvious, as symptoms are often non-specific or may be completely absent. Physical signs include crackles and/or a pleural rub, but auscultation may be entirely normal. Dullness to percussion and decreased breath sounds may be appreciated if a pleural effusion is present (occurs in about 10 per cent of all cases of radiation pneumonitis). Irradiated skin areas may demonstrate erythema outlining the radiation port. Radiation fibrosis may present with progressive dyspnoea and Velcro-like crackles on auscultation. Hypoxaemia, pulmonary hypertension and right heart strain may be present in more advanced cases.

Laboratory tests

There are currently no commonly available laboratory tests that predict the development of radiation pneumonitis reliably. The inflammatory state may be reflected by mild leucocytosis, elevated sedimentation rate and C-reactive protein. Elevated pro-calcitonin levels are highly specific for systemic bacterial infections which may help to distinguish radiation pneumonitis from bacterial pneumonia. Serum KL-6, a mucinous high-molecular-weight glycoprotein, is highly expressed in bronchial epithelial cells and type 2 pneumocytes. It has been found that elevated KL-6 levels correspond with the development of radiation pneumonitis.[30] So far, experience with KL-6 levels in this setting is still limited, and this test has not become part of the routine laboratory repertoire. Elevated KL-6 levels are also found in other interstitial lung diseases, but are usually normal in patients with bacterial pneumonia.

Radiographic patterns of radiation injury

Manifestations of radiation injury are usually confined to the radiation port. Occasionally, radiation pneumonitis and its radiographic sequelae can occur outside the irradiated area. Parallel to the clinical course of radiation injury, radiographic changes of radiation pneumonitis usually occur 8–12 weeks after completion of therapy. The radiographic pattern depends on the time interval after radiation therapy has been completed. The typical correlates of radiation pneumonitis are ground-glass opacities and consolidation within the treatment port (Fig. 16.3). However, radiation pneumonitis may also present as focal or nodular opacities. Occasionally, these changes are accompanied by an ipsilateral pleural effusion that is usually small to moderate in size and may cause compression atelectasis.[31]

Depending on the intensity of radiation injury, the opacities that correspond to radiation pneumonitis either resolve or they may progress to radiation fibrosis. Radiographically, radiation fibrosis is characterized by well-defined areas of increased interstitial markings, signs of volume loss, consolidation and traction bronchiectasis. Areas of consolidation typically have sharp borders that do not follow anatomic structures but rather correspond to the radiation port. The demarcation between normal and irradiated lung structures becomes more defined with progression of radiation-induced fibrosis.

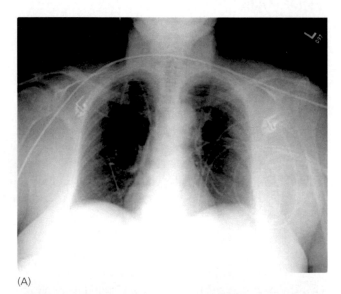

(A)

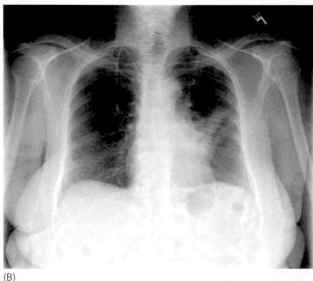

(B)

Fig. 16.3 (a) Baseline chest X-ray of a patient with a left hilar lung mass. (b) Acute pneumonitis following radiotherapy.

Ipsilateral mediastinal shift may occur when volume loss is substantial. Signs of pleural involvement may include pleural thickening and effusions that are usually small in size.[1,31]

With three-dimensional conformal radiation, the appearance of radiation injury may be different from what is commonly observed in conventional radiation therapy. Three distinct patterns have been described.

- In the first subset of patients, the radiographic changes reflective of radiation injury may be similar to those seen in conventional radiation therapy – consolidation, volume loss and traction bronchiectasis. These changes have been classified as 'modified conventional' and are usually less pronounced compared to two-dimensional radiation techniques.
- In the second subset of patients, changes of consolidation and volume loss may be very confined to the treatment port and present with a mass-like appearance.
- In the third subset of patients, radiographic changes will be quite subtle and present with linear opacities. This radiographic pattern has been classified as 'scar-like'.

It is unclear whether a more discrete pattern of radiation injury correlates with a better prognosis.[32]

In general, computed tomography is more sensitive in picking up radiographic changes of radiation injury, especially when high-resolution protocols are used. In addition, CT is more accurate when lesions need to be correlated with the irradiated field, especially when three-dimensional radiation techniques or unusual ports are used. For example, tangential-beam radiation, which is often used in the treatment of breast cancer, typically affects a small rim of lung tissue that is adjacent to the anterior pleura, which can easily be identified on CT.[33]

Nuclear medicine studies are only of limited value in the diagnosis of radiation pneumonitis. Increased uptake on gallium scanning has been reported in patients with radiation pneumonitis. However, this finding is non-specific and has also been associated with other inflammatory, infectious and malignant disorders of the lung. In addition, gallium uptake by the sternum hampers evaluation of lung areas that are close to the mediastinum. Indium scans offer the advantage of higher resolution and lacking interference with mediastinal structures. Indium scans can also be used for single photon emission tomography (SPECT) imaging which offers improved spatial discrimination. The activity on indium scanning has been shown to correlate with severity of radiation pneumonitis and response to therapy,[34] but experience with this technique has been limited to the research arena and this diagnostic modality has not become part of the standard nuclear medicine repertoire.

Positron emission tomography (PET) imaging has become a very useful and widely available diagnostic tool in the staging of cancer patients. Following radiation therapy, PET imaging was shown to demonstrate an early increase in uptake in non-irradiated lung areas, primarily involving subpleural lung zones. These changes were also present in patients who were clinically asymptomatic, suggesting that PET imaging may prove to be a very sensitive tool to pick up radiation-induced lung injury. Diffuse background activity on PET scanning was associated with significant radiation pneumonitis.[35] There is some evidence from a retrospective analysis performed in oesophageal cancer patients where post-treatment PET scans were used regularly to assess for recurrent disease that the degree of fluorodeoxyglucose (FDG) uptake correlates with symptoms of radiation pneumonitis.[36] It has to be pointed out, however, that an increase in FDG activity is not specific to radiation-induced lung injury but may also be seen in the setting of pulmonary infections or tumour recurrence. Experience with PET in diagnosing radiation-induced lung injury is limited, and it remains to be seen whether the additional information provided by this rather expensive diagnostic tool proves useful in the assessment of patients receiving radiation therapy.

Endobronchial brachytherapy

Endobronchial brachytherapy is mainly used as an adjunctive palliative treatment of symptoms related to endobronchial tumour growth. In the usual setting, continuous growth of a primary lung cancer causes endobronchial obstruction. Much less commonly, endobronchial obstruction is caused by metastatic disease. Symptoms are mainly related to tumour erosion of the endobronchial vasculature and atelectasis with possible post-obstructive pneumonia and include cough, dyspnoea, haemoptysis and fever. In cases of severe obstruction, endobronchial brachytherapy is usually preceded by procedures directed at immediate alleviation of the obstruction, such as endobronchial laser or cryotherapy, electrocauterization and endoluminal stent placement.[37] Endobronchial brachytherapy may then be employed to achieve long-term control of the obstruction.

The radiation is delivered to the target area with a radioactive application catheter that is endoscopically inserted into the bronchial tree. Usually, the total radiation dose does not exceed 25–30 Gy with fraction sizes between 5 and 10 Gy in weekly intervals. Larger fraction sizes have been associated with a higher incidence of complications like mucosal ulcerations, necrosis and haemorrhage. Overall symptomatic relief is achieved in more than two-thirds of the patients. Immediate procedure-related side-effects are reported to occur in about 3 per cent of all patients and include pneumothorax, bronchospasm, haemoptysis, cardiac arrhythmias and hypotension.[38]

Late toxicities occur weeks to months after completion of treatment and may be difficult to differentiate from complications of tumour progression. The reported rate of fatal haemoptysis varies between 0 and 32 per cent. In most cases this complication may not be treatment-related but occurs as a result of vascular tumour invasion.

Tracheoesophageal fistulas are another serious and potentially fatal complication, reported to occur in 5.3 per cent of all cases in one series.[39] In patients with centrally located tumours, the tracheal and oesophageal wall should be inspected carefully for pre-existing ulcerations to prevent fistula formation. Patients with tumour invasion of the oesophagus carry an even higher risk of developing fistulas.

In long-term survivors of endobronchial brachytherapy, late treatment complications like bronchial stenosis, radiation bronchitis and tracheomalacia have been observed. The reported incidences varied between 4 and 13 per cent. Higher fraction sizes, total radiation doses, dose volumes, and central location of the irradiated area have been associated with an increased risk of these late complications.[19]

Pulmonary function tests

Radiation-induced lung injury usually causes a decline in lung volumes (total lung capacity, vital capacity and residual volume) and compliance.[40] In patients with lung cancer where long-term follow-up pulmonary function test data beyond one year are available, those changes seem to be most pronounced at 6 months, with some recovery of lung function at 1 year and a slow, progressive decline thereafter, suggesting that radiation injury is an ongoing process that extends well beyond completion of therapy.[41] The gas transfer factor for carbon monoxide (DLCO) appears to be the most sensitive marker of radiation injury.[42] In general, the decline in lung function in patients irradiated for breast cancer seems to be less pronounced than in lung cancer patients, which in part could be related to lower radiation doses and tangential-beam radiation techniques that spare central regions of lung tissue from radiation exposure.

PREDICTING THE RISK OF RADIATION PNEUMONITIS

Clinical risk factors

Unfortunately, many studies that were conducted to identify independent risk factors that are associated with an increased risk of radiation pneumonitis produced inconclusive results. Age is a potential risk factor for radiation pneumonitis, but studies have not confirmed consistently that higher age is associated with increased risk. Since matching between ventilation and perfusion is different in the upper and lower lung zones, it was suspected that tumour location may influence the risk to develop radiation pneumonitis. However, it could not be demonstrated consistently that tumour location in the lower lobes carries a higher risk of radiation pneumonitis.

Patients with pre-existing pulmonary dysfunction may be predisposed to develop radiation pneumonitis. Hypoxaemia prior to radiation therapy was demonstrated to increase the risk of severe radiation pneumonitis from 5 to 19 per cent in a retrospective trial in patients with lung cancer.[43] There are also some preliminary data that good performance (i.e. walking distance) during a 6-minute walk-test is associated with a lower risk of radiation pneumonitis. Additional prognostic information may be gained if the 6-minute walking distance is correlated with the FEV_1 (i.e. lower risk in patients with high walking distance/FEV_1 ratio).[44] While poor baseline lung function has been demonstrated to increase the risk of radiation pneumonitis, there are data in patients with unresectable lung cancer to suggest that patients with good baseline lung function (FEV_1 at least 50 per cent) have larger decrements in lung function after radiation therapy compared to those with poor baseline lung function. Poor overall performance status and comorbid disease are other risk factors for the development of radiation pneumonitis.[45,46] The role of smoking in the development of radiation pneumonitis is paradoxical. A history of smoking increases the risk of radiation injury, probably as a result of smoking-induced lung disease.[46] On the other hand, active smoking at the time of radiation exposure seems to have a protective effect to prevent radiation-induced lung injury; the exact mechanism is unknown.[47,48]

Dose factors

The scheduling and dose intensity of radiation therapy may have an impact on the risk of toxicity. Across different studies, the risk of radiation pneumonitis has been shown to increase with the total radiation dose. In conventional radiotherapy, the incidence of radiation pneumonitis is 6 per cent for doses below 45 Gy, 9 per cent for doses of 45–54 Gy, and 12 per cent for doses above 55 Gy.[49] Larger individual fraction size (> 2.67 Gy once-daily) and the administration of once versus twice-daily radiotherapy have also been shown to increase the risk of radiation pneumonitis. The cumulative radiation dose that is administered to normal lung tissue has consistently been shown to be the most important predictor of radiation injury. However, it can be difficult to calculate these doses accurately, especially when complex three-dimensional radiation ports are used. Dose volume histograms that are derived from the treatment plan are commonly used to estimate the percentage of lung volume that will be exposed to a certain radiation dose. On this two-dimensional diagram, the cumulative radiation dose is plotted on the x-axis and the percentage of lung volume that will receive more than this dose is plotted on the y-axis. V_{20} refers to the percentage of lung volume that will receive a radiation dose of more than 20 Gy. This value has been shown to be an important predictor of radiation pneumonitis. In one series of 99 patients with inoperable non-small-cell lung cancer,[50] no grade 2 or higher radiation pneumonitis (defined as pneumonitis requiring either steroids, oxygen or diuretics) was observed if the V_{20} was below 22 per cent; 16 per cent grade 2–4 radiation pneumonitis was observed with V_{20} between 22 and 31 per cent. Higher-grade radiation pneumonitis occurred more frequently with higher V_{20} percentages, including some fatal cases with V_{20} exceeding 35 per cent. Both the Radiation Therapy Oncology Group (RTOG) and Southwest Oncology Group (SWOG) have specified upper limits of V_{20} in the range 30–35 percent for patients treated on recent group protocols.

Role of chemotherapy

The role of chemotherapy as a radiation sensitizer has been demonstrated most extensively for bleomycin. Other chemotherapeutic agents that have been shown to increase the risk of radiation injury include vincristine, mitomycin, recombinant interferon-α, taxanes, doxorubicin, dactinomycin, cyclophosphamide, gemcitabine and irinotecan.[45,51,52] The finding that these agents increase the risk of radiation-induced lung injury is not consistent across different studies. One has to keep in mind that chemotherapeutic agents such as bleomycin, mitomycin, taxanes, gemcitabine and cyclophosphamide are known to cause pulmonary toxicity, thus it may be difficult to ascertain which modality (radiation or chemotherapy) is the cause of the injury to the lung.

The timing of chemotherapy may influence the risk of radiation pneumonitis. Compared to sequential therapy regimens, most studies in lung cancer have demonstrated a slightly increased risk of radiation pneumonitis when chemotherapy and radiation are administered concurrently.[53] This difference appears to be more pronounced in women receiving concurrent radiation and chemotherapy for breast cancer, especially when anthracycline-based chemotherapy is used,[54] so that such therapy regimens are generally avoided.

Circulating radiotherapy

Therapy with radioactively labelled molecules has been used in a variety of clinical settings. The administration of radioactive iodine-131 to treat thyroid cancer and hyperthyroidism has been established since the 1940s. Radio-iodine therapy is also effective in patients with pulmonary metastasis of thyroid cancer. Patients with diffuse lung metastases in particular can be expected to have considerable intrapulmonary iodine uptake, which will inevitably expose normal lung tissue to significant amounts of radiation. Observations of severe and sometimes fatal radiation pneumonitis and fibrosis made in the late 1950s in patients receiving iodine-131 for the treatment of pulmonary metastases of thyroid cancer[55] resulted in dosing constraints that are still in effect today.

Selective internal radiation with yttrium-90 is used to treat inoperable hepatocellular carcinoma or hepatic metastasis. The radioisotope is usually injected into the hepatic artery during hepatic angiography. The development of radiation pneumonitis following selective internal radiation has been reported in several case series.[56] In some of these patients, radiation pneumonitis was attributed to increased hepatopulmonary shunting with subsequent intrapulmonary uptake of radioisotopes. Since the intraluminal diameter of the pulmonary vasculature decreases closer to the pleura, the characteristic pattern of radiographic opacities with sharp lateral margins that run parallel to the pleural surface[57] could be related to lodging of yttrium particles in small peripheral pulmonary arterioles.

THERAPY OF RADIATION-INDUCED LUNG INJURY

Even though efficacy has never been demonstrated in prospective randomized trials, radiation pneumonitis is commonly treated with corticosteroids. There is no consensus concerning dose, length of therapy and tapering schedule. Usually, at least 60 mg of prednisone per day are given for about 2 weeks and then tapered gradually over 3 months. Radiation-induced bronchiolitis obliterans organizing pneumonia (BOOP) usually responds favourably to corticosteroid therapy, but has a tendency to recur when treatment is discontinued.[17] The treatment effects of corticosteroids may be related to their anti-inflammatory properties, attenuation of TNF-α-mediated endothelial cell injury or direct lymphotoxic effects. Prophylactic steroid therapy,[58] antibiotics or heparin[59] do not appear to protect against radiation pneumonitis. Experience with other immunosuppressive agents including ciclosporin A and azathioprine in the treatment of radiation-induced lung injury is anecdotal.

Pentoxifylline is a xanthine derivative that is mainly used for platelet inhibition and improvement of microcirculation. The drug also displays immunomodulatory effects that are probably mediated via inhibition of TNF-α and interleukin-1. Combined therapy with pentoxifylline and tocopherol has been used successfully in the treatment of radiation-induced fibrosis of the skin and subcutaneous tissue,[60] and there are animal data suggesting pentoxifylline-mediated inhibition of bleomycin-induced lung injury. There is limited experience showing a beneficial effect of pentoxifylline in preventing the development of radiation-induced lung injury in patients irradiated for breast and lung cancer,[61] but this needs to be confirmed in a large prospective trial.

Promising results suggesting that amifostine, which displays cytoprotective effects by neutralizing free oxygen radicals, could reduce the incidence of radiation pneumonitis[62] could not be reproduced. Animal data suggesting that captopril reduced radiation-induced fibrosis in rats[63] could not be confirmed in humans. There is no experience with inhibitors of collagen synthesis like interferon-γ, penicillamine or prifenidone in the prevention of radiation-induced fibrosis.

KEY POINTS

- Radiation-induced lung injury is associated with significant morbidity and even mortality, so its prevention is a primary goal in patients that often have an already limited prognosis.

- The development of sophisticated computer-based three-dimensional radiation techniques that allow precise targeting of a tumour while sparing normal tissue structures has been a significant advancement.

- Likewise, the search needs to continue for agents that can be used as preventive therapy and have the potential to attenuate radiation-induced lung injury without compromising the efficacy of tumour-directed treatment.

- Radiation techniques will continue to evolve. Continuous efforts need to be made to identify potential clinical and diagnostic markers that allow the clinician to identify patients who are at increased risk for radiation injury.

- More research needs to be directed towards identifying therapeutic alternatives to corticosteroids in the treatment of radiation pneumonitis.

REFERENCES

1. Marks LB, Yu X, Vujaskovic Z, et al. Radiation-induced lung injury. Semin Radiat Oncol 2003;13:333–45.
2. Greenberger JS, Epperly MW, Gretton J, et al. Radioprotective gene therapy. Curr Gene Ther 2003;3:183–95.
3. Morgan GW, Breit SN. Radiation and the lung: a reevaluation of the mechanisms mediating pulmonary injury. Int J Radiat Oncol Biol Phys 1995;31:361–9.
4. Anscher MS, Kong FM, Andrews K, et al. Plasma transforming growth factor β₁ as a predictor of radiation pneumonitis. Int J Radiat Oncol Biol Phys 1998;41:1029–35.
5. Rubin P, Johnston CJ, Williams JP, McDonald S, Finkelstein JN. A perpetual cascade of cytokines postirradiation leads to pulmonary fibrosis. Int J Radiat Oncol Biol Phys 1995;33:99–109.
6. Coia LR, Myerson RJ, Tepper JE. Late effects of radiation therapy on the gastrointestinal tract. Int J Radiat Oncol Biol Phys 1995;31:1213–36.
7. Emami B, Lyman J, Brown A, et al. Tolerance of normal tissue to therapeutic irradiation. Int J Radiat Oncol Biol Phys 1991;21:109–22.
8. Singh AK, Lockett MA, Bradley JD. Predictors of radiation-induced esophageal toxicity in patients with non-small-cell lung cancer treated with three-dimensional conformal radiotherapy. Int J Radiat Oncol Biol Phys 2003;55:337–41.
9. Leigh BR, Lau DH. Severe esophageal toxicity after thoracic radiation therapy for lung cancer associated with the human immunodeficiency virus: a case report and review of the literature. Am J Clin Oncol 1998;21:479–81.
10. Gibson PG, Bryant DH, Morgan GW, et al. Radiation-induced lung injury: a hypersensitivity pneumonitis? Ann Intern Med 1988;109:288–91.
11. Azria D, Magne N, Zouhair A, et al. Radiation recall: a well recognized but neglected phenomenon. Cancer Treat Rev 2005;31:555–70.
12. Kitani H, Kosaka T, Fujihara T, Lindquist K, Elkind MM. The 'recall effect' in radiotherapy: is subeffective, reparable damage involved? Int J Radiat Oncol Biol Phys 1990;18:689–95.
13. Crestani B, et al. Groupe d'Etudes et de Recherche sur les Maladies Orphelines Pulmonaires. Bronchiolitis obliterans organizing pneumonia syndrome primed by radiation therapy to the breast. Am J Respir Crit Care Med 1998;158:1929–35.
14. Iijima M, Sakahara H. [Radiation pneumonitis resembling bronchiolitis obliterans organizing pneumonia after postoperative irradiation for lung cancer: a case report]. Nippon Igaku Hoshasen Gakkai Zasshi 2003;63:332–3.
15. Crestani B, Kambouchner M, Soler P, et al. Migratory bronchiolitis obliterans organizing pneumonia after unilateral radiation therapy for breast carcinoma. Eur Respir J 1995;8:318–21.
16. Tobias ME, Plit M. Bronchiolitis obliterans organizing pneumonia with migratory infiltrates: a late complication of radiation therapy. Am J Roentgenol 1993;160:205–6.
17. Arbetter KR, Prakash UB, Tazelaar HD, Douglas WW. Radiation-induced pneumonitis in the 'nonirradiated' lung. Mayo Clin Proc 1999;74:27–36.
18. Rowinsky EK, Abeloff MD, Wharam MD. Spontaneous pneumothorax following thoracic irradiation. Chest 1985;88:703–8.
19. Speiser BL, Spratling L. Radiation bronchitis and stenosis secondary to high dose rate endobronchial irradiation. Int J Radiat Oncol Biol Phys 1993;25:589–97.
20. Miller KL, Shafman TD, Anscher MS, et al. Bronchial stenosis: an underreported complication of high-dose external beam radiotherapy for lung cancer? Int J Radiat Oncol Biol Phys 2005;61:64–9.
21. Putterman C, Polliack A. Late cardiovascular and pulmonary complications of therapy in Hodgkin's disease: report of three unusual cases, with a review of relevant literature. Leuk Lymphoma 1992;7:109–15.

22. Applefeld MM, Cole JF, Pollock SH, *et al.* The late appearance of chronic pericardial disease in patients treated by radiotherapy for Hodgkin's disease. *Ann Intern Med* 1981; **94**:338–41.

23. Sagrista-Sauleda J, Angel J, Sanchez A, Permanyer-Miralda G, Soler-Soler J. Effusive-constrictive pericarditis. *New Engl J Med* 2004;**350**:469–75.

24. Lin YS, Jen YM, Lin JC. Radiation-related cranial nerve palsy in patients with nasopharyngeal carcinoma. *Cancer* 2002; **95**:404–9.

25. Cheng VS, Schultz MD. Unilateral hypoglossal nerve atrophy as a late complication of radiation therapy of head and neck carcinoma: a report of four cases and a review of the literature on peripheral and cranial nerve damages after radiation therapy. *Cancer* 1975;**35**:1537–44.

26. Nguyen NP, Frank C, Moltz CC, *et al.* Dysphagia severity and aspiration following postoperative radiation for locally advanced oropharyngeal cancer. *Anticancer Res* 2008; **28**:431–4.

27. Nguyen NP, Moltz CC, Frank C, *et al.* Effectiveness of the cough reflex in patients with aspiration following radiation for head and neck cancer. *Lung* 2007;**185**:243–8.

28. McWilliams A, Gabbay E. Chylothorax occurring 23 years post-irradiation: literature review and management strategies. *Respirology* 2000;**5**:301–3.

29. Dechambre S, Dorzee J, Fastrez J, *et al.* Bronchial stenosis and sclerosing mediastinitis: an uncommon complication of external thoracic radiotherapy. *Eur Respir J* 1998;**11**:1188–90.

30. Kohno N, Hamada H, Fujioka S, *et al.* Circulating antigen KL-6 and lactate dehydrogenase for monitoring irradiated patients with lung cancer. *Chest* 1992;**102**:117–22.

31. Libshitz HI, Southard ME. Complications of radiation therapy: the thorax. *Semin Roentgenol* 1974;**9**:41–9.

32. Koenig TR, Munden RF, Erasmus JJ, *et al.* Radiation injury of the lung after three-dimensional conformal radiation therapy. *Am J Roentgenol* 2002;**178**:1383–8.

33. Coscina WF, Arger PH, Mintz MC, Coleman BG. CT demonstration of pulmonary effects of tangential beam radiation. *J Comput Assist Tomogr* 1986;**10**:600–2.

34. Valdes Olmos RA, van Zandwijk N, Boersma LJ, *et al.* Radiation pneumonitis imaged with indium-111-pentetreotide. *J Nucl Med* 1996;**37**:584–8.

35. Hassaballa HA, Cohen ES, Khan AJ, *et al.* Positron emission tomography demonstrates radiation-induced changes to nonirradiated lungs in lung cancer patients treated with radiation and chemotherapy. *Chest* 2005;**128**:1448–52.

36. Hart JP, McCurdy MR, Ezhil M, *et al.* Radiation pneumonitis: correlation of toxicity with pulmonary metabolic radiation response. *Int J Radiat Oncol Biol Phys* 2008;**71**:967–71.

37. Baas P, van Zandwijk N. Endobronchial treatment modalities in thoracic oncology. *Ann Oncol* 1995;**6**:523–31.

38. Gollins SW, Burt PA, Barber PV, Stout R. Long-term survival and symptom palliation in small primary bronchial carcinomas following treatment with intraluminal radiotherapy alone. *Clin Oncol* (R Coll Radiol) 1996;**8**:239–46.

39. Macha HN, Wahlers B, Reichle C, von Zwehl D. Endobronchial radiation therapy for obstructing malignancies: ten years' experience with iridium-192 high-dose radiation brachytherapy afterloading technique in 365 patients. *Lung* 1995;**173**:271–80.

40. Mehta V. Radiation pneumonitis and pulmonary fibrosis in non-small-cell lung cancer: pulmonary function, prediction, and prevention. *Int J Radiat Oncol Biol Phys* 2005;**63**:5–24.

41. Miller KL, Zhou SM, Barrier RC, *et al.* Long-term changes in pulmonary function tests after definitive radiotherapy for lung cancer. *Int J Radiat Oncol Biol Phys* 2003;**56**:611–15.

42. Werner-Wasik M, Yu X, Marks LB, Schultheiss TE. Normal-tissue toxicities of thoracic radiation therapy: esophagus, lung, and spinal cord as organs at risk. *Hematol Oncol Clin North Am* 2004;**18**:131–60,x–xi.

43. Inoue A, Kunitoh H, Sekine I, *et al.* Radiation pneumonitis in lung cancer patients: a retrospective study of risk factors and the long-term prognosis. *Int J Radiat Oncol Biol Phys* 2001;**49**:649–55.

44. Miller KL, Kocak Z, Kahn D, *et al.* Preliminary report of the 6-minute walk test as a predictor of radiation-induced pulmonary toxicity. *Int J Radiat Oncol Biol Phys* 2005; **62**:1009–13.

45. Rancati T, Ceresoli GL, Gagliardi G, Schipani S, Cattaneo GM. Factors predicting radiation pneumonitis in lung cancer patients: a retrospective study. *Radiother Oncol* 2003;**67**: 275–83.

46. Robnett TJ, Machtay M, Vines EF, *et al.* Factors predicting severe radiation pneumonitis in patients receiving definitive chemoradiation for lung cancer. *Int J Radiat Oncol Biol Phys* 2000;**48**:89–94.

47. Hernando ML, Marks LB, Bentel GC, *et al.* Radiation-induced pulmonary toxicity: a dose-volume histogram analysis in 201 patients with lung cancer. *Int J Radiat Oncol Biol Phys* 2001;**51**:650–9.

48. Johansson S, Bjermer L, Franzen L, Henriksson R. Effects of ongoing smoking on the development of radiation-induced pneumonitis in breast cancer and oesophagus cancer patients. *Radiother Oncol* 1998;**49**:41–7.

49. Roach M, Gandara DR, Yuo HS, *et al.* Radiation pneumonitis following combined modality therapy for lung cancer: analysis of prognostic factors. *J Clin Oncol* 1995;**13**:2606–12.

50. Graham MV, Purdy JA, Emami B, *et al.* Clinical dose-volume histogram analysis for pneumonitis after 3D treatment for non-small cell lung cancer. *Int J Radiat Oncol Biol Phys* 1999;**45**:323–9.

51. Taghian AG, Assaad SI, Niemierko A, *et al.* Risk of pneumonitis in breast cancer patients treated with radiation therapy and combination chemotherapy with paclitaxel. *J Natl Cancer Inst* 2001;**93**:1806–11.

52. McDonald S, Rubin P, Phillips TL, Marks LB. Injury to the lung from cancer therapy: clinical syndromes, measurable endpoints, and potential scoring systems. *Int J Radiat Oncol Biol Phys* 1995;**31**:1187–203.

53. Socinski MA, Rosenman JG. Chemotherapeutic issues in the management of unresectable stage III non-small cell lung cancer. *Semin Oncol* 2005;**32**:S18–24.

54. Lingos TI, Recht A, Vicini F, *et al.* Radiation pneumonitis in breast cancer patients treated with conservative surgery and radiation therapy. *Int J Radiat Oncol Biol Phys* 1991;**21**:355–60.

55. Rall JE, Alpers JB, Lewallen CG, *et al.* Radiation pneumonitis and fibrosis: a complication of radioiodine treatment of pulmonary metastases from cancer of the thyroid. *J Clin Endocrinol Metab* 1957;**17**:1263–76.

56. Leung TW, Lau WY, Ho SK, *et al.* Radiation pneumonitis after selective internal radiation treatment with intraarterial [90]yttrium-microspheres for inoperable hepatic tumors. *Int J Radiat Oncol Biol Phys* 1995;**33**:919–24.

57. Lin M. Radiation pneumonitis caused by yttrium-90 microspheres: radiologic findings. *Am J Roentgenol* 1994;**162**:1300–2.

58. Kwok E, Chan CK. Corticosteroids and azathioprine do not prevent radiation-induced lung injury. *Can Respir J* 1998;**5**:211–14.

59. Moss WT, Haddy FJ, Sweany SK. Some factors altering the severity of acute radiation pneumonitis: variation with cortisone, heparin, and antibiotics. *Radiology* 1960; **75**:50–4.

60. Delanian S, Balla-Mekias S, Lefaix JL. Striking regression of chronic radiotherapy damage in a clinical trial of combined pentoxifylline and tocopherol. *J Clin Oncol* 1999;**17**:3283–90.

61. Ozturk B, Egehan I, Atavci S, Kitapci M. Pentoxifylline in prevention of radiation-induced lung toxicity in patients with breast and lung cancer: a double-blind randomized trial. *Int J Radiat Oncol Biol Phys* 2004;**58**:213–19.

62. Antonadou D. Radiotherapy or chemotherapy followed by radiotherapy with or without amifostine in locally advanced lung cancer. *Semin Radiat Oncol* 2002;**12**:50–8.

63. Ward WF, Molteni A, Ts'ao CH. Radiation-induced endothelial dysfunction and fibrosis in rat lung: modification by the angiotensin converting enzyme inhibitor CL242817. *Radiat Res* 1989;**117**:342–50.

17

Pulmonary complications of bone-marrow and stem-cell transplantation

BEKELE AFESSA, ANDREW D BADLEY, STEVE G PETERS

INTRODUCTION

Tens of thousands of patients undergo haematopoietic stem-cell transplantation (HSCT) annually, primarily for haematological and lymphoid cancers, but also for other disorders.[1] Intensive care support is provided for 15–40 per cent of HSCT recipients, most of whom require mechanical ventilation for respiratory failure.[2] Although recent studies have shown improvement in outcome, the mortality rate of HSCT recipients receiving invasive ventilation may exceed 90 per cent. This chapter describes the pulmonary complications that develop in HSCT recipients.

TIMING AND TYPES OF COMPLICATIONS

Appropriate diagnostic work-up and early treatment are essential for the management of HSCT recipients. The pre-transplant conditioning regimen virtually eliminates all pre-existing innate and acquired immunity.[3] After HSCT, the immune system recovers along predictable patterns depending on the underlying disorder, stem-cell source, and complications such as graft-versus-host disease (GVHD).[3] Recovery occurs faster in autologous recipients, in those who receive peripheral blood stem-cell grafts, and after a non-myeloablative conditioning regimen.

Pulmonary complications develop in 30–60 per cent of HSCT recipients.[4] They are the immediate cause of death in approximately 61 per cent.[5,6] Risk factors for the development of pulmonary complications include pre-transplant radiation and chemotherapy, total body irradiation, an allogeneic stem-cell source, and GVHD.[4,7] Both infectious and non-infectious pulmonary complications occur frequently in HSCT recipients (Fig. 17.1 and Box 17.1).[4,7,8]

The post-transplant period is divided into three phases: pre-engraftment, early post-transplant and late post-transplant.[9] The pre-engraftment phase (0–30 days) is characterized by neutropenia and breaks in the mucocutaneous barriers. During this phase, the most prevalent pathogens are bacteria and *Candida* species and, if neutropenia persists, *Aspergillus* species. During the period of neutropenia there is no significant difference in the type of infection between allogeneic and autologous HSCT recipients.[10] The early post-engraftment phase (30–100 days) is dominated by impaired cell-mediated immunity. The impact of this cell-mediated defect is determined by the development of GVHD and the immunosuppressant medications used to treat it. Cytomegalovirus (CMV), *Pneumocystis jiroveci* and *Aspergillus* species are the predominant pathogens during this phase. The late post-transplant phase (> 100 days) is characterized by defects in cell-mediated and humoral immunity, as well as function of the reticuloendothelial system in allogeneic transplant recipients. During this phase, allogeneic HSCT recipients are at risk for viral infection and infection by encapsulated bacteria such as *Haemophilus influenzae* and *Streptococcus pneumoniae*. In endemic areas of the world, pulmonary tuberculosis occurs during the late post-transplant phase.[1,11,12] After neutrophil engraftment, infections occur more frequently in allogeneic HSCT recipients.[10]

Non-infectious pulmonary complications also follow a characteristic temporal pattern.[7] The distinguishing features of the main non-infectious pulmonary complications are outlined in Table 17.1. Among the non-infectious pulmonary complications, pulmonary oedema, diffuse alveolar haemorrhage (DAH) and peri-engraftment respiratory distress syndrome (PERDS) usually occur during the first 30 days following transplant (see Fig. 17.1). Idiopathic pneumonia syndrome (IPS) can occur at any time following transplant.

APPROACH TO PULMONARY COMPLICATIONS

When HSCT recipients present with pulmonary infiltrates and signs of infection, most clinicians initiate empiric antibacterial

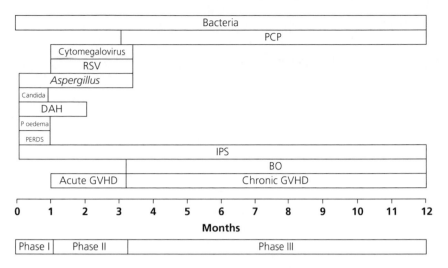

Fig. 17.1 Timing of the major non-infectious pulmonary complications following haematopoietic stem-cell transplantation. Phase I = pre-engraftment period; phase II = early post-engraftment period; phase III = late post-engraftment period. BO, bronchiolitis obliterans; DAH, diffuse alveolar haemorrhage; GVHD, graft-versus-host disease; IPS, idiopathic pneumonia syndrome; P oedema, pulmonary oedema; PCP, pneumocystic carinii pneumonia; PERDS, peri-engraftment respiratory distress syndrome; RSV, respiratory syncytial virus.

Box 17.1 Pulmonary complications in haematopoietic stem-cell transplant recipients

Infectious

Viral

- Cytomegalovirus
- Respiratory syncytial virus
- Influenza A and B
- Parainfluenza
- Measles
- Herpes simplex virus II

Bacterial

- Gram-positive
  - *Staphylococcus aureus*
  - *Streptococcus pneumoniae*
- Gram-negative
  - *Pseudomonas aeruginosa*

Mycobacterial

Fungal

- *Aspergillus* spp
- *Candida* spp
- *Pneumocystis jiroveci*

Protozoa

- *Toxoplasma gondii*

Non-infectious

- Isolated abnormality in pulmonary function
- Asthma
- Acute pulmonary oedema
- Diffuse alveolar haemorrhage
- Peri-engraftment respiratory distress syndrome
- Bronchiolitis obliterans
- Bronchiolitis obliterans organizing pneumonia
- Idiopathic pneumonia syndrome
- Delayed pulmonary toxicity syndrome
- Pulmonary cytolytic thrombi
- Pulmonary veno-occlusive disease
- Progressive pulmonary fibrosis
- Pulmonary hypertension
- Hepatopulmonary syndrome
- Pulmonary alveolar proteinosis
- Eosinophilic pneumonia

---

therapy, adding antifungal treatment if risk factors are present and there is no response to initial treatment.[13] The initial antibiotic therapy in HSCT recipients with neutropenia can be monotherapy with third- or fourth-generation cephalosporins or carbapenem. Combination antibiotic therapy usually includes an aminoglycoside and antipseudomonal penicillin or antipseudomonal cephalosporin or carbapenem. Since quinolones are widely used for prophylaxis in neutropenia, they are not recommended for empirical treatment of suspected infection.[14] For patients at high risk for methicillin-resistant *Staphylococcus aureus* infection (MRSI), the initial empirical antibiotic treatment should include vancomycin. If the patient remains septic after 5 days of antibacterial therapy and no cause is identified, empirical antifungal therapy should be

Table 17.1 Distinguishing features of the main non-infectious pulmonary complications in haematopoietic stem-cell transplant recipients

Pulmonary complication	Distinguishing features
BO	Absence of fever and pulmonary infiltrates, and presence of airway obstruction
BOOP	Fever, patchy pulmonary air-space consolidation, and typical lung histology
IPS	Mimics pneumonia of infectious aetiology, and diagnosed by exclusion of other causes
DAH	Diffuse pulmonary infiltrates, and bronchoalveolar lavage with progressively bloodier return and/or ≥ 20% haemosiderin-laden macrophages
PERDS	Onset within 5 days of neutrophil engraftment, and exclusion of cardiac and infectious causes
DPTS	In autologous haematopoietic stem cell recipients with breast cancer, and following high-dose pre-transplant chemotherapy; good prognosis
PCT	Fever and pulmonary nodules in children with GVHD, and typical lung histology

BO, bronchiolitis obliterans; BOOP, bronchiolitis obliterans organizing pneumonia; DAH, diffuse alveolar haemorrhage; DPTS, delayed pulmonary toxicity syndrome; GVHD, graft-versus-host disease; IPS, idiopathic pneumonia syndrome; PERDS, peri-engraftment respiratory distress syndrome; PCT, pulmonary cytolytic thrombi.

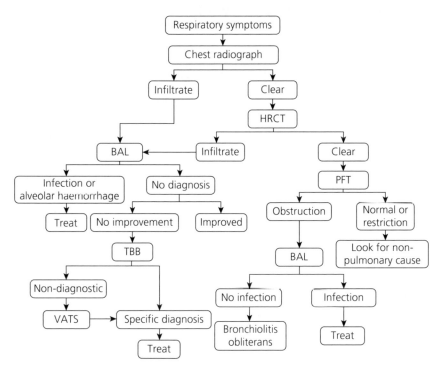

Fig. 17.2   Diagnostic approach to the haematopoietic stem-cell transplant recipient with pulmonary complication. BAL, bronchoalveolar lavage; HRCT, high-resolution computed tomography of the chest; PFT, pulmonary function test; TBB, transbronchial lung biopsy; VATS, video-assisted thoracoscopic surgery. Reproduced with permission from [185].

added.[13] Cultures of blood and respiratory secretions should be obtained. In the appropriate clinical setting, antigen and polymerase chain reaction (PCR) assays for *Aspergillus* and CMV may be helpful.

If a chest radiograph shows parenchymal infiltrates, the initial evaluation is non-diagnostic and the patient does not respond to empiric therapy, we advocate early bronchoalveolar lavage (BAL) with transbronchial lung biopsy if tolerated (Fig. 17.2).[7] For peripherally located nodules and masses, fluoroscopy or computed tomography (CT) guided transthoracic fine-needle aspiration may be helpful.[7] If these studies are non-diagnostic and there are no contraindications, we advocate video-assisted thoracoscopic lung biopsy, since specific

diagnoses may lead to appropriate treatment and avoidance of unnecessary or potentially harmful therapy.

If there are respiratory symptoms and the chest radiograph is normal or non-specific, high-resolution CT (HRCT) may suggest specific diagnoses.[7] Pulmonary function testing should be part of the evaluation in the non-acute setting, and when bronchiolitis obliterans is suspected.

VIRAL INFECTIONS

Viral pulmonary infections are a major cause of morbidity and mortality in HSCT recipients. Several studies have shown CMV to be the most common viral pathogen causing lower respiratory tract infection.[7] During endemic seasons, community respiratory viruses should be included in the differential diagnoses of respiratory symptoms in the HSCT recipient.

Cytomegalovirus

The frequency of CMV pneumonia in the early post-transplantation period has been substantially reduced by prophylaxis. However, CMV pneumonia continues to be a major cause of morbidity and mortality in the late post-transplant period. The clinical manifestations of CMV infection in HSCT recipients vary from completely asymptomatic to serious infection with multiple organ function abnormalities.

In allogeneic HSCT recipients, the reported incidence rate of CMV pneumonia had varied between 9 and 53 per cent.[7] With the use of early detection and application of prophylactic and pre-emptive treatment, this rate has declined to less than 5 per cent recently. In autologous HSCT recipients, the CMV pneumonia incidence rate is less than 5 per cent.

The risk factors for CMV pneumonia include old age, allogeneic transplant, and positive recipient or donor serology status, GVHD, systemic corticosteroid use and T-cell-depleted stem cells.[7] CMV viraemia and urinary excretion usually precede the development of pneumonia, but are not necessarily predictive of pneumonia. Although the existing data are conflicting, the incidence of CMV pneumonia may be reduced when allogeneic transplantation is performed with peripheral blood stem cells instead of bone marrow.[7]

CMV infection may result from primary infection or reactivation. The highest risk groups are 'donor CMV positive'/'recipient CMV negative', in whom the absence of CMV immunity allows aggressive primary infection.

Most CMV infections occur within the first 100 days following transplant, usually between 6 and 12 weeks.[7] Although rare, CMV pneumonia may develop prior to engraftment, especially in HSCT recipients with positive CMV serology. HSCT recipients with CMV pneumonia present with fever, nonproductive cough, dyspnoea and hypoxaemia. Chest radiographs usually show subtle patchy or diffuse ground-glass opacities, often with small concurrent multiple pulmonary nodules.[7] The plain chest radiograph may be normal. Pleural effusions are uncommon. HRCT demonstrates ground-glass opacities, air-space consolidation and pulmonary nodules.[15]

Viral culture, shell vial-based assay, antigen assays and PCR assays are used for the early detection of CMV infection in urine, blood and respiratory secretions. The viral culture method is too slow to be of any clinical utility to guide early treatment. In the appropriate clinical setting, the diagnosis of CMV pneumonia relies on the identification of CMV from transbronchial or surgical lung biopsy (Fig. 17.3). In cases where lung tissue cannot be safely obtained, CMV detection from BAL fluid in the appropriate clinical setting is used to make the diagnosis. The definitive diagnosis of CMV pneumonia requires positive culture results from BAL fluid samples in addition to identification of the characteristic cytopathic feature of intranuclear inclusions.[16]

The use of CMV seronegative blood products and leucocyte-depleted platelets can reduce the rate of CMV infection in CMV

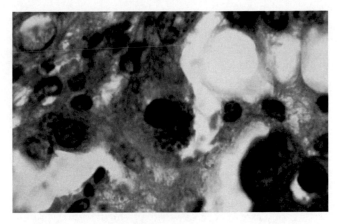

Fig. 17.3 Typical cytomegalovirus inclusion in lung tissue. Image courtesy of Thomas F Smith PhD, Mayo Clinic College of Medicine.

seronegative HSCT recipients.[7] However, these techniques will not prevent primary CMV infection. Because of the associated high morbidity and mortality, antiviral therapy with acyclovir, valacyclovir, valganciclovir, ganciclovir or foscarnet are used prophylactically or pre-emptively in HSCT recipients at risk for CMV disease.[7] Although ganciclovir and foscarnet are effective for CMV prophylaxis, their use is limited by bone marrow and renal toxicities, and they are best reserved for pre-emptive therapy or treatment. For the treatment of CMV pneumonia, intravenous immunoglobulin is usually used in combination with ganciclovir or foscarnet.[7] The combined use of foscarnet and ganciclovir is associated with increased toxicity without any therapeutic benefit.[17] If CMV pneumonia or blood stream infection fails to respond to therapy, drug resistance should be considered and tested.

The mortality rate associated with untreated CMV pneumonia may exceed 90 per cent. Even with treatment, the mortality exceeds 50 per cent.[7] Respiratory failure requiring mechanical ventilation, other concurrent infections and old age are poor prognostic indicators. Survivors of CMV pneumonitis may have persistent lymphocytosis in BAL fluid and peripheral blood and may develop restrictive lung disease and BOOP.[18]

Community respiratory viruses

Community respiratory virus infections are common causes of morbidity and mortality in HSCT recipients. They occur in similar frequency in both autologous and allogeneic HSCT recipients and mostly during the first 30 days following transplant.[19] The primary defence against the acquisition of a virus is pre-formed antibody, and the primary mechanisms for recovery are humoral and cell-mediated immunity.[20] HSCT recipients have both humoral and cell-mediated deficiencies. The loss of mucosal integrity associated with mucositis and reduced local IgA antibody increase the susceptibility to upper respiratory infection, and reduced serum antibody level leads to lower respiratory infection.[20]

Respiratory virus infections usually occur seasonally, except parainfluenza which occurs year-round.[20,21] Nosocomial respiratory virus infections may occur outside the usual seasons in HSCT recipients.[21]

In a multicentre prospective study from Europe, the frequency of documented respiratory virus infections was 3.5 per cent among allogeneic and 0.4 per cent among autologous HSCT recipients.[22] In patients with symptoms and signs suggestive of respiratory tract infection, the frequency may rise to 30 per cent.[19,23,24] Compared to immunocompetent patients, immunocompromised patients are more likely to manifest a non-seasonal pattern, to acquire nosocomial infection and to progress to pneumonia and death.[20]

The development of pneumonia is suggested by symptoms of respiratory tract infection and new radiographic infiltrates. The frequency of pneumonia ranges between 30 and 60 per cent.[19,21,22] The diagnosis requires viral cultures or antigen testing of nasopharyngeal washing, throat swabs, sputum, endotracheal aspirations or BAL fluid. Because of coexistent complications, it is difficult to determine the mortality rate attributable to respiratory virus infections after HSCT. Outcome varies by the type of virus and for some community-acquired infections mortality approaches 100 per cent if untreated.[19,21,24]

Hospital-acquired respiratory virus infections can be minimized by rigorous infection control measures including strict respiratory and contact precautions, restriction of visitors and hospital personnel with respiratory illness, vaccination of hospital personnel for the influenza season, and administration of antiviral agents to HSCT recipients during outbreaks and when the frequency of nosocomial transmission is high.[25]

Respiratory syncytial virus

Respiratory syncytial virus (RSV), the most common respiratory virus infection in HSCT recipients, is isolated in 30–50 per cent of HSCT recipients with community respiratory virus infection.[19,21] Seasonal outbreaks are common during the winter.[19,26]

RSV infection deteriorates to pneumonia, primarily viral, in over 50 per cent, usually in the first 30 days following HSCT.[19,21,23,27] Identification of RSV by cultures may take several days. Histopatholgy shows multinucleated syncytial cells or giant cells with intracytoplasmic viral inclusions in RSV pneumonia. Immunohistochemical studies with monoclonal antibodies confirm the diagnosis.

RSV-specific immune globulin prevents RSV disease in children.[28] Early treatment, before the development of respiratory failure, with aerosolized ribavirin and intravenous immunoglobulin has been associated with good outcome.[21,29]

RSV pneumonia is associated with a mortality of 30–80 per cent.[19,21,23,30] If not treated promptly, mortality approaches 100 per cent.

Influenza virus

Influenza accounts for 10–40 per cent of respiratory virus infections in HSCT recipients.[19,21,24] Although both influenza A and B cause severe infection, influenza A is more common.[24]

Pneumonia is reported to develop in between 10 and 70 per cent of HSCT recipients with influenza infection.[19,21,24,31]

However, unlike RSV, in at least half of the cases the pneumonia is secondary to bacteria or fungi.[19]

Since severe disease is more likely to occur near the time of transplantation, the procedure should be delayed in patients with respiratory virus infection.[32] Immunization with inactivated influenza is recommended. However, response to vaccine is lost for 6 months after transplant, and chronic GVHD is associated with poor response.[32] Amantidine or rimantidine prophylaxis may be used to provide partial protection when influenza A activity is detected in the community.[32] Although ribavirin is active in vitro against influenza A and B viruses, clinical data are not available.[32] Treatment with amantadine or rimantadine is not of benefit if initiated after the development of respiratory failure.[33] Oseltamivir may play a beneficial role for therapy and for prophylaxis following confirmed exposure to influenza A and B.[34] Pneumonia in patients with influenza infection is associated with 50 per cent mortality.[19]

Parainfluenza virus

Parainfluenza accounts for 10–30 per cent of respiratory virus infections in HSCT recipients.[19,21,24] Pneumonia is reported to develop in between 20 and 70 per cent of HSCT recipients with parainfluenza infection and is more common in allogeneic recipients.[35,36] There is no specific therapy for parainfluenza infection. The mortality rate associated with parainfluenza infection ranges between 0 and 40 per cent.[24,35,36] The mortality rate is higher in allogeneic recipients.[35]

Rhinovirus

Rhinovirus accounts for 6–25 per cent of respiratory virus infections in HSCT recipients.[19,21] Pneumonia develops in about 20 per cent.[19,21] Because there is no effective treatment, prevention is important.

Adenovirus

Adenoviruses are acquired by reactivation or exogenously.[33] They are isolated in 3–6 per cent of HSCT recipients with respiratory virus infection.[19,24] They can also cause haemorrhagic cystitis, enterocolitis, hepatitis, encephalitis and disseminated disease.[37,38] The clinical and histopathological features mimic CMV infection.[37] In a study of 85 HSCT recipients with adenovirus infection, 15 (18 per cent) had pneumonia.[38] Adenovirus pneumonia in HSCT recipients is associated with mortality in excess of 70 per cent.[23,38] Ribavirin has no effect on mortality.[23]

Herpes simplex virus

Herpes simplex virus (HSV)-1 is a common cause of vesicular lesions of the oral mucosa. HSV-2 commonly affects genital mucosa. In immunocompromised patients, HSV can also cause severe involvement of the genitalia, liver, lung, eye and central nervous system. Although HSV-1 is the more common, HSV-2 can also cause pneumonia in HSCT recipients.[39]

Without antiviral prophylaxis, about 70–80 per cent of HSCT recipients seropositive for HSV will develop reactivation.[40] With prophylaxis, the reactivation rate is reduced to 5 per cent. Before antiviral prophylaxis was widely applied, HSV pneumonia was commonly reported in autopsy series of HSCT recipients.[41] More recent studies have found HSV pneumonia to be uncommon.[5,6]

HSV reactivation occurs frequently during the first month following transplant. Although uncommon, HSV pneumonitis develops by extension from the upper airways secondary to tracheobronchitis.[40] HSV pneumonia presents with dyspnoea, cough and fever, and airway involvement may lead to haemoptysis.

The diagnosis of HSV infection can be made by a variety of techniques including viral culture, serology, immunofluorescence or a bedside Tzanck prep from BAL fluid or lung tissue. The histological pattern of HSV pneumonia includes parenchymal necrosis, haemorrhages and mononuclear cell inflammatory infiltrates.

HSCT recipients who are seronegative for HSV should avoid exposure to patients with HSV infection. Acyclovir and its derivatives (valacyclovir and famciclovir) are used for the prevention and treatment of HSV infection. Foscarnet is used if resistance to acyclovir exists.

Varicella zoster virus

Although varicella zoster virus infection is common in HSCT recipients after engraftment, it rarely causes disseminated disease.[12] Varicella zoster is an uncommon cause of interstitial pneumonia.[42] Pneumonia usually develops 3–7 days following rash. HSCT recipients should avoid exposure to persons with active varicella zoster infection. If they are exposed to patients with chickenpox or shingles and they are seronegative, they should receive varicella zoster immune globulin. HSCT recipients with varicella zoster rashes should be treated with acyclovir.

Epstein–Barr virus

Epstein–Barr virus (EBV) is distributed worldwide and 90–95 per cent of adults are seropositive for EBV. EBV-associated post-transplant lymphoproliferative disorder (PTLD) is more common in patients who have received T-cell-depleted stem cells and those who have been treated with anti-CD3 antibody.[43] GVHD is a risk factor for PTLD.[43] During the immunosuppression of HSCT recipients, loss of cytotoxic T-lymphocyte activity leads to uncontrolled Epstein–Barr driven B-cell proliferation, resulting in PTLD.[40,44,45]

EBV infection may present as uncomplicated infectious mononucleosis, benign B-cell hyperplasia or B-cell lymphoma. PTLD presents with a mononucleosis-like illness with fever, lymphadenopathy and leucopenia.[40] About 85 per cent of PTLD occurs within the first year of transplant. It may present with multiple organ involvement and fulminant clinical course.

EBV-associated infectious mononucleosis and benign B-cell hyperplasia are treated with reduction of immunosuppressive therapy and antivirals. Mortality of PTLD in HSCT recipients may exceed 80 per cent.[46] However, recent advances in treatment with anti-B-cell monoclonal antibodies may reduce PTLD associated mortality.[47,48]

Other viruses

Human herpes virus (HHV)-6 has been associated with pneumonia in HSCT recipients.[49] Since the virus is detected in organs with no pathology, the causal relationship is not clearly established.[40]

HHV-8-related Kaposi's sarcoma occurs frequently in immunocompromised patients. However, it is rarely reported in HSCT recipients.[50] Although Kaposi's sarcoma usually presents with skin lesions, pulmonary haemorrhaging is one of the manifestations.[50]

Human metapneumovirus, a recently discovered human respiratory virus, is found in 3 per cent of BAL fluid from HSCT recipients.[51] Clinical symptoms and signs include cough, nasal congestion, sore throat, low-grade fever and wheezing. Diffuse pulmonary infiltrates are seen in almost all patients and bronchoscopy reveals diffuse alveolar haemorrhaging.[51] The diagnosis is made by reverse transcriptase PCR.[51] Treatment with corticosteroids has no beneficial effect. In one report of five HSCT recipients with human metapneumovirus, progression to respiratory failure occurred at a median of 4 days and only one patient survived, with prolonged morbidity.[51]

FUNGAL INFECTIONS

With improved prevention and treatment of other opportunistic infections, especially CMV, invasive fungi have become the leading infectious cause of morbidity and mortality in HSCT recipients.[52] Candida and Aspergillus species are the most common fungal pathogens.[53] With the wider use of fluconazole prophylaxis, the incidence of Candida albicans infection after transplant has declined, and infections with non-albicans Candida species and Aspergillus have become the dominant cause of fungal pneumonia.[7] Even in endemic areas, HSCT recipients rarely develop infections with other fungi such as Cryptococcus neoformans, Coccidioides immitis or Histoplasma capsulatum.[53] The experience of pulmonary infections by other fungi such as mucormycosis[54,55] and Cunninghamella bertholletiae is limited to rare case reports.[56,57]

There is substantial controversy regarding the optimal criteria for the diagnosis of invasive fungal infections including pneumonia. The European Organization for Research and Treatment of Cancer/Invasive Fungal Infections Cooperative Group and the National Institute of Allergy and Infectious Diseases Mycoses Study Group have proposed the diagnoses of invasive fungal infections in the immunocompromised host to be labelled as 'proven', 'probable' or 'possible'.[58] These levels depend on histopathological examination and the presence of minor and major criteria derived from clinical, radiological and microbiological findings.

Pulmonary aspergillosis

The reported incidence of invasive aspergillosis in allogeneic recipients ranges between 3.6 and 28 per cent.[7] Because of its association with neutropenia and GVHD, the incidence has a bimodal distribution, early during the neutropenic phase and later with chronic GVHD.[53]

In addition to neutropenia and GVHD, damage to epithelial barriers, impaired monocyte function, altered cell-mediated immunity, prolonged corticosteroid therapy, and CMV disease increase the risk for *Aspergillus* infection.[7] In the presence of a normal graft and in the absence of GVHD, most immune impairment resolves by 12–15 months.[53] In the absence of neutropenia, invasive *Aspergillosis* is uncommon in autologous recipients.[10,59] *Aspergillus* infection is usually acquired through inhalation. Building construction increases the rates of *Aspergillus* colonization and invasive infection. In the absence of adequate immune response, the organism invades the sinuses or lung tissue and disseminates to other organs by hyphal invasion of blood vessels.[53]

In a recent publication of allogeneic HSCT recipients with invasive aspergillosis, the diagnosis was made within the first 40 days after transplant in 30 per cent, between 40 and 180 days in 53 per cent, and later than 180 days in 17 per cent.[60] The clinical presentations of pulmonary aspergillosis are non-specific and include fever, cough, sputum production and pleuritic chest pain.[61,62] Haemoptysis is uncommon but may occur when neutropenia resolves.[63] Pulmonary aspergillosis uncommonly may manifest with superior vena cava syndrome,[64] and irreversible airway obstruction mimicking bronchiolitis obliterans (BO).[65] Rhinocerebral aspergillosis is reported in 2–3 per cent of HSCT recipients.[66–68] However, since sinus aspirations are not commonly performed, the incidence of *Aspergillus* sinusitis is probably underestimated.[53]

The radiographic findings of pulmonary aspergillosis include nodules, diffuse pulmonary infiltrates and cavitation.[7] Nodules, masses and the 'halo sign' are more frequently seen in pulmonary aspergillosis compared to other pulmonary complications. The halo sign consists of a nodule surrounded by ground-glass attenuation due to coagulation necrosis and haemorrhagic infarction, whereas the 'air crescent sign' represents the development of necrosis. The halo sign is seen more often in neutropenic patients.[69] Cavitation is a late finding in invasive pulmonary aspergillosis (Fig. 17.4). A recent study has shown that the 'hypodense sign', the presence of hypodensity in nodules or consolidations, may have high specificity for the diagnosis of invasive pulmonary aspergillosis.[70] Since CT findings often precede plain radiography findings, chest CT should be performed in patients with suspected pulmonary aspergillosis.

In high-risk patients, galactomannan *Aspergillus* antigen and PCR assays can be used to screen for and monitor the response to treatment of *Aspergillus* infection.[71] However, both the antigen and PCR assays may have a high false-positive rate, and the clinical utility of these tests is not yet defined.[72]

The diagnostic work-up of HSCT recipients with suspected pulmonary aspergillosis should include a procedure intended to obtain tissue for histology and culture. The sensitivity of BAL is high for bronchopneumonia, but low in the setting of

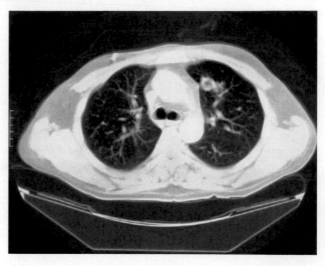

Fig. 17.4 Chest computed tomography of HSCT recipient with invasive pulmonary aspergillosis showing nodular infiltrate associated with cavitation.

isolated nodules.[53] The roles of *Aspergillus* antigen and PCR assays in BAL fluid are not well defined. The histology demonstrates angular (45°) dichotomously branching and septate hyphae.

Because of the high mortality associated with invasive aspergillosis, many clinicians use antifungals empirically, without confirming the diagnosis. If the histology of lung tissue shows *Aspergillus*, the diagnosis of *Aspergillus* pneumonia is considered definitive.[58] This definitive diagnosis usually requires surgical or radiologically guided transthoracic lung biopsy. In the absence of the typical histopathological findings, the presence of one microbiological criterion and one major or two minor clinical criteria is required for the diagnosis of probable, and one microbiological or one major or two minor criteria for possible, *Aspergillus* pneumonia.[58] The microbiological criteria include positive *Aspergillus* culture from sputum or BAL fluid or *Aspergillus* antigen from BAL fluid. Chest CT findings of halo sign, air-crescent sign or cavity within an area of consolidation are considered major criteria. The minor criteria include cough, chest pain, haemoptysis, dyspnoea, pleural rub, any pulmonary infiltrates not fulfilling the major criteria, and pleural effusion.

Historically, amphotericin B or related compounds have been the gold standard for therapy of *Aspergillus* infections. More recently, the success and tolerability of voriconazole have made it the choice of drug for invasive aspergillosis.[7] The appropriate dose is 6 mg/kg every 12 hours for 1 day followed by 4 mg/kg every 12 hours. Amphotericin B, caspofungin, micafungin and itraconazole are other alternatives. The optimal duration of antifungal therapy is unknown and depends on the extent of the infection, response to treatment, and improvement of the underlying immune deficit. Surgical resection combined with antifungal therapy may be useful in selected patients with localized pulmonary aspergillosis.

Since spore inhalation is the usual route of *Aspergillus* infection, environmental surveillance and strict preventive measures are important.[73] Nursing of patients undergoing stem-cell transplantation in high-efficiency particulate air (HEPA) filtered rooms is an effective protection, even under environmental conditions with increased air contamination.[74,75]

The addition of aerosolized amphotericin B inhalation to standard antifungal prophylaxis has not been shown to be effective in preventing invasive aspergillosis. During high-risk periods, granulocyte transfusion and oral itraconazole or intravenous amphotericin B have been used until neutropenia resolves or the *Aspergillus* colonization disappears. A randomized clinical trial has shown that itraconazole prophylaxis may decrease the mortality associated with invasive aspergillosis, although it does not reduce the incidence.[76]

The reported case fatality rate of invasive aspergillosis ranges between 29 and 87 per cent.[59,61,77] A recent study of allogeneic HSCT recipients with invasive aspergillosis has reported a 1-year survival rate of about 20 per cent.[60] In patients receiving mechanical ventilation, mortality exceeds 90 per cent.[78]

Pulmonary candidiasis

Candida infections, especially of the bloodstream, are common during the neutropenic phase. The portal of entry is usually the gastrointestinal tract or indwelling central venous catheter. Isolated *Candida* pneumonia is uncommon. In one study of 1359 HSCT recipients, only one late-onset pulmonary candidiasis was reported.[52] Autologous HSCT recipients are at low risk for fungal infection once their neutrophils recover. However, *Candida* infection remains a major threat to patients receiving allogeneic transplant. Risk factors include old age, prolonged duration of neutropenia, GVHD, and use of irradiation as part of the preparative regimen.[79,80] *Candida fungemia* is responsive to antifungal therapy if it is initiated early and accompanied by removal of foreign objects such as central venous catheters. Fluconazole prophylaxis will prevent invasive disease due to fluconazole-susceptible *Candida* species during neutropenia.[74]

Pneumocystis jiroveci pneumonia

The incidence of *P. jiroveci* pneumonia (PCP) in HSCT recipients has decreased sharply with trimethoprim–sulfamethoxazole (TMP–SMZ) prophylaxis. In four recent studies that included a total of 2875 HSCT recipients, PCP developed in 46 (1.6 per cent), all in patients who were not receiving TMP–SMZ prophylaxis.[12,52,81,82]

The radiographic findings of PCP consist of diffuse, bilateral reticular or granular pattern that can progress to alveolar consolidation.[82] The diagnosis usually requires BAL, with or without transbronchial lung biopsy.

The drug of choice for both prevention and treatment of PCP is TMP–SMZ. PCP prophylaxis should be provided for at least 6 months after autologous HSCT for treatment of haematological malignancies and throughout all periods of immunosuppression (use of prednisone or ciclosporin or the presence of GVHD) in allogeneic HSCT recipients.[52,74] For patients who do not tolerate TMP–SMZ, dapsone, aerosolized pentamidine and atovaquone are alternatives for prophylaxis – albeit less effective than TMP–SMZ.[74]

Other fungal infections

Fungi other than *Candida* and *Aspergillus* cause less than 10 per cent of fungal infections. Although *Fusarium* usually causes localized infection in the immunocompetent host, it may cause disseminated disease mimicking aspergillosis in HSCT recipients.[83] Rhinocerebral and pulmonary mucormycosis follow inhalation of the spores. Mucormycosis affects mostly allogeneic recipients.[84] Pulmonary mucormycosis is characterized by infarction, necrosis and dissemination to the heart and mediastinum.[54] Despite surgical resection and antifungal therapy, mortality is high.[85]

BACTERIAL INFECTIONS

Bacterial pneumonia

Bacterial infections are common during the neutropenic period in HSCT recipients. However, since most of the HSCT recipients receive prophylactic and empirical therapy during this period for suspected bacterial pneumonia, the exact incidence has not been well defined. In a recent study of 1359 HSCT recipients, 58 per cent of late-onset pneumonias were diagnosed clinically, without tissue or microbial documentation.[52]

The aetiology of bacterial infection in neutropenic patients is in flux.[86] Gram-negative bacteria, especially *Pseudomonas aeruginosa*, used to be the most common pathogens causing pneumonia.[87] However, owing to the use of prophylactic antibiotics to prevent Gram-negative infection, and the use of intravascular catheters, Gram-positive organisms are becoming more frequent. Patients who have received minimal or no prophylaxis develop infection predominantly with Gram-negative organisms, whereas those with antibiotic prophylaxis develop infection with Gram-positive bacteria. The use of narcotics for pain associated with mucositis increases the risk of aspiration pneumonia.

Persistent deficits in cellular and humoral immunity predispose HSCT recipients to intracellular pathogens such as *Listeria monocytogenes* and *Nocardia* species and to encapsulated organisms such as *Streptococcus pneumoniae* and *Haemophilus influenzae*. Invasive pneumococcal infection occurs usually in the late post-transplant phase in allogeneic HSCT recipients with GVHD.[88] Outbreaks of *Legionella* pneumonia in HSCT recipients may result from a colonized water supply.

Bacterial pneumonia may occur at any time during the pre-engraftment period. Late-onset bacterial pneumonia occurs at a median of 99 days (range 22–813) after transplant.[52] The initial clinical symptoms of bacterial pneumonia in the HSCT recipient may be subtle, but usually evolve quickly.

The radiographic appearance of bacterial pneumonia in HSCT recipients is similar to that of immunocompetent patients. Most present with focal consolidation.[15,82,87] *Pseudomonas* and *Legionella* tend to cause multifocal consolidation.[82] Some HSCT recipients with bacterial pneumonia may present with nodules and masses on chest radiograph.[15,82] Although uncommon, the halo sign (see above) and cavitations have been reported in bacterial pneumonia.[15,82] Blood and sputum, if possible, cultures are obtained and most HSCT recipients are treated empirically. If there is uncertainty in the diagnosis or lack of response to empirical treatment, BAL may be helpful.

If the causative organism is identified, antibiotic therapy should focus on the specific cause. However, no organism is identified in most of the HSCT recipients with suspected bacterial pneumonia. The initial empirical antibiotic regimen should include coverage of *Pseudomonas aeruginosa* and methicillin-resistant *Staphylococcus aureus*.

Because bacteria are carried on the hands, appropriate hand washing should be practised by all individuals in contact with HSCT recipients.[74] The data are insufficient to recommend routine use of gut decontamination and systemic antibiotics for afebrile, asymptomatic patients. However, some advocate using fluoroquinolone prophylaxis to reduce infection by Gram-negative rods in allogeneic HSCT recipients.[89] Although growth factors shorten the duration of neutropenia after HSCT, they do not reduce the attack rate of invasive bacterial disease. Some experts advise giving intravenous immunoglobulin (IVIG) to prevent bacterial infection in HSCT recipients with unrelated graft and severe hypogammaglobulinaemia within 100 days after transplant. During the late post-transplant phase, antibiotic prophylaxis is recommended to prevent infection with encapsulated organisms among allogeneic recipients with chronic GVHD. The 23-valent pneumococcal polysaccharide vaccine should be administered at 12–24 months after transplant.[87]

Pulmonary nocardiosis

Nocardiosis is a Gram-positive bacterial infection caused by aerobic Actinomycetes. The lungs are involved in the majority of patients with nocardiosis. *Nocardia* infection can affect multiple organs including the central nervous system and skin and may also cause bacteraemia. Nocardiosis has been reported in less than 2 per cent of allogeneic HSCT recipients. It usually develops during the late post-transplant period in patients with GVHD. Sulfonamides are used for treatment. Mortality attributable to *Nocardia* infection in HSCT recipients is about 16 per cent.[90]

Mycobacterial infection

The reported incidence of mycobacterial tuberculosis infection in HSCT recipients depends on geographical location. In a study from the University of Minnesota, *Mycobacterium tuberculosis* was reported only in 2 of 2241 HSCT recipients (< 0.1 per cent) compared to 2–6 per cent in Asia.[11,12,91,92] Most of the *Mycobacterium* tuberculosis infections occur during the late post-transplant period, typically in patients with GVHD. The clinical manifestations in HSCT recipients include fever and pulmonary infiltrates.[92] Acid-fast smears and mycobacterial cultures of sputum, BAL fluid and lung tissue are used for diagnosis.[91] Most of the tuberculous infections respond to standard anti-mycobacterial medications.[12,91,92]

Atypical mycobacteria infections by *M. chelonae/fortuitum* and *M. avium-intracellulare* are reported rarely in HSCT recipients.[92,93]

NON-INFECTIOUS PULMONARY COMPLICATIONS

Pulmonary function abnormalities

Restrictive and obstructive ventilatory defects, and gas transfer abnormalities, are frequent long-term sequelae following allogeneic transplantation. A systematic review of 20 publications published between 1996 and 2001 found decreased carbon monoxide diffusion (DLCO) in 83 per cent, restriction in 35 per cent and obstruction in 23 per cent of allogeneic HSCT recipients.[94] A more recent study of over 500 HSCT recipients documented somewhat lower frequencies: impaired diffusion in 35 per cent, restriction in 12 per cent, and obstruction in only 6 per cent of long-term survivors.[95] While the frequency of restrictive defect and of impaired DLCO remained constant over time, this study suggested a declining frequency of airflow obstruction. The development of airflow obstruction increases the risk of mortality following transplantation.[95-97]

Risk factors for the development of PFT abnormalities following HSCT include smoking history, pre-transplant pulmonary infection, viral infection in the early post-transplant period, older age, underlying disease, pre-transplant chemotherapy and conditioning regimen, graft-versus-host disease and HLA-mismatch.[4] Respiratory muscle weakness due to GVHD-associated myositis has been described following allogeneic HSCT and may rarely be a contributing factor to restrictive pulmonary function abnormalities.

Although there are conflicting studies, abnormal pre-transplant pulmonary function may be a risk factor for pulmonary complications in HSCT recipients.[4,98] There is considerable variability among published studies in the pulmonary parameters deemed to be predictive, the complications that they predict, and the strength of the association. In patients with systemic lupus, the pulmonary impairment associated with the disease may improve following HSCT.[99] In light of these limitations, baseline PFT cannot be used to identify recipients at risk for pulmonary complications, and its role to guide preventive and therapeutic interventions needs further studies.

Upper airway injury

Significant mucosal barrier injury occurs in about 75 per cent of HSCT recipients.[100] Total body irradiation, allogeneic transplant, leukaemia and delayed neutrophil engraftment are risk factors for mucositis in HSCT recipients.[101] Upper airway inflammation due to mucositis may lead to laryngeal oedema, dysphagia and aspiration pneumonia. Life-threatening upper airway complications are more common in children.[102,103] The severity of mucositis is associated with secondary infection, requirement for narcotics, longer duration of parenteral nutrition and hospital length of stay, and overall mortality.[101,104]

Recent studies have shown beneficial effects of cryotherapy, keratinocyte growth factor, laser therapy, glutamine supplementation and calcium phosphate mouth rinse with fluoride in decreasing the severity and duration of mucositis.[105-108] However, upper airway injury in HSCT recipients is usually

Colour plates

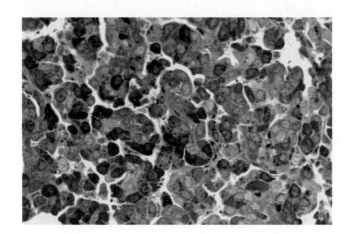

Fig. 2.5a

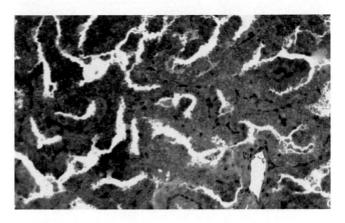

Fig. 2.5b

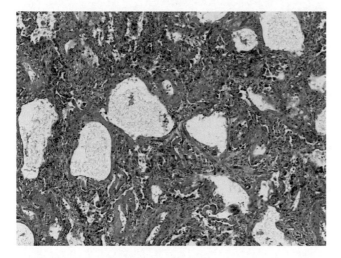

Fig. 5.1

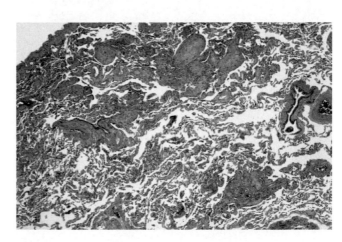

Fig. 5.2

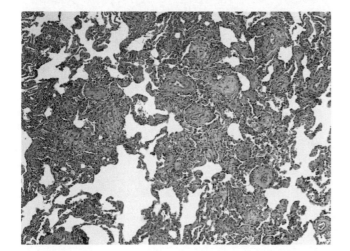

Fig. 5.3

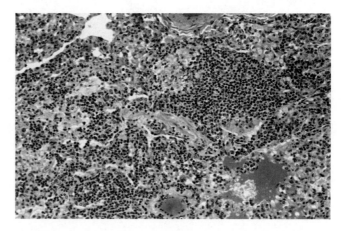

Fig. 5.4

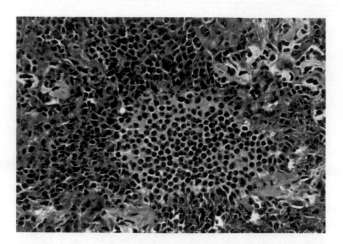

Fig. 5.5

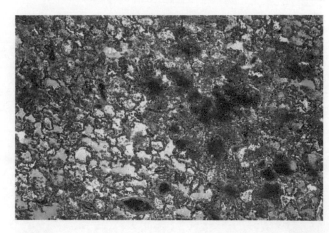

Fig. 5.6

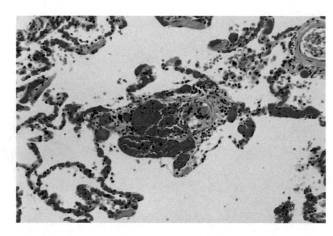

Fig. 5.7

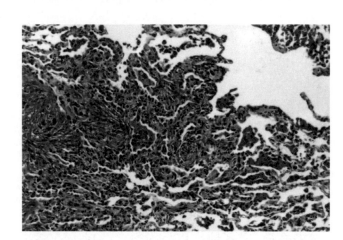

Fig. 11.1

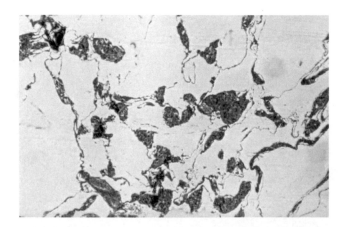

Fig. 12.1

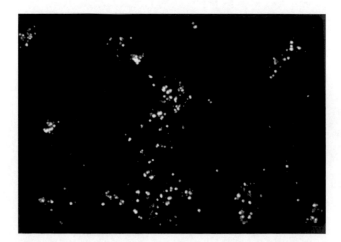

Fig. 12.2

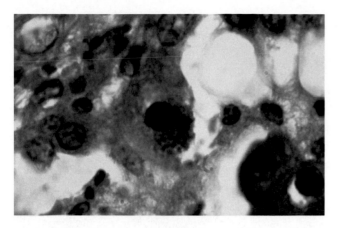

Fig. 17.3

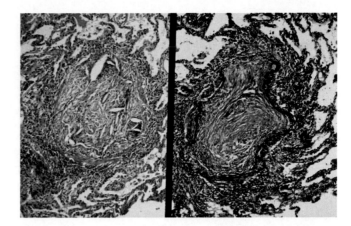

Fig. 17.5

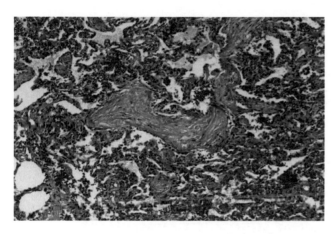

Fig. 17.6

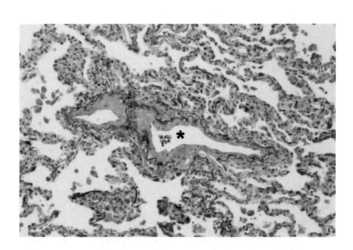

Fig. 22.1d

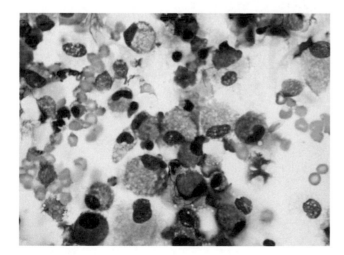

Fig. 22.3a

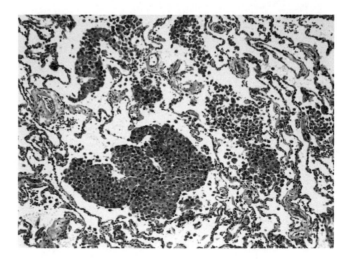

Fig. 22.3b

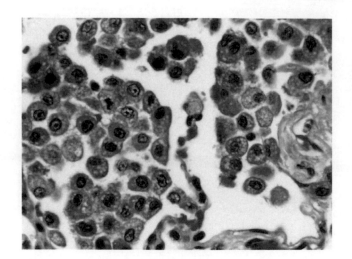

Fig. 22.3c

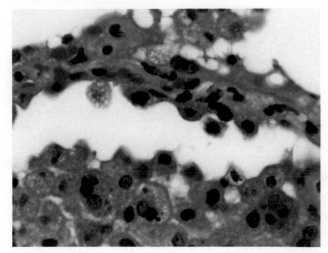

Fig. 22.3d

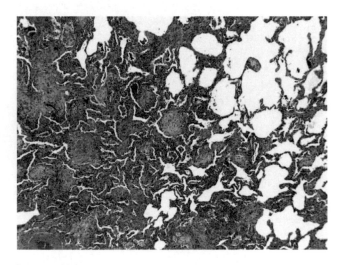

Fig. 22.3e

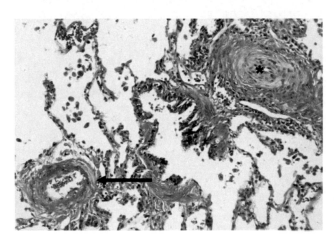

Fig. 29.1c

managed with supportive care, including local symptomatic therapy and endotracheal intubation in severe cases.

Asthma

There are case reports of asthma developing after allogeneic HSCT.[4] Limited data suggest that the serum IgE of donor origin is elevated following transplantation. Allergen-specific IgE-mediated hypersensitivity can be transferred from donor to HSCT recipient by B-cells with allergen-specific memory leading to atopic dermatitis, allergic rhinitis and asthma. The management of asthma in HSCT recipients is similar to that in the general population.

Acute pulmonary oedema

Although pulmonary oedema has been described as part of the capillary leak syndrome, it usually results from increased hydrostatic pressure due to large volumes of fluid and blood products infused during conditioning and the immediate post-transplant period.[109,110] Diagnostic criteria include acute onset of dyspnoea, crackles on physical examination, reduced arterial oxygen tension and chest radiograph showing vascular redistribution and increased interstitial markings in the absence of systemic or pulmonary infection.[110] Chest radiographic findings are typical of pulmonary oedema with bilateral diffuse ground-glass pulmonary opacities often accompanied by Kerley B lines.[111] Cardiomegaly is generally absent. Pulmonary oedema can be prevented by fluid restriction and diuretic therapy.[110]

Idiopathic pneumonia syndrome

No infectious aetiology is identified in many HSCT recipients with suspected pneumonia. In a 1985 review of pulmonary complications of 4500 HSCT recipients, Krowka et al. reported a 35 per cent frequency of idiopathic pneumonia.[112] A more recent study found this frequency to be 7.3 per cent among 1164 HSCT recipients.[113] The wide variation in the reported incidence of idiopathic pneumonia syndrome (IPS) was partly due to the lack of uniform definition and diagnostic criteria.

A workshop sponsored by the National Heart, Lung, and Blood Institute (NHLBI) defined IPS as the presence of widespread alveolar injury in the absence of lower respiratory tract infection, characterized by acute, bilateral pulmonary infiltrates, associated symptoms of cough and dyspnoea, hypoxaemia, and restrictive physiology, in the absence of infection or heart failure (Box 17.2).[114] Absence of infection is typically established by negative BAL and by the lack of clinical response to antimicrobial therapy. DAH and PERDS also fulfil the diagnostic criteria of IPS.[115,116] Despite some overlap in the clinical features of IPS with DAH and PERDS, their responses to treatment and clinical courses are different.[115]

The overall incidence of IPS is 10 per cent (range 2–17).[52,117,118] The median time of onset of IPS is between 21 and 87 days, with a range 0 to 1653 days after transplant.

Box 17.2 Criteria for the diagnosis of idiopathic pneumonia syndrome in haematopoietic stem-cell transplant recipients (from [114])

I. Evidence of widespread alveolar injury
   a Multilobar infiltrate
   b Symptoms and signs of pneumonia
   c Abnormal pulmonary physiology with increased alveolar to arterial oxygen gradient and increased restrictive defect
II. Absence of lower respiratory tract infection after appropriate evaluation with:
   a Bronchoalveolar lavage negative for bacterial and non-bacterial pathogens
   b Lack of improvement with broad-spectrum antibiotics
   c Transbronchial lung biopsy if tolerated
   d A second confirmatory test for infection within 2–14 days

---

Compared to patients with GVHD, IPS may present earlier in patients without GVHD.[113]

Risk factors for the development of IPS include old age, transplant for malignancy other than leukaemia, pre-transplant chemotherapy, total body irradiation, GVHD, and positive donor cytomegalovirus serology.[117,119]

The pathogenesis of IPS is not well defined.[120–122] Lung tissue injury, inflammation and cytokine release have been implicated. Pre-transplant radiation and chemotherapy and undocumented infections may be responsible for the initial injury. The increased levels of interleukin-6, interleukin-8 and tumour necrosis factor (TNF)-α in the serum and BAL of HSCT recipients with IPS, and the clinical improvement following the administration of etanercept, suggest the role of these cytokines in the pathogenesis.[120–123]

The clinical presentation of IPS includes dyspnoea, dry cough, hypoxaemia and non-lobar radiographic infiltrates.[114] The clinical spectrum is broad, ranging from acute respiratory failure to incidental radiographic abnormalities.[114]

The NHLBI workshop requires BAL or lung biopsy to exclude an infectious aetiology before entertaining the diagnosis of IPS.[114] In our diagnostic approach to patients with suspected IPS, we perform BAL and, if there are no contraindications, transbronchial lung biopsy. We resort to video-assisted thoracoscopic lung biopsy if transbronchial lung biopsy is contraindicated or the transbronchial lung biopsy specimen is inadequate. Lung biopsies of patients with IPS may show diffuse alveolar damage, organizing or acute pneumonia, and interstitial lymphocytic inflammation.[113,124]

Despite case reports of patients with IPS responding to treatment with corticosteroids, larger studies have not shown any outcome benefit.[113,124,125] Currently, the only accepted treatment regimens are supportive care and prevention and treatment of infection.

The pneumonitis resolves in about 31 per cent of patients with IPS.[113,125] However, the clinical course may often be complicated by viral and fungal infections, pneumothorax, pneumomediastinum, subcutaneous emphysema, pulmonary fibrosis, and autoimmune polyserositis.[117] The case fatality of IPS is about 74 per cent (range 60–86).[117] The reported 1-year survival rate of IPS is less than 15 per cent.[113,126] For those who require mechanical ventilation, the hospital mortality exceeds 95 per cent.[113] However, more recent studies show a higher survival rate.[52,118]

Diffuse alveolar haemorrhage

DAH occurs in approximately 5 per cent of autologous and allogeneic HSCT recipients, with a range of approximately 2 to 21 per cent.[115] The reported frequency is higher in series of patients identified in the intensive care unit or at autopsy.

Risk factors for DAH include age above 40 years, intensive chemotherapy and total body irradiation, and the presence of inflammatory cells in BAL fluid.[115] Following transplant, the development of high fever and severe mucositis, and acute GVHD, are also risk factors for DAH.

There are no associations between DAH and prolonged prothrombin or partial thromboplastin time or a low platelet count.[115] Although most patients with DAH have thrombocytopenia, DAH is not corrected with platelet transfusion.[127] Pre-transplant pulmonary function is not predictive of DAH following HSCT, but evidence of airway inflammation, defined by BAL neutrophils above 20 per cent and any eosinophils, has been associated with subsequent DAH.[128]

Despite recognition of the risk factors, the aetiology and pathogenesis of DAH in the HSCT recipient have not been clearly established. Lung tissue injury, inflammation and cytokine release are implicated in the pathogenesis of DAH.[115]

Symptoms of DAH typically include dyspnoea, fever and cough.[127] Haemoptysis is rare, reported in fewer than 20 per cent.[115,129] The onset of DAH is usually within the first 30 days (median between 11 and 24) following HSCT.

Measurements of spirometry, lung volume and DLCO are usually not available since HSCT recipients with DAH are too ill to perform pulmonary function testing. Arterial blood gas studies show hypoxaemia.

Chest radiographs usually show alveolar and interstitial infiltrates involving middle and lower lung zones.[130] In early phases, the radiographic changes may be subtle, unilateral or asymmetric.[130] Although CT of the chest may be helpful in patients in whom a focal abnormality is suspected, it has a limited role in DAH. The most common CT findings in DAH are bilateral areas of ground-glass attenuation or consolidation.[131] The criteria for the diagnosis DAH are outlined in Box 17.3. The visual description of the BAL fluid and the presence of haemosiderin-laden macrophages play complementary roles in the diagnosis of DAH.[129] Lung tissues in DAH show diffuse alveolar damage.[5,127]

Based on retrospective studies, HSCT recipients with DAH are treated with systemic corticosteroids.[115,129] We commonly use intravenous methylprednisolone approximately 1 g daily in four divided doses for 5 days, followed by 1 mg/kg for 3 days and tapering off over 2–4 weeks.[115] Although a retrospective study showed no outcome benefit,[132] there are case reports of allogeneic HSCT recipients with DAH successfully treated with recombinant factor VIIa.[133,134]

The majority of HSCT recipients with DAH require mechanical ventilator support for respiratory failure.[115,129] HSCT recipients with DAH are at high risk for subsequent

---

> **Box 17.3** Criteria for the diagnosis of diffuse alveolar haemorrhage (from [115])
>
> 1. Signs and symptoms of pneumonitis:
>   a Hypoxaemia and/or restrictive ventilatory defects, and
>   b Radiographic infiltrates involving multiple lobes
> 2. No evidence of infection
> 3. Bronchoalveolar lavage showing:
>   a Progressively bloodier return from separate subsegmental bronchi, OR
>   b >20% haemosiderin-laden alveolar macrophages

---

infectious complications.[135,136] The reported mortality rate of DAH in HSCT recipients ranges between 48 and 100 per cent.[115,129] Although the initial presentation of DAH is usually respiratory failure, the two most common reported causes of death had been multiple organ failure and sepsis.[136,137] However, in two recent studies, respiratory failure was the most common cause of death in HSCT recipients with DAH.[129,132]

Peri-engraftment respiratory distress syndrome

PERDS refers to the pulmonary component of the engraftment syndrome that may also present with diarrhoea and skin rash. It responds to treatment and has a low case fatality rate.[116] There is overlap between PERDS and DAH, about one-third of DAH occurring during the peri-engraftment period and about one-third of patients with PERDS having DAH.[116,129]

The engraftment syndrome, including non-pulmonary presentations, has been reported in up to 20 per cent of autologous HSCT recipients.[116,138,139] Capizzi et al. reported the incidence of PERDS to be about 5 per cent in autologous HSCT recipients.[116]

The pathogenesis of PERDS is not well defined but is believed to be a complex interaction between the conditioning-related endothelial damage and the cytokine release associated with the neutrophil and lymphocyte recovery.[116]

The diagnostic criteria of PERDS include fever (> 38.3°C) and pulmonary injury evidenced by hypoxia ($SaO_2 < 90$ per cent) and/or pulmonary infiltrates, in the absence of cardiac dysfunction and infection, within 5 days of neutrophil engraftment.[116] In the study by Capizzi et al. the median time to onset of PERDS was 11 days (range 4–25) after transplant.[116] Dyspnoea was the initial symptom in all the patients. Fever was present in 12 patients (63 per cent) at the onset of symptoms. Bilateral pulmonary infiltrates may not be present on plain chest radiography at the onset of symptoms.[116]

BAL may show neutrophilic inflammation.[116] Transbronchial lung biopsy is usually contraindicated during the peri-engraftment period because of thrombocytopenia. Surgical lung biopsy may show diffuse alveolar damage but is rarely necessary.[116]

HSCT recipients with PERDS are treated with high-dose corticosteroids that usually lead to rapid clinical improvement.[116] A short course of corticosteroid therapy has been used effectively to reduce engraftment syndrome in autologous HSCT recipients.[140]

Unlike DAH and IPS, only about one-third of HSCT recipients with PERDS require intensive care and mechanical ventilation. The reported mortality rate in PERDS is about 26 per cent.[116]

Bronchiolitis obliterans

About 45 per cent of long-term surviving allogeneic HSCT recipients develop chronic GVHD.[141] The histological pulmonary manifestations of GVHD include diffuse alveolar damage, lymphocytic bronchitis/bronchiolitis with interstitial pneumonitis, bronchiolitis obliterans organizing pneumonia (BOOP), and bronchiolitis obliterans (BO).[142] BO is a severe manifestation of chronic GVHD, characterized by airflow limitation.

With the exception of rare case reports in autologous recipients, BO almost exclusively affects allogeneic HSCT recipients with GVHD.[143] The overall frequency of BO in allogeneic HSCT recipients is about 3.9 per cent.[4] In long-term survivors with GVHD, the incidence of BO may reach 35 per cent.[4,117]

Risk factors for BO include GVHD, older donor and recipient age, myeloablative conditioning, methotrexate use, antecedent respiratory infection, and serum immunoglobulin deficiency.[4,117] T-cell depletion may prevent BO in allogeneic HSCT recipients.[144]

The pathogenesis of BO in HSCT recipients is not well understood. GVHD, recurrent aspiration, abnormal local immunoglobulin secretory function in the lungs, and unrecognized infections are implicated.[4,117]

Bronchiolitis obliterans is reported to have occurred at between 2 months and 9 years after transplantation.[4,117] The clinical presentation includes dry cough and dyspnoea in most, wheezing in about 40 per cent and antecedent cold symptoms in 20 per cent.[117] Twenty per cent of the patients with BO had no respiratory symptoms at the time of the abnormal pulmonary function testing.[145] Since the presenting respiratory symptoms are non-specific, the diagnostic work-up should focus on multiple organs likely to be affected by GVHD, including the liver, sinuses and oesophagus.[117]

PFTs show irreversible airway obstruction. The chest radiograph may show hyperinflation.[4,117] HRCT of the chest may show decreased lung attenuation, bronchial dilatation, centrilobular nodules and non-homogeneous air trapping.[4,117]

BAL may show neutrophilic and/or lymphocytic inflammation.[146] Transbronchial lung biopsy is usually non-diagnostic. Video-assisted thoracoscopic lung biopsy, showing fibrinous obliteration of the small airway lumen, is required to make a definitive histological diagnosis of BO (Fig. 17.5).[4,117] However, surgical biopsies are rarely indicated since the diagnosis can usually be made clinically, by the presence of irreversible airflow obstruction and the exclusion of other causes of this functional abnormality, in allogeneic HSCT recipients with chronic GVHD.

The treatment of BO consists of corticosteroids and augmented immunosuppression, targeting chronic GVHD. However, only a minority of the patients show clinical improvement.[117] In a recent study, Khalid and colleagues reported on eight HSCT patients with BO whose pulmonary function improved after treatment with azithromycin at an

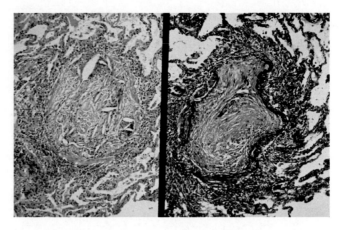

Fig. 17.5 Lung pathology in bronchiolitis obliterans showing bronchial inflammation and luminal obliteration associated with excess fibrous connective tissue. Alveoli and their ducts are spared. *Haematoxylin and eosin and Verhoeff–Van Gieson elastic tissue stain.* Reproduced with permission from [186].

initial dose of 500 mg daily for 3 days followed by 250 mg three times weekly for 12 weeks.[147] A randomized clinical trial is warranted to define the role of macrolide therapy for the treatment of BO in HSCT recipients. Although not confirmed by prospective clinical trials, inhaled budesonide/formoterol, a combination of an inhaled steroid and long-acting bronchodilatator, has been shown to improve the lung function of HSCT recipients with mild to moderately severe BO.[148] The role of blocking TNF-α awaits further trial.[149] In a recent pilot study of 19 patients with chronic GVHD, the addition of montelukast to the standard immunosuppressant regimen improved the PFT of three of the five patients with lung involvement, and no lung involvement occurred in the other 14 patients.[150] One study reported improvement of lung function in one of the three patients with chronic GVHD using extracorporeal photochemotherapy.[151] In addition to immunosuppression and anti-inflammatory therapy, prophylaxis against PCP and *Streptococcus pneumoniae* should be provided.

In selected HSCT recipients with respiratory failure secondary to BO, lung transplantation is an option.[4,117]

The airflow limitation in HSCT recipients with BO improves in only 8–20 per cent.[4,117] The reported overall case fatality rate is 59 per cent (range 14–100).[4,117] In a recent study, the 5-year survival rate of 47 HSCT recipients with BO was 10 per cent compared to 40 per cent for those without BO.[152]

Bronchiolitis obliterans organizing pneumonia

The published medical literature on BOOP in HSCT recipients is limited to case reports, with a maximum number of five patients.[4,117,153]

The occurrence of BOOP mostly in allogeneic HSCT recipients with GVHD suggests that it may represent a form of allo-immune injury by the transplanted stem cell.[154] However, it has also been reported in autologous HSCT recipients, suggesting additional mechanisms.[155] Unrecognized infections may also play a role.[156] T-cell depletion may prevent BOOP in allogeneic HSCT recipients.[144]

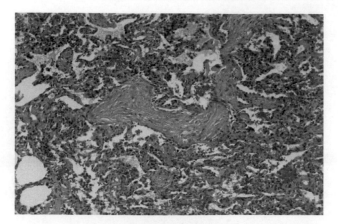

Fig. 17.6 Lung pathology in bronchiolitis obliterans organizing pneumonia showing the presence of intraluminal granulation tissue in bronchioli alveolar ducts, and alveoli. There is also interstitial infiltration with mononuclear cells and foamy macrophages. *Haematoxylin and eosin stains.* Reproduced with permission from [186].

The onset of BOOP has been reported to be 1 month to 2 years after transplant.[117] Presenting symptoms include dry cough, dyspnoea and fever. PFT shows a restrictive defect, decreased DLCO, normal airflow flow and hypoxaemia.[4,117] Chest radiographs and CT show patchy air-space consolidation, ground-glass attenuation and nodular opacities. Although usually bilateral, the radiographic abnormalities can also be unilateral.[155] Exhaled nitric oxide concentration may be increased in HSCT recipients with BOOP and may decline as a response to treatment.[157]

The definitive diagnosis of BOOP in the HSCT recipient requires transbronchial or, more commonly, surgical lung biopsy.[154] Typical findings include patchy intraluminal fibrosis, with polypoid plugs of immature fibroblasts resembling granulation tissue obliterating the distal airways, alveolar ducts, and peribronchial alveolar space (Fig. 17.6).[158]

About 80 per cent of HSCT recipients with BOOP respond favourably to treatment with corticosteroids.[4,117] The duration and dosage of corticosteroid therapy have not been clearly defined. Radiographic abnormalities usually clear within 1–3 months of initiating corticosteroid therapy. Erythromycin has been used in conjunction with corticosteroid in one allogeneic HSCT recipient with BOOP with favourable outcome.[159]

The case fatality rate of BOOP in HSCT recipients is about 19 per cent.[4,117]

Delayed pulmonary toxicity syndrome

Chemotherapeutic agents and radiation therapy have been associated with pulmonary toxicities that manifest weeks to years later.[160,161] In the 1990s, many patients with breast cancer were treated with a high-dose chemotherapy regimen consisting of cyclophosphamide, cisplatin and bischloroethylinitrosurea (BCNU) followed by autologous HSCT.[162] A significant number of these patients developed pulmonary complications, including delayed pulmonary toxicity syndrome (DPTS).[4]

DPTS develops in up to 72 per cent of autologous HSCT recipients who have received high-dose chemotherapy for breast cancer.[4] The relatively high frequency, low mortality, and good response to corticosteroid treatment distinguish DPTS from idiopathic pneumonia syndrome. Because recent studies have not shown a survival benefit, the use of autologous HSCT following high-dose chemotherapy for breast cancer has declined.

The pathogenesis of DPTS is not known. The depletion of reduced glutathione and impaired antioxidant defences caused by cyclophosphamide and BCNU have been implicated.[163]

Patients with DPTS present with cough, dyspnoea and fever.[4] The onset of symptoms ranges from 2 weeks to 4 months following transplantation. In the context of prior breast cancer treated with high-dose chemotherapy and autologous HSCT, DPTS is diagnosed by demonstration of a decline in DLCO and exclusion of infectious causes.[4] In patients with DPTS, the median absolute DLCO decrement is 26 per cent (range 10–73) and a nadir is reached in 15–18 weeks following transplant. The most common findings on CT of the chest are ground-glass opacities.[164] Because of the typical clinical presentation and response to therapy, invasive procedures such as bronchoscopy are not usually required.[4]

Corticosteroid therapy for DPTS usually results in resolution of symptoms and improvement in DLCO without long-term pulmonary sequelae.[4] One case of DPTS refractory to steroids was treated successfully with interferon-γ.[165] Prophylactic inhaled corticosteroids may reduce the frequency of DPTS.[166] No deaths attributable to DPTS have been reported.[4]

Pulmonary cytolytic thrombi

PCT is a non-infectious pulmonary complication of unknown aetiology. It occurs exclusively after allogeneic procedures, typically in the setting of GVHD. All but one of the 17 HSCT recipients with PCT reported in the medical literature are from a single institution.[167-169] Sixteen of the 17 patients were under age 18 years at the time of diagnosis. Despite the seemingly rare and previously unrecognized nature of PCT, it was found in 15 of 33 HSCT recipients (45 per cent) who underwent surgical lung biopsy for diagnosis of pulmonary nodules at the University of Minnesota.[167]

The pathogenesis of PCT is not known. Although the haemorrhagic infarcts in PCT are similar to those seen in angio-invasive fungal infections, none of the lung biopsies in the reported PCT cases had evidence of infection.[167,168] The development of PCT exclusively in allogeneic HSCT recipients, chiefly in those with GVHD, suggests that it may be a manifestation of GVHD targeting the endothelium of the lungs.[167]

Most HSCT recipients with PCT have active GVHD at the time of presentation.[168,170,171] The onset of PCT has been reported at between 8 and 343 days (median 72) after transplantation.[168,170,171] All patients are febrile and some have cough at presentation, but dyspnoea has not been noted.

Chest radiographs may be normal in 25 per cent of the patients with PCT.[171] Abnormal chest radiographic findings include nodules, interstitial prominence and atelectasis.[171] Chest CT shows multiple peripheral pulmonary nodules, ranging in size from a few millimetres to 4 cm.

Bronchoscopy with BAL is used to exclude infection. Because of the peripheral and intravascular location of the nodules in PCT, transbronchial lung biopsy is unlikely to yield a diagnosis. Histological demonstration of PCT requires surgical lung biopsy or necropsy.[168,169,171] Features of PCT include occlusive vascular lesions and haemorrhagic infarcts due to thrombi that consist of intensely basophilic, amorphous material that may extend into the adjacent tissue through the vascular wall.[167] The amorphous material suggests cellular breakdown products.[170] Immunohistochemical studies show a discontinuous endothelial cell layer. The cells that make up PCT are exclusively monocytes.[169]

In one recent case report, treatment with intravenous methylprednisolone 2 mg/kg per day and ciclosporin 3 mg/kg three times daily for one week followed by oral prednisone 30 mg per day and 125 mg ciclosporin twice daily resulted in improvement.[168] Most of the patients with PCT described in the literature improved clinically within 1–2 weeks and radiographically over weeks to months.[171] There has been no reported death attributed to PCT.[167,168] Of the 15 HSCT recipients with PCT reported from the University of Minnesota, 10 were still alive at an average of 13 months after diagnosis and five died, one from GVHD and four from infectious complications.[167]

Pulmonary veno–occlusive disease

PVOD is a rare cause of pulmonary hypertension that has been associated with various conditions including HSCT. Pulmonary hypertension in PVOD results from intimal fibrosis obstructing the pulmonary veins and venules.

The incidence of PVOD in HSCT recipients is not known. Wingard et al. reported PVOD in 19 of 154 autopsies of allogeneic HSCT recipients (12 per cent).[172] However, in an autopsy review of 71 adult HSCT recipients from our institution (39 allogeneic), we did not identify any case of PVOD.[6] There are about 28 HSCT recipients with PVOD in the published literature.[4] Most underwent HSCT for haematological malignancy, and only 2 of the 28 patients with PVOD had autologous grafts.

The aetiology of PVOD is not well defined. Infection, radiation and chemotherapy are hypothesized to be contributing factors.[4]

Presenting symptoms and signs of PVOD are non-specific.[173] Radiographic findings include Kerley B lines, prominent central pulmonary arteries, and scattered patchy opacities.[173,174] Unlike other forms of primary pulmonary hypertension, pleural effusions are common in PVOD.[173] PFT demonstrates a decreased DLCO and restrictive ventilatory defect.[4]

The triad of pulmonary arterial hypertension, radiographic evidence of pulmonary oedema, and normal pulmonary artery occlusion pressure suggests PVOD. However, many patients with PVOD do not have this triad, and it is often difficult to get an accurate pulmonary artery occlusion pressure in these patients.[173] PVOD is characterized by a high percentage of hemosiderin-laden macrophages in BAL fluid.[175] The definitive diagnosis of PVOD requires surgical lung biopsy. In the absence of this, most cases are diagnosed at post-mortem examination.[4] The pathologic hallmark of PVOD is the extensive and diffuse occlusion of pulmonary veins and venules by fibrous tissue.

There is no proven therapy for PVOD. Some advocate treating PVOD with long-term warfarin.[173] There are case reports suggesting that steroids may be beneficial in HSCT recipients with PVOD.[4] Despite case reports of successful therapy with oral vasodilators or intravenous administration of prostacyclin in PVOD, these agents can dilate the arterial vessels without concomitant venodilation, leading to increased transcapillary hydrostatic pressures, acute pulmonary oedema and death.[4] Lung transplantation is a consideration in patients free of other comorbidities. Owing to the limited treatment options and the generally progressive nature of the disease, most patients with PVOD die within two years of diagnosis.[173]

Other pulmonary complications

Several cases of pulmonary arterial hypertension have been reported in HSCT recipients.[176–178] The pathogenesis is not clearly defined, but radiation and chemotherapy-associated endothelial damage may be contributing mechanisms. Pulmonary hypertension has also been reported in an autologous HSCT recipient as part of pulmonary–renal syndrome associated with thrombotic microangiopathy.[179] Prostacyclin infusion and calcium-channel antagonists have been used for the treatment of pulmonary arterial hypertension in HSCT recipients.[177,178]

Pulmonary alveolar proteinosis is characterized by excessive accumulation of surfactant lipoprotein in the alveoli, leading to abnormal gas exchange.[180] The diagnosis of pulmonary alveolar proteinosis is made by the presence of periodic acid Schiff proteineceaous material in BAL fluid.[181] In an autopsy study of 71 HSCT recipients, pulmonary alveolar proteinosis was found in one.[6] Cordonnier et al. described three HSCT recipients with pulmonary alveolar proteinosis.[181] All three had received allogeneic HSCT for leukaemia between 12 and 90 days before the onset of pulmonary symptoms. Chest radiographs showed diffuse infiltrates. The clinicians had suspected infection in two and alveolar haemorrhaging in one. Only one of the three survived. Because of the small number of cases reported in the literature, the roles of BAL and aerosolized granulocyte–macrophage colony-stimulating factor in HSCT recipients with pulmonary alveolar proteinosis are unknown.

Three cases of chronic eosinophilic pneumonia have been reported in HSCT recipients, one autologous and two allogeneic.[182–184] Despite initial response to steroid therapy, one patient had a fatal course.[182]

KEY POINTS

- Pulmonary complications are the immediate cause of death in the majority of recipients of haematopoietic stem-cell transplantation. Both infectious and non-infectious pulmonary complications are more common in allogeneic than autologous HSCT recipients.[185,186]

- With the application of prophylactic and pre-emptive therapy, the incidences of PCP and CMV pulmonary infections have declined and invasive fungi have become the leading infectious cause of morbidity and mortality.

(Contd)

- The main non-infectious pulmonary complications include BO, IPS and DAH. BO occurs exclusively in allogeneic recipients and is characterized by the presence of airway obstruction in the absence of fever and pulmonary infiltrates. IPS mimics pneumonia with no infectious aetiology. DAH presents with diffuse pulmonary infiltrates on chest radiographs with BAL showing progressively bloodier return and/or ≥ 20% haemosiderin-laden macrophages.

- PERDS differs from IPS by its occurrence during the neutrophil peri-engraftment period and favourable response to corticosteroid therapy.

- DPTS occurs in autologous haematopoietic stem-cell recipients with breast cancer, and following high-dose pre-transplant chemotherapy. Since the use of HSCT for the treatment of metastatic breast cancer has declined, DPTS has become rare.

- Knowledge of the immune status, appropriate diagnostic evaluation and early treatment are essential to improve the prognosis associated with these complications.

REFERENCES

1. Copelan EA. Hematopoietic stem-cell transplantation. *New Engl J Med* 2006;**354**:1813–26.
2. Afessa B, Tefferi A, Dunn WF, *et al.* Intensive care unit support and Acute Physiology and Chronic Health Evaluation III performance in hematopoietic stem cell transplant recipients. *Crit Care Med* 2003;**31**:1715–21.
3. Matulis M, High KP. Immune reconstitution after hematopoietic stem-cell transplantation and its influence on respiratory infections. *Semin Respir Infect* 2002;**17**:130–9.
4. Afessa B, Peters SG. Chronic lung disease after hematopoietic stem cell transplantation. *Clin Chest Med* 2005;**26**:571–86.
5. Roychowdhury M, Pambuccian SE, Aslan DL, *et al.* Pulmonary complications after bone marrow transplantation: an autopsy study from a large transplantation center. *Arch Pathol Lab Med* 2005;**129**:366–71.
6. Sharma S, Nadrous HF, Peters SG, *et al.* Pulmonary complications in adult blood and marrow transplant recipients: autopsy findings. *Chest* 2005;**128**:1385–92.
7. Afessa B, Peters SG. Major complications following hematopoietic stem cell transplantation. *Semin Respir Crit Care Med* 2006;**27**:297–309.
8. Peters SG, Afessa B. Acute lung injury after hematopoietic stem cell transplantation. *Clin Chest Med* 2005;**26**:561–9.
9. Gosselin MV, Adams RH. Pulmonary complications in bone marrow transplantation. *J Thorac Imaging* 2002;**17**:132–44.
10. Ninin E, Milpied N, Moreau P, *et al.* Longitudinal study of bacterial, viral, and fungal infections in adult recipients of bone marrow transplants. *Clin Infect Dis* 2001;**33**:41–7.
11. George B, Mathews V, Srivastava A, *et al.* Infections among allogeneic bone marrow transplant recipients in India. *Bone Marrow Transplant* 2004;**33**:311–15.
12. Yoo JH, Lee DG, Choi SM, *et al.* Infectious complications and outcomes after allogeneic hematopoietic stem cell transplantation in Korea. *Bone Marrow Transplant* 2004;**34**:497–504.
13. Hughes WT, Armstrong D, Bodey GP, *et al.* 2002 guidelines for the use of antimicrobial agents in neutropenic patients with cancer. *Clin Infect Dis* 2002;**34**:730–51.
14. Mitchell AE, Derrington P, Turner P, *et al.* Gram-negative bacteraemia after 428 unrelated donor bone marrow transplants (UD-BMT): risk factors, prophylaxis, therapy and outcome. *Bone Marrow Transplant* 2004;**33**:303–10.
15. Escuissato DL, Gasparetto EL, Marchiori E, *et al.* Pulmonary infections after bone marrow transplantation: high-resolution CT findings in 111 patients. *Am J Roentgenol* 2005;**185**:608–15.
16. Shelhamer JH, Gill VJ, Quinn TC, *et al.* The laboratory evaluation of opportunistic pulmonary infections. *Ann Intern Med* 1996;**124**:585–99.
17. Mattes FM, Hainsworth EG, Geretti AM, *et al.* A randomized, controlled trial comparing ganciclovir to ganciclovir plus foscarnet (each at half dose) for preemptive therapy of cytomegalovirus infection in transplant recipients. *J Infect Dis* 2004;**189**:1355–61.
18. Chien SM, Chan CK, Kasupski G, *et al.* Long-term sequelae after recovery from cytomegalovirus pneumonia in allogeneic bone marrow transplant recipients. *Chest* 1992;**101**:1000–4.
19. Whimbey E, Champlin RE, Couch RB, *et al.* Community respiratory virus infections among hospitalized adult bone marrow transplant recipients. *Clin Infect Dis* 1996;**22**:778–82.
20. Couch RB, Englund JA, Whimbey E. Respiratory viral infections in immunocompetent and immunocompromised persons. *Am J Med* 1997;**102**:2–9.
21. Bowden RA. Respiratory virus infections after marrow transplant: the Fred Hutchinson Cancer Research Center experience. *Am J Med* 1997;**102**:27–30.
22. Ljungman P, Ward KN, Crooks BN, *et al.* Respiratory virus infections after stem cell transplantation: a prospective study from the Infectious Diseases Working Party of the European Group for Blood and Marrow Transplantation. *Bone Marrow Transplant* 2001;**28**:479–84.
23. Ljungman P. Respiratory virus infections in bone marrow transplant recipients: the European perspective. *Am J Med* 1997;**102**:44–7.
24. Raboni SM, Nogueira MB, Tsuchiya LR, *et al.* Respiratory tract viral infections in bone marrow transplant patients. *Transplantation* 2003;**76**:142–6.
25. Raad I, Abbas J, Whimbey E. Infection control of nosocomial respiratory viral disease in the immunocompromised host. *Am J Med* 1997;**102**:48–52.
26. Abdallah A, Rowland KE, Schepetiuk SK, *et al.* An outbreak of respiratory syncytial virus infection in a bone marrow transplant unit: effect on engraftment and outcome of pneumonia without specific antiviral treatment. *Bone Marrow Transplant* 2003;**32**:195–203.
27. Small TN, Casson A, Malak SF, *et al.* Respiratory syncytial virus infection following hematopoietic stem cell transplantation. *Bone Marrow Transplant* 2002;**29**:321–7.
28. Englund JA, Piedra PA, Whimbey E. Prevention and treatment of respiratory syncytial virus and parainfluenza viruses in immunocompromised patients. *Am J Med* 1997;**102**:61–70.
29. Whimbey E, Champlin RE, Englund JA, *et al.* Combination therapy with aerosolized ribavirin and intravenous immunoglobulin for respiratory syncytial virus disease in adult

bone marrow transplant recipients. *Bone Marrow Transplant* 1995;**16**:393-9.

30. Harrington RD, Hooton TM, Hackman RC, *et al.* An outbreak of respiratory syncytial virus in a bone marrow transplant center. *J Infect Dis* 1992;**165**:987-93.

31. Whimbey E, Elting LS, Couch RB, *et al.* Influenza A virus infections among hospitalized adult bone marrow transplant recipients. *Bone Marrow Transplant* 1994;**13**:437-40.

32. Hayden FG. Prevention and treatment of influenza in immunocompromised patients. *Am J Med* 1997;**102**:55-60.

33. Whimbey E, Englund JA, Couch RB. Community respiratory virus infections in immunocompromised patients with cancer. *Am J Med* 1997;**102**:10-18.

34. Machado CM, Boas LS, Mendes AV, *et al.* Use of oseltamivir to control influenza complications after bone marrow transplantation. *Bone Marrow Transplant* 2004;**34**:111-14.

35. Lewis VA, Champlin R, Englund J, *et al.* Respiratory disease due to parainfluenza virus in adult bone marrow transplant recipients. *Clin Infect Dis* 1996;**23**:1033-7.

36. Wendt CH, Weisdorf DJ, Jordan MC, *et al.* Parainfluenza virus respiratory infection after bone marrow transplantation. *New Engl J Med* 1992;**326**:921-6.

37. Carrigan DR. Adenovirus infections in immunocompromised patients. *Am J Med* 1997;**102**:71-4.

38. La Rosa AM, Champlin RE, Mirza N, *et al.* Adenovirus infections in adult recipients of blood and marrow transplants. *Clin Infect Dis* 2001;**32**:871-6.

39. Gasparetto EL, Escuissato DL, Inoue C, *et al.* Herpes simplex virus type 2 pneumonia after bone marrow transplantation: high-resolution CT findings in 3 patients. *J Thorac Imaging* 2005;**20**:71-3.

40. Taplitz RA, Jordan MC. Pneumonia caused by herpes viruses in recipients of hematopoietic cell transplants. *Semin Respir Infect* 2002;**17**:121-9.

41. Ramsey PG, Fife KH, Hackman RC, *et al.* Herpes simplex virus pneumonia: clinical, virologic, and pathologic features in 20 patients. *Ann Intern Med* 1982;**97**:813-20.

42. Sullivan KM, Meyers JD, Flournoy N, *et al.* Early and late interstitial pneumonia following human bone marrow transplantation. *Int J Cell Cloning* 1986;**4**(Suppl 1):107-21.

43. Curtis RE, Travis LB, Rowlings PA, *et al.* Risk of lymphoproliferative disorders after bone marrow transplantation: a multi-institutional study. *Blood* 1999;**94**:2208-16.

44. Hopwood P, Crawford DH. The role of EBV in post-transplant malignancies: a review. *J Clin Pathol* 2000;**53**:248-54.

45. van Esser JW, van der HB, Meijer E, *et al.* Epstein-Barr virus reactivation is a frequent event after allogeneic stem cell transplantation and quantitatively predicts EBV-lymphoproliferative disease following T-cell-depleted SCT. *Blood* 2001;**98**:972-8.

46. Shapiro RS, McClain K, Frizzera G, *et al.* Epstein-Barr virus associated B-cell lymphoproliferative disorders following bone marrow transplantation. *Blood* 1988;**71**:1234-43.

47. Benkerrou M, Jais JP, Leblond V, *et al.* Anti-B-cell monoclonal antibody treatment of severe posttransplant B-lymphoproliferative disorder: prognostic factors and long-term outcome. *Blood* 1998;**92**:3137-47.

48. Kuehnle I, Huls MH, Liu Z, *et al.* CD20 monoclonal antibody (rituximab) for therapy of Epstein-Barr virus lymphoma after hemopoietic stem-cell transplantation. *Blood* 2000;**95**:1502-5.

49. Buchbinder S, Elmaagacli AH, Schaefer UW, *et al.* Human herpes virus 6 is an important pathogen in infectious lung disease after allogeneic bone marrow transplantation. *Bone Marrow Transplant* 2000;**26**:639-44.

50. Tamariz-Martel R, Maldonado MS, Carrillo R, *et al.* Kaposi's sarcoma after allogeneic bone marrow transplantation in a child. *Haematologica* 2000;**85**:884-5.

51. Englund JA, Boeckh M, Kuypers J, *et al.* Brief communication: fatal human metapneumovirus infection in stem-cell transplant recipients. *Ann Intern Med* 2006;**144**:344-9.

52. Chen CS, Boeckh M, Seidel K, *et al.* Incidence, risk factors, and mortality from pneumonia developing late after hematopoietic stem cell transplantation. *Bone Marrow Transplant* 2003;**32**:515-22.

53. Marr KA, Bowden RA. Fungal infections in patients undergoing blood and marrow transplantation. *Transpl Infect Dis* 1999;**1**:237-46.

54. Lee DG, Choi JH, Choi SM, *et al.* Two cases of disseminated mucormycosis in patients following allogeneic bone marrow transplantation. *J Korean Med Sci* 2002;**17**:403-6.

55. Pavie J, Lafaurie M, Lacroix C, *et al.* Successful treatment of pulmonary mucormycosis in an allogenic bone-marrow transplant recipient with combined medical and surgical therapy. *Scand J Infect Dis* 2004;**36**:767-9.

56. Garey KW, Pendland SL, Huynh VT, *et al. Cunninghamella bertholletiae* infection in a bone marrow transplant patient: amphotericin lung penetration, MIC determinations, and review of the literature. *Pharmacotherapy* 2001;**21**:855-60.

57. Sivakumar S, Mathews MS, George B. *Cunninghamella* pneumonia in postbone marrow transplant patient: first case report from India. *Mycoses* 2005;**48**:360-2.

58. Ascioglu S, Rex JH, de Pauw B, *et al.* Defining opportunistic invasive fungal infections in immunocompromised patients with cancer and hematopoietic stem cell transplants: an international consensus. *Clin Infect Dis* 2002;**34**:7-14.

59. Grow WB, Moreb JS, Roque D, *et al.* Late onset of invasive *Aspergillus* infection in bone marrow transplant patients at a university hospital. *Bone Marrow Transplant* 2002;**29**:15-19.

60. Marr KA, Carter RA, Boeckh M, *et al.* Invasive aspergillosis in allogeneic stem cell transplant recipients: changes in epidemiology and risk factors. *Blood* 2002;**100**:4358-66.

61. Soubani AO, Qureshi MA. Invasive pulmonary aspergillosis following bone marrow transplantation: risk factors and diagnostic aspect. *Haematologia* (Budapest) 2002;**32**: 427-37.

62. Soubani AO, Chandrasekar PH. The clinical spectrum of pulmonary aspergillosis. *Chest* 2002;**121**:1988-99.

63. Pagano L, Ricci P, Nosari A, *et al.* Fatal haemoptysis in pulmonary filamentous mycosis: an underevaluated cause of death in patients with acute leukaemia in haematological complete remission. A retrospective study and review of the literature. Gimema Infection Program (Gruppo Italiano Malattie Ematologiche dell'Adulto). *Br J Haematol* 1995;**89**:500-5.

64. Takatsuka H, Wakae T, Mori A, *et al.* Superior vena cava syndrome after bone marrow transplantation caused by aspergillosis: a case report. *Hematology* 2002;**7**:169-72.

65. Nusair S, Amir G, Or R, *et al.* Invasive airway aspergillosis with new airflow obstruction mimicking post-BMT bronchiolitis obliterans. *Bone Marrow Transplant* 2002;**29**:711-13.

66. Guermazi A, Gluckman E, Tabti B, *et al*. Invasive central nervous system aspergillosis in bone marrow transplantation recipients: an overview. *Eur Radiol* 2003;**13**:377–88.

67. Jantunen E, Volin L, Salonen O, *et al*. Central nervous system aspergillosis in allogeneic stem cell transplant recipients. *Bone Marrow Transplant* 2003;**31**:191–6.

68. Saah D, Sichel JY, Schwartz A, *et al*. CT assessment of bone marrow transplant patients with rhinocerebral aspergillosis. *Am J Otolaryngol* 2002;**23**:328–31.

69. Bruno C, Minniti S, Vassanelli A, *et al*. Comparison of CT features of *Aspergillus* and bacterial pneumonia in severely neutropenic patients. *J Thorac Imaging* 2007;**22**:160–5.

70. Horger M, Einsele H, Schumacher U, *et al*. Invasive pulmonary aspergillosis: frequency and meaning of the 'hypodense sign' on unenhanced CT. *Br J Radiol* 2005;**78**:697–703.

71. Verweij PE, Meis JF. Microbiological diagnosis of invasive fungal infections in transplant recipients. *Transpl Infect Dis* 2000;**2**:80–7.

72. Bialek R, Moshous D, Casanova JL, *et al*. *Aspergillus* antigen and PCR assays in bone marrow transplanted children. *Eur J Med Res* 2002;**7**:177–80.

73. Mossad SB. Prevention and treatment of fungal infections in bone marrow transplantation. *Curr Hematol Rep* 2003;**2**:302–9.

74. Guidelines for preventing opportunistic infections among hematopoietic stem cell transplant recipients. *MMWR Recomm Rep* 2000;**49**:1–7.

75. Kruger WH, Zollner B, Kaulfers PM, *et al*. Effective protection of allogeneic stem cell recipients against aspergillosis by HEPA air filtration during a period of construction: a prospective survey. *J Hematother Stem Cell Res* 2003;**12**:301–7.

76. Oren I, Rowe JM, Sprecher H, *et al*. A prospective randomized trial of itraconazole vs. fluconazole for the prevention of fungal infections in patients with acute leukemia and hematopoietic stem cell transplant recipients. *Bone Marrow Transplant* 2006;**38**:127–34.

77. Lin SJ, Schranz J, Teutsch SM. Aspergillosis case-fatality rate: systematic review of the literature. *Clin Infect Dis* 2001;**32**:358–66.

78. Trullas JC, Cervera C, Benito N, *et al*. Invasive pulmonary aspergillosis in solid organ and bone marrow transplant recipients. *Transplant Proc* 2005;**37**:4091–3.

79. Goodrich JM, Reed EC, Mori M, *et al*. Clinical features and analysis of risk factors for invasive candidal infection after marrow transplantation. *J Infect Dis* 1991;**164**:731–40.

80. Verfaillie C, Weisdorf D, Haake R, *et al*. *Candida* infections in bone marrow transplant recipients. *Bone Marrow Transplant* 1991;**8**:177–84.

81. De Castro N, Neuville S, Sarfati C, *et al*. Occurrence of *Pneumocystis jiroveci* pneumonia after allogeneic stem cell transplantation: a 6-year retrospective study. *Bone Marrow Transplant* 2005;**36**:879–83.

82. Leung AN, Gosselin MV, Napper CH, *et al*. Pulmonary infections after bone marrow transplantation: clinical and radiographic findings. *Radiology* 1999;**210**:699–710.

83. Boutati EI, Anaissie EJ. Fusarium, a significant emerging pathogen in patients with hematologic malignancy: ten years' experience at a cancer center and implications for management. *Blood* 1997;**90**:999–1008.

84. Maertens J, Demuynck H, Verbeken EK, *et al*. Mucormycosis in allogeneic bone marrow transplant recipients: report of five cases and review of the role of iron overload in the pathogenesis. *Bone Marrow Transplant* 1999;**24**:307–12.

85. Leleu X, Sendid B, Fruit J, *et al*. Combined anti-fungal therapy and surgical resection as treatment of pulmonary zygomycosis in allogeneic bone marrow transplantation. *Bone Marrow Transplant* 1999;**24**:417–20.

86. Collin BA, Leather HL, Wingard JR, *et al*. Evolution, incidence, and susceptibility of bacterial bloodstream isolates from 519 bone marrow transplant patients. *Clin Infect Dis* 2001;**33**:947–53.

87. Lossos IS, Breuer R, Or R, *et al*. Bacterial pneumonia in recipients of bone marrow transplantation: a five-year prospective study. *Transplantation* 1995;**60**:672–8.

88. Engelhard D, Cordonnier C, Shaw PJ, *et al*. Early and late invasive pneumococcal infection following stem cell transplantation: a European Bone Marrow Transplantation survey. *Br J Haematol* 2002;**117**:444–50.

89. Kruger WH, Bohlius J, Cornely OA, *et al*. Antimicrobial prophylaxis in allogeneic bone marrow transplantation. Guidelines of the infectious diseases working party (AGIHO) of the German Society of Haematology and Oncology. *Ann Oncol* 2005;**16**:1381–90.

90. Van Burik JA, Hackman RC, Nadeem SQ, *et al*. Nocardiosis after bone marrow transplantation: a retrospective study. *Clin Infect Dis* 1997;**24**:1154–60.

91. Ip MS, Yuen KY, Woo PC, *et al*. Risk factors for pulmonary tuberculosis in bone marrow transplant recipients. *Am J Respir Crit Care Med* 1998;**158**:1173–7.

92. Roy V, Weisdorf D. Mycobacterial infections following bone marrow transplantation: a 20 year retrospective review. *Bone Marrow Transplant* 1997;**19**:467–70.

93. Mohite U, Das M, Saikia T, *et al*. Mycobacterial pulmonary infection post allogeneic bone marrow transplantation. *Leuk Lymphoma* 2001;**40**:675–8.

94. Marras TK, Szalai JP, Chan CK, *et al*. Pulmonary function abnormalities after allogeneic marrow transplantation: a systematic review and assessment of an existing predictive instrument. *Bone Marrow Transplant* 2002;**30**:599–607.

95. Marras TK, Chan CK, Lipton JH, *et al*. Long-term pulmonary function abnormalities and survival after allogeneic marrow transplantation. *Bone Marrow Transplant* 2004;**33**:509–17.

96. Chien JW, Martin PJ, Flowers ME, *et al*. Implications of early airflow decline after myeloablative allogeneic stem cell transplantation. *Bone Marrow Transplant* 2004;**33**:759–64.

97. Chien JW, Martin PJ, Gooley TA, *et al*. Airflow obstruction after myeloablative allogeneic hematopoietic stem cell transplantation. *Am J Respir Crit Care Med* 2003;**168**:208–14.

98. Afessa B. Pretransplant pulmonary evaluation of the blood and marrow transplant recipient. *Chest* 2005;**128**:8–10.

99. Traynor AE, Corbridge TC, Eagan AE, *et al*. Prevalence and reversibility of pulmonary dysfunction in refractory systemic lupus: improvement correlates with disease remission following hematopoietic stem cell transplantation. *Chest* 2005;**127**:1680–9.

100. Stiff P. Mucositis associated with stem cell transplantation: current status and innovative approaches to management. *Bone Marrow Transplant* 2001;**27**(Suppl 2):S3–11.

101. Rapoport AP, Miller Watelet LF, Linder T, *et al.* Analysis of factors that correlate with mucositis in recipients of autologous and allogeneic stem-cell transplants. *J Clin Oncol* 1999;**17**:2446–53.

102. Drew B, Peters C, Rimell F. Upper airway complications in children after bone marrow transplantation. *Laryngoscope* 2000;**110**:1446–51.

103. Murray JC, Chiu JK, Dorfman SR, *et al.* Epiglottitis following preparation for allogeneic bone marrow transplantation. *Bone Marrow Transplant* 1995;**15**:997–8.

104. Robien K, Schubert MM, Bruemmer B, *et al.* Predictors of oral mucositis in patients receiving hematopoietic cell transplants for chronic myelogenous leukemia. *J Clin Oncol* 2004;**22**:1268–75.

105. Aquino VM, Harvey AR, Garvin JH, *et al.* A double-blind randomized placebo-controlled study of oral glutamine in the prevention of mucositis in children undergoing hematopoietic stem cell transplantation: a pediatric blood and marrow transplant consortium study. *Bone Marrow Transplant* 2005;**36**:611–16.

106. Lilleby K, Garcia P, Gooley T, *et al.* A prospective, randomized study of cryotherapy during administration of high-dose melphalan to decrease the severity and duration of oral mucositis in patients with multiple myeloma undergoing autologous peripheral blood stem cell transplantation. *Bone Marrow Transplant* 2006;**37**:1031–5.

107. Papas AS, Clark RE, Martuscelli G, *et al.* A prospective, randomized trial for the prevention of mucositis in patients undergoing hematopoietic stem cell transplantation. *Bone Marrow Transplant* 2003;**31**:705–12.

108. Spielberger R, Stiff P, Bensinger W, *et al.* Palifermin for oral mucositis after intensive therapy for hematologic cancers. *New Engl J Med* 2004;**351**:2590–8.

109. Cahill RA, Spitzer TR, Mazumder A. Marrow engraftment and clinical manifestations of capillary leak syndrome. *Bone Marrow Transplant* 1996;**18**:177–84.

110. Dickout WJ, Chan CK, Hyland RH, *et al.* Prevention of acute pulmonary edema after bone marrow transplantation. *Chest* 1987;**92**:303–9.

111. Wah TM, Moss HA, Robertson RJ, *et al.* Pulmonary complications following bone marrow transplantation. *Br J Radiol* 2003;**76**:373–9.

112. Krowka MJ, Rosenow EC, Hoagland HC. Pulmonary complications of bone marrow transplantation. *Chest* 1985;**87**:237–46.

113. Kantrow SP, Hackman RC, Boeckh M, *et al.* Idiopathic pneumonia syndrome: changing spectrum of lung injury after marrow transplantation. *Transplantation* 1997;**63**:1079–86.

114. Clark JG, Hansen JA, Hertz MI, *et al.* NHLBI workshop summary. Idiopathic pneumonia syndrome after bone marrow transplantation. *Am Rev Respir Dis* 1993;**147**:1601–6.

115. Afessa B, Tefferi A, Litzow MR, *et al.* Diffuse alveolar hemorrhage in hematopoietic stem cell transplant recipients. *Am J Respir Crit Care Med* 2002;**166**:641–5.

116. Capizzi SA, Kumar S, Huneke NE, *et al.* Peri-engraftment respiratory distress syndrome during autologous hematopoietic stem cell transplantation. *Bone Marrow Transplant* 2001;**27**:1299–303.

117. Afessa B, Litzow MR, Tefferi A. Bronchiolitis obliterans and other late onset non-infectious pulmonary complications in hematopoietic stem cell transplantation. *Bone Marrow Transplant* 2001;**28**:425–34.

118. Wong R, Rondon G, Saliba RM, *et al.* Idiopathic pneumonia syndrome after high-dose chemotherapy and autologous hematopoietic stem cell transplantation for high-risk breast cancer. *Bone Marrow Transplant* 2003;**31**:1157–63.

119. Cooke KR, Yanik G. Acute lung injury after allogeneic stem cell transplantation: is the lung a target of acute graft-versus-host disease? *Bone Marrow Transplant* 2004;**34**:753–65.

120. Hauber HP, Mikkila A, Erich JM, *et al.* TNF-α, interleukin-10 and interleukin-18 expression in cells of the bronchoalveolar lavage in patients with pulmonary complications following bone marrow or peripheral stem cell transplantation: a preliminary study. *Bone Marrow Transplant* 2002;**30**:485–90.

121. Schots R, Kaufman L, Van RI, *et al.* Proinflammatory cytokines and their role in the development of major transplant-related complications in the early phase after allogeneic bone marrow transplantation. *Leukemia* 2003;**17**:1150–6.

122. Yanik G, Hellerstedt B, Custer J, *et al.* Etanercept (Enbrel) administration for idiopathic pneumonia syndrome after allogeneic hematopoietic stem cell transplantation. *Biol Blood Marrow Transplant* 2002;**8**:395–400.

123. Uberti JP, Ayash L, Ratanatharathorn V, *et al.* Pilot trial on the use of etanercept and methylprednisolone as primary treatment for acute graft-versus-host disease. *Biol Blood Marrow Transplant* 2005;**11**:680–7.

124. Griese M, Rampf U, Hofmann D, *et al.* Pulmonary complications after bone marrow transplantation in children: twenty-four years of experience in a single pediatric center. *Pediatr Pulmonol* 2000;**30**:393–401.

125. Crawford SW, Longton G, Storb R. Acute graft-versus-host disease and the risks for idiopathic pneumonia after marrow transplantation for severe aplastic anemia. *Bone Marrow Transplant* 1993;**12**:225–31.

126. Crawford SW, Hackman RC. Clinical course of idiopathic pneumonia after bone marrow transplantation. *Am Rev Respir Dis* 1993;**147**:1393–400.

127. Robbins RA, Linder J, Stahl MG, *et al.* Diffuse alveolar hemorrhage in autologous bone marrow transplant recipients. *Am J Med* 1989;**87**:511–18.

128. Sisson JH, Thompson AB, Anderson JR, *et al.* Airway inflammation predicts diffuse alveolar hemorrhage during bone marrow transplantation in patients with Hodgkin disease. *Am Rev Respir Dis* 1992;**146**:439–43.

129. Afessa B, Tefferi A, Litzow MR, *et al.* Outcome of diffuse alveolar hemorrhage in hematopoietic stem cell transplant recipients. *Am J Respir Crit Care Med* 2002;**166**:1364–8.

130. Witte RJ, Gurney JW, Robbins RA, *et al.* Diffuse pulmonary alveolar hemorrhage after bone marrow transplantation: radiographic findings in 39 patients. *Am J Roentgenol* 1991;**157**:461–4.

131. Worthy SA, Flint JD, Muller NL. Pulmonary complications after bone marrow transplantation: high-resolution CT and pathologic findings. *Radiographics* 1997;**17**:1359–71.

132. Gupta S, Jain A, Warneke CL, *et al.* Outcome of alveolar hemorrhage in hematopoietic stem cell transplant recipients. *Bone Marrow Transplant* 2007;**40**:71–8.

133. Hicks K, Peng D, Gajewski JL. Treatment of diffuse alveolar hemorrhage after allogeneic bone marrow transplant with

recombinant factor VIIa. *Bone Marrow Transplant* 2002;**30**: 975–8.

134. Pastores SM, Papadopoulos E, Voigt L, *et al.* Diffuse alveolar hemorrhage after allogeneic hematopoietic stem-cell transplantation: treatment with recombinant factor VIIa. *Chest* 2003;**124**:2400–3.

135. Mandanas RA, Saez RA, Selby GB, *et al.* Cytomegalovirus surveillance and prevention in allogeneic bone marrow transplantation: examination of a preemptive plan of ganciclovir therapy. *Am J Hematol* 1996;**51**:104–11.

136. Metcalf JP, *et al.*, University of Nebraska Medical Center Bone Marrow Transplant Group. Corticosteroids as adjunctive therapy for diffuse alveolar hemorrhage associated with bone marrow transplantation. *Am J Med* 1994;**96**:327–34.

137. Lewis ID, DeFor T, Weisdorf DJ. Increasing incidence of diffuse alveolar hemorrhage following allogeneic bone marrow transplantation: cryptic etiology and uncertain therapy. *Bone Marrow Transplant* 2000;**26**:539–43.

138. Akasheh M, Eastwood D, Vesole DH. Engraftment syndrome after autologous hematopoietic stem cell transplant supported by granulocyte-colony-stimulating factor (G-CSF) versus granulocyte-macrophage colony-stimulating factor (GM-CSF). *Bone Marrow Transplant* 2003;**31**:113–16.

139. Maiolino A, Biasoli I, Lima J, *et al.* Engraftment syndrome following autologous hematopoietic stem cell transplantation: definition of diagnostic criteria. *Bone Marrow Transplant* 2003;**31**:393–7.

140. Mossad S, Kalaycio M, Sobecks R, *et al.* Steroids prevent engraftment syndrome after autologous hematopoietic stem cell transplantation without increasing the risk of infection. *Bone Marrow Transplant* 2005;**35**:375–81.

141. Carlens S, Ringden O, Remberger M, *et al.* Risk factors for chronic graft-versus-host disease after bone marrow transplantation: a retrospective single centre analysis. *Bone Marrow Transplant* 1998;**22**:755–61.

142. Yousem SA. The histological spectrum of pulmonary graft-versus-host disease in bone marrow transplant recipients. *Hum Pathol* 1995;**26**:668–75.

143. Paz HL, Crilley P, Patchefsky A, *et al.* Bronchiolitis obliterans after autologous bone marrow transplantation. *Chest* 1992;**101**:775–8.

144. Ditschkowski M, Elmaagacli AH, Trenschel R, *et al.* T-cell depletion prevents bronchiolitis obliterans and bronchiolitis obliterans with organizing pneumonia after allogeneic hematopoietic stem cell transplantation with related donors. *Haematologica* 2007;**92**:558–61.

145. Clark JG, Crawford SW, Madtes DK, *et al.* Obstructive lung disease after allogeneic marrow transplantation: clinical presentation and course. *Ann Intern Med* 1989;**111**:368–76.

146. St John RC, Gadek JE, Tutschka PJ, *et al.* Analysis of airflow obstruction by bronchoalveolar lavage following bone marrow transplantation: implications for pathogenesis and treatment. *Chest* 1990;**98**:600–7.

147. Khalid M, Al Saghir A, Saleemi S, *et al.* Azithromycin in bronchiolitis obliterans complicating bone marrow transplantation: a preliminary study. *Eur Respir J* 2005;**25**:490–3.

148. Bergeron A, Belle A, Chevret S, *et al.* Combined inhaled steroids and bronchodilators in obstructive airway disease

after allogeneic stem cell transplantation. *Bone Marrow Transplant* 2007;**39**:547–53.

149. Fullmer JJ, Fan LL, Dishop MK, *et al.* Successful treatment of bronchiolitis obliterans in a bone marrow transplant patient with tumor necrosis factor-alpha blockade. *Pediatrics* 2005;**116**:767–70.

150. Or R, Gesundheit B, Resnick I, *et al.* Sparing effect by montelukast treatment for chronic graft versus host disease: a pilot study. *Transplantation* 2007;**83**:577–81.

151. Bisaccia E, Palangio M, Gonzalez J, *et al.* Treatment of extensive chronic graft-versus-host disease with extracorporeal photochemotherapy. *J Clin Apher* 2006;**21**:181–7.

152. Dudek AZ, Mahaseth H, DeFor TE, *et al.* Bronchiolitis obliterans in chronic graft-versus-host disease: analysis of risk factors and treatment outcomes. *Biol Blood Marrow Transplant* 2003;**9**:657–66.

153. Dodd JD, Muller NL. Bronchiolitis obliterans organizing pneumonia after bone marrow transplantation: high-resolution computed tomography findings in 4 patients. *J Comput Assist Tomogr* 2005;**29**:540–3.

154. Alasaly K, Muller N, Ostrow DN, *et al.* Cryptogenic organizing pneumonia: a report of 25 cases and a review of the literature. *Medicine* (Baltimore) 1995;**74**:201–11.

155. Hayes-Jordan A, Benaim E, Richardson S, *et al.* Open lung biopsy in pediatric bone marrow transplant patients. *J Pediatr Surg* 2002;**37**:446–52.

156. Yata K, Nakajima M, Takemoto Y, *et al.* [Pneumonitis with a bronchiolitis obliterans organizing pneumonia-like shadow in a patient with human herpes virus-6 viremia after allogeneic bone marrow transplantation]. *Kansenshogaku Zasshi* 2002;**76**:385–90.

157. Kanamori H, Fujisawa S, Tsuburai T, *et al.* Increased exhaled nitric oxide in bronchiolitis obliterans organizing pneumonia after allogeneic bone marrow transplantation. *Transplantation* 2002;**74**:1356–8.

158. Myers JL, Colby TV. Pathologic manifestations of bronchiolitis, constrictive bronchiolitis, cryptogenic organizing pneumonia, and diffuse panbronchiolitis. *Clin Chest Med* 1993;**14**:611–22.

159. Ishii T, Manabe A, Ebihara Y, *et al.* Improvement in bronchiolitis obliterans organizing pneumonia in a child after allogeneic bone marrow transplantation by a combination of oral prednisolone and low dose erythromycin. *Bone Marrow Transplant* 2000;**26**:907–10.

160. Abratt RP, Morgan GW, Silvestri G, *et al.* Pulmonary complications of radiation therapy. *Clin Chest Med* 2004; **25**:167–77.

161. Limper AH. Chemotherapy-induced lung disease. *Clin Chest Med* 2004;**25**:53–64.

162. Pedrazzoli P, Ferrante P, Kulekci A, *et al.* Autologous hematopoietic stem cell transplantation for breast cancer in Europe: critical evaluation of data from the European Group for Blood and Marrow Transplantation (EBMT) Registry 1990–99. *Bone Marrow Transplant* 2003;**32**:489–94.

163. Todd NW, Peters WP, Ost AH, *et al.* Pulmonary drug toxicity in patients with primary breast cancer treated with high-dose combination chemotherapy and autologous bone marrow transplantation. *Am Rev Respir Dis* 1993;**147**: 1264–70.

164. Wilczynski SW, Erasmus JJ, Petros WP, et al. Delayed pulmonary toxicity syndrome following high-dose chemotherapy and bone marrow transplantation for breast cancer. Am J Respir Crit Care Med 1998;**157**:565-73.

165. Suratt BT, Lynch DA, Cool CD, et al. Interferon-gamma for delayed pulmonary toxicity syndrome resistant to steroids. Bone Marrow Transplant 2003;**31**:939-41.

166. McGaughey DS, Nikcevich DA, Long GD, et al. Inhaled steroids as prophylaxis for delayed pulmonary toxicity syndrome in breast cancer patients undergoing high-dose chemotherapy and autologous stem cell transplantation. Biol Blood Marrow Transplant 2001;**7**:274-8.

167. Gulbahce HE, Parnbuccian SE, Jessurun J, et al. Pulmonary nodular lesions in bone marrow transplant recipients: impact of histologic diagnosis on patient management and prognosis. Am J Clin Pathol 2004;**121**:205-10.

168. Morales IJ, Anderson PM, Tazelaar HD, et al. Pulmonary cytolytic thrombi: unusual complication of hematopoietic stem cell transplantation. J Pediatr Hematol Oncol 2003;**25**:89-92.

169. Peters A, Manivel JC, Dolan M, et al. Pulmonary cytolytic thrombi after allogeneic hematopoietic cell transplantation: a further histologic description. Biol Blood Marrow Transplant 2005;**11**:484-5.

170. Gulbahce HE, Manivel JC, Jessurun J. Pulmonary cytolytic thrombi: a previously unrecognized complication of bone marrow transplantation. Am J Surg Pathol 2000;**24**: 1147-52.

171. Woodard JP, Gulbahce E, Shreve M, et al. Pulmonary cytolytic thrombi: a newly recognized complication of stem cell transplantation. Bone Marrow Transplant 2000;**25**: 293-300.

172. Wingard JR, Mellits ED, Jones RJ, et al. Association of hepatic veno-occlusive disease with interstitial pneumonitis in bone marrow transplant recipients. Bone Marrow Transplant 1989;**4**:685-9.

173. Mandel J, Mark EJ, Hales CA. Pulmonary veno-occlusive disease. Am J Respir Crit Care Med 2000;**162**:1964-73.

174. Williams LM, Fussell S, Veith RW, et al. Pulmonary veno-occlusive disease in an adult following bone marrow transplantation: case report and review of the literature. Chest 1996;**109**:1388-91.

175. Rabiller A, Jais X, Hamid A, et al. Occult alveolar haemorrhage in pulmonary veno-occlusive disease. Eur Respir J 2006;**27**:108-13.

176. Seguchi M, Hirabayashi N, Fujii Y, et al. Pulmonary hypertension associated with pulmonary occlusive vasculopathy after allogeneic bone marrow transplantation. Transplantation 2000;**69**:177-9.

177. Shankar S, Choi JK, Dermody TS, et al. Pulmonary hypertension complicating bone marrow transplantation for idiopathic myelofibrosis. J Pediatr Hematol Oncol 2004;**26**: 393-7.

178. Vaksmann G, Nelken B, Deshildre A, et al. Pulmonary arterial occlusive disease following chemotherapy and bone marrow transplantation for leukaemia. Eur J Pediatr 2002;**161**:247-9.

179. Perkowska-Ptasinska A, Sulikowska-Rowinska A, Pazik J, et al. Thrombotic nephropathy and pulmonary hypertension following autologous bone marrow transplantation in a patient with acute lymphoblastic leukemia: case report. Transplant Proc 2006;**38**:295-6.

180. Presneill JJ, Nakata K, Inoue Y, et al. Pulmonary alveolar proteinosis. Clin Chest Med 2004;**25**:593-613.

181. Cordonnier C, Fleury-Feith J, Escudier E, et al. Secondary alveolar proteinosis is a reversible cause of respiratory failure in leukemic patients. Am J Respir Crit Care Med 1994;**149**:788-94.

182. Gross TG, Hoge FJ, Jackson JD, et al. Fatal eosinophilic disease following autologous bone marrow transplantation. Bone Marrow Transplant 1994;**14**:333-7.

183. Brunet S, Muniz-Diaz E, Baiget M. [Chronic eosinophilic pneumonia in a patient treated with allogeneic bone marrow transplantation]. Med Clin (Barcelona) 1994;**103**:677.

184. Richard C, Calavia J, Loyola I, et al. [Chronic eosinophilic pneumonia in a patient treated with allogenic bone marrow transplantation]. Med Clin (Barcelona) 1994;**102**:462-4.

185. Afessa B, Peters SG. Major complications following hematopoietic stem cell transplantation. Semin Respir Crit Care Med 2006;**27**:297-309.

186. Afessa B, Litzow MR, Tefferi A. Bronchiolitis obliterans and other late onset non-infectious pulmonary complications in hematopoietic stem cell transplantation. Bone Marrow Transplant 2001;**28**:425-34.

18

Pulmonary complications of solid–organ transplantation

ROBERT M KOTLOFF

INTRODUCTION

The era of human solid-organ transplantation was inaugurated a half century ago when a kidney harvested from a healthy donor was implanted into the body of his identical twin brother suffering from renal failure. Over the ensuing decades, organ transplantation has evolved to include not only kidney but also heart, liver, lung and pancreas replacement. While offering extended survival and enhanced quality of life to many patients with lethal and debilitating conditions, these procedures are accompanied by a multitude of complications.

The lungs are particularly vulnerable, subject to the traumatic insults of the transplant surgery as well as to the infectious, neoplastic and toxic consequences of the immunosuppressive agents required to maintain the health of the allograft. In keeping with the general theme of this book, this chapter will focus on pulmonary complications that can be viewed as iatrogenic – that is, resulting in some way from the surgical and pharmacological manipulation of the recipient of a solid-organ transplant.

INFECTIOUS PULMONARY COMPLICATIONS

The pharmacological agents currently employed to prevent rejection following organ transplantation induce a state of profound global immunosuppression. This predisposes the transplant recipient to a multitude of infectious complications. Pulmonary infections in particular are common, represent the leading site of infection in lung and heart transplant recipients and the second most common site (following intra-abdominal infection) in liver transplant recipients. The incidence of pulmonary infection is lowest in kidney transplant recipients, reflecting the less rigorous surgical procedure required to implant the allograft and the decreased level of immunosuppression employed.

The spectrum of microorganisms responsible for post-transplantation pulmonary infections is similar among the various solid-organ transplantation populations and is discussed in detail in the sections that follow. The sequence in which these different organisms appear in the post-transplantation course is fairly characteristic.[1] The first month is influenced predominantly by infectious risks posed by surgery and intensive care, and to a lesser extent by the initiation of immunosuppressive agents. Nosocomial bacterial infections predominate, similarly to the general surgical population. The second stage extends from 1 to 6 months, a period of maximum sustained immunosuppression that is characterized by the emergence of opportunistic pathogens. Beyond 6 months, allograft functioning in the majority of patients is sufficiently stable to permit reduction in the level of immunosuppression. Consequently, infections are largely due to common community-acquired pathogens. Opportunistic infections occur less frequently in this later period but remain especially prevalent among the subset of patients requiring augmentation of immunosuppression for treatment of chronic rejection or recurrent episodes of acute rejection.

Bacterial pneumonia

Bacterial pneumonia may be either nosocomial or community-acquired. The time of onset, responsible pathogens and prognosis are distinct for these two modes of infection. Nosocomial pneumonia is almost exclusively a perioperative complication. Gram-negative pathogens predominate, but *Staphylococcus aureus* and, in some centres, *Legionella* species are also encountered. An increase in the prevalence of methicillin-resistant *Staphylococcus aureus* infections (MRSI) has been documented and must be considered when initiating empirical antibiotic therapy. The need for prolonged postoperative mechanical ventilation is a major risk factor for nosocomial pneumonia following transplantation. Impairment in cough that accompanies extensive surgical manipulation of the thorax or upper abdomen also contributes to the risk. Among lung transplant recipients, additional factors can potentially compromise local pulmonary defences: narrowing of the bronchial anastomosis, diminished cough reflex due to lung denervation, disruption of pulmonary lymphatics, and impairment in the mucociliary 'escalator' resulting from ischaemic injury to the bronchial mucosa. Passive transfer of occult pneumonia initially acquired by the donor is another circumstance unique to lung transplantation, though the presence of organisms on Gram stain of donor bronchial washings is not predictive of subsequent pneumonia in the recipient.[2] Although the incidence of nosocomial pneumonia has declined to less than 10 per cent in liver and heart transplant recipients and to approximately 15 per cent in lung transplant recipients, mortality remains high.

Community-acquired bacterial pneumonia occurs later in the post-transplantation period. *Hemophilus influenzae*, *Streptococcus pneumoniae* and *Legionella* species are among the commonly identified organisms. Response to therapy is generally excellent, with reported mortality rates of 0–33 per cent.[3,4] Lower respiratory tract infections occurring in the later phases after transplantation are particularly prevalent among lung transplant recipients who have developed bronchiolitis obliterans syndrome (BOS). These patients typically present with recurrent episodes of purulent bronchitis and pneumonia, most commonly due to *Pseudomonas aeruginosa*. Bronchiectasis can be demonstrated on high-resolution computed tomography (HRCT) in up to one-third of these patients.

Once relatively common among solid-organ transplant recipients, the prevalence of *Nocardia* infections has more recently declined to 0.2–2.1 per cent.[3,5] This trend reflects the introduction of ciclosporin-based immunosuppressive regimens that permitted use of reduced doses of corticosteroids and the widespread use of sulfonamides for *Pneumocystis jiroveci* pneumonia (PCP) prophylaxis. Clinicians must remain particularly vigilant for this infection in patients in whom trimethoprim-sulfamethoxazole has either not been administered due to allergy or has been discontinued after the first year. Infection due to this aerobic, Gram-positive filamentous rod is most common beyond the first month after transplantation.

Patients may be asymptomatic or may present subacutely with fever, non-productive cough, pleuritic chest pain, dyspnoea, haemoptysis and weight loss. Dissemination to brain, skin and soft tissue occurs in up to one-third of infected patients. Chest radiographs and CT scans typically demonstrate one or several nodules that may be cavitary.

Sulfonamides are the treatment of choice; minocycline, amikacin, imipenem and ceftriaxone are alternative agents for sulfa-allergic patients. Treatment for at least 3 months is recommended for pulmonary infection, and for up to 12 months for disseminated disease. Mortality directly attributable to *Nocardia* infection ranges from 0 to 30 per cent among the various solid-organ transplant populations.[5,6]

Tuberculosis

Tuberculosis has been reported in approximately 0.5–2 per cent of organ transplant recipients in the United States and Europe.[7] Although a relatively uncommon post-transplantation infection in developed countries, the annualized rate of infection is 30–100-fold higher than that of the general population. In areas where tuberculosis remains endemic, such as India, rates of infection in transplant recipients approach 15 per cent.[8]

Reactivation of latent infection is postulated to be the predominant mechanism for development of active tuberculosis following transplantation, but this assumption is difficult to verify based on available data. In published series, information on pre-transplantation PPD status is available for less than half of organ transplant patients who developed active tuberculosis following transplantation. Among those who were tested, less than one-quarter demonstrated a positive response, though one must assume that there is a high false-negative skin-test rate due to multiple factors present in patients with chronic organ dysfunction.[7] Additionally, only 5 per cent of transplant recipients with tuberculosis have a pre-transplantation history of infection and only 12 per cent demonstrate chest radiographic evidence of prior infection.[7] Less common modes of acquisition include nosocomial outbreaks and donor transmission through infected kidney, lung and liver allografts.

Onset of infection occurs a median of 9 months following transplantation and nearly two-thirds of cases develop within the first year.[7] Approximately half of patients present with tuberculosis restricted to the lungs, 15 per cent have focal extrapulmonary infection, and the remaining one-third present with disseminated disease. Counting both patients with pulmonary tuberculosis and those with lung involvement as part of disseminated disease, the overall rate of lung involvement is about 70 per cent.

Fever is the most common presenting symptom, occurring in over 90 per cent of patients with disseminated disease but in only 66 per cent of those with pulmonary tuberculosis. Chest radiographic abnormalities include focal infiltrates in 40 per cent, a miliary pattern in 22 per cent, pleural effusions in 13 per cent, diffuse interstitial infiltrates in 5 per cent, and cavitary lung disease in only 4 per cent.

Treatment of active tuberculosis in organ transplant recipients involves the use of combination therapy as per standard guidelines for the general population. Administration of this regimen to transplant recipients can be challenging, however. The risk of isoniazid-induced hepatotoxicity is markedly enhanced in liver transplant recipients, necessitating discontinuation of this agent in 41–83 per cent of patients.[7] Other recipient populations tolerate this agent better, with reported discontinuation rates of less than 5 per cent.[7] Administration of rifampin, a potent inducer of the hepatic P450 microsomal enzyme system, dramatically increases the clearance of ciclosporin and tacrolimus, consequently lowering blood levels of these drugs and enhancing the risk of rejection. Concurrent use of rifampin mandates frequent monitoring of immunosuppressive drug levels and appropriate adjustment in dosing to maintain therapeutic levels. Because of difficulties in maintaining therapeutic drug levels, some clinicians have advocated avoidance of rifampin in favour of alternative agents.

Even in contemporary series, mortality among transplant recipients with tuberculosis remains in the range 25–40 per cent.[7] This high rate reflects not only the direct consequences of the infection but also the impact of enhanced rejection and graft loss among treated but suboptimally immunosuppressed patients and the adverse impact of comorbid conditions that may have contributed to development of tuberculosis. For patients who complete a full course of treatment, response is highly favourable and mortality rates are low.

Given the significant morbidity and mortality associated with active tuberculosis in the transplant population, routine skin-testing, novel interferon γ release assay and pre-emptive treatment of latent infection are recommended. Ideally, skin-testing should be performed prior to transplantation, when there is a greater likelihood that the latently infected patient will mount a delayed hypersensitivity response to antigenic challenge.

When therapy is initiated at the time of listing, most or all of the recommended 9-month course of isoniazid can often be completed prior to transplantation. In this way, interactions with immunosuppressive drugs that could lead to an enhanced risk of hepatotoxicity can be avoided. The approach to liver transplant candidates and recipients with latent infection is more problematic. There is an understandable reluctance to use isoniazid prior to transplantation because of the presence of severe liver disease, though the drug can be administered safety under close observation.[9] The use of isoniazid after liver transplantation is associated with an increased risk of hepatotoxicity, but the risk may be lower when the drug is administered as a single agent rather than as part of a multi-drug antituberculous regimen. If treatment of latent infection with isoniazid is attempted following liver transplantation, liver biopsy should be considered in the setting of elevated liver enzymes, since aetiologies other than drug toxicity (e.g. allograft rejection) may be responsible in up to half of these instances.[7] A 4-month course of rifampin is an acceptable alternative to isoniazid and is ideally initiated prior to transplantation in order to avoid interactions with the calcineurin inhibitors.

The other approved regimen for treatment of latent infection – a 2-month course of rifampin and pyrazinamide – has been associated with frequent and severe hepatotoxicity. For this reason, this regimen is absolutely contraindicated in patients with chronic liver disease awaiting transplantation and must be used with extreme caution in liver transplant recipients.

Non-tuberculous mycobacterial infection

Among lung transplant recipients, the non-tuberculous mycobacteria may be more common than *Mycobacterium tuberculosis* as a cause of pulmonary infections. In the largest published series encompassing 261 patients, pulmonary infection due to non-tuberculous mycobacteria was documented in 16 patients (6.1 per cent) compared to only two cases of pulmonary tuberculosis (0.8 per cent).[10] Thirteen of the 16 cases were due to *Mycobacterium avium complex*; *M. kansasii*, *M. abscessus* and *M. asiaticum* each accounted for one case. Pulmonary infection tended to occur late in the post-transplantation course and was associated with pre-existent chronic rejection in over half of the cases. Treatment resulted in clinical improvement in approximately half of treated patients and there were no deaths directly attributable to these infections. Pulmonary infection due to non-tuberculous mycobacterial species is considerably less common in other solid-organ transplant populations.

Cytomegalovirus

Cytomegalovirus (CMV) is the most common viral pathogen encountered in all solid-organ recipient populations. Infection can occur through passive transfer of virus with the allograft or by reactivation of latent virus remotely acquired by the recipient. Seronegative recipients who acquire organs from seropositive donors are at greatest risk for developing infection, and these primary infections tend to be the most severe. The use of antilymphocyte antibody therapy for immunosuppression also enhances the likelihood and severity of infection in susceptible recipients.

CMV infection typically emerges 1–3 months following transplantation, though onset may be delayed in patients receiving prophylaxis. Infection is often subclinical, marked only by asymptomatic viraemia or shedding of virus in the respiratory tract or urine. Clinically apparent disease can assume a number of forms including a mononucleosis-like 'CMV syndrome' with fever, malaise and leucopenia, and organ-specific involvement of the lungs, gastrointestinal tract, liver, myocardium and central nervous system. In addition to the direct effect of tissue invasion in producing symptoms and dysfunction of involved organs, CMV infection leads to an enhanced state of host immunosuppression, accounting for the frequent emergence of other opportunistic infections in its wake. CMV infection has also been linked to the subsequent development of chronic allograft dysfunction and graft loss.

Contemporary series employing a variety of antiviral prophylactic strategies document an incidence of CMV pneumonitis of 0–9.2 per cent among liver transplant recipients,[11] 0.8–6.6 per cent among heart transplant recipients,[12] and less than 1 per cent among renal transplant recipients.[13] In contrast, an incidence of 15–55 per cent has been reported following lung transplantation.[14] The higher frequency of CMV pneumonitis in the lung transplant population is consistent with the notion that the lung is a major site of CMV latency and, therefore, that large quantities of CMV can be transmitted in the allograft. The high incidence likely also reflects the widespread performance of surveillance bronchoscopy in this population, facilitating detection of subclinical disease. Indeed, surveillance bronchoscopy has demonstrated that approximately 10–15 per cent of cases of CMV pneumonitis are asymptomatic.[14]

For the majority of transplant recipients, a prodrome of fever, malaise and myalgias frequently precedes the onset of pneumonitis, which is heralded by non-productive cough and dyspnoea. Associated laboratory findings of leucopenia, thrombocytopenia and elevated liver transaminases provide important clues to the presence of CMV. Radiographically, CMV pneumonitis is associated with a number of non-specific findings including ground-glass opacities, air-space consolidation and nodules.

A diagnosis of CMV pneumonitis is unequivocally established only by demonstration of characteristic viral inclusions on histological or cytological specimens. Unfortunately, the yield of transbronchial lung biopsy and bronchoalveolar lavage (BAL) in demonstrating these changes is relatively low, and surgical lung biopsy, while definitive, is invasive. This has prompted the development of alternative diagnostic techniques but the clinical limitations of these newer tools must be appreciated. Use of the rapid shell vial culture of BAL fluid is extremely efficient in detecting virus in the lung. However, since shedding of virus into the respiratory tract can occur in the absence of tissue invasion, a positive culture in the appropriate clinical setting at best provides only a presumptive diagnosis. Precise quantification of viral load in peripheral blood can now be achieved by means of the pp65 antigenaemia assay and by polymerase chain reaction (PCR) techniques. Preliminary data suggest that the use of these assays as a surrogate marker of CMV pneumonitis has a low positive predictive value but high negative predictive value. There is a larger body of literature on the use of peripheral blood viral load quantification not as a diagnostic tool *per se* but as a means of identifying patients most likely to progress from subclinical infection to clinical disease.[11,12] These studies suggest that a high or rapidly increasing peripheral blood viral load is a sensitive but non-specific marker of imminent progression to symptomatic disease, with a positive predictive value of only 49–64 per cent and a negative predictive value in excess of 95 per cent. In the final analysis, viral load quantification is most appropriately employed in targeting patients for pre-emptive prophylaxis (see below), but its role in diagnosing established or incipient CMV disease (including pneumonitis) is limited.

The efficacy of ganciclovir in the treatment of CMV disease in various solid-organ transplant populations has been suggested in uncontrolled clinical series and has become the standard of care. In renal transplant recipients, for example, ganciclovir reduced the overall mortality associated with CMV pneumonitis from 50 to 20 per cent, and mortality in the subset requiring mechanical ventilation from over 90 to 60 per cent.[15] Standard treatment consists of a 2- to 3-week course of intravenous ganciclovir at a dose of 5 mg/kg twice daily, adjusted for renal insufficiency. Some experts advocate the addition of CMV hyperimmune globulin in treatment of severe disease, but evidence supporting this practice is scant. Although treatment is effective, relapse rates of up to 60 per cent in primary infection and 20 per cent in previously exposed recipients have been reported.[16] Administration of a 3-month course of oral ganciclovir following definitive intravenous therapy modestly reduces the risk of relapse.[17]

In an attempt to minimize the adverse impact of CMV infection on the post-transplantation course, emphasis has shifted to preventive strategies. Numerous prospective, randomized trials have documented the efficacy of antiviral prophylaxis in diminishing the risk of CMV infection and invasive disease.[18,19] Ganciclovir, administered orally or intravenously, has been the drug most commonly employed for the various solid-organ recipient populations; valacyclovir is an alternative agent whose benefits have been demonstrated principally following kidney transplantation.[20] In current practice, oral valganciclovir is emerging as an attractive option owing to its excellent bioavailability after oral administration and once-daily prophylactic dosing requirement. A recent prospective, randomized trial suggests that this agent is at least as effective as oral ganciclovir in the prevention of CMV disease in solid-organ transplant recipients.[21]

Universal prophylaxis of all donor seropositive/recipient seronegative patients is recommended as the risk of CMV disease is high. Since the risk of disease is significantly lower in seropositive recipients (independent of donor status), it has been argued that universal prophylaxis of this group leads to over-treatment, increasing costs and unduly exposing patients to the risk of drug toxicity. In this population, pre-emptive strategies targeting antiviral therapy exclusively to patients demonstrating a rising viral load in peripheral blood are being developed.

Emergence of ganciclovir-resistant strains of CMV has been documented in 15.2 per cent of lung, 5.3 per cent of heart, 5.6 per cent of liver, and 2.2 per cent of kidney recipients.[22] Across all populations, donor positive/recipient negative patients are at greatest risk of developing resistance, likely due to the high viral load associated with primary infection.[23,24] Other risk factors that have been identified include the use of potent immunosuppressive agents such as antilymphocyte antibodies and daclizumab, and prolonged exposure to ganciclovir.[23,24] There is particular concern that the low serum levels achieved with use of oral ganciclovir may predispose to development of resistance. Foscarnet is the agent of choice for treatment of ganciclovir-resistant disease but its use may be limited by associated nephrotoxicity. The presence of ganciclovir-resistant disease is associated with decreased survival in lung transplant recipients.[24]

Community respiratory viruses

Infections due to the community respiratory viruses influenza, parainfluenza, adenovirus and respiratory syncytial virus (RSV) are common in the general population, typically presenting as mild, self-limited upper respiratory tract illnesses. It is unclear whether the risk of acquiring these viral infections is increased among solid-organ transplant recipients, but there is a greater propensity for these pathogens to involve the lower respiratory tract and, therefore, to result in a more severe spectrum of illness. Of the various organ transplant populations, the highest rates of infection have been reported in lung transplant recipients, up to 21 per cent of whom develop respiratory viral infections.[25,26] Lung transplant recipients appear to be particularly predisposed to a newly recognized community viral pathogen, human metapneumovirus. In one recent series of lung transplant recipients presenting with signs and symptoms of a viral respiratory tract infection, human metapneumovirus was the most common agent identified, accounting for one-third of all cases in which a viral aetiology could be defined.[27]

RSV, human metapneumovirus and influenza virus infections occur seasonally, with epidemics in the winter and spring months, while adenovirus and parainfluenza infections are seen throughout the year. Patients with lower respiratory tract involvement, in the form of bronchiolitis or pneumonitis, typically present with fever, dyspnoea, cough and wheezing. Chest radiographs may be normal or may show only subtle interstitial changes. CT of the chest is more sensitive and findings of ground-glass, air-space consolidation, nodules, and tree-in-bud opacities are seen. The diagnosis rests on demonstration of virus in respiratory secretions obtained by nasopharyngeal swabbing, nasal wash or bronchoalveolar lavage. Viral culture represents the gold standard for diagnosis but typically entails a period of 3–14 days before results are available. Rapid diagnostic tests utilizing enzyme-linked immunosorbent assays or immunofluorescence techniques to identify viral antigens are now available but results should be corroborated by standard viral culture.

Mortality rates in the range 0–20 per cent have been reported in association with respiratory viral infections in the various solid-organ transplant populations. In addition to their immediate consequences, there is a suggestion that respiratory viral infections may increase the risk of chronic rejection in lung transplant recipients, possibly through stimulation of alloimmune mechanisms targeting the bronchial epithelium.[28]

Treatment options are limited and largely supportive. The efficacy of antiviral therapy has been convincingly demonstrated only in the treatment of influenza in immunocompetent hosts. In this context, initiation of amantadine or rimantidine within 48 hours of symptom onset shortens the severity and duration of illness due to influenza A. Similar results are achieved with the early administration of the neuraminidase inhibitors zanamivir and osteltamivir, which have the added advantage of efficacy against both influenza A and B. Studies confirming benefit of these agents in organ transplant recipients are notably lacking. There is no established treatment for infections due to the other respiratory viruses, although aerosolized or intravenous ribavirin has been advocated as treatment for RSV, human metapneumovirus and parainfluenza virus infections based on anecdotal reports of success.

The limited treatment options have led to an increased focus on prevention. Respiratory viruses are highly contagious and infection control measures that emphasize minimizing contact with infected individuals and frequent hand washing may reduce risk. The inactivated influenza vaccine should be administered to all transplant recipients and close contacts. Although the antibody response appears to be attenuated in the solid-organ transplant recipient, many are still able to mount protective antibody responses. Case reports suggesting that vaccination may augment the allo-immune response and increase the risk of allograft rejection have not been confirmed in recent studies and should not prevent routine vaccination. Chemoprophylaxis of transplant recipients with amantadine, rimantidine, or one of the neuraminidase inhibitors should be considered in the setting of a major influenza outbreak or following exposure to a sick contact with presumed influenza.

Aspergillosis

Although a number of opportunistic and endemic fungi have been reported to cause pulmonary infections in organ transplant recipients, *Aspergillus* species are by far the most frequent and lethal fungal pathogens encountered. The incidence of invasive aspergillosis is about 5 per cent among the liver, heart and lung transplant populations but occurs considerably less frequently following kidney transplantation. Invasive disease is most commonly diagnosed within the first 6 months and nearly always involves the lung. Dissemination to distant sites, particularly the brain, occurs in a considerable minority of patients. Symptoms of invasive aspergillosis are non-specific and include fever, cough, pleuritic chest pain and haemoptysis. Radiographically, pulmonary aspergillosis may appear as single or multiple nodular opacities, cavities, or alveolar consolidation (Fig. 18.1). The 'halo sign', considered a highly characteristic radiographic feature of invasive aspergillosis in the haematopoietic stem-cell transplant population, is infrequently encountered and considerably less specific in the solid-organ transplant populations.

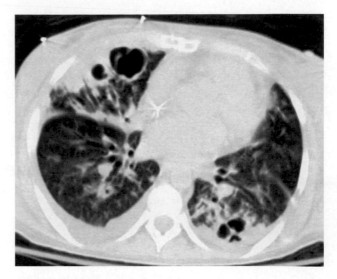

Fig. 18.1 Pulmonary aspergillosis. Computed tomography scan demonstrating areas of consolidation and cavitation in a transplant recipient with invasive aspergillosis.

Diagnosis of invasive aspergillosis can be problematic. *Aspergillus* is cultured from sputum in only 8–34 per cent and from BAL fluid in 45–62 per cent of patients with invasive disease.[29] Conversely, 28–55 per cent of organ transplant recipients demonstrate airway colonization in the absence of invasive disease, with the highest rates of airway colonization seen following lung transplantation.[30,31] In a patient with compatible clinical and radiographic features and/or demonstration of *Aspergillus* in respiratory secretions by culture or cytology, the clinician must exercise judgement in deciding whether to initiate an empirical trial of antifungal therapy or pursue more definitive proof by means of transthoracic needle biopsy or surgical lung biopsy.

Amphotericin B has traditionally been the mainstay of therapy for invasive aspergillosis but its many side-effects and need for intravenous administration have made it a rather unappealing choice. More recently, liposomal amphotericin preparations have been introduced that are less nephrotoxic than the parent compound, an important feature for a drug used concurrently with calcineurin inhibitors (ciclosporin, tacrolimus). The triazoles itraconazole and voriconazole offer the advantages of availability in both oral and intravenous formulations and absence of nephrotoxicity. Voriconazole absorption from the gastrointestinal tract is much more reliable than that of itraconazole, making it the preferred oral agent. Voriconazole recently was shown to have superior efficacy and less toxicity than amphotericin B in the treatment of invasive asp(e)rillosis, and it is emerging as first-line therapy in many centres.[32] Both of the triazoles are potent inhibitors of the P450 hepatic enzyme system and can lead to dangerously high blood levels of ciclosporin and tacrolimus if appropriate adjustments in the dosing of these agents are not made. Because of particularly profound interactions, the concurrent use of voriconazole and sirolimus is contraindicated. Caspofungin, the first of a new class of echinocandins, has been approved in the United States for treatment of invasive aspergillosis in patients who fail or are intolerant of first-line therapy. Despite the availability of antifungal therapy, mortality is in the range of

30–90 per cent, with the highest mortality rates associated with disseminated disease. The therapeutic role of surgical resection remains uncertain but surgery has been advocated in cases of localized infection refractory to medical therapy.[33]

Endobronchial aspergillosis is uniquely encountered in the lung transplant population, with an observed frequency of about 5 per cent.[31] In most cases, infection is localized to the bronchial anastomosis, where devitalized cartilage and foreign suture material create a nurturing environment. Less commonly, infection may present as a more diffuse ulcerative bronchitis with formation of pseudomembranes, typically following in the wake of a severe ischaemic injury to the bronchial mucosa. Clustered within the first 6 months after transplantation, these airway infections are usually asymptomatic and detected only by surveillance bronchoscopy. Although usually responsive to oral azoles or to inhaled or intravenous amphotericin, airway infections have rarely progressed to invasive pneumonia or have resulted in fatal erosion into the adjacent pulmonary artery.[31] An increased risk of subsequent bronchial stenosis or bronchomalacia has also been reported but it is unclear whether this is a consequence of the infection or of an underlying ischaemic injury to the bronchus that predisposed to infection.[34]

Pneumocystis jiroveci pneumonia

Prior to the widespread introduction of chemoprophylaxis, PCP was a common opportunistic infection among solid-organ transplant recipients. As documented in older series, organ-specific prevalence rates for at-risk patients (i.e. not receiving prophylaxis) were 4 per cent for kidney and heart transplant recipients, 11 per cent for liver transplant recipients, and up to 33 per cent for heart–lung recipients.[35] With the administration of low-dose trimethoprim–sulfamethoxazole or an alternative prophylactic agent, PCP can be effectively prevented. In a large contemporary series from the Cleveland Clinic, PCP was documented in only 25 of 1299 solid-organ transplant recipients and it occurred exclusively in patients who were not receiving prophylaxis.[35] The greatest risk of PCP falls between the second and sixth post-transplantation months. The risk declines significantly beyond the first year for all groups except lung transplant recipients,[35] prompting many non-lung programmes to discontinue prophylaxis beyond this point. Likely because of the need for augmented immunosuppression, patients with refractory acute allograft rejection or with chronic allograft rejection appear to be at increased risk for the late development of PCP and continuation or resumption of PCP prophylaxis may be warranted under these circumstances.[35] Indefinite prophylaxis is advocated for lung transplant recipients.

PCP in solid-organ transplant recipients typically presents in a subacute fashion, with a mean duration of symptoms prior to diagnosis of 14 days.[35] Dyspnoea, fever and cough are the most common presenting symptoms. Radiographic abnormalities are typically bilateral and may appear as interstitial, alveolar or ground-glass opacities, the latter best demonstrated on CT scan. The diagnosis can be established by bronchoalveolar lavage alone in approximately 90 per cent of cases; performance of transbronchial biopsies modestly enhances

the yield.[35] High-dose trimethoprim–sulfamethoxazole represents first-line therapy, with intravenous pentamidine typically reserved for patients who fail or cannot tolerate trimethaprim–sulfamethoxazole. Mortality directly attributable to PCP is about 20 per cent in contemporary series.[35]

Miscellaneous fungal infections

While *Aspergillus* species account for the majority of invasive fungal infections of the lung in solid-organ transplant recipients, myriad other fungal organisms can cause pulmonary disease. These include *Cryptococcus neoformans*, the zygomycetes (especially *Muror* species), and the geographically restricted endemic fungi (*Histoplasma capsulatum, Coccidioides immitus* and *Blastomyces dermatitidis*). In general, the presenting features of these fungal infections are neither clinically nor radiographically distinct and all have the potential to cause disseminated disease in addition to pneumonia. *Candida* species are responsible for a number of serious post-transplant infections, including sepsis, intra-abdominal abscesses and urinary tract infections, but involvement of the lungs is conspicuously rare. The one notable exception to this is infection of the bronchial anastomosis that occasionally occurs following lung transplantation. An emerging pulmonary pathogen is *Scedosporium apiospermum* (previously known as *Pseudallescheria boydii*). This organism has been documented to cause infection in all of the major solid-organ transplant populations.[36] Invasive pulmonary disease is a feature in approximately 50 per cent of cases; central nervous system and endovascular involvement as well as widespread dissemination are also common.[36] In distinction from the other fungal pathogens discussed above, *Scedosporium* is frequently resistant to amphotericin B; voriconazole is the preferred agent.

NON-INFECTIOUS PULMONARY COMPLICATIONS

Liver transplantation

RESPIRATORY FAILURE

Ventilatory support beyond 24 hours is required by approximately 10 per cent of recipients immediately following liver transplantation.[37] For those who are successfully extubated, reintubation at some point in the perioperative period occurs in one-third of patients who had initially required greater than 24 hours of ventilatory support and 10 per cent of those initially ventilated for less than 24 hours.[37] The need for reintubation is associated with significantly poorer survival.

A number of factors contribute to respiratory failure, including impaired respiratory muscle function from extensive upper abdominal surgery, malnutrition, and debility; marked intravascular volume shifts and volume overload; aggressive blood product support; and a relatively high frequency of postoperative pneumonia. Additionally, patients undergoing liver transplantation are often critically ill at the time of transplantation and it is not unusual for patients to have been supported on mechanical ventilation for varying periods prior to transplantation.

Acute respiratory distress syndrome (ARDS) is a particularly lethal cause of postoperative respiratory failure following liver transplantation. The reported incidence ranges from 4 to 16 per cent, with a mortality rate as high as 80–100 per cent.[38] Sepsis is the most common risk factor reported but other potential risk factors include massive blood transfusions, transfusion-related acute lung injury, aspiration, and the use of OKT3 antilymphocyte therapy. Recipients with fulminant hepatic failure prior to transplantation are predisposed to non-cardiogenic pulmonary oedema as a component of their liver failure.

PLEURAL EFFUSIONS

Perioperative pleural effusions are present in 40–100 per cent of liver transplant recipients.[39,40] Effusions are transudative and are typically right-sided or bilateral but are rarely exclusively on the left. Disruption of diaphragmatic lymphatics during hepatectomy is postulated to be the principal mechanism of fluid accumulation.[39] Other contributing mechanisms include volume overload, hypoalbuminaemia and atelectasis. Effusions may enlarge over the first postoperative week but typically resolve by the third week. The need for drainage because of perceived respiratory compromise has been reported in up to 31 per cent of patients.[38–40] Effusions that continue to enlarge beyond the first week, persist beyond 3 weeks, or involve only the left hemithorax should be sampled to rule out other causes. Persistent or enlarging effusions should also prompt consideration of subdiaphragmatic processes including haematoma, biloma or subphrenic abscess.

DIAPHRAGMATIC DYSFUNCTION

Right-sided diaphragmatic dysfunction is a common complication of liver transplantation. It is postulated to result from crush injury to the right phrenic nerve by the suprahepatic vena caval clamp placed during surgery. In one series of 48 patients, evidence of delayed or absent right-sided phrenic nerve conduction was found in 79 per cent of patients while the left phrenic conducted normally in all cases.[41] In 38 per cent of patients, there was associated right diaphragmatic paralysis documented by ultrasound. Phrenic nerve injury was not associated with increased duration of mechanical ventilatory support or hospital stay. In a subset of patients followed with serial testing, abnormalities in phrenic nerve conduction and diaphragmatic excursion normalized by 9 months after surgery.

Lung transplantation

PRIMARY GRAFT DYSFUNCTION

Mild, transient pulmonary oedema is a nearly universal feature of the freshly transplanted lung allograft. It is presumed to be a consequence of ischaemia–reperfusion injury and the attendant increase in microvascular permeability, but surgical trauma and lymphatic disruption may be contributing factors. In approximately 12–22 per cent of cases, injury is sufficiently severe to cause a form of acute respiratory distress syndrome termed 'primary graft dysfunction'.[42]

Risk factors for development of primary graft dysfunction include donor female gender, donor African American race, donor age (< 21 or > 45 years) and a recipient diagnosis of primary pulmonary hypertension.[42] An association between prolonged graft ischaemic time and primary graft dysfunction has been observed in some studies,[43] but not in others.[44,45] The diagnosis of primary graft dysfunction rests on the development of widespread radiographic infiltrates and markedly impaired oxygenation within the initial 72 hours following transplantation, and the exclusion of other causes of early graft dysfunction such as volume overload, pneumonia, rejection, atelectasis and pulmonary venous outflow obstruction. Histology of lung tissue from patients with primary graft dysfunction reveals a prevailing pattern of diffuse alveolar damage.[44]

Treatment is supportive, relying on conventional mechanical ventilation utilizing low tidal volume strategies, as well as on such adjunct measures as independent lung ventilation and extracorporeal life support for select patients who otherwise cannot be stabilized. The use of nitric oxide in patients with established graft injury has been associated with sustained reduction in pulmonary artery pressures and improvement in oxygenation. However, the prophylactic administration of nitric oxide to all recipients at the time of reperfusion does not reduce the incidence of primary graft dysfunction.[46]

With an associated in-hospital mortality rate of approximately 40 per cent, severe primary graft dysfunction is a leading cause of perioperative deaths among transplant recipients.[47] The risk of death remains excessive even beyond the first year, suggesting that primary graft dysfunction has lingering adverse consequences well after resolution of the acute event.[47] Recovery in survivors is often protracted and incomplete, though attainment of normal lung function and exercise tolerance is possible. Results of emergent retransplantation in this setting have been poor.

BRONCHIAL ANASTOMOTIC DEHISCENCE

With advances in surgical technique, extensive life-threatening dehiscence of the bronchial anastomosis now occurs in fewer than 2 per cent of lung transplant recipients. Surgical repair of extensive airway dehiscence using an intercostal muscle flap or omentopexy is often unsuccessful or deemed to be too risky. Bronchoscopic placement of an uncovered metal stent across the anastomosis can promote growth of granulation tissue; this technique has been reported to result in successful resolution of the dehisced airway in a small case series.[48] Focal areas of dehiscence, incidentally detected on bronchoscopy or heralded by the appearance of a spontaneous pneumothorax, are encountered in 1–6 per cent of patients.[49] Tube thoracostomy may be required for evacuation of pneumothoraces but focal dehiscence usually heals without other intervention.

BRONCHIAL ANASTOMOTIC STENOSIS

The most common form of airway complication currently encountered is anastomotic narrowing, with a reported frequency of 12–24 per cent in contemporary series.[49,50] Some centres have reported a higher rate of anastomotic stenosis associated with the use of the telescoped anastomosis compared to the end-to-end technique.[51] Narrowing of the anastomosis can result from ischaemia-induced stricture or bronchomalacia, or from the formation of excessive granulation tissue.

Independent of the mechanism, anastomotic narrowing typically develops within several weeks following transplantation. Clues to its presence include focal wheezing on the involved side, recurrent bouts of pneumonia or purulent bronchitis, and suboptimal pulmonary function studies demonstrating airflow obstruction. Diagnosis is based on direct bronchoscopic visualization.

Techniques to address anastomotic narrowing include balloon dilatation, laser debridement and stent placement, all readily performed through a flexible bronchoscope.[49] Endobronchial brachytherapy has been successfully employed as an adjunct in managing recurrent episodes of stenosis due to hyperplastic granulation tissue.

NATIVE LUNG HYPERINFLATION

Acute hyperinflation of the native lung leading to respiratory and haemodynamic compromise in the immediate postoperative period has been reported in 15–30 per cent of emphysema patients undergoing single lung transplantation.[52,53] The combination of positive-pressure ventilation and significant allograft oedema serves to magnify the compliance differential between the two lungs and may predispose to this complication. Acute hyperinflation can be rapidly addressed by insertion of a double-lumen endotracheal tube and initiation of independent lung ventilation, ventilating the native lung with a low respiratory rate and prolonged expiratory time to facilitate complete emptying.

Beyond the perioperative period, some single-lung transplant recipients with underlying emphysema demonstrate exaggerated or progressive native lung hyperinflation that more insidiously compromises the function of the allograft. In this setting, surgical volume reduction of the native lung can result in significant functional improvement.

PHRENIC NERVE INJURY

Phrenic nerve injury following lung transplantation results from intraoperative traction or from division of the nerve in the setting of extensive fibrous adhesions and difficult hilar dissection. In two prospective studies, phrenic nerve injury was documented in 7.4 and 29.6 per cent of patients.[54,55] The significantly higher frequency found in the latter study likely relates to the more sensitive screening technique utilized, involving phrenic nerve electrophysiological testing in addition to radiographic visualization of diaphragmatic motion.

Phrenic nerve injury has been associated with increases in ventilator days, tracheostomy rates and duration of intensive care.[56] Achievement of a normal functional outcome is ultimately possible for those with reversible injury but recovery in some cases may be protracted or incomplete. For severely impaired patients, nocturnal non-invasive ventilatory support and diaphragmatic plication have been successfully employed.

Heart transplantation

GENERAL COMPLICATIONS

Heart transplant recipients are subject to the same generic perioperative pulmonary complications encountered in the general cardiac surgical population. These include atelectasis, pulmonary oedema, pleural effusions and mediastinitis. The risk of respiratory failure and acute lung injury following heart transplantation is relatively low. In a series of 200 procedures, prolonged respiratory failure requiring tracheostomy was reported in only 7 cases (3.5 per cent); 5 of the 7 cases occurred within the first 6 months following transplantation.

PHRENIC NERVE INJURY

Diaphragmatic dysfunction due to phrenic nerve injury occurs in up to 12 per cent of heart transplant recipients.[54] The right phrenic nerve is most prone to injury. Diaphragmatic dysfunction has been associated with an increased risk of postoperative pneumonia and a trend towards increased length of intubation.

DIAPHRAGMATIC HERNIA

Diaphragmatic hernias have been reported in heart transplant recipients who had left ventricular assist devices (LVADs) implanted prior to transplantation. These devices are placed either pre- or intraperitoneally in the left upper quadrant. The inflow cannula penetrates the left hemidiaphragm and attaches to the left ventricle while the outflow cannula emerges from the ascending aorta and crosses anterior to the diaphragm near the midline. At the time of heart transplantation, the LVAD is explanted and the left-sided diaphragmatic defect is routinely repaired. However, repair of the anterior diaphragmatic defect is not necessarily standard practice, as it has been reasoned this would add to the operative time and that midline scarring would naturally close the rent.

In the largest published series, diaphragmatic hernias were detected in 8 of 67 heart transplant recipients with prior LVADs.[57] The prevalence was 16 per cent among the subset of patients in whom closure of the anterior midline diaphragmatic defect was not performed, and in all cases the hernia arose at the site of the unrepaired midline defect. In contrast, only 1 of 22 patients (4 per cent) experienced this complication when the surgical procedure was modified to include routine closure of the anterior defect. That case, as well as others reported in the literature, involved herniation through a previously closed left-sided defect.

Patients with diaphragmatic hernias can be entirely asymptomatic, experience subacute gastrointestinal symptoms (abdominal pain, nausea, vomiting), or present emergently with colonic incarceration. Chest radiographs can be non-specific, demonstrating only an ill-defined opacity at the base of the right or left lung (Fig. 18.2a). More suggestive findings include the presence of air within the opacity or actual visualization of colonic haustra. The diagnosis can usually be established definitively with CT of the chest and abdomen, after administration of oral contrast (Fig. 18.2b). Surgical repair, either via conventional laparotomy or performed laparoscopically, is indicated even in asymptomatic cases because of risk of incarceration.

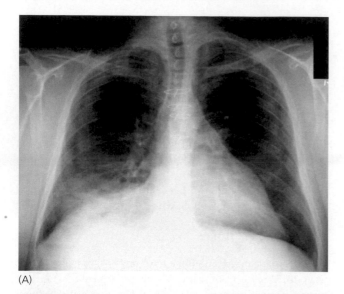

(A)

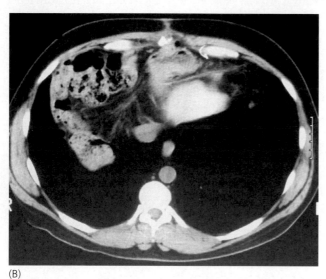

(B)

Fig. 18.2   Diaphragmatic hernia following heart transplantation. (a) Chest radiograph showing an opacity at the right base with ill-defined borders. (b) Chest computed tomography after administration of oral contrast demonstrates loops of bowel within the right hemithorax. Reproduced with permission from [91].

KIDNEY TRANSPLANTATION

Kidney transplantation is carried out with relatively few perioperative pulmonary complications, reflecting the use of a lower abdominal incision and the comparative good health of the recipients. The vast majority of patients are extubated in the operating room. Perioperative respiratory failure was documented in 4 per cent of 178 kidney transplant recipients from the University of Pittsburgh.[58] The most common non-infectious pulmonary complication is hydrostatic pulmonary oedema due to impaired salt and water excretion in the setting of early allograft dysfunction or rejection. ARDS, on the other hand, occurs infrequently. In a retrospective review of a national kidney transplant database encompassing over 42 000 transplants, ARDS was documented in only 86 patients (0.2 per cent) and only 1 of these cases occurred within the perioperative period.[59]

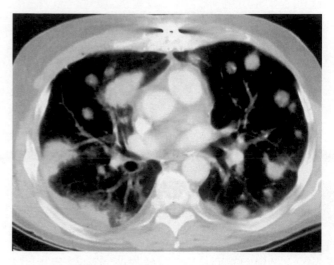

Fig. 18.3   Post-transplantation lymphoproliferative disorder. Chest computed tomography scan from a bilateral lung transplant recipient demonstrating multiple nodules and masses. Surgical lung biopsy revealed a B-cell lymphoma. Special stains for Epstein–Barr virus were positive.

NEOPLASTIC DISORDERS

Post-transplant lymphoproliferative disorder (PTLD) is second only to non-melanoma skin cancers as a leading neoplastic complication of organ transplantation. PTLD encompasses a spectrum of abnormal B-cell proliferative responses ranging from benign polyclonal hyperplasia to malignant lymphomas. Epstein–Barr virus (EBV) has been identified as the stimulus for B-cell proliferation, which proceeds in an unchecked fashion due to the inadequate cytotoxic T-cell response in the immunosuppressed host. EBV-naive recipients who acquire primary infection at the time of organ transplantation are at greatest risk of developing PTLD. A higher intensity of immunosuppression and, in particular, the use of antilymphocyte antibody preparations have also been implicated as risk factors.

Likely reflecting differences in the magnitude of immunosuppression employed, the incidence of PTLD is only 1–2 per cent among kidney and liver transplant recipients but is in the range of 5–7 per cent among recipients of hearts and lungs.[60,61] The incidence is greatest within the first year after transplantation. Lung transplant recipients and heart transplant recipients are the most likely to present with intrathoracic involvement, which typically assumes the form of one or multiple pulmonary nodules (Fig. 18.3) occasionally accompanied by regional adenopathy or pleural effusions.

Initial treatment involves reduction in the level of immunosuppression to permit partial restoration of host cellular immunity. Such a strategy can lead to regression of tumour in up to two-thirds of solid-organ recipients but carries the risk of precipitating acute or chronic allograft rejection.[62] For patients who fail to achieve a complete remission, cannot tolerate reduced immunosuppression, or have widespread disease, immunotherapy with anti-CD20 monoclonal antibodies (rituximab) is emerging as the preferred option.[63] Experience with standard chemotherapy has been poor owing to the high risk of infection during periods of neutropenia. Antiviral

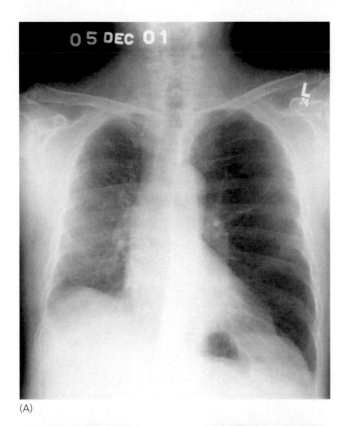

(A)

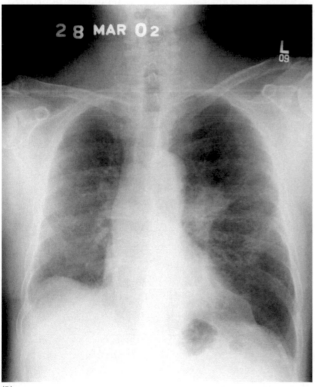

(B)

Fig. 18.4   Lung cancer. Chest radiographs taken approximately 4 months apart demonstrating a rapidly enlarging perihilar mass in the native emphysematous left lung of a single lung transplant recipient. Biopsies demonstrated non-small-cell lung cancer.

therapy has not been demonstrated to be beneficial in established PTLD but may have a role in prophylaxis.[64]

The mortality attributable to PTLD is not well defined. In multivariate analysis, increased age, elevated lactate dehydrogenase levels, severe organ dysfunction, presence of B symptoms, and multiorgan involvement are independent markers of poor prognosis, but the type of transplant is not. There are insufficient data to ascertain the impact of the various treatment regimens on survival.

Other malignancies occasionally present in the lungs following transplantation. Bronchogenic carcinoma develops in 2–4 per cent of heart transplant recipients.[65] It has been reported with similar frequency in the native lung of single-lung transplant recipients with underlying chronic obstructive pulmonary disease or pulmonary fibrosis (Fig. 18.4), the vast majority of whom are also smokers.[66] It is unclear whether the development of lung cancer in the heart and lung populations reflects an increased risk or simply represents the expected occurrence rate in a population with a high prevalence of cigarette smoking. Lung cancer in the transplant recipient can progress in a rapid fashion over a short period, mimicking an infectious process.[66] This aggressive behaviour may reflect loss of anti-tumour immune surveillance in the immunosuppressed host or may be due to a more specific effect of the calcineurin inhibitors in promoting tumour growth.[67]

Among liver transplant recipients with a pre-transplantation history of hepatocellular carcinoma, the lung is the most common site of recurrence. Recurrence usually occurs within two years of transplantation and appears radiographically as single or multiple lung nodules. An elevated α-fetoprotein level provides an important clue to the possibility of recurrent disease.

DRUG-INDUCED LUNG DISEASE

Sirolimus, also known as rapamycin, is a potent immunosuppressive agent recently introduced into clinical practice. Since its release, numerous cases of interstitial pneumonitis developing in association with sirolimus administration have been reported.[68,69] The incidence of this complication remains unknown. Initial reports suggested that interstitial pneumonitis was largely a result of excessive sirolimus blood concentrations but more recent reports describe this complication in the setting of therapeutic drug levels.[70] Patients typically present with dry cough, progressive dyspnoea, fatigue and weakness; fever and haemoptysis are less commonly present.[68] Radiographic abnormalities include interstitial infiltrates, alveolar consolidation and nodular opacities (Fig. 18.5).[68,69] Bronchoalveolar lavage reveals evidence of a lymphocytic alveolitis and, less commonly, of alveolar haemorrhage. Reported findings on transbronchial lung biopsies include bronchiolitis obliterans with organizing pneumonia, interstitial lymphocytic infiltrates, and non-necrotizing granulomas.[68,69] Discontinuation of the drug leads to prompt clinical improvement while radiographic abnormalities may take several months to fully resolve. Conversion from sirolimus to everolimus, an agent that shares similar biochemical features and mechanism of action, has led to resolution of pneumonitis in some cases,[71,72] but case reports of everolimus-induced pneumonitis have also recently appeared.[73,74]

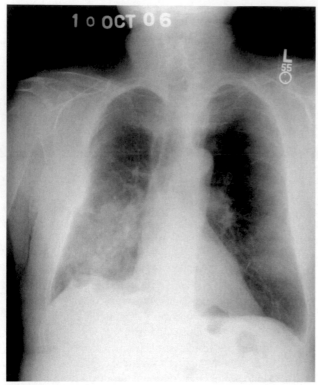

(A)

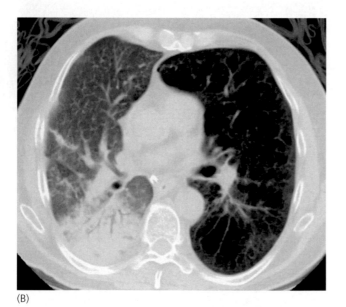

(B)

Fig. 18.5 Sirolimus-induced pneumonitis. Chest radiograph and computed tomography scan demonstrating areas of consolidation in the right lung allograft of a single lung transplant recipient who was receiving sirolimus as part of his immunosuppressive regimen. Lung biopsies demonstrated lymphocytic interstitial infiltrates with scattered non-caseating granulomas; special stains and cultures for microorganisms were negative. The radiographic abnormalities resolved with discontinuation of sirolimus and a tapering course of prednisone.

Sirolimus has the potential to impair wound healing as a consequence of its inhibitory effects on fibroblast proliferation. Surgical wound dehiscence has been reported in association with the use of sirolimus following kidney and liver transplantation. Among lung transplant recipients, an unusually high incidence of bronchial anastomotic dehiscence has been reported.[75,76] Because of this, sirolimus should not be initiated prior to complete healing of the airway.

The murine monoclonal anti-CD3 antibody OKT3 was commonly used in the past for induction immunosuppression and treatment of refractory acute rejection. This agent is associated with a 'cytokine release syndrome' that is most pronounced with the first dose and that clinically presents with fever, rigors, nausea, hypotension and dyspnoea. A small number of patients have been reported to develop non-cardiogenic pulmonary oedema, which rarely can be fatal. Because of its toxicity and the availability of alternative agents, the use of OKT3 has diminished in recent years. The interleukin-2 receptor antagonists basiliximab and daclizumab are being used with increasing frequency as induction agents, in part because of a generally favourable side-effect profile. However, several cases of non-cardiogenic pulmonary oedema in renal transplant recipients have been reported in association with basiliximab infusion.[77] The mechanism underlying this reaction has not been determined.

PULMONARY ALLOGRAFT REJECTION

Among the solid-organ transplant populations, lung transplant recipients are uniquely predisposed to pulmonary injury mediated by alloreactive T-lymphocytes in the form of acute allograft rejection. Approximately 55–75 per cent of transplant recipients experience at least one episode of acute rejection within the first year, highlighting the failings of current immunosuppressive strategies. Factors that influence the likelihood of developing acute rejection remain poorly defined. The degree of HLA discordance between donor and recipient has been inconsistently identified as a risk factor.[78,79] Polymorphisms in toll-like receptor 4 that down-regulate recipient innate immune responsiveness may be associated with a lower incidence of acute rejection.[79]

Symptoms of acute rejection are non-specific and include malaise, low-grade fever, dyspnoea and cough. Radiographic infiltrates, a decline in arterial oxygenation at rest or with exercise, and an abrupt fall of greater than 10 per cent in spirometric values are important clues to the possible presence of rejection, but similar findings accompany infectious episodes. Published series employing surveillance lung biopsies have demonstrated that histologically significant acute rejection (≥ grade A2) may be clinically silent in 14–40 per cent of patients.[14,80,81]

Transbronchial lung biopsy is the procedure of choice for diagnosing acute rejection. The procedure is safe, can be performed in serial fashion over time, and has a sensitivity of 61–94 per cent and specificity in excess of 90 per cent for the diagnosis of acute rejection.[82] The histological hallmark of acute rejection is the presence of perivascular lymphocytic infiltrates that, in more severe cases, spill over into the adjacent interstitium and alveolar air-spaces. Lymphocytic bronchitis or bronchiolitis may accompany the parenchymal involvement or may be an independent feature. An histology classification system has been universally adopted to categorize and grade the severity of acute cellular rejection.[83]

Conventional treatment of acute rejection consists of a three-day pulse of high-dose intravenous methylprednisolone. In most cases, this results in rapid improvement in symptoms, pulmonary function and radiographic abnormalities, but follow-up biopsies show histological evidence of persistent rejection in 30 per cent of patients with prior mild acute rejection and 44 per cent of patients with prior moderate acute rejection.[81] A transient increase in oral prednisone and subsequent taper over several weeks is advocated by some clinicians following the intravenous steroid pulse or as primary treatment for histologically mild episodes.

Chronic rejection, in the form of bronchiolitis obliterans, develops in up to two-thirds of lung transplant recipients and represents the major impediment to long-term graft and patient survival.[84] Bronchiolitis obliterans is a fibroproliferative process characterized by submucosal inflammation and fibrosis of the bronchiolar walls, ultimately leading to complete obliteration of the airway lumen. The functional consequence of this process is progressive and irreversible airflow obstruction. Because the characteristic histology is difficult to demonstrate by transbronchial lung biopsy, a surrogate diagnostic schema based on the magnitude of decline in FEV_1 has evolved, termed 'bronchiolitis obliterans syndrome' (BOS).

Acute rejection, particularly when recurrent or severe, and lymphocytic bronchiolitis have been consistently identified as major risk factors for development of BOS, supporting the view that BOS is a consequence of alloimmune injury. Non-immune factors may also be important in initiating or perpetuating injury but have been more variably substantiated. These factors include CMV and other respiratory viral infections, prolonged ischaemic times, and gastroesophageal reflux with occult aspiration.

The incidence of BOS is greatest within the first 2 years but the risk remains considerable and steady beyond this time point. Onset of disease is typically insidious but may be abrupt in more aggressive cases. Dyspnoea, cough and recurrent bouts of purulent tracheobronchitis, with recovery of *Pseudomonas aeruginosa* from sputum cultures, are highly characteristic features. While chest radiographs are usually unremarkable, HRCT reveals evidence of air trapping on expiratory images in the majority of patients and evidence of bronchiectasis in some. Progressive airflow obstruction is the rule, though the pace of decline is highly variable and the course may be interrupted by periods of functional stability. The prognosis is generally poor, with a 40 per cent mortality rate within 2 years of onset.[85] Patients with early-onset of BOS (i.e. within the first 3 years) appear to experience more rapid decline in lung function and higher mortality.

Myriad immunosuppressive modalities have been employed in the treatment of BOS, including pharmacological agents, antilymphocyte antibodies, photopheresis and total lymphoid irradiation, but consensus is lacking on the optimal approach. At best, immunosuppressive measures appear to slow the rate of decline rather than to fully arrest or reverse the process. A potential role for macrolide therapy is suggested by recent case series demonstrating an average increase in FEV_1 of 11–18 per cent in patients with BOS treated with azithromycin.[86,87] Macrolides possess anti-inflammatory properties and have been shown to decrease airway neutrophilia and IL-8 production in patients with BOS, possibly accounting for their observed beneficial effects.[87] The only definitive treatment is retransplantation but this strategy remains controversial in the context of a scarce donor organ pool.

The development of strategies to prevent BOS is an area of intense interest but, to date, little substantive progress. In recognition of the established link between acute rejection and BOS, many transplant centres routinely perform surveillance lung biopsies to detect and treat clinically silent acute rejection, but the impact of this strategy on risk of BOS remains uncertain.[88] The use of statins may be associated with a reduction in the number and severity of acute rejection episodes and a lower incidence of BOS.[89] Inhaled ciclosporin, initiated within 6 weeks of transplantation, has also been associated with a marked reduction in the incidence of BOS and with improved survival.[90,91]

FUTURE DIRECTIONS

Although solid-organ transplantation remains a risky endeavour, substantial inroads have been made in mitigating these risks. Advances in surgical technique and postoperative pain control have lessened the immediate decrement in lung function associated with the transplant operation, thereby reducing the risk of respiratory failure and pneumonia. Development of effective prophylactic strategies has dramatically reduced the lethal impact that CMV pneumonia and PCP once had on transplant recipients. In order for the field to advance further, however, there is a dire need to improve on current immunosuppressive strategies, which involve the use of relatively toxic agents and expose the recipient to the infectious and neoplastic consequences of global immunosuppression. The answer to this vexing problem lies in the ability to induce a state of graft-specific immune tolerance while preserving host immune responses to infection and neoplasm. Ongoing research in this area holds the promise of ultimately achieving this 'holy grail' of organ transplantation.

KEY POINTS

- Pulmonary infections are a major complication following solid-organ transplantation. Nosocomial bacterial pneumonias predominate in the initial month after transplantation, while opportunistic infections emerge as a major concern beyond this point.

- CMV remains the major viral threat, though the widespread use of prophylactic regimens has dramatically reduced the incidence and severity of infections.

- Aspergillosis is the major fungal pathogen responsible for pulmonary and disseminated infections. Voriconazole has become the agent of choice, though mortality remains high.

- Non-infectious pulmonary complications related to the surgery itself include respiratory failure, phrenic nerve injury/diaphragmatic dysfunction (heart, liver, lung), pleural effusions (liver), and airway anastomotic dehiscence and stricture (lung).

(Contd)

- Post-transplant lymphoproliferative disorder, mediated by EBV, is the most common non-cutaneous neoplastic disorder encountered. Treatment options include reduction in immunosuppression and the use of rituximab.

- Sirolimus, one of the newer immunosuppressive agents, has been associated with a spectrum of pulmonary toxicities including interstitial pneumonitis, alveolar haemorrhage and BOOP. Among lung transplant recipients, it has been associated with bronchial dehiscence.

- Lung transplant recipients are uniquely predisposed to alloimmune lung injury mediated by T-cells (i.e. acute rejection). Chronic allograft dysfunction in this patient population is a result of both alloimmune (prior acute rejection episodes) and non-immune (airway ischaemia, viral infection, aspiration) processes.

REFERENCES

1. Fishman JA, Rubin RH. Infection in organ-transplant recipients. *New Engl J Med* 1998;**338**:1741–51.
2. Weill D, Dey GC, Hicks RA, Young KR, *et al.* A positive donor Gram stain does not predict outcome following lung transplantation. *J Heart Lung Transplant* 2002;**21**:555–8.
3. Cisneros JM, *et al.*, Spanish Transplantation Infection Study Group. Pneumonia after heart transplantation: a multi-institutional study. *Clin Infect Dis* 1998;**27**:324–31.
4. Singh N, Gayowski T, Wagener M, *et al.* Pulmonary infections in liver transplant recipients receiving tacrolimus: changing pattern of microbial etiologies. *Transplantation* 1996;**61**:396–401.
5. Husain S, McCurry K, Dauber J, *et al. Nocardia* infection in lung transplant recipients. *J Heart Lung Transplant* 2002;**21**:354–9.
6. Montoya JG, Giraldo LF, Efron B, *et al.* Infectious complications among 620 consecutive heart transplant patients at Stanford University Medical Center. *Clin Infect Dis* 2001;**33**:629–40.
7. Singh N, Paterson DL. *Mycobacterium tuberculosis* infection in solid-organ transplant recipients: impact and implications for management. *Clin Infect Dis* 1998;**27**:1266–77.
8. John GT, Shankar V, Abraham AM, *et al.* Risk factors for post-transplant tuberculosis. *Kidney Int* 2001;**60**:1148–53.
9. Singh N, Wagener MM, Gayowski T. Safety and efficacy of isoniazid chemoprophylaxis administered during liver transplant candidacy for the prevention of posttransplant tuberculosis. *Transplantation* 2002;**74**:892–5.
10. Malouf MA, Glanville AR. The spectrum of mycobacterial infection after lung transplantation. *Am J Respir Crit Care Med* 1999;**160**:1611–16.
11. Rayes N, Seehofer D, Schmidt CA, *et al.* Prospective randomized trial to assess the value of preemptive oral therapy for CMV infection following liver transplantation. *Transplantation* 2001;**72**:881–5.
12. Senechal M, Dorent R, du Montcel ST, *et al.* Monitoring of human cytomegalovirus infections in heart transplant recipients by pp65 antigenemia. *Clin Transplant* 2003;**17**:423–7.
13. Akalin E, Sehgal V, Ames S, *et al.* Cytomegalovirus disease in high-risk transplant recipients despite ganciclovir or valganciclovir prophylaxis. *Am J Transplant* 2003;**3**:731–5.
14. Hopkins PM, Aboyoun CL, Chhajed PN, *et al.* Prospective analysis of 1235 transbronchial lung biopsies in lung transplant recipients. *J Heart Lung Transplant* 2002;**21**:1062–7.
15. Ettinger NA, Trulock EP. Pulmonary considerations of organ transplantation. *Am Rev Respir Dis* 1991;**143**:1386–405.
16. Rubin RH. Prevention and treatment of cytomegalovirus disease in heart transplant patients. *J Heart Lung Transplant* 2000;**19**:731–5.
17. Turgeon N, Fishman JA, Doran M, *et al.* Prevention of recurrent cytomegalovirus disease in renal and liver transplant recipients: effect of oral ganciclovir. *Transpl Infect Dis* 2000;**2**:2–10.
18. Sia IG, Patel R. New strategies for prevention and therapy of cytomegalovirus infection and disease in solid-organ transplant recipients. *Clin Microbiol Rev* 2000;**13**:83–121.
19. Couchoud C, Cucherat M, Haugh M, Pouteil-Noble C. Cytomegalovirus prophylaxis with antiviral agents in solid organ transplantation: a meta-analysis. *Transplantation* 1998;**65**:641–7.
20. Lowance D, *et al.*, International Valacyclovir Cytomegalovirus Prophylaxis Transplantation Study Group. Valacyclovir for the prevention of cytomegalovirus disease after renal transplantation. *New Engl J Med* 1999;**340**:1462–70.
21. Rubin RH, Kemmerly SA, Conti D, *et al.* Prevention of primary cytomegalovirus disease in organ transplant recipients with oral ganciclovir or oral acyclovir prophylaxis. *Transpl Infect Dis* 2000;**2**:112–17.
22. Lurain NS, Bhorade SM, Pursell KJ, *et al.* Analysis and characterization of antiviral drug-resistant cytomegalovirus isolates from solid organ transplant recipients. *J Infect Dis* 2002;**186**:760–8.
23. Limaye AP, Corey L, Koelle DM, *et al.* Emergence of ganciclovir-resistant cytomegalovirus disease among recipients of solid-organ transplants. *Lancet* 2000;**356**:645–9.
24. Bhorade SM, Lurain NS, Jordan A, *et al.* Emergence of ganciclovir-resistant cytomegalovirus in lung transplant recipients. *J Heart Lung Transplant* 2002;**21**:1274–82.
25. Billings JL, Hertz MI, Savik K, Wendt CH. Respiratory viruses and chronic rejection in lung transplant recipients. *J Heart Lung Transplant* 2002;**21**:559–66.
26. Palmer SM, Jr., Henshaw NG, Howell DN, *et al.* Community respiratory viral infection in adult lung transplant recipients. *Chest* 1998;**113**:944–50.
27. Hopkins P, McNeil K, Kermeen F, *et al.* Human metapneumovirus in lung transplant recipients and comparison to respiratory syncytial virus. *Am J Respir Crit Care Med* 2008;**178**:876–81.
28. Khalifah AP, Hachem RR, Chakinala MM, *et al.* Respiratory viral infections are a distinct risk for bronchiolitis obliterans syndrome and death. *Am J Respir Crit Care Med* 2004;**170**:181–7.
29. Paterson DL, Singh N. Invasive aspergillosis in transplant recipients. *Medicine* (Baltimore) 1999;**78**:123–38.
30. Horvath JA, Dummer S. The use of respiratory-tract cultures in the diagnosis of invasive pulmonary aspergillosis. *Am J Med* 1996;**100**:171–8.

31. Mehrad B, Paciocco G, Martinez FJ, *et al.* Spectrum of *Aspergillus* infection in lung transplant recipients: case series and review of the literature. *Chest* 2001;**119**:169–75.

32. Herbrecht R, Denning DW, Patterson TF, *et al.* Voriconazole versus amphotericin B for primary therapy of invasive aspergillosis. *New Engl J Med* 2002;**347**:408–15.

33. Robinson LA, Reed EC, Galbraith TA, *et al.* Pulmonary resection for invasive *Aspergillus* infections in immunocompromised patients. *J Thorac Cardiovasc Surg* 1995;**109**:1182–96; discussion 96–7.

34. Nunley DR, Gal AA, Vega JD, *et al.* Saprophytic fungal infections and complications involving the bronchial anastomosis following human lung transplantation. *Chest* 2002;**122**:1185–91.

35. Gordon SM, LaRosa SP, Kalmadi S, *et al.* Should prophylaxis for *Pneumocystis carinii* pneumonia in solid organ transplant recipients ever be discontinued? *Clin Infect Dis* 1999;**28**: 240–6.

36. Castiglioni B, Sutton DA, Rinaldi MG, *et al. Pseudallescheria boydii* (*Anamorph Scedosporium apiospermum*): infection in solid organ transplant recipients in a tertiary medical center and review of the literature. *Medicine* (Baltimore) 2002;**81**: 333–48.

37. Glanemann M, Langrehr J, Kaisers U, *et al.* Postoperative tracheal extubation after orthotopic liver transplantation. *Acta Anaesthesiol Scand* 2001;**45**:333–9.

38. Golfieri R, Giampalma E, Morselli Labate AM, *et al.* Pulmonary complications of liver transplantation: radiological appearance and statistical evaluation of risk factors in 300 cases. *Eur Radiol* 2000;**10**:1169–83.

39. Judson MA, Sahn SA. The pleural space and organ transplantation. *Am J Respir Crit Care Med* 1996;**153**:1153–65.

40. Pirat A, Ozgur S, Torgay A, *et al.* Risk factors for postoperative respiratory complications in adult liver transplant recipients. *Transplant Proc* 2004;**36**:218–20.

41. McAlister VC, Grant DR, Roy A, *et al.* Right phrenic nerve injury in orthotopic liver transplantation. *Transplantation* 1993;**55**:826–30.

42. Christie JD, Kotloff RM, Pochettino A, *et al.* Clinical risk factors for primary graft failure following lung transplantation. *Chest* 2003;**124**:1232–41.

43. Thabut G, Vinatier I, Stern JB, *et al.* Primary graft failure following lung transplantation: predictive factors of mortality. *Chest* 2002;**121**:1876–82.

44. Christie JD, Bavaria JE, Palevsky HI, *et al.* Primary graft failure following lung transplantation. *Chest* 1998;**114**:51–60.

45. King RC, Binns OA, Rodriguez F, *et al.* Reperfusion injury significantly impacts clinical outcome after pulmonary transplantation. *Ann Thorac Surg* 2000;**69**:1681–5.

46. Meade MO, Granton JT, Matte-Martyn A, *et al.* A randomized trial of inhaled nitric oxide to prevent ischemia-reperfusion injury after lung transplantation. *Am J Respir Crit Care Med* 2003;**167**:1483–9.

47. Christie JD, Kotloff RM, Ahya VN, *et al.* The effect of primary graft dysfunction on survival after lung transplantation. *Am J Respir Crit Care Med* 2005;**171**:1312–16.

48. Mughal MM, Gildea TR, Murthy S, *et al.* Short-term deployment of self-expanding metallic stents facilitates healing of bronchial dehiscence. *Am J Respir Crit Care Med* 2005;**172**:768–71.

49. Chhajed PN, Malouf MA, Tamm M, *et al.* Interventional bronchoscopy for the management of airway complications following lung transplantation. *Chest* 2001;**120**:1894–9.

50. Herrera JM, McNeil KD, Higgins RS, *et al.* Airway complications after lung transplantation: treatment and long-term outcome. *Ann Thorac Surg* 2001;**71**:989–93; discussion 93–4.

51. Garfein ES, Ginsberg ME, Gorenstein L, *et al.* Superiority of end-to-end versus telescoped bronchial anastomosis in single lung transplantation for pulmonary emphysema. *J Thorac Cardiovasc Surg* 2001;**121**:149–54.

52. Weill D, Torres F, Hodges TN, *et al.* Acute native lung hyperinflation is not associated with poor outcomes after single lung transplant for emphysema. *J Heart Lung Transplant* 1999;**18**:1080–7.

53. Yonan NA, el-Gamel A, Egan J, *et al.* Single lung transplantation for emphysema: predictors for native lung hyperinflation. *J Heart Lung Transplant* 1998;**17**:192–201.

54. Dorffner R, Eibenberger K, Youssefzadeh S, *et al.* Diaphragmatic dysfunction after heart or lung transplantation. *J Heart Lung Transplant* 1997;**16**:566–9.

55. Sheridan PH, Cheriyan A, Doud J, *et al.* Incidence of phrenic neuropathy after isolated lung transplantation. *J Heart Lung Transplant* 1995;**14**:684–91.

56. Ferdinande P, Bruyninckx F, Van Raemdonck D, *et al.* Phrenic nerve dysfunction after heart-lung and lung transplantation. *J Heart Lung Transplant* 2004;**23**:105–9.

57. Chatterjee S, Williams NN, Ohara ML, *et al.* Diaphragmatic hernias associated with ventricular assist devices and heart transplantation. *Ann Thorac Surg* 2004;**77**:2111–14.

58. Sadaghdar H, Chelluri L, Bowles SA, Shapiro R. Outcome of renal transplant recipients in the ICU. *Chest* 1995;**107**:1402–5.

59. Shorr AF, Abbott KC, Agadoa LY. Acute respiratory distress syndrome after kidney transplantation: epidemiology, risk factors, and outcomes. *Crit Care Med* 2003;**31**:1325–30.

60. Gao SZ, Chaparro SV, Perlroth M, *et al.* Post-transplantation lymphoproliferative disease in heart and heart-lung transplant recipients: 30-year experience at Stanford University. *J Heart Lung Transplant* 2003;**22**:505–14.

61. Paranjothi S, Yusen RD, Kraus MD, *et al.* Lymphoproliferative disease after lung transplantation: comparison of presentation and outcome of early and late cases. *J Heart Lung Transplant* 2001;**20**:1054–63.

62. Tsai DE, Hardy CL, Tomaszewski JE, *et al.* Reduction in immunosuppression as initial therapy for posttransplant lymphoproliferative disorder: analysis of prognostic variables and long-term follow-up of 42 adult patients. *Transplantation* 2001;**71**:1076–88.

63. Reams BD, McAdams HP, Howell DN, *et al.* Posttransplant lymphoproliferative disorder: incidence, presentation, and response to treatment in lung transplant recipients. *Chest* 2003;**124**:1242–9.

64. Levine SM, Angel L, Anzueto A, *et al.* A low incidence of posttransplant lymphoproliferative disorder in 109 lung transplant recipients. *Chest* 1999;**116**:1273–7.

65. Dorent R, Mohammadi S, Tezenas S, *et al.* Lung cancer in heart transplant patients: a 16-year survey. *Transplant Proc* 2000;**32**:2752–4.

66. Arcasoy SM, Hersh C, Christie JD, *et al.* Bronchogenic carcinoma complicating lung transplantation. *J Heart Lung Transplant* 2001;**20**:1044–53.

67. Hojo M, Morimoto T, Maluccio M, *et al.* Cyclosporine induces cancer progression by a cell-autonomous mechanism. *Nature* 1999;**397**:530–4.

68. Morelon E, Stern M, Israel-Biet D, *et al.* Characteristics of sirolimus-associated interstitial pneumonitis in renal transplant patients. *Transplantation* 2001;**72**:787–90.

69. McWilliams TJ, Levvey BJ, Russell PA, *et al.* Interstitial pneumonitis associated with sirolimus: a dilemma for lung transplantation. *J Heart Lung Transplant* 2003;**22**:210–13.

70. Haydar AA, Denton M, West A, *et al.* Sirolimus-induced pneumonitis: three cases and a review of the literature. *Am J Transplant* 2004;**4**:137–9.

71. Rehm B, Keller F, Mayer J, Stracke S. Resolution of sirolimus-induced pneumonitis after conversion to everolimus. *Transplant Proc* 2006;**38**:711–13.

72. De Simone P, Petruccelli S, Precisi A, *et al.* Switch to everolimus for sirolimus-induced pneumonitis in a liver transplant recipient: not all proliferation signal inhibitors are the same. A case report. *Transplant Proc* 2007;**39**:3500–1.

73. Alexandru S, Ortiz A, Baldovi S, *et al.* Severe everolimus-associated pneumonitis in a renal transplant recipient. *Nephrol Dial Transplant* 2008;**23**:3353–5.

74. Otton J, Hayward CS, Keogh AM, *et al.* Everolimus-associated pneumonitis in 3 heart transplant recipients. *J Heart Lung Transplant* 2009;**28**:104–6.

75. King-Biggs MB, Dunitz JM, Park SJ, *et al.* Airway anastomotic dehiscence associated with use of sirolimus immediately after lung transplantation. *Transplantation* 2003;**75**:1437–43.

76. Groetzner J, Kur F, Spelsberg F, *et al.* Airway anastomosis complications in de-novo lung transplantation with sirolimus-based immunosuppression. *J Heart Lung Transplant* 2004;**23**:632–8.

77. Bamgbola FO, Del Rio M, Kaskel FJ, Flynn JT. Non-cardiogenic pulmonary edema during basiliximab induction in three adolescent renal transplant patients. *Pediatr Transplant* 2003;**7**:315–20.

78. Schulman II, Weinberg AD, McGregor C, *et al.* Mismatches at the HLA-DR and HLA-B loci are risk factors for acute rejection after lung transplantation. *Am J Respir Crit Care Med* 1998;**157**:1833–7.

79. Palmer SM, Burch LH, Davis RD, *et al.* The role of innate immunity in acute allograft rejection after lung transplantation. *Am J Respir Crit Care Med* 2003;**168**:628–32.

80. Trulock EP, Ettinger NA, Brunt EM, *et al.* The role of transbronchial lung biopsy in the treatment of lung transplant recipients. *Chest* 1992;**102**:1049–54.

81. Guilinger RA, Paradis IL, Dauber JH, *et al.* The importance of bronchoscopy with transbronchial biopsy and bronchoalveolar lavage in the management of lung transplant recipients. *Am J Respir Crit Care Med* 1995;**152**:2037–43.

82. Trulock EP. Lung transplantation. *Am J Respir Crit Care Med* 1997;**155**:789–818.

83. Yousem SA, *et al.*, Lung Rejection Study Group. Revision of the 1990 working formulation for the classification of pulmonary allograft rejection. *J Heart Lung Transplant* 1996;**15**:1–15.

84. Estenne M, Hertz MI. Bronchiolitis obliterans after human lung transplantation. *Am J Respir Crit Care Med* 2002;**166**:440–4.

85. Date H, Lynch JP, Sundaresan S, *et al.* The impact of cytolytic therapy on bronchiolitis obliterans syndrome. *J Heart Lung Transplant* 1998;**17**:869–75.

86. Gerhardt SG, McDyer JF, Girgis RE, *et al.* Maintenance azithromycin therapy for bronchiolitis obliterans syndrome: results of a pilot study. *Am J Respir Crit Care Med* 2003;**168**:121–5.

87. Verleden GM, Vanaudenaerde BM, Dupont LJ, Van Raemdonck DE. Azithromycin reduces airway neutrophilia and interleukin-8 in patients with bronchiolitis obliterans syndrome. *Am J Respir Crit Care Med* 2006;**174**:566–70.

88. Tamm M, Sharples LD, Higenbottam TW, *et al.* Bronchiolitis obliterans syndrome in heart-lung transplantation: surveillance biopsies. *Am J Respir Crit Care Med* 1997;**155**:1705–10.

89. Johnson BA, Iacono AT, Zeevi A, *et al.* Statin use is associated with improved function and survival of lung allografts. *Am J Respir Crit Care Med* 2003;**167**:1271–8.

90. Iacono AT, Johnson BA, Grgurich WF, *et al.* A randomized trial of inhaled cyclosporine in lung-transplant recipients. *New Engl J Med* 2006;**354**:141–50.

91. Kotloff RM. Noninfectious pulmonary complications of liver, heart, and kidney transplantation. *Clin Chest Med* 2005;**26**:623–9.

19

Pulmonary infection induced by drugs

MARC B FEINSTEIN, DOROTHY A WHITE

INTRODUCTION

Immunosuppressive agents are commonly used to treat a variety of clinical disorders characterized by alteration in the immune system. These include collagen vascular diseases, inflammatory bowel diseases, dermatological conditions, graft-versus-host disease (GVHD) following haematopoietic stem-cell transplantation, and graft rejection following solid-organ transplantation. Over the last few years, drugs that specifically target lymphocytes have been developed and are now widely used against lymphoid haematologic malignancies. Some immunosuppressive agents have also been found to have antiproliferative effects and are now being studied in patients with solid tumours.

The use of immunosuppressive drugs can be expected to increase the risk of infections. Many of these infections can involve the lungs and it is therefore important for the pulmonologist to be aware of the most commonly used agents and their associated infections. This chapter will not discuss agents that cause neutropenia and the well-known infectious complications described in this setting. Rather, it will focus on drugs with other immune targets. Some drugs (e.g. glucocorticoids) have broad effects while others may have more specific and unique effects, such as tumour necrosis factor (TNF) antagonists. The drugs discussed in this chapter and associated infections are listed in Box 19.1.

The association of a drug with any specific infection is difficult to determine with accuracy except in specifically designed studies. Compounding factors include underlying immune defects from the disease being treated, other immunosuppressive agents used concurrently, as well as the dose and duration of treatment. We here consider a relationship to exist between a drug and a specific infection only where sufficient evidence supporting a link exists based on theoretical considerations and supporting epidemiological or clinical data.

INCIDENCE AND EPIDEMIOLOGY

The incidence of a drug-associated pulmonary infection is extremely variable depending on the clinical setting, such as the disease being treated and the dose of drug, as well as whether prophylactic agents are used. Ciclosporin (cyclosporin, cyclosporine) and tacrolimus, for instance, used to treat GVHD and solid-organ rejection, are associated with a risk of serious infection in up to 40 per cent of patients treated.[1] Alemtuzumab, a drug employed in lymphoid leukaemia, is associated with infection in up to 80 per cent, roughly 43 per cent having an opportunistic infection.[2] Corticosteroids have been the most widely studied class of drug, and are associated with a much lower but still significant incidence of infection, at least in those receiving lower doses. One large meta-analysis found the relative risk (RR) of infection among patients taking steroids was 1.6 (95% confidence interval (CI) 1.3–1.9).[3] Other agents, such as rituximab, TNF-α inhibitors, temozolomide and methotrexate, are associated with infection but less regularly.

The risk of specific infections associated with these drugs will be discussed in a later section of this chapter.

AETIOLOGY

Drugs can suppress the immune system by a number of mechanisms, including both broad and narrow effects. When the mechanism specific to a drug is known, it can help predict the types of infection that may occur. For instance, drugs that interfere with normal T-cell numbers or function, including ciclosporin, fludarabine and alemtuzumab, are associated with infections from such pathogens as *Pneumocystis jiroveci* (formerly *Pneumocystis carinii*), Cytomegalovirus (CMV) and *Legionella* species, against which T-cells are typically protective. Drugs that impair humoral immunity, such as rituximab,

---

Box 19.1   Medications and their associated pulmonary pathogens. Owing to immunological suppression, many medications also increase the susceptibility non-specifically to Gram-positive and Gram-negative bacteria. For purposes of simplicity, these infections are not listed here

Glucocorticoids

- *Mycobacterium* species, including *M. tuberculosis*
- *Pneumocystis jiroveci*
- *Aspergillus* species
- *Rhizopus oryzae* (mucormycosis)
- Herpes virus
- *Candida*
- *Strongyloides stercoralis*

Ciclosporin and tacrolimus

- CMV

Purine analogues

- *Listeria monocytogenes*
- *Nocardia asteroides*
- *Mycobacterium* species
- *Cryptococcus neoformans*

- *Pneumocystis jiroveci*
- *Aspergillus* species
- *Candida albicans*
- CMV
- *Varicella zoster*
- *Herpes zoster*

Alemtuzumab

- *Pneumocystis jiroveci*
- *Aspergillus* species
- *Rhizopus oryzae* (mucormycosis)
- *Histoplasma capsulatum*
- *Cryptococcus neoformans*
- CMV

Rituximab

- Encapsulated bacteria

TNF antagonists

- *Mycobacterium tuberculosis*
- *Pneumocystis jiroveci*
- *Histoplasma capsulatum*
- *Aspergillus* species
- *Cryptococcus neoformans*

Temozolomide

- *Pneumocystis jiroveci*
- *Aspergillus* species
- *Herpes simplex*
- *Herpes zoster*

Methotrexate

- *Nocardia asteroides*
- *Pneumocystis jiroveci*
- *Cryptococcus neoformans*
- CMV

---

are expected to increase the risk of bacterial infections, especially with encapsulated organisms. However, in some settings, the relationship of immune defect and infection remains elusive, such as the well-described association of TNF antagonists with tuberculosis.

Glucocorticoids are thought to exert their effects via three separate pathways. First, glucocorticoid molecules may bind to the glucocorticoid receptor in the cytoplasm. The resultant glucocorticoid–glucocorticoid receptor complex then migrates to the nucleus where it attaches to glucocorticoid response elements on the DNA, inducing transcription. Production of annexin I, an inhibitor of cytosolic phospholipase A2α (and therefore an indirect inhibitor of eicosanoid production), is thought to operate by this mechanism.[4,5] Second, the glucocorticoid–glucocorticoid receptor complex may directly modulate the activity of other inflammatory transcription factors. The transcriptional activity of nuclear factor-κB, for instance, a pro-inflammatory transcription factor that stimulates production of numerous cytokines, chemokines, adhesion molecules and receptors, is thought to be blocked by the glucocorticoid–glucocorticoid receptor complex.[6] Finally, glucocorticoids may bind directly to receptors on the cell surface, without entering the cytoplasm, thereby activating secondary messenger systems. Glucocorticoids also lead to a reduction in all circulating leucocytes, with the exception of neutrophils, which demarginate from the vessel wall and are found in increased numbers in the blood. Detachment of neutrophils from the vessel walls diminishes the ability of these cells to exit the blood stream and enter sites of infection.

Ciclosporin and tacrolimus exert their effects predominantly via helper T-lymphocytes, without leading to lymphopenia or neutropenia. These drugs bind to specific families of cytoplasmic proteins (cyclophilins for ciclosporin and FK binding proteins for tacrolimus) which inhibit calcineurin, a calcium- and calmodulin-dependent phosphatase. The result is decreased translocation to the nucleus of the NF-AT family of transcription factors, and consequently decreased transcription of IL-2, TNF-α, IL-3, IL-4, granulocyte–macrophage colony stimulating factor, and interferon-γ. These drugs also inhibit the AP-1 and NF-κB transcription factors by different mechanisms.

Purine analogues, including fludarabine, cladribine and pentostatin, modulate their effects by interrupting DNA synthesis. It has also been found that fludarabine inhibits the cytokine-induced activation of STAT1, and STAT1-dependent transcription in normal resting or activated lymphocytes. STAT1 is a molecule essential for cell-mediated immunity and the ability to control viral infections.[7] These drugs are noted for the depth of immunosuppression they cause and the prolonged duration of specific effects, which can sometimes last for years. They are highly effective against lymphocytes, and their use is associated with profound lymphocytopenias. For patients taking fludarabine, for instance, there exists a marked decrease in CD4 cells. There is also an effect on B-cells, but these cells generally recover more quickly, often within a year, and the effect on immunoglobulin levels is very variable. Neutropenia can occur in some patients. Other potential immunological consequences include transient monocytopenia, transient reduction in NK cells, and variable effect on lymphocyte and NK-cell function.[8] Most infections occur early in the course of treatment, even though the CD4 cell may remain very low for extended periods. This is felt to be due to improvement in monocyte function and other immune parameters CLL.[8]

Alemtuzumab (Campath-1H) is a humanized monoclonal antigen against the CD52 antigen. This antigen is expressed on virtually all lymphocytes at various stages of differentiation, as well as eosinophils, macrophages and monocytes. The highest levels are seen on T-prolymphocytic leukaemia cells, followed by B-cells of CLL; the lowest levels are on normal B-cells. Rituximab is chimeric human–mouse monoclonal antibody directed against the CD20 antigen found on the surface of most normal and malignant B-cells. It induces lymphopenia in most patients and low immunoglobulin levels, with recovery usually within a year.

TNF-α antagonists inhibit the cytokine TNF, a key modulator of the inflammatory response. These drugs represent a family of monoclonal antibodies that blocks binding to the TNF receptor and has proven effective in a number of autoimmune diseases, such as rheumatoid arthritis and inflammatory bowel disease. Temozolomide is an oral compound metabolized into an active alkylating agent. Prolonged administration has been associated with lymphopenia, CD4 T-cells being preferentially affected, which can be persistent. In a majority of treated patients, lymphopenia was still present 2 months after drug discontinuation.[9] Methotrexate (MTX) is an inhibitor of the enzyme dihydrofolate reductase. It indirectly impedes folinic acid-dependent metabolic pathways within the cell, including purine and pyrimidine metabolism and, at high doses, amino acid synthesis. Its mechanism of action as an anti-inflammatory agent at low doses is less well known, but may result from the clonal deletion or apoptosis of T-cells, decreased cytokine production, or effects on adenosine metabolism.[10,11]

CLINICAL PRESENTATIONS

Specific infections reported in association with the drugs listed in Box 19.1 are described below.

Glucocorticoids

Glucocorticoids are a class of drugs widely employed for their anti-inflammatory and immunosuppressive properties. Potential uses include asthma, collagen vascular disease, glomerulonephritis, inflammatory bowel disease, and allergic skin diseases. They are also used among transplant recipients, for the prophylaxis and treatment of rejection, as well as some patients with lymphoma.

The relatively diffuse immunosuppressive qualities of glucocorticoids have associated them with numerous forms of infection. Stuck et al. performed a meta-analysis encompassing approximately 4200 patients and 71 clinical trials.[3] They found the relative risk of infection among patients taking steroids to be 1.6 (95% CI: 1.3–1.9). Risk was highest among those with neurologic disease (RR = 2.8). It was not elevated among those taking the equivalent of 10 mg prednisone daily or a cumulative dose below 700 mg. The most common opportunistic pathogens associated with corticosteroid in this meta-analysis were Pneumocystis jiroveci, Aspergillus species and mycobacteria. Infections with bacteria, herpes virus and Candida species also occurred with greater frequency with steroids.

Studies have shown a clear association between glucocorticoid use and Pneumocystis pneumonia in the HIV-negative host. Yale and Limper reviewed the history of 116 consecutive HIV-negative patients diagnosed with Pneumocystis pneumonia at the Mayo Clinic between 1985 and 1991.[12] Of these, 30 per cent suffered from a haematological malignancy, 25 per cent had received organ transplants, and most of the remainder had inflammatory diseases, such as collagen vascular disease. Over 90 per cent of patients had received systemic corticosteroids within a month of developing pneumonia. The median dose of steroid was equivalent to 30 mg of prednisone daily; however, one-quarter of patients were receiving 16 mg or less. The median length of steroid treatment prior to pneumonia diagnosis was 12 weeks. In 1992, Sepkowitz et al. described their 12-year experience with HIV-negative patients diagnosed with Pneumocystis pneumonia at a cancer hospital, and found that 87 per cent of patients had taken systemic steroids.[13] The median dose was the equivalent of 80 mg of prednisone, and the median length of therapy was 3 months. In a German study, 87 per cent of HIV-negative patients with Pneumocystis pneumonia had taken steroids.[14]

There is no well-established mechanism by which corticosteroids are known to predispose to Pneumocystis pneumonia. Some have proposed an inhibitory effect on CD4 lymphocytes[15,16] or alveolar macrophages.[12] One study in non-HIV-infected patients on steroid therapy found a wide range of CD4 counts, from 37 to 1397 cells/μL. Those who developed Pneumocystis pneumonia had a mean CD4 count of 64.9 ± 9.5 cells/μL. Most had CD4 counts less than 300 cells/μL. The patients likely to develop low CD4 counts on corticosteroid therapy were those with haematological malignancy or recent transplants.[17] The mortality rate for Pneumocystis pneumonia in this population ranges from 33 to 50 per cent and exceeds the rate of HIV-infected patients.[12,13]

The genus Aspergillus encompasses a group of saprophytic filamentous fungi found in most environments. This fungus spreads through the air via small conidia, whose diameter of 2–3 μm enables them to reach the alveoli. In the immunocompetent host, defence against this organism rests with the destruction of hyphae by reactive oxygen species produced by granulocytes, while resident pulmonary macrophages ingest conidia, kill germinating cells and secrete cytokines and chemokines to coordinate secondary defence. There is now also evidence that T-helper cell cytokines play a role in adaptive immunity against Aspergillus. Th1 cytokines, such as TNF-α, interferon-γ, IL-12 and IL-15, augment superoxide production and enhance the antifungal activity of granulocytes and mononuclear phagocytes.[18]

Invasive pulmonary aspergillosis (IPA) can occur in those who receive high cumulative doses of steroids. Unfortunately, suspicion of IPA may be delayed by the fact that fever, the most common symptom, may be masked by steroids. Many patients are completely asymptomatic, and infection is not evident until radiographic abnormalities develop. Patients can then rapidly develop cough, dyspnoea or respiratory failure (Fig. 19.1).

Animal studies suggest that patterns of Aspergillus infection may differ between patients taking steroid and those with granulocytopenia. Berenguer et al. challenged two populations of rabbits with the same intratracheal inoculum of Aspergillus.[19] The first group was granulocytopenic, while the second group was immunosuppressed with ciclosporin A and methylprednisolone. Granulocytic animals had more coagulative necrosis

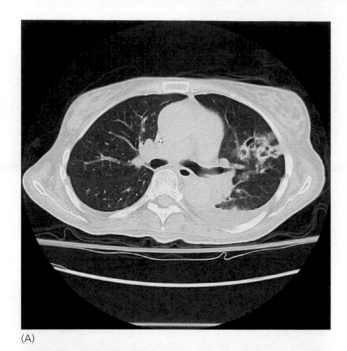

(A)

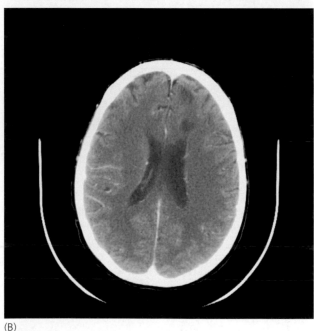

(B)

Fig. 19.1 A 58-year-old woman who received an allogeneic stem-cell transplant for acute lymphoblastic leukaemia required alemtuzumab for graft-versus-host disease that was difficult to control. She developed overwhelming disease with *Aspergillus fumigatus*. (a) Extensive cavitary lung disease; (b) area of lucency in the left frontal lobe of the brain related to *Aspergillus* infection.

and histological evidence of angioinvasion by hyphae. Mortality among granulocytopenic rabbits was 100 per cent while none of the rabbits in the ciclosporin/methylprednisolone group died. In another study, Balloy *et al.* administered *Aspergillus fumigatus conidia* intratracheally to mice exposed to either a glucocorticoid, vinblastine or no pre-treatment.[20] Whereas all animals receiving steroid or vinblastine died, bronchoalveolar lavage (BAL) fluid from steroid-treated mice contained

significantly more granulocytes and higher haemoglobin levels (a measure of alveolocapillary injury) but lower levels of TNF-α than vinblastine-treated animals. Gross pathology of lung tissue in steroid-treated animals showed large foci of pneumonia and exudative bronchiolitis with destruction of bronchi and alveoli. In contrast, vinblastine-treated animals showed less inflammatory exudate, but dissemination of fungal organisms. The implication, according to the authors, was that *Aspergillus* morbidity in the setting of corticosteroids, in contrast to granulocytopenia, is more associated with dysregulation of the host response, as opposed to direct invasion by the organism.

An association between corticosteroid use and tuberculosis reactivation is well known.[21] Evidence suggests that patients on steroids are at higher risk both of reactivation and of disseminated disease. In a recent case–control study, the adjusted odds ratio of tuberculosis among current users of glucocorticoids was 4.9.[22] This rose to 7.7 if the steroid dose was the equivalent of 15 mg prednisone daily or higher. Tam *et al.* reviewed 526 patients with systemic lupus erythematosus treated at a Hong Kong hospital, and found that receiving either intravenous 'pulse' methylprednisolone or high cumulative doses of prednisolone was an independent risk factor for developing tuberculosis.[23] Among those with tuberculosis, 67 per cent had miliary or extrapulmonary disease. In one report of cancer patients where tuberculosis reactivated in the setting of glucocorticoids or chemotherapy, infection was more severe and was associated with 17 per cent mortality (Fig. 19.2).

Ciclosporin and tacrolimus

Ciclosporin, developed during the 1980s, and tacrolimus (formerly FK506), developed during the 1990s, exert their effects predominantly via helper T-lymphocytes by suppressing the transcription of IL-2 and a number of other inflammatory cytokines. These drugs have greatly improved the survival of solid-organ transplant recipients, and have also been effective against GVHD in allogeneic bone marrow or peripheral blood stem-cell transplant recipients as well as a number of rheumatologic and inflammatory conditions listed in Box 19.2.

Ciclosporin and tacrolimus are each associated with a significant risk of infection. In a large European randomized trial, 40 per cent of 545 liver transplant recipients from eight academic centres receiving either ciclosporin or tacrolimus developed significant infections, and 20 per cent developed sepsis. These patients were, however, on multiple immunosuppressive drugs and the exact role of ciclosporin or tacrolimus cannot be established. CMV was a particularly common pathogen, occurring in 15–20 per cent of patients.[1] CMV pneumonia is already common in lung transplant patients, occurring with greater frequency than in renal or cardiac transplant recipients.[24] Whether the risk of CMV infection differs between patients taking ciclosporin and tacrolimus is unclear. One study of lung transplant recipients showed no difference in the incidence of bacterial or CMV infections,[25] while a study in renal transplant patients found a higher rate in those receiving tacrolimus rather than ciclosporin.[26]

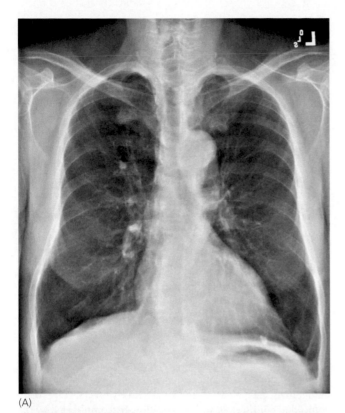

(A)

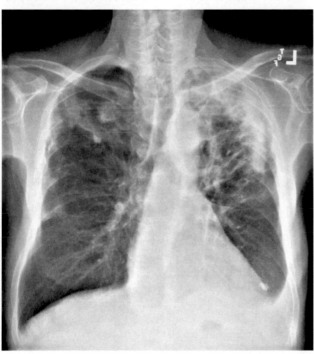

(B)

Fig. 19.2 Extensive upper lobe changes of tuberculosis rapidly developed in this 72-year-old man with metastatic colorectal cancer receiving chemotherapy. He was premedicated every 2 weeks with high-dose dexamethasone. His weight loss, failure to thrive and radiographic findings were initially attributed to the underlying cancer. (a) Baseline radiograph; (b) radiograph after 3 months of intermittent corticosteroids.

Box 19.2 Conditions treated with ciclosporin (cyclosporin, cyclosporine) and tacrolimus

- Allogeneic bone-marrow/stem-cell transplantation
- Behçet's disease
- Dermatomyositis
- Dry eye disease (administered as eye drops)
- Inflammatory bowel disease
- Juvenile rheumatoid arthritis
- Lupus membranous nephritis
- Pemphigus vulgaris
- Polymyositis
- Psoriasis
- Sjögren's syndrome
- Solid-organ transplantation
- Steroid-resistant nephrotic syndrome
- Systemic sclerosis
- Wegener's granulomatosis

Purine analogues

Fludarabine, cladribine and pentostatin are purine nucleoside analogues that represent a major advance in the treatment of indolent lymphoid malignancies, including chronic lymphocytic leukaemia (CLL), hairy-cell leukaemia, Waldenström's macroglobulinaemia, and non-Hodgkin's lymphoma. Fludarabine causes myelosuppression, particularly leucopenia with a profound decrease in all lymphocyte subsets. Of the three drugs, cladribine is the most myelosuppressive and pentostatin the least.[7]

Among patients with untreated or conventionally treated CLL, pulmonary infections most commonly result from B-cell dysfunction and involvement by encapsulated bacteria. However, in the setting of fludarabine treatment and resultant lymphocytopenia, the spectrum of pulmonary pathogens changes. Cases of *Pneumocystis* pneumonia and sepsis from *Listeria monocytogenes*, for example, have been reported. In a retrospective analysis of 402 patients treated with fludarabine at the MD Anderson Cancer Center, infection with *Pneumocystis jiroveci* or *Listeria* occurred in 14 patients.[27] Although all patients with these infections were also receiving prednisone, those taking steroids were also more likely to have refractory disease and lower CD4 counts. A relationship between opportunistic infections and the use of fludarabine and prednisone was reported in another series of 2269 patients.[28] Most infections occur in the first three cycles of therapy, but data continue to accumulate on the significant late morbidity caused by herpes viruses, such as CMV, *Varicella zoster* and *Herpes simplex*.[29] Other pathogens that have been associated with fludarabine include *Nocardia asteroides*, *Cryptococcus neoformans*, *Aspergillus fumigatus* and the *Mycobacterium* genus of organisms.[28,30–32] Some of these infections may be related to the presence of CLL itself.

A number of factors, in association with the use of purine analogues, are thought to additionally predispose patients to infection, including prior chemotherapy, advanced cancer stage, older age, poor performance status, low baseline CD4 counts, pre-treatment pancytopenia, hypogammaglobulinaemia, concurrent treatment with steroids and other immunosuppressive agents, as well as prior purine analogue therapy.[27,30]

The spectrum of infections associated with cladribine and pentostatin is thought to be similar to that of fludarabine. In a phase II trial of 39 patients taking pentostatin, severe, life-threatening or lethal bacterial infections developed in 34 per cent of patients. Cases included pneumonia, cellulitis, bronchitis and sepsis. Moreover, opportunistic pathogens were identified in 26 per cent of patients, including *Herpes simplex*, *H. zoster*, *Candida albicans* and *Pneumocystis jiroveci*. All infections occurred within 6 weeks of initiating therapy.[33]

Alemtuzumab

Alemtuzumab is used for fludarabine-refractory CLL; it is also effective at preventing GVHD among allogeneic transplant recipients receiving non-myeloablative conditioning regimens.[34] Although concurrent treatment with other forms of immunosuppression complicates any assessment of the prevalence of infectious complications, available data suggest the risk of infection is high. In a study of 93 patients treated with alemtuzumab at 21 centres, infection developed in 55 per cent of patients, and was severe in 27 per cent. Opportunistic pulmonary infections included *Pneumocystis* pneumonia, aspergillosis, mucormycosis and cryptococcosis.[35] Other infections of concern are invasive non-candidal fungi, such as *Histoplasma capsulatum*, as well as such parasitic pathogens as *Toxoplasma gondii*. CMV is thought to be a particularly significant problem. Among 101 patients receiving a non-myeloablative regimen containing alemtuzumab, GVHD – usually a predictor of CMV infection – occurred in only 4 per cent of recipients. But CMV infection, as determined by PCR positivity on two consecutive assays, occurred among 85 per cent of patients at risk.[36] All patients who developed clinically apparent CMV disease died; infection was common 100 or more days after transplant. Taken in sum, this is much higher than the 40–50 per cent infection rate of CMV expected among recipients of unmanipulated stem-cell grafts. In another study with use of alemtuzumab for lymphoproliferative disease, the incidence of CMV reactivation among recipients taking alemtuzumab was 66.7 per cent, versus 37 per cent in the non-alemtuzumab group.[2]

Rituximab

Rituximab, like alemtuzumab, is useful for haematological malignancies but is targeted towards those with abnormal B-cells. It is used widely, both alone and in combination with other chemotherapy drugs, against such disorders as B-cell non-Hodgkin's lymphoma, B-cell CLL, and post-transplant lymphoproliferative disorders. Rituximab has also been used to treat idiopathic thrombocytopenic purpura (ITP), systemic lupus erythematosus, rheumatoid arthritis and other autoimmune conditions associated with abnormal antibody production.

In the setting of treatment against B-cells, the predicted pulmonary pathogens include those associated with abnormal humoral immunity. Encapsulated bacteria and herpes virus can cause infections several months following therapy.[37]

It is notable, however, that rituximab is relatively well tolerated. In a trial of 35 patients receiving rituximab for ITP, one patient died from bacterial pneumonia 13 weeks following his last treatment. However, this patient suffered from significant underlying lung disease and no other notable infections were encountered; consequently, the authors concluded that the pneumonia was most likely coincidental.[38] In some patients, however, repeated sinusitis, bronchitis and pneumonitis have been seen.

TNF antagonists

Tumour necrosis factor α is a key cytokine in the inflammatory response. TNF antagonists represent an important recent development in the treatment of several inflammatory conditions, including rheumatoid arthritis and inflammatory bowel disease. Three drugs are currently in use: etanercept, infliximab and adalimumab.

As TNF is ordinarily considered crucial in the body's defences, the risk of infection among patients taking TNF antagonists has been studied extensively. Data have been somewhat conflicting, and are additionally complicated by the fact that rheumatoid arthritis patients may already be at somewhat increased risk for infection relative to the general population. In one randomized double-blind placebo-controlled trial, 636 patients with rheumatoid arthritis were randomized to receive adalimumab or a placebo. Over 24 weeks, there were no statistically significant differences between the two groups regarding the rates of infections (52.2 vs. 49.4 per cent for adalimumab and placebo, respectively) or serious infections (1.3 vs. 1.9 per cent).[39] However, another study comparing a single patient cohort before and after anti-TNF therapy found a 20-fold increase in the incidence rates of serious infection during treatment with this class of medication.[40] Most of the infections were bacterial.

The pulmonary infection most closely associated with TNF antagonists has been tuberculosis. In 2001, Keane *et al.* reported 70 cases of tuberculosis occurring during infliximab use.[41] Of these, 64 cases occurred in countries with a low incidence of TB, although this finding may relate to the availability of these costly medications. Twenty-two patients (31 per cent) had pulmonary tuberculosis, and 56 patients (80 per cent) had evidence of extrapulmonary disease. The median duration of anti-TNF therapy at the time of diagnosis was 12 weeks. Since then, numerous other case reports have cumulatively described over 300 episodes of tuberculosis occurring in the setting of these medications. Interestingly, the risk of tuberculosis among patients taking infliximab (41 cases per 100 000 person-years) is higher than that among patients taking etanercept (20.7 per 100 000 person-years). However, the rate for both drugs is clearly higher than the baseline rate among patients not taking TNF antagonists (6.2 per 100 000 person-years).[42] The TB risk associated with adalimumab is less well known, but is likely also increased; the increased risk is mentioned on the package insert. In most cases, tuberculosis develops early, suggesting reactivation rather than acquired disease.

A number of authors have implicated other pulmonary opportunistic infections among patients taking TNF antagonists, including *Aspergillus fumigatus* pneumonia, histoplasmosis, *Pneumocystis* pneumonia and pulmonary cryptococcosis.[43–46] Histoplasmosis is an infection by *Histoplasma capsulatum*, a fungus endemic to the Ohio River Valley, the Caribbean and Central America. It has been estimated that 20 per cent of the US population has been exposed to *H. capsulatum*. Clinical manifestation may result from an acute infection, reactivation of a latent infection, or reinfection. Lee *et al.* reported 10 cases of *H. capsulatum* infection occurring within 6 months of beginning TNF blockade, one of which was fatal.[45] Nine of these patients had been treated with infliximab and one etanercept, but all had been taking other forms of immunosuppression concurrent with the TNF antagonists. The rate of clinically significant *Histoplasma* infections among patients receiving TNF antagonists has been estimated to be 16.7 per 100 000 patients – less than that seen with tuberculosis, but higher than in the general population.[47]

The exact relationship between TNF blockade and these granulomatous pulmonary infections is unclear. It is known that *M. tuberculosis*-infected tissues express TNF during the quiescent phase of infection.[48] Mice deficient in functional TNF are more susceptible to fulminant tuberculosis[49] and *H. capsulatum* infection.[50] Therefore, these suggest that this cytokine may help suppress the spread of pathogens from the primary site of infection.

Temozolomide

Temozolomide is an alkylating agent widely used for treatment of melanoma and brain tumours. During prolonged use, lymphopenia and very low CD4 counts have been noted. This has been associated with documented opportunistic infection with *Pneumocystis* and *Aspergillus* pneumonia. In addition, cases of *H. simplex* and *H. zoster* have been seen.[9]

Methotrexate

MTX is used against a variety of medical conditions, and is prescribed in both high and low doses depending on the condition targeted. At high doses it is used as a chemotherapy agent against such malignancies as head and neck, breast, bladder and gastric cancer, as well as acute leukaemia, lymphoma and sarcoma. At lower doses, MTX serves as an anti-inflammatory agent and is employed, often concurrently with steroids, in the treatment of rheumatoid arthritis, psoriasis and dermatomyositis. MTX has also been used (although fewer data are available) against primary biliary cirrhosis, sarcoid and asthma.

At high doses, MTX precipitates lung infection predominantly through bone marrow suppression, with resultant sepsis and/or bacterial infection. At low doses, the risk of infection is low and complicated by the fact that MTX is often used with steroids. MTX has been linked to those organisms associated with abnormal lymphocyte function, including *Pneumocystis jiroveci*, CMV, *Nocardia asteroides* and *Cryptococcus neoformans*. van der Veen *et al.* compared 77 rheumatoid arthritis patients treated with MTX to 151 patients not

receiving this drug. The relative risk for respiratory infections was 1.43 (95% CI: 0.96–2.14), and was independent of prednisone use.[51] In another prospective study, the risk of infection was associated with rheumatoid arthritis severity and the proximity to the initiation of therapy; patients were most vulnerable to opportunistic infection during the first 1.5 years of treatment.[52]

INVESTIGATIONS

Immunocompromised patients who develop pulmonary infiltrates frequently undergo empirical treatment against common bacterial infection prior to performing invasive diagnostic procedures, such as fibreoptic bronchoscopy, fine-needle aspiration and surgical lung biopsy. This had particularly been the case with CLL, prior to purine analogue therapy, where encapsulated bacterial infection was most likely. However, when immunosuppressed patients have received purine antagonists or alemtuzumab, the spectrum of potential pathogens is considerably broader. Consideration should be given to proceeding towards early invasive testing to establish a diagnosis. It is also important to recognize that multiple pathogens may cause infection simultaneously. Several non-invasive serological tests are now available, such as the CMV serum antigen, which may be helpful for disease surveillance or for detecting early infection, and should be used in this setting.

TREATMENT/MANAGEMENT OF COMPLICATIONS

Given the high incidence and severity of pulmonary infections associated with some drugs with significant consequence, prophylaxis is advised in certain clinical contexts. However, specific patient populations who would most benefit from various prophylactic strategies remain controversial, and no standard approaches have yet been developed.

Glucocorticoid effects

There exist no universally accepted guidelines regarding *Pneumocystis* prophylaxis among HIV-negative patients receiving steroids. The incidence of *Pneumocystis* pneumonia is generally low, except for the transplant populations where prophylaxis is already given in most centres. In one study, *Pneumocystis* infection developed in only 11 of 180 patients (6 per cent) treated for Wegener's granulomatosis over a 24-year period.[53] Moreover, trimethoprim–sulfamethoxazole (TMP–SMZ), the drug of choice, is associated with a long list of potential side-effects, including rash, fever, hepatitis, renal dysfunction and myelosuppression. At our institution we commonly prescribe prophylaxis to any patient with an underlying malignancy, usually haematologic, receiving the equivalent of 20 mg daily of prednisone for longer than 4 weeks. Following CD4 counts and treating when under 300 cells/μL has also been suggested.

Patients with a positive purified protein derivative (PPD) skin-test being treated with glucocorticoids should receive prophylaxis against tuberculosis reactivation. Decisions regarding the need for prophylaxis are complicated by the fact that patients taking steroids are less likely to express a reaction to a positive reaction regardless of prior exposure. Therefore, treatment is typically recommended not only for patients having a positive PPD, but also for those with a history of previous tuberculosis or radiographic features of prior tuberculosis that were not treated adequately. For those taking the equivalent of 15 mg prednisone daily or more, a positive PPD is considered to be 5 mm induration.

Ciclosporin and tacrolimus effects

Given the increased risk and severity of CMV among transplant recipients taking ciclosporin or tacrolimus, combinations of prophylactic and pre-emptive strategies are recommended. CMV prophylaxis has already been shown effective in some specific patient groups. In kidney transplant recipients who received antithymocyte lymphocytic antibodies, the incidence of CMV was 13 per cent in those with prophylaxis and 43 per cent in those without.[54] For haematological transplants, recipients and donors are typically tested for serum anti-CMV IgG antibodies prior to transplant. Those considered at high risk for developing CMV infection may receive prophylaxis with ganciclovir after engraftment.[55] Other antiviral agents, such as valacyclovir and foscarnet, as well as intravenous immunoglobulin may be effective either alone or in combination with other drugs.[56] Early or pre-emptive therapy is contingent on a number of available tests designed to detect early CMV infection, such as viral culture, shell vial assays, antigenaemia assays and polymerase chain reaction, which may have reduced the incidence of CMV infection among haematologic transplant patients. For solid-organ transplants, the approach to prophylactic therapy is specific to the organ type and risk of the patient.

Purine analogue effects

For patients receiving these drugs, who are particularly susceptible to encapsulated bacteria, vaccination can be effective in decreasing the incidence of *Haemophilus influenzae* type b, as well as pneumococcal and meningococcal infections. Vaccines must be administered prior to treatment. As patients, even under optimal conditions, may not manifest an adequate response, the vaccination of household contacts may also be helpful in limiting exposure. The majority of opportunistic infections occur among patients taking glucocorticoids, so it is prudent to avoid treatment concurrently with systemic steroids if possible. A low threshold to antibiotic treatment (i.e. amoxicillin–clavulanate) should exist when patients manifest signs and symptoms of respiratory infections. Finally, prophylaxis with TMP–SMX and acyclovir should be considered among patients who have advanced disease, previous cytotoxic therapy, or CD4 counts below 200 cells/μL.

Alemtuzumab effects

For patients on this drug, prophylaxis against CMV and *Pneumocystis* is advised.[57] Prophylaxis against *Pneumocystis* should continue for a minimum of 2 months from the last dose of alemtuzumab, until the CD4 count exceeds 200 cells/μL. One study found a lower rate of CMV infection among patients receiving prophylaxis with pre-transplant ganciclovir and post-transplant valacyclovir.[56] The German Society of Haematology and Oncology recommends that all patients receiving alemtuzumab receive acyclovir or valacyclovir prophylaxis against herpes, but did not recommend CMV prophylaxis in the absence of sufficient clinical data.[58]

For those patients with CLL who have already had frequent pulmonary infections, antibiotic chemoprophylaxis may be considered. Antibiotic prophylaxis may take many forms; one popular method is to administer antibiotics of different classes for the first 10 days of every month (i.e. sulfa medication in the first month, fluoroquinolone in the second month, and β-lactam in the third month). Intravenous immunoglobulin therapy may be helpful among patients with diminished IgG levels and frequent bacterial infections, but it is expensive and entails a side-effect profile similar to blood transfusions.

TNF antagonist effects

Drug inserts for this medication class now recommend screening for tuberculosis before beginning therapy. Caution, however, is advised in interpreting results of PPD testing. Many patients may have received the BCG vaccine or are anergic secondary to their underlying disease and/or other immunosuppressive agents, raising the possibility of a false-negative result. Also, a non-reactive finding would not be predictive of novel cases of tuberculosis. Consequently, *ex-vivo* assays have been developed that quantify the release of interferon-γ from T-cells responding to TB-specific mycobacterial antigens. Several studies suggest these assays to be more helpful in detecting latent tuberculosis infection than PPD skin-testing alone.[59,60] Any patient who develops tuberculosis should have their TNF antagonist stopped. For those with latent infection, prophylaxis with isoniazid should be prescribed for 9 months. There is some debate regarding when patients receiving treatment for latent infections may begin TNF antagonists. Some have suggested a brief course of a few weeks or 2 months of isoniazid first, while others have advised completing a full course of prophylaxis.[61] Data are currently not sufficient to warrant definite guidelines.

Patients with active *H. capsulatum* infections should not receive TNF antagonists. Because some cases may represent reactivation, histoplasmosis may occur outside of endemic areas. Nevertheless, there are insufficient data to currently support screening all patients with a consistent travel history. Confounding issues include the large percentage of the population already exposed to *H. capsulatum*, as well as uncertainty regarding whether patients taking TNF antagonists are more susceptible to develop primary infection or reactivation of a previous exposure. Future research is needed to determine the most appropriate and cost-effective strategies for endemic regions.

COMPLICATIONS

Drug-induced pulmonary infections often require multiple anti-infection treatments administered simultaneously, even if invasive procedures have been carried out. It must be assumed that more than one infection can be present and extreme vigilance has to be used. Toxicity of these multiple medications must be carefully managed, particularly with the effects of many immunosuppressives on metabolism and disposition of drugs.

FUTURE DIRECTIONS

Over the past 20 years, immunosuppressive and anti-inflammatory medications have evolved from non-selective agents, which inhibit multiple cell lines, to newer more specific agents, which target individual cell lines or markers. While these targeted drugs are generally better tolerated than corticosteroids, the level of immunosuppression can nevertheless be profound and long-lasting, leading to serious infection. Knowledge of the types of infections associated with each drug may teach us more about the mechanisms leading to opportunistic infections in specific patients. This information might be helpful in determining which groups to target for prophylaxis. Optimization of dosing and length of therapy, as well as mechanisms to enhance the immune response to specific antigens via cytokines and dendritic cells, is still in development and may cut the risk of infections while preserving the therapeutic benefit.

KEY POINTS

- Currently used immunosuppressive drugs increase the risk of serious infections, depending on the nature of the drug, its dose and duration, and the underlying disease of the patient.

- Defects induced by newer targeted drugs, such as the purine antagonists and monoclonal antibodies directed against lymphocytes, may be profound and last for prolonged periods after the drug is discontinued.

- The type of immunologic deficit induced by any specific drug is helpful in predicting the types of infection with which it is associated, but relationships can be obscure and vigilance is needed.

- Early diagnostic intervention to establish the nature of any infection should be considered given the wide spectrum of infections that can be seen in some settings.

- Prophylactic regimens should be used when high-risk situations can be identified and regimens exist.

REFERENCES

1. European FK506 Multicentre Liver Study Group. Randomised trial comparing tacrolimus and cyclosporin in prevention of liver allograft rejection. *Lancet* 1994;**344**:423–8.
2. Martin SISI, Marty FMFM, Fiumara KK, *et al.* Infectious complications associated with alemtuzumab use for lymphoproliferative disorders. *Clin Infect Dis* 2006;**43**:16–24.
3. Stuck AE, Minder CE, Frey FJ. Risk of infectious complications in patients taking glucocorticosteroids. *Rev Infect Dis* 1989;**11**:954–63.
4. Kim SW, Rhee HJ, Ko J, *et al.* Inhibition of cytosolic phospholipase A2 by annexin I: specific interaction model and mapping of the interaction site. *J Biol Chem* 2001;**276**:15712–19.
5. Antonicelli F, De Coupade C, Russo-Marie F, Le Garrec Y. CREB is involved in mouse annexin A1 regulation by cAMP and glucocorticoids. *Eur J Biochem* 2001;**268**:62–9.
6. De Bosscher K, Vanden Berghe W, Haegeman G. The interplay between the glucocorticoid receptor and nuclear factor-κB or activator protein-1: molecular mechanisms for gene repression. *Endocr Rev* 2003;**24**:488–522.
7. Samonis GG, Kontoyiannis DDP. Infectious complications of purine analog therapy. *Curr Opin Infect Dis* 2001;**14**:409–13.
8. Cheson BBD. Infectious and immunosuppressive complications of purine analog therapy. *J Clin Oncol* 1995;**13**:2431–48.
9. Su YYB, Sohn SS, Krown SESE, *et al.* Selective CD4+ lymphopenia in melanoma patients treated with temozolomide: a toxicity with therapeutic implications. *J Clin Oncol* 2004;**22**:610–16.
10. Genestier L, Paillot R, Fournel S, *et al.* Immunosuppressive properties of methotrexate: apoptosis and clonal deletion of activated peripheral T cells. *J Clin Invest* 1998;**102**:322–8.
11. Gerards AH, de Lathouder S, de Groot ER, Dijkmans BA, Aarden LA. Inhibition of cytokine production by methotrexate: studies in healthy volunteers and patients with rheumatoid arthritis. *Rheumatology* (Oxford) 2003;**42**:1189–96.
12. Yale SH, Limper AH. *Pneumocystis carinii* pneumonia in patients without acquired immunodeficiency syndrome: associated illness and prior corticosteroid therapy. *Mayo Clin Proc* 1996;**71**:5–13.
13. Sepkowitz KA, Brown AE, Telzak EE, Gottlieb S, Armstrong D. *Pneumocystis carinii* pneumonia among patients without AIDS at a cancer hospital. *J Am Med Assoc* 1992;**267**:832–7.
14. Ewig S, Bauer T, Schneider C, *et al.* Clinical characteristics and outcome of *Pneumocystis carinii* pneumonia in HIV-infected and otherwise immunosuppressed patients. *Eur Respir J* 1995;**8**:1548–53.
15. Masur H, Lane HC, Kovacs JA, Allegra CJ, Edman JC. NIH Conference. *Pneumocystis* pneumonia: from bench to clinic. *Ann Intern Med* 1989;**111**:813–26.
16. Walzer PD, LaBine M, Redington TJ, Cushion MT. Lymphocyte changes during chronic administration of and withdrawal from corticosteroids: relation to *Pneumocystis carinii* pneumonia. *J Immunol* 1984;**133**:2502–8.
17. Mansharamani NG, Balachandran D, Vernovsky I, Garland R, Koziel H. Peripheral blood CD4+ T-lymphocyte counts during *Pneumocystis carinii* pneumonia in immunocompromised patients without HIV infection. *Chest* 2000;**118**:712–20.
18. Mencacci A, Cenci E, Bacci A, *et al.* Cytokines in candidiasis and aspergillosis. *Curr Pharm Biotech* 2000;**1**:235–51.
19. Berenguer J, Allende MC, Lee JW, *et al.* Pathogenesis of pulmonary aspergillosis: granulocytopenia versus cyclosporine and methylprednisolone-induced immunosuppression. *Am J Respir Crit Care Med* 1995;**152**:1079–86.
20. Balloy V, Huerre M, Latge JP, Chignard M. Differences in patterns of infection and inflammation for corticosteroid treatment and chemotherapy in experimental invasive pulmonary aspergillosis. *Infect Immun* 2005;**73**:494–503.

21. Sahn SA, Lakshminarayan S. Tuberculosis after corticosteroid therapy. *Br J Dis Chest* 1976;**70**:195–205.

22. Jick SS, Lieberman ES, Rahman MU, Choi HK. Glucocorticoid use, other associated factors, and the risk of tuberculosis. *Arthritis Rheum* 2006;**55**:19–26.

23. Tam LS, Li EK, Wong SM, Szeto CC. Risk factors and clinical features for tuberculosis among patients with systemic lupus erythematosus in Hong Kong. *Scand J Rheumatol* 2002;**31**: 296–300.

24. Ison MGMG, Fishman JAJA. Cytomegalovirus pneumonia in transplant recipients. *Clin Chest Med* 2005;**26**:691–705,viii.

25. Treede H, Klepetko W, Reichenspurner H, *et al.* Tacrolimus versus cyclosporine after lung transplantation: a prospective, open, randomized two-center trial comparing two different immunosuppressive protocols. *J Heart Lung Transplant* 2001;**20**:511–17.

26. Reichenberger F, Dickenmann M, Binet I, *et al.* Diagnostic yield of bronchoalveolar lavage following renal transplantation. *Transpl Infect Dis* 2001;**3**:2–7.

27. Anaissie EJ, Kontoyiannis DP, O'Brien S, *et al.* Infections in patients with chronic lymphocytic leukemia treated with fludarabine. *Ann Intern Med* 1998;**129**:559–66.

28. Byrd JC, Hargis JB, Kester KE, *et al.* Opportunistic pulmonary infections with fludarabine in previously treated patients with low-grade lymphoid malignancies: a role for *Pneumocystis carinii* pneumonia prophylaxis. *Am J Hematol* 1995;**49**: 135–42.

29. Weiss MMA. A phase I and II study of pentostatin (Nipent) with cyclophosphamide for previously treated patients with chronic lymphocytic leukemia. *Semin Oncol* 2000;**27**(2 Suppl 5): 41–3.

30. Bergmann L, Fenchel K, Jahn B, Mitrou PS, Hoelzer D. Immunosuppressive effects and clinical response of fludarabine in refractory chronic lymphocytic leukemia. *Ann Oncol* 1993;**4**:371–5.

31. Deane M, Gor D, Macmahon ME, *et al.* Quantification of CMV viraemia in a case of transfusion-related qraft-versus-host disease associated with purine analogue treatment. *Br J Haematol* 1997;**99**:162–4.

32. Wijermans PW, Gerrits WB, Haak HL. Severe immunodeficiency in patients treated with fludarabine monophosphate. *Eur J Haematol* 1993;**50**:292–6.

33. Dillman RO, Mick R, McIntyre OR. Pentostatin in chronic lymphocytic leukemia: a phase II trial of Cancer and Leukemia group B. *J Clin Oncol* 1989;**7**:433–8.

34. Kottaridis PD, Milligan DW, Chopra R, *et al.* In-vivo CAMPATH-1H prevents graft-versus-host disease following nonmyeloablative stem cell transplantation. *Blood* 2000;**96**: 2419–25.

35. Keating MJMJ, Flinn II, Jain VV, *et al.* Therapeutic role of alemtuzumab (Campath-1H) in patients who have failed fludarabine: results of a large international study. *Blood* 2002;**99**:3554–61.

36. Chakrabarti S, Mackinnon S, Chopra R, *et al.* High incidence of cytomegalovirus infection after nonmyeloablative stem cell transplantation: potential role of Campath-1H in delaying immune reconstitution. *Blood* 2002;**99**:4357–63.

37. Plosker GL, Figgitt DP. Rituximab: a review of its use in non-Hodgkin's lymphoma and chronic lymphocytic leukaemia. *Drugs* 2003;**63**:803–43.

38. Braendstrup PP, Bjerrum OWOW, Nielsen OJOJ, *et al.* Rituximab chimeric anti-CD20 monoclonal antibody treatment for adult refractory idiopathic thrombocytopenic purpura. *Am J Hematol* 2005;**78**:275–80.

39. Furst DE, Schiff MH, Fleischmann RM, *et al.* Adalimumab, a fully human anti tumor necrosis factor-alpha monoclonal antibody, and concomitant standard antirheumatic therapy for the treatment of rheumatoid arthritis: results of STAR (Safety Trial of Adalimumab in Rheumatoid Arthritis). *J Rheumatol* 2003;**30**:2563–71.

40. Phillips K, Husni ME, Karlson EW, Coblyn JS. Experience with etanercept in an academic medical center: are infection rates increased? *Arthritis Rheum* 2002;**47**:17–21.

41. Keane JJ, Gershon SS, Wise RRP, *et al.* Tuberculosis associated with infliximab, a tumor necrosis factor alpha-neutralizing agent. *New Engl J Med* 2001;**345**:1098–104.

42. Crum NFNF, Lederman ERER, Wallace MRMR. Infections associated with tumor necrosis factor-alpha antagonists. *Medicine* 2005;**84**:291–302.

43. De Rosa FG, Shaz D, Campagna AC, *et al.* Invasive pulmonary aspergillosis soon after therapy with infliximab, a tumor necrosis factor-alpha-neutralizing antibody: a possible healthcare-associated case? *Infect Control Hosp Epidemiol* 2003;**24**:477–82.

44. Hage CA, Wood KL, Winer-Muram HT, *et al.* Pulmonary cryptococcosis after initiation of anti-tumor necrosis factor-alpha therapy. *Chest* 2003;**124**:2395–7.

45. Lee JH, Slifman NR, Gershon SK, *et al.* Life-threatening histoplasmosis complicating immunotherapy with tumor necrosis factor alpha antagonists infliximab and etanercept. *Arthritis Rheum* 2002;**46**:2565–70.

46. Tai TL, O'Rourke KP, McWeeney M, *et al. Pneumocystis carinii* pneumonia following a second infusion of infliximab. *Rheumatology* (Oxford) 2002;**41**:951–2.

47. Wallis RS, Broder MS, Wong JY, Hanson ME, Beenhouwer DO. Granulomatous infectious diseases associated with tumor necrosis factor antagonists. *Clin Infect Dis* 2004;**38**:1261–5.

48. Flynn JL, Scanga CA, Tanaka KE, Chan J. Effects of aminoguanidine on latent murine tuberculosis. *J Immunol* 1998;**160**:1796–803.

49. Flynn JJL, Goldstein MMM, Chan JJ, *et al.* Tumor necrosis factor-alpha is required in the protective immune response against *Mycobacterium tuberculosis* in mice. *Immunity* 1995;**2**:561–72.

50. Allendoerfer RR, Deepe GGS. Blockade of endogenous TNF-alpha exacerbates primary and secondary pulmonary histoplasmosis by differential mechanisms. *J Immunol* 1998;**160**:6072–82.

51. van der Veen MJ, van der Heide A, Kruize AA, Bijlsma JW. Infection rate and use of antibiotics in patients with rheumatoid arthritis treated with methotrexate. *Ann Rheum Dis* 1994;**53**:224–8.

52. Boerbooms AAM, Kerstens PPJ, van Loenhout JJW, Mulder JJ, van de Putte LLB. Infections during low-dose methotrexate treatment in rheumatoid arthritis. *Semin Arthritis Rheum* 1995;**24**:411–21.

53. Ognibene FP, Shelhamer JH, Hoffman GS, *et al. Pneumocystis carinii* pneumonia: a major complication of immunosuppressive therapy in patients with Wegener's granulomatosis. *Am J Respir Crit Care Med* 1995;**151**(3 Pt 1):795–9.

54. Said TT, Nampoory MMRN, Johny KKV, *et al.* Cytomegalovirus prophylaxis with ganciclovir in kidney transplant recipients receiving induction antilymphocyte antibodies. *Transplant Proc* 2004;**36**:1847–9.

55. Verkruyse LA, Storch GA, Devine SM, Dipersio JF, Vij R. Once daily ganciclovir as initial pre-emptive therapy delayed until threshold CMV load > or = 10 000 copies/mL: a safe and effective strategy for allogeneic stem cell transplant patients. *Bone Marrow Transplant* 2006;**37**:51–6.

56. Kline J, Pollyea DA, Stock W, *et al.* Pre-transplant ganciclovir and post transplant high-dose valacyclovir reduce CMV infections after alemtuzumab-based conditioning. *Bone Marrow Transplant* 2006;**37**:307–10.

57. Thursky KA, Worth LJ, Seymour JF, Miles Prince H, Slavin MA. Spectrum of infection, risk and recommendations for prophylaxis and screening among patients with lymphoproliferative disorders treated with alemtuzumab. *Br J Haematol* 2006;**132**:3–12.

58. Sandherr M, Einsele H, Hebart H, *et al.* Antiviral prophylaxis in patients with haematological malignancies and solid tumours: Guidelines of the Infectious Diseases Working Party (AGIHO) of the German Society of Haematology and Oncology (DGHO). *Ann Oncol* 2006;**17**:1051–9.

59. Bocchino M, Matarese A, Bellofiore B, *et al.* Performance of two commercial blood IFN-gamma release assays for the detection of *Mycobacterium tuberculosis* infection in patient candidates for anti-TNF-alpha treatment. *Eur J Clin Microbiol Infect Dis* 2008;**27**:907–13.

60. Ponce de Leon D, Acevedo-Vasquez E, Alvizuri S, *et al.* Comparison of an interferon-gamma assay with tuberculin skin testing for detection of tuberculosis infection in patients with rheumatoid arthritis in a TB-endemic population. *J Rheumatol* 2008;**35**:776–81.

61. Gardam MAMA, Keystone ECEC, Menzies RR, *et al.* Anti-tumour necrosis factor agents and tuberculosis risk: mechanisms of action and clinical management. *Lancet Infect Dis* 2003;**3**:148–55.

20

Therapy-induced neoplasms

LOIS B TRAVIS, NOREEN AZIZ

INTRODUCTION

Given continuing advances in early detection, supportive care and therapy, the number of cancer survivors in the United States has tripled since 1971, and is increasing by 2 per cent each year.[1] The almost 10 million cancer survivors in 2001 comprised approximately 3.5 per cent of the US population, and the 5-year relative survival rate after a diagnosis of all cancers taken together now approaches 66 per cent.[2] As survival statistics continue to improve, ascertainment and quantification of the late sequelae of cancer and its treatment become essential, with the occurrence of second primary cancers representing one of the most severe complications.[3,4] The number of survivors with multiple primary malignancies is rising, with independent cancers representing about one in six incident cancers in 2003, as registered by the National Cancer Institute's (NCI) Surveillance, Epidemiology, and End Results (SEER) programme.[2]

Further, second solid tumours are the most common cause of death among several groups of long-term survivors, including Hodgkin's lymphoma (HL) patients.[5] It should always be recognized that second solid tumours can reflect not only the late effects of treatment, but also the contribution of lifestyle influences, environmental exposures, host factors, and combinations of factors, including gene–environment and gene–gene interactions.[6] Second primary cancers were recently categorized into three major groups,[7] according to major aetiological factors (i.e. treatment-related), syndromic, and those due to shared aetiological factors – with the non-exclusivity of these groups underscored.

In the present chapter we will focus on selected highlights and findings in treatment-associated lung tumours, concentrating on survivors of adult cancer. Given the high baseline incidence rates of lung cancer in the general population,[2] even a small increase in the relative risk can translate into large absolute risk, which is considered the optimal measure of disease burden in a population. The overall magnitude of lung cancer incidence, and the high associated mortality rate,[2] emphasizes the importance of its occurrence as a second primary cancer in cancer survivors. Given the known role of tobacco use in cancers which occur among lung cancer survivors, the occurrence of second primary lung cancers will not be addressed in this patient population; instead the reader is referred to the summary by Caporaso and colleagues.[8]

LUNG CANCER AFTER BREAST CANCER

Breast cancer is the leading incident cancer among women in the United States.[2] As the relative survival rates (at 5, 10 and 20 years) continue to increase,[2] the impact of long-term treatment-related effects, including second primary cancers, assumes greater clinical significance. It is estimated that approximately 12 per cent of women diagnosed with breast cancer will develop a second primary neoplasm.[9]

In a recent analysis that included 335 191 females registered in the National Cancer Institute's SEER Programme, observed numbers of subsequent cancers in women diagnosed with a first primary breast cancer between 1973 and 2000 were compared with the expected numbers based on age-adjusted incidence rates.[10] Investigators found that, while most second primary cancer risks decreased with age, lung cancer risk increased. In this analysis, second cancer incidence during the first 8 years after breast cancer diagnosis was calculated by designing three inception cohorts derived from SEER Programme data: 1975–77 ($n = 25\,920$), 1983–85 ($n = 32\,722$) and 1991–93 ($n = 40\,819$). It was reported that, between the 1970s and the 1990s, the incidence rate of malignant second cancers significantly increased among female breast cancer patients; lung cancer increased by 50 per cent, while non-Hodgkin's lymphoma and kidney cancer increased by about 150 per cent, and cancers of thyroid, uterine corpus and malignant melanoma by 80 per cent. In the 1990s, the risk ratios (RR) of cancers at all sites were found to be:

- 5.5 for ages 20–49 (95% confidence interval: 5.0–6.1);
- 1.3 for ages 50–64 (95% CI: 1.3–1.4);
- 1.2 for ages 65 and over (95% CI: 1.1–1.2).

In another series of 1884 breast cancer patients treated with conservative surgery and radiotherapy, 147 (8 per cent) developed a second non-breast neoplasm or malignancy (SNBM) compared with 127.7 expected based on general population rates (SEER Programme), corresponding to an absolute excess of 1 per cent of the study population and a relative increase of 15 per cent greater than that expected ($p = 0.05$).[11] Within the first 5 years, the observed and expected rates of SNBMs were identical (47 vs. 46.9). After 5 years, 24 per cent more SNBMs were observed than expected (100 vs. 80.8; $p = 0.02$). Among patients under 50 years old at breast cancer diagnosis, 43 per cent more observed SNBMs occurred than expected (40 vs. 28; $p = 0.02$). For patients 50 or more years of age, 7 per cent more SNBMs were observed than expected (107 vs. 99.7; $p = 0.25$). Lung SNBMs were observed in 33 women, 52 per cent more than the 21.67 predicted by SEER ($p = 0.01$). Most of the lung SNBMs occurred more than 5 years after treatment ($n = 23$) and in women who were over 50 years at the time of breast cancer diagnosis ($n = 27$). Information with regard to tobacco use was not available. Thus, in this series, SNBMs appeared to occur in a minority (8 per cent) of patients treated with conservative surgery and radiotherapy, the absolute excess risk compared with the general population was quite low (1 per cent), and this excess risk was only evident after 5 years.

Adjuvant radiotherapy for breast cancer, especially as delivered decades ago, is associated with the development of second primary lung cancers. A study of women diagnosed with primary breast or lung cancer between 1972 and 1989 identified 4123 lung cancer patients, 3537 breast cancer patients, and 42 patients with both diagnoses from tumour registry records.[12] The two malignancies were diagnosed simultaneously in five patients, lung cancer was diagnosed first in six patients and breast cancer was diagnosed first in 31 ($p < 0.001$). Nineteen of the 31 patients had received adjuvant radiotherapy for breast cancer and developed lung cancer a median of 17 years later. Of the 19 irradiated patients who subsequently developed lung cancer, 15 did so in the ipsilateral lung and only 4 had lung cancer contralateral to the previously irradiated site ($p < 0.001$).

In another analysis, SEER Programme data for 1973–86 were utilized to investigate whether radiation therapy for breast cancer affected the risk of subsequent lung cancer.[13] These investigators found that the risk of lung cancer overall was increased (RR = 2.0; 95% CI: 1.0–4.3) in women who underwent irradiation (compared with those who were not irradiated) 10 years after the initial breast cancer diagnosis. The elevated risk of lung cancer was in the ipsilateral lung, compared with the contralateral lung for irradiated women, and was observed after 10 years of diagnosis both for lung cancer overall and for the three major histological subgroups (small cell, squamous cell and adenocarcinoma). Specific information on radiation doses, treatment plans and cigarette smoking was not available.

In a subsequent case–referent study,[14] the Connecticut Tumor Registry was used to identify 8976 women diagnosed with histologically confirmed invasive breast cancer between 1935 and 1971 who had survived for at least 10 years, in order to ascertain subsequent lung cancers occurring in this group between 1945 and 1981 and to quantify the risk in relation to radiation dose. Sixty-one women with a second primary lung cancer met study eligibility criteria. For these patients and 120 reference subjects selected from the same registry and matched according to race, age at breast cancer diagnosis, year of breast cancer diagnosis, and survival without a second primary tumour, hospital charts were reviewed to collect medical history and radiotherapy information. A medical physicist estimated radiation dose to different lung segments on the basis of radiotherapy reports and experimental simulations of treatments. The overall relative risk of lung cancer associated with initial radiotherapy for breast cancer for these 10-year survivors was 1.8 (95% CI: 0.8–3.8). The risk estimates increased with time since treatment. The relative risk for periods of 15 years or more after radiotherapy was 2.8 (95% CI: 1.0–8.2). The mean dose to the ipsilateral lung was 15.2 Gy, 4.6 Gy to the contralateral lung, and 9.8 Gy for both lungs combined. The excess RR was calculated to be 0.08 per Gy, based on average dose to both lungs, and 0.20 per Gy to the affected (cancerous) lung. Thus, these investigators demonstrated that breast cancer radiotherapy regimens in use before the 1970s were associated with an elevated lung cancer risk many years following treatment. Given the unavailability of smoking data, the contribution of tobacco use could not be assessed.

In a British study comparing radiotherapy and non-radiotherapy regimens, elevated relative risks were observed for lung cancer at 10–14 years and at 15+ years after initial breast cancer diagnosis (RR = 1.62, 95% CI: 1.05–2.54; and RR = 1.49, 95% CI: 1.05–2.14, for the two groups, respectively).[15]

While studies have demonstrated that women who receive adjuvant radiation therapy after mastectomy for breast carcinoma have an increased risk of a second primary lung carcinoma after 10 years,[16,17] the risk associated with adjuvant radiation therapy after breast-conserving surgery (lumpectomy) is also of key importance. In an analysis of 182 122 women with breast cancer reported to the SEER Programme registries (1973–91), Travis et al. showed that breast-conserving surgery accompanied by radiotherapy was not associated with an increased risk of lung cancer for up to 10 years of follow-up, noting that too few women were followed for longer periods to determine subsequent risk; in contrast, those women who received radiotherapy following non-breast-conserving surgery had a significantly increased 2-fold risk of lung cancer after one decade of follow-up.[18] In an extension of this analysis, Zablotska and Neugut examined SEER Programme data up to 1998.[19] Of women with non-metastatic invasive breast carcinoma, 194 981 had been treated with mastectomy and 65 560 were treated with lumpectomy. Although no statistically significant elevation in risk for second primary lung carcinoma prior to 10 years was observed, the authors estimated relative risks of 2.06 (95% CI: 1.53–2.78) and 2.09 (95% CI: 1.50–2.90) for ipsilateral lung carcinoma at 10–14 years and at 15+ years after postmastectomy radiation therapy, respectively. No increased risk was observed for the contralateral lung. This excess risk of ipsilateral lung carcinoma after postmastectomy radiation was found for all three major histological subtypes of lung carcinoma (adenocarcinoma, squamous cell carcinoma and small-cell carcinoma). No increased risk of lung carcinoma was observed 10–14 years after postlumpectomy radiation therapy for either lung. Thus, postmastectomy radiation was found to confer a moderate increase in risk for ipsilateral lung carcinoma starting 10 years after exposure, this increased risk potentially persisting to at least 20 years. Postlumpectomy radiation did not appear to lead to an increased risk, confirming the results of the earlier SEER Programme-based study.[18]

Comparing the incidence of subsequent primary lung carcinoma in patients with breast carcinoma participating in two large NSABP trials (B-04 and B-06), investigators compared women who received radiotherapy as part of their treatment with those patients who did not.[16] Records of all patients who developed a recurrence in the lung or a new primary lung tumour were reviewed to determine the incidence and laterality of confirmed and probable primary lung carcinomas. Of the 1665 evaluable patients in the NSABP B-04 trial (mean follow-up 21.4 years), a total of 23 subsequent confirmed and probable ipsilateral or contralateral primary lung carcinomas was observed. Among patients who had received comprehensive postmastectomy radiotherapy, a statistically significant ($p = 0.029$) increase in the incidence of these new primary tumours was found. The incidence of confirmed new primary ipsilateral lung carcinoma alone was elevated ($p = 0.013$) in patients who had received radiotherapy as part of their treatment. When confirmed and probable ipsilateral lung carcinomas were analysed, a marginally significant excess ($p = 0.066$) was observed in those patients who had received radiotherapy. For the 1850 evaluable patients in NSABP trial B-06 (mean follow-up 19.0 years), a total of 30 second primary lung carcinomas were found, but no increase in either ipsilateral or contralateral primary tumours of the lung was observed among patients who had received radiotherapy.

In a separate investigation, there was no elevated risk of second malignancies when women treated with lumpectomy and radiation therapy for early-stage breast cancer were compared with a similar cohort of patients undergoing mastectomy without radiation.[17] The 15-year cumulative risk of any second malignancy was nearly identical for both cohorts (17.5 and 19.0 per cent, respectively). The second breast malignancy rate at 15 years was 10 per cent for both groups and the 15-year risk of a second non-breast malignancy was 11 vs. 10 per cent. In the subset of patients 45 years of age or younger at the time of treatment, the second breast and non-breast malignancy rates at 15 years were 10 and 5 per cent for patients undergoing lumpectomy and radiation vs. 7 and 4 per cent for patients undergoing mastectomy without radiation. A detailed analysis of patients managed with lumpectomy plus radiation therapy showed that second lung malignancies were associated with a history of tobacco use. The adjuvant use of chemotherapy did not significantly affect the risk of second malignancies.

It is important to assess the risk factors for and patterns of second malignancy among women with breast cancer, and to identify the subgroup(s) at increased risk for a second cancer. Factors that may be associated with second malignancies in general include age, menopausal status, race, family history, obesity, smoking, tumour size, location, histology, pathological nodal status, region(s) treated with radiation, and the use and type of adjuvant therapy. In a study of 1253 women with breast cancer who were treated with lumpectomy and breast irradiation and followed for a median of 8.9 years, Fowble et al. found a 1 per cent cumulative incidence of second lung cancer at 10 years (2 per cent in smokers and 0.3 per cent in non-smokers; p for difference = 0.16). Overall 176 patients developed a second malignancy: 87 contralateral breast cancers (median 5.8 years), 98 solid tumours (median 7.2 years), and 9 patients with both a contralateral breast cancer and solid tumour.[20] Young age was associated with an increased risk of contralateral breast cancer, while older age was associated with an increased risk of a second non-breast cancer second malignancy. A family history of breast cancer increased the risk of contralateral breast cancer, but not non-breast cancer malignancies.[21] There was no significant effect of chemotherapy or the regions treated with radiation on either contralateral breast cancer or non-breast cancer second malignancy.

Cigarette smoking is thought to have a multiplicative effect on the risk of lung cancer following breast cancer radiotherapy.[22] In a case–control study of women registered with primary breast cancer in the Connecticut Tumor Registry who developed a second malignancy between 1986 and 1989, those diagnosed with a subsequent lung cancer were compared with those diagnosed with a subsequent non-smoking, non-radiation-related second malignancy.[22] No radiation effects were observed within 10 years of initial primary breast cancer. Among both smokers and non-smokers diagnosed with second primary cancers more than 10 years after an initial primary breast cancer, radiation therapy was associated with a 3-fold increased risk of lung cancer. The authors did observe a multiplicative effect, with women exposed to both cigarette smoking and breast cancer radiation therapy having a relative risk of 32.7 (95% CI: 6.9–154). The radiation carcinogenic effect was observed only for the ipsilateral lung and not for the contralateral lung both in smokers and in non-smokers. Modern techniques, however, may significantly decrease the radiation dose to the lungs, which may decrease the risk of lung cancer. Nonetheless, young breast cancer patients who smoke remain a subgroup of concern.

A more recent analysis examined the risk of lung cancer in women treated with radiotherapy for breast cancer by evaluating the lung dose in relation to different radiotherapy techniques.[23] Investigators provided the excess relative risk estimate for radiation-associated lung cancer and evaluated the influence of tobacco use by using data from the nationwide Swedish Cancer Registry. They identified 182 women diagnosed with breast and subsequent lung cancers between 1958 and 2000. Radiotherapy was administered to 116 patients. Radiation dose was estimated from the original treatment charts, and information on smoking history was retrieved from case records and through interviews with relatives. The risk of lung cancer was assessed in a case-only approach, where each woman contributed a pair of lungs. The average mean lung dose to the ipsilateral lung was 17.2 Gy (range 7.1–32.0). A significantly increased relative risk of a subsequent ipsilateral lung cancer was observed at 10 or more years of follow-up (RR = 2.04; 95% CI: 1.24–3.36). Squamous cell carcinoma was the histopathological subgroup most closely related to ionizing radiation (RR = 4.00; 95% CI: 1.50–10.66). The effect of radiotherapy was restricted to smokers (RR = 3.08;

95% CI: 1.61–5.91). The excess relative risk per Gy for lung cancer that developed 10 or more years after exposure was 0.11 (95% CI: 0.02–0.44). Thus, these data indicate that radiotherapy for breast cancer significantly increases the risk of lung carcinoma more than 10 years after exposure in those women who smoke at the time of breast cancer diagnosis.

Breast cancer survivors may also be at increased risk of death from second cancers, especially when treated with radiotherapy at a young age. A recent study examined mortality in 7425 patients treated for early-stage breast cancer between 1970 and 1986.[24] Treatment-specific mortality was evaluated by calculating standardized mortality ratios (SMRs) based on comparison with general population rates and by using Cox proportional hazards regression. After a median follow-up of 13.8 years, 4160 deaths were observed, of which 76 per cent were due to breast cancer. Second malignancies showed a slightly increased SMR of 1.2 (95% CI: 1.0–1.3). Patients treated before age 45 experienced a higher SMR (2.0) for solid tumours (95% CI: 1.6–2.7). The 10- and 25-year actuarial risks of death were 34.8 and 49.0 per cent for breast cancer, and 1.8 and 10.2 per cent for all second cancers (excluding breast). The SMR for a second lung malignancy was 1.72 (95% CI: 1.31–2.23) and the absolute excess risk (AER) was 2.8.

LUNG CANCER AFTER HODGKIN'S LYMPHOMA

Lung cancer is one of the most frequent tumours following HL,[25] with risks elevated as early as 1–4 years after treatment and persistent for several decades.[3,26,27] In the largest population-based survey of HL patients to date, which encompassed 32 591 subjects, lung cancer accounted for the largest absolute excess risk of second primary cancers among men (12.7 excess cases per 10 000 men per year),[26] and the second largest excess risk among women (5.8 excess cases per 10 000 women per year), superseded only by breast cancer (10.5 excess cases per 10 000 women per year).[26]

In the past, HL was frequently treated with large-dose, wide-field chest radiotherapy, which resulted in exposure of the lung to a diverse spectrum of radiation doses. Smaller doses were received by tissues under the lung blocks (which shielded parts of the lung, and were conformed to the individual patient) whereas lung parenchyma in the radiotherapy field was exposed to full doses, which commonly reached 35–40 Gy.[28]

Ionizing radiation is an established lung carcinogen,[29] while the effect of chemotherapy for HL is more controversial.[30–32] Boice recently summarized evidence for the carcinogenic influence of lower-linear energy transfer irradiation of the lung.[21] A number of case–control studies to date have sought to clarify the relative importance of chemotherapy and radiotherapy for HL in the subsequent occurrence of lung cancer, but results have been inconsistent, based on sparse numbers, and some were inadequately adjusted for tobacco use.[30,31] Based on 30 cases, Van Leeuwen et al. found a significant dose–response relation between radiation for HL and lung cancer risk.[32] However, radiation dose was averaged across entire lung lobes, with only six cases given mean doses of 9 Gy or more. In larger studies[30,31] a quantitative relation between radiation dose and lung cancer risk was not observed. An association between total number of cycles of chemotherapy for HL and lung cancer risk was not evident in one investigation of 98 cases,[30] nor with

chemotherapy in the Netherlands series.[32] Swerdlow et al. noted a relative risk of 1.66 for lung cancer among 45 British HL patients given mechlorethamine, procarbazine, vincristine and prednisone (MOPP); a difference in lung cancer risk after 1–6 compared with 7 or more cycles, however, was not apparent, and data on cumulative dose were not collected.[31] Of most analytical investigations only the Dutch study had detailed data on smoking habits.[32] Information on tobacco use was gathered for 39 per cent of subjects in the British series,[31] and for 59 per cent of patients in the investigation by Kaldor et al.[30]

Only one large analytical study to date has addressed the interaction between radiotherapy, chemotherapy and tobacco use on lung cancer risk and assessed long-term temporal patterns, utilizing sophisticated radiation dosimetry to estimate the dose received by the specific site in which lung cancer later developed.[28] In collaboration with cancer registries in Iowa, Denmark, Finland, Sweden, Ontario, New Jersey and the Netherlands, the NCI conducted a case–control study of lung cancer among 19 046 HL patients (1965–94).[28] Cumulative dose of cytotoxic drugs, radiation dose to the specific area in the lung where cancer developed, and smoking habits were compared for 222 patients in whom lung cancer was diagnosed and 444 matched controls who did not develop lung cancer. Radiation dose of 5 Gy or more was associated with significant overall 6-fold excesses of lung cancer, and increased with doses of up to 40 Gy or more (p of trend < 0.001). The risk estimates for the highest dose categories of 30.0–39.9 Gy and ≥ 40 Gy were 8.5 (95% CI: 3.3–24) and 6.3 (95% CI: 2.2–19), compared to no radiotherapy, with a linear model providing a good fit to the data.[33] Neither the addition of a dose-squared term nor adding a term to reflect a decline in risk due to cell killing significantly improved the model fit ($p > 0.5$ in both instances), again indicating the good fit of the linear model. Alkylating agent chemotherapy, given without radiotherapy, was associated with significantly increased 4-fold risks of lung cancer, and risk increased with increasing number of cycles (p of trend < 0.001). The risk for patients who received nine or more cycles was 13-fold, with all estimates adjusted for radiation dose and tobacco use. Among patients given MOPP, risk of lung cancer increased with increasing cumulative dose of both mechlorethamine and procarbazine ($p < 0.001$) when evaluated separately. Significant lung cancer excesses also followed therapy with other alkylating agents (RR = 6.3). After treatment with both radiotherapy and alkylating agents, the risk of lung cancer was as expected if individual excess relative risks were summed (RR = 8.0). Smoking was associated with a 20-fold increased risk of lung cancer and multiplied risks from both chemotherapy and radiotherapy. The largest risks for lung cancer (RR = 49.1) were observed among moderate-to-heavy smokers given both radiotherapy and alkylating agents. Significant excesses of lung cancer occurred as early as 1–4 years after alkylating agent treatment and remained significantly elevated for 5–9 years, with marginally significantly increased risks (RR = 3.0) in the 10–14 year interval after HL treatment. In contrast, increased risks after radiotherapy began after 5 years and remained elevated for over 20 years. Significantly increased 6-fold and 9-fold risks of squamous cell and small-cell lung cancers occurred after alkylating agent therapy, and non-significant 3- to 4-fold excesses for adenocarcinoma and large-cell carcinoma were also noted. Following radiotherapy for HL, significantly elevated 8- to 14-fold risks for all major histological types of lung cancer occurred.

Prior to that large international study, several investigations had explored the effect of chemotherapy for HL on the development of lung cancer.[30-32] Kaldor et al. indicated that the risk of lung cancer following chemotherapy for HL was 2-fold compared with patients who received either radiotherapy alone or combined modality therapy.[30] Importantly, these investigators were among the first to point out that chemotherapy for HL might be at least as carcinogenic to lung as radiotherapy, even though risk did not increase with number of treatment cycles. Mechlorethamine, procarbazine and chlorambucil cause lung cancer in rodents,[34] and mechlorethamine is structurally related to sulphur mustard, a human lung carcinogen.[35]

The molecular pathways which might link the administration of alkylating agents to subsequently increased risks of lung cancer are not known, but various hypotheses were outlined by Travis and colleagues.[28] Alkylating agents achieve antitumour effects by reactions with DNA bases.[36] Methylating agents, such as procarbazine, can form the same DNA adduct (06-methylguanine) that is produced by 4-(methylnitrosamino)-1-(3-pyridyl)-1-butanone (NNK). 06-methylguanine is mutagenic and carcinogenic,[37] with concentrations of this DNA adduct linearly correlated ($p < 0.01$) with cumulative amount of procarbazine in lymphoma patients.[38] NNK is a tobacco metabolite which is a potent lung-specific carcinogen in laboratory animals.[39] Additional studies are needed to clarify the carcinogenic pathways associated with the elevated risks of lung cancer after alkylating agent therapy for HL.[28] It is not clear, however, whether results in HL patients can be extrapolated to patients with other cancers, given the immune defects inherent to this lymphoma,[40] and the high prevalence of tobacco use among case patients and control subjects in the international investigation.[28] Further, it was estimated that 9.6 per cent of all lung cancers in the above series were due to treatment, 24 per cent were due to smoking, but that 63 per cent were due to treatment and smoking in combination; the remainder (3 per cent) represented tumours in which neither smoking nor treatment played a role. It is noteworthy that of the 198 patients who developed lung cancer after HL for whom smoking status was known, 191 (96 per cent) had a history of tobacco use.[28]

To our knowledge, the above study is the first to demonstrate that excess lung cancers after alkylating agents for HL develop considerably earlier than the latency period associated with radiotherapy associated solid tumours, in which significantly elevated risks are typically not reported until 10 or more years after exposure.[21] These disparate latencies may be reflective of different mechanisms for cancer induction. The peak latency of 5–9 years for lung cancer after radiotherapy for HL observed in the international study[28] is consistent with prior reports.[41,42] The shortened latency of radiation-associated lung cancer may indicate a susceptibility state associated with the defects in cellular immunity,[40] or the genomic instability[43,44] observed in HL patients. As discussed below, it is notable that lung cancer excesses have also been found in patients with other lymphopoietic malignancies,[45,46] as well as in organ transplant recipients.[47] Whether the immunological defects inherent to HL also contribute to alkylating agent-related lung cancer should be explored. It seems plausible that

the immunosuppressive effects of tobacco smoking[48] may also heighten the immune alterations associated with HL.[40] A comprehensive review of lung cancer after HL was recent published by Lorigan and colleagues.[49]

LUNG CANCER AFTER NON-HODGKIN'S LYMPHOMA

Significantly increased risks of lung cancer have been reported in several surveys of NHL patients,[45,50-52] although analytical studies to examine the effects of radiotherapy, chemotherapy and tobacco use have not been undertaken. In a survey of NHL patients reported to population-based cancer registries that participate in the SEER Programme (1973–2001), Tward et al.[50] recently confirmed the significantly increased risks of lung cancer (standardized incidence ratio, SIR = 1.31; observed = 1095) that had been reported in an earlier review of these data (1973–87).[51] Lung cancer accounted for the largest absolute excess risk (6.76 excess cases per 10 000 person-years) of second primary cancers in both men and women (6.36 and 7.19, respectively).[50] Data with regard to the latency period of excess lung cancer risk were not presented. Comparable excesses of lung cancers were observed in NHL patients who initially received radiotherapy and those who did not receive irradiation. It should be recognized, however, that irradiation for NHL may not necessarily encompass the chest, since only about 20 per cent of patients present with mediastinal involvement. The effect on lung cancer risk of initial treatment for NHL with chemotherapy could not be evaluated,[50] because these data are not available in the public-use SEER Programme data file. In the earlier review of the SEER Programme data,[51] significantly increased risks of lung cancer were observed following radiation alone, regimens which included any chemotherapy, and the other/no treatment group. In neither of the above studies could the role of smoking be taken into consideration, since these data are not collected by the SEER Programme.

In a recent analysis of 2456 patients with NHL less than 60 years of age who were treated from 1973 to 2000 in centres participating in the British National Lymphoma Investigation (BNLI), the risk of lung cancer was significantly increased (SIR = 1.6; 95% CI: 1.1–2.3; observed = 28 cases) compared to the general population.[52] Lung cancer accounted for the largest absolute risk of solid cancers (5.8 excess cases per 10 000 person-years). The risk for lung cancer was significantly increased among 1219 patients who received chemotherapy with cyclophosphamide, doxorubicin, vincristine and prednisone (CHOP) (SIR = 2.1; 95% CI: 1.1–3.7), but not among 752 patients who received chorambucil (SIR = 1.1; 95% CI: 0.4–2.4). Data on radiotherapy fields or smoking habits were not available. Excess lung cancers were observed for up to 15 years after NHL treatment, but then declined to below expectation; similar latency patterns were reported in a study of 6171 1-year survivors of NHL reported to population-based registries in North America and Europe.[45] In both studies, however, only a relatively small number of patients (336 and 977, respectively) were followed for 15 or more years, thus, sufficient statistical power to detect elevated risks may not have existed. In neither investigation were data on tobacco habits available. The 15-year cumulative risk of lung cancer in the BNLI survey was 2.8 per cent. The early excesses of lung cancer after NHL are remarkably similar to the short latency period evident for lung cancers which follow alkylating agent therapy for HL.[28] Whether the short latency period reflects in part shared aetiological factors is not known. One survey[53] reported a significant relation between number of cigarettes smoked and NHL mortality, although other investigations[54-56] in which tobacco use has been linked with NHL have not found dose–response gradients. Associations between tobacco use and NHL have not been found in other studies.[57-59]

LUNG CANCER AFTER CHRONIC LYMPHOCYTIC LEUKAEMIA

Among 16 367 patients with CLL reported to the SEER Programme (1973–96), a significantly increased risk of lung cancer was reported (SIR = 1.66; observed = 421), with comparable elevations in men and women.[60] Although the impact of treatment on lung cancer risk was not directly addressed, it was noted that significantly increased risks of all solid tumours combined (23 per cent of which were lung cancer) were found in patients either initially given chemotherapy (O/E = 1.21) or in those who received no treatment (SIR = 1.19).[60] In an earlier survey of 9456 CLL patients reported to the SEER Programme (1973–88), significantly increased risks of lung cancer followed initial management with either chemotherapy (SIR = 2.19) or no treatment (SIR = 1.89).[46] In neither survey were data on tobacco use available. Parekh et al., however, observed that 85 per cent of patients with CLL (n = 1329) who also had lung cancer (n − 22) were smokers.[61]

It has been suggested that the immunodeficiency inherent to CLL may play a role in the subsequently increased risks for lung cancer as well as malignant melanoma.[62] Similarly, increased risks of melanoma have been reported in patients with HL[26] and NHL[45,50,51] – two other lymphopoietic malignancies that are often associated with immune defects.[41,63] Whether iatrogenic immunosuppression may enhance lung cancer risk in CLL patients was suggested in one study. Robak et al. reported lung cancer in 7 of 251 CLL patients (2.8 per cent) treated with cladribine and in 7 of 323 CLL patients (2.2 per cent) treated with cladribine and alkylating agents, compared with 3 of 913 CLL patients (0.3 per cent) who received alkyating agents alone ($p < 0.001$ and $p < 0.01$, respectively).[64] It should be noted, however, that the calculated percentages, which are based on sparse numbers, represent crude estimates. Although the prevalence of tobacco use was comparable in the three described treatment groups, the estimates do not take into account any intergroup differences in age and gender distribution, which were not presented. It has been postulated that cladribine, a nucleoside analogue that has been associated with severe immunosuppressive and infectious complications,[65] may add to the inherent immunodeficiency of CLL. However, Cheson et al. reported no increased risk of secondary malignancies among CLL patients given nucleoside analogue therapies.[65] Hisada et al. examined the risk of second cancers among CLL patients treated during 1973–89 and 1990–96 in the SEER Programme database, with the former interval representing an era before the widespread

use of nucleoside analogues.[60] This surrogate measure was used, since data on specific drugs were not available. Risks for all solid tumours combined were comparable in the two calendar year intervals (SIR = 1.19 and 1.29, respectively). Lung cancer risk, however, was not specifically addressed.

To our knowledge, only one large study has described the latency of lung cancer excesses after CLL.[46] In the survey of 9456 CLL patients described above, lung cancer risks were significantly increased in the periods of < 1 year, 1–4 years, 5–9 years and 10 or more years after CLL diagnosis (SIR = 1.60, 1.87, 1.98 and 2.80, respectively), akin to the overall temporal patterns described after both HL and NHL.[3]

Few data describe survival after diagnosis of a second primary lung cancer among patients with lymphopoietic malignancies. The majority of information to date derives from patients with HL. Survival according to type of second solid tumour (n = 131) was comprehensively described among 1319 HL patients by Ng et al.[66] The poorest prognosis was observed for HL patients who developed lung cancer (n = 22 cases), with a median survival of 1 year. Travis et al. anecdotally noted that the median survival following a diagnosis of lung cancer (n = 222) among over 19 000 HL patients was 5.0 months (212 deaths).[28] Parekh et al. showed that among 26 CLL patients who developed lung cancer, all died of lung cancer.[61] Thus, survival after a diagnosis of a second primary lung cancer is poor, consistent with the 5-year relative survival rate of 16 per cent observed for de-novo lung cancer.[67]

SECOND PRIMARY MALIGNANT MESOTHELIOMA

An increasing body of evidence points to a possible association between high-dose radiation and the subsequent development of malignant mesothelioma. In 2005, Travis et al. reported for the first time a significantly elevated relative risk for malignant mesothelioma in a large, international population-based study of over 13 000 testicular cancer patients initially given radiotherapy (RR = 4.0; 95% CI = 2.0–8.1; 14 cases). These results should be viewed in relation to the supradiaphragmatic radiotherapy fields administered in the past to treat testicular cancer,[68-70] which delivered high-dose radiation of the order of 30 Gy to mediastinum.[69] In a subsequent population-based study of 21 111 irradiated NHL survivors reported to the SEER Programme (1973–2001), Tward et al.[50] reported a significantly elevated risk of malignant mesothelioma (SIR = 2.26; 9 cases); in contrast, the risk among over 55 000 NHL patients who did not initially receive radiotherapy did not exceed expectation (SIR = 0.86; 9 cases). In an international survey of 376 825 1-year breast cancer survivors reported to nationwide, population-based cancer registries in Sweden, Denmark, Finland and Norway (1943–2002), Brown et al. recently found a significantly elevated risk of malignant mesothelioma (SIR 1.42; 95% CI: 1.02–1.94).[71] Excesses were not apparent until 10 or more years after breast cancer diagnosis, with the largest risks (SIR 5.60; 95% CI: 2.05–12.21; 6 cases) observed at 30 or more years of follow-up – a temporal pattern consistent with the late carcinogenic sequelae of radiation. The SIRs for the subgroup of 22 malignant mesotheliomas that occurred among breast cancer patients reported to registries which routinely record treatment data were 1.78 (95% CI: 1.0–2.95; 15 cases) for women initially given radiotherapy compared to 0.95 (95% CI: 0.38–1.97; 7 cases) for those who did not receive radiation.

In addition to the large studies described above, reports compiled by Cavazza et al. document the occurrence of additional cases of pleural mesothelioma following chest radiotherapy for HL, breast cancer and testicular cancer.[72] Neugut et al., in an early analysis of breast cancer patients reported to the SEER Programme (1973–93), found a slight non-significant risk of mesothelioma following radiotherapy, based on two cases.[73]

In a study of 2384 Danish and Swedish patients injected with radioactive Thorotrast (Th-232) during radiographic procedures, increasing cumulative dose of radiation was associated with an increasing risk of cancers of the peritoneum and other digestive sites (p of trend = 0.01).[74] The overall standardized incidence ratio was 14.6 (5 cases), whereas no cases occurred in 1180 patients injected with non-radioactive contrast agents during radiographic procedures (RR = ∞; 95% CI: 1.7–∞). Mortality due to cancer at this site was also significantly increased in the Japanese Thorotrast study,[75] with non-significant excesses of peritoneal mesothelioma noted in the survey of German Thorotrast patients.[76] The uniformity of this observation between various international investigations and the close proximity of parts of the peritoneum to the predominant areas of Thorotrast deposition add support to the radiogenic aetiology of cancers at this site.

Given the low population incidence rates for malignant mesothelioma,[2] it has only been possible to detect elevated risks when large numbers of patients are followed following exposure to high doses of radiotherapy. Even then, population-based surveys of cancer patients reported to cancer registries do not collect data on workplace exposures, such as asbestos, which is the main cause of malignant mesothelioma, with no effect apparent from smoking.[77] Thus, although the above studies add to a growing body of evidence that very high doses of radiation might be associated with malignant pleural mesothelioma, any role of asbestos exposure as well as any interaction between radiation and asbestos must be carefully examined in analytical studies. Prognosis following a diagnosis of malignant mesothelioma is poor, with a 5-year relative survival rate of 8.4 per cent.[2]

COMMENT ON SMOKING

The importance of tobacco use in the subsequent risk of lung cancer among HL patients has been established through analytical studies.[28,32,33] This finding is not unexpected, given that 90 per cent of patients who develop de-novo lung cancer use tobacco.[77] Studies of patients with HL[28,32,33] and breast cancer[22] show that patients who smoke will have a considerably greater risk of lung cancer after chest radiotherapy than those who do not smoke, an observation that supports findings in other radiation-exposed groups.[78] Thus, smokers who have received chest radiotherapy should be strongly advised to abstain from tobacco use. Similarly, smoking appeared to multiply the risk of lung cancer associated with past alkylating agent chemotherapy regimens for HL. It is not yet known whether more commonly used current chemotherapy regimens

for HL might also increase the risk of lung cancer. The efficacy of lung cancer screening in HL patients has not been addressed, but an international investigation is planned to evaluate the efficacy of screening with low-dose spiral computed tomography.[49] Thus, although second primary lung tumours in cancer survivors might represent in part an iatrogenic effect, it is noteworthy that tobacco use still plays a pivotal role, as it does in de-novo lung cancer.

Additional examination of the relation between breast cancer, CLL, NHL and subsequent lung cancer risk should be undertaken in analytical studies, in which detailed data on administered cytotoxic drugs, radiation dose, patient tobacco use and other factors are collected. Future investigations should incorporate molecular markers to further understand the underlying pathways that are operant in treatment-associated second cancers.[7] In the interim, cancer survivors who smoke would be well advised to quit.

KEY POINTS

- Advances in early detection, supportive care and therapy have allowed the number of cancer survivors to increase dramatically since the 1970s.

- The number of survivors with multiple primary malignancies is rising, with independent cancers representing about one in six incident cancers.

- As survival statistics continue to improve, ascertainment and quantification of the late sequelae and their treatment become essential, with the occurrence of second primary cancers representing one of the most severe complications.

- Second solid tumours are the most common cause of death among several groups of long-term survivors. These tumours can reflect not only the late effects of treatment, but also the contribution of lifestyle influences, environmental exposures, host factors and combinations of factors, including gene–environment and gene–gene interactions.

- Given the high baseline incidence rates of lung cancer in the general population, even a small increase in the relative risk can translate into a large absolute risk, which is considered the optimal measure of disease burden in a population.

- Exposure to tobacco smoke should be strongly discouraged, particularly in patients with a history of treated malignant lymphoma or lung cancer.

REFERENCES

1. Editorial. Cancer survivors: living longer, and now, better. Lancet 2004;**364**:2153–4.
2. Ries LAG, Harkins D, Krapcho M, et al. (eds) SEER Cancer Statistics Review, 1975–2003 [based on November 2005 SEER data submission]. Bethesda, MD: National Cancer Institute, 2006.
3. Van Leeuwen FE, Travis LB. Second cancers. In: DeVita VT et al. (eds) Cancer: Principles and Practice of Oncology, 7th edn, pp 2575–602. Philadelphia, PA: Lippincott Williams & Wilkins, 2005.

4. Aziz NM. Late effects of cancer treatments: surgery, radiation therapy, chemotherapy. In: Chang AE *et al.* (eds) *Oncology: An Evidence-based Approach*, ch 101. New York: Springer-Verlag, 2006.

5. Dores G, Schonfeld S, Chen J, *et al.* Long-term cause-specific mortality among 41,146 one-year survivors of Hodgkin lymphoma. *Proc Am Soc Clin Oncol* 2005;**23**:562S.

6. Travis LB. Therapy-associated solid tumors. *Acta Oncol* 2002;**41**:323–33.

7. Travis LB, Rabkin CS, Brown LM, *et al.* Cancer survivorship. Genetic susceptibility and second primary cancers: research strategies and recommendations. *J Natl Cancer Inst* 2006;**98**:15–25.

8. Caporaso N, Dodd K, Tucker MA. New malignancies following cancer of the respiratory tract. In: Curtis RE *et al.* (eds) *New Malignancies among Cancer Survivors: SEER Cancer Registries, 1973–2000*. NIH Publication 05-5302, 2006.

9. Raymond JS, Hogue CJ. Multiple primary tumours in women following breast cancer, 1973–2000. *Br J Cancer* 2006;**94**:1745–50.

10. Yu GP, Schantz SP, Neugut AI, Zhang ZF. Incidences and trends of second cancers in female breast cancer patients: a fixed inception cohort-based analysis (United States). *Cancer Causes Control* 2006;**17**:411–20.

11. Galper S, Gelman R, Recht A, *et al.* Second nonbreast malignancies after conservative surgery and radiation therapy for early-stage breast cancer. *Int J Radiat Oncol Biol Phys* 2002;**52**:406–14.

12. Wiernik PH, Sklarin NT, Dutcher JP, Sparano JA, Greenwald ES. Adjuvant radiotherapy for breast cancer as a risk factor for the development of lung cancer. *Med Oncol* 1994;**11**(3/4):121–5.

13. Neugut AI, Robinson E, Lee WC, *et al.* Lung cancer after radiation therapy for breast cancer. *Cancer* 1993;**71**:3054–7.

14. Inskip PD, Stovall M, Flannery JT. Lung cancer risk and radiation dose among women treated for breast cancer. *J Natl Cancer Inst* 1994;**86**:983–8.

15. Roychoudhuri R, Evans H, Robinson D, Moller H. Radiation-induced malignancies following radiotherapy for breast cancer. *Br J Cancer* 2004;**91**:868–72.

16. Deutsch M, Land SR, Begovic M, *et al.* The incidence of lung carcinoma after surgery for breast carcinoma with and without postoperative radiotherapy: results of National Surgical Adjuvant Breast and Bowel Project (NSABP) clinical trials B-04 and B-06. *Cancer* 2003;**98**:1362–8.

17. Obedian E, Fischer DB, Haffty BG. Second malignancies after treatment of early-stage breast cancer: lumpectomy and radiation therapy versus mastectomy. *J Clin Oncol* 2000;**18**:2406–12.

18. Travis LB, Curtis RE, Inskip PD, Hankey BF. Re: Lung cancer risk and radiation dose among women treated for breast cancer. *J Natl Cancer Inst* 1995;**87**:60–1.

19. Zablotska LB, Neugut AI. Lung carcinoma after radiation therapy in women treated with lumpectomy or mastectomy for primary breast carcinoma. *Cancer* 2003;**97**:1404–11.

20. Fowble B, Hanlon A, Freedman G, Nicolaou N, Anderson P. Second cancers after conservative surgery and radiation for stages I/II breast cancer: identifying a subset of women at increased risk. *Int J Radiat Oncol Biol Phys* 2001;**51**:679–90.

21. Boice JD. Ionizing radiation. In: Schottenfeld D, Fraumeni JF (eds) *Cancer Epidemiology and Prevention*, 3rd edn, pp 259–93. New York: Oxford University Press, 2006.

22. Neugut AI, Murray T, Santos J, *et al.* Increased risk of lung cancer after breast cancer radiation therapy in cigarette smokers. *Cancer* 1994;**73**:1615–20.

23. Prochazka M, Hall P, Gagliardi G, *et al.* Ionizing radiation and tobacco use increases the risk of a subsequent lung carcinoma in women with breast cancer: case-only design. *J Clin Oncol* 2005;**23**:7467–74.

24. Hooning MJ, Aleman BM, van Rosmalen AJ, *et al.* Cause-specific mortality in long-term survivors of breast cancer: a 25-year follow-up study. *Int J Radiat Oncol Biol Phys* 2006;**64**:1081–91.

25. Hoppe RT. Hodgkin's disease: complications of therapy and excess mortality. *Ann Oncol* 1997;**8**(Suppl 1):115–18.

26. Dores GM, Metayer C, Curtis RE, *et al.* Second malignant neoplasms among long-term survivors of Hodgkin's disease: a population-based evaluation over 25 years. *J Clin Oncol* 2002;**20**:3484–94.

27. Travis LB, Curtis RE, Bennett WP, *et al.* Lung cancer after Hodgkin's disease. *J Natl Cancer Inst* 1995;**87**:1324–7.

28. Travis LB, Gospodarowicz M, Curtis RE, *et al.* Lung cancer following chemotherapy and radiotherapy for Hodgkin's disease. *J Natl Cancer Inst* 2002;**94**:182–92.

29. UNSCEAR. Reports available at www.unscear.org/unscear/en/publications.html.

30. Kaldor JM, Day NE, Bell J, *et al.* Lung cancer following Hodgkin's disease: a case–control study. *Int J Cancer* 1992;**52**:677–81.

31. Swerdlow AJ, Schoemaker MJ, Allerton R, *et al.* Lung cancer after Hodgkin's disease: a nested case–control study of the relation to treatment. *J Clin Oncol* 2001;**19**:1610–18.

32. Van Leeuwen FE, Klokman WJ, Stovall M, *et al.* Roles of radiotherapy and smoking in lung cancer following Hodgkin's disease. *J Natl Cancer Inst* 1995;**87**:1530–7.

33. Gilbert ES, Stovall M, Gospodarowicz M, *et al.* Lung cancer after treatment for Hodgkin's disease: focus on radiation effects. *Radiation Res* 2003;**159**:161–73.

34. International Agency for Research on Cancer (IARC). Overall evaluations of carcinogenicity: an updating of IARC Monographs volumes 1 to 42. *IARC Monographs on the Evaluation of Carcinogenic Risks to Humans*, Suppl 7. Albany, NY: WHO Press, 1987.

35. Blair A, Kazerouni N. Reactive chemicals and cancer. *Cancer Causes Control* 1997;**8**:473–90.

36. Colvin OM. Antitumor alkylating agents. In: DeVita VT *et al.* (eds) *Cancer: Principles and Practice of Oncology*, 6th edn, pp 363–76. Philadelphia, PA: Lippincott Williams & Wilkins, 2001.

37. Pegg AE. Methylation of the O6 position of guanine in DNA is the most likely initiating event in carcinogenesis by methylating agents. *Cancer Invest* 1984;**2**:223–31.

38. Souliotis VL, Kaila S, Boussiotis VA, Pangalis GA, Kyrtopoulos SA. Accumulation of O6-methylguanine in human blood leukocyte DNA during exposure to procarbazine and its relationships with dose and repair. *Cancer Res* 1990;**50**:2759–64.

39. Hecht SS. Tobacco smoke carcinogens and lung cancer. *J Natl Cancer Inst* 1999;**91**:1194–210.

40. Mauch PM, Armitage JO, Diehl V, Hoppe RT, Weiss LM (eds). *Hodgkin's Disease*. Philadelphia, PA: Lippincott Williams & Wilkins, 1999.

41. Tucker MA, Coleman CN, Cox RS, Varghese A, Rosenberg SA. Risk of second cancers after treatment for Hodgkin's disease. *New Engl J Med* 1988;**318**:76–81.

42. Van Leeuwen FE, Klokman WJ, Hagenbeek A, *et al.* Second cancer risk following Hodgkin's disease: a 20-year follow-up study. *J Clin Oncol* 1994;**12**:312–25.

43. Behrens C, Travis LB, Wistuba, II *et al.* Molecular changes in second primary lung and breast cancers after therapy for Hodgkin's disease. *Cancer Epidemiol Biomarkers Prev* 2000;**9**:1027–35.

44. Falzetti D, Crescenzi B, Matteuci C, *et al.* Genomic instability and recurrent breakpoints are main cytogenetic findings in Hodgkin's disease. *Haematologica* 1999;**84**:298–305.

45. Travis LB, Curtis RE, Glimelius B, *et al.* Second cancers among long-term survivors of non-Hodgkin's lymphoma. *J Natl Cancer Inst* 1993;**85**:1932–7.

46. Travis LB, Curtis RE, Hankey BF, Fraumeni JF. Second cancers in patients with chronic lymphocytic leukemia. *J Natl Cancer Inst* 1992;**84**:1422–7.

47. Birkeland SA, Storm HH, Lamm LU, *et al.* Cancer risk after renal transplantation in the Nordic countries, 1964–86. *Int J Cancer* 1995;**60**:183–9.

48. Sopori ML, Kozak W. Immunomodulatory effects of cigarette smoke. *J Neuroimmunol* 1998;**83**(1/2):148–56.

49. Lorigan P, Radford J, Howell A, Thatcher N. Lung cancer after treatment for Hodgkin's lymphoma: a systematic review. *Lancet Oncol* 2005;**6**:773–9.

50. Tward JD, Wendland MM, Shrieve DC, Szabo A, Gaffney DK. The risk of secondary malignancies over 30 years after the treatment of non-Hodgkin lymphoma. *Cancer* 2006;**107**:108–15.

51. Travis LB, Curtis RE, Boice JD, Hankey BF, Fraumeni JF. Second cancers following non-Hodgkin's lymphoma. *Cancer* 1991;**67**: 2002–9.

52. Mudie NY, Swerdlow AJ, Higgins CD, *et al.* Risk of second malignancy after non-Hodgkin's lymphoma: a British Cohort Study. *J Clin Oncol* 2006;**24**:1568–74.

53. Linet MS, McLaughlin JK, Hsing AW, *et al.* Is cigarette smoking a risk factor for non-Hodgkin's lymphoma or multiple myeloma? Results from the Lutheran Brotherhood Cohort Study. *Leuk Res* 1992;**16**(6/7):621–4.

54. Williams RR, Horm JW. Association of cancer sites with tobacco and alcohol consumption and socioeconomic status of patients: interview study from the Third National Cancer Survey. *J Natl Cancer Inst* 1977;**58**:525–47.

55. Franceschi S, Serraino D, Bidoli E, *et al.* The epidemiology of non-Hodgkin's lymphoma in the north-east of Italy: a hospital-based case-control study. *Leuk Res* 1989;**13**:465–72.

56. Brown LM, Everett GD, Gibson R, *et al.* Smoking and risk of non-Hodgkin's lymphoma and multiple myeloma. *Cancer Causes Control* 1992;**3**:49–55.

57. Cartwright RA, McKinney PA, O'Brien C, *et al.* Non-Hodgkin's lymphoma: case control epidemiological study in Yorkshire. *Leuk Res* 1988;**12**:81–8.

58. Hoar SK, Blair A, Holmes FF, *et al.* Agricultural herbicide use and risk of lymphoma and soft-tissue sarcoma. *J Am Med Assoc* 1986;**256**:1141–7.

59. Paffenbarger RS, Wing AL, Hyde RT. Characteristics in youth predictive of adult-onset malignant lymphomas, melanomas, and leukemias: brief communication. *J Natl Cancer Inst* 1978;**60**:89–92.

60. Hisada M, Garber JE, Fung CY, Fraumeni JF, Li FP. Multiple primary cancers in families with Li–Fraumeni syndrome. *J Natl Cancer Inst* 1998;**90**:606–11.

61. Parekh K, Rusch V, Kris M. The clinical course of lung carcinoma in patients with chronic lymphocytic leukemia. *Cancer* 1999;**86**:1720–3.

62. Greene MH, Hoover RN, Fraumeni JF. Subsequent cancer in patients with chronic lymphocytic leukemia: a possible immunologic mechanism. *J Natl Cancer Inst* 1978;**61**: 337–40.

63. Advani RH, Horning SJ. Treatment of early-stage Hodgkin's disease. *Semin Hematol* 1999;**36**:270–81.

64. Robak T, Blonski JZ, Gora-Tybor J, *et al.* Second malignancies and Richter's syndrome in patients with chronic lymphocytic leukaemia treated with cladribine. *Eur J Cancer* 2004;**40**:383–9.

65. Cheson BD, Vena DA, Barrett J, Freidlin B. Second malignancies as a consequence of nucleoside analog therapy for chronic lymphoid leukemias. *J Clin Oncol* 1999;**17**:2454–60.

66. Ng AK, Bernardo MV, Weller E, *et al.* Second malignancy after Hodgkin disease treated with radiation therapy with or without chemotherapy: long-term risks and risk factors. *Blood* 2002;**100**:1989–96.

67. Jemal A, Siegel R, Ward E, *et al.* Cancer statistics, 2006. *CA Cancer J Clin* 2006;**56**:106–30.

68. Maier JG, Mittemeyer B. Carcinoma of the testis. *Cancer* 1977;**39**(2 Suppl):981–6.

69. White RL, Maier JG. Testis tumors. In: Perez CA, Brady LW (eds) *Principles and Practice of Radiation Oncology*, pp 899–944. Philadelphia, PA: JB Lippincott, 1987.

70. Zagars GK, Ballo MT, Lee AK, Strom SS. Mortality after cure of testicular seminoma. *J Clin Oncol* 2004;**22**:640–7.

71. Tward JD, Wendland MM, Shrieve DC, Szabo A, Gaffney DK. The risk of secondary malignancies over 30 years after the treatment of non-Hodgkin lymphoma. *Cancer* 2006;**107**: 108–15.

72. Cavazza A, Travis LB, Travis WD, *et al.* Post-irradiation malignant mesothelioma. *Cancer* 1996;**77**:1379–85.

73. Neugut AI, Ahsan H, Antman KH. Incidence of malignant pleural mesothelioma after thoracic radiotherapy. *Cancer* 1997;**80**:948–50.

74. Travis LB, Hauptmann M, Gaul LK, *et al.* Site-specific cancer incidence and mortality after cerebral angiography with radioactive thorotrast. *Radiation Res* 2003;**160**:691–706.

75. Mori T, Fukutomi K, Kato Y, *et al.* 1998 results of the first series of follow-up studies on Japanese thorotrast patients and their relationships to an autopsy series. *Radiation Res* 1999;**152**(6 Suppl):S72–80.

76. van Kaick G, Dalheimer A, Hornik S, *et al.* The German thorotrast study: recent results and assessment of risks. *Radiation Res* 1999;**152**(6 Suppl):S64–71.

77. Blot WJ, Fraumeni JF. Cancers of the lung and pleura. In: Schottenfeld D, Fraumeni JF (eds) *Cancer Epidemiology and Prevention*, 2nd edn, pp 637–65. New York: Oxford University Press, 1996.

78. Committee on the Biological Effects of Ionizing Radiations (BEIR III), National Research Council (US). *The Effects on Populations of Exposure to Low Levels of Ionizing Radiation*. Washington, DC: National Academy Press, 1980.

NON-ONCOLOGICAL CONDITIONS

21

Pneumonitis induced by non-cytotoxic agents

UMAIR A GAUHAR, J ALLEN D COOPER JR

INTRODUCTION

Certain non-cytoxic drugs used in the treatment of benign disorders can cause pulmonary toxicities. Pulmonary adverse effects of amiodarone are the best studied. However, there are many other therapeutic agents in this class that are relatively commonly used and cause various pulmonary toxicities. In this chapter, we will discuss the pulmonary adverse effects of antirheumatic drugs as well as other selected non-cytotoxic medications (Box 21.1). Pulmonary reactions to amiodarone are discussed in Chapter 22.

METHOTREXATE

Methotrexate (MTX) is an antimetabolite used in the treatment of many benign and malignant disorders. It is one of the disease-modifying antirheumatic drugs (DMARDs) used for three decades in the treatment of rheumatoid arthritis (RA) and has also found its use in the treatment of patients with sarcoidosis and psoriasis. Combined with other agents, methotrexate is used in the treatment of various malignant disorders such as acute lymphocytic leukaemia, choriocarcinoma, Burkitt's lymphoma in children, breast cancer, osteogenic carcinoma and head and neck cancers.[1]

Methotrexate is structurally related to the vitamin folic acid and acts as an antagonist of that vitamin by inhibiting dihydrofolate reductase, the enzyme involved in converting folic acid to its active coenzyme form, tetrahydrofolic acid. This leads to decreased biosynthesis of thymidylic acid, methionine, serine and the purines and thus eventually to decreased DNA, RNA and protein synthesis causing cell death.[1] The dose of methotrexate used in RA is much smaller (15–25 mg per week administered orally, intramuscularly or subcutaneously) as compared to doses used in malignant diseases (100–1000 mg/m^2).[2]

Box 21.1 Non-cytotoxic drugs causing interstitial pneumonitis (discussed in this chapter)

Drug class	*Specific drugs*
• Antirheumatics	• Methotrexate
	• Gold salts
	• D-penicillamine
	• Leflunomide
• Antimicrobials	• Nitrofurantoin
	• Nalidixic acid
	• Ceftizoxime
	• Trimethoprim
• Antipyretics/analgesics	• Acetaminophen
• Antineoplastic agents	• Nilutamide
• Beta-adrenergic antagonists	• Propranolol
	• Acebutolol
	• Nadolol
• Antiarrhythmics	• Flecainide
	• Tocainide
• Anti-inflammatory agents	• Mesalamine
	• Sulfasalazine
• Tumour necrosis factor-α antagonists	• Infliximab
	• Etanercept
	• Adalimumab
• Immunosuppressant agents	• Azathioprine

EPIDEMIOLOGY

There are five pulmonary syndromes associated with MTX therapy:

• acute interstitial pneumonitis (hypersensitivity pneumonitis);

Table 21.1   Risk factors for methotrexate-induced lung toxicity in rheumatoid arthritis (data from [9])

Risk factor	Odds ratio
Diabetes mellitus	35.6
Hypoalbuminaemia	19.5
Rheumatoid pleuropulmonary involvement	7.1
Previous use of disease-modifying agents (gold, sulfasalazine, penicillamine)	5.6
Older age	5.1

- less commonly interstitial fibrosis;
- pleuritis with or without pleural effusions;
- pulmonary nodules;
- non-cardiogenic pulmonary oedema, mainly seen after intrathecal methotrexate.[3]

This chapter will focus on interstitial pneumonitis which is by far the most common complication.

The first case of MTX-induced pulmonary toxicity was described in 1969 in a young patient treated for childhood leukaemia.[4] However, for several years it was believed that MTX causes lung toxicity only when used in doses greater than 20 mg.[5]

The first reports of lung toxicity associated with low-dose MTX therapy in rheumatoid arthritis appeared in 1983.[6,7] Many retrospective and prospective studies of MTX-treated RA patients, published from 1976 to 1992, have shown the prevalence of low-dose MTX-induced lung toxicity to range from 0.3 to 11.6 per cent.[8] The clinical characteristics and risk factors for low-dose MTX-related lung disease were well described in two large publications from the Methotrexate Lung Study Group in 1997.[9,10] The first was a case–control study with 29 RA cases and 82 MTX-treated RA controls and identified several risk factors for MTX-induced pulmonary reactions in RA patients (Table 21.1).[9] The second publication by this group described the clinical, laboratory, radiological and histopathological features of this condition in the same cohort of patients. The mean duration of therapy was about 18 months before the occurrence of MTX-induced lung injury, with 11 per cent developing adverse effects within the first 10 weeks and 48 per cent within 32 weeks.[10] Three smaller studies also found age, pleuropulmonary disease and previous DMARD therapy to increase risk for MTX-induced lung toxicity.[11–13] Renal dysfunction also increases the risk for MTX-associated pneumonitis.[14,15] Gender, disease duration, route of administration of the drug, length of therapy and cumulative dose are not associated with an increased risk of MTX pulmonary toxicity,[8] although, in one case, switching from the oral to the parenteral route was temporally associated with the clinical onset of methotrexate pneumonitis.

Rooney et al. did not find an increased risk of MTX-associated pneumonitis with concomitant administration of non-steroidal anti-inflammatory drugs.[16] One small study found a greater proportion of males and smokers among RA patients with MTX-induced pneumonitis.[17] Methotrexate-induced pneumonitis can occur after a single dose of the drug.[18,19] Interestingly, MTX-induced pneumonitis can also develop several weeks after discontinuation of treatment, although this is quite unusual.[20–22] Association of MTX-induced pneumonitis with certain HLA haplotypes has been noted, such as DRW3 in one case and DR4 in two cases.[18,23] In a review of 189 cases, Zisman et al. found RA to be the most common underlying disease in patients with MTX-induced pneumonitis (53 per cent).[24] A few cases have developed shortly after institution of therapy against tumour necrosis factor (TNF).[25]

PATHOGENESIS

The underlying mechanism of MTX-induced pneumonitis is thought to be acute hypersensitivity reaction.[26] This idea is supported by the presence of fever, peripheral eosinophilia, granulomatous changes on biopsy in some cases, and response to corticosteroids.[8] However, there are reported cases where methotrexate was successfully reintroduced after development of MTX-induced pneumonitis.[14,27,28] Also, the pneumonitis has been shown to resolve despite continued use of methotrexate.[29] These cases suggest that there may be other mechanisms (possible idiosyncratic) involved in the pathogenesis of MTX-induced lung disease rather than true hypersensitivity. The presence of leucocyte inhibiting factor in the peripheral blood of patients with pneumonitis and lymphocytic alveolitis in the bronchoalveolar lavage (BAL) fluid supports an immunological cell-mediated damage mechanism.[30,31] Recently, however, the lymphocyte transformation test was not found to be useful to the diagnosis of methotrexate pneumonitis.[32] The pathogenesis of methotrexate pneumonitis may differ according to dosage and time to onset, with early-onset and low-dose cases being characterized by neutrophilic alveolitis in the BAL and increased mortality compared to late high-dose cases which are characterized by lymphocytes in the BAL and lower mortality.[33]

CLINICAL PRESENTATION

The clinical presentation of MTX-induced pneumonitis is similar whether it is used for benign or malignant diseases.[5] In a review of 123 reported cases of MTX-induced pneumonitis, shortness of breath (82.1 per cent), cough (81.3 per cent) and fever (76.4 per cent) were found to be the most common presenting symptoms.[34] Other manifestations included chest pain (9.8 per cent), tachypnoea (42.3 per cent) and crackles (52 per cent). Average age of presentation was 49.3 years. Out of 123 reported cases, 47 (38.2 per cent) were males and 76 (61.8 per cent) were females. In most cases the onset is subacute over several days to weeks but more acute presentations have been described.[6,17,18,35,36] Bedrossian et al. also described a more chronic case with progression over several months.[37] There may be a reporting bias towards more severe cases.

DIAGNOSIS

The diagnosis of MTX-induced pneumonitis can be challenging because there are no pathognomonic findings and the disease may be difficult to separate from an infection. Diagnostic criteria have been proposed in three different studies (Boxes 21.2 to 21.4).[14,17,38] Diagnosis is often made retrospectively by exclusion of other aetiologies, especially infection.

Box 21.2   Diagnostic criteria for methotrexate-induced interstitial pneumonitis (from [14])

Classification: *probable*, > 3/4 criteria; *possible*, 2/4 criteria; *unlikely*, 1/4 criteria

- Clinical course consistent with a hypersensitivity reaction
- Exclusion of infection and other pulmonary disease
- Resolution of infiltrates on chest radiograph after discontinuation of methotrexate
- Histopathology consistent with drug-induced injury or hypersensitivity pneumonitis

---

Box 21.3   Diagnostic criteria for methotrexate-induced interstitial pneumonitis (data from [17])

Classification: *definite*, 6/9 criteria; *probable*, 5/9 criteria; *possible*, 4/9 criteria

- Acute onset of dyspnoea
- Fever > 38°C
- Tachypnoea > 28/min + non-productive cough
- WBC < 15k (+ eosinophilia)
- Room air PO_2 < 55 mmHg on admission
- Negative blood and sputum cultures (obligatory)
- Radiographic evidence of interstitial or alveolar infiltrates
- Pulmonary function testing showing restrictive pattern with decreased diffusion capacity for CO (DLCO)
- Histopathology showing bronchiolitis or interstitial pneumonitis with giant cells and no evidence of pathogenic microorganisms

---

Box 21.4   Diagnostic criteria for methotrexate-induced interstitial pneumonitis (from [38])

- Patient receiving methotrexate prior to onset of pulmonary symptoms
- New bilateral infiltrates on chest radiograph
- Exclusion of an infectious aetiology with fibre-optic bronchoscopy with BAL samples and protected brush specimens, sputum and blood cultures and serological tests
- Exclusion of congestive heart failure by forced diuresis, cardiac echocardiogram and absence of alveolar haemorrhage on BAL cytological analysis
- Exclusion of underlying pulmonary disease by BAL cytological analysis and bronchial biopsy
- Improvement of chest radiograph after methotrexate withdrawal

---

Since these patients are immunosuppressed it is important to consider opportunistic infections such as *Pneumocystis jiroveci* or viral pneumonia. Laboratory abnormalities are non-specific and can include moderate leucocytosis, mild eosinophilia (up to 41 per cent of cases) and elevation of LDH.[8] Hypoxaemia is observed in 90–95 per cent of cases.[8] Most common chest X-ray abnormalities are bilateral, mixed interstitial and alveolar infiltrates. Unilateral infiltrates, reticulonodular opacities, pleural effusions and transient hilar lymphadenopathy are other reported abnormalities.[39] These findings must be differentiated from imaging features of the underlying rheumatic condition, and review of earlier films is warranted. High-resolution chest CT usually shows ground-glass parenchymal opacities, and occasionally features suggesting fibrosis.[8] Gallium-67 and Tc-99 diethylene-triamine-pentacetate (DTPA) scans show increased uptake which is non-specific and of undetermined clinical utility.[40–42] Pulmonary function testing (PFT) demonstrates hypoxaemia, restriction, airflow obstruction and decreased lung diffusing capacity (DLCO).[5,43] It is to be noted that uncomplicated long-term therapy with methotrexate does not alter PFTs, and periodic PFT monitoring during methotrexate therapy has not been found to be helpful in detecting pneumonitis before the onset of clinical symptoms.[44–47]

Bronchoalveolar lavage may be needed to rule out infections. Although BAL may show neutrophilia with methotrexate treatment, typically there is lymphocytosis (with increased CD4 and/or CD8 cells) with an increase in CD4/CD8.[48,49] Fuhrman *et al.* found a variable CD4/CD8 ratio (ranging from 0.4 to 9.6) but CD4/CD8 ratio was found to correlate negatively with the time elapsed from last methotrexate dose as well as steroid cumulative dose received by the patients.[38] BAL eosinophilia has also been reported in methotrexate pneumonitis.[49] It should be kept in mind that BAL lymphocytosis with an increased CD4/CD8 ratio may be found in RA patients not taking methotrexate and in the absence of detectable underlying pathology.[50] Also, a similar BAL fluid profile can be seen in pneumonitis caused by other antirheumatic drugs such as gold, D-penicillamine, non-steroidal anti-inflammatory drugs and azathioprine.[50,51] Transbronchial biopsy or surgical lung biopsy may be required to establish definitive diagnosis and exclude infections.

In a review of 49 biopsy specimens, the most frequent histopathological changes were interstitial inflammation (71.4 per cent), interstitial fibrosis (59.2 per cent), type II pneumocyte hyperplasia (38.8 per cent) and granuloma formation (34.7 per cent).[34] Other findings included giant cells (26.5 per cent), increased intra-alveolar macrophages (26.5 per cent) and increased tissue eosinophils (18.4 per cent). Less common changes were intra-alveolar organization (10.2 per cent), hyaline membranes (8.2 per cent), bronchiolitis obliterans (8.2 per cent) and bronchial epithelial cell atypia (2 per cent).

TREATMENT

There are no controlled trials to establish definitive guidelines. Withdrawal of methotrexate seems appropriate, although pneumonitis has been shown to resolve despite its continued use.[28] Traditionally high-dose corticosteroids have been used and can hasten recovery.[6,7,14,52] Corticosteroids should be continued until clinical improvement is seen. There is no evidence that symptoms recur with withdrawal of steroid therapy.[8] In some cases improvement can occur just by discontinuation of methotrexate without the need for corticosteroids. There are case reports where methotrexate therapy was successfully restarted without recurrence of pneumonitis.[14,27,28] In a review by Imokawa *et al.* methotrexate was reintroduced in 16 patients and only 4 had recurrence.[34] Corticosteroids or folinic acid do not prevent the development of MTX-induced pneumonitis.[8] Most patients with MTX-induced pneumonitis improve. Out of the 121 patients with available follow-up data summarized in the Imokawa study, 99 improved, 1 had progressive disease and 16 died from respiratory failure. Approximate mortality rate from MTX-induced pneumonitis is 10 per cent, with many patients dying due to superimposed bacterial infections.[5,26]

GOLD SALTS

Oral and parenteral gold salts (auranofin, gold sodium thiomalate, and gold aurothioglucose) have been used in the treatment of rheumatoid arthritis for over 50 years but are rarely used today because of the availability of safer and more effective therapeutic modalities. However, parenteral gold is a therapeutic option for patients refractory to other antirheumatic drugs. Gold salts are also used in selected cases of juvenile rheumatoid arthritis, ankylosing spondylitis, psoriatic arthritis and pemphigus.[53,54] Gold may also be used in Felty's syndrome (associated with anaemia and thrombocytopenia), since in such cases methotrexate may cause further bone marrow suppression.[39] Gold salts exert an immunomodulatory effect through a decrease in neutrophil chemotaxis, decreased monocyte responsiveness and reduction in the release of proteolytic enzymes in the synovial fluid.[53] They have also been shown to have a dose-dependent effect on cell proliferation by affecting thymidine incorporation and collagen synthesis.[55]

EPIDEMIOLOGY

Gold therapy can cause interstitial pneumonitis, bronchiolitis obliterans (BO) and bronchiolitis obliterans organizing pneumonia (BOOP). Gold-induced pneumonitis is seen in about 1 per cent of patients. It was first described by Savilahti in 1948 but brought to greater attention in 1976 by Winterbauer.[56,57] Sixty cases had been reported before 1986.[58] Tomioka and King in 1997 analysed a cohort of 140 patients with suspected chrysotherapy-induced pulmonary toxicity (from 1966 to December 1994).[54] Their analysis showed that there was no clear relationship between the total dose of gold received and the onset of symptoms or their severity. Women outnumbered men four to one and mean age of onset was 53.1 ± 12.3 years. Interestingly, fewer than 4 per cent of the patients had simultaneous acute rheumatoid flare and gold-induced pulmonary toxicity. In a study of 85 gold-treated RA patients (compared to 283 healthy controls), Hakala *et al.* found a positive correlation between gold-induced interstitial pneumonitis and the presence of HLA-B40 (relative risk 10.5) and HLA-Dw1 (relative risk 6.2).[59] RA patients with gold-induced pneumonitis have also been found to express two extended HLA haplotypes, HLA-A3 B35 Dw1 BfF C4A3,2(BO) and HLA-B40 with a C4 null allele at higher frequency than controls.[60] Adverse effects are more commonly seen with gold sodium thiomalate compared with auranofin.[61] Symptoms usually begin within 2–6 months after therapy, but Hafejee and Burke have reported a case

of acute pneumonitis occurring within 2 hours of intramuscular gold administration.[62]

PATHOGENESIS

The underlying mechanism of gold-associated pulmonary toxicity is thought to be a cell-mediated hypersensitivity reaction.[39,63] McCormick *et al.* demonstrated the elaboration of lymphokines, migration inhibition factor (MIF) and macrophage chemotactic factor (MCF) from the peripheral lymphocytes of patients with gold-induced pulmonary toxicity after incubation with gold salts.[63] A cell-mediated hypersensitivity mechanism is also suggested by the presence of lymphocytosis and CD8+ lymphocytes in BAL fluid of these patients.[54]

CLINICAL PRESENTATION

In the study by Tomioka and King, dyspnoea was the most common symptom (92 per cent) followed by cough (67 per cent).[54] About one-half of the patients have fever and over 35 per cent may have an associated skin rash. Inspiratory crackles are heard in 75 per cent of the patients. Subcutaneous nodules and finger clubbing are absent.

DIAGNOSIS

Diagnostic criteria were suggested by Tomioka and King in their study (Box 21.5).[54] It is important to rule out other aetiologies, especially infections (including opportunistic ones). This is especially important since mild panhypogammaglobulinaemia can occur in 50 per cent of the patients on chrysotherapy.[64] Severe antibody deficiency can be seen with gold therapy and can predispose to opportunistic infections.[65,66] Hypoxaemia,

---

Box 21.5 Diagnostic criteria for gold-induced interstitial pneumonitis (from [54])

Classification: *probable,* $> 9/13$ criteria; *possible,* 7–8/13 criteria

- Acute onset of dyspnoea
- Recent onset of dry cough
- Fever $> 38°C$
- Skin rash
- Absence of finger clubbing
- Crackles on chest examination
- Peripheral blood eosinophilia
- Positive lymphocyte stimulation test to gold salts
- BAL fluid lymphocytosis of $> 25\%$ with a CD4/CD8 ratio of < 1
- Restrictive pattern on pulmonary function testing or decreased diffusion capacity for CO (DLCO)
- Interstitial and alveolar opacities on chest radiograph
- Peribronchovascular alveolar opacities on chest CT
- Non-specific histopathological changes (may be useful in excluding other causes)

hypocarbia, peripheral blood eosinophilia (up to 42 per cent of cases), elevated serum IgE and lactate dehydrogenase may be seen.[39,67,68] Titres of rheumatoid factor are usually low.[54] Chest X-ray shows interstitial and/or alveolar opacities.[68] CT findings include peribronchovascular alveolar opacities.[54] BAL fluid demonstrates lymphocytosis (with inversion of CD4/CD8 ratio).[69,70] This is in contrast to RA-associated lung disease which has mainly neutrophils in BAL.[71] Drug lymphocyte stimulation testing using lymphocytes from BAL can assist in diagnosis.[72,73] Pulmonary function testing reveals reduced vital capacity, restrictive ventilatory defect and decreased diffusing capacity.[68] Pulmonary functions may take up to 12 months after discontinuation of drug to improve.[74] Pathological changes are non-specific and include interstitial infiltrates of plasma cells, lymphocytes, interstitial and alveolar wall fibrosis and intra-alveolar desquamation of mononuclear cells.[67]

TREATMENT

Gold therapy has to be discontinued in all cases. Corticosteroids are used for treatment but there is no consensus as to the dose and duration of therapy and therapy is usually guided by clinical response. Relapses can occur upon discontinuation of steroid treatment.[57] Most patients respond well to steroids and the overall prognosis is good. Interestingly, McFadden and colleagues described a patient with gold-induced pneumonitis who did not improve until naproxen was discontinued as well, suggesting that naproxen may unmask or exacerbate gold-induced pneumonitis.[75]

D–PENICILLAMINE

D-penicillamine is produced from hydrolysis of penicillin. It is used in many diseases including rheumatoid arthritis, scleroderma, primary biliary cirrhosis, Wilson's disease and heavy metal chelation.[76] D-penicillamine exerts an anti-inflammatory effect by down-regulation of T-lymphocyte functions, decreased rheumatic factor and immune complex levels and impaired fibroblast proliferation.[39]

EPIDEMIOLOGY

D-penicillamine can cause myriad adverse effects with several distinct pulmonary syndromes such as pulmonary–renal syndrome (similar to Goodpasture's syndrome), bronchiolitis obliterans, acute interstitial pneumonitis, chronic fibrosing alveolitis and drug-induced lupus.[75] A study by Chakravarty showed that 21 per cent of D-penicillamine-treated RA patients developed a restrictive pulmonary defect when followed for 2 years.[74] HLA-B8, DR3, poor sulphoxidizers and higher age are risk factors for penicillamine-associated adverse effects.[77]

PATHOGENESIS

Though not completely understood, D-penicillamine-induced lung disease is likely a hypersensitivity pneumonitis.[77] Other proposed mechanisms include a direct toxic effect of the drug, and interference with collagen and elastin synthesis may also contribute.[78]

CLINICAL PRESENTATION AND DIAGNOSIS

Typically clinical manifestations include dyspnoea, cough, fever, tachypnoea and inspiratory rales, hypoxaemia and leucocytosis.[78] Chest imaging shows mixed interstitial and alveolar opacities. Miliary pulmonary infiltrates have been described as well.[79] Pulmonary function testing reveals a restrictive ventilatory pattern.[74] However, pulmonary function abnormalities are also seen in asymptomatic patients receiving D-penicillamine.

TREATMENT

D-penicillamine should be withdrawn in patients suspected of having penicillamine-induced pneumonitis. In some cases of persistent symptoms, corticosteroids can be used.[39] In contrast to the very high mortality in penicillamine-induced pulmonary–renal syndrome, the outcome for pneumonitis is better once the drug is stopped.

LEFLUNOMIDE

Leflunomide is an immunomodulatory disease-modifying antirheumatic drug used in the treatment of rheumatoid arthritis. It inhibits T-lymphocyte proliferation by inhibiting the enzyme dihydroorotate dehydrogenase involved in pyrimidine biosynthesis.[80] Leflunomide-induced interstitial pneumonitis is extremely rare, with an estimated incidence of 0.02 per cent in Western countries.[81] However, the incidence was found to be higher in Japan (0.48 per cent), posing the question whether there are any genetic and/or environmental risk factors involved.[82] In a study by Suissa et al. the risk of ILD was increased about 2-fold in RA patients treated with leflunomide compared with those not receiving the drug.[83] However, the risk was only increased in the subgroup of RA patients who had prior ILD or had received prior methotrexate. There was no increased risk in the absence of prior ILD or MTX therapy. The study concluded that the apparent increase in ILD due to leflunomide was due to channelling bias as RA patients with prior ILD were found twice as likely to receive leflunomide as first DMARD when compared to MTX.

The symptoms of leflunomide-induced pneumonitis are non-specific and treated with withdrawal of the drug and a trial of corticosteroids.[81] Cholestyramine can be used to lower serum leflunomide levels. The mortality rate of leflunomide-induced pneumonitis is about 30 per cent.[81]

NITROFURANTOIN

Nitrofurantoin is an antimicrobial commonly used to treat urinary tract infections. It is also used for chronic suppressive therapy in patients with recurrent urinary tract infections. Nitrofurantoin is converted by bacterial nitroreductases into highly electrophilic intermediates that attack bacterial ribosomes inhibiting protein synthesis.[84]

EPIDEMIOLOGY

Nitrofurantoin can cause acute and subacute interstitial pneumonitis as well as chronic fibrosing alveolitis. Fulminant pulmonary haemorrhaging with haemoptysis and respiratory failure has been reported.[85–87] Nitrofurantoin-induced pulmonary toxicity (non-cardiogenic pulmonary oedema) was first described in 1965 by Murray.[88] In 1980, Holmberg et al. published a review of 921 cases of nitrofurantoin adverse effects from 1966 to 1976.[89] The two most common adverse effects recorded were acute pulmonary reactions (43 per cent) and allergic reactions (42 per cent). The incidence of nitrofurantoin-induced lung toxicity is less than 1 in 1000 exposures, with acute pneumonitis being 10 times more common than chronic interstitial fibrosis.[90] Jick et al. found only three hospitalizations out of 16 101 patients receiving their first course of nitrofurantoin.[91] In a review of 398 cases of acute and 49 cases of chronic nitrofurantoin-associated pneumonitis, Holmberg and Boman found 85 per cent of the patients to be women – likely representing the increased susceptibility to recurrent UTIs in women.[92] Patients with acute pneumonitis were generally younger (mean 59 years) than those with chronic lung toxicity (mean 68 years).[92] Toxicity does not appear to be dose-related and can occur with small doses.[90,93]

Symptoms of acute pneumonitis usually occur within 4 weeks of starting treatment, with about 8 per cent of patients developing symptoms within 8–9 days.[93] There are also reports of acute pneumonitis developing within 24 hours of treatment.[89,94,95] Acute toxicity has been described in patients who had no reaction to prior treatment with the drug.[87] Subacute pneumonitis usually develops within 1–6 months of starting treatment, while chronic interstitial fibrosis is seen after long-term suppressive therapy (several months to years) with nitrofurantoin.[93]

PATHOGENESIS

Several mechanisms have been proposed by different authors (Box 21.6).

CLINICAL PRESENTATION

Clinical manifestations of nitrofurantoin-induced acute pneumonitis include fever, chills, dry cough, dyspnoea, chest pain, myalgias, arthralgias and rash. Physical examination shows tachypnoea, tachycardia, wheezes, rales, rhonchi or decreased breath sounds. Chronic alveolitis with the distinctive pattern of desquamative interstitial pneumonia or fibrosis presents as weight loss, fatigue, dry cough, dyspnoea on exertion, tachypnoea and rales. Fever is uncommon in the chronic form.[96] Combined

---

Box 21.6 Proposed mechanisms for nitrofurantoin-associated lung toxicity

- Hypersensitivity reaction[96, 97]
- Direct oxidant-mediated injury to lung parenchyma[98, 99]
- Immune complex-mediated injury[100]
- Alveolar epithelial cell injury mediated by lymphocytes[101, 102]
- Exacerbation of nitrofurantoin lung damage by hyperoxia[99, 103]
- Direct pulmonary endothelial damage[104]

---

lung and liver toxicity with elevated antinuclear antibodies (ANA) and anti-smooth muscle antibodies can occur.[105,106]

DIAGNOSIS

Laboratory studies show peripheral eosinophilia in about 80 per cent of the acute cases but only 44 per cent of the chronic cases.[93] Eosinophilia may be initially absent and appear with continued or repeat treatment.[93] BAL fluid also shows eosinophils. The chronic form is associated with elevated IgG (80 per cent) and elevated ANA titres (60 per cent).[93] Chest imaging in acute pneumonitis shows interstitial and/or alveolar opacities with slight basal predominance. Chronic lung toxicity shows reticular opacities, traction bronchiectasis and honeycombing which are usually irreversible but can improve over several weeks after stopping the drug.[97] CT scanning shows ground-glass opacities, patchy consolidation and linear opacities in a subpleural and peripheral pattern. Pulmonary function testing can show isolated reduction in diffusion capacity for carbon monoxide (DLCO) to restrictive ventilatory defect characterized by reduction in forced vital capacity (FVC) and lung volume.[97,100,107] Histopathological changes on biopsy are non-specific.

TREATMENT

Treatment includes withdrawal of the drug and supportive care. Corticosteroids may be helpful in some cases.[89,91,92,94,95] Most patients with acute toxicity have rapid improvement in symptoms within 24–72 hours, but symptoms can persist for several weeks in the subacute and chronic forms.[93] Holmberg and Boman reported a mortality of 0.5 per cent in acute pneumonitis and 8 per cent in chronic nitrofurantoin lung toxicity.[92]

NALIDIXIC ACID

Nalidixic acid is a first-generation quinolone and is rarely, if at all, used these days. Dan et al. described a case of hypersensitivity pneumonitis in a patient, which developed 1 week after starting therapy with nalidixic acid.[108] Peripheral blood and BAL fluid showed eosinophilia. There was also a positive migration inhibitory factor test with nalidixic acid.

CEFTIZOXIME

Ceftizoxime is a third-generation cephalosporin. A case of ceftizoxime-induced pneumonitis was reported by Suzuki et al.[109]

TRIMETHOPRIM

Higgins and Niklasson described a case of a 43-year-old woman who developed respiratory symptoms and chest infiltrates after taking an extended course of trimethoprim.[110] Symptoms improved on discontinuation of the drug but reappeared with a repeat course 3 days later. Symptoms included fever, cough, myalgias, arthralgias, leucocytosis with 11 per cent eosinophils, moderate transaminitis. Clinical and laboratory improvement was seen after discontinuation of trimethoprim. Serological studies for mycoplasma, viruses, toxocara and ornithosis were negative and stool studies did not reveal any ova or cysts.

ACETAMINOPHEN

Kitaguchi *et al.* described a case of a 20-year-old female who developed cough, fever, dyspnoea and reticulonodular infiltrates on chest X-ray after taking 1.2 g of acetaminophen for toothache.[111] BAL fluid showed an increase in lymphocytes, eosinophils and neutrophils. The lymphocyte stimulation test was positive for acetaminophen component of Norshin. All medications were stopped and she was treated with methylprednisolone resulting in complete improvement of symptoms, pulmonary functions and chest X-ray.

NILUTAMIDE

Nilutamide is a non-steroidal, synthetic antiandrogen used in the treatment of metastatic prostate cancer. Pfitzenmeyer *et al.* described eight patients with metastatic prostate cancer who were on nilutamide therapy and developed respiratory symptoms, hypoxaemia with bilateral infiltrates on chest imaging.[112] Pulmonary function testing showed restrictive parameters in six patients. Average duration of therapy was 113 days and mean cumulative exposure was 21.8 g. Six patients underwent bronchoscopy with BAL revealing lymphocytosis in four and neutrophilia in two patients. Nilutamide therapy was stopped in seven patients and decreased from 300 mg to 150 mg per day in one. Two patients were also given corticosteroids. All patients showed improvement in symptoms, pulmonary functions and PaO_2. This study estimated an incidence of 1 per cent for nilutamide-induced interstitial pneumonitis. Akoun *et al.* demonstrated BAL lymphocytosis with a reversed CD4/CD8 ratio and the release of leucocyte inhibition factor by sensitized peripheral blood T-cells in the presence of nilutamide, suggesting that the underlying mechanism for the lung damage may be cell-mediated hypersensitivity reaction.[113] Simultaneous lung and liver toxicity that improved on discontinuation of the drug has been reported.[114] The congener bicalutamide may also cause ILD.[115]

BETA-ADRENERGIC ANTAGONISTS

Cases of interstitial pneumonitis have been described with propranolol, acebutolol and nadolol.[116–118] BAL fluid showed lymphocytic alveolitis with an inverted CD4/CD8 ratio.[116] Clinical improvement was seen after discontinuation of the beta-blocker with normalization of the BAL lymphocyte subset ratio. The underlying mechanism is thought to be cell-mediated hypersensitivity.[116]

ANTIARRHYTHMIC AGENTS

Flecainide is a class 1C antiarrhythmic agent used for the treatment of ventricular and supraventricular arrhythmias. Dyspnoea and chest pain can occur in 4–7 per cent of the patients on flecainide therapy but pulmonary toxicity is much less rare.[119] Five cases of suspected flecainide-induced pneumonitis have been described.[120–123] Cell-mediated immune responses are thought to be the causative mechanism.[120] Withdrawal of the drug and corticosteroid therapy has been helpful. Tocainide, another class 1C antiarrhythmic with a similar structure, has also been reported to cause interstitial pneumonitis.[124,125] Amiodarone toxicity is discussed in Chapter 22.

MESALAMINE (5-AMINOSALICYLIC ACID)

Mesalamine, used in the treatment of inflammatory bowel disease, has been reported to cause interstitial pneumonitis.[126]

TUMOUR NECROSIS FACTOR-α ANTAGONISTS

TNF-α blockers (infliximab, etanercept and adalimumab) are a group of newer immunomodulatory agents used in a variety of inflammatory conditions including rheumatoid arthritis, psoriatic arthritis, juvenile rheumatoid arthritis, systemic sclerosis, ankylosing spondylitis and psoriasis. Infliximab is also used to treat fistulous Crohn's disease and ulcerative colitis.

There have been several cases reported of interstitial pneumonitis associated with TNF-α blocker therapy.[127–134] Most of these occurred after the second or third dose of anti-TNF α therapy. The most common underlying disease was rheumatoid arthritis, but pulmonary toxicity of these drugs has also been reported in systemic sclerosis, Crohn's disease and ankylosing spondylitis.[134–136] The underlying mechanism of infliximab-associated pneumonitis may be deficient apoptosis of infiltrating inflammatory cells.[127] Many patients who developed pneumonitis related to TNF α had pre-existing interstitial lung disease (ILD) and had been exposed to methotrexate in the past. It is believed that TNF-α blockers may cause pulmonary toxicity by themselves and may also exacerbate methotrexate-induced pneumonitis.[127,137] Prognosis depends on the underlying histopathological process. Several fatal cases were found to have usual interstitial pneumonitis (UIP).[135]

This topic is covered in more detail in Chapter 15.

AZATHIOPRINE

Azathioprine is an immunosuppressant agent used in organ transplantation and inflammatory disorders such as rheumatoid arthritis, inflammatory bowel disease, sarcoidosis and bullous pemphigoid. Several cases of associated interstitial pneumonitis have been described in renal transplant patients.[138,139]

KEY POINTS

- A number of non-cytotoxic drugs can cause interstitial pneumonitis, and the list is growing (see www.pneumotox.com).
- Although the underlying pathogenesis is not clearly understood, there is evidence that it is related to immune-mediated mechanisms.

(Contd)

- Since the clinical manifestations and histopathological changes are non-specific, diagnosis rests on exclusion of other aetiologies, especially infections.

- In most cases, BAL fluid shows increased cellularity, often with a reversed lymphocyte subset ratio. Eosinophilia in peripheral blood and BAL fluid is suggestive but not always present.

- The most important aspect of management is prompt recognition and withdrawal of the suspected drug. Corticosteroids can be helpful in some cases.

- Prognosis is usually good but fatal cases have been described.

REFERENCES

1. Anticancer drugs. In: Harvey RA *et al.* (eds) *Lippincott's Illustrated Reviews: Pharmacology*, 2nd edn, pp 373–400. Philadelphia, PA: Lippincott–Raven, 2000.

2. Perry MC, Anderson CM, Donehower RC. Chemotherapy. In: Abeloff MD *et al.* (eds) *Clinical Oncology*, p 494. Edinburgh: Churchill Livingstone, 2004.

3. Iikuni N, Iwami S, Kasai S, Tokuda H. Noncardiogenic pulmonary edema in low-dose oral methotrexate therapy. *Intern Med* 2004;**43**:846–51.

4. Acute Leukemia Group B. Acute lymphocytic leukemia in children: maintenance therapy with methotrexate administered intermittently. *J Am Med Assoc* 1969;**207**:923–8.

5. Sostman HD, Mathay RA, Putman CE, Smith GJ. Methotrexate-induced pneumonitis. *Medicine* 1976;**55**:371–88.

6. Cannon GW, Ward JR, Clegg DO, Samuelson CO, Abott TM. Acute lung disease associated with low-dose pulse methotrexate therapy in patients with rheumatoid arthritis. *Arthritis Rheum* 1983;**26**:1269–74.

7. Engelbrecht JA, Calhoon SL, Scherrer JJ. Methotrexate pneumonitis after low-dose therapy for rheumatoid arthritis. *Arthritis Rheum* 1983;**26**:1275–8.

8. Barrera P, Laan RF, van Riel PL, *et al.* Methotrexate-related pulmonary complications in rheumatoid arthritis. *Ann Rheum Dis* 1994;**53**:434–9.

9. Alarcon GS, *et al.*, Methotrexate Lung Study Group. Risk factors for methotrexate-induced lung injury in patients with rheumatoid arthritis: a multicenter, case-control study. *Ann Intern Med* 1997;**127**:356–64.

10. Kremer JM, Alarcon GS, Weinblatt ME, *et al.* Clinical, laboratory, radiographic and histopathologic features of methotrexate-associated lung injury in patients with rheumatoid arthritis: a multicenter study with literature review. *Arthritis Rheum* 1997;**40**:1829–37.

11. Carroll GJ, Thomas R, Phatouros CC, *et al.* Incidence, prevalence and possible risk factors for pneumonitis in patients with rheumatoid arthritis receiving methotrexate. *J Rheumatol* 1994;**21**:51–4.

12. Golden MR, Katz RS, Balk RA, Golden HE. The relationship of preexisting lung disease to the development of methotrexate pneumonitis in patients with rheumatoid arthritis. *J Rheumatol* 1995;**22**:1043–7.

13. Ohosone Y, Okano Y, Kameda H, *et al.* Clinical characteristics of patients with rheumatoid arthritis and methotrexate-induced pneumonitis. *J Rheumatol* 1997;**24**:2299–303.

14. Carson CW, Cannon GW, Egger MJ, Ward JR, Clegg DO. Pulmonary disease during treatment of rheumatoid arthritis with low dose pulse methotrexate. *Semin Arthritis Rheum* 1987;**16**:186–95.

15. McKendry RJR, Cyr M. Toxicity to methotrexate compared with azathioprine in the treatment of rheumatoid arthritis: a case–control study of 131 patients. *Arch Int Med* 1989;**149**:685–9.

16. Rooney TW, Furst DE. Comparison of toxicity in methotrexate treated rheumatoid arthritis patients also taking aspirin or other NSAID. *Arthritis Rheum* 1986;**29**:S76.

17. Searles G, McKendry RJR. Methotrexate pneumonitis in rheumatoid arthritis: potential risk factors. Four case reports and a review of the literature. *J Rheumatol* 1987;**14**:1164–71.

18. Ridley MG, Wolfe CS, Mathews JA. Life threatening acute pneumonitis during low dose methotrexate treatment for rheumatoid arthritis: a case report and review of the literature. *Ann Rheum Dis* 1988;**47**:784–8.

19. Schoenfeld A, Mashiach R, Vardy M, Ovadia J. Methotrexate pneumonitis in nonsurgical treatment of ectopic pregnancy. *Obstet Gynaecol* 1992;**80**:520–1.

20. Elasser S, Dalquen P, Soler M, Perruchoud AP. Methotrexate-induced pneumonitis: appearance four weeks after discontinuation of treatment. *Am Rev Respir Dis* 1989;**140**:1089–92.

21. Pourel J, Guillemin F, Fener P, *et al.* Delayed methotrexate pneumonitis in rheumatoid arthritis (letter). *J Rheumatol* 1991;**18**:303–4.

22. De Bandt M, Rat AC, Palazzo E, Kahn MF. Delayed methotrexate pneumonitis (letter). *J Rheumatol* 1991;**18**:1943.

23. Kremer JM, Phelps C. Long-term prospective study of the use of methotrexate in the treatment of rheumatoid arthritis: update after a mean of 90 months. *Arthritis Rheum* 1992;**35**:138–45.

24. Zisman DA, McCune WJ, Tino G, Lynch JP. Drug-induced pneumonitis: the role of methotrexate. *Sarcoidosis Vasc Diffuse Lung Dis* 2001;**18**:243–52.

25. Hagiwara K, Sato T, Takagi Kobayashi S, *et al.* Acute exacerbation of preexisting interstitial lung disease after administration of etanercept for rheumatoid arthritis. *J Rheumatol* 2007;**34**:1151–4.

26. Cooper JAD, White DA, Mathay RA. Drug-induced pulmonary disease. I: Cytotoxic drugs. *Am Rev Respir Dis* 1986;**133**:321–40.

27. Cook NJ, Carroll GI. Successful reintroduction of methotrexate after pneumonitis in two patients with rheumatoid arthritis. *Ann Rheum Dis* 1992;**51**:272–4.

28. Fehr T, Jacky E, Bachii EB. Successful reintroduction of methotrexate after acute pneumonitis in a patient with acute lymphoblastic leukemia. *Ann Hematol* 2003;**82**:193–6.

29. Clarysse AH, Cathey WJ, Cartwright GE, Wintrobe HH. Pulmonary disease complicating intermittent therapy with methotrexate. *J Am Med Assoc* 1969;**209**:1861–4.

30. Akoun GM, Gauthier-Rahman S, Mayaud CB, *et al.* Leukocyte migration inhibition in methotrexate induced

pneumonitis: evidence for an immunologic cell-mediated mechanism. *Chest* 1987;**91**:96–8.

31. Schnabel A, Richter C, Bauerfeind S, Gross WL. Bronchoalveolar lavage cell profile in methotrexate induced pneumonitis. *Thorax* 1997;**52**:377–9.

32. Hirata S, Hattori N, Kumagai K, *et al.* Lymphocyte transformation test is not helpful for the diagnosis of methotrexate-induced pneumonitis in patients with rheumatoid arthritis. *Clin Chim Acta* 2009;**407**:25–9.

33. Chikura B, Sathi N, Lane S, Dawson JK. Variation of immunological response in methotrexate-induced pneumonitis. *Rheumatology* (Oxford) 2008;**47**:1647–50.

34. Imokawa S, Colby TV, Leslie KO, Helmers RA. Methotrexate pneumonitis: review of the literature and histopathological findings in nine patients. *Eur Respir J* 2000;**15**:373–81.

35. Hargreaves MR, Mowat AG, Benson MK. Acute pneumonitis associated with low dose methotrexate treatment for rheumatoid arthritis: report of five cases and review of published reports. *Thorax* 1992;**47**:628–33.

36. Horrigan TJ, Fanning J, Marcotte MP. Methotrexate pneumonitis after systemic treatment for ectopic pregnancy. *Am J Obstet Gynecol* 1997;**176**:714–15.

37. Bedrossian CWM, Miller WC, Luna MA. Methotrexate-induced diffuse interstitial pulmonary fibrosis. *South Med J* 1979;**72**:313–18.

38. Fuhrman C, Parrot A, Wislez M, *et al.* Spectrum of CD4 to CD8 T-cell ratios in lymphocytic alveolitis associated with methotrexate-induced pneumonitis. *Am J Respir Crit Care Med* 2001;**164**:1186–91.

39. Lock BJ, Eggert M, Cooper JA. Infiltrative lung disease due to noncytotoxic agents. *Clin Chest Med* 2004;**25**:47–52.

40. Bisson G, Drapeau G, Lamoureux G, *et al.* Computer based quantitative analysis of gallium-67 uptake in normal and diseased lungs. *Chest* 1983;**84**:513–17.

41. Elstad MR. Lung biopsy, bronchoalveolar lavage and gallium scanning. In: Cannon GW, Zimmerman GA (eds) *The Lung and Rheumatic Diseases*, pp 117–42. New York: Marcel Dekker, 1990.

42. King TE. Idiopathic pulmonary fibrosis. In: King TE, Decker BC (eds) *Interstitial Lung Disease*, pp 139–69. Toronto: Schwartz MI, 1988.

43. Bell MJ, Geddie WR, Gordon DA, Reynolds WJ. Pre-existing lung disease in patients with rheumatoid arthritis may predispose to methotrexate lung. *Arthritis Rheum* 1986;**29**:575.

44. Jeurissen MEC, Boerbooms AM, Festen J, van de Putte LBA, Doesburg W. Serial pulmonary function tests during a randomized, double-blind trial of azathioprine versus methotrexate in rheumatoid arthritis. *Arthritis Rheum* 1991;**34**:S90.

45. Croock AD, Furst DF, Helmers RA, *et al.* Methotrexate does not alter pulmonary function in patients with rheumatoid arthritis. *Arthritis Rheum* 1980;**32**:S60.

46. Velay B, Lamboley L, Massonnet B. Prospective study of respiratory function in rheumatoid arthritis treated with methotrexate. *Eur Respir J* 1988;**1**(Suppl 2):371S.

47. Cottin V, Tebib J, Massonnet B, Souquet PJ, Bernard JP. Pulmonary function in patients receiving long term low-dose methotrexate. *Chest* 1996;**109**:933–8.

48. Akoun GM, Maynaud CM, Touboul JL, *et al.* Use of bronchoalveolar lavage in the evaluation of methotrexate lung disease. *Thorax* 1987;**42**:652–5.

49. White DA, Rankin JA, Stover DE, Gellene RA, Gupta S. Methotrexate pneumonitis: bronchoalveolar lavage findings suggest an immunologic disorder. *Am Rev Respir Dis* 1989;**139**:18–21.

50. Kolarz G, Scherak O, Popp W, *et al.* Bronchoalveolar lavage in rheumatoid arthritis. *Br J Rheumatol* 1993;**32**:556–61.

51. Cannon GW. Pulmonary complications of antirheumatic drug therapy. *Semin Arthritis Rheum* 1990;**19**:353–64.

52. St Clair EW, Rice JR, Snyderman R. Pneumonitis complicating low-dose methotrexate therapy in rheumatoid arthritis. *Arch Intern Med* 1985;**145**:2035–8.

53. Libby D, White DA. Pulmonary toxicity of drugs used to treat systemic autoimmune diseases. *Clin Chest Med* 1998;**19**:809–21.

54. Tomioka R, King TE. Gold-induced pulmonary diseases: clinical features, outcome, and differentiation from rheumatoid lung disease. *Am J Respir Crit Care Med* 1997;**155**:1011–20.

55. Goldberg RL, Parrott DP, Kaplan SR, Fuller GC. A mechanism of action of gold sodium thiomalate in diseases characterized by proliferative synovitis: reversible changes in collagen production in cultured human synovial cells. *J Pharmacol Exp Ther* 1981;**218**:395–403.

56. Savilahti M. Pulmonary complications following use of gold salts. *Ann Med Intern Fenn* 1948;**37**:263–6.

57. Winterbauer RH, Wilske KR, Wheelis RF. Diffuse pulmonary injury associated with gold treatment. *New Engl J Med* 1976;**294**:919–21.

58. Fernandez CM, Casas HA, Leczycki H. Pulmonary toxicity of gold salts. *Medicine* (B Aires) 1981;**51**(1):59–61.

59. Hakala M, Van Assendelft AHW, Ilonen J, Jalava S, Tiilikainen A. Association of different HLA antigens with various toxic effects of gold salts in rheumatoid arthritis. *Ann Rheum Dis* 1986;**45**:177–82.

60. Partanen J, Van Assendelft AHW, Koskimies S, *et al.* Patients with rheumatoid arthritis and gold-induced pneumonitis express two high-risk major histocompatibility complex patterns. *Chest* 1987;**92**:277–81.

61. Davis P, Menard H, Thompson J, Harth M, Beaudet F. One-year comparative study of gold sodium thiomalate and auranofin in the treatment of rheumatoid arthritis. *J Rheumatol* 1985;**12**:60–7.

62. Hafejee A, Burke MJ. Acute pneumonitis starting 2 hours after intramuscular gold administration in a patient with rheumatoid arthritis. *Ann Rheum Dis* 2004;**63**:1525–6.

63. McCormick J, Cole S, Lahirir S, *et al.* Pneumonitis caused by gold salts therapy: evidence for role of cell-mediated immunity in its pathogenesis. *Am Rev Respir Dis* 1980;**122**:14552.

64. Lorber A, Simon T, Leeb J, Peter A, Wilcox S. Chrysotherapy: suppression of immunoglobulin synthesis. *Arthritis Rheum* 1978;**21**:785–91.

65. Snowden N, Dietch DM, The LS, Hilton RC, Haeney MR. Antibody deficiency associated with gold treatment: natural history and management in 22 patients. *Ann Rheum Dis* 1996;**55**:616–21.

66. Lawson TM, Bevan M, Linton S, Williams BD. Serious opportunistic infection associated with gold-induced panhypogammaglobulinemia. *Br J Rheumatol* 1998;**37**:914–16.

67. Sinha A, Silverstone EJ, O'Sullivan MM. Gold-induced pneumonitis: computed tomography findings in a patient

with rheumatoid arthritis. *Rheumatology* (Oxford) 2001;**40**:712–14.

68. Blancas R, Moreno JL, Martin E, *et al.* Alveolar-interstitial pneumopathy after gold-salts compounds administration, requiring mechanical ventilation. *Intens Care Med* 1998;**24**:1110–12.

69. Evans RB, Ettensohn DB, Fawaz-Estrup F, Lally EV, Kaplan SR. Gold lung: recent developments in pathogenesis, diagnosis and therapy. *Semin Arthritis Rheum* 1987;**16**:196–205.

70. Zitnik RJ, Cooper JA. Pulmonary disease due to antirheumatic agents. *Clin Chest Med* 1990;**11**:139–50.

71. Ettensohn DB, Roberts NJ, Condemi JJ. Bronchoalveolar lavage in gold lung. *Chest* 1984;**85**:569–70.

72. Matsumura Y, Miyake A, Ishida T. A case of gold lung with positive lymphocyte stimulation test to gold, using bronchoalveolar lymphocytes. *Nihon Kyobu Shikkan Gakkai Zasshi* 1992;**30**:472–7.

73. Bando M, Takishita Y, Bando H, Hashimoto Y, Sano T. A case of gold-induced pneumonitis showing a positive reaction in the drug lymphocyte stimulation (DLST) for gold. *Nihon Kyobu Shikkan Gakkai Zasshi* 1992;**30**:128–32.

74. Chakravarty K, Webley M. A longitudinal study of pulmonary function in patients with rheumatoid arthritis treated with gold and D-penicillamine. *Br J Rheumatol* 1992;**31**:829–33.

75. McFadden RG, Fraher LJ, Thompson JM. Gold–naproxen pneumonitis: a toxic drug interaction? *Chest* 1989;**96**:216–18.

76. Frankel SK. Drug-induced lung disease. In: Hanley ME, Welsh CH (eds) *Current Diagnosis and Treatment in Pulmonary Medicine*, pp 337–47. Denver, CO: McGraw-Hill, 2003.

77. Grasedyck K. D-penicillamine: side effects, pathogenesis and decreasing the risks. *Z Rheumatol* 1988;**47**(Suppl 1):17–19.

78. Camus P, Degat OR, Justrabo E, Jeannin L. D-penicillamine-induced severe pneumonitis. *Chest* 1982;**81**:376–8.

79. Petersen J, Moller I. Miliary pulmonary infiltrates and penicillamine. *Br J Radiol* 1978;**51**:915–16.

80. Fox RI. Mechanism of action of leflunomide in rheumatoid arthritis. *J Rheumtol Suppl* 1998;**53**:20–6.

81. Kamata Y, Nara H, Kamimura T, *et al.* Rheumatoid arthritis complicated with acute interstitial pneumonia induced by leflunomide as an adverse reaction. *Intern Med* 2004;**43**:1201–4.

82. Ito S, Sumida T. Interstitial lung disease associated with leflunomide. *Intern Med* 2004;**43**:1103–4.

83. Suissa S, Hudson M, Ernst P. Leflunomide use and the risk of interstitial lung disease in rheumatoid arthritis. *Arthritis Rheum* 2006;**54**:1435–9.

84. McOsker CC, Fitzpatrick PM. Nitrofurantoin: mechanism of action and implications for resistance development in common uropathogens. *J Antimicrob Chemother* 1994;**33**(Suppl A):23–30.

85. Averbuch SD, Yungbluth P. Fatal pulmonary hemorrhage due to nitrofurantoin. *Arch Intern Med* 1980;**140**:271–3.

86. Meyer MM, Meyer RJ. Nitrofurantoin-induced pulmonary hemorrhage in a renal transplant recipient receiving immunosuppressive therapy: a case report and review of the literature. *J Urol* 1994;**152**:938–40.

87. Bucknall CE, Adamson MR, Banham SW. Nonfatal pulmonary hemorrhage associated with nitrofurantoin. *Thorax* 1987;**42**:475–6.

88. Murray MJ, Kronenburg R. Pulmonary reactions simulating cardiac pulmonary edema caused by nitrofurantoin. *New Engl J Med* 1965;**273**:1185–7.

89. Holmberg L, Boman G, Bottinger LE, *et al.* Adverse reactions to nitrofurantoin: analysis of 921 reports. *Am J Med* 1980;**69**:733–8.

90. D'Arcy PF. Nitrofurantoin. *Drug Intell Clin Pharm* 1985;**19**:540–7.

91. Jick SS, Jick H, Walker AM, Hunter JR. Hospitalization for pulmonary reactions following nitrofurantoin use. *Chest* 1989;**96**:512–15.

92. Holmberg L, Boman G. Pulmonary reactions to nitrofurantoin. 447 cases reported to the Swedish Adverse Drug Reaction Committee 1966–76. *Eur Respir Dis* 1981;**62**:180–9.

93. Vahid B, Wildemore BMM. Nitrofurantoin pulmonary toxicity: a brief review. *Internet J Pulmon Med*, 2006 (online). Available at www.ispub.com/ostia/index.php?xmlFilePath = journals/ijpm/vol6n2/nitro.xml (accessed October 2007).

94. Chundnofsky CR, Otten EJ. Acute pulmonary toxicity to nitrofurantoin. *J Emerg Med* 1989;**7**(1):15–19.

95. Sovijarvi AR, Lemola M, Stenius B, Idanpaan-Heikkila J. Nitrofurantoin induced acute, subacute and chronic pulmonary reactions. *Scand J Respir Dis* 1977;**58**:41–50.

96. Ozkan M, Dweik RA, Ahmad M. Drug-induced lung disease. *Cleve Clin J Med* 2001;**68**:782–95.

97. Sheehan RE, Wells AU, Milne DG, Hansell DM. Nitrofurantoin-induced lung disease: two cases demonstrating resolution of apparently irreversible CT abnormalities. *J Comput Assist Tomogr* 2000;**24**:259–61.

98. Martin WJ. Nitrofurantoin: evidence for oxidant injury of lung parenchymal cells. *Am Rev Respir Dis* 1983;**127**:482–6.

99. Dunbar JR, DeLucia AJ, Bryant LR. Glutathione status of isolated rabbit lungs: effects of nitrofurantoin and paraquat perfusion with normoxic and hyperoxic ventilation. *Biochem Pharmacol* 1984;**33**:1343–8.

100. Taskinen E, Tukiainen P, Sovijarvi AR. Nitrofurantoin-induced alterations in pulmonary tissue: a report on five patients with acute or subacute reactions. *Acta Pathol Microbiol Scan A* 1977;**85**:713–20.

101. Brutinel WM, Martin WJ. Chronic nitrofurantoin reaction associated with T-lymphocyte alveolitis. *Chest* 1986;**89**:150–2.

102. Pearsall HR, Ewalt J, Tsoi MS, *et al.* Nitrofurantoin lung sensitivity: report of a case with prolonged nitrofurantoin lymphocyte sensitivity and interaction of nitrofurantoin-stimulated lymphocytes with alveolar cells. *Lab Clin Med* 1974;**83**:728–37.

103. Suntres ZE, Shek PN. Nitrofurantoin-induced pulmonary toxicity: in-vivo evidence for oxidative stress-mediated mechanisms. *Biochem Pharmacol* 1992;**43**:1127–35.

104. Martin WJ, Powis GW, Kachel DL. Nitrofurantoin-stimulated oxidant production in pulmonary endothelial cell. *J Lab Clin Med* 1985;**105**:23–9.

105. Reinhart HH, Reinhart E, Korlipara P, Peleman R. Combined nitrofurantoin toxicity to liver and lung. *Gastroenterology* 1992;**102**(4 Pt 1):1396–9.

106. Schattner A, Von der Walde J, Kozak N, Sokolovskaya N, Knobler H. Nitrofurantoin-induced immune-mediated lung and liver disease. *Am J Med Sci* 1999;**317**:336–40.

107. Cameron RJ, Kolbe J, Wilshner ML, Lambie N. Bronchiolitis obliterans organizing pneumonia associated with the use of nitrofurantoin. *Thorax* 2000;**55**:249–51.

108. Dan M, Aderka D, Topilsky M, Livni E, Levo Y. Hypersensitivity pneumonitis induced by nalidixic acid. *Arch Intern Med* 1986;**146**:1423–4.

109. Suzuki K, Yamamoto K, Kishimoto A, Hayakawa T, Yamamoto T. A case of ceftizoxime-induced pneumonitis. *Nihon Kyobu Shikkan Gakkai Zasshi* 1985;**23**:1357–61.

110. Higgins T, Niklasson PM. Hypersensitivity pneumonitis induced by trimethoprim. *Br Med J* 1990;**300**:1344.

111. Kitaguchi S, Miyazawa T, Minesita M, *et al.* A case of acetaminophen-induced pneumonitis. *Nihon Kyobu Shikkan Gakkai Zasshi* 1992;**30**:1322–6.

112. Pfitzenmeyer P, Foucher P, Piard F, *et al.* Nilutamide pneumonitis: report on eight patients. *Thorax* 1992;**47**:622–7.

113. Akoun GM, Liote HA, Liote F, Gauthier-Rahman S, Kurtz D. Provocation test coupled with bronchoalveolar lavage in diagnosis of drug (nilutamide)-induced hypersensitivity pneumonitis. *Chest* 1990;**97**:495–8.

114. Gomez JL, Dupont A, Cusan L, *et al.* Simultaneous liver and lung toxicity related to the nonsteroidal antiandrogen nilutamide (Anadron): a case report. *Am J Med* 1992;**92**:563–6.

115. Bennett CL, Raisch DW, Sartor O. Pneumonitis associated with nonsteroidal antiandrogens: presumptive evidence of a class effect. *Ann Intern Med* 2002;**137**:625.

116. Akoun GM, Milleron BJ, Maynaud CM, Tholoniat D. Provocation test coupled with bronchoalveolar lavage in the diagnosis of propranolol-induced hypersensitivity pneumonitis. *Am Rev Respir Dis* 1989;**139**:247–9.

117. Thompson RN, Grennan DM. Acebutolol induced hypersensitivity pneumonitis. *Br Med J* (Clin Res) 1983;**286**:894.

118. Levy MB, Fink JN, Guzzetta PA. Nadolol and hypersensitivity pneumonitis. *Ann Intern Med* 1986;**105**:806–7.

119. Gentzkow GD, Sullivan JY. Extracardiac adverse effects of flecainide. *Am J Cardiol* 1984;**53**:101B–105B.

120. Pesenti S, Lauque D, Daste G, *et al.* Diffuse infiltrative lung disease associated with flecainide: report of two cases. *Respiration* 2002;**69**:182–5.

121. Akoun GM, Cadranel JL, Israel-Biet D, Gauthier-Rahman S. Flecainide-associated pneumonitis. *Lancet* 1991;**337**:49.

122. Hanston P, Evrand P, Mahieu P, *et al.* Flecainide-associated interstitial pneumonitis. *Lancet* 1991;**337**:371–2.

123. Robain A, Perchet H, Fuhrman C. Flecainide-associated pneumonitis with acute respiratory failure in a patient with the leopard syndrome. *Acta Cardiol* 2000;**55**:45–7.

124. Feinberg L, Travis WD, Ferrans V, Sato N, Bernton HF. Pulmonary fibrosis associated with tocainide: report of a case with literature review. *Am Rev Respir Dis* 1990;**141**:505–8.

125. Stein MG, Demarco T, Gamsu G, Finkbeiner W, Golden JA. Computed tomography: pathologic correlation in lung disease due to tocainide. *Am Rev Respir Dis* 1988;**137**:458–60.

126. Sviri S, Gafanovich I, Kramer MR, Tsvang E, Ben-Chetrit E. Mesalamine-induced hypersensitivity pneumonitis: a case report and review of literature. *J Clin Gastroenterol* 1997;**24**:34–6.

127. Villeneuve E, St-Pierre A, Haraoui B. Interstitial pneumonitis associated with infliximab therapy. *J Rheumatol* 2006;**33**:1189–93.

128. Tengstrand B, Ernestam S, Engvall IL, Rydvald Y, Hafstrom I. TNF blockade in rheumatoid arthritis can cause severe fibrosing alveolitis: six case reports. *Lakartidningen* 2005;**102**:3788–90, 3793.

129. Chatterjee S. Severe interstitial pneumonitis associated with infliximab therapy. *Scand J Rheumatol* 2002;**33**:276–7.

130. Cairns AP, Taggart AJ. Anti-tumour necrosis factor therapy for severe inflammatory arthritis: two years of experience in Northern Ireland. *Ulster Med J* 2002;**71**:101–5.

131. Kramer N, Chuzhin Y, Kaufman LD, Ritter JM, Rosenstein ED. Methotrexate pneumonitis after initiation of infliximab therapy for rheumatoid arthritis. *Arthritis Rheum* 2002;**47**:670–1.

132. Courtney PA, Alderdice J, Whitehead EM. Comment on methotrexate pneumonitis after initiation of infliximab therapy for rheumatoid arthritis. *Arthritis Rheum* 2003;**49**:617 (author reply 617–18).

133. Ostor AJ, Chilvers ER, Somerville MF, *et al.* Pulmonary complications of infliximab therapy in patients with rheumatoid arthritis. *J Rheumatol* 2006;**33**:622–8.

134. Allanore Y, Devos-Fracois G, Caramella C, *et al.* Fatal exacerbation of fibrosing alveolitis associated with systemic sclerosis in a patient treated with adalimumab. *Ann Rheum Dis* 2006;**65**:834–5.

135. Dotan I, Yeshurun D, Hallak, *et al.* Treatment of Crohn's disease with anti TNF alpha antibodies: the experience in the Tel Aviv Medical Center. *Harefuah* 2001;**140**:289–93, 368.

136. Braun J, Brandt J, Listing J, *et al.* Treatment of active ankylosing spondylitis with infliximab: a randomised controlled multicentre trial. *Lancet* 2002;**359**:1187–93.

137. Roos JC, Chilvers ER, Ostor AJ. Interstitial pneumonitis and anti-tumor necrosis factor-alpha therapy. *J Rheumatol* 2007;**34**:238–9 (author reply 239).

138. Bedrossian CW, Sussman J, Conklin RH, Kahan B. Azathioprine-associated interstitial pneumonitis. *Am J Clin Pathol* 1984;**82**:148–54.

139. Brown AL, Corris PA, Ashcroft T, Winkinson R. Azathioprine-related interstitial pneumonitis in a renal transplant recipient. *Nephrol Dial Transplant* 1992;**7**:362–364.

22

Amiodarone pulmonary toxicity

PHILIPPE CAMUS, THOMAS V COLBY, EDWARD C ROSENOW III

INTRODUCTION

Amiodarone is a coronary vasodilator and class 3 antiarrhythmic agent used to treat or prevent recurrences of supraventricular and ventricular arrhythmias. Amiodarone is also used prophylactically in the perioperative period after thoracic, cardiac and other surgery, and in patients who received an implantable defibrillator to reduce the incidence and discomfort associated with automatic electric shocks. Amiodarone is mostly prescribed by cardiologists, primary-care physicians, physicians caring for the elderly and critical-care physicians. Optimal communication with physicians of these disciplines and with anaesthetists is of critical importance to prevent and detect amiodarone pulmonary toxicity (APT) in a timely fashion. Notwithstanding the fact that the use of amiodarone is associated with a plethora of adverse effects, the drug is popular among cardiologists and intensivists (e.g. 500 000 prescriptions filled in Canada annually), and sales are on the rise.[1]

APT is also referred to as 'amiodarone lung disease' or 'amiodarone pneumonitis'. The condition has been known for 30 years, has an incidence rate of about one case per 100 person-years, and its clinical presentation is multifaceted. Studies are under way to determine whether the amiodarone congener dronedarone, which lacks iodine and may not effectuate pulmonary toxicity, is as efficacious as amiodarone.[2]

Amiodarone is a *lipophilic* drug that sequesters in fat and tissues, notably in the lungs. Fat may act as a reservoir, so amiodarone effluxes slowly upon drug discontinuance.[3] This may explain two cardinal features of APT:

- Improvement in symptoms and radiographic abnormalities may not follow drug withdrawal as the sole intervention unless corticosteroids are given.
- Relapse of APT may occur despite drug discontinuance weeks or months earlier, at a time when corticosteroids are tapered.

Amiodarone is also *amphiphilic* and *lysosomotropic*. The drug sequesters in cell lysosomes, where it impairs phospholipid turnover. This explains the distinctive pattern of dyslipidosis that can be seen in cells of the lung and other tissues that are the main targets for the adverse effects of amiodarone. Foamy alveolar macrophages in the bronchoalveolar lavage (BAL) fluid of patients exposed to the drug are a marker of pulmonary phospholipidosis, and are a useful diagnostic adjunct.

The two iodines present on each amiodarone molecule account for the thyroid and pulmonary adverse effects of the drug, and are thought to account for the increased attenuation on computed tomography (CT) of tissues where amiodarone sequesters, notably the liver and condensed areas of amiodarone pneumonitis in amiodarone toxic patients.

The drawback of amiodarone is its extensive adverse-effect profile, which includes (in decreasing order of frequency) thyroid disorders, skin reactions, pulmonary toxicity, arrhythmias, hepatotoxicity, corneal deposits and neurologic impairment. Therefore, appropriate monitoring of patients who are exposed to the drug on a continual basis is required. As a consequence of amiodarone's adverse effects, the number of patients who can tolerate the drug gradually decreases with time, with approximately one-half of patients remaining on the drug by 3 years of follow-up, for reasons of inefficacy, intolerance or toxicity. Among the latter group, which accounts for about 20 per cent of the reasons for withdrawal,[4] pulmonary toxicity can be devastating and is the most worrisome adverse effect of the drug.

Amiodarone pulmonary toxicity is a distinctive form of drug-induced lung injury clinically, on imaging and pathology. Cases of APT were recognized in the late 1970s[5] but the condition was first described as a discrete entity in 1980.[6] Along with nitrofurantoin pulmonary toxicity, APT is the most common form of drug-induced pneumonitis, with approximately 700 papers and more than 1500 cases reported worldwide.[7] Classic and novel patterns of APT and reviews are still being published.[1,8] Although areas of uncertainty remain, the diagnosis and management of APT rest on a strong evidence base.

The probability of developing APT is related to several factors, among which drug dosage ranks first.[9] Although the lower maintenance dose advised nowadays has certainly translated into a decreased overall incidence, APT can occur with low maintenance doses of the drug.[10] The majority of cases develop 6–12 months into treatment, corresponding to a total dosage of 100–150 g. In fact, time to onset is variable, from a few days[11] to more than 10 years. Patients receiving a high maintenance dose tend to develop the disease earlier than those who are taking a low dose. Severity at presentation varies from the asymptomatic state to respiratory failure, acute lung injury (ALI) or acute respiratory distress syndrome (ARDS). Imaging findings are manifold and may quickly change over time in a given patient.

The most distinguishing feature of APT is its histological appearance. This includes a foamy cytoplasm or vacuolation of histiocytes, pneumocytes and other resident lung cells, and foamy alveolar macrophages in the BAL fluid which, although not entirely specific, are routinely used as a supportive diagnostic tool. The absence of lipid-laden macrophages in the BAL fluid will often exclude the likelihood of APT, although APT should not be entirely rejected on that basis. In any case, the diagnosis of APT remains one of exclusion, in that other pulmonary conditions and interstitial lung diseases need to be carefully ruled out. The management of APT includes drug discontinuation and, often, corticosteroid therapy is required as well. Mortality is 5–10 per cent and may be as high as 30 per cent in the more severely ill patients.

Appropriate monitoring of the patient on amiodarone, early detection, work-up of the patient suspected of having APT in the most conservative yet efficient way, diagnosis and management are important. Although history-taking, imaging and pulmonary physiology can detect APT cases early, the frequency at which these tests ought to be done may not meet the requirements for patient acceptance and cost-effectiveness. This probably accounts for the wide variance in clinical practices with regard to monitoring.

PATHOPHYSIOLOGY

The amiodarone molecule ($M = 643.3$) is notable for the following reasons.

- It has a benzofuran ring which gives the drug high lipid solubility – hence the storage of amiodarone in tissues, notably the lung, liver and fat. The latter acts as a reservoir for drug bidirectional storage and efflux. The benzofuran ring *per se* may not explain the drug's pulmonary toxicity.
- It has a cationic hydrophilic side-chain, which renders the drug amphiphilic and lysosomotropic. This explains its sequestration in the acidic milieu of lysosomes, where amiodarone blocks phospholipid processing, turnover and clearance, causing phospholipids to build up in cells of lung, liver, nerves and other organs.
- It has two iodine atoms, which play a mechanistic role in the thyroid and pulmonary adverse effects of the drug, and account for the high Hounsfield numbers on imaging of tissues where amiodarone sequesters, notably liver and lung.

Pharmacokinetics and metabolism

Amiodarone plasma concentration (about 1.3 μg/mL on average) increases with increasing drug dosage, albeit with a wide variability between patients for any given dose level. No correlation is found between amiodarone plasma concentration and the presence or type of adverse effects. Owing to its lipophilic and amphiphilic nature, amiodarone sequesters in tissues, as reflected by a large volume of distribution of 6.8–21 L/kg in normal subjects, resulting in delayed onset of action. This is why intravenous amiodarone is often required whenever rapid control of arrhythmia is to be achieved. Amiodarone infiltrates particularly into the liver, lung, the discoloured skin of amiodarone toxic patients, thyroid and adrenals. The amiodarone tissue-to-plasma concentration ratios in the human lung and liver are about 100–150 and 300, respectively, and the ratios tend to increase further with increasing duration of treatment. Calculations indicate that the concentration of amiodarone prevailing in lung tissues *in vivo* is 6- to 20-fold the concentration that is toxic to lung cells *in vitro*.[12,13] Uptake of amiodarone in tissues goes hand in hand with the development of lysosomal inclusion bodies in cells, which reflect phospholipidosis (see below). Uptake of amiodarone and its chief metabolite desethylamiodarone in tissues increases with time, requiring reduction in drug dosage or episodic administration of the drug. Since most of the tissues where amiodarone sequesters are common targets for the adverse effects of the drug, it is considered that amiodarone toxicity results at least in part from sequestration of the drug and metabolite.

Amiodarone is chiefly metabolized in the liver into its chief metabolite mono-desethylamiodarone (usually refered to as 'desethylamiodarone'), under cytochrome P450-1A1- and 3A4-driven dealkylation. Desethylamiodarone plasma concentration is 30–120 per cent that of amiodarone,[14] and it may be more in amiodarone toxic patients. Desethylamiodarone participates in the antiarrhythmic efficacy of amiodarone, and is also capable of producing phospholipidosis. Further dealkylation of desethylamiodarone leads to minute amounts of *bis*-desethylamiodarone. Metabolism of amiodarone takes place to a lesser extent in the gut. There is insignificant biotransformation of amiodarone in the lungs. Upon metabolism, amiodarone and its metabolite are excreted into the bile and eliminated via the faecal route. Although desethylamiodarone is more water-soluble and more readily excretable than is amiodarone, it sequesters about 5-fold more in tissues compared to amiodarone, particularly in liver and lung. Further, desethylamiodarone is more toxic to lung cells than is amiodarone, with a LD50 2.5- to 18-fold less than that for amiodarone. Thus, desethylamiodarone may play a pathogenic role in APT. There is no significant correlation of amiodarone or desethylamiodarone concentration in plasma and the likelihood of developing pulmonary toxicity. Likewise, plasma concentration of amiodarone and/or desethylamiodarone or their ratio may not reliably separate amiodarone-toxic from non-toxic patients in clinical practice.

The half-life of elimination of amiodarone and desethylamiodarone is up to 60 days in chronically treated patients. Deep-seated tissue stores, mostly fat, serve as reservoirs from which amiodarone is slowly released, and measurable drug levels can be found in tissues several months after drug discontinuance.[15] This may explain why merely stopping the drug does not translate into abatement of the signs and symptoms of APT. This may also explain why early withdrawal of corticosteroid therapy can be followed by relapse of APT, a feature nearly unique to amiodarone.

Amiodarone interacts with the pharmacokinetics and metabolism of many cardiovascular and non-cardiovascular drugs, including statins with an augmented risk of myositis or rhabdomyolysis, warfarin with a consequent increase in the INR, digoxin, beta-blockers, calcium-channel blockers, clozapine and sirolimus. Vassallo and Trohman have provided a review.[16]

Drug-induced phospholipidosis

The cationic amphiphilic anorectic drug chlorphentermine is the prototypical phospholipidosis-inducing agent in rats. Upon chronic administration, chlorphentermine sequesters in cell lysosomes in the lung and other organs, where the drug blocks lysosomal functions. Massive accumulation of foam cells in alveolar spaces ensues.[17] Phospholipidosis is a distinctive drug-induced, as opposed to inherited, lysosomal storage disease. The changes seen in the chlorphentermine rat model are also found in rats fed amiodarone and, to a certain extent, in human APT as well. As in the chlorphentermine model, phospholipids increase in lung and BAL cells of amiodarone-toxic patients, and significant amounts of the drug and metabolite are present in BAL subcellular fractions.[18,19] To date, amiodarone is the only drug whose toxicity expresses itself with histological and cytological features of dyslipidosis in humans (limited evidence exists for perhexilline and statins). The derangement in intracellular phospholipid processing correlates with the pathological finding of fine granular or linear appearance of the cytoplasm of lung cells including macrophages on microscopy, and the whorled lamellar membranous inclusions and myelinic figures seen on transmission electron microscopy. Although distinctive,[20] these changes are reflective of the amiodarone effect[21] and do not necessarily indicate toxicity, except when they are associated with pulmonary infiltrates on imaging, subnormal pulmonary physiology, and interstitial inflammation, organizing pneumonia and/or fibrosis on pathology. In the absence of these changes, foam cells are a routine finding and indicate chronic exposure to amiodarone.

Animals (mainly Fisher 344 rats and the hamster) chronically fed amiodarone may exhibit some of the changes that characterize the human form of APT, including interstitial inflammation, an increased number and/or proportion of inflammatory cells in the BAL, and decreased lung compliance. Grossly and on microscopy, pulmonary fibrosis is generally absent in the rat, except after intratracheal administration of amiodarone, a model that markedly differs from oral administration of the drug. Changes in rodents following exposure to amiodarone are less than in humans, despite dosages of the drug up to 500 mg/kg per day; i.e. 170-fold the usual dose used in humans.

Other amiodarone-induced effects

Other candidate contributors to explain APT include the production of an unstable aryl radical during deiodination of amiodarone, release of reactive oxygen species, caspase activation followed by apoptosis and cell death,[22] augmented TGF production, uncoupling of mitochondrial oxidative phosphorylation followed by mitochondrial toxicity, oxidant radical production and stimulation of the angiotensin converting enzyme pathway. In a series of studies, Uhal's group confirmed that amiodarone and desethylamiodarone are toxic to lung cells, leading to apoptosis and cell death at or near the therapeutic concentration.[23] Epithelial cell injury may initiate immunomodulatory events and inflammation in the lung.[24] This can be followed by lung fibrosis.[25] As mentioned above, the concentration of amiodarone that prevails in lung and other tissues is more than two orders of magnitude greater than in plasma. Thus, cell toxicity and death is plausible in tissues *in vivo*. Uhal's group showed in addition that the angiotensin-converting enzyme inhibitor captopril abrogated amiodarone toxicity *in vitro*, and that fibrosis of the lung was less in amiodarone-exposed animals given captopril, compared to those merely exposed to amiodarone. Interestingly, two recent clinical studies support their theory.[26,27] *In vivo* in humans, APT can be boosted by concomitant exposure to high fractional concentration of oxygen. Recent studies also indicate that tissue and/or urinary metabolic fingerprinting may identify phospholipidosis in amiodarone-exposed animals. The ability of this technique to separate amiodarone-toxic from non-toxic patients has not been investigated clinically.

EPIDEMIOLOGY AND RISK FACTORS

According to the classification of Hill, the signal for amiodarone pulmonary toxicity is *strong*, with many cases reported around the world, is *consistent*, since most cases have a resembling clinical, imaging and pathologic background, and is *distinctive*, with a characteristic BAL cell and pathological pattern of involvement. *Temporality* also supports the drug aetiology, with suggestive dechallenge and/or re-challenge in many patients. A *biological gradient* exists, in that greater dosages result in an increased risk of developing APT. In addition, amiodarone lung disease is *mechanistically plausible* and *coherent*, since the drug blocks phospholipid catabolism, evidence for phospholipidosis is present on pathology, and several of the features of APT in humans can be consistently reproduced in animals. Lastly, there is *statistical evidence* for the condition, since randomized studies showed that the incidence rate of 'interstitial lung disease' (ILD) in patients on amiodarone is about twice the background rate of 'idiopathic' ILD in controls unexposed to the drug, a statistically significant difference.[28,29]

Pooled incidence data from double-blind placebo-controlled trials involving 3439 patients receiving daily amiodarone doses of less than 400 mg indicate an incidence rate of 1.6 per cent, a figure consistent with the 2 per cent mentioned by Roy[30] and by Goldschlager et al.,[4] respectively, and the 2.2 per cent found in Kosseifi et al.'s study.[27] Chandhok et al.[31] found that 7 per cent of patients had developed APT at 3 years, a figure consistent with the results by Yamada et al.[32] who reported cumulative incidences of 4.2, 7.8 and 10.6 per cent at 1, 3 and 5 years, respectively. Olshansky reported an incidence of 3.5 per cent by 4 years into treatment.[33]

The incidence of APT depends on the daily maintenance dose of amiodarone. On average, APT develops in 0.1–0.5 per cent of patients treated with less than 200 mg, compared with 5–15 per cent of patients on more than 400 mg,[34,35] and up to 50 per cent of those on more than 1200 mg. An intravenous loading dose of 1000 mg per day increases the risk of developing APT.[36] One patient developed APT following a stepwise increase in daily dose. Thus, amiodarone should be titrated to the lowest possible effective dosage level.[37] Some even advise episodic use of amiodarone to reduce build-up of stores in the body, and this translated into a decrease in the incidence of some adverse effects, including pulmonary toxicity.[38] While low-dose amiodarone regimens have decreased the incidence of APT, there is no current evidence that severity has changed.[10]

Amiodarone pulmonary toxicity is preferentially described in males in a 20/1 ratio to females. The condition is rarely reported before 35 years of age, possibly reflecting the distribution of indications for administering the drug.[39] There are a few reports in infants, children and adolescents, with features similar to the adult form of the disease, suggesting that no age or gender is immune to the adverse effects of amiodarone.[39–41] The risk of developing APT increases in patients over the age of 60 years.[9,42] Having an abnormal chest radiograph or poor pulmonary reserve prior to the commencement of treatment may increase the risk of developing APT several-fold.[43] At 4 years into treatment, 5.9 versus 3.1 per cent of patients with or without pre-existing lung disease developed APT,[44] a significant difference. However, this has not been confirmed in all studies.[42,45] Evidently, this reinforces the need for pretherapy evaluation of pulmonary physiology. To the extent that pre-existing ILD or low baseline diffusing capacity constitutes a greater risk of developing APT, closer monitoring and follow-up may be needed in this subset of patients.[34] Evidence for APT prior to thoracic surgery or implantation of an automatic defibrillator was predictive of severe postoperative pulmonary complication in one study.[46]

A recent study of 237 APT cases showed a peak incidence for the clinical onset of the disease between 6 and 12 months into treatment, corresponding to a 100–150 g cumulative dose of amiodarone.[9] However, amiodarone pneumonitis can develop just a few days into treatment (particularly if the patient receives an intravenous loading dose or there is a history of recent thoracic surgery or exposure to oxygen),[36,42,47] and up to a decade or more into treatment. Time to onset of amiodarone pneumonitis is shorter in patients exposed to a high maintenance dose.[48] APT cases with an acute presentation develop the disease after an average of 3 months, as opposed to patients with a more indolent presentation. Rarely does APT develop after cessation of exposure, generally 3 months at most after drug discontinuance.[49]

Exposure to high fractional concentration of oxygen in the perioperative period may trigger the onset of acute severe APT.[50,51] This was reflected in a study of 288 patients who were receiving 400 mg amiodarone daily, of whom 33 underwent surgery (cardiac surgery in 22, thoracic surgery in 5, abdominal or orthopaedic surgery in 6).[50] Acute postoperative APT developed within 72 hours of starting intravenous amiodarone in four patients, an incidence of 12 per cent, and was fatal in all four. All of the patients had received a high intraoperative concentration of oxygen. None of the patients had evidence

suggesting pre-existing interstitial lung disease, and no distinguishing variables could predict the development of acute APT. Left ventricular failure was ruled out on the basis of a normal pulmonary capillary wedge pressure. Autopsy showed the characteristic features of APT. A study of prophylactic intravenous amiodarone immediately following pulmonary resection had to be halted, because ARDS developed with an incidence of 11 per cent in patients receiving the drug, a 6-fold higher rate than that in controls unexposed to the drug. Therapy with amiodarone in the context of resectional lung surgery for cancer poses a particular risk, with the combination of diminished ventilatory reserve, a frequent background of chronic obstructive pulmonary disease (COPD) or interstitial lung disease, and the need for oxygen supplementation which is known to be deleterious.[52] The triggering role of oxygen was even more specifically suggested in three other cases. In one patient, left-sided pulmonary shadowing thought to reflect acute APT developed after selective ventilation of the left lung with 100 per cent O_2 for the 90 minutes required for contra-lateral resectional surgery for lung cancer.[53] Post-mortem studies indicated diffuse alveolar damage in the left lung (DAD). A similar reaction occurred in two other patients whose right lung was selectively ventilated.[54] Preventative measures as regards oxygen therapy are critically needed in this setting.

Isolated reports have described fatal ARDS in amiodarone-treated patients following injection of iodinated contrast media[55] or cyclophosphamide.[56,57] APT has been reported in a patient awaiting heart transplantation[58] and following heart or lung transplantation.[59] Indirect evidence for the noxious effect of amiodarone in this setting is that a cohort of patients on the drug had more postoperative days on the ventilator and decreased survival in the first year after transplantation.[60] In heart and lung transplant candidates or recipients, APT needs to be distinguished from heart failure, heart and/or lung rejection, sirolimus pneumonitis and opportunistic infections, using appropriate tests.

One study showed that the rate of APT increased 2-fold in patients with repeated episodes of congestive heart failure compared to cardiac patients with a stable heart condition.[26] Another study of 883 patients showed that the incidence of APT was 3.9 per cent in patients who were not on an angiotensin-converting enzyme inhibitor (ACEI) or an angiotensin-receptor blocker (ARB), versus 1 per cent in those so treated, suggesting a role for these drugs in the prevention of, or protection from, APT.[27]

CLASSIC AMIODARONE PULMONARY TOXICITY

Signs and symptoms

The most common pattern of APT is a subacute infiltrative lung disease with new restrictive lung dysfunction and hypoxaemia and arterial desaturation on exertion.[1,61–65] Onset of APT is insidious or it can be rapidly progressive.[66] Signs and symptoms are non-specific and include malaise, weakness, moderate fever, bodyweight loss, dyspnoea on exertion progressing over weeks to months, an unproductive cough that can be at the forefront and may predate the clinical or imaging onset of APT, crackles (a non-specific finding in the aged[67]), and pleuritic chest pain. Patient and doctor factors combine so that it takes about 2 months on average for APT to be diagnosed in a symptomatic patient, a delay that negatively affects outcome. The perception of dyspnoea can be exaggerated in the presence of the hyperkinetic state from amiodarone-induced hyperthyroidism, should that adverse effect of amiodarone be present simultaneously. Other adverse effects of amiodarone are infrequently present, and no adverse effects have shown a significant association with APT, be it at the time of diagnosis or later. Nevertheless, a search for adverse non-pulmonary effects of the drug is warranted. Amiodarone can produce a bluish skin discolouration causing pseudocyanosis.

Imaging

See Fig. 22.1 showing multiple variations of amiodarone toxicity on a chest X-ray and Fig. 22.2 also showing the many variations on a CT including the liver and spleen.

On the chest radiograph a pacemaker can be present. APT manifests with localized, patchy, disseminated or diffuse discreet or well-formed infiltrates, reticular opacities, haze, ground- or frosted-glass shadowing, consolidation, fuzzy or flocculent mottling involving the lung bilaterally, albeit often in an asymmetric fashion. It is not unusual for a specific patient to offer a distinct presentation. The opacities of APT may localize in the bases, mid-lung regions or apices, or they can be diffuse, sparing virtually no area of the lung.[68] Patients with basilar disease tend to have denser shadowing.[69] A useful finding in some patients is that volume loss may be less than it is in lung fibrosis.[70] Inexplicably when the disease is mild, there is an impression that the right lung, mainly the right upper lobe, is more frequently or more densely involved than the left lung.[71-73] Right upper lobe shadowing and a concave adjacent minor fissure suggesting volume loss in a patient taking amiodarone is suggestive. A few cases present with unilateral involvement. Kerley B lines, reticular opacities, alveolar shadows and pleural effusion may be present. These features may overlap with those of hydrostatic pulmonary oedema, although then they are more likely to clear following forced diuresis.[74] Multiple flocculent ill-defined alveolar shadowing indicates more acute presentations characterized by a shorter time to onset, significant lymphocytosis in the BAL, and response to corticosteroid therapy.[75]

Although features are sometimes largely unilateral on the chest radiograph, high-resolution CT (HRCT) generally indicates bilateral disease,[65] in the form of localized or scattered discrete or dense haze or ground-glass opacities, with or without a recognizable segmental or lobar distribution, inter- or intralobular, localized or diffuse interstitial septal thickening, disseminated patchy shadows, and/or X-ray dense mass or masses with a random distribution.

In a study of 20 patients with moderate APT, all patients had ground-glass opacities. Areas of consolidation were seen in four and intralobular reticulation in five. A subpleural distribution of the opacities was more common (18 cases) than a central one (2 cases), and high attenuation in the area of involvement was present in eight patients.[76] CT-pathology correlates showed that subclinical intralobular septal thickening in patients on amiodarone indicates the presence of mural and intra-alveolar foam cells. Nevertheless, this finding may not dictate withdrawal of amiodarone, as the outcome of such changes is variable, even though amiodarone is continued.

Some opacities will disappear with time, while others stabilize and yet others progress. Increased attenuation numbers of the pulmonary shadowing with a cut-off limit of > 70 Hounsfield units (HU) seems a more consistent and diagnostically useful feature in subacute APT.[77] This measurement should be routine in the evaluation of patients suspected of having APT, particularly if the imaging presentation is ambiguous, if the patient cannot undergo BAL or an invasive procedure, or when areas of condensation are discovered incidentally on an HRCT done for other reasons in a patient taking amiodarone.

Iodine-induced high liver attenuation can be found in patients chronically treated with amiodarone, and is a clue to the diagnosis of APT. The normal range for liver attenuation is 30–70 HU, and in one study in seven patients chronically exposed to amiodarone all displayed increased liver density averaging 114 HU (range 95–145). In practice, this finding is found in 60–100 per cent of patients taking amiodarone on a regular basis. Smooth-edged visceral pleural thickening *en face* the pulmonary opacities of APT, and/or pleural effusion (usually an exudate), can be present in association with the pulmonary opacities in up to 30 per cent of patients with APT.[78] In a few, pleural changes occur in isolation.

The expression of APT on imaging, but not its clinical expression, can be less in patients with a background of emphysema. The organizing pneumonia pattern of APT or BOOP, an expressive acronym denoting pathological evidence of interstitial inflammation and ductal fibrosis (discussed later; see Fig. 22.4, page 254), is characterized by triangular opacities containing air bronchograms, which classically fluctuate or wander on sequential chest radiographs.

No systematic study of gallium-67 scanning is available. The test initially showed promise in the diagnosis of APT,[66] but negative results have been reported in authentic APT as well.[79] Irradiation is not inconsiderable and the test is not widely available.

Di-deoxy-18F glucose is taken up in focal areas and masses of APT, although reported cases are still scarce. Thus, the test may not clearly distinguish an amiodarone induced mass from malignancy.[8,80,81]

Laboratory investigations

There is no single biological abnormality or concatenation of findings that is specific for APT. An increased erythrocyte sedimentation rate (ESR) is classically present at the time of diagnosis, and may predate the clinical onset of APT. However, an increased ESR can be present in patients with left ventricular failure.[73] Leucocytosis, a raised C-reactive protein (CRP) or plasma lactate dehydrogenase (LDH) have been reported, and the latter may predate the clinical onset of APT, as does the ESR.[82] Hyperglobulinaemia can develop in patients on amiodarone; in a recent study, 11 out of 15 patients with amiodarone-associated hyperglobulinaemia developed adverse amiodarone effect, pulmonary toxicity in nine.[83] An increased KL-6, a human MUC1 mucin, has been found in patients with APT,[84] as in several other forms of lung injury. Cases of APT with a normal KL-6 level (< 500 U/L) have also been reported.[85] Although some investigators use KL-6 to diagnose APT, this test should be considered non-specific, and using it as a surrogate marker is likely to flaw the diagnosis of APT. Increased

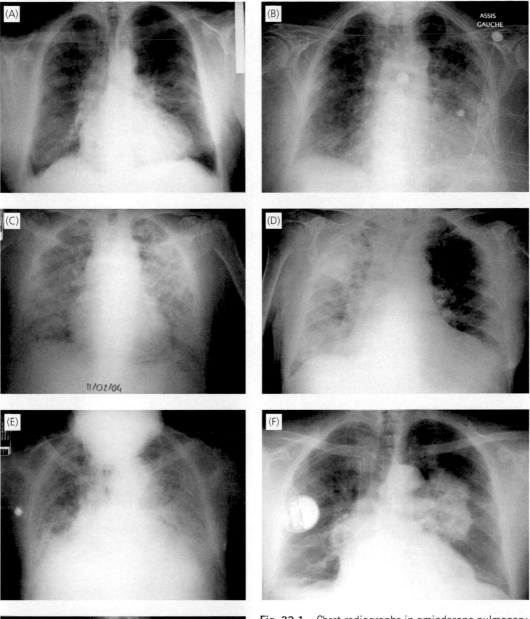

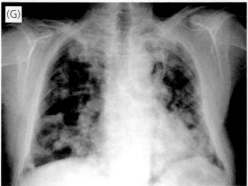

Fig. 22.1 Chest radiographs in amiodarone pulmonary toxicity (APT). Almost every patient with APT displays a distinct pattern, simulating a host of other pulmonary conditions. Cardiomegaly is often present, owing to the underlying cardiac condition. Diuresis with post-test chest radiograph is indicated to rule out interstitial pulmonary oedema. (a and b) Generally, APT is in the form of bilateral, disseminated, asymmetrical ill-defined interstitial or alveolar opacities. (c) Acute presentations tend to occur earlier, are associated with diffuse alveolar shadowing, BAL shows lymphocytosis, and respiratory compromise is present. APT responds to drug withdrawal and corticosteroid therapy in the vast majority of cases. (d) Unilateral predominant involvement, particularly of the right upper lobe, is common in APT. (e) Acute lung injury or acute respiratory distress syndrome (ARDS) can occur in the postoperative period, especially in patients exposed to a high fractional concentration of oxygen during surgery. APT may present with (f) a mass lesion or, less often, (g) in the form of multiple lung nodules mimicking malignancy. Pleural effusion, usually an exudate, is a common finding in patients with substantial involvement, and may occur as an isolated finding.

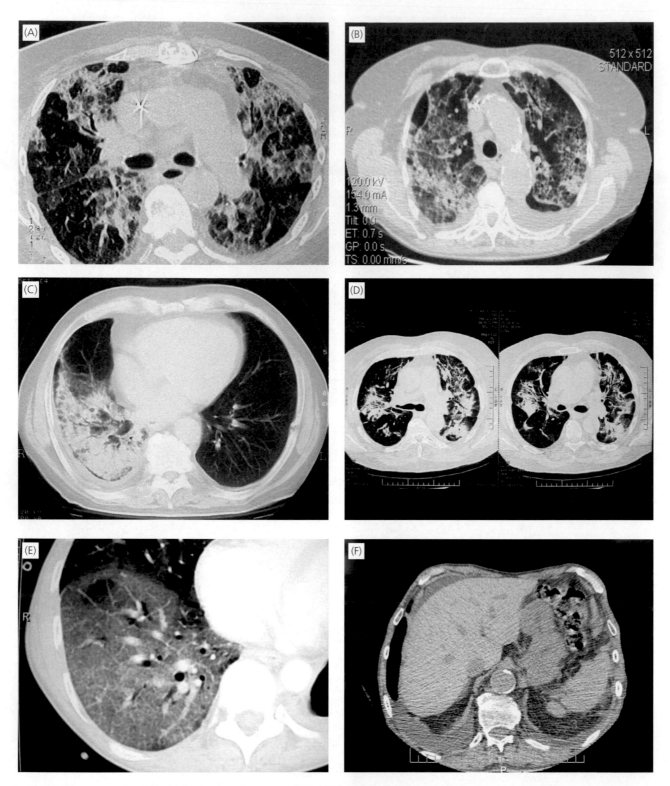

Fig. 22.2 High-resolution CT in amiodarone pulmonary toxicity (APT). (a and b) Patterns include disseminated or diffuse intralobular filling with or without pleural thickening or effusion, or (c) localized alveolar shadowing with a recognizable lobar distribution. Often, pleural thickening is present, as in (c). Other presentations include (d) retractile mass lesions that contain air bronchograms and produce parenchymal distortion, and (e) localized or diffuse haze or ground-glass with increased intralobular markings. High attenuation of (f) the spleen and/or liver, and of (g) the pulmonary opacities (left upper lobe in this particular patient) on unenhanced CT is a clue to the diagnosis of APT. (h) Multiple shaggy nodules, or (i and j) large nodules with or without a centre of low attenuation, are less common presentations. (k) Small areas of intralobular thickening may be detected in asymptomatic patients taking amiodarone. They correspond to mural foam cells on pathology, and dictate amiodarone withdrawal or watchful waiting, as not all will evolve to frank APT. *Note:* Patients with a history of pneumonectomy, such as the patient in (e), are at risk owing to their diminished ventilatory reserve.

(Contd)

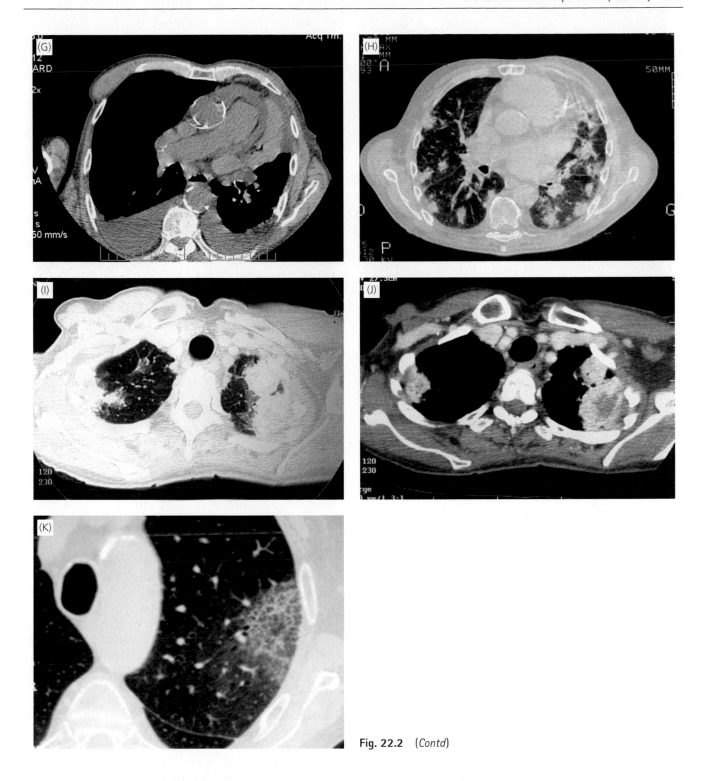

Fig. 22.2   (Contd)

serum surfactant-associated protein D has been reported in two patients with APT.[85] The role of serum brain natriuretic peptide (BNP) in distinguishing APT from left ventricular failure remains imprecise,[73] in as much as patients may present with both conditions concomitantly or in sequence. Adding to the difficulty, amiodarone may decrease plasma BNP.[86] Liver chemistry and thyroid function should be evaluated. In light of rare cases of amiodarone-induced lupus, the antinuclear antibody (ANA) test is advised in patients taking amiodarone who develop pleural or pericardial effusion as an isolated finding.

Pulmonary physiology

PRE–THERAPY AND SERIAL MONITORING

Patients on amiodarone may be active or reformed smokers and may present with a background of smoking-related pulmonary disease such as COPD/emphysema, smoking-related interstitial lung disease including pulmonary fibrosis, heart failure and inspiratory muscle dysfunction.[87] Patients with coronary artery disease have an increased background incidence of ILD.[88] Since all these conditions have an impact on pulmonary function and do so in unpredictable ways, pulmonary physiology should be evaluated before commencing treatment with amiodarone.[89] This will serve as a baseline to which any further change on amiodarone can be compared.

Pulmonary function testing and the chest radiograph are now recommended in the initial work-up of patients who will be treated with amiodarone.[4,90] Some authors advocate three or four repetitions of pulmonary function and diffusing capacity (DLCO) testing in the first few months of treatment with amiodarone, to establish a baseline free from the intrinsic variability of the tests,[42] but this may not be practical and may not have patient compliance. For instance, a study in the United States showed that a single baseline pulmonary function and DLCO measurement were performed in only 52 per cent of the patients intended for treatment with amiodarone.[90] Likewise, a recent audit study at one institution in the United Kingdom showed that, among 50 eligible patients, adequacy of baseline pulmonary function testing was a mere 4 per cent.[91]

With regard to serial follow-up of pulmonary function tests in patients taking amiodarone in the long term, the diffusing capacity tends to decrease about 15 per cent in patients on maintenance amiodarone, to then stabilize or improve back to or near pre-therapy values, even though there is no evidence for pulmonary toxicity developing.[42] An isolated decrease of the diffusing capacity *per se* does not necessarily indicate clinically recognizable disease and should not prompt discontinuation of amiodarone unless there is clinical or imaging evidence of disease,[92] since overt APT will develop in only about a third of such patients.[42] Nevertheless, closer follow-up and repeat measurements may be indicated in this subset of patients. Conversely, a stable diffusing capacity is reassuring and indicates the lack of clinically meaningful APT.[93] An isolated measurement showing a low diffusing capacity during treatment with amiodarone in the absence of pre-therapy evaluation should be interpreted with caution, as this may simply reflect a previous background of emphysema or ILD. Although systematic follow-up of pulmonary function and the diffusing capacity in all patients on a low-dose amiodarone regimen may not be economic, many will err on the safe side and have them performed. One of the authors (ECR) performs an initial complete pulmonary function test, and serially follows patients with forced vital capacity (FVC), the diffusing capacity (DLCO) and O_2 saturation with exercise.

ESTABLISHED APT

A restrictive lung function defect is classically present at the time of diagnosis in APT, with a vital capacity averaging 75 per cent of predicted and hypoxaemia (58 mmHg on average). A reduced FEV_1/FVC ratio is not uncommon, is generally ascribed to smoking-related airway obstruction, and heralds increased mortality. In one patient with emphysema, APT manifested with further reduction of the diffusing capacity and a transient 'improvement' of airflow obstruction, which was deemed to result from APT-related increase in lung elastic recoil.[94] The earliest consistent physiological abnormality in APT is a 15–20 per cent decrease in the diffusing capacity.[32,42] Of note, left heart failure does not significantly alter this test.[73,95] Using the diffusion as a surrogate test for APT, Magro *et al.* have established that the optimal sensitivity and specificity

were, respectively, falls of ≥ 15 per cent and ≥ 30 per cent in this measurement.[42] Patients who develop APT may demonstrate a precipitous decrease in this measurement in a few weeks, which may be missed if the test is done episodically.

No study is available of cost-effectiveness of repeat measurements of the diffusing capacity in the context of therapy with amiodarone. It is recommended to perform the test before treatment, and then annually in patients taking more than 200 mg per day, and if new symptoms or changes on imaging occur in a patient taking amiodarone.

FOLLOW-UP AFTER DRUG WITHDRAWAL

Pulmonary function testing and measurement of the diffusing capacity are traditionally performed at the time of drug withdrawal and at each follow-up visit of patients who developed APT, until recovery, to confirm functional improvement and titrate corticosteroid dosage.

Bronchoalveolar lavage

Figure 22.3 demonstrates some of the multiple variations of histologic presentation of amiodarone toxicity.

The contribution of BAL is to rule out other conditions and to support the diagnosis of APT when abundant foam cells are present. An increased proportion of lymphocytes, neutrophils or any other cell type is not a prerequisite for diagnosing APT.[96] Generally, total cells are increased in BAL fluid, averaging 550 cells/μL. Pooled results indicate that about 25 per cent of patients with APT exhibit a neutrophilic BAL pattern, 20 per cent display a lymphocytic pattern and 30 per cent have a mixed lymphocytic and neutrophilic pattern; in the remainder, the differential is normal.[93,96,97] The percentage of lymphocytes in the BAL fluid can be as high as 68 per cent.[98] A high lymphocyte count indicates symptomatic cases and a shorter time to onset of amiodarone pneumonitis, compared to patients lacking this feature. The BAL fluid in amiodarone-induced ALI/ARDS may show lymphocytes or neutrophils, in addition to the characteristic foam cells that may be detectable after a few days on the drug. An increased percentage of CD8+ lymphocytes relative to CD4+ cells was once considered a possible diagnostic clue, but time has not confirmed that. Eosinophilia as an isolated finding in the BAL fluid is an extremely unusual finding.[99] A few patients present with diffuse antineutrophil cytoplasmic antibody (ANCA)-negative alveolar haemorrhage, with haemorrhagic BAL return and haemosiderin-laden macrophages on microscopy, superimposed on a background of foam cells (reviewed by Iskandar *et al.*[100]).

Foam cells with a finely granular cytoplasm on light microscopy and lamellar inclusions on electron microscopy (now rarely performed) in the BAL fluid or induced sputum[79] are considered a valuable diagnostic adjunct for APT, when there is suggestive clinical and/or imaging evidence for the disease. However, this finding is not entirely specific for APT, being also found in patients with *Rhodococcus equi* pneumonia, diffuse panbronchiolitis, and other reactive and inflammatory lung conditions.[101] Importantly, foam cells are a routine finding in amiodarone-treated patients, and do not firmly establish toxicity. For instance, Bedrossian *et al.* found foam cells in

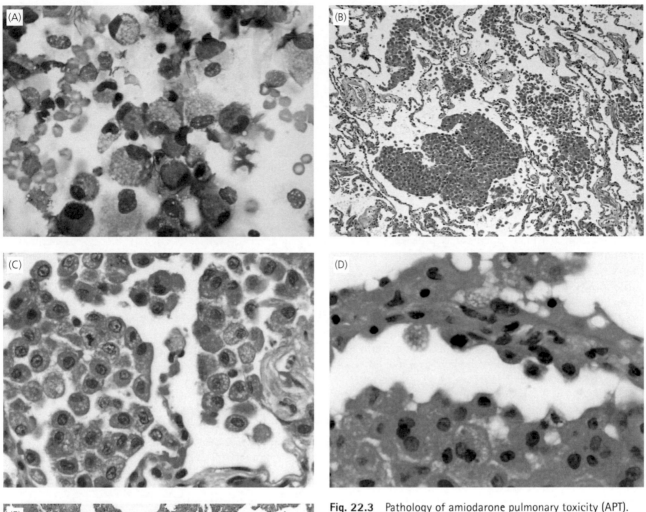

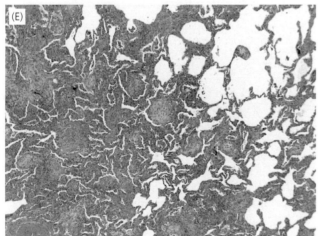

Fig. 22.3 Pathology of amiodarone pulmonary toxicity (APT). Foam cells in the BAL fluid are a sign of exposure to amiodarone. These are reminiscent of those seen on pathology sections, and are found in nearly every APT case. (a) There is prominent increase in air-space macrophages, in some regions forming sheets as in (b). On this section, interstitial tissue is preserved. This appearance may simulate desquamative interstitial pneumonia at low power. (c) Higher power evaluation of the case shown in (b) shows macrophages with prominent vacuolization in the cytoplasm. These changes are non-specific but are consistent with a lipid storage disorder. (d) Detail of macrophages (from the case shown in (e)) shows vacuolization in the macrophages as well as some vacuolization in the Type II cells along the alveolar septum in the upper half of the figure. In addition, some of the macrophages show haemosiderin (5 o'clock), a common finding in amiodarone toxicity, related to passive congestion from coexisting cardiac disease. (e) Amiodarone toxicity illustrating a BOOP (organizing pneumonia) pattern with patchy regions of the lung involved by air-space organization with intraluminal polypoid connective tissue plugs. In such cases the increase in macrophages may not be readily apparent in all fields, as shown in the field illustrated.

all their amiodarone-treated patients, regardless of whether they had APT.[20] Similarly, Jolly *et al.* reported BAL findings in 11 patients with no evidence of disease taking an average of 345 mg amiodarone per day.[21] Foamy macrophages were evidenced with light microscopy in all 11 patients, and lamellar bodies were detected by electron microscopy. Amiodarone was continued with no harm to the patients. It thus appears that foam cells in the BAL fluid may not indicate toxicity and are not necessarily the harbinger for overt disease if amiodarone is continued. The absence of foam cells, however, challenges the diagnosis of APT, although a reporting bias may have favoured only those cases with foam cells in the BAL fluid. Foam cells in APT may stain positive with Oil Red O.[102]

Pathology

Pathology is the gold standard for diagnosing APT, but a lung biopsy is not required in every patient. A number of cases have been diagnosed using the transbronchial approach (TBLB), which has a reasonably low rate of complications. An open or video-assisted thoracic surgery (VATS) lung biopsy is arguably better in this setting, although the lung sample is definitely larger with this technique compared to TBLB. It is still unclear whether the lung biopsy is justified and cost-effective in APT, and/or in which patients it is indicated. Preferred indications are in cases with unusual presentations such as the solitary nodule (also called amiodaronoma), in candidates to heart or lung transplantation where it is deemed important to definitely and quickly rule out malignancy, and in patients in whom amiodarone cannot be withdrawn without undue risk of arrhythmia. In contrast, classic APT tends to be managed in a more conservative way, using drug withdrawal and corticosteroid therapy. The fine-needle transthoracic aspiration approach has been used successfully in a few cases.

Histopathological appearances of classic APT include septal inflammation, oedema and fibrosis, lipids or vacuoles in such resident cells of the lung as histiocytes, endothelial and epithelial cells, or lying free in alveolar spaces, and a distinctive increase in air-space macrophages which pack densely in the alveolar space forming sheets and showing foam cell changes.[43,71,73,93,103] The degree of interstitial changes versus alveolar foam cells varies from patient to patient, and from one area of the lung specimen to the other. Packed foam cells in alveolar spaces is sometimes the sole pathological finding while the interstitium is preserved, producing a desquamative interstitial pneumonia (DIP) pattern.[104] Other histopathological findings include a reactive epithelium, active or resolving diffuse alveolar damage and hyaline membranes, alveolar haemorrhaging, and luminal fibrosis suggesting organizing pneumonia. All these coexist with more classic features of APT on the same or on contiguous lung samples. Indeed, according to autopsy studies, the pathological picture is heterogeneous throughout the lung, so what is seen on a lung section should be generalized to the entire lung only with caution.

Acute APT cases are characterized by more severe involvement, in the form of interstitial and/or alveolar oedema, diffuse alveolar damage and/or alveolar haemorrhaging on a background of foam cells. Rarely, APT manifests in the form of acute fibrinous organizing pneumonia (AFOP), which corresponds to the association of organizing pneumonia and intra-alveolar fibrin.[105]

Patient work-up and diagnosis

Although no consensual guidelines exist, proposals for the work-up of patients suspected of having APT may include the steps described in Boxes 22.1 and 22.2. It is advisable to document any completed steps in the patient's record.

Once a presumptive diagnosis of APT is made, the requirement for continued amiodarone therapy should be quickly evaluated and the drug should be discontinued whenever possible. Continued amiodarone therapy is almost a warrant for deterioration of APT. In a few cases, amiodarone could not be withheld and was continued at a lower (halved) dosage along with corticosteroid therapy and the pulmonary infiltrates did not progress or cleared.

Deciding on a lung biopsy remains an area of uncertainty, especially in patients in whom amiodarone therapy can be withheld without undue risk of recurrence of life-threatening arrhythmias. Indeed, patients may deteriorate after the lung biopsy and fatal complications have been reported post-biopsy.[71] A risk–benefit evaluation of transbronchial versus VATS, or biopsy versus a conservative approach, is not and may never be available.[73] The lung biopsy is sometimes considered in patients who show no convincing improvement after a few months of drug discontinuation and corticosteroid therapy. However, resolution of the pulmonary infiltrates of APT may be extremely slow and might take more than a year. With the accumulation of literature on APT, not relying on the lung biopsy and instituting empirical treatment with corticosteroids does not seem to pose undue obstacles to the management of APT. It turns out that the concept of 'pulmonary infiltrates without another cause clearing upon discontinuation of amiodarone' may progressively replace the undisputable academic diagnosis of APT. The gallium-97 scan may provide inconsistent results in APT. FDG–positron-emission tomography is reserved for nodular or mass-like APT, and may not clearly distinguish focal APT from malignancy.

Differential diagnosis

History-taking, the presence of dyspnoea when lying flat, the presence or absence of cardiomegaly or valvular heart disease, a history of definite heart failure, the diuresis test, cardiac ultrasound, review of previous imaging and pulmonary function and, in selected cases, cardiac catheterization with measurement of the pulmonary capillary wedge pressure should make it possible to weigh the respective likelihoods of hydrostatic pulmonary oedema or APT.[42] If the pulmonary infiltrates clear suboptimally following a course of diuretics, continued consideration of amiodarone-induced lung disease is warranted.[58] Other competing diagnoses that need to be particularly ruled out in this setting include interstitial lung disease of other cause, including those due to cardiovascular drugs, an infection, aspiration pneumonia, pulmonary embolism/infarction, idiopathic organizing-, non-specific- or eosinophilic interstitial pneumonia, exogenous lipoid pneumonia and, less commonly, bronchioloalveolar carcinoma or pulmonary lymphoma. With regard to cardiovascular drugs, patients may develop adverse pulmonary reactions to such drugs as aspirin, beta-blockers, ACEIs, oral or intravenous anticoagulants, platelet glycoprotein IIb/IIIa inhibitors (abciximab, clopidogrel), HMG-CoA reductase inhibitors (statins), hydrochlorothiazide, thiazolinediones, flecainide, procainamide and tocainide, in the form of interstitial lung disease, eosinophilic pneumonia, organizing pneumonia, alveolar haemorrhage or pulmonary oedema, with few features that might definitely separate these adverse reactions from APT.[7]

Box 22.1   Work-up for a patient suspected of having amiodarone pulmonary toxicity (APT)

- Record detailed history.
- Record detailed drug history (see www.pneumotox.com).
- Retrieve/review all previous imaging studies.
- Retrieve/review all previous pulmonary function tests, and diffusing capacity of the lung for carbon monoxide (DLCO) and diffusion capacity corrected for alveolar ventilation (KCO).
- Perform physical examination, especially a cardiac exam to listen for an S3 gallop.
- Order usual lab tests (haematology, blood chemistry, globulins), erythrocyte sedimentation rate (ESR), C-reactive protein, liver chemistry, tests of thyroid function.
- Perform new pulmonary function tests and DLCO/KCO, and compare with earlier values.
- Obtain new chest radiograph and high-resolution CT, and compare with earlier. (Manage for density measurement of liver and areas of lung involvement on high-resolution CT.)
- Serum amiodarone or desethylamiodarone concentration has limited diagnostic contribution.
- Discuss other possible causes for pulmonary infiltrates.
- Discuss diuresis, with pre-/post-test chest radiograph.
- Perform bronchoalveolar lavage, patient status permitting. Have foamy macrophages formally evaluated. Quantitate the proportion of inflammatory cells, red blood cells and haemosiderin-laden macrophages. (Measurement of amiodarone/desethylamiodarone levels in BAL fluid has no established diagnostic contribution.)

- Obtain cardiology opinion as regards:
  - heart ultrasound
  - cardiac catheterization and measurement of capillary wedge pressure
  - preparation for amiodarone discontinuation and institution of other antiarrhythmic therapy or implantation of a rhythm assist device.
- Discuss benefit/risk of transbronchial or video-assisted lung biopsy.
- Discuss whether drug withdrawal should be the sole intervention, or whether corticosteroids are indicated (this depends on symptoms, impact on lung functions and extent of involvement on imaging in the patient).
- Withdraw amiodarone and explain the reason to the patient. Document in chart.
- Write an action plan to the patient (appointments for visits, new pulmonary function tests and imaging; how to self-detect adverse effects of corticosteroids).
- Ensure adequacy of follow-up visits.
- Plan corticosteroid tapering after opacities of APT begin to clear (usually corticosteroid therapy needs to be given for 6–12 months or more). Ensure corticosteroid therapy is not discontinued by the patient or other care provider, otherwise relapse may occur.
- Inform all referring physicians. Inform the anaesthetist not to expose patient to high concentrations of oxygen.
- Make sure the patient is not unnecessarily re-challenged with amiodarone.
- Notify the local agency for adverse reactions to medicines.

Follow-up and outcome

On stopping amiodarone the patient is observed in the hospital and followed as an outpatient, using clinical evaluation of dyspnoea, ESR, body weight, chest radiograph, HRCT, pulmonary function and blood gases at serial time points. The frequency at which a patient is evaluated is dictated by clinical severity, extent of involvement, and by how much pulmonary function was initially impaired. One possibility is checks at 1, 2, 4 and 8 weeks, then at 4, 8 and 12 months, and then 6-monthly or yearly afterwards – depending on the response and whether pulmonary infiltrates persist. It is crucial that prevention of arrhythmia recurrence is organized, as cardiac death has been described following amiodarone withdrawal.

APT being a phospholipid storage disease and an inflammatory lung condition, adjunctive corticosteroid therapy is often required, except in patients with minimal/moderate disease who may improve quickly on discontinuing the drug. There is no evidence suggesting that observing such patients for a few days or for 1–2 weeks after drug withdrawal without initial recourse to corticosteroid therapy is in any way deleterious. However, despite the lack of controlled studies, evidence has accumulated supporting the use of corticosteroid therapy in patients with APT and significant pulmonary involvement. Indeed, early mortality was observed at a time when corticosteroids were not known to be efficacious,[106,107] or were given late in the course of APT.[42,108] In other patients, amiodarone therapy withdrawal was not followed by measurable improvement and/or their status deteriorated until corticosteroids were eventually given.[109] Corticosteroid therapy is considered efficacious in controlling the symptoms of APT more rapidly, and these drugs probably accelerate the resolution of pulmonary opacities as well.[66] Whether corticosteroids reduce the likelihood of long-term fibrotic changes in the lung is unknown. About 40–50 mg equivalent prednisolone per day is the typical starting dose. Corticosteroids are particularly indicated in patients with disseminated or diffuse involvement, and/or in those with a new restriction and/or hypoxaemia. There is no evidence that augmenting the daily dose of corticosteroids will alter the rate at which pulmonary infiltrates resolve. However, if no clinical and radiographic response is obtained following conventional doses, increased corticosteroid dosage should be considered. Corticosteroids are tapered once pulmonary infiltrates begin to clear. Improvement of imaging and pulmonary function generally lags behind clinical improvement.[110] Clearing of the pulmonary opacities typically occurs within 1–4 months.[71]

Box 22.2   Diagnostic features of amiodarone pulmonary toxicity (APT)

Key features

- Compatible temporality (new onset of the non-specific symptoms of dyspnoea, cough and body weight loss after commencement of therapy with amiodarone).

- New onset of pulmonary infiltrates.

- Confirmatory pattern and density measurements on imaging (chest radiograph and high-resolution CT; hyperdense liver and/or lung).

- Negative diuresis test (chest radiograph and weight unchanged following a 5-day course of increased diuresis).

- Exclusion of other possible diagnoses.

- Presence of foam cells in the bronchoalveolar lavage fluid. (Lamellar inclusion bodies visible by electron microscopy probably not justified in all patients because of cost. It may be merely a sign of exposure to the drug, not toxicity.)

- Pathological features (if biopsy indicated) consistent with APT (chronic interstitial inflammation or pneumonitis, organizing pneumonia, fibrosis, phospholipidosis, diffuse alveolar damage).

- Dechallenge followed by abatement of signs and symptoms and lower ESR.

- Re-challenge would prove the diagnosis, but it is not generally recommended.

Supportive features

- Age > 60 years.

- Daily dosage > 200 mg/day (incidence and dosage increase hand in hand, but APT can occur in patients taking only 200 mg 5 days a week).

- Time on the drug: most cases occur within the first year, particularly the 6–12 months window, but early and late cases (a few days to > 10 years) have been reported.

- Pulmonary function test results: decline in total lung capacity and vital capacity (> 15%) or of the diffusing capacity (> 20%) from baseline.

- Beneficial effect of corticosteroid therapy (to be given once other causes are ruled out; generally, corticosteroid therapy is extremely efficacious, even in some but not all advanced or severe cases).

Incidental findings of low intrinsic value for the diagnosis of APT

- Evidence of amiodarone toxicity in other organ systems (e.g. skin discolouration, photosensitivity, hyperthyroidism or hypothyroidism, corneal deposits, hepatitis, ataxia, peripheral neuropathy).

Undetermined or of no value

- KL-6.
- Drug and metabolite levels in blood.
- Gallium-67 scanning.
- 18FDG-PET scan.

Note: Causality is more often inferred rather than proven.

During corticosteroid therapy, patients need to be instructed and monitored with regard to the development of adverse effects and complications of corticosteroid therapy, including salt and water retention, hypertension, glucose intolerance, weight gain, infection, muscle wasting/myopathy, bruising, gastric ulcers and cataracts. Appropriate tests should be included in the follow-up plan to monitor these complications. A few cases of opportunistic pulmonary infections (but not *Pneumocystis* pneumonia) have been reported.

Corticosteroid tapering must be slow, to avoid relapse of the pulmonary infiltrates (up to three relapses in one case[111]). This occurs in about 5 per cent of patients with classic APT, and can occur up to 8 months after amiodarone withdrawal. Relapse is generally controlled by augmented corticosteroid dosage, but sometimes relapse is in the form of accelerated lung disease with uncontrollable respiratory failure.[112,113] Careful follow-up is needed to avoid relapses. Owing to the persistence of amiodarone in tissues whence it is slowly released, and the consequent risk of relapse, corticosteroids are typically given for 6–12 months. Difficult APT patients are those with life-threatening arrhythmias and a major contraindication to corticosteroid therapy present or developing, as corticosteroids are the only form of therapy that is capable of controlling APT at this time.

Imaging studies show clearing of the pulmonary opacities in about 85 per cent of patients. Opacities wane gradually, or

they move slowly from one area of the lung to a contigous one, like 'evaporating' before disappearing completely. Residual disease persists in the form of streaks which presumably correspond to non-progressive lung fibrosis in 10–15 per cent of cases.[76] In a small subset of patients, the disease progressed in spite of corticosteroid therapy, leading to respiratory death.[47,96] A decrease in the diffusing capacity may persist, especially in those patients with residual pulmonary infiltrates. In some patients who develop a residual picture consistent with lung fibrosis, prolonged low-dose corticosteroid therapy may be indicated.

In a few APT patients in whom amiodarone was deemed the only means to treat their underlying cardiac condition, the dosage of amiodarone was titrated downward (e.g. halved) and corticosteroid therapy was given, enabling control of both conditions.[114]

Overall mortality is an estimated 5–10 per cent, or 0.2 per cent of all patients on amiodarone. Outcome is worse if amiodarone pneumonitis is on a background of COPD.[92] Mortality from APT can be as high as 21–33 per cent in the more severely ill patients who are admitted to hospital.[48,66,96,115]

Unwitting continuation of treatment with amiodarone in the patient with APT leads to an increase in extent, severity and rate of progression of the disease. Thus early diagnosis is encouraged. Resumption of amiodarone after a drug holiday, if indicated for the control of refractory or life-threatening arrhythmia, is followed by relapse in about two-thirds of patients, within a short period.[116,117]

OTHER PATTERNS OF AMIODARONE-INDUCED INVOLVEMENT

Amiodarone pulmonary toxicity is not a single clinical entity, and several other patterns have been catalogued.

Mass, masses and nodules

A single mass (sometimes called amiodaronoma[118]) or multiple mass lesions with, generally, increased attenuation on HRCT is/are present in 6–12 per cent of patients as the sole manifestation of APT.[69,119-121] On imaging, mass or masses may be irregularly shaped and/or lobulated, mimic cavitary lung disease, be ill-defined, contain air bronchograms, or abut the pleura, in which case the pleural membrane is often thickened *en face* the mass causing the clinical sign and symptom of pleuritic chest pain and friction rub. Less often, mass lesions are deep-seated in the lung, where they can be mistaken for pulmonary infarction, resolving or organizing pneumonia, or malignancy.[121] Patients may be symptomatic, or the mass is discovered incidentally on CT. On HRCT the high density of the mass is sometimes noted, owing to the iodinated chemistry of the drug. On pathology of the lung or fine-needle aspirate, classic features of APT – chronic interstitial inflammation, myofibroblasts, organizing pneumonia and foam cells – are present.

Other patients present with round nodules, which are solitary or multiple, shaggy, lobulated or well delineated, and can be surrounded by a halo of decreasing attenuation. Yet other patients present with one or more large nodules up to 3.5 cm in diameter, with a decreased attenuation in the centre, and with or without a background of interstitial lung disease in the remaining lung. The nodules may correspond to classic changes of APT which attenuate peripherally,[122] to a lympho-plasmacytic infiltrate,[81] or to a novel pattern.[8] Ruangchira-Urai *et al.* described the pathological features of excised nodular APT in four patients treated with amiodarone and reviewed eleven earlier cases. None of the four cases had a preoperative diagnosis of APT, and the initial suspicion was malignancy, and the PET-scan was positive in one case so tested. Histopathology was strikingly similar in all these four patients, with 'vacuolated histiocytes massed within alveoli to form macroscopic nodules with tissue breakdown'.[8] Ultrastructural examination of the lung showed characteristic inclusions in the cytoplasm of swollen histiocytes. In addition to these changes, there was necrotizing tissue destruction and necrotizing bronchiolitis in the absence of infection or obstruction. Three of the four patients had received a high maintenance dose of 800 mg amiodarone daily for 7 months or more. Vacuolation may represent more advanced toxicity in patients on that high dosage, compared to the usual foamy cytoplasmic changes of classic APT occurring in patients taking the conventional low-dosage regimen. The four above patients responded to drug withdrawal and/or corticosteroid therapy. It is important to keep this distinctive pattern of APT in mind, since this may

obviate the need for lung biopsy. Also, drug history should be made available to the pathologist in similar cases, as the drug history of none of these four cases was made available to the pathologist at the time of examination of the lung sample.

Pulmonary fibrosis

Amiodarone-induced pulmonary fibrosis is an uncommon, severe and disabling pattern of involvement in APT, with an estimated incidence of 0.1 per cent.[123] In 5–7 per cent of patients with classic APT, pulmonary fibrosis follows an episode of classic APT and is diagnosed during follow-up in spite of treatment with corticosteroids, with no predictive or annunciating features.[69] Pulmonary fibrosis may also appear as a *de-novo* phenomenon in a few weeks in patients taking amiodarone, with no warning signs, usually leading to respiratory death,[47] or the fibrosis develops insidiously in a patient on chronic amiodarone treatment, possibly reflecting earlier undiagnosed classic APT. A normal pre-therapy chest imaging and pulmonary function, and the absence of another cause, point to the causal role of amiodarone in such cases of pulmonary fibrosis. Coarse crackles are present on chest auscultation. On imaging, there are dense bilateral reticular shadows and, on CT, the characteristic findings of pulmonary fibrosis are present, including coarse reticulation, traction bronchiectasis and, later, honeycombing. Pulmonary physiology indicates significant restrictive lung dysfunction and resting hypoxaemia. Histology findings include interstitial fibrosis with septal thickening and type 2 cell hyperplasia. Characteristic foam cells are present, depending on the time from discontinuation of amiodarone to the biopsy.[37] The condition is largely irreversible, and response to corticosteroids is limited or transient.

Organizing pneumonia

Amiodarone-induced organizing pneumonia (BOOP) has been reported as a manifestation of APT in about 8 per cent of APT patients, and it is estimated that about 3 per cent of all organizing pneumonia cases occur as the result of therapy with amiodarone.[124] The disease manifests distinctively with stationary or migratory opacities on sequential imaging (Fig. 22.4; see also Chapter 24).[72,105,125]

The clinical and imaging features of BOOP associated with amiodarone are indistiguishable from those of idiopathic organizing pneumonia. The pathological picture of amiodarone-induced BOOP is of foam cells superimposed on a background of interstitial inflammation and endoluminal fibrosis. Withdrawal of amiodarone and often corticosteroid therapy are required. Several relapses may occur over a period of months, until the drug is discontinued. Outcome is good.

ALI/ARDS

An ALI or ARDS picture can occur in patients treated with amiodarone (in the long or short term) who undergo thoracic or cardiac surgery and are given a high perioperative fractional concentration of oxygen in association with amiodarone.

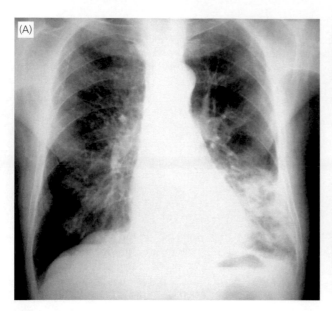
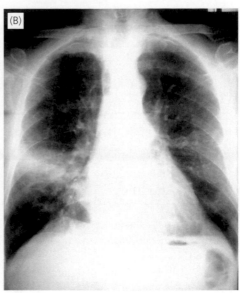

Fig. 22.4 (a and b) Bronchiolitis obliterans organizing pneumonia (BOOP) as a manifestation of amiodarone toxicity. Alveolar shadowing wanders from one area of the lung to the other over a period of a few weeks. Manifestations of the disease generally do not remit until amiodarone is withdrawn or corticosteroids are added while tapering the amiodarone.

Anaesthetists and critical-care personnel must know about the hazard of this dreadful association. Very short-term postoperative amiodarone may not expose the patient to this risk.[126]

The clinical picture is of rapidly progressive respiratory failure in the context of diffuse alveolar opacities and volume loss. Mechanical ventilation is indicated. Pathology shows the typical features of diffuse alveolar damage and, in some cases, foam cells were present, in spite of very short-term exposure to the drug, and interstitial inflammation. In a blinded postmortem study of seven ARDS cases in patients who had received amiodarone postoperatively, the three patients with > 2 days amiodarone already exhibited changes suggestive of APT, including endogenous lipoid pneumonia and foam cells superimposed on a non-specific background of diffuse alveolar damage and desquamated alveolar cells.[127] Whether ARDS in these three patients specifically corresponds to acute APT, or merely to the combination of early phospholipidosis and ARDS of other cause as a chance association, is unclear. Importantly, foam cells can be retrieved in the BAL fluid after a few days on amiodarone therapy. ARDS has also been reported in up to 10 per cent of patients on amiodarone who receive an automatic defibrillator.[11]

Although the response of ALI/ARDS to treatment with corticosteroids is uncertain, a trial of corticosteroid therapy is indicated. The fatality rate is up to 50 per cent.[128] The ALI/ARDS picture can develop in patients with amiodarone-induced fibrosis as a terminal exacerbation, in a manner similar to fibrosis of other causes.

Subclinical amiodarone pulmonary toxicity

Subclinical amiodarone pulmonary toxicity was evidenced in patients on chronic amiodarone therapy and a normal chest radiograph, after HRCT indicated ground-glass density, diminutive alveolar opacities[119] or increased septal lines.[129] These abnormalities correspond histologically to aggregates of mural or alveolar foam cells or alveolitis.[20,129] The significance of subclinical changes in terms of possible progression to overt APT is unknown. Increased proportions of lymphocytes or neutrophils have been detected in the BAL fluid of patients on amiodarone who had normal imaging findings. Those patients with alveolitis had a lower PaO_2.[21] A positive gallium-67 lung scan has been reported in asymptomatic patients chronically exposed to amiodarone, and was taken as evidence of silent APT.[130] Dyspnoea on exertion developed in a 65-year-old man who erroneously took 1000 mg amiodarone daily for 6 weeks. No infiltrates were present on imaging. DLCO and the diffusing capacity dropped from a pre-therapy value of 98 per cent to 55 per cent of predicted. BAL lymphocytes were slightly above normal and the proportion of neutrophils was 10 per cent. Symptoms resolved and the diffusing capacity improved in 2.5 months on drug discontinuation.

Diffuse alveolar haemorrhage

ANCA-negative diffuse alveolar haemorrhage as the primary manifestation of APT has been reported in seven patients.[131–133] Patients presented rather acutely, with diffuse pulmonary opacities, haemoptysis, a drop in haemoglobin, and haemorrhagic BAL return. On pathology, there is alveolar haemorrhaging on a background of foam cells. Drug withdrawal and corticosteroid therapy are followed by improvement. The main competing diagnoses include haemorrhagic pulmonary oedema and anticoagulant-induced and platelet glycoprotein IIB/IIA inhibitor-induced alveolar haemorrhage.[7]

Pleural effusion

A moderate, uni- or bilateral pleural exudate develops in up to a third of patients with the features of parenchymal APT.[77] A lone free-flowing effusion is unusual. Pericardial effusion can be present as well. One patient on high-dose amiodarone (1.6 g, followed by 1.2 g daily) developed bilateral exudates having the same biochemical characteristics; the effusions resolved on drug discontinuation.[134] In another patient, a large bilateral serosanguinous effusion predated the onset of classic APT.[135] The pleural fluid in amiodarone-associated effusions is a lymphocyte-rich[136] or lymphocyte- and neutrophil-rich exudate, with a range of protein concentration of 2.8–5.5 g/dL.[134] Foam cells resembling those in the BAL fluid can be present in pleural fluid.[135] On pathology in one case, there was pleural thickening and foam cells were present in pleural tissue.[134] When pleural effusion develops during treatment with amiodarone, this is often against the background of heart failure which can also manifest with pleural transudate or exudate.[137]

A few cases of amiodarone-induced lupus have been reported, two of which manifested with pleural[138] or pleuropericardial[139] effusion in the context of positive ANA. Pleural fluid was examined in one of the two cases, and was a lymphocytic exudate.[138] Patients with amiodarone-induced lupus improve and ANA titres decrease slowly on discontinuing the drug.

Pleural thickening

Imaging studies often indicate smooth-edged pleural thickening in APT patients,[119,140] especially those with a large infiltrate or a mass abutting the pleura in a proportion of 65 per cent.[76] Upon discontinuation of the drug there is improvement in pleural thickening, although some pleural thickening may persist after resolution of the pulmonary opacities and participate in the late restrictive physiology.

Angioedema

A single report described angioedema without upper airway obstruction or breathing difficulties during treatment with amiodarone. Symptoms relapsed on re-challenge with amiodarone.[141]

MONITORING, PREVENTION AND EARLY DETECTION OF APT

Given the frequency and potential severity of amiodarone pneumonitis, early detection is desirable but no consensus exists. A go-low dosage and episodic treatment with amiodarone are indicated whenever possible. Early diagnosis of APT may simplify management and improve prognosis.

There is some discrepancy between the guidelines from manufacturers, authoritative bodies and experts. However, Box 22.3 shows the corpus that emerges for prescribing physicians and pulmonologists.

Box 22.3   Pulmonary monitoring of patients taking amiodarone

1. Baseline (pre-therapy) evaluation:
   - Chest radiograph
   - Pulmonary function tests (PFTs), including diffusing capacity
   - Sedimentation rate
   - Refer to pulmonologist and have high-resolution CT done if chest radiograph and/or PFT abnormal, or if unexplained dyspnoea.
2. Ensure communication with patient, primary-care provider, cardiologist and pharmacist. Record in history.
3. Check implementation and adequacy of non-respiratory monitoring (chiefly liver and thyroid function tests).
4. Ask patient to report any new and otherwise unexplained dyspnoea, malaise, fever or body weight loss.
5. Consider imaging with or without PFTs (i.e. imaging without PFT, or imaging and PFT if imaging abnormal) once within 6–12 months into treatment, as most cases of APT develop during this timeframe. There is no firm evidence for cost-effectiveness of yearly chest radiograph or PFT thereafter. Consider radiation exposure. A simple diffusing capacity is inexpensive and may be rewarding, to the extent that the considerations under (7) below are taken into account).
6. Restrict serial tests (every 3 to 12 months, depending on perceived risk): sedimentation rate, chest radiograph and lung function evaluation to higher-risk patients (those aged > 60, those with poor pre-therapy lung function, and/or those exposed to high intravenous or oral dosages of the drug). Particular attention to be given to patients with underlying lung disease such as chronic obstructive disease and/or fibrosis, or those with antecedent pneumonectomy and recipients of heart, lung or heart and lung transplant.
7. If PFTs are done in the patient on amiodarone, an isolated decrease of the diffusing capacity from baseline does not necessarily indicate disease and should not prompt drug discontinuation unless there is evidence of interstitial lung disease (ILD) or other adverse effects of amiodarone. It could represent early congestive heart failure, although CHF alters this test less than does APT. A stable diffusing capacity indicates clinically and radiographically meaningful APT is unlikely.
8. Should the patient on amiodarone undergo a surgical procedure, communication with the anaesthetist, responsible cardiologist and critical-care specialists should be optimized to keep fractional concentration of oxygen as low as possible.
9. Reasons to refer to a pulmonologist:
   - abnormal chest imaging or pulmonary function prior to initiating treatment or at any time later
   - to separate the role of other drugs the patient might be on from amiodarone
   - for unexplained respiratory symptoms (cough, dyspnoea), imaging or pulmonary physiology at any time point while on the drug
   - before deciding on a lung biopsy, as the possibility of conservative management of APT should not be overlooked.

KEY POINTS

- Amiodarone is prescribed by cardiologists, anaesthesiologists, primary-care physicians, physicians caring for the elderly and critical-care physicians.

- It is a coronary vasodilator and class 3 antiarrhythmic agent used to treat or prevent recurrences of supraventricular and ventricular arrhythmias. It is used prophylactically in the perioperative period after thoracic, cardiac and other surgery, and in patients who receive an implantable defibrillator.

- The drawback of amiodarone is its extensive adverse-effect profile, among which is APT.

- Amiodarone sequesters in fat and tissues, notably in the lungs. Fat may act as a reservoir, so amiodarone effluxes slowly upon drug discontinuance. Improvement in symptoms of APT and radiographic abnormalities may not follow drug withdrawal unless corticosteroids are given. Relapse may occur despite drug discontinuance weeks or months earlier, at a time when corticosteroids are tapered.

- The two iodines present on each amiodarone molecule account for the thyroid and pulmonary adverse effects of the drug, and are thought to account for the increased attenuation on CT of tissues where amiodarone sequesters, notably the liver and condensed areas of amiodarone pneumonitis in amiodarone-toxic patients.

- The distinguishing feature of APT is its histological appearance. This includes a foamy cytoplasm or vacuolation of histiocytes, pneumocytes and other resident lung cells, and foamy alveolar macrophages in the BAL fluid.

- The probability of developing APT is related to several factors, the principal one being amiodarone dosage. The lower maintenance dose advised nowadays has led to reduced incidence of APT, but it occurs with low maintenance doses.

REFERENCES

1. Wolkove N, Baltzan M. Amiodarone pulmonary toxicity. Can Respir J 2009;16:43–8.
2. Singh BN. Amiodarone as paradigm for developing new drugs for atrial fibrillation. J Cardiovasc Pharmacol 2008;52:300–5.
3. Camus P, Coudert B, d'Athis P, et al. Pharmacokinetics of amiodarone in the isolated rat lung. J Pharmacol Exp Ther 1990;254:336–43.
4. Goldschlager N, Epstein AE, Naccarelli GV, et al. A practical guide for clinicians who treat patients with amiodarone. Heart Rhythm 2007;4:1250–9.
5. ter Maaten JC, Strack van Schijndel RJ, Jukema GJ, Kerkhof SC. Respiratory insufficiency due to pulmonary toxicity of amiodarone. Ned Tijdschr Geneeskd 1995;139:1501–4.
6. Rotmensch HH, Liron M, Tupilsky M, Laniado S. Possible association of pneumonitis with amiodarone therapy. Am Heart J 1980;100:412–13.
7. www.pneumotox.com: Pneumotox® website 1997. Produced by P Foucher – Ph Camus. Last update June 2010.
8. Ruangchira-Urai R, Colby TV, Klein J, et al. Nodular amiodarone lung disease. Am J Surg Pathol 2008;32:1654–60.
9. Ernawati DK, Stafford L, Hughes JD. Amiodarone-induced pulmonary toxicity. Br J Clin Pharmacol 2008;66:82–7.
10. Fung RC, Chan WK, Chu CM, Yue CS. Low dose amiodarone-induced lung injury. Int J Cardiol 2006;113:144–5.
11. Liverani E, Armuzzi A, Mormile F, et al. Amiodarone-induced adult respiratory distress syndrome after nonthoracotomy subcutaneous defibrillator implantation. J Intern Med 2001;249:565–6.
12. Bargout R, Jankov A, Dincer E, et al. Amiodarone induces apoptosis of human and rat alveolar epithelial cells in vitro. Am J Physiol Lung Cell Mol Physiol 2000;278:L1039–44.
13. Quaglino D, Ha HR, Duner E, et al. Effects of metabolites and analogs of amiodarone on alveolar macrophages: structure-activity relationship. Am J Physiol Lung Cell Mol Physiol 2004;287:L438–47.
14. Lafuente-Lafuente C, Alvarez JC, Leenhardt A, et al. Amiodarone concentrations in plasma and fat tissue during chronic treatment and related toxicity. Br J Clin Pharmacol 2009;67:511–19.
15. Maggioni AP, Maggi A, Volpi A, et al. Amiodarone distribution in human tissues after sudden death during Holter recording. Am J Cardiol 1983;52:217–18.
16. Vassallo P, Trohman RG. Prescribing amiodarone: an evidence-based review of clinical indications. J Am Med Assoc 2007;298:1312–22.
17. Camus P, Joshi UM, Lockard VG, et al. Effects of drug-induced pulmonary phospholipidosis on lung mechanics in rats. J Appl Physiol 1989;66:2437–45.
18. Martin WJI, Rosenow EC. Amiodarone pulmonary toxicity: recognition and pathogenesis, part 1. Chest 1988;93:1067–75.
19. Martin WJI, Rosenow EC. Amiodarone pulmonary toxicity: recognition and pathogenesis, part 2. Chest 1988;93:1242–8.
20. Bedrossian CW, Warren CJ, Ohar J, Bhan R. Amiodarone pulmonary toxicity: cytopathology, ultrastructure, and immunocytochemistry. Ann Diagn Pathol 1997;1:47–56.
21. Jolly E, Belloti MS, Lowenstein J, et al. Cytologic changes in bronchoalveolar lavage in amiodarone treated patients. Medicina (Buenos Aires) 1991;51:19–25.
22. Yano T, Itoh Y, Yamada M, Egashira N, Oishi R. Combined treatment with l-carnitine and a pan-caspase inhibitor effectively reverses amiodarone-induced injury in cultured human lung epithelial cells. Apoptosis 2008;13:543–52.
23. Uhal BD, Wang R, Lauuka J, et al. Inhibition of amiodarone-induced lung fibrosis but not alveolitis by angiotensin system antagonists. Basic Clin Pharacol Toxicol 2003;92:81–7.
24. Hotchkiss RS, Strasser A, McDunn JE, Swanson PE. Cell death. New Engl J Med 2009;361:1570–83.
25. Sisson TH, Mendez M, Choi K, et al. Targeted injury of type II alveolar epithelial cells induces pulmonary fibrosis. Am J Respir Crit Care Med 2010;181:254–63.
26. Nikaido A, Tada T, Nakamura K, et al. Clinical features of and effects of angiotensin system antagonists on amiodarone-induced pulmonary toxicity. Int J Cardiol 2010;143:328–35.
27. Kosseifi SG, Halawa A, Bailey B, et al. Reduction of amiodarone pulmonary toxicity in patients treated with angiotensin-converting enzyme inhibitors and angiotensin receptor blockers. Ther Adv Respir Dis 2009;3:289–94.

28. Vorperian VR, Havighurst TC, Miller S, January CT. Adverse effects of low dose amiodarone: a meta-analysis. *J Am Coll Cardiol* 1997;**30**:791–8.

29. Piccini JP, Berger JS, O'Connor CM. Amiodarone for the prevention of sudden cardiac death: a meta-analysis of randomized controlled trials. *Eur Heart J* 2009;**30**:1245–53.

30. Roy D, *et al*., Canadian Trial of Atrial Fibrillation Investigators. Amiodarone to prevent recurrence of atrial fibrillation. *New Engl J Med* 2000;**342**:913–20.

31. Chandhok S, Schwartzman D. Amiodarone therapy for atrial rhythm control: insights gained from a single center experience. *J Cardiovasc Electrophysiol* 2007;**18**:714–18.

32. Yamada Y, Shiga T, Matsuda N, Hagiwara N, Kasanuki H. Incidence and predictors of pulmonary toxicity in Japanese patients receiving low-dose amiodarone. *Circ J* 2007;**71**:1610–16.

33. Olshansky B. Amiodarone-induced pulmonary toxicity. *New Engl J Med* 1997;**337**:1814.

34. Greene HL, Graham EL, Werner JA, *et al*. Toxic and therapeutic effects of amiodarone in the treatment of cardiac arrhythmias. *J Am Coll Cardiol* 1983;**2**:1114–28.

35. Morady F, Sauve MJ, Malone P, *et al*. Long-term efficacy and toxicity of high-dose amiodarone therapy for ventricular tachycardia or ventricular fibrillation. *Am J Cardiol* 1983;**52**:975–9.

36. Kaushik S, Hussain A, Clarke P, Lazar HL. Acute pulmonary toxicity after low-dose amiodarone therapy. *Ann Thorac Surg* 2001;**72**:1760–1.

37. Lengyel C, Boros I, Varkonyi T, Selmeczi A, Fazekas T. Amiodarone-induced pulmonary fibrosis. *Orv Hetil* 1996;**137**:1759–62.

38. Ahmed S, Rienstra M, Crijns H, *et al*. Continuous vs. episodic prophylactic treatment with amiodarone for the prevention of atrial fibrillation: a randomized trial. *J Am Med Assoc* 2008;**300**:1784–92.

39. Kothari SS, Balijepally S, Taneja K. Amiodarone-induced pulmonary toxicity in an adolescent. *Cardiol Young* 1999;**9**:194–6.

40. Daniels CJ, Schutte DA, Hammond S, Franklin WH. Acute pulmonary toxicity in an infant from intravenous amiodarone. *Am J Cardiol* 1997;**80**:1113–16.

41. Bowers PN, Fields J, Schwartz D, Rosenfeld LE, Nehgme R. Amiodarone induced pulmonary fibrosis in infancy. *Pace Pacing Clin Electrophysiol* 1998;**21**:1665–7.

42. Magro SA, Lawrence EC, Wheeler SH, *et al*. Amiodarone pulmonary toxicity: prospective evaluation of serial pulmonary function tests. *J Am Coll Cardiol* 1988;**12**:781–8.

43. Dean PJ, Groshart KD, Porterfield JG, Iansmith DH, Golden EB. Amiodarone-associated pulmonary toxicity: a clinical and pathologic study of eleven cases. *Am J Clin Pathol* 1987;**87**:7–13.

44. Olshansky B, Sami M, Rubin A, *et al*. Use of amiodarone for atrial fibrillation in patients with preexisting pulmonary disease in the AFFIRM study. *Am J Cardiol* 2005;**95**:404–5.

45. Kudenchuk PJ, Pierson DJ, Greene HL, *et al*. Prospective evaluation of amiodarone pulmonary toxicity. *Chest* 1984;**86**:541–8.

46. Nalos PC, Kass RM, Gang ES, *et al*. Life-threatening postoperative pulmonary complications in patients with previous amiodarone pulmonary toxicity undergoing cardiothoracic operations. *J Thorac Cardiovasc Surg* 1987;**93**:904–12.

47. Kharabsheh S, Abendroth CS, Kozak M. Fatal pulmonary toxicity occurring within two weeks of initiation of amiodarone. *Am J Cardiol* 2002;**89**:896–8.

48. Biour M, Hugues FC, Hamel JD, Cheymol G. Adverse pulmonary effects of amiodarone: analysis of 162 cases. *Therapie* 1985;**40**:343–8.

49. Wilhelm JM, Thannberger P, Derragui A, *et al*. Interstitial pneumonia two months after discontinuing amiodarone. *Presse Med* 1999;**28**:2040.

50. Kay GN, Epstein AE, Kirklin JK, *et al*. Fatal postoperative amiodarone pulmonary toxicity. *Am J Cardiol* 1988;**62**:490–2.

51. Handschin AE, Lardinois D, Schneiter D, Bloch KE, Weder W. Acute amiodarone-induced pulmonary toxicity following lung resection. *Respiration* 2003;**70**:310–12.

52. van Mieghem W, Coolen L, Malysse I, *et al*. Amiodarone and the development of ARDS after lung surgery. *Chest* 1994;**105**:1642–5.

53. Saussine M, Colson P, Alauzen M, Mary H. Postoperative acute respiratory distress syndrome: a complication of amiodarone associated with 100 percent oxygen ventilation. *Chest* 1992;**102**:980–1.

54. Herndon JC, Cook AO, Ramsay MAE, Swygert TH, Capehart J. Postoperative unilateral pulmonary edema: possible amiodarone pulmonary toxicity. *Anesthesiology* 1992;**76**:308–12.

55. Wood DL, Osborn MJ, Rooke J, Holmes DR. Amiodarone pulmonary toxicity: report of two cases associated with rapidly progressive fatal adult respiratory distress syndrome after pulmonary angiography. *Mayo Clin Proc* 1985;**60**:601–3.

56. Bhagat R, Sporn TA, Long GD, Folz RJ. Amiodarone and cyclophosphamide: potential for enhanced lung toxicity. *Bone Marrow Transplant* 2001;**27**:1109–11.

57. Gupta S, Mahipal A. Fatal pulmonary toxicity after a single dose of cyclophosphamide. *Pharmacotherapy* 2007;**27**:616–18.

58. Silva CP, Bacal F, Pires PV, *et al*. The importance of amiodarone pulmonary toxicity in the differential diagnosis of a patient with dyspnea awaiting a heart transplant. *Arq Bras Cardiol* 2006;**87**:e4–7.

59. Diaz-Guzman E, Mireles-Cabodevila E, Arrossi A, Kanne JP, Budev M. Amiodarone pulmonary toxicity after lung transplantation. *J Heart Lung Transplant* 2008;**27**:1059–63.

60. Blomberg PJ, Feingold AD, Denofrio D, *et al*. Comparison of survival and other complications after heart transplantation in patients taking amiodarone before surgery versus those not taking amiodarone. *Am J Cardiol* 2004;**93**:379–81.

61. Zitnik RJ. Drug-induced lung disease: cardiovascular agents. *J Respir Dis* 1996;**17**:293–8.

62. Zitnik RJ. Drug-induced lung disease: antiarrhythmic agents. *J Respir Dis* 1996;**17**:254–70.

63. Cooper JAJ. Drug-induced lung disease. *Adv Intern Med* 1997;**42**:231–8.

64. Kanji Z, Sunderji R, Gin K. Amiodarone-induced pulmonary toxicity. *Pharmacotherapy* 1999;**19**:1463–6.

65. Rossi SE, Erasmus JJ, McAdams P, Sporn TA, Goodman PC. Pulmonary drug toxicity: radiologic and pathologic manifestations. *RadioGraphics* 2000;**5**:1245–59.

66. Darmanata JI, Van Zandwijk N, Düren DR, et al. Amiodarone pneumonitis: three further cases with a review of published reports. Thorax 1984;**39**:57–64.

67. Kataoka H, Matuno O. Age-related pulmonary crackles (rales) in asymptomatic cardiovascular patients. Ann Fam Med 2008;**6**:239–45.

68. Gefter WB, Epstein DM, Pietra GG, Miller WT. Lung disease caused by amiodarone, a new antiarrhythmic agent. Radiology 1983;**147**:339–44.

69. Polverosi R, Zanellato E, Doroldi C. Thoracic radiography and high resolution computerized tomography in the diagnosis of pulmonary disorders caused by amiodarone. Radiol Med 1996;**92**:58–62.

70. Emby DJ. Amiodarone-induced lung disease. S Afr Med J 2008;**98**:9.

71. Marchlinski FE, Gansler TS, Waxman HL, Josephson ME. Amiodarone pulmonary toxicity. Ann Intern Med 1982;**97**:839–45.

72. Camus P, Lombard JN, Perrichon M, et al. Bronchiolitis obliterans organising pneumonia in patients taking acebutolol or amiodarone. Thorax 1989;**44**:711–15.

73. Malhotra A, Muse VV, Mark EJ. An 82-year-old man with dyspnea and pulmonary abnormalities. New Engl J Med 2003;**348**:1574–85.

74. Cunha Ribeiro CM, Marchiori E, Rodrigues R, et al. Hydrostatic pulmonary edema: high-resolution computed tomography aspects. J Bras Pneumol 2006;**32**:515–22.

75. Torres-Romero J, Asin-Guillén JM, Peral-Martinez JI, Caro-Paton Gomez T. Pulmonary fibrosis induced by amiodarone. Rev Esp Cardiol 1984;**37**:142–4.

76. Vernhet H, Bousquet C, Durand G, Giron J, Senac JP. Reversible amiodarone-induced lung disease: HRCT findings. Eur Radiol 2001;**11**:1697–703.

77. Siniakowicz RM, Narula D, Suster B, Steinberg JS. Diagnosis of amiodarone pulmonary toxicity with high-resolution computerized tomographic scan. J Cardiovasc Electrophysiol 2001;**12**:431–6.

78. Clarke B, Ward DE, Honey M. Pneumonitis with pleural and pericardial effusion and neuropathy during amiodarone therapy. Int J Cardiol 1985;**8**:81–8.

79. Fireman E, Topilsky I, Viskin S, Priel IE. The role of induced sputum in amiodarone-associated interstitial lung diseases. Cardiology 2007;**108**:223–7.

80. Azzam I, Tov N, Elias N, Naschitz JE. Amiodarone toxicity presenting as pulmonary mass and peripheral neuropathy: the continuing diagnostic challenge. Postgrad Med J 2006;**82**:73–5.

81. Kimura T, Kuramochi S, Katayama T, et al. Amiodarone-related pulmonary mass and unique membranous glomerulonephritis in a patient with valvular heart disease: diagnostic pitfall and new findings. Pathol Int 2008;**58**:657–63.

82. Drent M, Cobben NAM, Dieijen-Visser MP, et al. Serum lactate dehydrogenase activity: indicator of the development of pneumonitis induced by amiodarone. Eur Heart J 1998;**19**:969–70.

83. Mouallem M, Antipov N, Mayan H, Sela BA, Farfel Z. Hyperglobulinemia in amiodarone-induced pneumonitis. Cardiovasc Drugs Ther 2007;**21**:63–7.

84. Endoh Y, Hanai R, Uto K, et al. Diagnostic accuracy of KL-6 as a marker of amiodarone-induced pulmonary toxicity. Pace Pacing Clin Electrophysiol 2000;**23**:2010–13.

85. Umetani K, Abe M, Kawabata K, et al. SP-D as a marker of amiodarone-induced pulmonary toxicity. Intern Med 2002;**41**:709–12.

86. Troughton RW, Richards AM, Yandle TG, Frampton CM, Nicholls MG. The effects of medications on circulating levels of cardiac natriuretic peptides. Ann Med 2007;**39**:242–60.

87. Lavietes MH, Gerula CM, Fless KG, Cherniack NS, Arora RR. Inspiratory muscle weakness in diastolic dysfunction. Chest 2004;**126**:838–44.

88. Ponnuswamy A, Manikandan R, Sabetpour A, Keeping IM, Finnerty JP. Association between ischaemic heart disease and interstitial lung disease: a case-control study. Respir Med 2009;**103**:503–07.

89. Horowitz LN. Detection of amiodarone pulmonary toxicity: to screen or not to screen, that is the question! J Am Coll Cardiol 1988;**12**:789–90.

90. Stelfox HT, Ahmed SB, Fiskio J, Bates DW. Monitoring amiodarone's toxicities: recommendations, evidence, and clinical practice. Clin Pharmacol Ther 2004;**75**:110–22.

91. Toms LD, Wordsworth MW, Bloom C, Mann B. Amiodarone: an audit to assess monitoring practices. Eur Respir J Suppl 2009; abstract.

92. Ohar JA, Jackson F, Redd RM, Evans GR, Bedrossian CW. Usefulness of serial pulmonary function testing as an indicator of amiodarone toxicity. Am J Cardiol 1989;**64**:1322–6.

93. Adams PC, Gibson GJ, Morley AR, et al. Amiodarone pulmonary toxicity: clinical and subclinical features. Q J Med 1986;**59**:449–71.

94. Cockcroft DW, Fisher KL. Near normalization of spirometry in a subject with severe emphysema complicated by amiodarone lung. Respir Med 1999;**93**:597–600.

95. Mettauer B, Lampert E, Charloux A, et al. Lung membrane diffusing capacity, heart failure, and heart transplantation. Am J Cardiol 1999;**83**:62–7.

96. Coudert B, Bailly F, André F, Lombard JN, Camus P. Amiodarone pneumonitis: bronchoalveolar lavage findings in 15 patients and review of the literature. Chest 1992;**102**:1005–12.

97. Xaubet A, Roca J, Rodriguez-Roisin R, et al. Bronchoalveolar lavage cellular analysis and gallium lung scan in the assessment of patients with amiodarone-induced pneumonitis. Respiration 1987;**52**:272–80.

98. Meurice JC, Debacque I, Boita F, et al. Interstitial pneumopathies during amiodarone treatment: determination of serum amiodarone, typing of lymphocytes from bronchiolo-alveolar lavage. Pathol Biol 1986;**34**:1074–80.

99. Grabczak EM, Zielonka TM, Wiwala J, et al. Amiodarone induced pneumonitis and hyperthyroidism: case report. Pol Arch Med Wewn 2008;**118**:524–9.

100. Iskandar SB, Abi-Saleh B, Keith RL, Byrd RP, Roy TM. Amiodarone-induced alveolar hemorrhage. South Med J 2006;**99**:383–7.

101. Wang CW, Colby TV. Histiocytic lesions and proliferations in the lung. Semin Diagn Pathol 2007;**24**:162–82.

102. Basset-Leobon C, Lacoste-Collin L, Aziza J, et al. Cut-off values and significance of Oil Red O-positive cells in bronchoalveolar lavage fluid. Cytopathology 2010;in press.

103. Myers JL, Kennedy JI, Plumb VJ. Amiodarone lung: pathologic findings in clinically toxic patients. *Hum Pathol* 1987;**18**: 349–54.

104. Morera J, Vidal R, Morell F, *et al*. Pulmonary fibrosis and amiodarone. *Br Med J* 1982;**285**:895.

105. Beasley MB, Franks TJ, Galvin JR, Gochuico B, Travis WD. Acute fibrinous and organizing pneumonia: a histologic pattern of lung injury and possible variant of diffuse alveolar damage. *Arch Pathol Lab Med* 2002;**126**:1064–70.

106. Sobol SM, Rakita L. Pneumonitis and pulmonary fibrosis associated with amiodarone treatment: a possible complication of a new antiarrhythmic drug. *Circulation* 1982;**65**:819–24.

107. Bates N, Dawson J. Suspected amiodarone-hypersensitivity pneumonitis. *Can J Hosp Pharm* 1998;**51**:117–21.

108. Wang T, Charette S, Smith ML. An unintended consequence: fatal amiodarone pulmonary toxicity in an older woman. *J Am Med Dir Assoc* 2006;**7**:510–13.

109. Butland RJA, Milliard FJC. Fibrosing alveolitis associated with amiodarone. *Eur J Respir Dis* 1984;**65**:616–19.

110. Olson LK, Forrest JV, Friedman PJ, Kiser PE, Henschke CI. Pneumonitis after amiodarone therapy. *Radiology* 1984;**150**:327–30.

111. Okayasu K, Takeda Y, Kojima J, *et al*. Amiodarone pulmonary toxicity: a patient with three recurrences of pulmonary toxicity and consideration of the probable risk for relapse. *Intern Med* 2006;**45**:1303–7.

112. Parra O, Ruiz J, Ojanguren I, Navas JJ, Morera J. Amiodarone toxicity: recurrence of interstitial pneumonitis after withdrawal of the drug. *Eur Respir J* 1989;**2**:905–7.

113. Chendrasekhar A, Barke RA, Druck P. Recurrent amiodarone pulmonary toxicity. *South Med J* 1996;**89**:85–6.

114. Zaher C, Hamer A, Peter T, Mandel W. Low-dose steroid therapy for prophylaxis of amiodarone-induced pulmonary infiltrates. *New Engl J Med* 1983;**308**:779

115. Dunn M, Glassroth J. Pulmonary complications of amiodarone toxicity. *Progr Cardiovasc Dis* 1989;**31**:447–53.

116. van Zandwijk N, Darmanata JI, Düren DR, *et al*. Amiodarone pneumonitis. *Eur Respir J* 1983;**64**:313–17.

117. Veltri EP, Reid PR. Amiodarone pulmonary toxicity: early changes in pulmonary function tests during amiodarone rechallenge. *J Am Coll Cardiol* 1985;**6**:802–5.

118. Jarand J, Lee A, Leigh R. Amiodaronoma: an unusual form of amiodarone-induced pulmonary toxicity. *Can Med Assoc J* 2007;**176**:1411–13.

119. Standertskjöld-Nordenstam CG, Wandtke JC, Hood WJ, Zugibe FT, Butler L. Amiodarone pulmonary toxicity: chest radiography and CT in asymptomatic patients. *Chest* 1985;**88**:143–5.

120. Kuhlman JE, Scatarige JC, Fishman EK, Siegelman SS. CT demonstration of high attenuation pleural-parenchymal lesions due to amiodarone therapy. *J Comput Assist Tomogr* 1987;**11**:160–2.

121. Rodriguez-Garcia JL, Garcia-Nieto JC, Ballesta F, *et al*. Pulmonary mass and multiple lung nodules mimicking a lung neoplasm as amiodarone-induced pulmonary toxicity. *Eur J Intern Med* 2001;**12**:372–6.

122. Chouri N, Langin T, Lantuejoul S, Coulomb M, Brambilla C. Pulmonary nodules with the CT halo sign. *Respiration* 2002;**69**:103–6.

123. Morera J, Vidal R, Morell F, *et al*. Amiodarone and pulmonary fibrosis. *Eur J Clin Pharmacol* 1983;**24**:591–3.

124. Barroso E, Hernandez L, Gil J, *et al*. Idiopathic organizing pneumonia: a relapsing disease. 19 years of experience in a hospital setting. *Respiration* 2007;**74**:624–31.

125. Anton Aranda E, Basanez RA, Jimenez YL. Bronchiolitis obliterans organising pneumonia secondary to amiodarone treatment. *Neth J Med* 1998;**53**:109–12.

126. Carrio ML, Fortia C, Javierre C, *et al*. Is short-term amiodarone use post cardiac surgery a cause of acute respiratory failure? *J Cardiovasc Surg* (Torino) 2007; **48**:509–12.

127. Donaldson L, Grant IS, Naysmith MR, Thomas JS. Acute amiodarone-induced lung toxicity. *Intens Care Med* 1998;**24**:626–30.

128. Greenspon AJ, Kidwell GA, Hurley W, Mannion J. Amiodarone-related postoperative adult respiratory distress syndrome. *Circulation* 1991;**84**(Suppl III):407–15.

129. Ren H, Kuhlman JE, Hruban RH, *et al*. CT-pathology correlation of amiodarone lung. *J Comput Assist Tomogr* 1990;**14**:760–5.

130. van Rooij WJ, van der Meer SC, van Royen EA, van Zandwijk N, Darmanata JI. Pulmonary gallium-67 uptake in amiodarone pneumonitis. *J Nucl Med* 1984;**25**:211–13.

131. Ravishankar R, Samuels LE, Kaufman MS, *et al*. Amiodarone-associated hemoptysis. *Am J Med Sci* 1998;**316**:390–2.

132. Iskander S, Raible DG, Brozena SC, *et al*. Acute alveolar hemorrhage and orthodeoxia induced by intravenous amiodarone. *Catheter Cardiovasc Interv* 1999;**47**:61–3.

133. Singh S. Amiodarone-induced pulmonary hemorrhage. *South Med J* 2006;**99**:329–30.

134. Gonzalez-Rothi RJ, Hannan SE, Hood I, Franzini DA. Amiodarone pulmonary toxicity presenting as bilateral exudative pleural effusion. *Chest* 1987;**92**:179–82.

135. Stein B, Zaatari GS, Pine JR. Amiodarone pulmonary toxicity: clinical, cytologic and ultrastructural findings. *Acta Cytol* 1987;**31**:357–61.

136. Akoun G, Milleron BJ, Badaro DM, *et al*. Pleural T-lymphocyte subsets in amiodarone-associated pleuropneumonitis. *Chest* 1989;**95**:596–7.

137. Gotsman I, Fridlender Z, Meirovitz A, Dratva D, Muszkat M. The evaluation of pleural effusions in patients with heart failure. *Am J Med* 2001;**111**:375–8.

138. Susano R, Caminal L, Ramos D, Diaz B. Amiodarone induced lupus. *Ann Rheum Dis* 1999;**58**:655–6.

139. Sheikhzadeh A, Schafer U, Schnabel A. Drug-induced lupus erythematosus by amiodarone. *Arch Intern Med* 2002;**162**: 834–6.

140. Kuhlman JE, Teigen C, Ren H, *et al*. Amiodarone pulmonary toxicity: CT findings in symptomatic patients. *Radiology* 1990;**177**:121–5.

141. Burches E, Garcia-Verdegay F, Ferrer M, Pelaez A. Amiodarone-induced angioedema. *Allergy* 2000;**55**:1199–200.

23

Eosinophilic pneumonia induced by drugs

JAMES N ALLEN

INTRODUCTION

The eosinophilic lung diseases are a diverse group of disorders that are accompanied by eosinophilic infiltration of the airways, alveoli, or interstitium of the lung. There are a large number of lung diseases in which the lung contains abnormally increased eosinophils and it is important to exclude these diseases during the clinical evaluation of suspected eosinophilic lung disease. The most common eosinophilic lung diseases are listed in Box 23.1.

Among the most common causes of eosinophilic lung disease are drugs, including prescribed medications, over-the-counter medications and illicit drugs. Many drugs can cause eosinophilic lung disease and it can often be difficult for the clinician to determine whether a given drug is responsible for a patient's symptoms. This chapter will review the mechanism of drug-induced eosinophilia, medications commonly causing drug-induced eosinophilic lung disease, the spectrum of clinical presentation, the management of eosinophilic lung disease, and long-term complications of drug-induced eosinophilic lung disease.

Box 23.1   Eosinophilic lung diseases

- Simple pulmonary eosinophilia
- Chronic eosinophilic pneumonia
- Acute eosinophilic pneumonia
- Churg–Strauss syndrome
- Idiopathic hypereosinophilic syndrome
- Allergic bronchopulmonary aspergillosis
- Parasite-induced pulmonary eosinophilia
- Fungus-induced pulmonary eosinophilia
- Drug-induced pulmonary eosinophilia

DEFINITIONS

Eosinophilic lung disease can be defined as occurring in one of three ways:

- increased blood eosinophils in combination with X-ray abnormalities;
- increased bronchoalveolar lavage (BAL) eosinophils;
- increased tissue eosinophils observed on histology sections.

Eosinophilic lung disease was historically defined as 'pulmonary infiltrates with eosinophilia' (PIE) syndromes based on the finding of increased blood eosinophils with infiltrates on chest radiographs. Blood eosinophilia is generally defined not in terms of the percentage of eosinophils on a white blood cell count but as an increase in the absolute number of eosinophils, generally > 350 per cubic millimetre.

The application of BAL to the diagnosis of lung disease has resulted in the realization that some lung diseases accompanied by increased eosinophils in BAL fluid are not always accompanied by increased blood eosinophils. BAL mainly samples the alveoli and distal airways that in some diseases can be the main site of eosinophil infiltration. In this regard, BAL fluid in normal subjects contains fewer than 2 per cent eosinophils. More than 5 per cent eosinophils in the BAL is often considered clinically significant.

Identification of increased tissue eosinophils can on occasion first occur during microscopic examination of lung pathological specimens. Although more difficult to express quantitatively, this can be an important means of identifying eosinophil involvement in the interstitium, especially if the alveolar space and the blood are unaccompanied by increased eosinophils.

Drug-induced lung disease can pose a challenge in diagnosis. Many patients presenting with otherwise unexplained pulmonary eosinophilia may have underlying diseases that can be associated with eosinophilia or may be taking more than one medication that can potentially cause eosinophilia. Although the most certain way to determine whether a patient has drug-induced disease is for the eosinophilia to resolve after discontinuing the medication and then recur after re-challenging the patient, this can be risky and should be avoided in most cases. More practically, five criteria for eosinophilic drug-induced lung disease should be met. The person should:

- have no other likely cause of lung disease;
- have symptoms consistent with the effects of the suspected drug;
- have a time course compatible with drug-induced lung disease;
- have tissue or bronchoalveolar lavage finding compatible with drug-induced lung disease;
- improve after the drug is discontinued.

Patients meeting all five of these criteria can be considered as having definite drug-induced eosinophilic lung disease, those with four considered as probable, and those with three considered as suspected.[1]

Sometimes, entire classes of drugs should be considered as candidates for drug-induced eosinophilic lung disease. For example, many penicillins, cephalosporins, angiotensin-converting enzyme inhibitors, statins, sulfa-containing antibiotics, and selective serotonin reuptake inhibitors have been associated with pulmonary eosinophilia. Therefore, a newly released drug from one of these classes should also be a suspect even if no reports yet exist associating the new drug with lung disease.

Lymphocyte transformation tests have been used to determine whether a given drug is responsible for eosinophilia. Although these studies may help support the diagnosis, they can be negative even when there is strong clinical evidence that a drug is responsible for eosinophilic lung disease – and so their positive and negative predictive values are uncertain.[2] Moreover, these tests are not widely available for routine clinical use and are largely relegated to experimental use.

In summary, the diagnosis is still best established based on the finding of increased blood or lung eosinophils and a clinical history of drug exposure in the context of the current literature regarding drugs known to cause eosinophilic lung disease.

INCIDENCE AND EPIDEMIOLOGY

The incidence of drug-induced eosinophilic lung disease is unknown. It is likely that many patients with allergic drug reactions who present with concurrent rash or peripheral blood eosinophilia are never recognized as having lung involvement, especially if the offending drug is discontinued on the basis of the skin or blood findings before the lung abnormalities reach the symptomatic threshold or before chest radiographs are obtained.

The list of drugs reported to cause eosinophilic lung disease continues to grow as new medications are released to market and as old medications are used in larger numbers of patients. For any given drug, there are varying degrees of certainty that it can be responsible for pulmonary eosinophilia. For example, a single case report of a drug in the medical literature provides less certainty than multiple case series of eosinophilic

lung disease associated with the same drug. Boxes 23.2 to 23.5 list drugs reported to cause eosinophilic lung disease.

One of the best resources for clinicians is the website www.pneumotox.com, which is maintained by Drs Pascal Foucher and Philippe Camus, and the Groupe d'Etudes de la Pathologie Pulmonaire Iatrogène. This site grades evidence that a given drug is responsible for a specific lung disease in four categories:

- 1–5 isolated case reports;
- about 10 cases;
- 20–100 cases;
- more than 100 cases.

It should be remembered that a drug used only very rarely (e.g a chemotherapy medication) has fewer opportunities to cause iatrogenic effects than a drug used frequently (e.g. an antibiotic). At the time of preparation of this chapter, www.pneumotox.com listed 118 drugs reported to cause pulmonary infiltrates with eosinophilia.

Box 23.2 Drugs commonly reported to cause pulmonary eosinophilia

- Amiodarone
- Angiotensin-converting enzyme (ACE) inhibitors
- Beta-blockers
- Bleomycin
- Captopril
- Gold salts
- Iodine, radiographic contrast media
- L-tryptophan
- Methotrexate
- Nitrofurantoin
- Phenytoin

Box 23.3 Drugs occasionally reported to cause pulmonary eosinophilia

- Acetyl salicylic acid
- Carbamazepine
- Fenfluromine/dexfenfluramine
- G(M)-CSF
- Hydrochlorothiazide
- Minocycline
- Nilutamide
- Non-steroidal anti-inflammatory drugs
- Penicillamine
- Propylthiouracil
- Sulfa-containing antibiotics
- Sulfasalazine

Box 23.4　Drugs rarely reported to cause pulmonary eosinophilia

- Beclomethasone
- Chloroquine
- Cocaine
- Cotrimoxazole
- Dapsone
- Desipramine
- Diclofenac
- Erythromycin
- Heroin

- Imipramine
- Interleukin-2
- Isoniazid
- Isotretinoin
- Mesalamine
- Methylphenidate
- Paclitaxel
- Para-(4)-aminosalicylic acid
- Penicillins

- Phenylbutazone
- Procarbazine
- Propranolol
- Ranitidine
- Simvastin
- Streptomycin
- Trimipramine
- Zafirlukast

Box 23.5　Drugs very rarely reported to cause pulmonary eosinophilia

- Amitriptyline
- Ampicillin
- wAzithromycin
- Bicalutamide
- Bucillamine
- Camptothecin
- Cephalosporins
- Chlorambucil
- Chlorpromazine
- Chlorpropamide
- Cladribine
- Clarithromycin
- Clindamycin
- Clofibrate
- Cromolyn
- Cyproterone acetate
- Daptomycin
- Diflunisal
- Duloxetine
- Enalapril
- Ethambutol
- Febarbamate
- Fenbufen
- Fenoprofen

- Fludarabine
- Fosinopril
- Furazolidone
- Glafenine
- Ibuprofen
- Ifenprodil
- Indomethacin
- Infliximab
- Interferon alpha
- Labetalol
- Leflunomide
- Levofloxacin
- Lovastatin
- Loxoprofen
- Maprotiline
- Meloxicam
- Mephenesin
- Metronidazole
- Montelukast
- Nalidixic acid
- Naproxen
- Niflumic acid
- Nimesulide
- Niridazol

- Nomifensine
- Oxaliplatin
- Pentamidine
- Perindopril
- Piroxicam
- Pranoprofen
- Progesterone
- Pyrimethamine
- Rifampicin
- Roxithromycin
- Serrapeptase
- Sertraline
- Sulindac
- Tamoxifen
- Tenidap
- Tetracycline
- Tiaprofenic acid
- Ticlopidine
- Tolazamide
- Tolfenamic acid
- Tosufloxacin
- Trazodone
- Troleandomycin
- Venlafaxine

AETIOLOGY

In most cases of drug-induced eosinophilia, the pivotal cell involved in the initiation of the host response is the T-lymphocyte, specifically the Th2 (T-helper type 2) lymphocyte. Interleukin-5 (IL-5) produced by Th2 cells is the main cytokine regulating eosinophil production by the bone marrow (Fig. 23.1). Other cytokines promoting eosinophil production include IL-3 and granulocyte-macrophage colony stimulating factor.[3]

Once produced, eosinophil release into the blood stream is regulated by cytokines such as IL-5 and eotaxin. Most eosinophils rapidly migrate to the tissues, especially the lung, gastrointestinal tract and skin. Eosinophils can preferentially accumulate in the lung by local production of eosinophil chemotactic factors such as IL-5 produced by pulmonary lymphocytes and eotaxin produced by alveolar macrophages, pulmonary endothelial cells, airway smooth muscle cells and alveolar epithelial cells.[4]

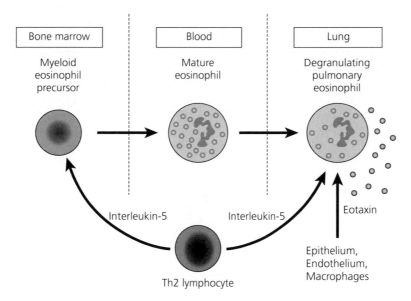

Fig. 23.1 Regulation of eosinophil production, chemotaxis and degranulation in the lung.

The eosinophil tissue/blood ratio is approximately 100 in most normal individuals. The eosinophil contains granules composed of toxic chemicals used for host defence, especially against parasites. Release of these granule contents in response to a drug can result in inadvertent tissue injury and the development of the clinical abnormalities associated with eosinophilic lung disease.[1]

Depending on the disease process, eosinophils in the lung may be predominately located in the airway, alveoli, interstitium or blood vessels. As tissue eosinophils age and degranulate within the lung, they can develop more than the usual two nuclear lobes and may lack cytoplasmic granules. Thus, eosinophils seen in BAL fluid or in histology specimens can sometimes be mistaken for neutrophils.

In drug-induced eosinophilic lung disease, antigen-presenting cells, such as alveolar macrophages and dendritic cells, consume drugs and then present them as antigens via their MHC receptors. This antigen–receptor complex is recognized by the Th2 lymphocyte T-cell receptor in conjunction with the lymphocyte CD4 receptor. This in turn activates the Th2 lymphocyte to release IL-5 and results in eosinophil production, chemotaxis to the lung, and degranulation (Fig. 23.2). The relative abundance of macrophages and dendritic cells in the lung could result in relatively large local production of IL-5 and eotaxin with the consequence of preferential accumulation of eosinophils within the lung.

CLINICAL PRESENTATIONS

Because drug-induced eosinophilic lung disease can mimic several other causes of the disease, brief mention of some of these is warranted. A more detailed discussion of these diseases may be found in other publications.[5]

Simple pulmonary eosinophilia, or Löffler's syndrome, causes minimal or no pulmonary symptoms and transient, often migratory, radiographic infiltrates. Chronic eosinophilic pneumonia presents with weeks or months of progressive

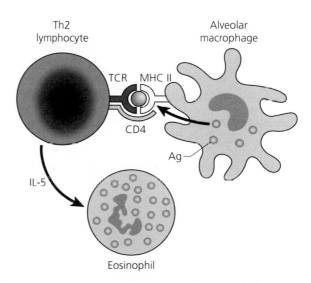

Fig. 23.2 Alveolar macrophages present antigens (Ag) to Th2 lymphocytes via their major histocompatibility complexes (MHC II). Lymphocyte T-cell receptors (TCR) in conjunction with CD4 receptors then recognize the antigen and stimulate the lymphocyte to produce interleukin-5 (IL-5) which activates eosinophils.

respiratory symptoms, diffuse or peripheral radiographic infiltrates, and increased blood eosinophils.[6] Acute eosinophilic pneumonia presents with sudden onset of respiratory failure and usually lacks blood eosinophilia.[7] Churg–Strauss syndrome presents with asthma, blood eosinophilia and granulomatous vasculitis involving the lung and other organs.[8] Idiopathic hypereosinophilic syndrome presents with profound blood eosinophilia (generally greater than 1500/mm^3) for more than 6 months; many organs, including the lung, may become infiltrated with eosinophils.[9] Other diseases that can present as eosinophilic lung disease include fungal infections (especially aspergillosis and coccidioidomycosis), parasitic infections, interstitial lung diseases, allergic bronchopulmonary aspergillosis, bronchocentric granulomatosis and malignancy.

Drug-induced eosinophilic lung disease can have several different clinical presentations including simple pulmonary eosinophilia, chronic eosinophilic pneumonia, acute eosinophilic pneumonia, and Churg–Strauss syndrome. Isolated eosinophilic pleural effusions are less common causes of pulmonary manifestations of drug allergy; however, pleural effusions frequently occur as part of a generalized pulmonary reaction that includes parenchymal infiltrates, especially when drug-induced eosinophilic lung disease manifests as an acute eosinophilic pneumonia-like presentation.

Perhaps the most common form of drug-induced eosinophilic lung disease is simple pulmonary eosinophilia or Löffler's syndrome. Patients may be asymptomatic or may have relatively mild cough or dyspnoea. Chest radiographs typically show patchy segmental or subsegmental pulmonary infiltrates that are often migratory over time. Peripheral blood eosinophil counts are usually elevated. Signs and symptoms resolve with discontinuing the drug and typically recur if it is resumed. Treatment with corticosteroids is rarely required. For many patients with simple pulmonary eosinophilia, the cause cannot be identified. The most important cause of simple pulmonary eosinophilia worldwide is parasitic infection. In Löffler's original series, occult ascaris infection may have been responsible for the majority of the patients.

In some patients, a subacute presentation can occur. These patients may present with weeks or even months of cough and dyspnoea. In contrast to patients with idiopathic chronic eosinophilic pneumonia who have symptoms limited to the lung, drug-induced eosinophilic lung disease is often accompanied by systemic symptoms such as rash or fever. Peripheral blood eosinophilia is usually present. Infiltrates may be patchy or confluent, unilateral or bilateral. The main diagnostic consideration is idiopathic chronic eosinophilic pneumonia. Like chronic eosinophilic pneumonia, patients presenting with this form of drug-induced eosinophilic lung disease will improve promptly with corticosteroids. Unlike idiopathic chronic eosinophilic pneumonia, drug-induced eosinophilic lung disease can also resolve with discontinuation of the offending medication without the use of steroids. Patients with idiopathic chronic eosinophilic pneumonia require months or years of steroids or relapse will occur. On the other hand, patients with drug-induced lung disease will not relapse after a brief course of steroids if the offending medication is discontinued.

Many drugs have been reported to present as acute eosinophilic pneumonia. Minocycline has been particularly frequently reported.[10–12] It is likely that drug allergy is just one kind of antigenic stimulation that culminates in acute eosinophilic pneumonia. For example, new exposure to tobacco smoke has also been linked to acute eosinophilic pneumonia.[13] Patients present with just a few days of severe dyspnoea and hypoxaemia. Other symptoms such as fever, myalgias and pleuritic chest pain are frequent. Patients with idiopathic acute eosinophilic pneumonia have normal or only slightly elevated blood eosinophil counts, whereas those with drug-induced acute eosinophilic pneumonia often have moderately or highly elevated blood eosinophil counts. The finding of an eosinophil count in excess of 1000 per cubic millimetre should heighten the suspicion of drug-induced as opposed to idiopathic acute eosinophilic pneumonia. Chest radiograph findings in idiopathic and drug-induced acute eosinophilic

pneumonia are similar.[14] The earliest findings are patchy interstitial infiltrates with Kerley B lines which later progress to diffuse alveolar infiltrates with small or moderate sized pleural effusions. Patients can improve simply by discontinuing the offending drug, but steroids are usually given to accelerate resolution of hypoxaemia and dyspnoea. Patients with idiopathic acute eosinophilic pneumonia resolve completely following a 4- to 6-week course of corticosteroids; however, those with drug-induced acute eosinophilic pneumonia will recur if the offending drug is re-administered.[10]

Churg–Strauss syndrome has been linked to leukotriene inhibitors including zafirlukast[15,16] and montelukast.[17,18] The clinical presentation of leukotriene inhibitor-associated Churg–Strauss syndrome cannot be distinguished from idiopathic Churg–Strauss syndrome. Patients present with antecedent asthma and allergic rhinitis followed later by high blood eosinophil counts and vasculitis involving many organs including the skin, peripheral nerves, heart, kidney and gastrointestinal system. It remains uncertain whether leukotriene inhibitors actually cause Churg–Strauss syndrome or are just associated with it.[19,20] These drugs are often used to treat asthma symptoms that may be the initial symptoms of Churg–Strauss syndrome. Furthermore, their use as steroid-sparing agents in patients with asthma may merely permit reduction of the steroid dose and development of symptoms of occult Churg–Strauss syndrome previously controlled by the steroids.[21] However, there is some evidence that the relationship between leukotriene inhibitors and Churg–Strauss syndrome is in fact causal. Treatment requires long-term corticosteroids plus cytotoxic medications such as cyclophosphamide.

Drug-induced lung disease can also present as the DRESS (drug rash with eosinophilia and systemic symptoms) syndrome. In this setting, pulmonary eosinophilia can be a component of the systemic eosinophilia. The condition can resemble a systemic vasculitis and has been associated with a large number of medications including minocycline[22] and sulfasalazine.[23]

The history and physical examination remains the single most important tool in diagnosis. Patients must be questioned carefully as they may not be forthcoming about non-prescription drugs, herbal preparations and street drugs. Many over-the-counter medications and herbal preparations have multiple ingredients and identification of the individual components can be difficult. Patients may be taking more than one drug known to cause eosinophilic lung disease that can make incrimination of a single drug difficult. In outpatients, the temporal association of eosinophilia and pulmonary abnormalities with initiation of a given medication can be an important clue. Inpatients are frequently started on many medications over a short period of time, eliminating these clues.

Often, drug-induced lung disease can be diagnosed only after excluding other causes. Thus, a history of asthma should raise the possibility of allergic bronchopulmonary aspergillosis, bronchocentric granulomatosis or Churg–Strauss syndrome.

Travel history may suggest parasitic infestation. Travel to relevant areas of the world should raise the possibility of filarial infection (tropical pulmonary eosinophilia), *Schistosoma* or *Paragonimus westermani*. In the United States, *Ascaris*, *Strongyloides*, *Toxacara* and *Ancylostoma* are the most common culprits.[4] *Strongyloides* is a human intestinal worm that can be particularly difficult to diagnose since it can exist for years or decades before causing respiratory symptoms, pulmonary infiltrates and blood eosinophilia. It can be diagnosed by finding larva in sputum, BAL fluid or stool. However, stool examination can be negative and so diagnosis often requires the use of anti-*Strongyloides* antibodies in the blood. *Strongyloides* is particularly important to identify and distinguish from drug-induced lung disease because treatment with steroids can worsen the condition and precipitate the *Strongyloides* hyperinfestation syndrome. *Ascaris* infection causes pulmonary infiltrates and high blood eosinophil counts. Skin rash is common. The stool examination is usually negative at the time of pulmonary symptoms and does not reveal ova until the worms mature in the intestine approximately 8 weeks later. *Toxacara canis* ('visceral larval migrans') is the dog roundworm and adult worms do not develop in humans. It causes marked blood eosinophilia with pulmonary infiltrates as the larvae migrate through the internal organs. Stool examinations are negative and definitive diagnosis usually requires serology. *Ancylostoma brasiliense* is the dog hookworm and in humans causes 'cutaneous larval migrans' also known as 'creeping eruption'. Adult worms do not develop in humans so stool examinations are negative. The diagnosis is established by finding the characteristic serpiginous, red rash on the extremities in association with blood eosinophilia and, occasionally, pulmonary infiltrates.

Two fungal infections can present as eosinophilic lung disease and these are particularly important to recognize since administration of steroids for presumptive drug-induced lung disease can result in progressive infection and even death. Coccidioidomycosis can occur after travel to the southwestern United States and can result in pulmonary infiltrates with peripheral blood eosinophilia.[24] The diagnosis can be suggested by the presence of antibodies against *Coccidioides immitis* or by growth of *Coccidioides* in sputum or BAL cultures. Aspergillosis can occur after exposure to compost or decomposing organic materials[25] and from dust created by building renovation or demolition.[26] It can be more difficult to identify than *Coccidioides immitis* as it tends to be a more tissue-invasive fungus and does not always grow in sputum or BAL cultures. Furthermore, many normal people may have airway colonization with *Aspergillus* species and so growth of *Aspergillus* from respiratory specimens does not necessarily indicate true infection. Serological studies can be suggestive but are rarely diagnostic. Lung biopsy may be required in difficult cases.

Laboratory evaluation is rarely diagnostic for drug-induced eosinophilic lung disease but may be useful to diagnose other diseases in the differential diagnosis. The blood eosinophil count is frequently elevated in drug-induced eosinophilic lung disease but is also elevated in many other forms of the disease. The IgE level is non-specific, but a markedly elevated IgE (greater than 2000 ng/mL) is mainly seen with allergic bronchopulmonary aspergillosis and some cases of Churg–Strauss syndrome. Chest X-ray and CT findings are not specific. In a study of 11 patients with eosinophilic lung disease undergoing thin-section CT, radiologists correctly diagnosed only 27 per cent of the 11 with drug-induced lung disease.[27] Radiographs are most useful to diagnose other lung diseases that may be confused with drug-induced lung disease. For example, bronchiectasis in the setting of increased blood eosinophils suggests allergic bronchopulmonary aspergillosis.

Bronchoalveolar lavage is often performed to exclude infection or other lung diseases. Although bacterial pneumonias are not associated with increases in either blood or BAL eosinophil counts, such increases can be seen in the setting of pneumonia if the patient has a coexistent drug allergy, for example to an antibiotic. In the intensive-care unit, the differentiation between drug-induced lung disease and nosocomial pneumonia can be particularly vexing since allergy to drugs prescribed in the unit can develop coincident with pulmonary infiltrates for a variety of other reasons, including pneumonia. BAL can also be useful to help exclude fungal or parasitic infection as a cause of pulmonary eosinophilia.

BAL in drug-induced eosinophilic lung disease demonstrates elevated percentages of eosinophils, usually in association with elevated percentages of lymphocytes. In one study of 19 patients with drug-induced lung disease, increased lymphocytes were the most common finding in 95 per cent of patients but increased eosinophils were noted in 42 per cent.[28] On the other hand, drug allergy is a relatively common cause of increased BAL eosinophils. In a series of 48 patients with more than 5 per cent BAL eosinophils, drugs were responsible in 12 per cent of patients.[29]

Lung biopsy is not required to diagnose drug-induced eosinophilic lung disease. However, biopsy may be required to diagnose other lung diseases such as Churg–Strauss syndrome, *Aspergillus* infection or interstitial lung disease. When performed in drug-induced eosinophilic lung disease, lung biopsy is non-specific and typically shows lymphocyte and eosinophil infiltration in the interstitium and alveoli. In some drug exposures (e.g. L-tryptophan) vasculitis may be seen.[30]

TWO ILLUSTRATIVE AND CAUTIONARY EVENTS

Two epidemics of drug-induced eosinophilic lung disease warrant special mention because of their historical significance: the eosinophilia–myalgia syndrome and the 'toxic oil' syndrome.

Eosinophilia–myalgia syndrome occurred in the 1980s and was caused by a contaminant(s) found in L-tryptophan manufactured by a single company.[31,32] At the time, L-tryptophan was commonly prescribed for insomnia and depression; it was also available in the United States without prescription and was often sold in health-food stores. In one series, approximately half of persons ingesting the contaminated drug developed acute peripheral blood eosinophilia accompanied by severe myalgias and multiple organ involvement.[33] Respiratory findings occurred in more than 50 per cent of patients[34] and included chest X-ray infiltrates, pleural effusions, dyspnoea, cough, pulmonary hypertension and respiratory muscle weakness. The IgE and creatine kinase levels were normal. Pulmonary function tests usually revealed a reduced diffusing capacity, often with restriction. Lung histology demonstrated interstitial infiltration by lymphocytes and eosinophils, alveolar eosinophils, and small to medium vessel vasculitis.[35] Although pulmonary involvement was generally self-limited, patients frequently had lasting abnormalities of the musculoskeletal system, skin and nervous system.

The 'toxic oil' syndrome occurred in 1981–82 when about 20 000 cases of lung disease accompanied by peripheral blood eosinophilia developed in Spain.[36,37] This epidemic was ultimately traced to rapeseed oil contaminated with oleoanilide fraudulently marketed as olive oil – hence the name of the syndrome. Symptoms typically included fever, respiratory distress, nausea, vomiting, pruritus, abdominal pain, headache and cough. Signs included rash, pulmonary crackles, generalized lymphadenopathy and hepatosplenomegaly. Chest X-rays showed interstitial and/or alveolar infiltrates, often with Kerley B lines. Corticosteroids were usually effective in the acute stage of the disease but not the later stages. Pulmonary fibrosis and pulmonary hypertension were common late sequelae; more than 300 fatalities were recorded.

Although the causative agents for eosinophilia–myalgia syndrome and the 'toxic oil' syndrome are no longer commercially available, these two epidemics illustrate the speed and extent to which contaminants in medications or foods can result in epidemics of eosinophilic lung disease. It is likely that other large- or small-scale epidemics will occur in the future as new medications, chemicals and food additives are produced.

TREATMENT OR MANAGEMENT

The primary treatment of drug-induced lung disease is removal of the offending drug. In many cases, particularly those resulting in a Löffler's-like condition, drug discontinuation is all that is required. One of the challenges is to determine whether a specific drug is, in fact, responsible for the eosinophilic process, especially in patients taking more than one candidate drug. In patients with mild symptoms, the clinician has the luxury of sequentially stopping different drugs to determine which specific medication is responsible. In this setting, it is reasonable to start with the medication that has been most commonly reported to cause eosinophilic lung disease (see Boxes 23.2 to 23.5) or the medication that was most recently started.

In patients with more severe symptoms, it may be necessary to discontinue all medications that are known to cause eosinophilic lung disease as it can be too risky to continue any one that could cause the patient to deteriorate. After clinical resolution, if there is a compelling clinical reason to resume some of these medications, then sequential reinstitution of the least likely medications to have caused the drug reaction can be attempted under careful clinical supervision, while following blood eosinophil counts, serial chest X-rays or other markers of drug reaction.

In more severe or prolonged cases, simply discontinuing the drug may be insufficient and the addition of oral or parenteral corticosteroids may be necessary. In patients who are critically ill – for example with an acute eosinophilic pneumonia-like syndrome – and in patients who have vomiting, intravenous corticosteroids may be necessary. Although some patients with idiopathic acute eosinophilic pneumonia (and presumably drug-induced acute eosinophilic pneumonia) will recover without treatment, deaths have been reported, so corticosteroids are generally necessary. The optimal dose is uncertain, but patients with idiopathic acute eosinophilic pneumonia are often treated with 1–2 mg/kg of methylprednisolone every 6–12 hours. In patients with less severe reactions, doses of 0.25–1 mg/kg of prednisone (or equivalent oral corticosteroid) are generally effective. Fortunately, patients with drug-induced eosinophilic lung disease recover promptly and a short course of corticosteroids is usually sufficient. Once a causative drug is identified, patients should be warned to not use the drug or other drugs in the same category in the future.

COMPLICATIONS

Long-term effects of drug-induced eosinophilic lung disease are variable. Fortunately, most patients recover completely without any residual pulmonary impairment. Patients presenting with an acute eosinophilic pneumonia-like syndrome can progress to respiratory failure and even death if the offending drug is not discontinued. Patients with chronic eosinophilic pneumonia-like presentation can have persistent interstitial lung disease following discontinuation of the offending drug, especially if the pulmonary process has been persistent for a long time. In this setting, use of supplemental oxygen and pulmonary rehabilitation can be useful. In patients with the Spanish 'toxic oil' syndrome (see above), chronic fibrosis resulted in many patients.

FUTURE DIRECTIONS

New medications are brought to the clinical market each year, not only prescribed medications but also over-the-counter and herbal preparations. As new drugs become available, there will inevitably be more medications identified as causing drug-induced eosinophilic lung disease. There is a great need to develop better tests to determine whether a specific drug is responsible, and future research efforts in the immunological mechanism(s) of drug-induced lung disease should permit new and better clinical laboratory techniques in order to more rapidly diagnose patients.

KEY POINTS

- Many drugs are capable of causing eosinophilic lung disease. A high index of suspicion is necessary to recognize the offending drug. There is no easy definitive diagnostic test. Prescribed, over-the-counter and illegal drugs should all be considered.

- Common patterns of clinical presentation include simple pulmonary eosinophilia, chronic eosinophilic pneumonia and acute eosinophilic pneumonia.

- In difficult cases, bronchoalveolar lavage may be required for diagnosis. Surgical lung biopsy is rarely required.

- Mild forms of drug-induced pulmonary eosinophilia may resolve simply by discontinuing the offending medication, while more severe cases may require systemic corticosteroids.

- Other causes of pulmonary eosinophilia must be excluded, especially parasitic and fungal infections.

REFERENCES

1. Allen JN. Drug-induced eosinophilic lung disease. *Clin Chest Med* 2004;**25**:77–88.
2. Primeau MN, Adkinson NF. Recent advances in the diagnosis of drug allergy. *Curr Opin Allergy Clin Immunol* 2001;**1**:337–41.
3. Allen JN, Davis WB. Eosinophils and the lung. In: Crystal RG *et al.* (eds) *The Lung: Scientific Foundations.* New York: Raven Press, 1996.
4. Conroy DM, Williams TJ. Eotaxin and the attraction of eosinophils to the asthmatic lung. *Respir Res* 2001;**2**:150–6.
5. Allen JN, Davis WB. State of the art: the eosinophilic lung diseases. *Am J Respir Crit Care Med* 1994;**150**:1423–38.
6. Marchand E, Reynaud-Gaubert M, Lauque D, *et al.* Idiopathic chronic eosinophilic pneumonia: a clinical and follow up study of 62 cases. *Medicine* 1998;**77**:299–312.
7. Pope-Harman AL, Davis WB, Christoforidis AJ, Allen ED, Allen JN. Acute eosinophilic pneumonia: a summary of fifteen cases and review of the literature. *Medicine* 1966;**75**:334–42.
8. Guillevin L, Cohen P, Gayraud M, *et al.* Churg–Strauss syndrome: clinical study and long-term follow-up of 96 patients. *Medicine* 1999;**78**:26–37.
9. Winn RE, Kollef MH, Meyer JI. Pulmonary involvement in the hypereosinophilic syndrome. *Chest* 1994;**105**:656–60.
10. Toyoshima M, Sato A, Hayakawa H, *et al.* A clinical study of minocycline-induced pneumonitis. *Intern Med* 1996;**35**:176–9.
11. Dussopt C, Mornex JF, Cordier JF, Brune J. Acute eosinophilic lung after a course of minocycline. *Rev Mal Respir* 1994;**11**:67–70.
12. Liegeon MN, De Blay F, Jaeger A, Pauli G. A cause of respiratory distress: eosinophilic pneumopathy due to minocycline. *Rev Mal Respir* 1996;**13**:517–19.
13. Shintani H, Fujimura M, Ishiura Y, Noto M. A case of cigarette smoking-induced acute eosinophilic pneumonia showing tolerance. *Chest* 2000;**117**:277–9.
14. Yokoyama A, Mizushima Y, Suzuki H, *et al.* Acute eosinophilic pneumonia induced by minocycline: prominent Kerley B lines as a feature of positive re-challenge test. *Jpn J Med* 1990;**29**:195–8.
15. Knoell DL, Lucas J, Allen JN. Churg–Strauss syndrome associated with zafirlukast. *Chest* 1998;**114**:332–4.
16. Wechsler ME, Garpestad E, Flier SR, *et al.* Pulmonary infiltrates, eosinophilia, and cardiomyopathy following corticosteroid withdrawal in patients with asthma receiving zafirlukast. *J Am Med Assoc* 1998;**279**:455–7.
17. Guilpain P, Viallard JF, Lagarde P, *et al.* Churg–Strauss syndrome in two patients receiving montelukast. *Rheumatology* 2002;**41**:535–9.
18. Franco J, Artes MJ. Pulmonary eosinophilia associated with montelukast. *Thorax* 1999;**54**:558–60.
19. Harrold LD, Patterson MK, Andrade Se, *et al.* Asthma drug use and the development of Churg–Strauss syndrome. *Pharmacoepidemiol Drug Safety* 2007;**16**:620–6.
20. Nathani N, Little MA, Kunsst H, Wilson D, Thickett DR. Churg–Strauss syndrome and leukotriene antagonist use: a respiratory perspective. *Thorax* 2008;**63**:883–8.
21. Weller PF, Plaut M, Taggart V, Trontell A. The relationship of asthma therapy and Churg–Strauss syndrome: NIH workshop summary report. *J Allergy Clin Immunol* 2001;**108**:175–83.
22. Favrolt N, Bonniaud P, Collet E, *et al.* Severe drug rash with eosinophilia and systemic symptoms after treatment with minocycline. *Rev Mal Respir* 2007;**334**:215–18.
23. Michel F, Navellou JC, Ferraud D, Toussirot E, Wendling D. DRESS syndrome in a patient on sulfasalazine for rheumatoid arthritis. *Joint Bone Spine* 2005;**72**:82–5.
24. Lombard CM, Tazelaar MD, Krasne DL. Pulmonary eosinophilia in coccidioidal infections. *Chest* 1987;**91**:734–6.
25. Ricker DH, Taylor SR, Gartner JC, Kurland G. Fatal pulmonary aspergillosis presenting as acute eosinophilic pneumonia in a previously healthy child. *Chest* 1991;**100**:875–7.
26. Bouza E, Pelaez T, Perez-Molina J, *et al.* Demolition of a hospital building by controlled explosion: the impact on filamentous fungal load in internal and external air. *J Hosp Infect* 2002;**52**:234–42.
27. Johkoh T, Muller NL, Akira M, *et al.* Eosinophilic lung diseases: diagnostic accuracy of thin-section CT in 111 patients. *Radiology* 2000;**216**:773–80.
28. Akoun GM, Cadranel JL, Milleron BJ, D'Ortho MP, Mayaud CM. Bronchoalveolar lavage cell data in 19 patients with drug-associated pneumonitis (except amiodarone). *Chest* 1991;**99**:98–104.
29. Allen JN, Davis WB, Pacht ER. Diagnostic significance of increased bronchoalveolar lavage fluid eosinophils. *Am Rev Respir Dis* 1990;**142**:642–7.
30. Tazelaar HD, Myers JL, Strickler JG, Colby TV, Duffy J. Tryptophan-induced lung disease: an immunophenotypic, immunofluorescent, and electron microscopic study. *Mod Pathol* 1993;**6**:56–60.
31. Kaufman LD, Seidman RJ, Gruber BL. L-tryptophan-associated eosinophilic perimyositis, neuritis, and fasciitis: a clinicopathologic and laboratory study of 25 patients. *Medicine* 1990;**69**:187–99.
32. Philen RM, Hill RH, Flanders WD, *et al.* Tryptophan contaminants associated with eosinophilia–myalgia syndrome. *Am J Epidemiol* 1993;**138**:154–9.
33. Kamb ML, Murphy JJ, Jones JL, *et al.* Eosinophilia–myalgia syndrome in L-tryptophan-exposed patients. *J Am Med Assoc* 1992;**267**:77–82.
34. Swygert LA, Maes EF, Sewell LE, *et al.* Eosinophilia–myalgia syndrome: results of national surveillance. *J Am Med Assoc* 1990;**264**:1698–703.
35. Tazelaar HD, Myers JL, Drage CW, *et al.* Pulmonary disease associated with L-tryptophan-induced eosinophilic myalgia syndrome: clinical and pathologic features. *Chest* 1990;**97**:1032–6.
36. Kilbourne EM, Rigau-Perez JG, Heath CW, *et al.* Clinical epidemiology of toxic oil syndrome: manifestations of a new illness. *New Engl J Med* 1983;**309**:1408–14.
37. Alonso-Ruiz A, Calabozo M, Perez-Ruiz F, Mancebo L. Toxic oil syndrome: a long-term follow-up of a cohort of 332 patients. *Medicine* 1993;**72**:285–95.

Bronchiolitis obliterans organizing pneumonia induced by drugs or radiotherapy

GARY R EPLER

INTRODUCTION

Bronchiolitis obliterans organizing pneumonia (BOOP) is generally idiopathic; however, this lesion may be caused by more than 35 medications and from breast radiotherapy. These drugs include the cancer chemotherapeutic agents, antibiotics, cardiovascular drugs and immunosuppressive agents noted in Table 24.1.[1,2] Unproductive cough is a common early symptom and shortness of breath occurs later. Fever is also often seen early. Eosinophila is unusual. The radiograph shows bilateral patchy infiltrates, and the chest CT shows ground-glass opacities, often peripheral and pleural-based (Fig. 24.1).

As this is an inflammatory process, the prognosis is excellent, with almost all patients responding to withdrawal of the medication with or without a brief course of corticosteroid therapy.

DEFINITION AND OVERALL ASSESSMENT OF BOOP

DEFINITION

BOOP is defined as organized granulation tissue in the distal airways extending into the alveolar ducts and alveoli.[3] Sometimes organizing pneumonia is seen without the organized process in the bronchioles, although a thorough search of a sufficiently large size of tissue usually shows the typical BOOP lesion. The term 'cryptogenic organizing pneumonia' (COP)[4] is sometimes used for the idiopathic BOOP, while BOOP is used for known causes such as the drug-induced lesions, or systemic associated conditions such as connective tissue disorders or after organ transplantation. The term 'BOOP' is preferred because this is a distinct simultaneously involved lesion of distal airways and alveoli with internationally well-established causes, clinical course and outcome.

INCIDENCE AND EPIDEMIOLOGY

Drug-induced BOOP may occur as frequently as 15 per cent among individuals taking high-dose amiodarone, but it is generally rare, occurring in less than 0.1 per cent of patients receiving a medication. The process occurs in men and women equally and at all ages. Thus far, there have been no specific risk factors such as genetic pleomorphisms established for the development of the drug-related BOOP lesion.

CLINICAL PRESENTATION

The clinical presentation is usually subacute with a non-specific cough and gradual onset of shortness of breath. There usually is no sputum production although a small amount of clear sputum may occur. Haemoptysis occurs in rare situations if alveolar haemorrhaging is part of the BOOP process.

Physical examination indicates bilateral end-inspiratory fine crackles, and there is no wheezing, rhonchi or finger clubbing. Pulmonary function tests show a decreased vital capacity and total lung capacity with no airflow obstruction. Virtually all patients have a decreased diffusing capacity.

The chest X-ray shows bilateral patchy infiltrates. Computed tomography is helpful in the diagnosis of drug-induced BOOP because the scans show ground-glass opacities located in all regions, often peripherally and pleural-based. During the early stage of drug-related BOOP, the ground-glass opacities may be subtle and sometimes unilateral. Focal nodular BOOP may occur. Honeycombing and traction bronchiectasis do not occur in drug-induced BOOP, but these findings may occur with agents such as amiodarone that are capable of producing concurrent pulmonary fibrosis. Pleural-based 'triangle' infiltrates are a common and specific chest CT finding for BOOP with the base of the triangle located along the pleura and the tip of the triangle towards the central mediastinum.

Table 24.1 Drug-related bronchiolitis obliterans organizing pneumonia (BOOP)

Name of drug	Strength of association[a]
Antimicrobials	
Minocycline	+++
Nitrofurantoin	+++
Cephalosporin	+++
Amphotericin B	+++
Daptomycin	++
Abacavir	+
Anticancer agents	
Bulsulfan	++
Methotrexate	++
Bleomycin	+++
Doxorubicin	++
Thalidomide	+
Cytosine-arabinoside (ARA-C)	++
Interferon-α, cytarabine ocfosfate	++
Chlorambucil	++
Rituximab	++
Oxaliplatin	+
Cardiovascular agents	
Amiodarone	+++
Acebutolol	++
Anti-inflammatory agents	
Gold	++
Sulfasalazine	+++
Mesalamine (mesalazine)	++
Bucillamine	++
Infliximab	+
Immunosuppressive agents	
Azathioprine	+++
6-mercaptopurine	+
Tacrolimus	+
Sirolimus	+++
Everolimus	+
Anticonvulsants	
Carbamazepine	+++
Phenytoin	++
Miscellaneous agents	
Interferons	
– alpha	+++
– beta	++
– gamma	+
Ticlopidine	++
L-tryptophan	+++
Cocaine (illicit use)	+++
Heroin	+
Fluvastatin	+
Risedronate	+++

[a] *Strength of association:* +minimal; ++moderate; +++strong.

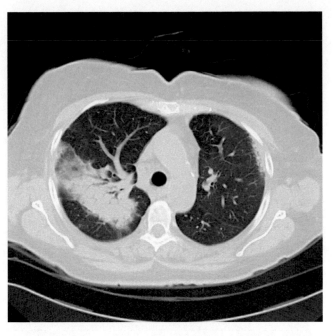

Fig. 24.1 Mid-lung computed tomography scan showing BOOP as a posterior ground-glass opacity and consolidation with an air bronchogram that is triangular, with the apex at the mediastinum and the base at the posterior pleura.

TREATMENT AND MANAGEMENT

The BOOP lesion is inflammation, so discontinuation of the drug often results in resolution. In some situations, such as with amiodarone which has a continual effect for several days to weeks, a course of corticosteroid therapy may be needed. Patients have improved symptoms in a few days and the radiograph begins to improve over a period of weeks. In some situations, such as with amiodarone, symptoms begin to improve during a 2- to 4-week period and the radiographs improve in 3 months and may continue to improve for as long as 18 months.

OUTCOME AND COMPLICATIONS

The outcome of drug-induced BOOP is favourable as this is an inflammatory lesion without residual scarring. However, the BOOP lesion may occur secondarily in association with other primary drug-related lung manifestations such as usual interstitial pneumonia (UIP) or constrictive bronchiolitis with obliteration. In this situation, the BOOP lesion may resolve or respond to corticosteroid therapy, yet the fibrosing UIP process or fibrotic bronchiolar lesion does not respond and persists as the primary cause of pulmonary dysfunction. There have been rare reports of a fatal outcome from drug-related BOOP.

ANTIMICROBIAL DRUGS AND BOOP

Minocycline

There have been two reports of women taking minocycline for acne who developed BOOP. The first was a 20-year-old woman who had taken minocycline for 3 months and developed progressive shortness of breath, a chest radiograph showing alveolar opacities in the upper lobes, and a lung biopsy showing BOOP.[5] The chest radiograph showed partial improvement 3 weeks after drug cessation. An 8-week course of corticosteroid was given for total symptomatic and radiographic resolution.

The second report was of a 39-year-old woman who was treated with minocycline for 5 months and developed a cough and radiographic infiltrates.[6] Bronchoalveolar lavage (BAL) showed increased lymphocytes, and a transbronchial biopsy showed BOOP. The minocycline was stopped, resulting in rapid remission without treatment.

Nitrofurantoin

There have been two reports of three women taking nitrofurantoin who developed BOOP. A 34-year-old woman developed cough and progressive shortness of breath beginning 2 years after she had started taking nitrofurantoin.[7] She had no crackles and the high-resolution chest CT scan showed bilateral patchy ground-glass opacities. The illness responded to corticosteroid therapy.

The second case was a 50-year-old woman who developed cough, shortness of breath, fatigue, anorexia, fever and night sweats. She had bilateral crackles. She had severe pulmonary impairment with a vital capacity of 0.84 litres (22 per cent of predicted); however, she responded to prednisone treatment with resolution of symptoms. The initial high-resolution chest CT scanned showed patchy infiltrates that responded to treatment although there were some residual abnormalities.

The third individual was an 82-year-old woman who had taken nitrofurantoin for 2 years and developed cough with gradual increase in breathlessness.[8] High-resolution chest CT showed areas of patchy ground-glass opacification. Prednisone treatment at 30 mg was given for 6 weeks and decreased during an additional 4-month period. Within 1 month of beginning treatment there was a decrease in cough and breathlessness. The patient returned to her baseline in 8 months and was well 3 years later.

Nitrofurantoin BOOP was reported in a 71-year-old woman who had taken nitrofurantoin for 6 months prior to developing unproductive cough, fever and shortness of breath.[9] This represented the seventh reported case. She had bilateral inspiratory crackles, and the chest CT showed peripheral infiltrates and effusions. A video-assisted thoracoscopy (VATS) biopsy showed proliferating fibromyomatous connective tissue with organization within the air-spaces consistent with BOOP. There was complete resolution of symptoms and lung volumes with prednisone therapy which was discontinued after 16 months; however, the diffusing capacity remained abnormal at 53 per cent predicted.

Cephalosporin

A 61-year-old patient was receiving cephradine for a urinary tract infection and developed a red, pruritic rash on her trunk during the fourth day of therapy.[10] Several days later she developed shortness of breath, an unproductive cough and malaise. One year previously she had been given cephalexin for an infected cat scratch and developed a rash without respiratory symptoms. This time there were bilateral crackles, and a chest radiograph showed patchy segmental infiltrates in the right mid-lung and left lower lung. A lung biopsy showed intra-alveolar myxomatous connective tissue consistent with BOOP. The radiograph cleared after discontinuation of the medication. Six months later she was re-challenged with cephradine 500 mg four times daily. On the third hospital day she developed a lingular infiltrate and progressive cough. The drug was discontinued. Corticosteroid therapy resulted in a prompt disappearance of symptoms and return of normal lung function.

Amphotericin B

A 1990 report described a 52-year-old woman who developed a fever and a right lower lung infiltrate after receiving amphotericin B.[11] The drug was continued and the patient developed shortness of breath and hypoxaemia. The chest radiograph showed bilateral diffuse infiltrates. Transbronchial biopsy showed respiratory bronchioles and alveolar ducts filled with polypoid plugs consistent with BOOP. The medication was stopped. The radiograph cleared. Four months later, amphotericin B was given empirically but was stopped because of shortness of breath.

Daptomycin

There have been two case reports of daptomycin-related BOOP. The first individual was an 84-year-old man who was treated for 4 weeks with intravenous daptomycin for an infected knee prosthesis.[12] The chest CT showed bilateral peripheral ground-glass opacities, one of which was triangular. BOOP was the primary histological pattern in this patient, and the transthoracic needle biopsy showed organizing pneumonia with eosinophils. His symptoms and CT scan improved after cessation of daptomycin.

The second individual was a 54-year-old man who developed shortness of breath 14 days after intravenous daptomycin treatment for a methicillin-resistant *Staphylococcus aureus* (MRSA) infection.[13] The chest CT scan showed bilateral patchy infiltrates and lung biopsy showed organizing pneumonia with eosinophils. After 4 weeks of corticosteroid therapy, the chest radiographic abnormalities resolved.

Abacavir for HIV infection

Abacavir is an HIV nucleoside analogue reverse transcriptase inhibitor and may cause hypersensitivity reactions in 3–5 per cent of patients within 6 weeks of treatment.[14] A case report described a 52-year-old woman infected with HIV who had been started on abacavir 2 weeks previously and developed progressive shortness of breath with cough. The chest radiograph showed bilateral diffuse infiltrates and the biopsy showed patchy areas of organizing pneumonia with alveolar fibrin deposition. The abacavir was discontinued and she showed a remarkable improvement with dramatic resolution of the diffuse bilateral infiltrates.

ANTICANCER AGENTS AND BOOP

Busulfan

For many years busulfan has been a commonly used drug for treatment of lymphoma and leukaemias. There have been reports of busulfan-related BOOP lesions.[4]

Methotrexate

There have been several types of lung lesion associated with methotrexate, including interstitial pneumonia, granulomatous lung lesions, diffuse alveolar damage and BOOP. In a study of 168 patients with rheumatoid arthritis, Carson *et al.* reported that three had a lung biopsy; one showed a bronchiolar lesion.[15] This 74-year-old woman with 7 years of rheumatoid arthritis developed shortness of breath, cough and fever after 1 year of methotrexate therapy. The chest radiograph showed bilateral interstitial infiltrates. The open lung biopsy showed ill-formed interstitial granulomas and bronchiolitis. The patient responded to discontinuation of methotrexate and a brief course of corticosteroid therapy.

In another report, a 67-year-old man with severe rheumatoid arthritis for 8 years had taken 10 mg of methotrexate weekly for 8 months when he developed fatigue, weight loss and an unproductive cough.[16] The chest X-ray showed normal lungs; however, 1 month later, he developed acute-onset right-sided chest pain after an episode of cough. Now, the chest radiograph showed a right upper lobe triangular consolidation in the axillary area with air bronchograms and an 8th posterior rib fracture. Transbronchial biopsy showed interalveolar septa infiltrated by mononuclear cells along with plugs of granulation tissue within bronchioles consistent with BOOP. The rib biopsy showed a rheumatoid nodule. Eight months later the patient had improved symptoms and the lung consolidation had partially resolved while continuing the methotrexate.

Bleomycin

Focal nodular BOOP has been reported in individuals receiving bleomycin.[17] This report described three patients who were 6, 19 and 19 years of age who received less than 200 mg of bleomycin for treatment of osteogenic sarcoma. The radiographs showed nodular ground-glass lesions, one of them pleural-based without major respiratory symptoms. The lung biopsies showed BOOP, and none progressed to diffuse BOOP.

Doxorubicin

Acute cyanotic respiratory failure and near fatal outcome of drug-related BOOP has been reported. A 56-year-old woman was treated with four cycles of doxorubicin for metastatic breast carcinoma, and prior to her fifth 21-day cycle of treatment she developed rapid-onset progressive shortness of breath and cyanosis.[18] Bilateral coarse inspiratory crackles were heard. The chest X-ray showed diffuse reticular opacities, and a high-resolution chest CT scan showed bilateral ground-glass opacities in a background of improved pulmonary metastases. Transbronchial biopsies showed intra-alveolar granulation tissue consistent with BOOP. Treatment with oxygen and dexamethasone resulted in rapid improvement.

Cytosine arabinoside (ARA-C)

There have been two reports of four individuals who developed BOOP from cytosine arabinoside. The first report described three children 6, 9 and 12 years of age with acute leukaemias who developed BOOP between 10 and 20 days after intravenous ARA-C for treatment.[19] The chest CT scans showed patchy infiltrates, two of them with the pleural-based triangular shape consistent with BOOP. The lung biopsies showed intraluminar buds of loose connective tissue consistent with BOOP. All three patients recovered completely.

In another report, a 59-year-old man with chronic myelogenous leukaemia was treated with subcutaneous interferon-α and ARA-C and developed a persistent cough 4 weeks later; the chest radiograph was normal.[20] Four weeks later he developed spiking fever, fatigue, 20-pound weight loss and shortness of breath. The chest radiograph now showed bilateral infiltrates, and the chest CT scan showed bilateral peripheral ground-glass opacities. The lung biopsy showed polypoid connective tissue plugs in airspaces consistent with BOOP. The patient was started on 60 mg of prednisone daily and the hydroxyurea was restarted. He showed dramatic clinical improvement and was discharged home within 2 days. One month later his performance status was normal and the radiographic infiltrates had resolved.

Fludarabine, cytosine arabinoside and mitoxantrone (FLAN)

FLAN-related BOOP has been reported in a 31-year-old man treated for acute myeloid leukaemia.[21] Seven days after cessation of treatment he developed progressive shortness of breath, and high-resolution CT showed ground-glass opacities. Transbronchial biopsy showed patchy interstitial cell inflammation and alveolar loose fibrotic buds. Treatment with intravenous corticosteroid therapy resulted in resolution of symptoms and physiological abnormalities.

Interferon–alpha and cytarabine ocfosfate

There has been a report of fulminant rapidly progressive BOOP from combined chemotherapy that resulted in death. A 65-year-old man was treated for chronic myelogenous leukaemia with hydroxyurea, oral cytarabine ocfosfate (YNK01) and subcutaneous injections of interferon-α.[22] Within 48 hours the patient developed fulminant respiratory failure. The chest radiograph showed bilateral patchy infiltrates, and CT scans showed ground-glass opacity and patchy consolidation with possible cavitation. The patient did not respond to 1 g per day of methylprednisolone and died 5 days after initiation of the chemotherapy regimen. The post-mortem lung tissue showed intraluminal buds of granulation tissue in the distal air-spaces and foci of organizing pneumonia consistent with BOOP. Interferon-α has been associated with BOOP,[20] and as cytarabine ocfosfate is a derivative of cytosine-arabinoside (ARA-C), the cytarabine was the probable causative agent.[19,20]

Chlorambucil

Haemoptysis and a unilateral focal lesion can occur in patients with drug-related BOOP. A 70-year-old man with chronic lymphocytic leukaemia had been treated with oral chlorambucil and methylprednisolone for 7 weeks and developed cough, chest pain and haemoptysis.[23] The chest CT scan showed a peripheral ground-glass opacity in the right mid-lung, and an open lung biopsy showed typical findings of BOOP with tufts of intraluminal organization and polypoid granulation tissue within alveolar ducts and alveoli. The patient had a compete clinical and radiographic recovery after surgery.

Thalidomide

Thalidomide is an angiogenesis inhibitor and decreases tumour necrosis factor. A 58-year-old man with multiple myeloma developed low-grade fever, cough and shortness of breath after two complete cycles of thalidomide at 200 mg per day.[24] He had bilateral crackles and the high-resolution chest CT scan showed diffuse ground-glass opacities peripherally. The thalidomide was discontinued. The bronchoalveolar lavage showed 31 per cent neutrophils, 40 per cent lymphocytes, 8 per cent monocytes and 20 per cent eosinophils. Transbronchial biopsy showed fibroblastic plugs in the alveolar spaces with surrounding interstitial inflammation consistent with BOOP. The chest CT scan 6 weeks later showed significant improvement and he had no recurrence of pulmonary symptoms after a 3-month course of corticosteroid therapy.

Rituximab

Rituximab is a chimeric monoclonal antibody that may induce apoptosis of B-cells and is used in treatment of lymphoma. There have been two reports of rituximab-related BOOP, one in combination with other chemotherapy and one as the sole treating agent.

In the first report, a 52-year-old man was treated with rituximab and CHOP (cyclophosphamide, vincristine, doxorubicin and prednisolone) for an abdominal lymphoma mass.[25] Ongoing medications included carbamazepine, pravastatin and warfarin. The R-CHOP regimen was given at 3-week intervals, and after three courses there was a more than 50 per cent decrease in tumour size with no abnormal lung parenchyma. Pegylated filgrastim, a neutrophil colony-stimulating factor, was given subcutaneously on day 2 of courses three and four. Two weeks after the fourth course, the patient developed acute-onset shortness of breath, pleuritic chest pain, hypoxaemia and fever. The chest X-ray showed bilateral patchy ground-glass opacities. Transbronchial biopsy showed fibroblastic tissue arising within alveoli and pigmented foamy macrophages consistent with BOOP. The radiographic appearance improved with prednisolone at 60 mg daily for 2 weeks. Additional courses of CHOP without rituximab or filgrastim were given for a complete lymphoma remission and no development of respiratory symptoms. The filgrastim may have been related to the BOOP although rituximab is the most likely cause.

In the second report,[26] a 61-year-old man developed gastric and duodenum lymphoma treated with four weekly intravenous doses of rituximab at 375 mg/m^2. Two months later the patient was asymptomatic, but the chest CT showed two new parenchymal opacities in the left lung, one in the left upper lobe and one in the left lower lobe. The patient was treated with an antibiotic and repeat CT scan 1 month later showed a stable left lower lobe nodule, but the left upper lobe nodule had increased to 3 cm and with spiculated margins. A wedge resection of this nodule showed BOOP. One month later the patient developed a cough and shortness of breath. Pulmonary function tests showed a decrease in function compared to preoperative values, and the chest CT scan showed increased size of the left lung nodules and a new right lower lobe nodule. Treatment with prednisone 40 mg daily was begun, and within 4 days the patient reported a dramatic improvement. By 1 month the cough and shortness of breath had resolved. The pulmonary function tests normalized.

Oxaliplatin

A 30-year-old woman with rectal cancer and a single liver metastasis received pelvic radiotherapy and 5-fluorouracil (5-FU).[27] After liver tumour resection, adjuvant therapy with 5-FU, leucovorin and oxaliplatin was given. After 10 weeks and the sixth cycle she developed cough and progressive shortness of breath. The chest CT scan showed ground-glass patchy infiltrates and the biopsy showed peribronchial thickening and alveolar plugs of fibrous tissue. The patient had resolution of shortness of breath and cough after 4 days of corticosteroid treatment and resolution of the chest CT scan 4 weeks later.

CARDIOVASCULAR AGENTS

Amiodarone

Amiodarone-related BOOP is the most commonly reported drug-related BOOP reaction. In a 2004 review of amiodarone pulmonary toxicity, Camus and Rosenow reported the occurrence is from 1 per cent to as high as 15 per cent.[28] There are several types of pulmonary lesions associated with amiodarone toxicity. BOOP is common as it was found in 6 out of 197 patients (3 per cent) in a study in Spain.[29]

Amiodarone and the metabolite desethylamiodarone are cationic amphiphilics that accumulate in tissues including the lungs. These compounds localize in cell lysosomes and block turnover of endogenous phospholipids, which explains the presence of foamy lipid-laden macrophages in lavage or lung tissue.

Clearance of these substances from tissues is very slow as autopsy studies have shown significant amounts of both compounds persisting in the lung 1 year after cessation of treatment.

The pathological findings show BOOP with foamy macrophages. High-resolution chest CT scan shows the typical BOOP pattern of ground-glass opacities, air bronchograms and subpleural opacities. These radiographic findings may wander from one lung region to other lung regions over time. The clinical and radiographic findings resolve from 3 months to as long as 18 months after cessation of amiodarone.[30] A small percentage of patients with amiodarone lung toxicity have subpleural honeycombing and traction bronchiectasis reflecting usual interstitial pneumonia (UIP) and fibrosis, which is a distinctly different lesion from BOOP.[28,30]

The occurrence of amiodarone-related BOOP is higher among patients receiving high doses; however, BOOP also occurs among patients taking 200 mg daily. Ott et al. reviewed eight patients receiving 200 mg daily.[31] The duration of therapy prior to symptoms was as short as 3 months, but most were 2 years and one patient had 5 years of treatment. Shortness of breath and cough were the most common symptoms. The light microscopy finding of vacuolization in pneumocyte cytoplasm and the electron microscopy findings of whorled, lamellar, membranous inclusions in alveolar pneumoctyes are characteristic of amiodarone-related BOOP. These findings were illustrated in the lung biopsy of an 82-year-old man who had taken amiodarone, 400 mg daily, reduced to 200 mg daily during a 15-month course.[32] The amiodarone was discontinued, prednisone was begun, his condition improved and within a few weeks he no longer needed supplemental oxygen. The prednisone was decreased, and he was well 15 months after discharge.

Adenopathy may occur in drug-related BOOP. A 67-year-old man had a pre-cardiac transplant evaluation that showed bilateral small parenchymal opacities in the lower lungs and a nodule in the right lower lobe.[33] A follow-up CT scan 3 months later showed persistence of the small opacities as well as bilateral ground-glass opacities, and an enlarged pre-carinal lymph node. The patient had shortness of breath, fatigue, cough and occasional chest pain that were attributed to his cardiac condition. A CT-guided fine-needle aspiration of the right lower lung nodule showed abundant foamy macrophages admixed with benign bronchial epithelial cells. A subsequent thoracoscopic biopsy showed typical BOOP as well as abundant foamy macrophages. The amiodarone was discontinued, and 2 months later a chest CT showed no new abnormalities and improvement in the lymphadenopathy. Four months later the patient had a successful heart transplantation. One year later a follow-up CT study showed significant improvement in the bilateral pulmonary opacities and lymphadenopathy.

Acebutolol

Acebutolol is a beta-blocker used for treatment of hypertension. There has been one report of acebutolol-related BOOP. A 59-year-old man had been taking acebutolol 200 mg daily for 1 month and developed a fever, cough and increasing shortness of breath.[34] The chest radiograph showed alveolar opacities. Bronchoalveolar lavage showed 15 per cent neutrophils.

A transbronchial biopsy showed organizing pneumonia and obstruction of small air passages by buds of connective tissue consistent with BOOP. After discontinuation of acebutolol, the fever remitted rapidly. Over several months the patient was symptom free and had a normal chest radiograph. The finding of lavage neutrophils is unusual because patients with BOOP usually have a high percentage of lymphocytes, not neutrophils. Whether the presence of neutrophils is suggestive of drug-related BOOP is not known.

ANTI-INFLAMMATORY AGENTS

Gold

Costabel et al. cited seven reports of possible gold-related BOOP.[35] One of these reports included a 49-year-old woman with rheumatoid arthritis who was treated with 2 months of weekly intramuscular gold injections and developed shortness of breath, cough, and a chest radiograph that showed nodular opacities and an air bronchogram.[36] The patient had also received naproxen 500 mg twice daily. The lung biopsy showed bronchiolitis obliterans with small bronchioles occluded by inflammatory cells, and their walls thickened by granulation tissue. The photomicrographs suggested BOOP although the authors note there was destruction of the alveolar structures, a finding usually not seen in BOOP. The patient had a prolonged post-biopsy course requiring assisted ventilation for 4 weeks. The patient gradually responded to corticosteroid and cyclophosphamide therapy. The chest radiograph was normal 18 months later.

Sulfasalazine

Sulfasalazine is commonly used for individuals with inflammatory bowel disease, and there have been several reports of sulfasalazine-related BOOP. In a review of inflammatory bowel disease by Camus et al. a 26-year-old woman with Crohn's disease developed BOOP after sulfasalazine treatment.[37] A previous report was of a 66-year-old man with ulcerative colitis who had taken sulfasalazine for 4 months and developed a cough with a chest X-ray showing patchy upper lung infiltrates.[38] The patient continued with progressive shortness of breath, cough and fever despite discontinuing the sulfasalazine. There was no eosinophilia. A biopsy showed BOOP. The patient was treated with prednisone 60 mg daily and began improving in 1 week. A follow-up chest radiograph showed residual apical fibrotic strands. This BOOP reaction may have been related to the inflammatory bowel disease since the process progressed after the sulfasalazine was stopped.

A report in 1988 described an individual with ulcerative colitis who developed shortness of breath, cough productive of white sputum, mild fever, bilateral crackles, and a chest radiograph showing patchy infiltrates.[39] There was associated eosinophilia. After cessation of the sulfasalazine, symptoms subsided and the chest radiograph returned to normal. This report was consistent with sulfasalazine-related BOOP although there was no biopsy confirmation of BOOP, and the patient had taken the drug for 6 years.

A 68-year-old woman with seronegative rheumatoid arthritis developed progressive shortness of breath and pruritus 4 months after initiation of sulfasalazine.[40] She had bilateral crackles and pulmonary function tests showed a vital capacity of 75 per cent of predicted, an FEV_1/FVC ratio of 75 per cent, and a decreased diffusing capacity to 49 per cent of predicted. The chest radiograph showed diffuse, bilateral reticular–nodular infiltrates in the middle and lower lobes, and the chest CT scan showed alveolar consolidation, pretracheal, carinal and prevascular bilateral axillary lymphadenopathy. Thoracoscopic lung biopsy showed plugs of granulation tissue within the lumina of small airways and alveolar accumulations of foamy macrophages and inflammation consistent with BOOP. The sulfasalazine was stopped and the patient was given prednisolone daily for 3 months, in decreasing doses. Respiratory symptoms improved and the previously elevated eosinophil count returned to normal within a few days. After 45 days, radiological improvement was seen on the chest CT scans.

Mesalamine

Mesalamine (mesalazine) is a preparation of 5-aminosalicylate (5-ASA), available in oral and rectal forms, which does not contain sulfonamide. It is used for the treatment of inflammatory bowel disease. BOOP has been reported in patients taking the oral as well as the rectal preparation. In the 1993 Camus review of inflammatory bowel disease, three patients with BOOP had received mesalamine treatment.[37] In a prior report, a 20-year-old woman with ulcerative colitis was initially treated with sulfasalazine that was changed to mesalazine, 800 mg three times daily, and 1 year later developed fever, anorexia, unproductive cough, central pleuritic chest pain and shortness of breath.[41] The chest radiograph showed a large peripheral infiltrate in the right mid-lung and right lower lung. Lung biopsy showed knots of granulation tissue in respiratory bronchioles, alveolar ducts and alveoli consistent with BOOP. Mesalamine was discontinued. There was rapid symptomatic and radiological improvement after 1 week of prednisolone 40 mg daily.

A 2001 report described an 18-year-old woman who developed ulcerative colitis and was treated with intravenous corticosteroids, oral sulfasalazine and parenteral nutrition.[42] She improved and 1 month later her medications were changed to oral mesalamine. Oral 6-mercaptopurine (6-MP) was added because of persistent bowel symptoms. Within 3 years she was in complete remission and all medications were stopped. She had an exacerbation of ulcerative colitis treated with oral mesalamine, oral 6-MP and mesalamine enemas. Two months later she developed fever, cough, pleuritic chest pain and shortness of breath. There were no increased eosinophils. The chest radiograph showed patchy peripheral densities with areas of consolidation in the left lung. All medications were stopped. Video-assisted thoracoscopy showed typical BOOP with a polyp of young connective tissue inside the bronchiolar lumen and organizing pneumonia. Treatment was begun with intravenous methylprednisolone 40 mg every 6 hours. The patient quickly improved during the ensuing 4 days. Within 3 weeks the patient had returned to almost all physical activity with almost complete clearing of the radiographic infiltrates. The symptoms of colitis were minimal and

6-MP was restarted without recurrence of the pulmonary disease.

Bucillamine

Bucillamine is an anti-inflammatory agent developed in Japan with a structure similar to D-penicillamine. A 74-year-old woman with rheumatoid arthritis for 10 years had been taking bucillamine at 100 mg per day for 3 months and developed unproductive cough and shortness of breath.[43] Fine crackles were detected at both lung bases. The chest X-ray showed patchy infiltrates bilaterally predominantly in the lower lungs. The chest CT scan showed ground-glass opacities adjoining bronchovascular bundles and air bronchograms with no honeycombing. The features were suggestive of BOOP. Because of severe hypoxaemia, a transbronchial biopsy and lavage were not performed. However, a drug lymphocyte stimulation test with bucillamine was positive. The bucillamine was discontinued and prednisolone 30 mg daily was given. She gradually improved, and by 2 months she had no cough or shortness of breath and the patchy infiltrates seen on the chest CT scan had nearly resolved.

Infliximab

Infliximab is an antitumour necrosis factor (anti-TNF) agent given by infusion for the treatment of rheumatoid arthritis. A 70-year-old man had been receiving methotrexate at 22.5 mg per week for 3 years and infliximab was added at 3 mg/kg. Nine days after the third infusion he was seen in the emergency room because of fever, fatigue and increasing shortness of breath.[44] He was febrile to 104.3°F (40.2°C) with oxygen saturation of 94 per cent. Lungs were clear to auscultation. The chest X-ray showed bilateral interstitial infiltrates in the upper two-thirds of the lungs. There were no eosinophils. High-resolution chest CT scan showed bilateral ground-glass opacities in the upper two-thirds of the lung without nodules, lymphadenopathy or fibrosis. Lavage revealed lymphocytic predominance. The methotrexate was discontinued and prednisone at 50 mg per day was begun and increased to 100 mg daily because of lack of improvement. The patient slowly improved and was discharged from the hospital 1 month later with home oxygen therapy. Methotrexate at 22.5 mg per week was reinstituted without respiratory symptoms. Six months later the chest X-ray showed improvement yet the chest CT continued to show the ground-glass opacities. The authors reported eight other patients who had received infliximab, some without methotrexate, who developed interstitial lung disease.

IMMUNOSUPPRESSIVE DRUGS

Azathioprine

Azathioprine is a purine analogue used as a corticosteroid-sparing agent for the treatment of inflammatory bowel disease and other diseases. A 71-year-old man with Crohn's disease for 12 years developed an exacerbation treated with mesalamine for 3 months without respiratory symptoms.[45] The mesalamine was discontinued, and azathioprine was added at 100 mg daily. Within 2 weeks the patient developed fever, with progressively worsening unproductive cough and shortness of breath; he required supplemental oxygen. A chest CT scan showed ground-glass opacities predominantly in the upper lungs bilaterally. An open lung biopsy showed histological features of BOOP. He was treated with intravenous corticosteroid therapy and the azathioprine was continued. The severe illness persisted with a white blood cell count of 27 000/mm^3 and a fever of 104°F (40°C). The patient was transferred to a new institution, and the azathioprine was immediately stopped. Within 3 days the WBC count had returned to normal with resolution of the fever. The corticosteroid therapy was gradually decreased over 6 months, with eventual complete recovery of pulmonary function.

Tacrolimus

Tacrolimus was one of the early macrolide preparations used as an immunosuppresive agent in organ transplant recipients. A 23-year-old woman who received a bone marrow transplantation for acute lymphocytic leukaemia developed BOOP after receiving FK506 (tacrolimus) for 7 weeks and died from respiratory failure 15 days later.[46] Autopsy showed BOOP without findings of graft-versus-host disease.

Sirolimus

Sirolimus is a macrolide that is used as an immunosuppressive agent in organ transplant recipients and causes inhibition of T-cell response to interleukin-2. In 2005, Lindenfeld et al. described seven patients who developed sirolimus-related BOOP.[47] The BOOP lesion developed 1–7 months after initiation of sirolimus, and fever, cough and shortness of breath were common presenting symptoms. The chest CT scans were consistent with BOOP, and the available lung biopsies showed BOOP. Two of the seven patients had complete resolution after withdrawal of the medication, four after corticosteroid therapy; one developed respiratory failure and died.

One year later, Champion et al. reported sirolimus-associated pneumonitis in 24 renal transplant recipients.[48] These patients had received sirolimus for a median of 5.5 months, and they developed fever, fatigue, cough and shortness of breath. Crackles occurred in ten patients (42 per cent). All patients had pulmonary infiltrates radiographically, and in four patients CT scans showed patchy bilateral asymmetrical peripheral consolidation and ground-glass opacities. The pulmonary infiltrates resolved after discontinuation of the medication.

Everolimus

Everolimus is a derivative of sirolimus. There has been a report of a 56-year-old woman with a renal transplant from polycystic kidney disease who developed focal nodular BOOP while receiving everolimus.[49] Whether there is a cause and effect relationship between the agent and BOOP is not known

as BOOP may occur in patients with polycystic kidney disease and has been reported in renal transplant recipients.

ANTICONVULSANT DRUGS

Carbamazepine

Carbamazepine is an established anticonvulsant medication used for treatment of epilepsy. A 72-year-old man was treated with cabamazepine for focal seizures and admitted to the hospital with fever, unproductive cough and progressive shortness of breath after 7 weeks of treatment.[50] He had bilateral crackles. The chest radiograph showed right upper lung consolidation. There was no eosinophilia. He was treated with antibiotics for community acquired pneumonia with worsening of symptoms. Transbronchial biopsy showed organizing tissue filling small bronchioles and alveolar spaces consistent with BOOP. Treatment with oral prednisolone at 60 mg daily was begun and the carbamazepine discontinued. He had a dramatic improvement and after 2 weeks was asymptomatic and the chest radiograph was normal.

Phenytoin

A 40-year-old man began taking phenytoin 300 mg daily after a craniotomy, and in 4 weeks developed fever, rash, unproductive cough, shortness of breath and bilateral crackles.[51] The chest radiograph showed bilateral patchy infiltrates, and the open lung biopsy showed myxoid granulation tissue in the bronchioles extending to alveolar ducts and alveoli consistent with BOOP. The patient had also developed new-onset cold agglutinin disease. Corticosteroid therapy was given for 1 year. The chest radiograph, liver function and haematological abnormalities resolved.

Sometimes, BOOP occurs in the context of a drug rash with eosinophilia and systemic symptoms (DRESS), a generalized syndrome that has been associated with anticonvulsant drugs such as carbamazepine and phenytoin.[52]

MISCELLANEOUS DRUGS

Interferons

The interferons have been used for treatment of hepatitis C, multiple sclerosis, malignant melanoma and idiopathic pulmonary fibrosis. In addition to the two reports that included interferon-α noted earlier,[20,22] a 1994 report described a 64-year-old man with hepatitis C who was treated with interferon-α for 11 weeks and developed fever, unproductive cough and shortness of breath.[53] The liver function tests had returned to normal. The chest radiograph showed bilateral patchy infiltrates. The biopsy showed BOOP. The interferon was stopped. Prednisolone at 40 mg per day was begun. The dyspnoea and radiographic infiltrates disappeared promptly within 1 week.

A 49-year-old man with multiple sclerosis for 8 years was treated with interferon-β.[54] Three months after beginning the interferon he developed a progressive unproductive cough and right-sided chest pain. The chest radiograph showed a single right basal infiltrate. Transbronchial biopsy showed BOOP. The symptoms and radiographic findings resolved after 2 months of prednisone treatment. A 77-year-old woman with malignant melanomas was treated with interferon-β and after a total cumulative dose of 5.4×10^7 units she developed shortness of breath.[55] Chest CT scan showed bilateral lower lung peripheral infiltrates, and a wedge biopsy showed BOOP. Prednisone at 30 mg daily resulted in improved symptoms and resolution of the radiographic abnormalities.

A 2003 report described four patients treated with interferon-γ for idiopathic pulmonary fibrosis who developed acute respiratory failure and new alveolar opacities radiographically.[56] The chest CT scan showed diffuse ground-glass opacities superimposed on the existing usual interstitial pneumonia pattern. A lung biopsy of one of the patients showed extensive interstitial fibrosis with honeycomb change and foci of BOOP. Whether the BOOP was related to the interferon-γ or was part of the inflammatory exacerbation component of UIP is not known.

Ticlopidine

Ticlopidine is a platelet aggregation inhibitor used as a preventive agent in patients with temporal arteritis, strokes or transient ischaemic attacks. A 76-year-old woman was taking prednisone 45 mg daily, plus ticlopidine 250 mg twice daily, for giant-cell temporal arteritis.[57] In 1 month she developed a pruritic skin rash, increasing shortness of breath and bilateral crackles. The chest radiograph showed diffuse bilateral peripheral infiltrates. Transbronchial biopsy showed air-spaces filled with organizing connective tissue plugs consistent with BOOP. The ticlopidine therapy was stopped. Prednisone was continued. The dyspnoea and blanching cutaneous lesions improved. Shortness of breath with exertional activities disappeared in 3 months and the interstitial radiographic pattern had resolved at 5 months.

L-tryptophan

L-tryptophan has been used as a dietary supplement usually with no adverse reactions at a small daily dose. A report described a 51-year-old woman who had taken 2668 mg daily for 2.5 months and was admitted to hospital because of fever, weakness, nausea, productive cough and bilateral crackles.[58] The chest radiograph showed bilateral infiltrates, and the open lung biopsy showed granulation tissue plugs in the bronchioles, alveolar ducts and adjacent alveoli consistent with BOOP. Treatment with intravenous methylprednisolone resulted in prompt resolution of symptoms and clearing of the chest radiographic abnormalities.

Illicit use of cocaine

The use of freebase cocaine was described as the cause of BOOP in a 1987 report.[59] A 32-year-old man had 10 days of

fever, unproductive cough and shortness of breath. He had bilateral crackles. The chest radiograph showed bilateral nodular opacities. An open lung biopsy showed granulation tissue in the bronchioles and alveoli consistent with BOOP. He received corticosteroid therapy for 6 months with resolution of the chest radiograph findings.

Intravenous use of heroin

A 24-year-old woman with 7 years of active intravenous heroin use developed progressive shortness of breath. A chest CT scan showed dense bilateral ground-glass opacities, and a lung biopsy showed patchy, temporally homogeneous pneumonia with intraluminal fibroblastic proliferation consistent with BOOP.[60] After cessation of heroin, respiratory symptoms subsided without corticosteroid therapy and follow-up chest CT scan showed marked improvement.

Fluvastatin

The statins may cause a BOOP reaction as shown by a report of a 66-year-old woman who had 3 weeks of progressive shortness of breath.[61] She had taken fluvastatin for 1 year prior to the onset of dyspnoea. She had bilateral crackles and a chest CT scan showing bilateral patchy consolidations. Transbronchial biopsy showed organizing pneumonia and proliferating fibroblasts extending along and branching within alveolar ducts. She had a favourable response with prednisone after cessation of the fluvastatin. She developed new-onset shortness of breath and linear opacities at the lung basis. A biopsy showed UIP. There was sustained improvement with azathioprine and prednisone therapy without a relapse.

Risedronate

The drug lymphocyte stimulation test (DLST) can be useful for the diagnosis of drug-related BOOP. Risedronate is used for the treatment of osteoporosis by inhibiting osteoclast resorption. A 66-year-old woman was given risedronate for osteoporosis and 2 months later developed fever, unproductive cough and a chest radiograph showing bilateral patchy infiltrates. The chest CT scan showed several ground-glass opacities and a large, pleural triangular-based infiltrate in the left mid lung.[62] A transbronchial lung biopsy showed cellular alveolitis with intraluminal polypoid organization consistent with BOOP. All of the drugs were stopped. The fever resolved after 5 days and the radiographic infiltrates disappeared 2 weeks later. A drug lymphocyte stimulation test on her peripheral lymphocytes gave a positive reaction to risedronate with a stimulation index of 265 per cent and negative reaction to the other four drugs, all of which had been administered to her for at least 4 years.

BOOP AFTER BREAST IRRADIATION

Post-breast-radiation BOOP has emerged as an important iatrogenic development as this lesion may occur in up to 2.3 per cent of women (Fig. 24.2).

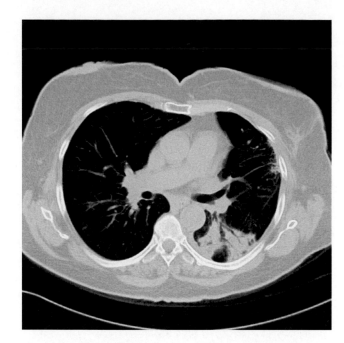

Fig. 24.2 Chest CT scan of a 46-year-old woman who had radiation for a left breast carcinoma. There is subpleural peripheral fibrosis in the left upper lung secondary to the radiation, and BOOP as a right mid-lung ground-glass opacity with air bronchograms. The BOOP resolved rapidly with prednisone therapy.

In a 1997 report, Van Laar *et al.* reported two women who developed BOOP after receiving adjuvant radiotherapy for breast carcinoma.[63] Both patients were treated with prednisone 60 mg daily, leading to rapid and dramatic clinical improvement of symptoms and infiltrative abnormalities with no subsequent relapse. A later report showed that symptoms may occur 1–12 months after completion of radiation therapy, and the chest X-ray shows peripheral patchy or alveolar infiltrates often outside of the radiation field.[64] In a 2000 report of 11 patients, Majori *et al.* reported spontaneous migration of infiltrates from the irradiated lung to the contralateral non-iradiated lung.[65] There can be a dramatic improvement with corticosteroid therapy, but relapses do occur.[64,66] Some investigators have suggested that radiation therapy may 'prime' the development of BOOP.[64,65] Bronchoalveolar lavage studies of these patients indicated an increase in lymphocytes, mast cells, CD3 cells and CD8 cells, and a decrease in CD4 cells and the CD4/CD8 ratio.[65]

In a 2008 Japanese study, Ogo *et al.* reported a 1.8–2.19 per cent incidence of BOOP among 2056 women receiving radiation therapy for breast carcinoma.[67] Most developed BOOP within 6 months of therapy. Cough, fever and sputum were the most common symptoms. They showed that the ground-glass opacities were unilateral in 70 per cent and often occurred outside of the radiation field. They found the BOOP resolved in 20 per cent of patients who received no medication. In another 2008 study of 702 women, 16 individuals (2.3 per cent) developed post-radiation BOOP.[68] Risk factors included age ≥ 50 and women who had received concurrent endocrine therapy, while tamoxifen was not found to be a risk factor. The mean latency of BOOP was 4.4 months after the end of radiation therapy (range 2.3–7.9). The lesion was radiographically unilateral in the majority, occurring in 10 patients. All 16 patients

improved, 5 without corticosteroid therapy; and among the others, corticosteroid therapy was used for between 1–3 weeks and 3.7 years, with a median of 1.1 years. Cataracts developed in two women on long-term therapy.

One case report showed a 71-year-old woman who completed right-sided breast radiation and developed a fever, sweats and unproductive cough 9 months later.[69] A chest CT scan showed a dense infiltrate with air bronchograms in the right mid-lung. Bronchoscopy lung biopsy showed BOOP. There was a dramatic resolution of symptoms within a week, and 3 weeks later the chest CT scan showed complete resolution.

A case report has described the recurrence of BOOP after low-dose breast cancer radiation in a woman with BOOP 1 year earlier.[70] Another 51-year-old woman developed idiopathic BOOP that was successfully treated with prednisone. Five months later she was found to have left-sided breast cancer, and intensity-modulated radiotherapy (IMRT) was begun to the left breast while she was taking prednisone 5 mg every other day and concluded after 25 treatments. Two months after completion of radiation therapy she developed cough. The chest CT scan showed a peripheral triangular ground-glass opacity with an air bronchogram in the left mid-lung. Prednisone was begun. No new respiratory symptoms developed and the CT scan showed improvement.

In conclusion, BOOP may occur as frequently as 2.3 per cent among women after breast radiotherapy and may resolve through monitoring without therapy or through treatment with corticosteroid therapy.

KEY POINTS

- BOOP is an inflammatory lung process and there are more than 35 medications associated with its development. Radiation therapy can also trigger BOOP.

- Fever, unproductive cough and progressive shortness of breath are common symptoms.

- Bilateral crackles occur in most individuals. Eosinophilia rarely occurs. The chest radiograph shows bilateral patchy infiltrates, and chest high-resolution CT (HRCT) shows ground-glass opacities and air bronchograms.

- There is a suggestion that drugs such as amiodarone or statins may cause the newly described entity name afop.[71]

- Although in very rare situations the outcome may be fatal, the prognosis is usually excellent, with complete resolution of the BOOP with cessation of the medication or a brief course of corticosteroid therapy.

REFERENCES

1. Hollingsworth HM. Drug-related bronchiolitis obliterans organizing pneumonia. In: Epler GR (ed.) *Diseases of the Bronchioles*, pp 367–76. New York: Raven Press, 1994.

2. Camus P, Bonniaud P, Fanton A, *et al.* Drug-induced and iatrogenic infiltrative lung disease. *Clin Chest Med* 2004;**25**:479–519.

3. Epler GR, Colby TV, McLoud TC, Carrington CB, Gaensler EA. Bronchiolitis obliterans organizing pneumonia. *New Engl J Med* 1985;**312**:152–8.

4. Cordier JF. Cryptogenic organizing pneumonia. *Eur Resp J* 2006;**28**:1272–85.

5. Piperno D, Donné C, Loire R, Cordier J–F. Bronchiolitis obliterans organizing pneumonia associated with minocycline therapy: a possible cause. *Eur Respir J* 1995;**8**:1018–20.

6. Kondo H, Fujita J, Inoue T, *et al.* Minocycline-induced pneumonitis presenting as multiple ring-shaped opacities on chest CT, pathologically diagnosed bronchiolitis obliterans organizing pneumonia. *Nihon Kokyuki Gakkai Zasshi* 2001;**39**:215–19.

7. Cameron, RJ, Kolbe J, Wilsher ML, Lambie N. Bronchiolitis obliterans organizing pneumonia associated with the use of nitrofurantoin. *Thorax* 2000;**55**:249–51.

8. Fawcett IW, Ibrahim NBN. BOOP associated with nitrofurantoin. *Thorax* 2001;**56**:161.

9. Fenton ME, Kanthan R, Cockcroft DW. Nitrofurantoin-associated bronchiolitis obliterans organizing pneumonia: report of a case. *Can Respir J* 2008;**15**:311–12.

10. Dreis DF, Winterbauer RH, Van Norman GA, Sullivan SL, Hammar SP. Cephalosporin-induced interstitial pneumonitis. *Chest* 1984;**86**:138–40.

11. Roncoroni AJ, Corrado C, Besushio S, Pavlovsky S, Narvaiz M. Bronchiolitis obliterans possibly associated with amphotericin B. *J Infect Dis* 1990;**161**:589.

12. Cobb E, Kimbrough RC, Nugent KM, Phy MP. Organizing pneumonia and pulmonary eosinophilic infiltration associated with daptomycin. *Ann Pharmacother* 2007;**41**:696–701.

13. Shinde A, Seifi A, Delre S, Moustafa Hussein WH, Ohebsion J. Daptomycin-induced pulmonary infiltrates with eosinophilia. *J Infect* 2009;**58**:173–4.

14. Yokogawa N, Alcid DV. Acute fibrinous and organizing pneumonia as a rare presentation of abacavir hypersensitivity reaction. *AIDS* 2007;**21**:2116–17.

15. Carson CW, Cannon GW, Egger MG, Ward JR, Clegg DO. Pulmonary disease during the treatment of rheumatoid arthritis with low dose pulse methotrexate. *Semin Arthritis Rheum* 1987;**16**:186–95.

16. García-Vicuña R, Díaz-González F, Castañeda S, Arranz M, López-Bote JP. Rheumatoid disease resembling lung neoplasia. *J Rheum* 1990;**17**:1686–8.

17. Santrach PJ, Askin FB, Wells RJ, Azizkhan RG, Merten DF. Nodular form of bleomycin-related pulmonary injury in patients with osteogenic sarcoma. *Cancer* 1989;**64**:806–11.

18. Jacobs C, Slade M, Lavery B. Doxorubicin and BOOP, a possible near fatal association. *Clin Oncol* 2002;**14**:262.

19. Battistini E, Dini G, Savioli C, *et al.* Bronchiolitis obliterans organizing pneumonia in three children with acute leukemias treated with cytosine arabinoside and anthracyclines. *Eur Respir J* 1997;**10**:1187–90.

20. Patel M, Ezzat W, Pauw KL, Lowsky R. Bronchiolitis obliterans organizing pneumonia in a patient with chronic myelogenous leukemia developing after initiation of interferon and cytosine arabinoside. *Eur J Haematol* 2001;**67**:318–21.

21. Salvucci M, Zanchini R, Molinari A, *et al.* Lung toxicity following fludarabine, cytosine arabinoside and mitoxantrone (FLAN) treatment for acute leukemia. *Haematologica* 2000;**85**:769–70.

22. Kalambokis G, Stefanou D, Arkoumani E, *et al.* Fulminant bronchiolitis obliterans organizing pneumonia following 2 days of treatment with hydroxyurea, interferon-alpha and oral cytarabine ocfosfate for chronic myelogenous leukemia. *Eur J Haematol* 2004;**73**:67–70.

23. Kalambokis G, Stefanou D, Arkoumani E, Tsianos E. Bronchiolitis obliterans organizing pneumonia following chlorambucil treatment for chronic lymphocytic leukemia. *Eur J Haematol* 2004;**73**:139–42.

24. Feaver AA, McCune DE, Mysliwiec AG, Mysliwiec V. Thalidomide-induced organizing pneumonia. *South Med J* 2006;**99**:1292–4.

25. Macartney C, Burke E, Elborn S, *et al.* Bronchiolitis obliterans organizing pneumonia in a patient with non-Hodgkin's lymphoma following R-CHOP and pegylated filgrastim. *Leuk Lymphoma* 2005;**46**:1523–6.

26. Biehn SE, Kirk D, Rivera MP, *et al.* Bronchiolitis obliterans with organizing pneumonia after rituximab therapy for non-Hodgkin's lymphoma. *Hematol Oncol* 2006;**24**:234–7.

27. Garrido M, O'Brien A, González S, Claverio JM, Orellana E. Cryptogenic organizing pneumonitis during oxaliplatin chemotherapy for colorectal cancer: case report. *Chest* 2007;**132**:1997–9.

28. Camus P, Martin WJ, Rosenow EC. Amiodarone pulmonary toxicity. *Clin Chest Med* 2004;**25**:65–75.

29. Barroso E, Hernandez L, Gil J, *et al.* Idiopathic organizing pneumonia: a relapsing disease. 19 years of experience in a hospital setting. *Respiration* 2007;**74**:624–31.

30. Vernhet H, Bousquet C, Durand G, Giron J, Senac JP. Reversible amiodarone-induced lung disease: HRCT findings. *Eur Radiol* 2001;**11**:1697–703.

31. Ott MC, Khoor A, Leventhal JP, Paterick TE, Burger CD. Pulmonary toxicity in patients receiving low-dose amiodarone. *Chest* 2003;**123**:646–51.

32. Malhotra A, Muse V, Mark EJ. An 82-year-old man with dyspnea and pulmonary abnormalities. *New Engl J Med* 2003;**348**:1574–85.

33. Omeroglu G, Kalugina Y, Ersahin C, Wojcik EM. Amiodarone lung toxicity in a cardiac transplant candidate initially diagnosed by fine-needle aspiration: cytologic, histologic, and electronmicroscopic findings. *Diagn Cytopathol* 2006;**34**:351–4.

34. Camus P, Lombard JN, Perrichon M, Piard F, *et al.* Bronchiolitis obliterans organizing pneumonia in patients taking acebutolol or amiodarone. *Thorax* 1989;**44**:711–15.

35. Costabel U, Teschler H, Schoenfeld B, *et al.* BOOP in Europe. *Chest* 1992;**102**:14S–20S.

36. Fort JG, Scovern H, Abruzzo JL. Intravenous cyclophosphamide and methylprednisolone for the treatment of bronchiolitis obliterans and interstitial fibrosis associated with crysotherapy. *J Rheumatol* 1988;**15**:850–4.

37. Camus P, Piard F, Ashcroft T, Gal AA, Colby TV. The lung in inflammatory bowel disease. *Medicine* 1993;**72**:151–83.

38. Williams T, Eidus L, Thomas P. Fibrosing alveolitis, bronchiolitis obliterans, and sulfasalazine therapy. *Chest* 1982;**81**:766–8.

39. Jordan A, Cowan RE. Reversible pulmonary disease and eosinophilia associated with sulphasalazine. *J Royal Soc Med* 1988;**81**:233–5.

40. Ulubas B, Sahin G, Ozer C, *et al.* Bronchiolitis obliterans organizing pneumonia associated with sulfasalazine in a patient with rheumatoid arthritis. *Clin Rheumatol* 2004;**23**:249–51.

41. Swinburn CR, Jackson GJ, Cobden I, *et al.* Bronchiolitis obliterans organizing pneumonia in a patient with ulcerative colitis. *Thorax* 1988;**43**:735–6.

42. Haralambou G, Teirstein AS, Gil J, Present DH. Bronchiolitis obliterans in a patient with ulcerative colitis receiving mesalamine. *Mt Sinai J Med* 2001;**68**:384–8.

43. Kajiya T, Kuroda A, Hokonohara D, Tei C. Radiographic appearance of bronchiolitis obliterans organizing pneumonia developing during bucillamine treatment for rheumatoid arthritis. *Am J Med Sci* 2006;**332**:39–42.

44. Villeneuve E, St-Pierre A, Haraqui B. Interstitial pneumonitis associated with infliximab therapy. *J Rheumatol* 2006;**33**:1189–93.

45. Ananthakrishnan AN, Attila T, Otterson MF, *et al.* Severe pulmonary toxicity after azathioprine/6-mercaptopurine initiation for the treatment of inflammatory bowel disease. *J Clin Gastroenterol* 2007;**41**:682–8.

46. Przepiorka D, Abu-Elmagd K, Huaringa A, *et al.* Bronchiolitis obliterans organizing pneumonia in a BMT patient receiving FK506. *Bone Marrow Transplant* 1993;**11**:502.

47. Lindenfeld JA, Simon SF, Zamora MR, *et al.* BOOP is common in cardiac transplant recipients switched from a calcineurin inhibitor to sirolimus. *Am J Transplant* 2005;**5**:1392–6.

48. Champion L, Stern M, Israël-Biet D, *et al.* Brief communication: sirolimus-associated pneumonitis: 24 cases in renal transplant recipients. *Ann Intern Med* 2006;**144**:505–9.

49. Carreño CA, Gadea M. Case report of a kidney transplant recipient converted to everolimus due to malignancy: resolution of bronchiolitis obliterans organizing pneumonia without everolimus discontinuation. *Transplant Proc* 2007;**39**:594–5.

50. Banka R, Ward MJ. Bronchiolitis obliterans and organizing pneumonia caused by carbamazepine and mimicking community acquired pneumonia. *Postgrad Med J* 2002;**78**:621–2.

51. Angle P, Thomas P, Chiu B, Freedman J. Bronchiolitis obliterans with organizing pneumonia and cold agglutinin disease associated with phenytoin hypersensitivity syndrome. *Chest* 1997;**112**:1697–9.

52. Allam JP, Paus T, Reichel C, Bieber T, Novak N. DRESS syndrome associated with carbamazepine and phenytoin. *Eur J Dermatol* 2004;**14**:339–42.

53. Ogata K, Koga T, Yagawa K. Interferon-related bronchiolitis obliterans organizing pneumonia. *Chest* 1994;**106**: 612–13.

54. Ferriby D, Stojkovic T. Clinical picture: bronchiolitis obliterans with organizing pneumonia during interferon-beta-1a treatment. *Lancet* 2001;**357**:751.

55. Kimura Y, Okuyama R, Watanabe H, *et al.* Development of BOOP with interferon-beta treatment for malignant melanoma. *J Eur Acad Dermatol Venereol* 2006;**20**: 999–1000.

56. Honoré I, Nunes H, Groussard O, *et al.* Acute respiratory failure after interferon-gamma therapy of end-stage pulmonary fibrosis. *Am J Respir Crit Care Med* 2003;**167**:953–7.

57. Alonso-Martinez JL, Elejalde-Guerra JI, Larrinaga-Linero D. Bronchiolitis obliterans organizing pneumonia caused by ticlopidine. *Ann Intern Med* 1998;**129**:71–2.

58. Mar KE, Sen P, Tan K, Krishnan R, Ratkalkar K. Bronchiolitis obliterans organizing pneumonia associated with massive L-trytophan ingestion. *Chest* 1993;**104**;1924–6.

59. Patel RC, Dutta D, Schonfeld SA. Free-base cocaine use associated with bronchiolitis obliterans organizing pneumonia. *Ann Intern Med* 1987;**107**:186.

60. Bishay A, Amchentsev A, Saleh A, *et al.* A hitherto unreported pulmonary complication in an IV heroin user. *Chest* 2008;**133**:549–51.

61. Naccache JM, Kambouchner M, Girard F, *et al.* Relapse of respiratory insufficiency one year after organizing pneumonia. *Eur Respir J* 2004;**24**:1062–5.

62. Arai T, Inoue Y, Hayashi S, Yamamoto S, Sakatani M. Risedronate induced BOOP complicated with sarcoidosis. *Thorax* 2005;**60**:613–14.

63. Van Laar JM, Holscher HC, van Krieken JHJM, Stolk J. Bronchiolitis obliterans organizing pneumonia after adjuvant radiotherapy for breast carcinoma. *Respir Med* 1997;**91**:241–4.

64. Crestani B, Valeyre D, Roden S, *et al.* Bronchiolitis obliterans organizing pneumonia syndrome primed by radiation therapy to the breast. *Am J Respir Crit Care Med* 1998;**158**:1929–35.

65. Majori M, Poletti V, Curti A, *et al.* Bronchoalveolar lavage in bronchiolitis obliterans organizing pneumonia primed by radiation therapy to the breast. *J Allerg Clin Immunol* 2000;**105**:239–44.

66. Arbetter KR, Prakash UBS, Tazelaar HD, Douglas WW. Radiation-induced pneumonitis in the 'nonirradiated' lung. *Mayo Clin Proc* 1999;**74**:27–36.

67. Ogo E, Komaki R, Fujimoto K, *et al.* A survey of radiation-induced bronchiolitis obliterans organizing pneumonia syndrome after breast-conserving therapy in Japan. *Int J Radiat Oncol Biol Phys* 2008;**71**:123–31.

68. Katayama N, Sato S, Katsui K, *et al.* Analysis of factors associated with radiation-induced bronchiolitis obliterans organizing pneumonia syndrome after breast-conserving therapy. *Int J Radiat Oncol Biol Phys* 2009;**73**:1049–54.

69. Bissoli L, Di Francesco V, Valbusa F, *et al.* A case of bronchiolitis obliterans organizing pneumonia after nine months post-operative irradiation for breast cancer. *Age Ageing* 2008;**37**:235.

70. Pierce L, Lewin A, Abdel-Wahab M, Elsayyad N. Early radiation-induced lung injury in a patient with prior diagnosis of bronchiolitis obliterans organizing pneumonitis. *J Natl Med Assoc* 2008;**100**:1474–6.

71. Beasley MB, Franks TJ, Galvin JR, Gochuico B, Travis WD. Acute Fibrinous and organizing pneumonia. A histologic pattern of lung injury and possible variant of diffuse alveolar damage. *Arch Pathol Lab Med* 2002;**126**:1064–70.

25

Coughing induced by drugs

ALYN H MORICE, HOSNIEH FATHI

INTRODUCTION

Coughing is the most common manifestation of respiratory tract symptoms. It both defends against respiratory pathogens and helps to clear the tracheobronchial tree for an optimal gas exchange. Drugs may cause cough by three basic mechanisms. First, there may be direct stimulation of cough by an inhaled medication. Second, they may give rise to a parenchymal or airway disease process which produces coughing as a major symptom. Third (and perhaps a more subtle mechanism), drugs may alter the intrinsic sensitivity of the cough reflex, thereby causing cough hypersensitivity making the patient prone to cough induced by other influences.

DRUGS INDUCING COUGH BY DIRECT MECHANISMS

Direct physicochemical effects may trigger coughing. This is particularly obvious with inhaled medications. The physical and chemical composition of the inhaled substance may precipitate coughing by direct stimulation of cough receptors. In understanding this mechanism of cough production, knowledge of the variety and complexity of cough receptor subtypes is required.

Over the past 20 years our understanding of the cough receptor mechanisms has developed considerably. Early attempts at categorizing cough receptors involved the electrophysiological examination of nerve fibre recordings.[1] This technique yielded valuable information about the wiring of the cough reflex (e.g. that cough is almost exclusively a vagal phenomenon), but was less helpful in determining the physicochemical stimuli provoking cough. With the advent of techniques such as patch clamping and calcium flurophor labelling, individual receptor channels have been identified, cloned and studied *in vitro*. Knowledge of these receptors, which respond to a variety of stimuli including mechanical and chemical irritants and disease processes, has enabled us to understand why certain inhaled medications give rise to cough.

The TRPV1 capsaicin receptor

The capsaicin receptor is one of a family of irritant receptors, known as the 'transient receptor potential' family.[2] The importance of this receptor in the production of cough in humans is demonstrated by the exquisite sensitivity of the human cough reflex to specific agonists of this receptor, such as capsaicin[3] and resiniferatoxin.[4] Indeed, resiniferatoxin is the most potent tussive agent known. The TRPV1 is a non-specific ion channel, which responds not only to vanilloids, such as capsaicin, but also to acidity and a change in temperature. The receptor may be up-regulated by inflammatory mediators,[5] and of particular interest it has been demonstrated that lipoxygenase products act intracellularly to increase the opening probability of the receptor.[6] Thus, in lung diseases such as asthma, TRPV1 activity is increased and may be responsible for the cough associated with inhaled antiasthmatic medication.[7]

However, the most likely mechanism whereby TRPV1 stimulation gives rise to drug side-effects is through its acid sensitivity. The pH of, particularly, nebulized solutions may be sufficiently low to precipitate coughing in sensitized individuals.

Other TRP receptors

Two other TRP receptors are important in the modulation of cough. The cold menthol receptor, TRPM8, has a dual effect with high concentrations of agonist causing a cough, whereas lower concentrations produce a cooling sensation which has been utilized to reduce cough.[8] This is particularly so in over-the-counter medications but also in the inhaler of nedocromil sodium, known as Tilade®.

The TRPA1 receptor plays an important modulating role in cough reflex sensitivity and may be particularly important in smokers through its potent agonist acreolin.

The TRPV4 receptor is sensitive to changes in osmotic pressure. Changes in the osmotic composition of the inhaled nebulized solution have important side-effects in terms of cough.[9] Thus, hypertonic saline recently introduced for the treatment of bronchiectasis may produce profound paroxysms of coughing.[10] In the 1980s the side-effects of bronchoconstriction and cough with nebulized ipratropium were narrowed down to the lack of the isosmolarity of the nebulized solution.[11,12] Subsequent production of an isosmolar solution has removed this problem.

Mechanosensitive receptors

For dry powder inhalation, perhaps the most important receptor is the recently discovered subepithelial receptor, described by Undem and colleagues at Johns Hopkins.[13] This receptor is exquisitely sensitive to light mechanical touch, and particles deposited in the airway may stimulate this receptor, leading to cough. The respirable fraction of any inhaled dry powder must aim, therefore, to minimize the deposition in the larger airways where this receptor predominates, particularly laryngeal deposition. The particle size of greater than $5\,\mu m$ should be avoided. In general, however, dry powder inhalers, particularly using drug without excipient, produces lower rates of cough than metered dose inhalers.[14]

DRUGS CAUSING CONDITIONS THAT PRECIPITATE COUGHING

Cough is the major symptom for which patients with lung disease seek medical advice. It is unsurprising that drugs that produce pathological changes within the lung will lead to a cough. These topics are covered extensively elsewhere in this book. The origin of the cough in such circumstances is by up-regulation of the cough receptors described above, through either increased production of receptor (demonstrated for TGF-β, interleukin-1α and lipopolysaccharide (LPS)), an increase in receptor sensitivity, or even an increase in density of sensory nerve fibres.[7]

Certain medical conditions in the process of the disease can produce cough as their major symptom by mechanically disrupting receptors. This is thought to be the primary mechanism whereby fibrotic lung disease produces cough. Drugs that can induce fibrosis – such as amiodarone,[15] ergots,[16] methotrexate,[7] and nitrofurantoin – will disrupt mechanoreceptors, whereas other drugs such as pergolide[18,19] may produce cough directly through TRPV receptors being up-regulated by inflammation.

DRUGS PRECIPITATING COUGH THAT ORIGINATES OUTSIDE THE LUNGS

There is one circumstance where the precipitant of drug-induced cough may not be located within the airways. This is

the case for cough induced by gastroesophageal reflux (GER).[20] Reflux cough is among the most common causes of chronic cough but is poorly recognized. This is because reflux cough is only occasionally associated with the classic features of GER, such as heartburn or dyspepsia. A higher index of suspicion is required coupled with knowledge of the characteristic features of reflux cough caused by transient opening of the lower oesophageal sphincter (LES), such as cough on phonation, postprandial cough and cough on rising in the morning.[21] Inappropriate relaxation of the LES, caused by drugs such as anticholinergics, calcium-channel blockers and ethanol, or reduction of the LES pressure by progesterone, estrogen,[22] theophylline, bisphosphonate[23] and tricyclic antidepressants,[24] can precipitate reflux and consequently cough.

ANGIOTENSIN–CONVERTING ENZYME INHIBITOR (ACEI) COUGH

Perhaps the most intriguing form of drug-induced cough arises from the antihypertensive drugs that inhibit ACE. Cough associated with ACE inhibitors was first reported with captopril in 1985.[25] It occurs at the rate of approximately 15 per cent, depending on the population studied,[26,27] and is independent of disease state with no increase in the incidence of cough in those with congestive heart failure or asthma. ACE inhibititor cough is twice as common in women as in men.[28] In this regard it is entirely typical of other types of chronic cough, which also have a female predominance. Cough occurs with all types of ACE inhibitors,[29] but not or to a far lesser extent with angiotensin II receptor antagonists, indicating that the property of ACE inhibitors that causes cough probably is not through the renin angiotensin system. ACE is a promiscuous enzyme that cleaves di- and tri-peptide bonds on a number of small peptides including the metabolism of bradykinin and substance P.[30] It is likely that the side-effect of cough is due to inhibition of the metabolism of these peptides.

The cough may come on at any time during therapy, starting sometimes with the first dose, but in other subjects there may be a gap of many months before cough is noted.[31] It does not seem to be dose-related.[32] The cough is usually a tickling sensation, dry, persistent, annoying and paroxysmal. It may get worse in the supine position and might be severe enough to cause hoarseness, vomiting or urinary incontinence.[33] Stopping the ACE inhibitor frequently does not bring immediate relief from the cough. Indeed, a delay of up to 18 months following the cessation of treatment may be needed for the cough to clear. The reason for this phenomenon is that ACE inhibitors do not directly cause cough, but act to enhance cough reflex sensitivity. This increased sensitivity has been shown in a randomized placebo-controlled study, where subjects were given captopril and cough reflex sensitivity determined by inhalational cough challenge.[34] Inhalation of the pungent extract of peppers, capsaicin, produced a concentration response curve in terms of coughing. This curve was shifted to the left by administration of captopril, indicating heightened cough reflex sensitivity. This study performed in humans tends to support the hypothesis that increase in substance P is the ultimate cause of the increased cough sensitivity since capsaicin is a highly specific stimulant of C-fibres, which contain substance P. In contrast, studies in guinea pigs suggest that bradykinin may be the sensitizing agent.[35] At present there is no definitive answer. Indeed, we do not understand whether the increased sensitivity arises from topical up-regulation of cough receptor sensitivity or an increase in neuronal traffic at the ganglion level, or whether the action is within the central pathways.

MISCELLANEOUS DRUG COUGH REACTIONS

Cough presumed to be drug-induced because the symptom disappeared on drug discontinuation has been described during treatments with morphine,[36] omeprazole,[37] sertraline,[38] sirolimus,[39] interferon,[40] mycophenolate mofetil[41] and topiramate.[42] Explosive coughing may develop abruptly following exposure to propofol,[43] fentanyl[44] or sufentanil.[45]

KEY POINTS

- Cough is a known side-effect of specific medications that can stimulate cough receptors either directly or indirectly. They can alter the sensitivity of these receptors.

- This effect can be on the respiratory system itself or on the related organs.

- To manage cough induced by drugs, the proposed option should be to stop the drug and replace it with one of a different class. The classical example would be stopping an ACE inhibitor and starting an angiotensin II receptor blocker.

- Cough suppressants can be considered.

REFERENCES

1. Widdicombe JG. Neurophysiology of the cough reflex. *Eur Respir J* 1995;**8**:1193–202.
2. Clapham DE. TRP channels as cellular sensors. *Nature* 2003;**426**:517–24.
3. Fuller RW, Karlsson JA, Choudry NB, Pride NB. Effect of inhaled and systemic opiates on responses to inhaled capsaicin in humans. *J Appl Physiol* 1988;**65**:1125–30.
4. Laude EA, Higgins KS, Morice AH. A comparative study of the effects of citric acid, capsaicin and resiniferatoxin on the cough challenge in guinea-pig and man. *Pulm Pharmacol* 1993;**6**:171–5.
5. Mitchell JE, Campbell AP, New NE, *et al.* Expression and characterization of the intracellular vanilloid receptor (TRPV1) in bronchi from patients with chronic cough. *Exp Lung Res* 2005;**31**:295–306.
6. Hwang SW, Cho H, Kwak J, *et al.* Direct activation of capsaicin receptors by products of lipoxygenases: endogenous capsaicin-like substances. *Proc Natl Acad Sci USA* 2000;**97**:6155–60.
7. Groneberg DA, Niimi A, Dinh QT, *et al.* Increased expression of transient receptor potential vanilloid-1 in airway nerves of chronic cough. *Am J Respir Crit Care Med* 2004;**170**:1276–80.

8. Morice AH, Marshall AE, Higgins KS, Grattan TJ. Effect of inhaled menthol on citric acid induced cough in normal subjects. *Thorax* 1994;**49**:1024–6.
9. Liedtke W, Choe Y, Marti-Renom MA, *et al.* Vanilloid receptor-related osmotically activated channel (VR-OAC), a candidate vertebrate osmoreceptor. *Cell* 2000;**103**:525–35.
10. Elkins MR, Robinson M, Rose BR, *et al.* A controlled trial of long-term inhaled hypertonic saline in patients with cystic fibrosis. *New Engl J Med* 2006;**354**:229–40.
11. Lowry RH, Wood AM, Higenbottam TW. Effects of pH and osmolarity on aerosol-induced cough in normal volunteers. *Clin Sci* 1988;**74**:373–6.
12. Portel L, de Lara JMT, Vernejoux JM, *et al.* The osmolarity of solutions used in nebulisation. *Rev Mal Respir* 1998;**15**:191–5.
13. Canning BJ, Mazzone SB, Meeker SN, *et al.* Identification of the tracheal and laryngeal afferent neurones mediating cough in anaesthetized guinea-pigs. *J Physiol* 2004;**557**(Pt 2):543–58.
14. Obyrne PM. Clinical comparisons of inhaler systems: what are the important aspects? *J Aerosol Med-Dep Clearance Effects Lung* 1995;**8**:S39–47.
15. Gardini A, D'Aloia A, Faggiano P. Amiodarone-induced adverse effects at the beginning of oral therapy: clinical implications. *Chest* 1998;**113**:848.
16. Oechsner M, Groenke L, Mueller D. Pleural fibrosis associated with dihydroergocryptine treatment. *Acta Neurol Scand* 2000;**101**:283–5.
17. Schnabel A, Dalhoff K, Bauerfeind S, *et al.* Sustained cough in methotrexate therapy for rheumatoid arthritis. *Clin Rheumatol* 1996;**15**:277–82.
18. Kastelik JA, Aziz I, Greenstone MA, *et al.* Pergolide-induced lung disease in patients with Parkinson's disease. *Resp Med* 2002;**96**:548–50.
19. Francis PC, Carlson KH, Owen NV, Adams ER. Preclinical toxicology studies with the new dopamine agonist pergolide: acute, subchronic, and chronic evaluations. *Arzneimittelforschung* 1994;**44**:278–84.
20. Irwin RS, Madison JM, Fraire AE. The cough reflex and its relation to gastroesophageal reflux. *Am J Med* 2000;**108**(4a):73S–78S.
21. Everett CF, Buckton GK, Kastelik JA, *et al.* Clinical history in the diagnosis of gastroesophageal reflux related cough. *Am J Respir Crit Care Med Suppl* 2003;**168**:A315.
22. Irwin RS. Chronic cough due to gastroesophageal reflux disease – ACCP evidence-based clinical practice guidelines. *Chest* 2006;**129**:80S–94S.
23. Ettinger B, Pressman A, Schein J. Clinic visits and hospital admissions for care of acid-related upper gastrointestinal disorders in women using alendronate for osteoporosis. *Am J Manag Care* 1998;**4**:1377–82.
24. Richter JE. Gastroesophageal reflux: diagnosis and management. *Hosp Pract* 1992;**27**(1):59–66.
25. Sesoko S, Kaneko Y. Cough associated with the use of captopril. *Arch Intern Med* 1985;**145**:1524.
26. Israili ZH, Hall WD. Cough and angioneurotic edema associated with angiotensin-converting enzyme inhibitor therapy: a review of the literature and pathophysiology. *Ann Intern Med* 1992;**117**:234–42.
27. Ravid D, Lishner M, Lang R, Ravid M. Angiotensin-converting enzyme inhibitors and cough: a prospective evaluation in

hypertension and in congestive heart failure. *J Clin Pharmacol* 1994;**34**:1116–20.

28. Coulter DM, Edwards IR. Cough associated with captopril and enalapril. *Br Med J* (Clin Res ed.) 1987;**294**:1521–3.

29. Wood R. Bronchospasm and cough as adverse reactions to the ACE inhibitors captopril, enalapril and lisinopril: a controlled retrospective cohort study. *Br J Clin Pharmacol* 1995;**39**: 265–70.

30. Hirata R, Nabe T, Kohno S. Augmentation of spontaneous cough by enalapril through up-regulation of bradykinin B1 receptors in guinea pigs. *Eur J Pharmacol* 2003;**474**(2/3): 255–60.

31. Yeo WW, Foster G, Ramsay LE. Prevalence of persistent cough during long-term enalapril treatment: controlled study versus nifedipine. *Q J Med* 1991;**80**:763–70.

32. Yeo WW, Higgins KS, Foster G, *et al.* Effect of dose adjustment on enalapril-induced cough and the response to inhaled capsaicin. *Br J Clin Pharmacol* 1995;**39**:271–6.

33. Morimoto T, Gandhi TK, Fiskio JM, *et al.* An evaluation of risk factors for adverse drug events associated with angiotensin-converting enzyme inhibitors. *J Eval Clin Pract* 2004;**10**: 499–509.

34. Morice AH, Lowry R, Brown MJ, Higenbottam T. Angiotensin converting enzyme and the cough reflex. *Lancet* 1987;**2**: 1116–18.

35. Ricciardolo FL, Rado V, Fabbri LM, *et al.* Bronchoconstriction induced by citric acid inhalation in guinea pigs: role of

36. Honan D, Jacob S, Malkan D, *et al.* The Cyclimorph® cough. *Anaesthesia* 2004;**59**:618–19.

37. Howaizi M, Delafosse C. Omeprazole-induced intractable cough. *Ann Pharmacother* 2003;**37**:1607–9.

38. Henry RE, Kadlee GJ. Pulmonary drug reaction secondary to sertraline (Zoloft). *J Allergy Clin Immunol* 1994;**93**:242.

39. Adibelli Z, Dilek M, Kocak B, *et al.* An unusual presentation of sirolimus associated cough in a renal transplant recipient. *Transplant Proc* 2007;**39**:3463–4.

40. Kartal ED, Alpat SN, Ozgunes I, Usluer G. Adverse effects of high-dose interferon-alpha-2a treatment for chronic hepatitis B. *Adv Ther* 2007;**24**:963–71.

41. Elli FN, Aroldi A, Montagnino G, Tarantino A, Ponticelli C. Mycophenolate mofetil and cough. *Transplantation* 1998;**66**:409.

42. Maggioni F, Mampreso E, Mainardi F, *et al.* Topiramate-induced intractable cough during migraine prophylaxis. *Headache* 2010;**50**:301–4.

43. Mitra S, Sinha PK, Anand LK, Gombar KK. Propofol-induced violent coughing. *Anaesthesia* 2000;**55**:707–8.

44. Tweed WA, Dakin D. Explosive coughing after bolus fentanyl injection. *Anesth Analg* 2001;**92**:1442–3.

45. Yemen TA. Small doses of sufentanil will produce violent coughing in young children. *Anesthesiology* 1998;**89**:271–2.

tachykinins, bradykinin, and nitric oxide. *Am J Respir Crit Care Med* 1999;**159**:557–62.

26

Pleural disease induced by drugs

STEVEN A SAHN

INTRODUCTION

In contrast to drug-induced parenchymal disease,[1] drug-induced pleural disease is relatively uncommon.[2,3] A pleural effusion is the most common manifestation, which can also cause chest pain, pleural fibrosis and pneumothorax. With the development of new drugs it is likely that many of these agents may also be associated with pleural complications. The implicated drugs may result in an isolated pleural effusion or concomitant lung and pleural disease.

The mechanisms responsible for drug-induced pleural disease include (in isolation or in association):[4]

- a hypersensitivity or allergic reaction;
- a direct toxic effect;
- enhanced oxygen free radical production;
- suppression of antioxidant defence mechanisms;
- chemical-induced inflammation.

Although there tends to be a temporal association between the initiation of the drug and the onset of pleural disease, some patients may present months to years later, suggesting a dose- or time-related drug effect.

CLINICAL PRESENTATIONS AND DIAGNOSIS

Dyspnoea, the most common presenting symptom, may be caused by large unilateral pleural or bilateral effusions, with or without concomitant parenchymal lung disease, which may be related to drug toxicity or be the target of drug therapy. Occasionally chest pain, either pleuritic or non-pleuritic, may be the initial presentation, which is typically associated with a pleural effusion. In contrast, patients may be asymptomatic with the pleural effusion discovered on a routine or unrelated chest radiograph.

A drug-induced pleural effusion should always be considered in the differential diagnosis of an exudative effusion when the cause is not apparent, and the patient's drug list should be scrutinized. In some instances the association may be obvious, such as when a patient has received chemotherapy for treatment of malignancy. On the other hand, an elderly woman may not report intermittent or long-term nitrofurantoin usage for urinary tract infection. It is imperative that the clinician exclude all drugs as a cause of an unknown exudative effusion before pursuing costly or invasive procedures.

Categories of drugs that are associated with pleural disease include cardiovascular agents, chemotherapeutic agents, ergoline drugs, sclerosing agents, and other classes of drugs that include cytokines, chelating agents, diabetes drugs, statins and antidepressants. A large number of drugs have been associated with drug-induced lupus pleural effusions, with a small group having a strong causal relationship (Box 26.1; see Chapter 27).

The presence of a pleural fluid eosinophilia (defined as > 10 per cent of the total nucleated cells being eosinophils) should raise the suspicion that an exudative effusion may be drug-related. There are a small number of currently used drugs that have been reported to cause pleural fluid eosinophilia on multiple occasions, with or without blood eosinophilia (Box 26.2). However, pleural fluid eosinophilia is more commonly associated with other diseases including haemothorax, pneumothorax, benign asbestos pleural effusion (BAPE), coccidiomycosis and histoplasmosis, Hodgkin's lymphoma, pulmonary embolism with infarction, and parasitic infection.

DRUGS ASSOCIATED WITH PLEURAL DISEASE

Acyclovir

Acyclovir, a nucleoside analogue that inhibits viral replication, is used for the treatment of herpetic infections. The adverse-effect profile includes gastrointestinal intolerance, renal insufficiency and fever. A single report of an elderly man who was treated for *Herpes zoster* ophthalmicus developed bilateral pulmonary

Box 26.1   Medications associated with drug-induced lupus

Strongly associated with lupus pleuritis

- Chlorpromazine
- Hydralazine (second most common)
- Isoniazid
- Methyldopa
- Penicillamine
- Procainamide (most common)
- Quinidine (third most common)

Anticonvulsants

- Carbamazepine
- Ethosuximide
- Primidone

Anti-inflammatory agents

- Diclofenac
- Ibuprofen
- Para-aminosalicyclic acid
- Sulindac
- Tolmetin

Antimicrobials

- Griseofulvin

- Nalidixic acid
- Nitrofurantoin
- Penicillin
- Streptomycin

Anti-TNF drugs

Cardiovascular agents

- Acebutolol
- Amiodarone
- Atenolol
- Captopril
- Clonidine
- Disopyramide
- Labetalol
- Lovastatin
- Minoxidil
- Practolol
- Prinolol
- Spironolactone

Endocrine agents

- Aminoglutethimide
- Methimazole
- Propylthiouracil

Gastrointestinal agents

- Promethazine
- Sulfasalazine

Gynaecological agents

- Danazol
- Oral contraceptives

Immune modulators

- Gold salts
- Interferon (alpha, beta)

Neurological agents

- Levodopa
- Methylsergide

Oncological agents

- Leuprolide acetate

Ophthamological agents

- Timolol eye drops

Psychiatric agents

- Lithium carbonate

Box 26.2   Drugs associated with an eosinophilic pleural effusion (information from www.pneumotox.com)

- Acyclovir
- Antidepressants
  - Amitriptyline
  - Desipramine
  - Imipramine
  - Tricyclic antidepressants
  - Trimipramine
- Beta-blockers
- Bromocriptine
- Clozapine

- Dantrolene
- Fenfluramine/dexfenfluramine
- Fluoxetine
- Gliclazide
- Imidapril
- Infliximab
- Isotretinoin
- Mesalamine
- Nevirapine

- Nitrofurantoin
- Praziquantel
- Progesterone
- Propylthiouracil
- Simvastin
- Sulfasalazine
- Tizanidine
- Valproic acid

---

infiltrates, fever, haemoptysis and a left pleural effusion after 4 days of therapy.[5] Acyclovir was discontinued on day 6 with a rapid resolution of fever; the pleural effusion and pulmonary infiltrates resolved by day 10.

All-*trans*-retinoic acid

ATRA is a normal plasma constituent that is derived from intracellular oxidation of plasma retinol that is absorbed by

the intestine. ATRA is accepted therapy for acute promyelocytic leukaemia.[6]

Based on the results of a phase II study of promyelocytic leukaemia patients treated with ATRA compared with historical controls, Frankel *et al.* described the ATRA syndrome, a life-threatening complication of unclear cause.[7] They found that 13 of 56 patients (23 per cent) treated with ATRA developed a distinct respiratory distress syndrome characterized by fever, pulmonary infiltrates, weight gain, pleural effusion, renal failure, pericardial effusion, heart failure and hypotension. In a retrospective study from South Korea, 15 of 69 patients (22 per cent) developed the ATRA syndrome during treatment of acute promyelocytic leukaemia.[8] Chest radiographs in these patients showed increased vascular pedicle (87 per cent), pulmonary congestion (87 per cent), pleural effusion (73 per cent), ground-glass opacities (60 per cent), septal lines (60 per cent), peribronchial cuffing (60 per cent), consolidation (47 per cent), nodule (47 per cent), and air bronchograms (33 per cent). Pleural effusions tended to be bilateral. De Botton et al. found no significant predictive factors for the ATRA syndrome, including pre-treatment leucocyte count.[9] While the ATRA syndrome was responsible for death in only 1.2 per cent of those treated with ATRA, it was associated with lower event-free survival and overall 2-year survival compared to non-ATRA-treated patients.

Amiodarone

The class III antiarrythmic drug amiodarone is an iodinated benzofuran, approved for the management of ventricular and supraventricular dysarrythmias. The adverse-effect profile is protean and virtually all organs can be affected. Approximately 6 per cent of patients receiving amiodarone have developed pulmonary toxicity that includes interstitial/alveolar infiltrates, pleural thickening, pulmonary nodules and pleural effusions.[10] Adverse effects are dose-dependent and occur less frequently with daily dosages below 400 mg; these patients can present from 2 to 30 days after the initiation of therapy. Pulmonary toxicity is manifested by the insidious onset of dyspnoea, unproductive cough, weight loss and, less commonly, fever. The CT scan shows ground-glass attenuation and at times increased brightness owing to the high iodine concentration of the drug. When a pleural effusion is present, there is usually concomitant parenchymal infiltrates; however, a single patient has been reported with amiodarone-associated bilateral pleural effusions without parenchymal involvement. The pleural fluid is typically unilateral and moderate in volume. A paucicellular exudate (total protein 2.8–5.5 g/dL) occurs in up to 33 per cent of those with amiodarone lung involvement. The pleural fluid is either lymphocyte or lymphocyte and neutrophil predominant. Foam cells resembling those found in bronchoalveolar lavage (BAL) fluid have been reported. Pleural thickening with foamy macrophages has been documented in pleural tissue. The effusion usually resolves on discontinuation of the drug. Those with persistent pleural effusions most likely would respond to corticosteroids based on data available from amiodarone pneumonitis.

In a single study, 13 of 20 patients (65 per cent) were noted to have pleural thickening, that is typically found in the area where pulmonary infiltrates are most prominent and abuts the pleura. Patients with pleural thickening may present with pleuritic chest pain. When amiodarone is discontinued, regression of the pleural thickening usually parallels the resolution of the pulmonary infiltrates.[10] The drug-induced lupus syndrome with pleuropericardial effusions appears to be a rare occurrence.

Bleomycin

Bleomycin, an antitumour antimicrobial agent, is used in the treatment of head and neck cancers, testicular carcinoma, and Hodgkin's and non-Hodgkin's lymphoma. Pleuropulmonary adverse effects occur in 6–10 per cent of patients with a mortality rate just under 2 per cent. The most common and devastating form of bleomycin-induced pulmonary disease is the development of interstitial pneumonitis and acute respiratory distress syndrome (ARDS) that may progress to pulmonary and pleural fibrosis enhanced by the use of high concentrations of oxygen, radiation therapy or other chemotherapeutic agents. Both pleural and lung toxicity are age- and dose-related: it is more common in patients older than 70 years and in those who receive a cumulative dose above 400 mg/m^2. Pleural effusions have been reported in a few patients with bleomycin lung toxicity.[11] The combination of drug cessation and corticosteroid therapy may result in resolution of the effusions and pneumonitis.

Bromocriptine

Bromocriptine, an ergot derivative with potent dopamine receptor agonist activity, activates postsynaptic dopamine receptors. The dopamine neurons modulate the secretion of prolactin and result in decreased prolactin levels. The drug is used in patients with acromegaly, as it lowers elevated levels of growth hormone. In addition, it has been used in Parkinson's disease, an entity in which there is progressive deficiency of dopamine synthesis in the substantia nigra. Chronic use of bromocriptine and other dopamine agonists (cabergoline, mesulergine and lisuride) or the ergoline methysergide can induce pleuropulmonary pathology.

Pleural effusions and pleural thickening have been reported in approximately 6 per cent of patients with Parkinson's disease treated with bromocriptine. The pleural effusion may or may not be associated with pulmonary infiltrates and/or rounded or folded lung. The development of pleural fibrosis appears to be related to the dose and the length of time that the patient has received the drug. Pleural effusions tend to occur between 12 and 48 months after the initiation of therapy. Pleural fluid analysis typically shows a lymphocyte predominant exudate, with rare reports of pleural fluid eosinophilia.[12] Withdrawal of the drug usually leads to complete resolution of the pleural effusions; however, pleural fibrosis and interstitial infiltrates may not resolve completely.[13] Corticosteroids may result in regression of pleural thickening and pulmonary infiltrates.

Clozapine

Clozapine, an atypical antipsychotic drug, has had success in the treatment of schizophrenia that is refractory to conventional neuroleptics. Adverse effects of clozapine include seizures, sialorrhea, agranulocytosis, hepatotoxicity, pancreatitis and hyperglycaemia. There have been a few reports of clozapine-induced pleural effusions, and there has been a single report of clozapine-induced polyserositis.[14] A pleural effusion appears to occur within the first 2–3 weeks of therapy. The effusions are bilateral or unilateral, usually accompanied by fever, and may be associated with haematuria and hepatocellular dysfunction. Pleural fluid analysis in a single patient revealed a neutrophil predominant exudate; pleural fluid eosinophilia has been reported on several occasions. Three cases of clozapine-induced polyserositis and three of isolated pleural effusions have been reported. In one patient, the erythrocyte sedimentation rate (ESR) was 90 mm/h without peripheral or pleural fluid eosinophilia. Pleural fluid in another patient was straw-coloured with 233 nucleated cells/μL with 74 per cent mononuclear cells and no eosinophilia; total protein was 2.4 g/dL with a serum/protein total ratio of 0.37, and the lactate dehydrogenase (LDH) was 88 IU/L. In the third patient, pleural fluid was straw-coloured with 1710 nucleated cells/μL, a total protein of 3.4 g/dL (protein/serum ratio of 0.52) and LDH of 149 IU/L; the patient had a moderate pericardial effusion without echocardiographic evidence of tamponade. In all reports the pleural effusions and fever resolved within 2 weeks following discontinuation of therapy. When re-challenged with the drug, the pleural effusion recurred but resolved with drug cessation.

Cabergoline

Cabergoline, a long-acting tetracyclic ergot derivative agonist, acts through dopamine antagonism, and is commonly used in the treatment of parkinsonism. Rarely, cabergoline and other ergoline drugs have been reported to cause a pleuropulmonary inflammatory–fibrotic reaction.[15,16] Subacute constrictive pericarditis requiring pericardiectomy has also been reported.[17] Drug dosage appears to be an important factor for the inflammation and fibrosis related to the ergoline drugs.[15,16] Long-term administration of ergolines appears to be a consistent finding in patients who develop the inflammatory–fibrotic reactions and appears to be unusual with less than 6 months of therapy. The mechanism of ergoline-induced pleural fibrosis is unclear; however, a serotonin-related mechanism has been suggested. Although the drug should be discontinued if pulmonary symptoms develop, worsening may still occur owing to the long half-life of the drug.

Cyclophosphamide

Cyclophosphamide, an immunosuppressive alkylating agent, is used in the treatment of a variety of non-malignant and malignant diseases. Adverse effects include bone marrow suppression, haemorrhagic cystitis, bladder cancer and gastrointestinal intolerance. The first case of cyclophosphamide-induced lung injury was reported approximately 40 years ago. Since the initial report, approximately 20 additional cases of pulmonary toxicity have been reported.[18]

The pulmonary toxicity from cyclophosphamide can occur either early or late. The early lung toxicity tends to occur 1–6 months following drug exposure. Typical symptoms in the early phase include dyspnoea, cough, fatigue and fever. The chest radiograph most commonly demonstrates bilateral reticular or reticulonodular infiltrates; acute pulmonary oedema has rarely been reported. Late-onset cyclophosphamide lung toxicity occurs months to years following drug initiation and has been reported as late as 6 years after discontinuation of the drug.[11] Patients with late toxicity present with the insidious onset of dyspnoea and unproductive cough. The radiographic features are similar to those seen with the early form of disease; however, pleural thickening is a useful radiographic finding of late-onset toxicity.[18] There has been a single report of a pleural effusion associated with cyclophosphamide; massive pleural effusions occurred 2 days following high-dose cyclophosphamide infusion prior to bone marrow transplantation.[19] Pleural fluid analysis revealed transudates (protein 2.7 g/dL and LDH 32 IU/L), which may have been related to myocardial injury that has been described as a complication of high-dose cyclophosphamide.

In early-onset cyclophosphamide toxicity, radiographic findings and symptoms typically remit with drug withdrawal and use of corticosteroids. Late-onset pulmonary toxicity is usually progressive and appears to be unresponsive to corticosteroids.[18]

Dantrolene

Dantrolene is used as a skeletal muscle relaxant. Common adverse effects include nausea, gastrointestinal atony, hepatitis and sedation. Dantrolene-associated pleural effusions have been reported in six patients who received doses between 100 and 400 mg per day.[20] The pleuropulmonary reaction typically presents with pleurodynia, fever and pleural effusion. Pleural effusion and pleural fibrosis are late sequelae with an onset 2 months to 12 years after initiation of the drug. The pleural effusion is typically unilateral, with pleural fluid eosinophilia being a prominent finding – ranging from 33 to 66 per cent of the total nucleated cell count. All effusions were exudates with normal glucose concentrations. Peripheral blood eosinophilia (7–18 per cent) is usually present. Parenchymal infiltrates have not been documented with dantrolene-induced pleural disease. Drug withdrawal is associated with a rapid decrease in symptoms and resolution of the pleural effusion.

Dapsone

Dapsone, a 4,4'-diaminodiphenylsulfone, has traditionally been used in the treatment of leprosy and dermatitis herpetiformis. Its usage has become more common in the treatment of diverse diseases including malaria, brown recluse spider bite, a number of dermatological diseases including acne, and as prophylaxis for *Pneumocystis jiroveci* infection.

The sulfone syndrome is a hypersensitivity reaction to dapsone manifested by fever, malaise, acute hepatitis, exfoliative dermatitis and haemolytic anaemia. The syndrome typically occurs 2–6 weeks after the initiation of dapsone and has been observed in patients taking as little as 50 mg daily for a short time. In addition to the previously described signs and symptoms of the sulfone syndrome, a teenager developed pancreatitis and a large right pleural effusion.[21] Pleural fluid analysis revealed an exudate with a total protein of 4.2 g/dL, a nucleated cell count of 345 cells/μL with 98 per cent mononuclear cells and 2 per cent neutrophils, glucose 99 mg/dL, amylase 69 IU/L and LDH 666 IU/L. Dapsone was discontinued and the patient was treated with corticosteroids, with resolution of hepatitis, pancreatitis, haemologic abnormalities and the pleural effusion. With a normal pleural fluid amylase, the pleural effusion was most likely caused by the underlying inflammatory process induced by dapsone.

Dasatinib

Dasatinib, an orally available Brc-Abl-kinase inhibitor, targets most imatinib-resistant BCR-ABL mutations in patients with chronic myelogenous leukaemia (CML) or Ph-positive acute lymphoblastic leukaemia (ALL). In a phase I study of 84 patients – 40 with chronic-phase CML, 11 with accelerated-phase CML, 23 with myeloid blast crisis, and 10 with lymphoid blast crisis or Ph-positive ALL – 15 patients (18 per cent) developed pleural effusions.[22] The rate of drug-related pleural effusion was 21 per cent in a series of five phase II trials involving 511 patients with chronic-phase, accelerated phase, blast crisis or Ph-positive ALL. Twenty-eight per cent of the patients had pleural effusions while receiving dasatinib at doses of under 140 mg per day, with twice-daily dosing. A randomized study is ongoing comparing once- with twice-daily dosing, at 100–140 mg daily, to determine whether the drug and the schedule affect the incidence of pleural effusion. A pleural effusion occurred in 48 patients (35 per cent: grade 3/4 in 23 patients), including 29 per cent of those treated in chronic phase, 50 per cent in accelerated phase, and 33 per cent in blast phase with dasatinib treatment. A history of cardiac disease, hypertension and use of twice-daily dosing appear to be risk factors associated with development of pleural effusions.[23]

The pleural fluid analysis characteristically shows a lymphocytic exudate.[24] We recently cared for two CML patients with dasatinib-induced pleural effusions. One patient had bilateral haemorrhagic effusions that were positive for chylomicrons and negative on flow cytometry; pleural fluid analysis on both sides showed lymphocyte-predominant, protein-discordant exudates with pH values of 7.33 and 7.37, glucose concentrations similar to serum, and triglycerides of 169 and 170 mg/dL. The second patient also had a lymphocyte-predominant, protein-discordant exudate with a triglyceride of 59 mg/dL and negative chylomicrons.

Docetaxel

Docetaxel is a chemotherapeutic agent approved for the treatment of ovarian and breast cancers. The intravenous administration of docetaxel has been reported to cause myelosuppression, peripheral neuropathy and a hypersensitivity reaction. A series of five patients was reported to have docetaxel-induced pleural effusions.[25] Pleural fluid analysis revealed that all the effusions were exudative with a glucose concentration similar to serum and pleural fluid pH in the alkaline range of 7.48–7.60. Cytological examination showed

predominantly reactive mesothelial cells without eosinophil predominance. The effusions appeared an average of 19 weeks following the initiation of drug therapy and resolved an average of 20 weeks following drug cessation.

Fluoxetine

Fluoxetine hydrochloride, a commonly used drug for depression, has been reported to cause a pleural effusion in a single patient.[26] Approximately 8 weeks into treatment the patient developed pleuritic chest pain and a right pleural effusion. Pleural fluid analysis revealed 22 500 nucleated cells/μL, with 3 per cent neutrophils, 56 per cent lymphocytes, 5 per cent monocytes and 36 per cent eosinophils. Total protein was 6.1 g/dL with an LDH of 2335 IU/L (the normal upper limit of serum LDH is 580 IU/L). Glucose was 98 mg/dL, the pH was 7.40, and amylase was 44 mg/dL. Pleural fluid cytology was negative. The blood leucocyte count was 6800/μL with 10 per cent eosinophils. Pleural biopsy showed inflammatory changes with mild fibrosis, infiltration of lymphocytes, and minimal mesothelial hyperplasia without eosinophils. Fluoxetine was discontinued with partial resolution of the effusion by 4 weeks and total resolution by 8 weeks. At the time of resolution of the effusion, blood eosinophilia had decreased to 2 per cent. The effusion did not recur. The pleural effusion was thought to be secondary to a hypersensitivity reaction. Fluoxetine has been reported to cause hypersensitivity pneumonitis and pulmonary phospholipidosis.

Gliclazide

Gliclazide, an oral hyperglycaemic agent, was described in an isolated case report as a cause of pleural fluid eosinophilia.[27] A 52-year-old man developed a unilateral effusion with parenchymal infiltrates 2 weeks following the administration of gliclazide. Blood eosinophilia (20 per cent) and pleural fluid eosinophilia (80 per cent) were documented. The drug was discontinued and 1 month later the patient became asymptomatic. The blood eosinophil count normalized, and chest radiograph showed complete resolution of the effusion.

Granulocyte colony-stimulating factor

GCSF, a haematopoietic growth factor, is used to stimulate bone marrow recovery following chemotherapy-induced myelosuppression. There is an isolated report of a 43-year-old woman with breast cancer who developed an exudative pleural effusion 10 days following the initiation of GCSF.[28] The pleural fluid analysis showed no evidence of infection or malignancy. Pleural fluid cytology demonstrated an increase of myeloid cells, both mature and immature forms. The pleural effusion resolved slowly following discontinuation of the GCSF.

Imatinib mesylate

Imatinib mesylate, a selective Ber-Abl tyrosine kinase inhibitor, has significant antileukaemic activity in patients with chronic myelogenous leukaemia. Imatinib has also been used in recurrent metastatic Ewing's sarcoma and metastatic primitive neuroectodermal tumours. The most frequent adverse events from imatinib are nausea, oedema, myalgias and rash. There have been a few reports of pleuropericarditis that has been described in both adults and children.[29] In the adult cases, the typical dose used was 600 mg daily; the effusions, which were bilateral or unilateral, occurred from 2 days to 54 weeks following initiation of imatinib with or without pericardial effusions. Some patients showed significant peripheral oedema and weight gain associated with pleuropericarditis. The effusion has only been described as an exudate without evidence of malignancy, infection or congestive heart failure. Only one case appeared to be related to cardiac tamponade. The effusions resolved with discontinuation of imatinib, with or without corticosteroids and diuretics. When patients were re-dosed at 400 mg per day, pleural effusion and pericardial effusion did not recur. It has been suggested that patients treated with imatinib at a dose above 600 mg per day should be more closely observed to see whether a prophylactic diuretic should be added to their regimen.

Interleukin-2

Recombinant human interleukin-2 (IL-2) has shown promise in the treatment of renal cell carcinoma and melanoma. Acute, but usually reversible, adverse effects include fever, chills, gastrointestinal distress, anaemia, thrombocytopenia, blood eosinophilia, and capillary leak syndrome; the latter has led to hypotension, renal insufficiency, hepatic dysfunction and thyroid dysfunction. Pleuropulmonary abnormalities occur in approximately 75 per cent of patients.[30,31] Chest radiograph has demonstrated bilateral perihilar alveolar/interstitial focal infiltrates and pleural effusions. Pleural effusions occur in approximately 50 per cent of patients and are associated with pulmonary oedema. Following cessation of therapy, the effusions typically improve; however, approximately 20 per cent have a persistent pleural effusion at 4 weeks following drug discontinuation. The effusions most likely result from capillary leak as occurs with any cause of ARDS. There has not been a description of the pleural fluid in these patients; however, it is likely that the effusions would be exudative based on the proposed pathogenesis.

Intravenous immunoglobulin

IVIG has been used as an immunomodulator for the treatment of autoimmune diseases, such as idiopathic thrombocytopenia (ITP) and myasthenia gravis, and as supplemental therapy for congenital immunodeficiency states. There is an isolated report of a recurrent lymphocytic pleural effusion in a young woman who was treated with IVIG for ITP.[32] The pleural fluid was described as a lymphocyte-predominant exudate (total protein 3.3 g/dL and LDH 274 IU/L) with a normal glucose. Microbiological and immunological studies of the pleural fluid were negative. The pleural effusion resolved within 2 weeks following discontinuation of the drug. The subject was re-challenged with the same preparation of

IVIG that had resulted in the recurrent pleural effusion that had a similar pleural fluid analysis. Drug withdrawal again resulted in resolution of the effusion. Interestingly, the patient was subsequently treated with a different preparation of IVIG, as she experienced recurrent ITP. However, there was no development of a pleural effusion, suggesting that the pleural injury from IVIG was not class-specific.

Isotretinoin

Isotretinoin, a retinoid compound used in the treatment of severe nodulocystic acne, has an adverse-effect profile that includes hypertriglyceridaemia, dry skin, cheilitis and teratogenicity. Two isolated reports of individuals in their 40s developed unilateral pleural effusions 1 and 7 months following the initiation of isotretinoin, one for cystic acne and the other for systemic sclerosis.[33] Both demonstrated pleural fluid eosinophilia without blood eosinophilia or interstitial lung disease. One patient had chronic eosinophilic infiltration on pleural biopsy. Following discontinuation of isotretinoin, the effusions resolved in 1 month and 3 months, respectively.

Itraconazole

Itraconazole, a triazole used to treat systemic mycosis, has been described in an isolated report to have caused a pleural effusion in a 49-year-old individual following 8 weeks of the drug at 200 mg twice-daily.[34] The pleural effusion was followed 1 week later by a pericardial effusion that required pericardiocentesis. Pleural fluid analysis was described as an exudate (total protein 4.3 g/dL); percutaneous pleural biopsy was not diagnostic. Following re-exposure to the drug 6 weeks later, parenchymal infiltrates and cardiomegaly developed without recurrent pericardial effusion. A cardiac evaluation showed no evidence of ischaemic or valvular heart disease, and the authors concluded that the cardiac dysfunction was induced by re-exposure to itraconazole.

Mesalamine

Mesalamine, a molecule composed of sulfasalazine linked to a sulfapyridine moiety, is used in the treatment of inflammatory bowel disease. Adverse effects, which are related to the sulfapyridine component, include gastrointestinal intolerance, urticaria, drug-induced lupus, hepatitis, blood dyscrasia and interstitial nephritis. There have been two reports of pleuropericarditis and an eosinophilic pleural effusion with parenchymal infiltrates associated with mesalamine therapy for ulcerative proctitis and presumed inflammatory bowel disease. After 2 weeks of mesalamine, a patient presented with fever, fatigue, chest pain, acute pericarditis and a small left pleural effusion without parenchymal infiltrates. The patient was re-challenged with mesalamine after the chest pain had resolved and recurrent symptoms developed which resolved after 1 week of systemic corticosteroids. Antinuclear antibodies and serological studies for viral pathogens were negative, suggesting that the clinical presentation was a mesalamine-induced hypersensitivity reaction. The second patient developed an eosinophilic pleural effusion with peripheral eosinophilia and parenchymal infiltrates following mesalamine therapy for presumed inflammatory bowel disease.[35] Drug cessation and systemic corticosteroids resulted in complete remission of the radiographic abnormalities over a few months.

Methotrexate

Methotrexate, an antimetabolite that acts as an immune modulator and steroid-sparing agent, is used in the treatment of autoimmune diseases, malignancy and psoriasis. The most significant adverse effects include bone marrow suppression, mucositis, hepatitis, gastrointestinal distress and pulmonary fibrosis. With high-dose methotrexate, approximately 3–4 per cent of patients develop pleuropulmonary disease. Of 317 patients who received methotrexate, 14 (4 per cent) developed pleuritic chest pain within 4 weeks of therapy.[36] Chest pain was described as severe and acute in onset and recurred with re-challenge of the drug. Four of the 14 patients (29 per cent) developed pleural effusions.

Lower dose methotrexate can lead to pleural thickening, pleural effusions and interstitial infiltrates with associated fever, cough and dyspnoea. A hypersensitivity reaction is thought to be the mechanism responsible for the pneumonitis and pleural disease.[37] Despite continuation of the drug, the pleural changes appear to be self-limited and rarely progress to significant respiratory impairment. Corticosteroid therapy appears to ameliorate symptoms rapidly; however, it may not affect radiographic abnormalities.

Methysergide

Methysergide, a serotonin antagonist, is used to treat migraine headaches. Significant adverse effects include alopecia, dermatitis, nausea, and coronary and valvular heart disease. Fibrosing mediastinitis and retroperitoneal fibrosis have been recorded in the medical literature since the 1960s. Subsequent reports have documented fibrosis of the pleura, lung, endocardium and pericardium. Pleuropulmonary complications occur in fewer than 1 per cent of patients taking the drug. Most patients present with fever, dyspnoea, weight loss and pleuritic chest pain 1–4 years after initiation of the drug.[38] Pleural fluid has been described as serous to bloody with total protein concentrations of 2.0–4.4 g/dL and a low nucleated cell count.[39] On pleural biopsy, there was evidence of chronic pleuritis with a mononuclear cell infiltrate with fibroblasts.

The proposed pathogenic mechanism for methysergide-induced fibrosis is thought to be stimulation of fibroblasts from the increased levels of serotonin. Treatment of pleuropulmonary injury is drug cessation which results in regression of the pleural thickening in some, but not all, patients.

Minoxidil

Minoxidil, an antihypertensive agent used to treat refractory hypertension, acts as a peripheral vasodilator. Significant adverse effects include excessive hair growth, leukopenia and thrombocytopenia, nausea, breast tenderness and skin rash. Minoxidil has been associated with the development of pericardial effusions. There is an isolated report of bilateral pleural effusions that developed in a patient following 3 months of minoxidil therapy.[40] Both thoracentesis and pericardiocentesis revealed an exudative effusion without evidence of infection, connective tissue disease or malignancy. Drug cessation resulted in resolution of the pericardial and pleural effusions. Re-challenge resulted in recurrence of only the pericardial effusion that subsequently resolved with drug cessation.

Mitomycin

Mitomycin, an alkylating agent used in the treatment of gastrointestinal, breast and lung cancers, is associated with an adverse-effect profile that consists of dermatitis, renal insufficiency, bone marrow suppression and gastrointestinal symptoms. Pleuropulmonary fibrosis has been reported.[41] Similar to bleomycin, ARDS has been described in patients receiving mitomycin and a high concentration of oxygen. Both pleural effusion and pleural thickening have been ascribed to mitomycin; however, pleural fluid analysis has not been reported. Biopsy of the pleural fibrotic lesions has shown lymphocytes and eosinophils in a single patient. Drug withdrawal and corticosteroids have resulted in partial resolution of the radiographic abnormalities in most patients.

Nitrofurantoin

Nitrofurantoin, an antibacterial agent, is used in the treatment of chronic urinary tract infections. The adverse-effect profile is extensive and includes dermatitis, urticaria, angioedema, gastrointestinal intolerance, hepatotoxicity, neuropathy and pleuropulmonary reactions. Nitrofurans are reduced by bacterial enzymes to highly reactive, short-lived intermediates that are believed to cause DNA strand damage and alter bacterial ribosomal proteins. This mechanism of nitrofurantoin action is a possible explanation for the hypersensitivity reaction involving the lung parenchyma and pleura.

The pleuropulmonary reactions from nitrofurantoin may be acute, subacute or chronic. Since the first report of nitrofurantoin-induced pleuropulmonary reaction, approximately 45 years ago, more than 2000 additional cases have been reported.[42] The acute pleuropulmonary syndrome occurs in 5–10 per cent of patients receiving the drug. The patient may present within days to 1 month following drug therapy with fever, dyspnoea and cough. Chest radiograph typically shows bibasilar alveolar/interstitial infiltrates. In about a third of cases, small-volume pleural effusions may be observed and rarely occur in the absence of parenchymal abnormalities.[43] In a series of 447 patients with pulmonary reactions to nitrofurantoin, 232 (52 per cent) had infiltrates alone, 68 (15 per cent) had infiltrates and effusions, and 14 (3 per cent) had only pleural effusions; 70 (16 per cent) had normal chest radiographs.[44] Blood eosinophilia (as high as 83 per cent) is a common finding with nitrofurantoin toxicity. In addition, pleural fluid eosinophilia has been reported. The pleural effusions are usually bilateral, and drug withdrawal generally results in resolution, which may be hastened with the use of corticosteroids.

Chronic nitrofurantoin lung toxicity typically presents with the insidious onset of progressive dyspnoea on exertion and unproductive cough with interstitial pneumonitis and fibrosis. Pleural effusions are rarely observed with the chronic form of the disease.[44]

Pleural thickening has been described with chronic nitrofurantoin administration. The severity of the chronic pulmonary reaction with nitrofurantoin appears to be related to dosage and duration of therapy. With the development of fibrosis, pulmonary function is impaired permanently and may continue to deteriorate despite discontinuation of the drug and corticosteroid therapy.

Penicillamine

Penicillamine, a chelating agent for copper, exhibits immunosuppressive effects and is approved for the treatment of Wilson's disease; however, it also has been used in the treatment of autoimmune diseases because of its immunosuppressive action. Adverse effects associated with D-penicillamine include blood dyscrasias (neutropenia, thrombocytopenia, haemolytic anaemia and aplastic anaemia), dermatitis (Stevens–Johnson syndrome and pemphigus), conditions that mimic other autoimmune diseases (rheumatoid arthritis, Goodpasture's syndrome, lupus erythematosus and myasthenia gravis), and nephrotic syndrome.

An isolated report described a young woman who developed a large, left-sided pleural effusion more than 12 years after receiving penicillamine therapy for Wilson's disease.[45] Pleural fluid analysis revealed straw-coloured fluid with negative cytology. Serological testing for connective tissue disease was negative. Multiple thoracenteses were required for this recurrent effusion. Video-assisted thoracic surgery demonstrated white plaques on the visceral and parietal pleura. Talc poudrage was unsuccessful with symptomatic re-accumulation of fluid. A thoracotomy and decortication were performed; pleural biopsy revealed chronic, non-specific inflammation without evidence of infection or malignancy. Successful lung re-expansion was accomplished with decortication. The patient was not re-challenged with penicillamine, and no further sequelae were reported.

Procarbazine

Procarbazine, a methylhydrazine, is used in the treatment of Hodgkin's disease. The most serious adverse effects include bone marrow suppression, neurological changes and gastrointestinal symptoms. Patients with procarbazine-induced pulmonary disease present with fever, cough, dyspnoea and blood eosinophilia.[46] Both unilateral and bilateral pleural effusions have been described with procarbazine treatment; these effusions may or may not be associated with concomitant parenchymal infiltrates. With discontinuation of the drug, symptoms and radiographic changes tend to resolve within a few days. Re-challenge with procarbazine has induced

the same clinical syndrome including pleural effusions. There have been no reports detailing pleural fluid analysis.

Propylthiouracil

Propylthiouracil, a thionamide approved for the treatment of Grave's disease, has been reported to cause an eosinophilic pleural effusion 3 weeks after starting the drug. The patient presented with pleuritic chest pain, and pleural fluid analysis showed an exudative effusion with an initial eosinophil count of 16 per cent that increased to 45 per cent on repeat thoracentesis.[47] A pleural biopsy showed chronic inflammation with an eosinophilic infiltrate. There was no concomitant blood eosinophilia or lung parenchymal abnormalities. Following discontinuation of the drug, the effusion resolved in 3 months.

Prostaglandin–E1

PGE1 is used to achieve vascular patency in occlusive arterial disease and to maintain vascular patency pending surgery in paediatric cardiac cases. It has also been used in Japan for reattachment surgeries to maintain circulation and revascularization.

An isolated report of a 75-year-old man documents the development of bilateral pleural effusions 12 days after daily PGE1 therapy at 120 μg per day following reattachment surgery.[48] Pleural fluid analysis revealed a total protein of 2.9 g/dL (pleural fluid/protein ratio 0.55), an LDH of 129 IU/L, and a normal glucose. PGE1 was discontinued following detection of the pleural effusions, and the effusions resolved 8 days later. The authors hypothesized that PGE1 induced the pleural effusions because of its propensity to increase capillary permeability; in addition, hypoalbuminaemia may have contributed to the pleural fluid volume.

Rosiglitazone

Rosiglitazone, a thiazolidinedione, is being increasingly used for glycaemic control in diabetes. Adverse effects of these drugs are fluid retention, peripheral oedema and congestive heart failure. These adverse effects have particularly been observed in patients with baseline left ventricular dysfunction and renal insufficiency when the drug is combined with insulin.

A 61-year-old man with diabetes mellitus developed a recurrent left pleural effusion 1 week following CABG surgery.[49] The effusion persisted following chest tube drainage, prednisone and indomethacin, while rosiglitazone and insulin were continued. Following discontinuation of rosiglitazone and prednisone, the effusion resolved. The authors speculated that the pleural effusion was caused by rosiglitazone, as it persisted despite treatment with a non-steroidal drug as well as a corticosteroid and did not resolve until rosiglitazone was discontinued. They hypothesized that thiazolidinediones cause increased vascular permeability leading to a 'vascular leak syndrome'. They also reasoned that increased vascular permeability induced by thiazolidinediones may have potentiated the vascular leak that was present from the inflammatory process associated with CABG surgery leading to persistent pleural effusions. It is well known, however, that persistent left-sided pleural effusions occur following CABG surgery owing to diaphragm dysfunction, harvesting of the internal mammary artery, and the post-cardiac-injury syndrome.

Simvastin

Simvastin, a co-enzyme A-reductase inhibitor, is used to treat certain forms of hypercholesterolaemia. The most serious adverse effect is drug-induced hepatitis. There is an isolated report of hypersensitivity pneumonitis in a 61-year-old man with BAL eosinophilia and concomitant interstitial lung disease.[50] The patient also demonstrated a large right-sided pleural effusion with pleural thickening in the left hemithorax. At thoracoscopy, the pleural fluid was dark brown without evidence of infection. Pleural fluid analysis was not reported. Following drug withdrawal and corticosteroid therapy, symptomatic improvement occurred within a few days.

Tizanidine

Tizanidine, an α_2-adrenergic receptor agonist, is used to treat chronic pain from muscle spasms and spasticity associated with central nervous system disorders, such as multiple sclerosis. The mechanism of action is related to its skeletal muscle relaxant and antinociceptive properties. Tizanidine has also been reported to control neuropathic pain, associated with complex regional pain syndrome, trigeminal neuralgia and post-herpetic neuralgia. Common adverse effects include asthenia, somnolence, sedation, dizziness, dry mouth, increased spasms or tone, hypotension and bradycardia.

An isolated report of a 42-year-old man treated with tizanidine (2 mg every 8 hours) developed a right pleural effusion 6 weeks later that was detected on a CT scan for evaluation of low back pain.[51] A few weeks after the tizanidine dose was doubled, the right pleural effusion progressed without concomitant parenchymal infiltrates. Pleural fluid analysis showed a cloudy, yellow fluid that was exudative (total protein 4.1 g/dL and LDH 235 IU/L) with a glucose of 117 mg/dL. Total nucleated cell count was 2731 cells with 26 per cent neutrophils, 21 per cent lymphocytes, 10 per cent eosinophils and 40 per cent other cells. Eight months following initiation of tizanidine, the drug was discontinued; 4 weeks following drug discontinuation the effusion had resolved.

Troglitazone

Troglitazone, used in the treatment of insulin-resistant diabetes mellitus, has the most serious adverse effect of hepatic dysfunction. There is an isolated report of a 47-year-old man treated with troglitazone who developed cough, dyspnoea and night sweats after a week of therapy.[52] Empirical antibiotics did not affect the symptoms. Chest radiograph showed small bilateral pleural effusions and basilar interstitial infiltrates. The patient's symptoms abated within 24 hours after discontinuing troglitazone; however, complete resolution of the pleural effusions and parenchymal infiltrates did not occur for 6 days. Because of the temporal relationship between drug initiation and occurrence of the pleuropulmonary disease, a causal relationship was implied.

L-tryptophan

L-tryptophan, an over-the-counter sleep-promoting agent, has been associated with pleuropulmonary reactions. It has also been implicated in the eosinophilia–myalgia syndrome; the pathogenesis may be related to a contaminant during the manufacturing process. Eosinophilia–myalgia syndrome often presents with fever, myalgias, arthralgias, dermatitis, neuropathy and marked blood eosinophilia. Dyspnoea is a common symptom. Chest radiographs have demonstrated bilateral alveolar/interstitial infiltrates and pleural effusions.[53] Acute eosinophilic pneumonitis, chronic interstitial lung disease and chronic pulmonary hypertension have also been reported; however, pleural fluid analysis has not been described. Although some individuals have improved following discontinuation of L-tryptophan or with corticosteroid therapy, the response is often incomplete.

Valproic acid

Valproic acid, a carboxylic acid derivative, is used for the treatment of specific seizure disorders and acute manic episodes in patients with bipolar disorder. It has also been used for migraine headache prophylaxis and other psychiatric conditions. Common adverse effects include somnolence, tremor, nausea and vomiting; hepatic failure has also been reported.

We recently reported a 25-year-old man with neurosarcoidosis who presented with an asymptomatic right pleural effusion and reviewed the five other reported cases.[54] Pleural fluid analysis showed a clear yellow fluid with total protein of 5.2 g/dL, LDH of 354 IU/L, 1200 nucleated cells with 75 per cent eosinophils, glucose 84 mg/dL, and a pH of 7.65. The pleural fluid findings of all six cases were consistent. The pleural fluid was exudative with an eosinophil percentage of at least 40 per cent (range 40–75). Blood eosinophilia was often, but not always, present. Patients may present with an asymptomatic pleural effusion or with fever, cough or chest pain. The effusions can be unilateral or bilateral. A surprising finding was that three of the four cases had a pleural pH above 7.56; a pH this alkaline has only been reported previously with proteus empyema, and there are reports of patients with docetaxel-induced pleural effusions with pH ranging from 7.48 to 7.60. When valproic acid was discontinued in the six patients, the pleural effusions resolved completely by 6 months. With the presence of blood and pleural fluid eosinophilia, valproic acid-induced pleural effusions are most likely caused by a hypersensitivity reaction.

Vitamin B5 and vitamin H

Vitamin B5 (pantothenic acid) and vitamin H (biotin) have been proposed as a treatment for hair loss. Vitamin H also has been used in biotin deficiency, multiple carboxylase deficiency, brittle nails and dermatitis. Vitamin B5 has been used in the treatment of acne and rheumatoid arthritis.

Only mild adverse events, including rash and diarrhoea, have been reported with vitamin B5. There is an isolated report of a 76-year-old woman who was hospitalized with chest pain and dyspnoea.[55] She had no history of atopic disease. She had been taking trimetazidine for 6 years, vitamin H (10 mg daily) and vitamin B5 (300 mg daily) for 2 months for the treatment of alopecia. Chest radiograph showed cardiac enlargement and a pleural effusion. Clinical examination and echocardiogram revealed cardiac tamponade. The absolute pleural fluid eosinophil count ranged from 1200 to 1500/μL. At pericardiectomy the fluid was a sterile exudate with protein 6.7 g/dL with 1500 eosinophils/μL. Histology showed an eosinophilic infiltrate. The pleural fluid also confirmed eosinophilia. Despite multiple thoracenteses the pleural effusion recurred, leading to thoracotomy. All studies for rheumatological disorders, infection and occult malignancy were negative. Vitamins B5 and H were discontinued, and 1 week later the patient improved dramatically with resolution of blood eosinophilia. Trimetazidine was also discontinued but readministered a few months later without symptoms or an increase in eosinophilia. The patient was symptom-free without relapse at a 2-year follow-up visit.

DRUG-INDUCED LUPUS PLEURITIS

Since the initial report of a lupus-like illness following sulfadiazine treatment approximately 60 years ago,[56] more than 80 drugs have been reported to cause a lupus-like reaction or exacerbation of pre-existing disease (see Box 26.1). However, only procainamide, hydralazine, chlorpromazine, isoniazid, D-penicillamine, methyldopa and quinidine have a strong association with drug-induced lupus (DIL).[57] Only procainamide and hydralazine have been studied both retrospectively and prospectively to determine their association with a lupus-like reaction.

Pleuropulmonary involvement is especially common in DIL. Almost 60 per cent of patients with procainamide-induced lupus have manifested pleural effusions and pleuritis.[57] Pleuropulmonary disease is reported in approximately 30 per cent of patients who have hydralazine-induced lupus.[57] The pleural fluid characteristics of DIL are similar to native lupus; the effusions are exudative with nucleated cell counts ranging from 200 to 15 000 cells/μL; pleural fluid pH and glucose are usually normal but may be low in 10–15 per cent of cases.[58] Pleural fluid antinuclear antibodies (ANA) are higher than concomitant serum values and frequently exceed titres of 1:250. Cytological examination may reveal the diagnostic presence of lupus erythematosus (LE) cells.[58] Antihistone antibodies are not specific for DIL and can be found in 50–80 per cent of the patients who have native lupus. For example, antihistone antibodies can be detected in patients with rheumatoid arthritis, Felty's syndrome, and mixed connective tissue disease.[57] Hypocomplementaemia and detectable anti-double stranded DNA are seen in fewer than 1 per cent of individuals with DIL, compared with native lupus where they occur at frequencies of, respectively, 66 and 44 per cent.[57] Intrinsic renal disease is extremely rare in DIL when compared with idiopathic lupus. Recently, interferon, TNF-α receptor

inhibitors and statins have been added to the list of drugs that induce lupus. *See also* Chapter 27.

Procainamide

Procainamide is the drug that is most commonly associated with DIL. More than 90 per cent of patients on procainamide will develop a positive serum ANA after 1 year of therapy, and a third will develop a lupus-like syndrome.[59] Procainamide-induced lupus can occur as early as 1 month or as late as 12 years from the initiation of therapy. The acetylator phenotype is an important determinant of symptom onset; slow acetylators developed a lupus-like syndrome at a lower dose of the drug and after a shorter duration of therapy.[59,60] Parenchymal infiltrates are noted in about 40 per cent of patients, and slightly more than 50 per cent have pleuritis and pleural effusions.[61] Antihistone antibodies associated with procainamide-induced lupus are predominantly IgG and are directed against (H2A-H2B)-DNA complex and chromatin.[62]

Hydralazine

Hydralazine-induced lupus occurs in 2–21 per cent of patients. The incidence is dependent on the dose, duration of treatment and acetylator phenotype.[57] The greatest risk is associated with dosages above 200 mg per day or more than a 100 g cumulative dose.[57] Pleuropulmonary disease has been reported in approximately 30 per cent of patients, whereas isolated parenchymal disease occurs in fewer than 5 per cent. The ANA is positive in half of the patients who receive 200 mg daily; the clinical syndrome develops in 10 per cent of these patients.[57]

Other drugs associated with drug-induced lupus

Quinidine was reported to induce a lupus-like syndrome in approximately 30 patients; serositis, arthritis, fever and skin rash were common presenting features.[57] Adult patients who received isoniazid (25 per cent) and chlorpromazine (16–52 per cent) have a positive ANA; however, only 1 per cent will develop a lupus-like reaction.[57]

Drug-induced lupus pleuritis typically improves rapidly after discontinuation of the drug and usually does not require specific therapy. Non-steroidal anti-inflammatory medications or short courses of low- to moderate-dose corticosteroids are beneficial in prolonged cases or in patients who are markedly symptomatic.

CHEMOTHERAPY-INDUCED SPONTANEOUS PNEUMOTHORAX

While secondary spontaneous pneumothorax is common in patients with lymphangioleiomyomatosis, Langerhans cell histiocytosis and chronic obstructive pulmonary disease (COPD), it appears to be rare in patients with malignancy. About 30 patients have been reported in the literature with chemotherapy-related secondary spontaneous pneumothoraces.[63] Lung cancer-related secondary spontaneous pneumothorax is a rare occurrence accounting for approximately 0.05 per cent of all pneumothoraces.[64] Although most malignancies may be associated with a risk of pneumothorax, this complication appears to occur most commonly in patients with metastatic osteogenic sarcoma and germ-cell tumours.[65] However, it has been described in patients with uterine cancer, breast carcinoma, small-cell lung cancer, uterine sarcoma, soft tissue sarcoma, Wilms' tumour, teratoma, synovial sarcoma, uterine leiomyosarcoma, endometrial cancer, lymphoma, and adenocarcinoma from an unknown primary (Box 26.3).

Malignancy-related spontaneous pneumothorax appears to occur in four distinct circumstances:

- at the time of diagnosis;
- during progression of disease;
- shortly following chemotherapy;
- late following chemotherapy or radiotherapy.

However, most malignancy-related spontaneous pneumothoraces are present at the time of diagnosis.

Secondary spontaneous pneumothorax may be a marker of effective chemotherapy, with a tendency to occur between 1 and 32 days following the initiation of treatment. Several mechanisms have been hypothesized to explain chemotherapy-associated spontaneous pneumothoraces:[65]

- underlying emphysema;
- rupture of a peripheral, chemosensitive tumour with direct leakage of air into the pleural space;
- enlargement of a rapidly necrotizing tumour;
- the combination of necrosis and chemotherapy-induced impairment of repair processes in the lung (e.g. doxorubicin);
- the elevation of intrathoracic pressure due to drug-related emesis (cisplatin).

Chemotherapy-induced pneumothorax from metastatic malignancy may be bilateral and occur simultaneously. Chemical pleurodesis through a chest tube should be considered in all patients with chemotherapy-induced secondary spontaneous pneumothorax, as the risk of recurrence with or without further chemotherapy appears to be considerable.

Box 26.3　Tumours associated with chemotherapy-induced pneumothorax

- Osteogenic sarcoma
- Germ-cell tumour
- Uterine leiomyosarcoma
- Endometrial carcinoma
- Synovial cell carcinoma
- Lymphoma
- Wilms' tumour
- Thymoma
- Small-cell lung cancer
- Breast cancer
- Adenocarcinoma of unknown primary

RECREATIONAL DRUGS ASSOCIATED WITH PNEUMOTHORAX AND PNEUMOMEDIASTINUM

See also Chapter 12.

Most of the cases of recreational drug-associated pneumothorax and pneumomediastinum are believed to be related to barotrauma from the technique of using the drug (Box 26.4). Possible mechanisms include vigorous inhalation with combined intentional or inadvertent Valsalva manoeuvre, a form of partner-induced positive-pressure ventilation used to enhance the inhaled drug experience, and a direct toxic effect on the lung parenchyma from inhalation of an irritating substance causing alveolar rupture.[66]

Pneumothorax has also been described following 'pocket shots', which are attempted intravenous injections into the internal jugular and subclavian vein. There are several reports of spontaneous pneumomediastinum and an isolated report of pneumothorax in an individual who performed 8 hours of vigorous whistle-blowing following the use of Ecstasy.[67]

Spontaneous pneumomediastinum most likely results from rupture of alveoli following a sudden increased bronchovascular pressure gradient. It is commonly associated with increased alveolar pressure during mechanical inhalation and exaggerated Valsalva manoeuvres occurring with emesis, coughing and parturition. It may also occur when there is decreased pulmonary interstitial pressure, as can occur with bronchiolitis. The increased bronchoalveolar pressure gradient promotes alveolar gas dissection along the perivascular sheaths into the mediastinum. The mediastinal air follows the path of least resistance, frequently into the neck along the contiguous layers of the cervical fascia, preventing tamponade and resulting in subcutaneous emphysema. If the mediastinal pressure increases to a critical level, a pneumothorax can develop.

While there is no evidence of a direct pharmacological effect of Ecstasy on the mediastinum or pleura, barotrauma may occur from the decreased interstitial pressure resulting in an increased bronchovascular gradient that occurs with enhanced physical exertion seen in some Ecstasy users.[68,69]

Recreational drug-related spontaneous pneumomediastinum is typically self-limited. These individuals can usually be managed as outpatients with supportive care and analgesia. Clinical and radiological resolution may be hastened by oxygen therapy that enhances the nitrogen washout effect from mediastinal air-pockets.

Box 26.4　Recreational drugs reported to induce pneumothorax and pneumomediastinum

- Amphetamines
- Alkaloid ('crack') cocaine
- Cocaine
- Ecstasy
- Heroin
- Marijuana
- Nitrous oxide

REFERENCES

1. Cooper JA, White DA, Matthay RA. Drug-induced pulmonary disease. I: Cytotoxic drugs. II: Noncytotoxic drugs. *Am Rev Respir Dis* 1986;**133**:488–505.

2. Morelock SY, Sahn SA. Drug-induced pleural disease. *Chest* 1999;**116**:212–21.

3. Huggins JT, Sahn SA. Drug-induced pleural disease. *Clin Chest Med* 2004;**25**:141–53.

4. Antony VB. Drug-induced pleural disease. *Clin Chest Med* 1998;**19**:331–40.

5. Pusateri DW, Muder RR. Fever, pulmonary infiltrates, and pleural effusion following acyclovir therapy for *Herpes zoster* ophthalmicus. *Chest* 1990;**98**:754–66.

6. Blomhoff A, Green MH, Green JB, Berg T, Norum KR. Vitamin A metabolism: new perspectives on absorption, transport, and storage. *Physiol Rev* 1991;**71**:951–90.

7. Frankel SR, Eardley A, Heller G, *et al.* All-*trans*-retinoic acid for acute promyelocytic leukemia: results of the New York study. *Ann Intern Med* 1994;**120**:278–86.

8. Jung JI, Choi JE, Hahn ST, *et al.* Radiologic features of all-*trans*-retinoic acid syndrome. *Am J Roentgenol* 2002;**178**: 475–80.

9. De Botton D, *et al.*, European APL Group. Incidence, clinical features, and outcome of *trans*-retinoic-acid syndrome in 413 cases of newly diagnosed acute promyelocytic leukemia. *Blood* 1998;**92**:2712–18.

10. Camus P, Martin WJ, Rosenow EC. Amiodarone pulmonary toxicity. *Clin Chest Med* 2004;**25**:65–75.

11. Bauer KA, Skarin AT, Balikian JP, *et al.* Pulmonary complications associated with combination chemotherapy programs containing bleomycin. *Am J Med* 1983;**74**:557–63.

12. McElvaney NG, Wilcox PG, Churg A, Fleetham JA. Pleuropulmonary disease during bromocriptine treatment of Parkinson's disease. *Arch Intern Med* 1988;**148**:2213–36.

13. LeWitt PA, Calne DB. Pleuropulmonary changes during long-term bromocriptine treatment for Parkinson's disease. *Lancet* 1981;**1**:44–5.

14. Thompson J, Shengappa KNR, Good CB, *et al.* Hepatitis, hyperglycemia, pleural effusion, eosinophilia, hematuria, and proteinuria occurring early in clozapine treatment. *Int Clin Psychopharmacol* 1998;**13**:95–8.

15. Bhatt MH, Keenan SP, Fletham JA, Caine DB. Pleuropulmonary disease associated with dopamine agonist therapy. *Ann Neurol* 1991;**30**:613–16.

16. Frans E, Dom R, Demedts M. Pleuropulmonary changes during treatment of Parkinson's disease with a long-acting ergot derivative cabergoline. *Eur Respir J* 1992;**5**:263–5.

17. Ling LH, Ahlskog JE, Munger TM, Limper AH, Oh JK. Constrictive pericarditis and pleuropulmonary disease linked to ergot dopamine agonist therapy (cabergoline) for Parkinson's disease. *Mayo Clin Proc* 1999;**74**:371–5.

18. Malik SW, Myers JL, DeRemee RA, Specks U. Lung toxicity associated with cyclophosphamide use. *Am Rev Respir Crit Care Med* 1996;**154**:1851–6.

19. Schaap N, Waymakers R, Schattenberg A, Ottevan JP, DeWitte T. Massive pleural effusion attributed to high-dose cyclophosphamide during conditioning. *Bone Marrow Transplant* 1996;**18**:247–8.

20. Mahoney JN, Bachtel ND. Pleural effusion associated with chronic dantrolene administration. *Ann Pharmacother* 1994;**28**:587–9.

21. Corp CC, Ghishan FK. The sulfone syndrome complicated by pancreatitis and pleural effusion in an adolescent receiving dapsone for treatment of acne vulgaris. *J Pediatr Gastroenterol Nutr* 1998;**26**:103–5.

22. Talpaz M, Shah NP, Kantarjian H, *et al.* Dasatinib in imatinib-resistant Philadelphia chromosome-positive leukemias. *New Engl J Med* 2000;**354**:2531–41.

23. Quintas-Cardama A, Kantarjian H, O'Brien S, *et al.* Pleural effusion in patients with chronic myelogenous leukemia treated with dasatinib after imatinib failure. *J Clin Oncol* 2007;**25**:3908–14.

24. Bergeron A, Rea D, Levy V, *et al.* Lung abnormalities after dasatinib treatment for chronic myeloid leukemia. *Am J Respir Crit Care Med* 2007;**176**:814–18.

25. Wohlrab JL, Liu MC, Anderson ED, *et al.* Docetaxel-induced pleural effusions. *Chest* 2002;**122**:94S–95S (abstract).

26. Behnia M, Dowdeswell I, Vakili S. Pleural fluid and serum eosinophilia: association with fluoxetine hydrochloride. *South Med J* 2000;**93**:611–13.

27. Tzanakis N, Bouros D, Siafakis N. Eosinophilic pleural effusion associated due to gliclazide. *Respir Med* 2000;**94**:94.

28. Busmanis IA, Beaty AE, Basser RL. Isolated pleural effusion with hematopoietic cells of mixed lineage in a patient receiving granulocyte-colony-stimulating factor and high-dose chemotherapy. *Diagn Cytopathol* 1988;**18**:204–97.

29. Breccia N, D'Elia MD, D'Andrea M, Latagliata R, Alimena G. Pleural–pericardic effusion as uncommon complication in CML patients treated with imatinib. *Eur J Haematol* 2005;**74**:89–90.

30. Vogelzang PJ, Bloom SM, Mier JW, Atkins MB. Chest roentgenography abnormalities in IL-2 recipients: incidence in correlation with clinical parameters. *Chest* 1992;**101**: 746–52.

31. Saxon RR, Klein JR, Bar MH, *et al.* Pathogenesis of pulmonary edema during interleukin-2 therapy: correlation of chest radiographic and clinical findings of 54 patients. *Am J Roentgenol* 1991;**156**:281–5.

32. Bolanos-Meade J, Keung YK, Cobos E. Recurrent lymphocytic pleural effusion after intravenous immunoglobulin. *Am J Hematol* 1999;**60**:248–9.

33. Bunker CB, Sheron N, Maurice PD, *et al.* Isotretinoin and eosinophilic pleural effusion. *Lancet* 1989;**1**:435–6.

34. Gunther J, Lode H, Raffenberg N, Scharberg T. Development of pleural and pericardial effusions during itraconazole therapy for pulmonary aspergillosis. *Eur J Clin Microbiol Infect Dis* 1992;**12**:723–4.

35. Trisolini R, Dore R, Biagi F, *et al.* Eosinophilic pleural effusion due to mesalamine: report of a rare occurrence. *Sarcoidosis Vasc Diffuse Lung Dis* 2000;**17**:288–91.

36. Walden PA, Michell-Weggs PF, Coppin C, Dent J, Bagshane KD. Pleurisy in methotrexate treatment. *Br Med J* 1972;**28**:67.

37. Everts CS, Westscott JL, Brag DG. Methotrexate therapy and pulmonary disease. *Radiology* 1973;**107**:539–43.

38. Dunn JM, Sloan H. Pleural effusions and fibrosis secondary to sansert administration. *Ann Thorac Surg* 1973;**15**:295–8.

39. Gefter WB, Epstein DN, Bonovita JA, Miller WT. Pleural thickening caused by sansert and ergotrate in the treatment of migraine. *Am J Roentgenol* 1980;**135**:375–7.

40. Webb DB, Whale RJ. Pleuropericardial effusion associated with minoxidil administration. *Postgrad Med J* 1982; **58**:319–20.

41. Ozols RF, Hochan WM, Ostchega Y, Young RC. MVP (mitomycin, vinblastine, and progesterone): a second-line regimen in ovarian cancer with a high incidence of pulmonary toxicity. *Cancer Treat Rep* 1983;**67**:721–2.

42. Israel HL, Diamond T. Recurrent pulmonary infiltration and pleural effusion due to nitrofurantoin sensitivity. *New Engl J Med* 1962;**266**:2024–6.

43. Rosenow EC, DeRemee RA, Dins DE. Chronic nitrofurantoin pulmonary reactions: report of five cases. *New Engl J Med* 1968;**279**:1258–62.

44. Holmberg L, Boman G. Pulmonary reactions to nitrofurantoin: 447 cases reported to the Swedish Adverse Drug Reaction Committee. *Eur J Respir Dis* 1981;**62**:180–9.

45. Karkos C, Moore H, Marche A, Thorpe JA. Pleural effusion associated with D-penicillamine therapy: a case report. *Clin Pharmacol Ther* 1996;**21**:15–17.

46. Jones SE, Moore M, Blank M, Castellino RA. Hypersensitivity to procarbazine manifested by fever and pleuropulmonary reaction. *Cancer* 1972;**29**:498–500.

47. Middleton KL, Santella R, Couser JI. Eosinophilic pleuritis due to propylthiouracil. *Chest* 1993;**103**:955–6.

48. Watanabe H, Anayama S, Horiuchi T, *et al.* Pleural effusion caused by prostaglandin E1 preparation. *Chest* 2003;**123**: 952–3.

49. Abdallah MH, Khalil PB, Jamaleddin G, Salti I, Dakik HA. Thiazolidinediones associated with recurrent pleural effusion post coronary artery bypass surgery. *Int J Cardiol* 2006; **106**:273–5.

50. DeGroot RE, Willems LN, Dijkman JH. Interstitial lung disease with pleural effusion caused by simvastin. *J Intern Med* 1996;**239**:361–3.

51. Moufarrege G, Frank E, Carstens DD. Eosinophilic exudative pleural effusion after initiation of tizanidine treatment: a case report. *Pain Med* 2003;**4**:85–90.

52. Koshida H, Shibata K, Kametani T. Pleuropulmonary disease in a man with diabetes who was treated with troglitazone. *New Engl J Med* 1998;**339**:1400–1.

53. Martin RW, Duffy J, Engel AG, *et al.* The clinical spectrum of the eosinophilia–myalgia syndrome associated with L-tryptophan ingestion. *Ann Intern Med* 1999;**113**:124–34.

54. Bullington W, Sahn SA, Judson M. Valproic acid-induced eosinophilic pleural effusion: a case report and review of the literature. *Am J Med Sci* 2007;**333**:290–2.

55. Debourdeau PM, Djezzar S, Estival JLF, *et al.* Life or death threatening eosinophilic pleuropericardial effusion related to vitamins B5 and H. *Ann Pharmacother* 2001;**35**: 424–6.

56. Hoffman BJ. Sensitivity of sulfadiazine resembling acute disseminated lupus erythematosus. *Arch Dermatol Syph* 1945;**51**:190–2.

57. Yung YL, Richardson BC. Drug-induced lupus. *Rheum Dis Clin N Amer* 1994;**20**:61–85.

58. Good JT, King TE, Antony VB, Sahn SA. Lupus pleuritis: clinical features and pleural fluid characteristics with special reference to pleural fluid antinuclear antibody titers. *Chest* 1984;**84**:714–18.

59. Henningsen NC, Cederberg A, Hanson A, Johansson BW. Effects of long-term treatment with procainamide: a prospective study with special regard to ANF and SEL fast and slow acetylators. *Acta Med Scand* 1975;**198**:475–82.

60. Woolsey RL, Drayer DE, Reidberg MM, *et al.* Effect of acetylator phenotype on the rate at which procainamide induces antinuclear antibodies in the lupus syndrome. *New Engl J Med* 1978;**298**:1157–9.

61. Cush JJ, Goldings EA. Drug-induced lupus: clinical spectrum and pathogenesis. *Am J Med Sci* 1985;**290**:36–45.

62. Burlingame LW, Ruben RL. Drug-induced anti-histone antibodies displayed two patterns of reactivity. *J Clin Invest* 1991;**88**:860–9.

63. Bini A, Zompatori M, Ansaloni L, *et al.* Bilateral recurrent pneumothorax complicating chemotherapy for pulmonary metastatic breast ductal carcinoma: report of a case. *Surg Today* 2000;**30**:469–72.

64. Steinhauslin CA, Cuttat JF. Spontaneous pneumothorax: a complication of lung cancer. *Chest* 1985;**88**:709–13.

65. Stein ME, Shklar Z, Drumea K, *et al.* Chemotherapy-induced spontaneous pneumothorax in a patient with bulky mediastinal lymphoma: a rare oncologic emergency. *Oncology* 1997;**54**:15–18.

66. Seaman ME. Barotrauma related to interventional inhalational drug abuse. *J Emerg Med* 1990;**8**:141–9.

67. Mazur S, Hitchcock T. Spontaneous pneumomediastinum, pneumothorax, and Ecstasy abuse. *Emerg Med* 2001;**13**: 121–3.

68. Pittman JAL, Pounsford JC. Spontaneous pneumomediastinum and Ecstasy abuse. *J Emerg Med* 1997;**14**:335–6.

69. Rezvani K, Kurbaan AS, Brenton D. Ecstasy-induced pneumomediastinum. *Thorax* 1996;**51**:960–1.

27

Lupus erythematosus syndrome induced by drugs

ROBERT L RUBIN

INTRODUCTION

Within 1 year after the introduction of hydralazine to control malignant hypertension in 1952, the first report appeared of a late-onset 'collagen disease' resembling systemic lupus erythematosus (SLE) in 17 out of 211 hydralazine-treated patients.[1] Although procainamide was introduced for the treatment of cardiac arrhythmia at about the same time, it was not until 1962 that Ladd reported a patient who developed lupus-like features after 6 months of procainamide therapy.[2] By 1966 a scattering of cases of lupus-like disease as a side-effect of therapy with isonizid, diphenylhydantoin, sulfamethoxypyradazine, primidone and tetracycline appeared. In the ensuing years a diverse array of more than 50 different drugs has been implicated in the *de-novo* development of autoantibodies and clinical features similar to those seen in patients with idiopathic SLE.

DEFINITIONS

It is important to distinguish drug-induced lupus-like disease from the more common condition of drug-induced autoimmunity. When patients receiving drug therapy develop autoantibodies or other laboratory features of autoimmunity such as elevated immunoglobulin levels, but are clinically asymptomatic, the term 'drug-induced autoimmunity' (DIA) should be used. In contrast, a minority of drug-treated patients develop *de-novo* clinical signs and symptoms similar to the spectrum of features associated with idiopathic SLE. This syndrome is referred to as 'drug-related lupus' or 'drug-induced lupus' (DIL). While DIA may be a more homeostatically regulated form of DIL as suggested by the quantitative differences in their overt autoimmune abnormalities, most asymptomatic drug-treated patients with persistently high levels of autoantibodies do not progress to develop lupus-like symptoms.[3]

There is an entrenched view that some cases of idiopathic SLE are 'unmasked' by drug therapy in patients with a lupus diathesis,[4] an idea difficult to discount or prove. Various drugs have been noted to have a temporal relationship with the exacerbation of SLE or with the onset of chronic SLE prior to diagnosis.[5] In the latter cases SLE remains after withdrawal of the implicated agent; this is the medical history of the patient sometimes claimed as the first case of drug-induced lupus.[6] Rash or dermatitis related to drugs typically has a history of rapid onset and behaves as a drug hypersensitivity-type reaction, sometimes triggered by exposure to ultraviolet light. The majority of adverse drug reactions in previously diagnosed SLE patients are of this category.[7] Whether or not an environmental or pharmaceutical agent might aggravate or unmask incipient SLE should be considered a clinical problem distinct from DIL because, by definition, symptoms of DIL resolve after discontinuation of therapy.

Adverse reactions to some drugs include symptoms and signs that overlap with those of DIL but do not necessarily have an autoimmune aetiology. For example, a variety of respiratory disorders symptomatically resembles the pleuritis associated with drug-induced lupus such as cough due to angiotensin-converting enzyme inhibitors and interstitial pneumonitis and/or pulmonary oedema associated with amiodarone, tocainidine, flecaindine and hydrochlorothiazide. In particular, nitrofurantoin can induce acute respiratory reactions including dyspnoea, cough, pleural effusions and pneumonitis within a week of initiating therapy, as well as a diffuse interstitial pneumonitis and lung fibrosis associated with chronic exposure.[8,9] With drugs such as nitrofurantoin that are also occasionally associated with DIL and antinuclear antibody (ANA) development, distinguishing DIL, a *bona fide* systemic autoimmune disease, from an adverse drug reaction limited to the lungs can be formally challenging although not of practical importance. While it is unlikely that nitrofurantoin-induced pneumonitis has an autoimmune aetiology, the coincidental induction of ANA in some patients can occur.

Even more diagnostically confounding is the development of only haemolytic anaemia after long-term therapy with certain drugs. Some drugs associated with DIA or DIL (methyldopa, L-dopa, procainamide, chlorpromazine and streptomycin) can induce autoantibodies specific for rhesus locus or other intrinsic red blood cell (RBC) antigens (Coombs' test positive) and occasionally haemolytic anaemia.[10] These autoantibodies, unlike the penicillin-type or the stibophen-type of anti-RBC antibodies, behave like the ANA associated with DIA and DIL in that the likelihood for autoantibody appearance is time- and drug-dose-dependent, but the drug is not required for antibody binding to its target antigen. While there is generally no correlation between positive Coombs' test and ANA, these autoantibodies coexist in some patients.[11,12] Drug-induced anti-RBC antibodies of the methyldopa type are also of historical relevance to DIL because, despite the drug-independence of anti-RBC binding, a drug-altered RBC model is commonly invoked[10] as the underlying mechanism and also incorrectly used as a paradigm for the origin of autoantibodies associated with DIA and DIL (see the later section, 'Investigations into mechanisms').

INCIDENCE AND EPIDEMIOLOGY

Box 27.1 lists currently used drugs reported to be associated with a lupus-like syndrome. In all cases drugs that induce lupus also induce autoantibodies in a much higher frequency. Excluded from this list are drugs implicated only in the exacerbation of SLE or with the onset of chronic SLE prior to diagnosis as discussed above. Also excluded are drugs that induce non-multisystem cell, tissue or organ abnormalities such as cytopenias or cutaneous manifestations, although these may have an autoimmune aetiology similar to DIL. Several of the 'statins' fall into this category owing to their causative association with autoantibody-positive cutaneous lupus erythematosus, but the statins included in Box 27.1 are associated with *bona fide* cases of drug-induced lupus. Some of the anticonvulsants listed (although probably not phenytoin and carbamazepine) may be unfairly implicated in DIL because manifestations of convulsive disorders may precede typical SLE by many years.[5] Several infrequently used drugs are included, in which only a single case of autoimmune-like disease has been reported, to heighten vigilance for these possible lupus-inducing drugs.

Box 27.1   Drugs commonly reported to cause pulmonary eosinophilia

Agent[a]	Risk[b]	Agent[a]	Risk[b]
Antiarrhythmics		*Anticonvulsants*	
Procainamide (Pronestyl)	High	Carbamazepine (Tegretol)	Low
Quinidine (Quinaglute)	Moderate	Phenytoin (Dilantin)	Very low
Disopyramide (Norpace)	Very low	Trimethadione (Tridone)	Very low
Propafenone (Rythmol)	Very low	Primidone (Mysoline)	Very low
Amiodarone (Cordarone)	Very low	Ethosuximide (Zarontin)	Very low
Antihypertensives		*Antithyroidals*	
Hydralazine (Apresoline)	High	Propylthiouracil (Propyl-thyracil)	Low
Methyldopa (Aldomet)	Low	*Anti-inflammatories*	
Captropril	Low	D-penicillamine (Cuprimine)	Low
Acebutolol (Sectral)	Low	Sulfasalazine (Azulfidine)	Low
Enalapril (Vasotec)	Very low	Phenylbutazone (Butazolidin)	Very low
Clonidine (Catapres)	Very low	TNF-α inhibitors[c] (Remicade)	Very low
Atenolol (Tenormin)	Very low	Mesalamine (Asacol)	Very low
Labetalol (Normodyne)	Very low	Zafirlukast (Accolate)	Very low
Pindolol (Visken)	Very low	*Diuretics*	
Minoxidil (Loniten)	Very low	Chlorthalidone (Hygroton)	Very low
Prazosin (Minipress)	Very low	Hydrochlorothiazide (Diuchlor H)	Very low
Lisinopril (Prinivil)	Very low	*Antihyperlipidaemics*	
Antipsychotics		Lovastatin (Mevacor)	Very low
Chlorpromazine (Thorazine)	Low	Simvastatin (Zocor)	Very low
Phenelzine (Nardil)	Very low	Atorvastatin (Lipitor)	Very low
Chlorprothixene (Taractan)	Very low	*Miscellaneous*	
Lithium carbonate (Eskalith)	Very low	Aminoglutethimide (Cytadren)	Very low
Perphenazine (Trilafon)	Very low	Interferon-a (Wellferon), -β, -γ	Very low
Antibiotics		Ticlopidine (Ticlid)	Very low
Isoniazid (INH)	Low	Timolol eye drops (Timoptic)	Very low
Minocycline (Minocin)	Low		
Nitrofurantoin (Macrodantin)	Very low		

[a] Commonly used brand name is enclosed in parentheses. [b] Risk refers to likelihood for lupus-like disease, not autoantibody induction, which is usually much more common. [c] Refers to a drug family possessing similar pharmacological properties.

Macromolecular immunotherapeutics that have potential for induction of autoimmunity such as cytokines and tumour necrosis factor-alpha (TNF-α) inhibitor – reviewed by Ioannou and Isenberg[13] – are included, although the mechanism underlying this phenomenon is probably different from classical DIL.

Thus, 46 drugs currently in use have a propensity for inducing autoantibodies and occasionally a lupus-like syndrome. Box 27.1 divides these drugs into therapeutic classes and indicates their approximate risk levels based on the number of reports. While listed separately, some drugs such as enalapril and lisinopril or simvastatin and atorvastatin are essentially the same chemical structure and could be considered a single agent. The drugs with by far the highest risk are procainamide and hydralazine, with approximately 20 per cent incidence for procainamide and 5–8 per cent for hydralazine during 1 year of therapy at currently used doses. The risk for developing lupus-like disease for the remainder of the drugs is much lower, considerably less than 1 per cent of treated patients. Quinidine can be considered moderate risk while sulfasalazine, chlorpromazine, penicillamine, methyldopa, carbamazepine, acebutalol, isoniazid, captopril, propylthiouracil and minocycline are relatively low risk. The remaining 33 drugs should be considered very low risk based on the paucity of reports in the literature, for some only one case report. Obviously, the perception of risk is not rigorous since it depends on dose and frequency of prescriptions as well as occasion to publish case reports and should not be equated with a fundamental, lupus-inducing propensity. New drugs such as amiodarone and ticlopidine are gaining suspicion as recent case reports accumulate. Some drugs listed as very low risk may be falsely implicated or are currently of negligible risk because customary treatment doses have been decreased, but most reports on drug-induced lupus are convincing because cessation of therapy usually results in prompt resolution of symptoms and eventually autoantibodies.

A broad variety of therapeutic purposes is encompassed by lupus-inducing drugs, including control of convulsion disorders, psychoses, hyperthyroidism, hypertension, fungal and bacterial infections, heart arrhythmias, oedema and even anti-inflammatory agents. Consequently, the structures of these drugs show wide disparity and this is reflected in their diverse physiochemical actions. Although some of these drugs are aromatic amines (procainamide, practolol and sulfapyridine, a metabolite of sulfasalazine) or aromatic hydrazines (hydralazine and isoniazid), there is no common denominator of a pharmacological, therapeutic or chemical nature that links the drugs with capacity to induce lupus-like disease. Nevertheless, the remarkable similarity in clinical features and laboratory findings in lupus induced by most drugs other than the biologics strongly suggests that the same mechanism underlies the process regardless of the inciting agent.

Twenty years ago the incidence of DIL was estimated to be 15 000 to 20 000 new cases annually in the United States, but most of these cases were probably due to procainamide, a drug that has now been largely replaced by other antiarrhythmics. Therefore, the incidence of DIL is currently probably considerably lower but is unknown. The incidence of DIL in other countries is also unknown but has been estimated to be 10 per cent that of idiopathic SLE.[5] Lee *et al.* reviewed the medical histories of 285 consecutive SLE case records and found that drugs were a possible causative factor in 12 per cent.[14] Fries and Holman identified 12 DIL cases in a population of 198 lupus patients.[15] In a more recent retrospective study of 30 patients with elevated anti-myeloperoxidase antibodies and vasculitis usually involving the kidney, lungs and/or skin, 60 per cent were exposed to a lupus-inducing drug for 1–10 years, usually hydralazine or propylthiouracil, suggesting that the majority of patients with high titre anti-myeloperoxidase antibodies may be drug-induced.[16] The frequency of DIL may be underestimated because most cases are mild, and only a small proportion are correctly diagnosed or seen by a rheumatologist.

The age of patients developing DIL reflects the age of the population undergoing treatment with the implicated drug, usually people of age 50 or higher. Two epidemiological features distinguish DIL from SLE. First, the high female-to-male predominance seen in SLE (9:1 vs. 7:1) is not seen in DIL largely because the majority of patients treated with the major lupus-inducing drugs are men. Nevertheless, procainamide- and hydralazine-induced lupus appear to be 2-fold to 4-fold more common in women than men. In the study of Totoritis *et al.*, the female/male ratio of procainamide-treated patients who developed lupus-like symptoms was 0.52 compared to a ratio of 0.19 for those who remained asymptomatic.[17] In the study of Cameron and Ramsay, the overall incidence of hydralazine-induced lupus during a 4-year observation period was 11.6 per cent in women and 2.8 per cent in men;[18] this is in contrast to the development of autoantibodies only (DIA) over a 3-year period in which no gender differences were noted.[19] Women were more than twice as likely as men to develop minocycline-induced lupus.[20] Further, unlike SLE, the frequency of hydralazine-induced lupus in blacks was reported to be 4- to 6-fold lower than in whites.[21,22] African Americans seem to be protected from lupus induced by procainamide as well.[3]

AETIOLOGY

By definition, DIL is caused by the ingested medication, and this is generally convincing because discontinuation of treatment generally results in prompt and permanent remission. However, it is unlikely that DIL is due to a direct action of the parent compound on some component of the immune system because of its idiosyncratic nature, not predicted by any of the diverse pharmacological properties of the various lupus-inducing drugs, and because DIA and DIL require many months of continual exposure to the implicated agent. However, it is possible that steady-state blood levels of the parent compound contribute to the immune perturbations. *In-vivo* metabolism of dissimilar drugs to a product with a common, reactive property may explain how compounds with widely different pharmacological and chemical characteristics could produce the same adverse reaction. Essentially all pharmacological classes of lupus-inducing drugs but not their non-lupus-inducing analogues have been demonstrated *in vitro* to undergo oxidative metabolism by activated neutrophils, as shown in Fig. 27.1.[23] Oxidative intermediates of lupus-inducing drugs are strong candidates for triggering autoimmunity. *N*-acetylation of hydralazine and procainamide competes with *N*-oxidation of these drugs, accounting for the lower probability for development of autoimmunity in people with the rapid acetylation phenotype.[3,24]

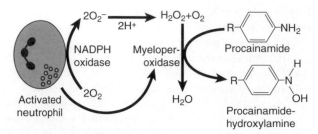

Fig. 27.1 Mechanism for transformation of drugs by activated neutrophils. Activation of neutrophils by opsonized particles or certain soluble factors triggers the ectoenzyme NADPH oxidase to produce superoxide anion (O_2^-) in the extracellular environment. O_2^- spontaneously dismutates to hydrogen peroxide (H_2O_2). Degranulation often follows, releasing myeloperoxidase (MPO). If a drug with an appropriate functional group is present, it will participate in electron transfer with the H_2O_2–MPO intermediate. Consequently, the functional group accepts an oxygen atom from H_2O_2, resulting in a new compound. In this example procainamide is undergoing oxidation to procainamide–hydroxylamine.

CLINICAL PRESENTATIONS

The clinical and laboratory features of procainamide- and hydralazine-induced lupus are shown in Table 27.1 and compared with SLE.[25] Musculoskeletal complaints are commonly observed, with arthralgia heading the list for both drugs. Approximately 50 per cent of patients have constitutional symptoms of fever, weight loss and fatigue. Arthritis is a less common feature with lupus induced by procainamide (20 per cent) than by hydralazine (50–100 per cent), whereas serositis (pleuritis and/or pericarditis) and/or myalgia are more common presenting features of procainamide-induced lupus. By contrast, hydralazine- and quinidine-induced lupus are associated with a higher frequency of cutaneous manifestations. However, in any one patient, lupus induced by procainamide, hydralazine or other drugs cannot be distinguished by clinical features. The onset of symptoms can be slow or acute, although an interval of 1–2 months typically passes before the diagnosis is made.

Some suggestion of a drug-specific symptomatology in addition to the usual musculoskeletal and constitutional symptoms is made by the literature, with pleuritis and pericarditis common to procainamide-induced lupus, polyarthritis common to quinidine and minocycline-induced lupus, glomerulonephritis and rash reported in hydralazine-induced lupus and autoimmune hepatitis in lupus related to minocycline and atorvastatin. Lung involvement in procainamide-induced lupus occurs in approximately 50 per cent of patients and consists of pleuritis, pleural effusions and/or pulmonary infiltrates; pericardial effusions are also common. Surprisingly, the clinical presentation of three patients with DIL associated with infliximab (a TNF-α blocking biological agent) also displayed pronounced pleural–pulmonary disease including bilateral effusions and pulmonary infiltrates,[26] similar to procainamide-induced lupus. Quinidine is sometimes associated with a mild drug reaction characterized only by polyarthalgias within 7 days to 3 months after initiation of therapy.[27] Lupus-like disease associated with minocycline is also atypical – patients frequently present with symmetrical polyarthritis and may have

evidence of hepatitis (elevated liver transaminases) and pneumonitis (due to pulmonary lymphocytic infiltrates), which are assumed to be autoimmune in nature. These patients may not have autoantibodies but frequently have perinuclear antineutrophil cytoplasmic antibodies (pANCA) due to anti-myeloperoxidase.[28] Autoimmune hepatitis in addition to systemic autoimmunity suggestive of DIL has also been associated with atorvastatin[29] and with mesalamine (5-aminosalicylic acid).[30]

The frequency of serological autoimmune abnormalities in lupus induced by procainamide and hydralazine is essentially identical (Box 27.2). The immune response in this setting is characteristic and restricted. The most commonly observed abnormality is positive antinuclear antibodies (ANA), which are largely due to histone-reactive antibodies, particularly IgG anti-[(H2A-H2B)-DNA] antibodies. These autoantibodies occur in over 90 per cent of patients who develop procainamide-induced lupus, and have been detected in individual patients with lupus induced by penicillamine, isoniazid, acebutalol, methyldopa, sulfasalazine, ophthalmic timolol and lisinopril. Although generally of low titre, antibodies to denatured (but not native) DNA are also common in DIL. Less common laboratory features of an autoimmune nature include rheumatoid factor (procainamide), circulating immune complexes, antineutrophil cytoplasmic antibodies (hydralazine, propylthiouracil, minocycline and sulfasalazine), positive Coombs' test (methyldopa, chlorpromazine and procainamide), complement activation (procainamide), and hypocomplementaemia (quinidine). The presence of the lupus anticoagulant (LAC) or of cardiolipin antibodies has been described in patients on hydralazine, procainamide, chlorpromazine, quinidine and quinine. Other laboratory features noted in a minority of patients include a mild anaemia, leukopenia, and thrombocytopenia, a hypergammaglobulinaemia and elevated erythrocyte sedimentation rate (ESR).

INVESTIGATIONS INTO MECHANISMS

Since symptoms and serological features of DIL overlap with those of idiopathic SLE, it is presumed that similar pathogenic factors underlie both syndromes. However, immune complex formation and deposition *in vital* organs, one of the mechanisms believed to operate in SLE, have not been well documented in DIL. Immune complexes have been reported in DIL,[31,32] but their composition, correlation with disease activity and pathogenic potential have not been evaluated. Evidence supporting the involvement of immune complexes in disease pathogenesis includes the observations that complement breakdown products (e.g. C3d and C4d) were detected in DIL.[33,34] Immune complexes involving IgG anti-[(H2A-H2B)-DNA] would be a candidate mediator of complement activation and subsequent inflammatory reactions. This possibility is consistent with the finding that these autoantibodies are predominantly IgG1 and IgG3,[35] immunoglobulin subclasses that are potent activators of the classical complement pathway when engaged by antigen. Detection of LE cells in procainamide-induced lupus[36] also suggests the presence of complement-fixing autoantibodies. However, in comparison to the complement-fixing activity of autoantibodies in SLE, autoantibodies from patients with DIL displayed much lower capacity

Table 27.1 Prevalence (percentage) of clinical and laboratory abnormalities in drug-induced lupus and systemic lupus erythematosus (data from [25])[a]

Feature	Hydralazine-induced lupus	Procainamide-induced lupus	Systemic lupus erythematosus
Symptom			
Arthralgia	80	85	80[b]
Arthritis	50–100	20	80[b]
Pleuritis, pleural effusion	<5	50	44
Fever, weight loss	40–50	45	48
Myalgia	<5	35	60
Hepatosplenomegaly	15	25	5–10
Pericarditis	<5	15	20
Rash	25	<5	71
Glomerulonephritis	5–10	<5	42
CNS disease	<5	<5	32
Sign			
Autoantibodies, including:	>95	>95	97
– LE cell	>50	80	71
– antihistone	>95	>95	54
– anti-[(H2A-H2B)-DNA]	43	96	70
– antidenatured DNA	50–90	50	82
– antinative DNA	<5	<5	28–67
– anticardiolipin	5–15	5–20	35
– rheumatoid factor	20	30	25–30
Anaemia	35	20	42
Elevated ESR	60	60–80	>50
Leucopenia	5–25	15	46
Positive Coombs' test	<5	25	25
Elevated gammaglobulins	10–50	25	32
Hypocomplementaemia	5	5	51

[a] Each prevalence represents a consensus value ± 5 percentage points. Abnormalities occurring in fewer than 5 per cent of patients are not listed. [b] Articular pain (arthralgia) and joint inflammation (arthritis) are the initial symptoms in >50 per cent of SLE patients, but frank arthritis is less commonly the presenting symptom in procainamide-induced lupus. ESR, erythrocyte sedimentation rate.

Box 27.2 Guidelines for diagnosis of drug-induced lupus

- Continuous treatment with a known lupus-inducing drug for at least 1 month, and usually much longer
- Presenting symptoms:
  - *Common:* arthralgias, myalgias, malaise, fever, serositis (pleuropericarditis, especially with procainamide), polyarthritis (especially with quinidine and minocycline)
  - *Rare:* rash or other dermatological problems, glomerulonephritis (primarily with hydralazine)
- Unrelated symptoms suggestive of systemic lupus erythrematosus: multisystem involvement especially neurological, renal and skin symptoms
- Laboratory profile:
  - *Common:* antinuclear antibodies (ANA) which is due to antihistone antibodies, especially IgG anti-[(H2A-H2B)-DNA], leukopenia, thrombocytopenia and mild anaemia, increased ESR
  - *Absent or rare:* antibodies to native DNA, Sm, RNP, SS-A/Ro, SS-B/La, hypocomplementaemia
- Improvement and permanent resolution of symptoms generally within days or weeks after discontinuation of therapy. However, serology findings, especially autoantibody levels, often require months to resolve.

---

to activate the complement system using an indirect immunofluorescence assay involving cell nuclei.[37–40] In addition to this weak complement-fixing activity which may be related to the relative homogeneity of autoantibodies in DIL compared to SLE,[39,40] overloading of the immune complex clearance machinery and pro-inflammatory signalling through complement

and Fc receptors may be necessary for the serious pathological features that are typical of SLE.

Regardless of the actual pathological mechanisms responsible for clinical symptoms, the origin of autoantibodies in DIL is a perplexing phenomenon that deserves an explanation, not only for the understanding of DIL but also because it may shed light on the origin of the similar autoantibodies arising in idiopathic diseases such as SLE. Studies attempting to determine which component(s) of the immune system are important in the initiation of DIL, and how autoimmunity is triggered by drugs or drug metabolites, are organized below into four general mechanistic hypotheses. While many of the older investigations into mechanisms, as well as some relatively recent ones,[41–43] report biological effects of the ingested, parent molecule, these drug effects are immediate and so do not readily account for the requirement of continuous and long-term drug exposure prior to manifestation of disease.

Hypothesis: Drug metabolites act as haptens for drug-specific T-cells

Most of the older experimental studies on DIL explored the presumed capacity of lupus-inducing drugs to form stable complexes with self-macromolecules or to directly stimulate lymphocytes. The premise underlying much of this work was apparently based on previously described immune reactions to xenobiotics such as the penicillin-type of drug hypersensitivity (allergic) reaction mediated by antibodies or the delayed-type hypersensitivity reaction associated with allergic contact dermatitis mediated by T-cells. Autoantibodies might develop if an immune response to the drug in the form of a hapten or to a self-antigen altered by the drug induces antibodies that cross-react with or cause spreading of the immune response to native self-macromolecules. For the most part these experiments have not been illuminating because non-pharmacological concentrations of drugs or artificial drug–macromolecular complexes together with immunological adjuvants were employed. An examination of the specificities of antibodies elicited in rabbits by immunization with drug–albumin complexes revealed no cross-reaction between denatured DNA or histones and antibodies to procainamide or its oxidative metabolites.[44]

A number of lupus-inducing drugs (hydralazine, chlorpromazine, carbamazepine, phenylbutazone, and nitrofurantoin) will cause significant enlargement of the draining popliteal lymph node when injected subcutaneously into the hind foot-pad of mice.[45,46] In this popliteal lymph node assay (PLNA), T-cells apparently respond to drug-altered self-proteins. Interestingly, procainamide, isoniazid and propylthiouracil were negative in this assay[45–47] unless oxidatively metabolized by rat liver microsomes[47] or peritoneal macrophage,[46] or if the metabolite itself such as PAHA or propylthiouracil 2-sulfonate was injected.[46] The requirement for a drug metabolite is in good agreement with the in-vitro studies demonstrating neutrophil- or macrophage-mediated metabolism of the same drugs (see Fig. 27.1).

T-cell immune responses require presentation of the antigen on the class II molecules of the major histocompatibility complex (MHC) expressed on antigen presenting cells, and most drugs are ignored by this peptide-based immune recognition machinery. However, if a drug or its metabolite can form a stable bond to self-molecules, a combined epitope generated by the drug–self-protein complex could be endocytosed, processed and presented on the MHC for recognition by receptors on T-cells. Presentation of a drug on the MHC could also occur by binding directly to the MHC through a non-covalent association.[48] B-cell receptors would be able to interact with drug-altered macromolecules directly. Alternatively, B-cells within the microenvironment of a drug-specific T-cell

response might become activated by cytokine-mediated bystander mechanisms.

The immunoreactivity of many drugs in the PLNA could be due to these types of process. Since monocytes, macrophages and Langerhans cells can present antigens to T-cells, they have received special attention as a potential source of drug biotransformation producing drug conjugates as well as immune presentation of the conjugate to initiate an immune response.[49] Oxidative metabolites of carbamazepine, chlorpromazine, hydralazine, phenytoin, procainamide and propylthiouracil have been demonstrated to form covalent bonds with cellular proteins. However, mice that developed enlarged lymph nodes in response to lupus-inducing drugs or their oxidative metabolites failed to develop autoreactive T-cells or autoantibodies, indicating that drug-specific T-cells do not typically lead to autoimmunity. In addition the autoantibodies that arise in people with DIL or drug-induced cytopenias are limited in specificity, a feature not consistent with a bystander activation scenario. However, if an incipient autoimmune response were independently under way and if the drug as a hapten becomes expressed on the MHC of autoreactive B-cells, it is possible that drug-specific T-cells could accelerate development of autoimmunity.

Hypothesis: Drug metabolites cause cell death

Lupus-inducing drugs are generally not cytotoxic at therapeutically relevant concentrations, a property which would have precluded their medicinal use, although activated lymphocytes were reported to be killed by chlorpromazine.[41] However, cytopenias associated with certain drugs may be related to the capacity of various reactive drug metabolites to directly cause cell death in in-vitro studies. This would be a non-immune-mediated process. Demonstration that procainamide–hydroxylamine (PAHA), the oxidative metabolite of procainamide (see Fig. 27.1), can under certain in-vitro conditions directly kill a wide variety of cells at pharmacologically relevant concentrations,[23,50–52] or enhance reactive oxygen species generation by murine macrophage[53] and human neutrophils,[52] is consistent with this view. Apoptosis rather than necrotic cell death was mediated by sulfamethoxazole–hydroxylamine[54] or by the nitrenium ion of clozapine.[55] Cell death may be initiated by damage to the plasma or mitochondrial membrane or by covalent binding to critical intracellular molecules or may be due to redox cycling with oxidized and reduced forms of nicotinamide adenine dinucleotide, depleting the cell of its energy stores as suggested by the correlation between cell reducing potential and sensitivity to PAHA cytotoxicity.[50] Cytotoxic drug metabolites generated by neutrophils in vitro have been demonstrated for amodiaquine, carbamazepine, chlorpromazine, clozapine, hydralazine, isoniazid, procainamide, propylthiouracil, quinidine and sulfonamides at therapeutically feasible concentrations. Cytopenia could occur by killing of stem cells if reactive drug metabolites were produced by myeloperoxidase released from immature promyelocytes undergoing granulopoiesis or from activated neutrophils recirculating into the bone marrow. In addition, certain haematopoietic cell lineages sensitized by an immune-mediated mechanism because of drug binding to the cell surface could be destroyed by an otherwise subtoxic concentration of a reactive drug metabolite. Lymphocytes from patients with a history of agranulocytosis secondary to clozapine therapy were somewhat more sensitive to the cytotoxic effects of oxidative metabolites of clozapine than normals or patients who did not develop agranulocytosis,[56,57] possibly related to differences in drug bioinactivation by intracellular glutathione or cysteine.[58] However, while it is possible that direct cytotoxicity of drug metabolites could be an independent pathogenic mechanism especially in certain susceptible populations, such a process cannot explain the bulk of the immune abnormalities in DIL.

It is also formally possible that drug toxicity alters degradation, clearance or processing of self-materials by antigen-presenting cells, producing abnormal macromolecular forms or unusual peptides presented on the MHC. These 'cryptic' T-cell autoepitopes could induce classical adaptive immune responses because immune tolerance to such unusual forms of self-materials was never established. Autoreactive B-cells pre-existing in the immune repertoire could present such cryptic epitopes to T-cells, resulting in B-cell activation and autoantibody secretion. Alternatively, repetitive macromolecular structures released from dying cells could theoretically elicit T-independent autoimmune responses. This type of scenario has been proposed to account for autoantibodies associated with diverse medications, environmental agents and viruses, but there remains little in the way of experimental support.

Hypothesis: Drugs non-specifically activate lymphocytes

Several lines of investigation suggest that lupus-inducing drugs can upregulate the immune system by causing non-specific lymphocyte activation. PAHA at 2 μM was reported to enhance pokeweed mitogen-mediated lymphocyte proliferation and increase the number of immunoglobulin-producing cells.[52] Micromolar levels of procainamide or hydralazine caused an approximately 2-fold increase in poly (ADP-ribosylation) levels in the lymphocyte Wil-2 cell line.[59] A lupus-like pathology including autoantibodies and glomerulonephritis has been produced in mice by injection monthly for 6 months of splenocytes treated with procainamide in vitro.[60] An antigen-specific T-cell clone treated with procainamide or hydralazine produced a similar in-vivo pathology,[61,62] suggesting that autoimmunity in this animal model was a non-specific process involving activated T-cells exposed to lupus-inducing drugs. The reactive nature of drug-treated cells was due to decreased DNA methyltransferase activity in CD4+ T-cells as a result of competitive inhibition by procainamide of DNA methyltransferase[63] or by hydralazine of the extracellular-regulated kinase pathway resulting in lower expression of DNA methyltransferase.[43] As a result, with each round of cell division undermethylation of deoxycytosine residues in CpG pairs occurs in the genomic DNA. Hypomethylation of promoter sequences is associated with enhanced gene transcription, and drug-treated T-cells showed increased expression of lymphocyte function antigen-1,[62] an important adhesion molecule that helps stabilize the interaction between T-cells and antigen-presenting cells, and enhanced CD70 expression,[64] a costimulatory ligand for CD27 on T- and B-cells. Longer contact between T-cells and antigen-presenting cells may promote T-cell activation and/or stimulate IgG production in B-cells.

One concern about these studies is that the immunopathological features of this mouse model do not resemble DIL but are more like the global autoimmune characteristics of a graft-versus-host reaction when adoptively transferred semi-allogeneic T-cells recognize histoincompatible MHC molecules in the host.[65] This syndrome is similar to SLE, while DIL displays much more limited autoimmune features (see Table 27.1). In addition, since multiple exposures to large numbers of dividing lymphocytes were required to produce the in-vivo phenomena, it is not clear what the natural counterpart of such a polyclonal T-cell activation would be. Immune responses to infectious agents that might occur coincident with medication with a lupus-inducing drug would be expected to be associated with a much more limited, oligoclonal T-cell activation. However, it is possible that, in DIL, autoreactive T-cells developing through another mechanism become more aggressive owing to such a drug-induced LFA-1 and/or CD70 over-expression process, thereby aggravating disease in DIL.

Hypothesis: Drug metabolites disrupt immune tolerance machinery

Injection of PAHA into the thymus of normal adult mice resulted in the delayed but long-lasting production of IgG anti-[(H2A-H2B)-DNA] antibodies, similar to those found in patients with DIL.[66] Transfer into naive mice of autoreactive peripheral T-cells derived from PAHA-injected animals elicited a similar autoantibody profile, indicating that autoreactive T-cells that emigrated from the thymus to the periphery accounted for autoantibody production in this system.[67] Observations in this animal model and in an in-vitro model of T-cell tolerance[68] indicated that PAHA did not reverse self-tolerance of the mature thymocyte and did not prevent deletion of high-affinity autoreactive T-cells in the thymus. Instead, PAHA apparently interfered with the establishment of tolerance to endogenous self-antigens that are normally presented by the MHC on thymic epithelial cells during the positive selection of thymocytes.[69] As a result, mature T-cells are produced that are capable of undergoing spontaneous activation when they encounter similar self-antigens in the periphery. These studies would imply that the restricted autoantibody production associated with DIL reflects a limited diversity in the selecting antigens encountered as T-cells undergo positive selection in the thymus, and that T-cell tolerance to these antigens is compromised in the presence of reactive drug metabolites.

By this mechanism, rather than stimulating mature T-cells, drug metabolites prevent the acquisition of central T-cell self-tolerance. T-cells originate in the thymus where their development ensures that only cells that are unresponsive to self are released into the periphery. The possibility that lupus-inducing drugs interfere in this process had been generally ignored because it was widely assumed that there is no thymic function in the adult, but several studies have demonstrated that T-cells are generated in the thymus at least up to the seventh decade of life in normals[70,71] and in patients with a history of DIL.[71] However, there is no evidence that reactive metabolites of procainamide can be produced in the thymus by circulating or resident phagocytic cells and that metabolites of other lupus-inducing drugs can also initiate autoimmunity by this mechanism. Ultimately, identification of the molecular target(s) in developing thymocytes that are compromised by reactive drug metabolites will be needed to make this a convincing story.

DIAGNOSTIC CONSIDERATIONS

Diagnostic criteria for drug-induced lupus

Specific criteria for the diagnosis of DIL have not been formally established. Although some of the criteria for the classification of SLE are applicable to DIL, the requirement for four manifestations as established by the American College of Rheumatology[72] is overly rigid for DIL and is obviously not useful for distinguishing between drug-induced and idiopathic SLE. Patients frequently present with mild or few lupus-like symptoms and often do not fulfil criteria for SLE, although symptoms typically worsen the longer the patient is maintained on the implicated drug. In particular, symptoms common to SLE such as malar or discoid rash, photosensitivity, oral ulcers, alopecia and renal or neurological disorders are very unusual in DIL.

Guidelines for a diagnosis of drug-induced lupus are shown in Box 27.2. DIL usually occurs after several months or years of continuous therapy and should not be confused with the short-term toxic side-effects that are often suffered by patients treated with pharmaceuticals as described under 'Definitions' above. In some patients, symptoms gradually appear and worsen over the course of many months of treatment with the implicated drug, whereas in others, symptom onset is rapid. Variation in time of onset of DIL is due in part to differences in the steady-state drug concentrations employed in order to maintain therapeutic control, but genetic factors may also be involved. The best documented genetic predisposing factor for procainamide- and hydralazine-induced lupus is low hepatic acetyltransferase activity,[24] presumably resulting in higher plasma drug concentration of the unacetylated, autoimmunity-inducing form. Human leucocyte antigen (HLA) allele association as well as certain complement-related protein alleles have been claimed to increase risk of DIL.

In most patients with DIL, symptoms are mild although indistinguishable from SLE and usually include fever, malaise, weight loss, polyarticular arthralgias and symmetric myalgias. Central nervous system (CNS) disease, while not a feature of lupus induced by most drugs (with the possible exception of quinidine), should not be an exclusion criterion because of possible independent neurotoxic effects of drugs or the occurrence of stroke, convulsions or dementia syndromes in the elderly commonly treated with these drugs. Finally, although a history of rheumatological disease independent of the suspected drug tends to negate a diagnosis of DIL, obviously a patient can have two diseases. This situation is characteristic of patients with various forms of arthritis who also develop DIL from penicillamine, sulfasalazine or minocycline therapy. In these difficult cases serological findings are especially informative.

Differential diagnosis

Anti-[(H2A-H2B)-DNA] antibodies (or anti-chromatin/nucleosome antibodies) are particularly useful for distinguishing asymptomatic drug-treated patients who develop benign autoantibodies (i.e. DIA) from patients with symptomatic DIL because only the latter have IgG anti-[(H2A-H2B)-DNA] antibodies.[73] Although antihistone and anti-[(H2A-H2B)-DNA]

antibodies are also common in SLE,[74] these patients are rarely monospecific for this activity. Therefore, when a diagnosis of SLE or DIL cannot be clearly distinguished on clinical grounds or history, the presence of antibodies to native DNA, Sm, RNP, SS-A/Ro, SS-B/La or other nuclear antigens should be considered as evidence against a diagnosis of DIL.

TREATMENT OR MANAGEMENT

Resolution of symptoms and laboratory abnormalities by withdrawing the offending drug is a defining feature of DIL. Symptoms usually resolve within 1–2 weeks after discontinuing the offending drug and, therefore, this action provides a key (although retrospective) diagnostic tool. Although autoantibodies and related laboratory abnormalities should also resolve after discontinuation of therapy, autoantibody disappearance often takes much longer than symptom resolution – autoantibodies can still be present 2 years after withdrawal of therapy. However, quantitative measurements of autoantibodies should show a systematic decline in activity once the causative agent is withdrawn.

COMPLICATIONS

Although there is a strong temptation to treat patients suspected of DIL with anti-inflammatory agents, this action may confound the diagnosis and should not be required for recovery from DIL. Treatment with anti-inflammatory agents, including corticosteroids, may be indicated for those with severe manifestations of the disease such as pericarditis with tamponade, inflammatory pleural effusions or debilitating polyarthritis. The judicious use of non-steroidal anti-inflammatory agents and corticosteroids in the elderly is important because of potential side-effects. The prolonged use of high doses of corticosteroids, or the use of chloroquine, hydroxychloroquine or immunosuppressive agents, is not indicated in the treatment of DIL. If DIL is associated with the rare feature of glomerulonephritis, as reported with hydralazine-induced lupus, corticosteroids can be used. Although some DIL patients have been restarted on procainamide or hydralazine without incident, this is not advisable if insufficient time (which may be as long as 1 year) has elapsed since the initial episode of DIL.

Clinicians are often consulted to consider the safety of drugs associated with DIL in the treatment of patients with idiopathic SLE. The use of hydralazine to treat hypertension in SLE patients has not been associated with exacerbations of the disease,[5] and procainamide has been used without incident in SLE.[75] Patients with established SLE have been treated with procainamide for long duration without exacerbation of symptoms.[15] The use of anticonvulsants to treat seizure disorders in SLE patients has not been associated with flares or acceleration of disease activity. Isoniazid has been given to SLE patients on corticosteroids without aggravating lupus.[76] These observations are contrary to the expectations of the hypothesis that drugs unmask a predisposition to SLE,[4] and help to establish the view that DIL is a fundamentally different disease from SLE. Despite the apparent safety of these drugs in the setting of SLE, the clinician should use the drugs judiciously and carefully document the clinical and serological status of the patient being considered for treatment.

KEY POINTS

- Drug-induced lupus is an idiosyncratic drug reaction associated with long-term exposure to several dozen medications currently in use. It is a rare or very rare adverse effect, although there is a much higher incidence of autoantibody induction in patients who remain asymptomatic.

- Symptoms and signs of drug-induced lupus resemble or overlap with those of SLE. However, diagnostic criteria for SLE should not be rigidly required in order to implicate a medication in causing lupus-like disease.

- Symptoms of drug-induced lupus should fully resolve after discontinuation of the implicated agent. Autoantibodies can persist for a year or more after discontinuation.

- The diversity of medications capable of inducing lupus-like disease is best explained by the *in-vivo* formation and action of drug metabolites with similar immune-perturbing properties.

- Drug-induced lupus does not behave like a drug hypersensitivity reaction, and development of drug-induced autoantibodies cannot be readily explained by mechanisms involving drug-altered self-materials or by possible cytotoxic effects of drug metabolites.

REFERENCES

1. Morrow JD, Schroeder HA, Perry HM. Studies on the control of hypertension by hyphex. II: Toxic reactions and side effects. *Circulation* 1953;**8**:829–39.
2. Ladd AT. Procainamide-induced lupus erythematosus. *New Engl J Med* 1962;**267**:1357–8.
3. Mongey AB, Sim E, Risch A, Hess E. Acetylation status is associated with serological changes but not clinically significant disease in patients receiving procainamide. *J Rheumatol* 1999;**26**:1721–6.
4. Alarcon-Segovia D. Drug-induced lupus syndrome. *Mayo Clin Proc* 1967;**44**:664–81.
5. Wallace DJ, Dubois EL. Drugs that exacerbate and induce systemic lupus erythematosus. In: Wallace DJ, Dubois EL (eds) *Dubois' Lupus Erythematosus*, pp 450–69. Philadelphia, PA: Lea & Febiger, 1987.
6. Hoffman BJ. Sensitivity to sulfadiazine resembling acute disseminated lupus erythematosus. *Arch Dermatol Syphilol* 1945;**51**:190–2.
7. Petri M, Allbritton J. Antibiotic allergy in systemic lupus erythematosus. *J Rheumatol* 1992;**19**:265–9.
8. Ben Noun L. Drug-induced respiratory disorders: incidence, prevention and management. *Drug Safety* 2000;**23**:143–64.
9. Sovijarvi AR, Lemola M, Stenius B, Idanpaan-Heikkila J. Nitrofurantoin-induced acute, subacute and chronic pulmonary reactions. *Scand J Respir Dis* 1977;**58**:41–50.

10. Petz LD. Autoimmune and drug-induced immune hemolytic anemia. In: Rose NR, *et al.* (eds) *Manual of Clinical Laboratory Immunology*, pp 325–43. Washington, DC: American Society for Microbiology, 1992.
11. Mackay IR, Cowling DC, Hurley TH. Drug-induced autoimmune disease: hemolytic anemia and lupus cells after treatment with methyldopa. *Med J Aust* 1968;**2**:1047–50.
12. Nordstrom DM, West SG, Rubin RL. Methyldopa-induced systemic lupus erythematosus. *Arthritis Rheum* 1989;**32**:205–8.
13. Ioannou Y, Isenberg DA. Current evidence for the induction of autoimmune rheumatic manifestions by cytokine therapy. *Arthritis Rheum* 2000;**43**:1431–42.
14. Lee SL, Rivero I, Siegel M. Activation of systemic lupus erythematosus by drugs. *Arch Intern Med* 1966;**117**:620–6.
15. Fries JF, Holman HR. Systemic lupus erythematosus: a clinical analysis. In: *Problems in Internal Medicine*. Philadelphia, PA: Saunders, 1975.
16. Choi HK, Merkel PA, Walker AM, Niles JL. Drug-associated antineutrophil cytoplasmic antibody-positive vasculitis: prevalence among patients with high titers of antimyeloperoxidase antibodies. *Arthritis Rheum* 2000;**43**:405–13.
17. Totoritis MC, Tan EM, McNally EM, Rubin RL. Association of antibody to histone complex H2A–H2B with symptomatic procainamide-induced lupus. *New Engl J Med* 1988;**318**:1431–6.
18. Cameron HA, Ramsay LE. The lupus syndrome induced by hydralazine: a common complication with low dose treatment. *Br Med J* 1984;**289**:410–12.
19. Mansilla-Tinoco R, Harland SJ, Ryan PJ, *et al.* Hydralazine, antinuclear antibodies, and the lupus syndrome. *Br Med J* 1982;**284**:936–9.
20. Sturkenboom MC, Meier CR, Jick H, Stricker BH. Minocycline and lupus-like syndrome in acne patients. *Arch Intern Med* 1999;**159**:493–7.
21. Condemi JJ, Moore-Jones D, Vaughan JH, Perry HM. Antinuclear antibodies following hydralazine toxicity. *New Engl J Med* 1967;**276**:486–90.
22. Perry HM. Late toxicity of hydralazine resembling systemic lupus erythematosus or rheumatoid arthritis. *Am J Med* 1973;**54**:58–72.
23. Jiang X, Khursigara G, Rubin RL. Transformation of lupus-inducing drugs to cytotoxic products by activated neutrophils. *Science* 1994;**266**:810–13.
24. Perry HMJ, Tan EM, Carmody S, Sakomoto A. Relationship of acetyl transferase activity to antinuclear antibodies and toxic symptoms in hypertensive patients treated with hydralazine. *J Lab Clin Med* 1970;**76**:114–25.
25. Rubin RL. Drug-induced lupus. In: Wallace DJ, Hahn BH (eds) *Dubois' Lupus Erythematosus*, 7th edn, pp 870–900. Philadelphia, PA: Lippincott Williams & Wilkins, 2007.
26. Diri E, Tello W, Ratnoff WD, Nugent K. Infliximab-induced SLE-like syndrome involving the lung and pleura. *Lupus* 2007;**16**:764–6.
27. Cohen MG, Kevat S, Prowse MV, Ahern MJ. Two distinct quinidine induced rheumatic syndromes. *Ann Intern Med* 1988;**108**:369–71.

28. Dunphy J, Oliver M, Rands AL, Lovell CR, McHugh NJ. Antineutrophil cytoplasmic antibodies and HLA class II alleles in minocycline-induced lupus-like syndrome. *Br J Dermatol* 2000;**142**:461–7.

29. Graziadei IW, Obermoser GE, Sepp NT, Erhart KH, Vogel W. Drug-induced lupus-like syndrome associated with severe autoimmune hepatitis. *Lupus* 2003;**12**:409–12.

30. Kirkpatrick AW, Bookman AA, Habal F. Lupus-like syndrome caused by 5-aminosalicylic acid in patients with inflammatory bowel disease. *Can J Gastroenterol* 1999; **13**:159–62.

31. Mitchell JA, Batchelor JR, Chapel H, Spiers CN, Sim E. Erythrocyte complement receptor type 1 (CR1) expression and circulating immune complex (CIC) levels in hydralazine-induced SLE. *Clin Exp Immunol* 1987;**68**:446–56.

32. Becker M, Klajman A, Moalem T, Yaretzky A, Ben-Efraim S. Circulating immune complexes in sera from patients receiving procainamide. *Clin Immunol Immunopathol* 1979;**12**:220–7.

33. Brandslund I, Ibsen HHW, Klitgaard NA, *et al.* Plasma concentrations of complement split product C3d and immune complexes after procainamide induced production of antinuclear antibodies. *Acta Med Scand* 1986;**220**:431–5.

34. Rubin RL, Nusinow SR, Johnson AD, *et al.* Serological changes during induction of lupus-like disease by procainamide. *Am J Med* 1986;**80**:999–1002.

35. Rubin RL, Tang F-L, Chan EKL, *et al.* IgG subclasses of autoantibodies in systemic lupus erythematosus, Sjogren's syndrome, and drug-induced autoimmunity. *J Immunol* 1986;**137**:2528–34.

36. Vivino FB, Schumacher HRJ. Synovial fluid characteristics and the lupus erythematosus cell phenomenon in drug-induced lupus. *Arthritis Rheum* 1989;**32**:560–8.

37. Fritzler M, Ryan P, Kinsella TD. Clinical features of systemic lupus erythematosus patients with antihistone antibodies. *J Rheumatol* 1982;**9**:46–51.

38. Klajman A, Farkas R, Ben-Efraim S. Complement-fixing activity of antinuclear antibodies induced by procainamide treatment. *Isr J Med Sci* 1973;**9**:627–30.

39. Kanayama Y, Peebles C, Tan EM, Curd JG. Complement activating abilities of defined antinuclear antibodies. *Arthritis Rheum* 1986;**29**:748–54.

40. Rubin RL, Teodorescu M, Beutner EH, Plunkett RW. Complement-fixing properties of antinuclear antibodies distinguish drug-induced lupus from systemic lupus erythematosus. *Lupus* 2004;**13**:249–56.

41. Hieronymus T, Grotsch P, Blank N, *et al.* Chlorpromazine induces apoptosis in activated human lymphoblasts: a mechanism supporting the induction of drug-induced lupus erythematosus? *Arthritis Rheum* 2000;**43**: 1994–2004.

42. Ablin J, Verbovetski I, Trahtemberg U, Metzger S, Mevorach D. Quinidine and procainamide inhibit murine macrophage uptake of apoptotic and necrotic cells: a novel contributing mechanism of drug-induced-lupus. *Apoptosis* 2005;**10**:1009–18.

43. Deng C, Lu Q, Zhang Z, *et al.* Hydralazine may induce autoimmunity by inhibiting extracellular signal-regulated kinase pathway signaling. *Arthritis Rheum* 2003;**48**:746–56.

44. Adams LE, Roberts SM, Donovan-Brand R, Zimmer H, Hess EV. Study of procainamide hapten-specific antibodies in rabbits and humans. *Int J Immunopharmacol* 1993;**15**:887–97.

45. Kammüller ME, Thomas C, De Bakker JM, Bloksma N, Seinen W. The popliteal lymph node assay in mice to screen for the immune disregulation potential of chemicals: a preliminary study. *Int J Immunopharmacol* 1989;**11**: 293–300.

46. Kubicka-Muranyi M, Goebels R, Goebel C, Uetrecht J, Gleichmann E. T-lymphocytes ignore procainamide, but respond to its reactive metabolites in peritoneal cells: demonstration by the adoptive transfer popliteal lymph node assay. *Toxicol Appl Pharmacol* 1993;**122**:88–94.

47. Katsutani N, Shionoya H. Popliteal lymph node enlargement induced by procainamide. *Int J Immunopharmacol* 1992; **14**:681–6.

48. von Greyerz S, Burkhart C, Pichler WJ. Molecular basis of drug recognition by specific T-cell receptors. *Intern Arch Allergy Immunol* 2000;**119**:173–80.

49. Griem P, Wulferink M, Sachs B, Gonzalez JB, Gleichmann E. Allergic and autoimmune reactions to xenobiotics: how do they arise? *Immunol Today* 1998;**19**:133–41.

50. Rubin RL, Uetrecht JP, Jones JE. Cytotoxicity of oxidative metabolites of procainamide. *J Pharmacol Exp Ther* 1987;**242**:833–41.

51. Wheeler JF, Lunte CE, Heineman WR, Adams L, Hess EV. Rapid communications: electrochemical determination of N-oxidized procainamide metabolites and functional assessment of effects on murine cells *in vitro*. *Proc Soc Exp Biol Med* 1988;**188**:381–6.

52. Adams LE, Sanders CE, Budinsky RA, *et al.* Immunomodulatory effects of procainamide metabolites: their implications in drug-related lupus. *J Lab Clin Med* 1989;**113**:482–92.

53. Adams LE, Roberts SM, Carter JM, *et al.* Effects of procainamide hydroxylamine on generation of reactive oxygen species by macrophages and production of cytokines. *Int J Immunopharmacol* 1990;**12**:809–19.

54. Hess DA, Sisson ME, Suria H, *et al.* Cytotoxicity of sulfonamide reactive metabolites: apoptosis and selective toxicity of CD8(+) cells by the hydroxylamine of sulfamethoxazole. *Faseb J* 1999;**13**:1688–98.

55. Williams DP, Pirmohamed M, Naisbitt DJ, Uetrecht JP, Park BK. Induction of metabolism-dependent and -independent neutrophil apoptosis by clozapine. *Mol Pharmacol* 2000;**58**:207–16.

56. Tschen AC, Rieder MJ, Oyewumi LK, Freeman DJ. The cytotoxicity of clozapine metabolites: implications for predicting clozapine-induced agranulocytosis. *Clin Pharmacol Ther* 1999;**65**:526–32.

57. Gardner I, Leeder JS, Chin T, Zahid N, Uetrecht JP. A comparison of the covalent binding of clozapine and olanzapine to human neutrophils *in vitro* and *in vivo*. *Mol Pharmacol* 1998;**53**:999–1008.

58. Williams DP, Pirmohamed M, Naisbitt DJ, Maggs JL, Park BK. Neutrophil cytotoxicity of the chemically reactive metabolite(s) of clozapine: possible role in agranulocytosis. *J Pharmacol Exp Ther* 1997;**283**:1375–82.

59. Ayer LM, Edworthy SM, Fritzler MJ. Effect of procainamide and hydralazine on poly (ADP-ribosylation) in cell lines. *Lupus* 1993;**2**:167–72.

60. Quddus J, Johnson KJ, Gavalchin J, *et al.* Treating activated CD4+ T-cells with either of two distinct DNA methyltransferase inhibitors, 5-azacytidine or procainamide, is sufficient to induce a lupus-like disease in syngeneic mice. *J Clin Invest* 1993;**92**:38–52.

61. Yung RL, Quddus J, Chrisp CE, Johnson KJ, Richardson BC. Mechanisms of drug-induced lupus. I: Cloned Th2 cells modified with DNA methylation inhibitors *in vitro* cause autoimmunity *in vivo. J Immunol* 1995;**154**:3025–35.

62. Yung R, Powers D, Johnson K, *et al.* Mechanisms of drug-induced lupus. II: T-cells overexpressing lymphocyte function-associated antigen 1 become autoreactive and cause a lupuslike disease in syngeneic mice. *J Clin Invest* 1996;**97**:2866–71.

63. Scheinbart LS, Johnson MA, Gross LA, Edelstein SR, Richardson BC. Procainamide inhibits DNA methyltransferase in a human T-cell line. *J Rheumatol* 1991;**18**:530–4.

64. Oelke K, Lu Q, Richardson D, *et al.* Overexpression of CD70 and overstimulation of IgG synthesis by lupus T cells and T cells treated with DNA methylation inhibitors. *Arthritis Rheum* 2004;**50**:1850–60.

65. Gleichmann E, Pals ST, Rolink AG, Radaskiewicz T, Gleichmann H. Graft-versus-host reactions: clues to the etiopathology of a spectrum of immunological diseases. *Immunol Today* 1984;**5**:324–32.

66. Kretz-Rommel A, Duncan SR, Rubin RL. Autoimmunity caused by disruption of central T cell tolerance: a murine model of drug-induced lupus. *J Clin Invest* 1997;**99**:1888–96.

67. Kretz-Rommel A, Rubin RL. Persistence of autoreactive T cell drive is required to elicit anti-chromatin antibodies in a murine model of drug-induced lupus. *J Immunol* 1999;**162**:813–20.

68. Kretz-Rommel A, Rubin RL. A metabolite of the lupus-inducing drug procainamide prevents anergy induction in T cell clones. *J Immunol* 1997;**158**:4465–70.

69. Kretz-Rommel A, Rubin RL. Disruption of positive selection of thymocytes causes autoimmunity. *Nature Med* 2000;**6**:298–305.

70. Douek DC, McFarland RD, Keiser PH, *et al.* Changes in thymic function with age and during the treatment of HIV infection. *Nature* 1998;**396**:690–5.

71. Rubin RL, Salomon DR, Guerrero RS. Thymus function in drug-induced lupus. *Lupus* 2001;**10**:795–801.

72. Tan EM, Cohen AS, Fries JF, *et al.* The 1982 revised criteria for the classification of systemic lupus erythematosus. *Arthritis Rheum* 1982;**25**:1271–7.

73. Rubin RL, Burlingame RW, Arnott JE, *et al.* IgG but not other classes of anti-[(H2A-H2B)-DNA] is an early sign of procainamide-induced lupus. *J Immunol* 1995;**154**:2483–93.

74. Burlingame RW, Boey ML, Starkebaum G, Rubin RL. The central role of chromatin in autoimmune responses to histones and DNA in systemic lupus erythematosus. *J Clin Invest* 1994;**94**:184–92.

75. Prockop LD. Myotonia, procaine amide, and lupus-like syndrome. *Arch Neurol* 1966;**14**:326–30.

76. Solinger AM. Drug-related lupus: clinical and etiological considerations. *Rheum Dis Clin North Am* 1988;**14**:187–202.

28

Vasculitis induced by drugs

MICHIEL DE VRIES, MARJOLEIN DRENT, JAN–WIL COHEN TERVAERT

INTRODUCTION

The term 'vasculitis' denotes a pathological process of inflammation, vessel destruction and tissue necrosis. Vasculitis can occur secondary to drugs, infection and/or other diseases such as rheumatoid arthritis, and may also occur without known underlying cause ('primary vasculitis'). These primary vasculitides are classified based on vessel size, and clinical manifestations as described in the Chapel Hill international consensus definitions. Recently a simple classification scheme was introduced by the EMEA study group. Based on these classification schemes, vasculitides can now be grouped as 'large-vessel vasculitides' (Takayasu disease and giant-cell arteritis), 'medium-vessel vasculitides' (polyarteritis nodosa and Kawasaki disease), and 'small-vessel vasculitides'.[1]

Wegener's granulomatosis (WG), Churg–Strauss syndrome (CSS), microscopic polyangiitis (MPA), Henoch–Schönlein purpura and mixed essential cryoglobulinaemic vasculitis all belong to the group of small-vessel vasculitides. Isolated pauci-immune pulmonary capillaritis and isolated renal vasculitis with pauci-immune necrotizing crescentic glomerulonephritis in the kidney biopsy can be considered as limited subsets of small-vessel vasculitis. In many patients with small-vessel vasculitis, antineutrophil cytoplasmic antibodies (ANCA) can be detected. Two major categories of ANCA can be recognized in vasculitis. These are PR3 ANCA that produce a cytoplasmic staining pattern in the indirect immunofluorescence (IF) technique and that react in the antigen specific ELISA test with proteinase,[2] and MPO-ANCA that produce a perinuclear staining pattern in the IF test and react with myeloperoxidase in the ELISA. PR3 ANCA positivity is mostly associated with WG, whereas CSS and MPA are mostly associated with MPO-ANCA. In primary forms of small-vessel vasculitis, generally, ANCA is directed to either PR3 or MPO. In drug-induced vasculitis, however, multiple ANCA specificities are frequently found. ANCA in these patients react not only with MPO and/or PR3 but also with elastase, lactoferrin and sometimes other myeloid proteins.[2]

Pulmonary vasculitis may occur as part of any systemic vasculitis or may be associated with another underlying disease such as a connective tissue disease, paraneoplastic, bronchocentric granulomatosis or inflammatory bowel disease.[3]

The presentation of vasculitis is highly variable. Only rarely does a patient present with 'classic' findings. This makes the delay in diagnosis most common.

Vasculitis frequently affects the lung. Pulmonary manifestations such as nodular shadows and/or infiltrative changes on the chest radiograph with or without haemoptysis (diffuse alveolar haemorrhage) are often present in the early phase of the disease process. Other symptoms like purpura, acute glomerulonephritis, destructive upper airway lesions and neurological symptoms (mononeuritis multiplex) should raise the suspicion for vasculitis. Awareness of physicians for related and unrelated symptoms cannot be overemphasized. Often combinations of clinical, lab and radiographic findings suggest the diagnosis of vasculitis.

Vasculitides are chronic diseases that present with relapses and remissions. In most cases of ANCA-associated vasculitis, relapses are preceded by increases of ANCA levels. Pathophysiological mechanisms in vasculitis are diverse.

In large-vessel vasculitis, invasion of the vessel walls with lymphoplasmacytic cells together with an autoimmune response is thought to play an important role in the pathophysiology. Medium-vessel vasculitis (i.e. PAN) and small-vessel vasculitis (non-ANCA associated) are thought to be immune-complex mediated diseases since lesions are characterized by the presence of immune deposits. In ANCA-associated small-vessel vasculitis, involvement of ANCA in the pathophysiology is well demonstrated.

In-vitro studies indicate that ANCA activate cytokine-primed neutrophils and monocytes through both direct Fab'2 binding and Fc receptor engagement. ANCA-activated neutrophils release oxygen radicals, lytic enzymes and inflammatory cytokines. Membrane expression of ANCA antigens on neutrophils is genetically regulated as been demonstrated for PR3. Leucocytes activated by ANCA have an increased capacity to adhere to endothelium and subsequently kill endothelial cells. This supports a direct pathogenic role of ANCA.

In-vivo models in which causality for PR3 ANCA can be tested have been lacking. In mice and rat models compelling evidence has been found that MPO-ANCA are a primary pathogenic factor in MPO-ANCA associated vasculitis by augmenting leucocyte–endothelial interaction and vascular wall damage.[4]

Furthermore, anti-endothelial cell antibodies (AECA) may play a pathophysiological role in vasculitides. In vitro, it has been demonstrated that AECA may directly cause EC injury and/or apoptosis of endothelial cells.

Therapy for vasculitis relies on immunosuppression, or in the case of drug-induced vasculitis cessation of the drug. Immunosuppression is generally titrated in accordance with the severity of the disease so that disease control is obtained while minimizing the adverse side-effects. Therapy of vasculitis is often divided into an initial 'remission-induction' phase in which more intensive immunosuppressive therapy is used, and a 'maintenance' phase in which less intensive therapy is used.[5]

This review will focus on clinical and therapeutic features of pulmonary vasculitis. Special attention will be paid to the role of medication in drug-induced pulmonary vasculitis.

ANCA-ASSOCIATED VASCULITIS

Churg–Strauss syndrome

Churg–Strauss syndrome (CSS) is a multisystem disorder characterized by allergic rhinitis, asthma and blood eosinophilia (> 10 per cent). In CSS, three sequential phases are described. First comes the 'prodromal phase', characterized by allergic rhinitis and asthma. The second phase is called the 'eosinophilic phase' with infiltration of eosinophils in multiple organs, especially in the respiratory and gastrointestinal tract. The third phase is the 'vasculitic phase', in which a systemic vasculitis of the small and medium vessels develops. This phase is often heralded by malaise, weight loss, fever, disabling muscle weakness and arthralgias. Other features of the vasculitic phase are purpura or nodular lesions, abdominal pain, pericarditis and glomerulonephritis. Involvement of the peripheral nervous system dominates the clinical picture of the vasculitic phase in most patients. With the onset of the vasculitic phase, the severity of asthma often increases. Although pulmonary haemorrhage and glomerulonephritis may occur, they are less common than in the other ANCA-associated small-vessel vasculitides (i.e. Wegener's granulomatosis and microscopic polyangiitis).[6] Cardiac involvement (pericarditis, myocarditis and pericardial effusion) is relatively common, causing 50 per cent of the deaths.[7]

Laboratory findings in CSS include a normochromic normocytic anaemia, leucocytosis with prominent eosinophilia. Most patients have elevated IgE levels. Approximately 50 per cent of the patients are ANCA-positive. In most of these patients ANCA is directed to myeloperoxidase (MPO-ANCA).

The pathophysiology of CSS is not completely clear. CSS is not associated with immune complex depositions. Activated eosinophils induce vessel wall damage. Variations in the balance between Th1 and Th2 cytokines at different disease stages could contribute to the distinct clinical courses seen in patients with CSS. In most cases a Th-2 cytokine profile may predominate. In serum and bronchoalveolar lavage (BAL) fluid, high levels of interleukin-5 and tumour necrosis factor-alpha (TNF-α) were found. Interleukin-5 is the most potent stimulator of eosinophil production and induces activation of mature eosinophils, the key effector cells in CSS. Seromarkers like soluble thrombomodulin, soluble interleukin-2 receptor and eosinophil cationic protein may predict a relapse of CSS.[8,9] Also, increasing MPO-ANCA levels in a patient in remission may predict relapse.

Radiographic abnormalities in patients with CSS are varied, with the chest X-ray revealing changes from transient focal areas of consolidation to more extensive peripheral opacities.[7] Pulmonary function tests reveal airflow obstruction (asthma) with either a reduced gas transfer factor or, when pulmonary haemorrhage is present, an increased gas transfer for carbon monoxide.[10]

Vaccinations and desensitization have been implicated as precipitating factors.[11,12] In addition, CSS is associated with use of drugs such as leukotriene receptor antagonists (LRAs). LRAs have an anti-inflammatory and a bronchodilatator effect. LRAs are zafirlukast, pranlukast and montelukast. Furthermore, a 5-lipoxygenase inhibitor (zileuton) is available which is also associated with the occurrence of CSS. A number of explanations may account for the association between LRAs and CSS:[13–18]

- There is coincidental development of a full-blown CSS with the introduction of a leukotriene receptor antagonist.
- LRAs prevent airway obstruction by blocking the bronchoconstrictor effects of leukotrienes and therefore decrease the need for corticosteroids which allow the manifestations of CSS to develop. Also CSS has been reported in several asthmatic patients not on LRAs while tapering corticosteroids.
- There is an imbalance in leukotriene receptor stimulation. The LRAs block the effect of cysteinyl leukotrienes LTC4, LTD4 and LTE4, but do not block the effect of LTB4 which is a chemoattractant for eosinophils and neutrophils. Since reports of LTB4 inhibition by zileuton, a strong 5-lipoxyoxigenase inhibitor, and CSS are also described, this hypothesis seems unlikely.
- CSS is a hypersensitivity reaction to LRA treatment.

The estimated incidence of CSS in the asthmatic population not receiving LRA therapy is 60 cases per million per year; including LRA therapy it is 64.4 cases per million. In the general population the incidence is 1.8–3.3 cases per million per year.

The diagnosis of CSS is based on the presence of asthma, peripheral eosinophilia, clinical signs of systemic vasculitis, histopathological findings and/or angiographic abnormalities. Biopsy of skin lesions, peripheral nerves and/or muscles reveals necrotizing vasculitis, often with many eosinophils in the infiltrate, but occasionally with granulomas. Kidney biopsies may show pauci-immune focal necrotizing and/or crescentic glomerulonephritis. Histological examination of the lungs typically reveals changes of asthma such as eosinophilic inflammation of the bronchiolar walls. Damage to the bronchi may cause bronchiectasis. In addition, infiltration of the alveolar septa with eosinophils and a considerable admixture of multinucleated giant cells can be found. Granulomas, vasculitis and/or necrotizing capillaritis may be present.[7]

Corticosteroids are the mainstay of treatment, and the response to treatment is often dramatic. The decision to use cytotoxic therapy should be made in the individual case based on disease extension (e.g. visceral or cardiac involvement or severe neuropathy).[8,19] However, the exact role of immunosuppressive agents and treatment options for refractory disease remain poorly studied.[20] In severe cases (e.g. life-threatening alveolar haemorrhage and/or severe kidney failure) plasma exchange is recommended.

Despite the strength of the association of CSS with LTRA, LTRA treatment still is a welcome addition in some patients according to the GINA guidelines.[21] Patients should be monitored for the possible development of symptoms consistent with CSS. CSS has also been observed in patients who were steroid-dependent asthmatics in whom inhaled corticosteroids allowed systemic corticosteroid withdrawal.[22] In addition, however, a few cases have been described in which fluticasone inhalation triggered CSS (or a relapse of CSS) that improved after discontinuation of the drug.[23,24] Several cases of CSS in association with omalizumab have been described, and the issues regarding pathophysiology are similar to those raised by LTRA drugs.[25]

Finally, other drugs such as penicillin, sulfonamides, macrolides, anticonvulsants and thiazides have been implicated to cause CSS in isolated case reports.

Wegener's granulomatosis

Wegener's granulomatosis (WG) is a necrotizing, granulomatous vasculitis that has a clinical predilection to involve the upper airways, lungs and kidneys. The prevalence of WG is 23.7 per million adult inhabitants in France. A decreasing north/south gradient is hypothesized with a prevalence of 53 per million adults in Norway.[26]

The upper airways are affected in more than 90 per cent of patients at some point of the disease's course.[27] Narrowing of the subglottic trachea occurs in approximately 20 per cent of patients with WG and can cause life-threatening situations.[27] WG also affects the lower airways. Common symptoms are cough, dyspnoea, haemoptysis and chest pain. In many cases, however, chest abnormalities are asymptomatic. On CT scans areas of consolidation and ground-glass attenuation with or without pulmonary nodules are seen in up to 50 per cent of the patients. In the series of Lohrmann et al., 15 per cent of the patients had pleural effusion at the initial CT scan.[28] Pleural irregularities were seen in 14 per cent. Hilar and mediastinal lymph node enlargement is rare in WG. In addition to the above-described organs, WG can affect virtually any organ system.[27]

WG is strongly associated with the presence of ANCA with specificity against proteinase 3 (PR3). In about 10–20 per cent of the patients PR3 ANCA is not found. In most of these patients ANCA of other specificity is found, i.e. MPO-ANCA or elastase-ANCA. In addition, the sequential measurement of PR3 ANCA levels is of value in diagnosing relapses of WG.[29-31] Most ANCA titre rises occur within 6 months before relapse is diagnosed. Genes and environmental factors influence the development of WG. Chronic nasal carriage of *Staphylococcus aureus* identifies a subgroup of WG patients who are prone to relapses.[32,33] The role of *S. aureus* in the pathogenesis of WG, however, remains controversial. Association between drugs and the onset of WG is rarely observed.[34] PTU, other antithyroid drugs, sulfasalazine and montelukast have been implicated.[34,35] Relapse of WG may be triggered by thyreostatic drugs or sulfasalazine.[36] Iatrogenic vasculitis in such cases should be suspected when a patient with WG that is PR3 ANCA-associated develops MPO-ANCA and/or elastase ANCA.

Recently, a patient has been described who developed a fatal relapse of WG after influenza vaccination.[37] Furthermore, two patients with WG have been described in whom the disease started after influenza vaccination, suggesting that vaccination might have a role in triggering the disease.[38] Cocaine may induce a PR3 ANCA associated generalized form of WG.[39,40] Interestingly, inhalation of nasal cocaine induces more often an elastase-ANCA associated midline destructive disease which mimics the limited form of WG.[41]

The diagnosis of WG is based on the presence of typical clinical symptoms, demonstration of antibodies to PR3 or MPO, and histopathological findings. The most characteristic pathological finding of WG is 'pathergic' necrosis and granulomatous inflammation in the absence of infectious or foreign-body granulomas, angio-invasive lymphomas and/or rheumatoid nodules.

Treatment consists of the combination of high-dose steroids and cyclophosphamide. Prednisone is tapered in 6–8 months, and cyclophosphamide is switched to azathioprine after 3–6 months. Azathioprine is subsequently stopped after 18–24 months. Alternative therapeutic options include methotrexate, the anti-CD-20 monoclonal antibody rituximab, or mycophenolate mofetil. In patients with life-threatening disease, pulse methylprednisolone and/or plasma exchange is added to the standard regimen.[42]

Microscopic polyangiitis

Microscopic polyangiitis (MPA) is another form of ANCA-associated systemic necrotizing vasculitis that histologically affects small vessels without granulomatous formation. Although MPA was originally considered a subclass of polyarteritis nodosa (PAN), the 1994 Chapel Hill consensus conference set a classification framework that differentiated PAN from MPA on the basis of the presence of vasculitis in small vessels. Most of the patients have MPO-ANCA, whereas in the remaining patients PR3 ANCA is found. Age at onset is 40–60 years. MPA seems more common in men than in women. Patients present with variable signs and symptoms such as microscopic haematuria and proteinuria, palpable purpura, abdominal pain, cough and haemoptysis. Histology reveals pauci-immune necrotizing glomerulonephritis with crescent formation and sometimes vasculitis involving arterioles and/or small interlobular arteries. Granulomatous lesions of the upper or lower respiratory tract are not found.[43] The kidney is the most commonly affected organ in 90 per cent of the patients.[44] Pulmonary involvement is seen in 30–50 per cent. Pulmonary features are cough, chest pain and haemoptysis. CT findings consisted of ground-glass attenuation in 94 per cent of cases. Other features are consolidations, thickening of bronchovascular bundles and honeycombing.[45] Lung biopsy specimens frequently demonstrate only necrotic tissue or show non-specific inflammation with or without haemorrhage. Alveolar haemorrhage, vasculitis with alveolar haemorrhage, pulmonary fibrosis, bronchitis, bronchiectasis and bronchiolitis obliterans organizing pneumonia (BOOP) have been reported in surgical biopsies.[45,46] Aggressive immunosuppressive therapy (corticosteroids and cyclophosphamide) is effective in inducing remission in about 80–90 per cent of cases. After 3–6 months cyclophosphamide is stopped and replaced by azathioprine. Other therapeutic options are mycophenolate mofetil or methotrexate.[47–49] Rituximab, a chimeric anti-CD20 monoclonal antibody, has been shown to be effective.[50] In severe lung haemorrhaging, pulse methylprednisolone and/or plasma exchange can be added.[42] Antithyroid drugs such as PTU may induce MPA.[51] Other drugs that are suspected to induce MPA are hydralazine and D-penicillamine. In these drug-induced cases, ANCA is frequently directed to two or more antigens including elastase.[52]

ANCA-associated pneumo-renal syndrome

This potentially life-threatening syndrome of diffuse alveolar haemorrhage (DAH) and glomerulonephritis is usually called Goodpasture's syndrome. The incidence is 1 per million. The syndrome affects two different age groups: persons in their mid 30s and in their late 50s.[53] This pneumo-renal syndrome (PRS) is due to pulmonary capillaritis in conjunction with rapidly progressive glomerulonephritis. Renal biopsy displays extracapillary proliferating glomerulonephritis and renal immunohistology facilitates detection of the underlying disease. A pneumo-renal syndrome has been reported to occur in ANCA-associated vasculites (MPO/PR3-ANCA), including WG, CSS and MPA, in patients with antibodies against glomerular and alveolar basement membranes (i.e. anti-GBM or Goodpasture's disease), systemic lupus erythematosus, and in infection-associated or drug-induced glomerulonephritis. Antiglomerular basement membrane (anti-GBM) antibodies attack the NC1 domain of the α_3 chain of type IV collagen located in the basement membranes of the kidneys and alveoli. ANCA-associated vasculitides account for about 60 per cent of the cases, anti-GBM disease syndrome for about 20 per cent.[54] There has been a tendency to believe that pulmonary disease takes a more aggressive course in PR3-positive ANCA patients,[55] although others did not confirm that.[56]

Respiratory involvement usually presents with life-threatening symptoms such as haemoptysis, or acute respiratory failure requiring mechanical ventilation. However, there can be more indolent initial presentations of the disease, associated with atypical symptoms such as cough, dyspnoea, fever and fatigue.

DAH should be suspected if the uptake of carbon monoxide (CO) by the lung is increased. The diffusing capacity for CO expressed as the carbon monoxide transfer coefficient KCO is increased. An increase of KCO of 30 per cent or more over baseline may indicate active DAH, but some patients with active haemorrhaging do not present with this feature.

Hydralazine, D-penicillamine, propylthiouracil (PTU) and other thyrestatic drugs are involved in MPO-ANCA associated pneumorenal syndrome.[51] In addition, MPO-ANCA associated pneumo-renal syndrome has been described during anti-TNF antibody therapy (i.e. etanercept) in a patient with rheumatoid arthritis.[57] ANCA-associated vasculitis has been reported in patients using the antithyroid drug PTU. These antibodies are directed against human neutrophil elastase, proteinase 3 and in the majority of cases against myeloperoxidase.[51,52,58]

PTU-induced vasculitis often occurs within weeks of treatment, but may develop after months or even years.[52] The vascular disorder is not necessarily dose-dependent. A possible mechanism for the production of ANCA is an interaction between PTU and neutrophils (MPO). MPO and hydrogen peroxide produced by neutrophils could metabolize the drug leading to reactive intermediates, that are immunogenic for T-cells and stimulate the immune system. These metabolites also have cytotoxic activity leading to cell death and abnormal degradation of cell material, as well as production of autoantibodies. PTU can also bind to MPO, changing the haem structure of the enzyme, which may then act as an antigen.[59]

Although PTU can induce ANCA in a substantial percentage of patients who receive the drug to treat their hyperthyroid state, most patients have no manifestations of vasculitis and drug withdrawal is not indicated in all cases. Slot et al. suggest that the amount and the binding capacity of MPO antibodies might be associated with the development of clinically manifest PTU-associated ANCA vasculitis.[52]

Therapy of choice is stopping the drug that is suspected to induce the vasculitis. However, in most cases immunosuppressive treatment is additionally needed. The goals of treatment in PRS are to remove the circulating antibodies, to stop further production of autoantibodies, and to remove any antigen that stimulates antibody production. Treatment is based on plasmapheresis, corticosteroids and cyclophosphamide.

LARGE-VESSEL VASCULITIS

Two forms of large-vessel vasculitis will be discussed: Takayasu arteritis (TA) and giant-cell arteritis (GCA). Takayasu arteritis (TA) is a form of vasculitis, mainly involving the aorta and its branches and the pulmonary and coronary arteries, causing various clinical conditions such as pulselessness, ischaemic congestive heart disease, aortic insufficiency, aneurysm formation, renovascular hypertension and/or cerebrovascular symptoms. The mechanism is unknown, but a complex interaction of multiple factors such as age, sex, ethnic background, immunogenetic mechanisms and environmental influences is probably involved.[60]

The estimated incidence in Western Europe and in the United States is approximately 2.6 cases per million per year. Women younger than 30 are frequently involved. Mortality rate is 3–6 per cent at 5 years after diagnosis. Diagnosis of TA arteritis is based on clinical and radiological features. Angiographic abnormalities in TA are usually seen as occlusion, stenoses, lumen irregularities and ectasias or aneurysms of large vessels. Pulmonary artery involvement is detected by angiography in 50–80 per cent of patients with TA. Beside stenosis and aneurysm formation, pulmonary hypertension is a common finding in TA. Pulmonary involvement in TA presents with dyspnoea, pleurisy and/or haemoptysis. Currently, treatment includes corticosteroids, to which 50 per cent of the patients respond, and methotrexate (MTX), to which a further 50 per cent respond. In addition, anti-TNF-based therapy is used for 'difficult' cases. To our knowledge, there are no case reports of drug-induced TA.

Giant-cell arteritis is also known as 'temporal arteritis' or Horton's disease. It is a disease of the elderly. GCA is a systemic granulomatous arteritis with a predilection for cranial vessels, especially the temporal arteries. The onset of symptoms may be gradual or abrupt. Generalized symptoms such as malaise, fatigue, weight loss, anorexia, low-grade fever and anaemia are usual findings. Visual manifestations are noted in approximately 25–50 per cent of the patients. Headache is the most common finding. The syndrome of polymyalgia rheumatica, estimated to occur in 50–75 per cent of these patients, is characterized by aching and stiffness in the proximal muscles. Jaw claudication, difficulty in swallowing and tongue pain are also familiar symptoms. Pulmonary involvement has been described in GCA.[61-63] A persistent dry cough is the most common symptom, together with dyspnoea, haemoptysis, pleural effusion, pulmonary infiltrates and/or interstitial lung disease. In many of these reported cases, a coexisting form of vasculitis such as Wegener's granulomatosis and/or Churg–Strauss syndrome was probably present. In addition, rare cases of pulmonary infarction due to giant cell arteritis are seen.[64-67]

Although the aetiology of GCA is unknown, data accumulated in recent years point to a T-cell-mediated immunity leading to granuloma formation. Emerging data support the hypothesis that GA is an antigen-driven disease.

Corticosteroids are the mainstay of therapy. In general it is started at 40–60 mg per day. After the induction period of 4 weeks it is tapered to zero in 6 months. Low-dose methotrexate reduces the incidence of disease relapses and is steroid-sparing. Azathioprine may be used as an alternative for patients in whom MTX is contraindicated.

Drug-induced large-vessel vasculitis has not been described. Drug-induced small-vessel vasculitis, however, may mimic large-vessel vasculitis. Lie and Dixit described a patient who developed NSAID-induced vasculitis that simulated GCA.[68]

POLYARTERITIS NODOSA

Polyarteritis nodosa (PAN) is a systemic vasculitis of medium-sized and small arteries in various organs without glomerulonephritis or vasculitis in arterioles, capillaries or venules.

PAN is a diagnosis of exclusion, so the diagnosis can be made only when the disease cannot fit into another recognized syndrome. At present two different types of PAN are recognized:

- systemic PAN – vasculitic involvement of visceral organs (kidneys, heart and gastrointestinal tract);
- limited PAN or cutaneous PAN – vasculitis restricted to the skin, nerves and/or musculoskeletal system.

The disease is seen at any age but the mean age of onset is 45–50 years. A substantial proportion of patients with PAN have a hepatitis B virus and/or human immunodeficiency virus (HIV) infection.

PAN is an ANCA-negative vasculitis. In systemic PAN, visceral angiograms demonstrate the presence of microaneurysms and/or stenoses in medium-sized vessels. These aneurysms form when the arterial wall is sufficiently weakened by the necrotizing process. Pulmonary involvement in PAN is very rare. The most common symptom is mononeuritis multiplex. Other organ systems frequently involved are kidneys (vascular nephropathy), gastrointestinal tract, skin, central nervous system and heart.

Initial therapy is with high-dose corticosteroids. In patients with visceral organ involvement and/or severe peripheral nerve disease, a cytotoxic (cyclophosphamide) agent has to be added. In less severe forms azathioprine or MTX can be used. In PAN associated with hepatitis B or HIV, antiviral therapy and plasma exchange is the recommended therapy.

Cutaneous PAN may be induced by drugs. Minocycline is one of the drugs described.[69] Remarkably, the reported cases were MPO-ANCA-positive,[69] whereas idiopathic cutaneous PAN is ANCA-negative.[70,71] The exact pathogenic mechanism involved remains to be elucidated. Both cell-mediated and humoral immune mechanisms appear to play a role. Several hypotheses have been suggested. One hypothesis is that minocycline is transformed into a reactive hydroxylated product. These reactive products then form complexes with MPO and subsequently induce an autoimmune response.[59]

Withdrawal of the offending agent alone is often sufficient to induce resolution of clinical manifestations.

KAWASAKI DISEASE

Kawasaki disease (KD) is an acute self-limited disease that occurs predominantly in infants and young children. KD is currently the leading cause of aquired cardiac disease in children in the United States and developed nations. KD presents with fever, rash, conjunctival injection, cervical lymphadenopathy, oral mucosal changes, and oedematous and erythematous hands and feet. Coronary artery aneurysms or ectasia develop in approximately 15–25 per cent of untreated patients and may lead to myocardial infarction, sudden death, or ischaemic heart disease.

The aetiology of KD remains unknown, although clinical and epidemiological features strongly suggest an infectious cause. The prominence of IgA plasma cells in the respiratory tract supports the theory of a respiratory portal of entry of an aetiologic agent(s). In the early phase of the disease the media of affected vessels demonstrate oedematous dissociation, endothelial cell swelling with subendothelial oedema and an influx of neutrophils. In a later phase, destruction of the internal elastic lamina and remodelling take place. Active inflammation is replaced over several weeks to months by progressive fibrosis, with scar formation.[72]

Pulmonary features are associated with acute KD. Cough is reported as a symptom in about 30 per cent and abnormal chest radiographs in 15 per cent of children with acute KD. As in fatal acute respiratory illnesses of childhood due to respiratory syncytial virus or influenza virus infections, IgA plasma cell infiltrations of the upper respiratory tract can be seen in KD. Pulmonary nodules can be seen on chest X-ray or CT scan and represent the more severe end of the spectrum.[73]

Treatment of KD consists of a combination of high-dose aspirin, intravenous gammaglobulin and, in severe cases, corticosteroids. The possible mechanisms of action of gammaglobulins include modulation of cytokine production, neutralization of bacterial superantigens, augmentation of T-cell suppressor activity, suppression of antibody synthesis, and provision of anti-idiotypic antibodies.

Drug-induced KD has been described. Drugs that are suspected to induce and/or mimic KD are carbamazepine, acetaminophen and abacavir.[74-77] It was stressed that a febrile illness with rash and constitutional symptoms ('Kawasaki-like syndrome') was a hypersensitivity reaction to the above drugs.

HENOCH–SCHÖNLEIN PURPURA

Henoch–Schönlein purpura (HSP) is a clinicopathological syndrome characterized clinically by palpable purpura, arthragias or arthritis, abdominal pain, gastrointestinal bleeding and renal involvement. It is a small-vessel vasculitis containing IgA-dominant immune complex deposits. HSP is generally an illness of childhood, although it can occur at any age. Pulmonary involvement is rare, but it does occur.[78] Pulmonary complications are haemoptysis and dyspnoea caused by diffuse alveolar haemorrhage or lung infarction.[79] Occasionally, pleural effusion and usual interstitial pneumonia are found.[80]

HSP is an ANCA-negative disease. The diagnosis is based on clinical findings in combination with the occurrence of IgA in skin and/or kidney biopsies. Skin biopsies show leucocytoclastic angiitis involving capillaries, arterioles and venules. Kidney biopsies show mesangial proliferative glomerulonephritis in most cases. Immunofluorescence studies of the lungs may show IgA deposits in alveolar septal vessels.

HSP is usually a self-limiting disease. If there is pulmonary haemorrhaging, however, aggressive immunosuppressive treatment is indicated.[81]

As many as 50 per cent of occurrences of HSP in paediatric patients are preceded by an upper respiratory infection. Several agents have been implicated, including group A streptococci, Varicella, hepatitis B, Epstein–Barr virus, parvovirus B19, Mycoplasma, Campylobacter and Yersinia. In adults, more commonly other factors are associated with HSP, including drugs, vaccination, malignancy, food, pregnancy, familial Mediterranean fever, liver cirrhosis, or exposure to cold.

Many drugs are reported to be associated with HSP,[82,83] particularly antibiotics such as amoxicillin, erythromycin and/or clarithromycin, analgetics such as acetaminophen/codeine and NSAIDs.[84-86] Interestingly, drugs such as acenocoumarol, etanercept and ranitidine are also reported to induce HSP.[86-89] These reactions often appear weeks after initiating the drug and withdrawal of the offending agent alone is often sufficient to induce resolution of clinical manifestations.

SECONDARY FORMS OF VASCULITIS

Vasculitis may be the primary pathological process or secondary to other conditions.

Lymphomatoid granulomatosis

Lymphomatoid granulomatosis (LG) is a disease characterized by infiltration of small and medium-sized blood vessels by lymphocytes. During the last 10 years it was suggested that LG is an angiocentric T-cell-rich B-cell lymphoproliferative disorder, and not strictly a form of vasculitis, since vessels are not inflamed but infiltrated. The lungs, skin and nervous system are involved frequently.

The chest radiograph typically shows multiple nodular densities, which may cavitate. Other findings include hilar adenopathy, reticular infiltrates and pleural effusions. A few cases of drug-induced LG have been reported. LG is considered to be induced by immunosuppressives such as methotrexate. The lesions may diminish after withdrawal of the immunosuppressive agent.[90]

Sarcoidosis

Sarcoidosis is a multisystemic granulomatous disorder of unknown cause characterized by frequent pulmonary involvement. It commonly affects young and middle-aged adults and frequently presents with bilateral hilar adenopathy, pulmonary infiltration, ocular and skin lesions. The diagnosis is established when clinico-radiographic findings are supported by histological evidence of non-caseating epithelioid cell granulomas. Frequently observed immunological features are depression of cutaneous delayed type-hypersensitivity and a heightened Th1 immune response at sites of disease. The Th1 immune response is felt to be a key immunological event in granuloma formation and the pathogenesis of sarcoidosis. Circulating immune complexes along with signs of B-cell hyperactivity may also be found. The disease occurs throughout the world, affecting both males and females.

An acute onset with erythema nodosum or asymptomatic bilateral hilar adenopathy usually heralds a self-limiting course.[91] Sarcoidosis is rarely associated with a second, distinct autoimmune disease. Six cases of small- and large-vessel vasculitis were described by Fernandes et al.[92] Clinical manifestations have ranged from mild cutaneous vasculitis to glomerulonephritis with renal failure, transient cerebral ischaemia and Takayasu-like disease. In three cases evidence of necrotizing sarcoid granulomatosis (NSG) was found. Liebow[93] described this mild disorder in 1973 with sarcoid-like granuloma, vasculitis of small vessels, and necrosis. Many of these autoimmune diseases are associated with dysregulated

Th1 immunity that would explain at least one common patho-genetic pathway shared between them.

Drug-induced sarcoidosis is very rare. Scattered reports of interferon-induced sarcoidosis have appeared in the literature.[94] Interferon (IFN) is widely used for the treatment of viral hepatitis C. The drug is capable of inducing, aggravating or unmasking a variety of autoimmune diaseases, for example autoimmune thyroid disease and SLE. The exact mechanism of interferon-induced sarcoidosis is not known. Administration of IFN-α is thought to increase IFN-γ production by CD4+ cells, which, along with interleukin-2 (IL-2) and IL-12, up-regulates the Th1 response.

Signs and symptoms of sarcoidosis develop between 2 weeks and 2 years after starting with IFN. Symptoms are fever, myalgias, arthralgias and in rare cases pulmonary symptoms. Clinical outcome is favourable on withdrawal of IFN. Sometimes, however, corticosteroid therapy is needed. Other drugs described in the literature that are suspected to induce sarcoidosis include allopurinol, mitomycin-C and anti-TNF-α antibody therapy.[95,96]

Mixed essential cryoglobulinaemia

Cryoglobulinaemia refers to the presence of one or more anti-bodies which precipitate at temperatures below 37°C and sol-ubilize on rewarming. Cryoglobulins are classified into three types based on whether they are monoclonal or not. Type II cryoglobulins are composed of monoclonal IgM and a poly-clonal IgG. The cryoglobulins have rheumatoid factor activity and can produce an immune-complex small-vessel vasculitis. Clinical manifestations of type II cryoglobulin-associated vasculitis are diverse. In many cases, this so-called 'essential' mixed cryoglobulinaemia is related to hepatitis C infection. Otherwise, cryoglobulinaemia in countries where hepatitis C prevalence is low is often a complication of a lymphoprolifera-tive disorder and occurs sometimes years after cytotoxic ther-apy is administered, either because of an autoimmune disease or as treatment of a malignancy.[97]

Classical clinical findings include purpura, arthralgias and glomerulonephritis, and alveolar haemorrhage due to pulmonary vasculitis. Antiviral treatment with an interferon mostly leads to improvement of vasculitic symptoms in hepatitis C-associated mixed cryoglobulinaemia, but vasculitis may also exacerbate dur-ing IFN treatment. In these cases, the most common symptom is vasculitic neuropathy. However, peripheral neuropathy in an HCV-infected patient treated with IFN may also be caused by direct neurotoxic or antiangiogenic effects of IFN itself.[98]

Collagen vascular diseases

Rheumatic disorders such as SLE, rheumatoid arthritis (RA), scleroderma, undifferentiated connective tissue disease and Sjögren's disease are often associated with pulmonary involve-ment. Documented vasculitis, however, is relatively uncom-mon in these diseases.

SLE can affect the skin, joint, kidneys, lung and other organs. In a retrospective cohort of 320 SLE patients, Shen *et al.* documented pleuropulmonary involvement in 142 patients (44.4 per cent).[99] Other pulmonary complications of SLE are

pulmonary hypertension, pulmonary embolism, and different forms of pulmonary infections. Drug-induced lupus erythema-tosus (DIL) (see Chapter 27) is a syndrome that shares symp-toms and laboratory characteristics with idiopathic systemic lupus erythematosus (SLE). The first case of DIL was reported in 1945 and associated with sulfadiazine. In 1953, it was reported that DIL was related to the use of hydralazine. Various drugs (minocycline, chlorpromazine, hydralazine, isoniazide, pro-cainamide, anti-TNF drugs and others) have been associated with DIL. Approximately half the patients with drug-induced SLE are women, compared with 90 per cent of patients with idiopathic SLE. Similarly to idiopathic lupus, DIL can be divided into systemic, subacute cutaneous (SCLE) and chronic cuta-neous lupus.

The syndrome is characterized by arthralgia, myalgia, pleurisy, rash and fever in association with antinuclear anti-bodies (ANA) in the serum. Central nervous system and renal involvement are rare in DIL.[100] Pulmonary infiltrates occur in isolation in a small minority of patients.

Statins are associated with drug-induced lupus. Statins are among the most prescribed drugs in adults, being taken by an estimated 25 million patients worldwide. Two types of drug-induced lupus are reported: SLE and SCLE. In most cases symptoms appeared many months or even years after starting statin therapy. Recovery after cessation was very slow. In many reported cases, ANAs were still present many months after clinical recovery. The exact mechanism is unclear. Cellular apoptosis with release of nuclear antigens may foster the pro-duction of pathogenic autoantibodies. Furthermore, a modu-lation of T-lymphocytes with a shift of T helper-1 (Th1) to Th2 response leading to B-cell reactivity and production of pathogenic autoantibodies may be important.[101]

In RA, nearly 50 per cent of patients demonstrate some type of extra-articular manifestation of the disease such as pleuritis, pleuropericarditis, vasculitis, pneumonitis, pulmo-nary fibrosis, scleritis or nodulosis. Pulmonary involvement in RA is common and can be due to the disease itself as well as to the therapies used to treat it. In fact, lung disease is the second most common cause of death (second to infection). RA-associated interstitial lung disease (ILD) is often subtle in onset, slowly progressive and of unclear aetiology and response to treatment.[102] Vasculitic lesions have been demonstrated in the so-called 'rheumatoid nodules' and are occasionally the basis of interstitial lung disease in these patients.[103] Novel anti-TNF agents have been shown to produce the lupus syn-drome. Anti-double strand DNA antibodies may be present, and this contrasts with DIL induced by most other drugs.

Diagnosis is complex in patients who receive the drug for the treatment of autoimmune conditions. Both clinical features and autoantibody profile of DIL can overlap with those of the underlying condition. A pre-therapy antibody panel is useful.

Gold-based drugs have long been used in the treatment of rheumatoid arthritis. The most common adverse effects appeared in skin, mucous and kidneys. Most gold-induced autoimmune manifestations are localized in the kidneys. Pulmonary manifestations are rare. Occasionally case reports are found in the literature about interstitial pulmonary fibrosis caused by gold therapy, possibly with an autoimmune patho-genesis. In addition, D-penicillamine has long been used for therapy in RA. D-penicillamine can induce an ANCA-associated small-vessel vasculitis.

REFERENCES

1. Watts R, Lane S, Hanslik T, et al. Development and validation of a consensus methodology for the classification of the ANCA-associated vasculitides and polyarteritis nodosa for epidemiological studies. Ann Rheum Dis 2007;66:222–7.

2. Cohen Tervaert JW, Damoiseaux J. Fifty years of antineutrophil cytoplasmic antibodies (ANCA) testing: do we need to revise the international consensus statement on testing and reporting on ANCA? APMIS Suppl 2009;127:55–9.

3. Rockall AG, Rickards D, Shaw PJ. Imaging of the pulmonary manifestations of systemic disease. Postgrad Med J 2001;77:621–38.

4. Heeringa P, Huugen D, Tervaert JW. Anti-neutrophil cytoplasmic autoantibodies and leukocyte–endothelial interactions: a sticky connection? Trends Immunol 2005;26:561–4.

5. Frankel SK, Cosgrove GP, Fischer A, Meehan RT, Brown KK. Update in the management of pulmonary vasculitis. Chest 2006;129:452–65.

6. Guilevin L, Cohen P, Gayraud M, et al. Churg–Strauss syndrome: clinical study and long-term follow-up of 96 patients. Medicine 1999;78:26–37.

7. Lanham JG, Churg J. Churg–Strauss syndrome. In: Churg A, Churg J (eds) Systemic Vasculitides, pp 101–20. New York: Igaku-Shoin, 1991.

8. Gross WL. Churg–Strauss syndrome: update on recent developments. Curr Opin Rheumatol 2002;14:11–14.

9. Kurosawa M, Nakagami R, Morioka J, et al. Interleukins in Churg–Strauss syndrome. Allergy 2000;55:785–6.

10. Thickett DR, Richter AG, Nathani N, Perkins GD, Harper L. Pulmonary manifestations of anti-neutrophil cytoplasmatic antibody (ANCA)-positive vasculitis. Rheumatology 2006;45:261–8.

11. Guillevin L, Guittard T, Bletry O, Godeau P, Rosenthal P. Systemic necrotizing angiitis with asthma: causes and precipitating factors in 43 cases. Lung 1987;165:165–72.

12. Vanoli M, Gambini D, Scorza R. A case of Churg–Strauss vasculitis after hepatitis B vaccination. Ann Rheum Dis 1998;57:256–7.

13. Jamaleddine G, Diab K, Tabbarah Z, Tawil A, Arayssi T. Leukotriene antagonists and the Churg–Strauss syndrome. Semin Arthritis Rheum 2002;31:218–27.

14. Donohue JF. Montelukast and Churg–Strauss syndrome. Chest 2001;119:668–77.

15. Conen D, Leuppi J, Bubendorf L, et al. Montelukast and Churg–Strauss syndrome. Swiss Med Wkly 2004;134:377–80.

16. Keogh KA, Specks U. Churg–Strauss syndrome: clinical presentation, antineutrophil cytoplasmic antibodies, and leukotriene receptor antagonists. Am J Med 2003;115:284–90.

17. Franco J, Artes MJ. Pulmonary eosinophilia associated with montelukast. Thorax 1999;54:558–60.

18. Noth I, Strek ME, Leff AR. Churg–Strauss syndrome. Lancet 2003;361:587–94.

19. Chumbley LC, Harrison EG, DeRemee RA. Allergic granulomatosis and angiitis (Churg–Strauss syndrome): report and analysis of 30 cases. Mayo Clin Proc 1977;52:477–84.

20. Keogh KA, Specks U. Churg–Strauss syndrome. Semin Respir Crit Care Med 2006;27:148–57.

21. Liard R, Leynaert B, Zureik M, Beguin FX, Neukirch F. Using global initiative for asthma guidelines to assess asthma severity in populations. Eur Respir J 2000;16:615–20.

22. Le Gall C, Pham S, Vighes S, et al. Inhaled corticosteroids and Churg–Strauss syndrome: a report of five cases. Eur Respir J 2000;15:978–81.

23. Termeer C, Simon JC, Schöpf E. Churg–Strauss syndrome associated with fluticasone therapy. Arch Dermatol 2001;137:1527–8.

24. Nepal M, Padma H. Fluticasone-associated cutaneous allergic granulomatous vasculitis: an underrecognized but important cause of drug-induced cutaneous Churg–Strauss syndrome. South Medical J 2008;101:761–3.

25. Wechsler ME, Wong DA, Miller MK, Lawrence-Miyasaki L. Churg–Strauss syndrome in patients treated with omalizumab. Chest 2009;136:507–18.

26. Mahr A, Guillevin L, Poissonnet M, Ayme S. Prevalences of polyarteritis nodosa, microscopic polyangiitis, Wegener's granulomatosis, and Churg–Strauss syndrome in a French urban multiethnic population in 2000: a capture–recapture estimate. Arthritis Rheum 2004;51:92–9.

27. Hoffman GS, Kerr GS, Leavitt RY, et al. Wegener granulomatosis: an analysis of 158 patients. Ann Intern Med 1992;116:488–94.

28. Lohrmann C, Uhl M, Kotter E, et al. Pulmonary manifestations of Wegener granulomatosis: CT findings in 57 patients and a review of the literature. Eur J Radiol 2005;53:471–7.

29. Boomsma MM, Stegeman CA, van der Leij MJ, et al. Prediction of relapses in Wegener's granulomatosis by measurement of

antineutrophil cytoplasmic antibody levels: a prospective study. *Arthritis Rheum* 2000;**43**:2025–33.

30. Cohen Tervaert JW, Huitema MG, Hené RJ, *et al.* Prevention of relapses in Wegener's granulomatosis by treatment based on antineutrophil cytoplasmic antibody titre. *Lancet* 1990; **336**:709–11.

31. Cohen Tervaert JW, van der Woude FJ, Fauci AS, *et al.* Association between active Wegener's granulomatosis and anticytoplasmic antibodies. *Arch Intern Med* 1989; **149**:2461–5.

32. Stegeman CA, Cohen Tervaert JW, Sluiter WJ, *et al.* Association of chronic nasal carriage of *Staphylococcus aureus* and higher relapse rates in Wegener granulomatosis. *Ann Intern Med* 1994;**120**:12–17.

33. Stegeman CA, Cohen Tervaert JW, de Jong PE, Kallenberg CGM. Trimethoprim–sulfamethoxazole (co-trimoxazol) for the prevention of relapses of Wegener's granulomatosis. *New Engl J Med* 1996;**335**:16–20.

34. Pillinger M, Staud R. Wegener's granulomatosis in a patient receiving propylthiouracil for Graves disease. *Semin Arthritis Rheum* 1998;**28**:124–9.

35. Denissen NHAM, Peters JGP, Masereeuw R, Barrera P. Can sulfasalazine therapy induce or exacerbate Wegener's granulomatosis? *Scand J Rheumatol* 2008;**37**:72–4.

36. Choi HK, Merkel PA, Cohen Tervaert JW, *et al.* Alternating antineutrophil cytoplasmic antibody specificity: drug-induced vasculitis in a patient with Wegener's granulomatosis. *Arthritis Rheum* 1999;**42**:384–8.

37. Spaetgens B, van Paassen P, Tervaert JW. Influenza vaccination in ANCA-associated vasculitis. *Nephrol Dial Transplant* 2009;**23**:654–8.

38. Birck R, Kaelsch I, Schnuelle P, Flores-Suárez LF, Nowack R. ANCA-associated vasculitis following influenza vaccination. *J Clin Rheumatol* 2009;**15**:289–91.

39. Cohen Tervaert JW, Stegeman CA. A difficult diagnosis. *Lancet* 2004;**364**:1313–14.

40. Neynaber S, Mistry-Burchardi N, Rust C, *et al.* PR3-ANCA-positive necrotizing multi-organ vasculitis following cocaine abuse. *Acta Derm Venereol* 2008;**88**:594–6.

41. Trimarchi M, Gregorini G, Facchetti F, *et al.* Cocaine-induced midline destructive lesions: clinical, radiographic, histopathologic, and serologic features and their differentiation from Wegener granulomatosis. *Medicine* 2001;**80**:391–404.

42. Jayne DR, *et al.*, European Vasculitis Study Group. Randomized trial of plasma exchange or high-dose methylprednisolone as adjunctive therapy for severe renal vasculitis. *J Am Soc Nephrol* 2007;**18**:2180–8.

43. Smyth L, Gaskin G, Pusey CD. Microscopic polyangiitis. *Semin Respir Crit Care Med* 2004;**25**:523–33.

44. Jennette JC, Thomas DB, Falk RJ. Microscopic polyangiitis (microscopic polyarteritis). *Semin Diagn Pathol* 2001;**18**: 3–13.

45. Ando Y, Okada F, Matsumoto S, Mori H. Thoracic manifestation of myeloperoxidase–antineutrophil cytoplasmic antibody (MPO-ANCA)-related disease: CT findings in 51 patients. *J Comput Assist Tomogr* 2004;**28**:710–16.

46. Collins CE, Quismorio FP. Pulmonary involvement in microscopic polyangiitis. *Curr Opin Pulm Med* 2005; **11**:447–51.

47. Stassen PM, Cohen Tervaert JW, Stegeman CA. Induction of remission in active anti-neutrophil cytoplasmic antibody-associated vasculitis with mycophenolate mofetil in patients who cannot be treated with cyclophosphamide. *Ann Rheum Dis* 2007;**66**:798–802.

48. Pagnoux C, Hamidou MA, Boffa JJ, *et al.* Azathioprine or methotrexate maintenance for ANCA-associated vasculitis. *New Engl J Med* 2008;**359**:2790–803.

49. Langfort CA, Talar-Williams C, Barron KS, Sneller MC. Use of a cyclophosphamide-induction methotrexate-maintenance regimen for the treatment of Wegener's granulomatosis: extended follow-up and rate of relapse. *Am J Med* 2003; **114**:463–9.

50. Stasi R, Stipa E, Poeta GD, *et al.* Long-term observation of patients with anti-neutrophil cytoplasmic antibody-associated vasculitis treated with rituximab. *Rheumatology* (Oxford) 2006;**45**:1432–6.

51. Bonaci-Nikolic B, Nikolic MM, Andrejevic S, Zoric S, Bukilica M. Antineutrophil cytoplasmic antibody (ANCA)-associated autoimmune diseases induced by antithyroid drugs: comparison with idiopathic ANCA vasculitides. *Arthritis Res Ther* 2005;**7**:R1072–81.

52. Slot MC, Links TP, Stegeman CA, Tervaert JW. Occurrence of antineutrophil cytoplasmic antibodies and associated vasculitis in patients with hyperthyroidism treated with antithyroid drugs: a long-term followup study. *Arthritis Rheum* 2005;**53**:108–13.

53. Bergs L. Goodpasture syndrome. *Crit Care Nurs* 2005;**25**:50–8.

54. de Groot K, Schnabel A. Pulmonary-renal syndrome. *Internist* (Berlin) 2005;**46**:769–81.

55. Stangou M, Asimaki A, Bamichas G, *et al.* Factors influencing patient survival and renal function outcome in pulmonary-renal syndrome with ANCA(+) vasculitis: a single-center experience. *J Nephrol* 2005;**18**:35–44.

56. Gal AA, Salinas FF, Staton GW. The clinical and pathological spectrum of ANCA related pulmonary disease: a comparison between perinuclear and cytoplasmic antineutrophil cytoplasmic antibodies. *Arch Pathol Lab Med* 1994; **118**:1209–14.

57. Stokes MB, Foster K, Markowitz GS, *et al.* Development of glomerulonephritis during anti-TNF-alpha therapy for rheumatoid arthritis. *Nephrol Dial Transplant* 2005; **20**:1400–6.

58. Dolman KM, Gans RO, Vervaat TJ, *et al.* Vasculitis and antineutrophil cytoplasmic autoantibodies associated with propylthiouracil therapy. *Lancet* 1993;**342**:651–2.

59. Jiang X, Khursigara G, Rubin RL. Transformation of lupus-inducing drugs to cytotoxic products by activated neutrophils. *Science* 1994;**266**:810–13.

60. Fietta P. Systemic vasculitis: immunogenetics and familial clustering. *Clin Exp Rheumatol* 2004;**22**:238–51.

61. Sonnenblick M, Nesher G, Rosin A. Nonclassical organ involvement in temporal arteritis. *Sem Arthritis Rheum* 1989;**19**:183–90.

62. Larson TS, Hall S, Hepper NGG, Hunder GG. Respiratory tract symptoms as a clue to giant cell arteritis. *Ann Intern Med* 1984;**101**:594–7.

63. Gur H, Ehrenfeld M, Izak E. Pleural effusion as a presenting manifestation of giant cell arteritis. *Clin Rheumatol* 1996;**15**:200–3.

64. Staudt LS, Silver RM. The lung and vasculitis. In: LeRoy EC (ed.) *Systemic Vasculitis: The Biologic Basis*, pp 273–301. New York: Marcel Dekker, 1992.

65. Chassagne P, Gligorov J, Dominique S. Pulmonary artery obstruction and giant-cell arteritis. *Ann Intern Med* 1995;**122**:732.

66. De Heide LJM, Pieterman H, Henneman G. Pulmonary infarction caused by giant-cell arteritis of the pulmonary arterie. *Neth J Med* 1995;**46**:36–40.

67. Andreas E, Kaltenbach G, Marcellin L, Imler M. Acute pulmonary embolism related to pulmonary giant cell arteritis. *Press Med* 2004;**33**:1328–9.

68. Lie JT, Dixit RK. Nonsteroidal anti-inflammatory drug induced hypersensitivity vasculitis clinically mimicking temporal arteritis. *J Rheumatol* 1996;**23**:183–5.

69. Culver B, Itkin A, Pischel K. Case report and review of minocycline-induced cutaneous polyarteritis nodosa. *Arthritis Rheum* 2005;**53**:468–70.

70. Cohen Tervaert JW, Limburg PC, Elema JD, *et al.* Detection of antibodies against myeloid lysosomal enzymes: a useful adjunct to classification of patients with biopsy-proven necrotizing arteritis. *Am J Med* 1991;**91**:59–66.

71. Kallenberg CG, Mulder AH, Tervaert JW. Antineutrophil cytoplasmic antibodies: a still-growing class of autoantibodies in inflammatory disorders. *Am J Med* 1992;**93**:675–82.

72. Newburger JW, Takahashi M, Gerber MA, *et al.* Diagnosis, treatment, and long-term management of Kawasaki disease: a statement for health professionals from the committee on rheumatic fever, endocarditis and Kawasaki disease. *Circulation* 2004;**110**:2747–71.

73. Freeman AF, Crawford SE, Finn LS, *et al.* Inflammatory pulmonary nodules in Kawasaki disease. *Pediatric Pulmonology* 2003;**36**:102–16.

74. Chinen J, Piecuch S. Anticonvulsant hypersensitivity syndrome vs Kawasaki disease: a challenging clinical diagnosis with therapeutic implications. *Clin Pediatr* 2000;**39**:109–11.

75. Toerner JG, Cvetkovich T. Kawasaki-like syndrome: abacavir hypersensitivity? *Clin Infect Dis* 2002;**34**:131–3.

76. Hurvitz H, Branski D, Gross-Kieselstein E, Klar A, Abrahamov A. Acetaminophen hypersensitivity resembling Kawasaki disease. *Isr J Med Sci* 1984;**20**:145–7.

77. Hicks RA, Murphy JV, Jackson MA. Kawasaki-like syndrome caused by carbamazepine. *Pediatr Infect Dis J* 1988;**7**:525–6.

78. Cohen Tervaert JW, van der Werf T, Stegeman CA, *et al.* Pulmonary manifestation of systemic vasculitides. In: Isenberg DA, Spiro SG (eds) *Autoimmune Aspects of Lung Disease*, pp 53–85. Birkhäuser, 1998.

79. Green RJ, Ruoss SJ, Kraft SA, *et al.* Pulmonary capillaritis and alveolar hemorrhage: update on diagnosis and management. *Chest* 1996;**110**:1305–16.

80. Nadrous HF, Yu AC, Specks U, Ryu JH. Pulmonary involvement in Henoch–Schönlein purpura. *Mayo Clin Proc* 2004;**79**:1151–7.

81. Leatherman JW. Immune alveolar hemorrhage. *Chest* 1987;**91**:891–7.

82. Carcia-Porrúa C, Conzález-Gay MA, López-Lázaro L. Drug associated cutaneous vasculitis in adults in northwestern Spain. *J Rheumatol* 1999;**26**:1942–4.

83. Chen KR, Carlson JA. Clinical approach to cutaneous vasculitis. *Am J Clin Dermatol* 2008;**9**:71–92.

84. Goldberg EI, Shoji T, Sapadin AN. Henoch–Schönlein purpura induced by clarithromycin. *Int J Dermatol* 1999;**38**:700–11.

85. Borras-Blasco J, Enriquez R, Amoros F, *et al.* Henoch–Schönlein purpura associated with clarithromycin: case report and review of literature. *Int J Clin Pharmacol Ther* 2003;**41**:213–16.

86. Santoro D, Stella M, Castellino S. Henoch–Schönlein purpura associated with acetaminophen and codeine. *Clin Nephrol* 2006;**66**:131–4.

87. Borras-Blasco J, Girona E, Navarro-Ruiz A, *et al.* Acenocoumarol-induced Henoch–Schönlein purpura. *Ann Pharmacother* 2004;**38**:261–4.

88. Duffy TN, Genta M, Moll S, Martin PY, Gabay C. Henoch–Schönlein purpura following etanercept treatment of rheumatoid arthritis. *Clin Exp Rheumatol* 2006;**24**(2 Suppl 41):S106.

89. Prajapati C, Casson IF. Henoch–Schönlein purpura associated with ranitidine. *Int J Clin Pract* 1997;**51**:251.

90. Shimada K, Matsui T, Kawakami M, *et al.* Methotrexate-related lymphoid granulomatosis: a case report of spontaneous regression of large tumours in multiple organs after cessation of methotrexate therapy in rheumatoid arthritis. *Scand J Rheumatol* 2007;**36**:64–7.

91. Drent M, Costabel U (eds). *Sarcoidosis. Eur Respir Mon* 2005;**32**(10):341.

92. Fernandes SRM, Singsen BH, Hoffman GS. Sarcoidosis and systemic vasculitis. *Semin Arthritis Rheum* 2000;**30**;33–46.

93. Liebow AA. Pulmonary angiitis and granulomatosis. *Am Rev Respir Dis* 1973:**108**;1–18.

94. Doyle MK, Berggren R, Magnus JH. Interferon induced sarcoidosis. *J Clin Rheumatol* 2006;**129**:241–8.

95. Cuervo Oinna MA, Cuervo Pinna C, Macias Castillo S, *et al.* Lung sarcoidosis following installation of mitomycin C in the urinary bladder. *Ann Med Intern* 2001;**12**:641–3.

96. Almirall J, Orellana R, Martinez Ocana JC, *et al.* Allopurinol-induced chronic granulomatous interstitial nephritis. *Nefrologia* 2006;**26**:741–4.

97. Cohen Tervaert JW, Van Paassen P, Damoiseaux J. Type II cryoglobulinemia is not associated with hepatitis C infection: the Dutch experience. *Ann NY Acad Sci* 2007;**1107**:251–8.

98. Beuthien W, Mellinghof HU, Kempis J. Vasculitic complications of interferon-alpha treatment for chronic hepatitis C virus infection: case report and review of the literature. *Clin Rheumatol* 2005;**24**:507–15.

99. Shen M, Wang Y, Xu WB, Zeng XJ, Zhang FC. Pleuropulmonary manifestations of systemic lupus erythematosus. *Zhonghua Yi Xue Za Zhi* 2005;**85**:3392–5.

100. Frankel SK, Cosgrove GP, Fischer A, Meehan RT, Brown KK. Update in the management of pulmonary vasculitis. *Chest* 2006;**129**:452–65.

101. Noel B. Lupus erythematosus and other autoimmune diseases related to statin therapy: a systematic review. *J Eur Acad Dermatol Venereol* 2007;**21**:17–24.

102. Horton MR. Rheumatoid arthritis associated interstitial lung disease. *Crit Rev Comput Tomogr* 2004;**45**:429–40.

103. Staudt LS, Silver RM. The lung and vasculitis. In: LeRoy EC (ed.) *Systemic Vasculitis: The Biologic Basis*, pp 273–301. New York: Marcel Dekker, 1992.

29

Pulmonary hypertension induced by drugs and toxins

KIM BOUILLON, YOLA MORIDE, LUCIEN ABENHAIM, MARC HUMBERT

INTRODUCTION

Pulmonary arterial hypertension (PAH) is a disease of the small pulmonary arteries, characterized by intense remodelling resulting in a progressive increase in pulmonary vascular resistance.[1] PAH is a progressive condition that can occur in an idiopathic form or in association with other disease states or exposures and is believed to result from environmental or disease-inciting factors coupled with genetically determined susceptibilities.[1] By definition, PAH patients do not have significant left heart disease, lung disease, or acute or chronic thromboembolic disease.[2] A diagnosis of PAH requires invasive haemodynamic criteria including a mean pulmonary artery pressure above 25 mmHg at rest and a normal pulmonary capillary wedge or left ventricular end-diastolic pressure under 15 mmHg.[1]

Multiple risk factors and associated conditions that trigger and/or worsen the progression of the disease have been recognized (Box 29.1).[2] Among them, several drugs and toxins have been shown to be associated with the development of PAH.

Box 29.1 Updated clinical classification of pulmonary hypertension (adapted from [2])

1. Pulmonary arterial hypertension
   1.1. Idiopathic pulmonary arterial hypertension
   1.2. Heritable
       1.2.1. BMPR2 mutations
       1.2.2. ALK1 or endoglin mutations with or without hereditary haemorrhagic telangiectasia
       1.2.3. Unknown genetic basis
   1.3. Drug- and toxin-induced
   1.4. Associated with:
       1.4.1. Connective tissue diseases
       1.4.2. Human immunodeficiency virus infection
       1.4.3. Portal hypertension
       1.4.4. Congenital heart diseases
       1.4.5. Schistosomiasis
       1.4.6. Chronic haemolytic anaemia
   1.5. Persistent pulmonary hypertension of the newborn
   1.6. Pulmonary veno-occlusive disease and/or pulmonary capillary haemangiomatosis
2. Pulmonary hypertension due to left heart disease
   2.1. Systolic dysfunction
   2.2. Diastolic dysfunction
   2.3. Valvular disease
3. Pulmonary hypertension due to lung diseases and/or hypoxia
   3.1. Chronic obstructive pulmonary disease
   3.2. Interstitial lung disease
   3.3. Other pulmonary diseases with mixed restrictive and obstructive pattern
   3.4. Sleep-disordered breathing
   3.5. Alveolar hypoventilation disorders
   3.6. Chronic exposure to high altitude
   3.7. Developmental abnormalities
4. Chronic thromboembolic pulmonary hypertension
5. Pulmonary hypertension with unclear multifactorial mechanisms
   5.1. Haematological disorders: myeloproliferative disorders, splenectomy
   5.2. Systemic disorders: sarcoidosis, pulmonary Langerhans cell histiocytosis, lymphangioleiomyomatosis, neurofibromatosis, vasculitis
   5.3. Metabolic disorders: glycogen storage disease, Gaucher's disease, thyroid disorders
   5.4. Others: tumoral obstruction, fibrosing mediastinitis, chronic renal failure on dialysis

PAH related to these factors has been classified as 'drug- and toxin-induced PAH' in the updated clinical classification of pulmonary hypertension.[2] Some drugs and toxins are associated with the development of isolated small pulmonary artery remodelling, while others can be associated with predominant pulmonary veno-occlusive disease (PVOD), with persistent pulmonary hypertension of the newborn (PPHN), and with newly recognized systemic diseases. These diseases are classified in Groups 1 and 1' according to the updated clinical classification of pulmonary hypertension.[2] They have many similarities in terms of clinical manifestations, haemodynamic measures, pathological changes and treatment. The common features in histopathology are characterized by pulmonary artery changes including medial hypertrophy, intimal thickening, adventitial fibrosis and plexiform lesions (in PVOD, the vascular changes predominate in small postcapillary pulmonary venules) (Fig. 29.1). As a consequence of elevated pulmonary vascular resistance, right ventricular hypertrophy and failure will subsequently develop.

The identification of drugs and toxins as risk factors for PAH poses a great challenge to both the physician and the epidemiologist. Drugs and toxins are categorized according to the strength of their association with PAH and the probability of a causal role.[2] In this chapter, only drugs or toxins associated with Groups 1 and 1' of PAH are reviewed. Table 29.1 summarizes the drugs and toxins considered in this chapter.

DRUGS ASSOCIATED WITH ISOLATED PAH

Appetite suppressants

AMINOREX

Aminorex resembles adrenaline and ephedrine in its chemical structure (Fig. 29.2) and acts as a potent appetite depressing agent and as a central stimulant. It has an adrenergic effect. In

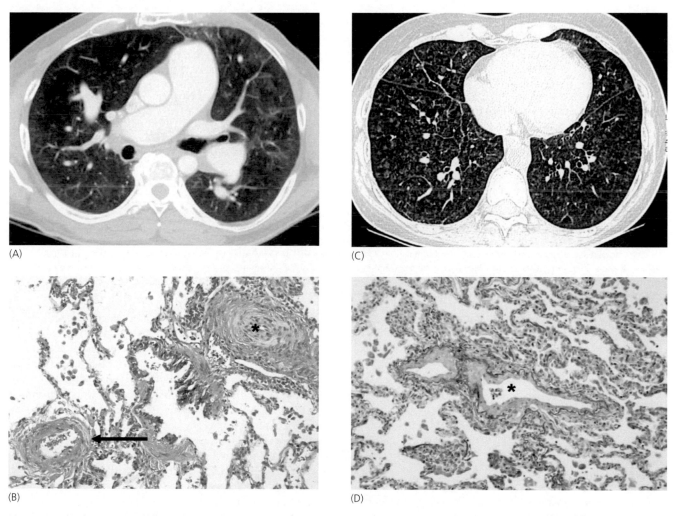

(A)

(B)

(C)

(D)

Fig. 29.1   Presentations of idiopathic pulmonary arterial hypertension and pulmonary veno-occlusive disease. (a) High-resolution computed tomography with contrast of the chest showing normal pulmonary parenchyma in idiopathic pulmonary arterial hypertension. (b) Pathological examination of the lung shows marked medial hypertrophy and plexiform lesions in idiopathic pulmonary arterial hypertension (arrow). (c) High-resolution CT of the chest shows nodular centrilobular ground-glass opacities and septal lines in pulmonary veno-occlusive disease. (d) Pathological examination of the lung shows obstruction of the small pulmonary veins. Images supplied by Drs David Montani, Peter Dorfmüller, Sophie Maître, Dominique Musset, Gérald Simonneau and Marc Humbert, Université Paris Sud 11, Hôpital Antoine Béclère, Clamart, France. * = the lumens of the pulmonary veins.

Table 29.1 Drugs and toxins related to pulmonary hypertension

Name	Updated risk factors for and associated conditions of PAH[a]	Name	Updated risk factors for and associated conditions of PAH[a]
Drugs and toxins associated with PAH		Methamphetamine	Likely
Appetite suppressants		Cocaine	Possible
Aminorex	Definite	Toluene	NC
Fenfluramine	Definite	**Herbal preparations**	
Phentermine	Definite	*Hypericum perforatum* (St John's wort)	Possible
Dexfenfluramine	Definite	Pyrrolizidine alkaloids	NC
Propylhexedrine	NC	**Drugs associated with PVOD**	
Phendimetrazine	NC	Mitomycin-C	Possible
Mazindol	NC	BCNU	Possible
Phenylpropanolamine	Possible	Cyclophosphamide with BMT	Possible
Diethylpropion	NC	Bleomycin	Possible
Other drugs		MOPP/COPP	Possible
Benfluorex	NC		
Pergolide	NC	**Drugs associated with PPHN**	
Phenformin	NC	*Non-steroidal anti-inflammatories:*	
Oral contraceptives and hormone replacement therapy	Unlikely	Indomethacin	NC
		Naproxen	NC
Leflunomide	NC	Ibuprofen	NC
Interferon-α_2	NC	Diclofenac	NC
Alglucerase	NC	Nimesulide	NC
Thalidomide	Possible	Fluoxetine	Possible
Cyclophosphamide without BMT	Possible	**Substances associated with systemic diseases**	
Bevacizumab	NC		
Dasatinib	NC	Toxic rapeseed oil	Definite
Substance abuse		L-tryptophan	Likely
4-methyl-aminorex	NC		

[a] According to a consensus of an international group of experts (Dana Point, California, 2008). A 'definite' association is defined as an epidemic, such as occurred with appetite suppressants in the 1960s, or large, multicentre epidemiological studies demonstrating an association between a drug and pulmonary arterial disease. A 'likely' association is defined as a single-centre, case–control study demonstrating an association or as multiple-case series. 'Possible' is defined as drugs with similar mechanisms of action as those in the 'definite' or 'likely' categories but which have not yet been studied. An 'unlikely' association is defined as one in which a drug has been studied in epidemiological studies and an association with PAH has not been demonstrated. BMT, bone marrow transplant; NC, drugs and toxins that have not yet been categorized by the group of experts; PAH, pulmonary arterial hypertension; PVOD, pulmonary veno-occlusive disease; PPHN, persistent pulmonary hypertension of the newborn.

the 1960s there was an outbreak of chronic pulmonary hypertension in Europe. In 1967, a sudden 20-fold increase in the number of cases in adults (known at that time as primary pulmonary hypertension, according to the former nomenclature), mostly women, subjected to cardiac catheterization was observed in a Swiss medical clinic. The disease showed rapid progression, and the interval between the onset of symptoms and diagnosis was 10 months on average. The onset of the symptoms followed the beginning of the anorectic treatment by an average of 8.6 months (range 6–12 months). There were no apparent changes in either size or composition of the population, or of diagnostic procedures. About half of the patients were to some degree overweight, with a female preponderance. It was noticed that a considerable number of these patients had taken aminorex to reduce weight.[3] A similar increase in the number of PAH cases observed was reported in other clinics in Switzerland,[4] and in Austria and Germany, where aminorex was also available. Subsequently, a cooperative retrospective study has been conducted with participation of 23 centres in these countries. A total of 582 cases of chronic pulmonary

hypertension were identified, among which 68 per cent had claimed they had used aminorex, either alone or in combination with other anorectic agents.[5] Aminorex was introduced on the Swiss market in November 1965, and was withdrawn in October 1968, a few months after the communications of these findings.

The overall incidence of PAH in patients who used aminorex was reported to be about 0.2 per cent.[6] The number of cases varied geographically, consistent with the marketing area of the product, and temporally with the intake of aminorex.

In patients with PAH, 10-year survival rate after diagnosis was better among the aminorex-exposed groups than among aminorex-free groups (respectively, 53 vs. 34 per cent[3] and 63 vs. 31 per cent[7]).

The prognosis in patients with PAH due to aminorex intake is critically dependent on the interruption of drug intake. If aminorex had been discontinued early enough, a regression of PAH and right ventricular hypertrophy could be observed.[7]

Although one study documented that the acute administration of aminorex increased pulmonary vascular resistance in anaesthetized dogs in a dose-dependent way, attempts to develop

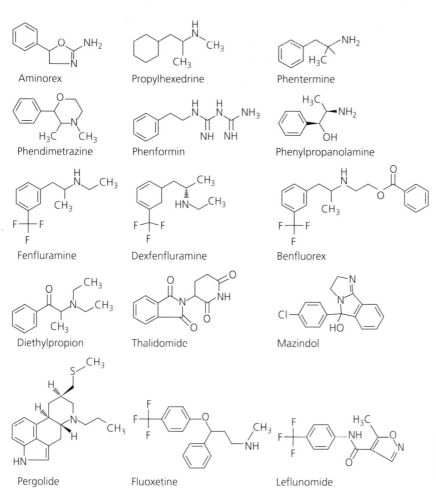

Fig. 29.2 Chemical structures of drugs associated with pulmonary arterial hypertension (PAH) and persistent pulmonary hypertension of the newborn (PPHN).

a chronic aminorex–pulmonary hypertension animal model failed.[8] Although a dose–response effect between aminorex intake and severity of pulmonary hypertension has not been established, and evidence has not been substantiated by an animal model, the epidemiological evidence strongly suggests a cause-and-effect relationship between this drug and the disease.

Even though aminorex was withdrawn from the market in 1968, a case series of PAH has been associated with ingestion of its analogue 4-methyl-aminorex in the United States in the context of illicit use as an alternative to methamphetamine.[9]

FENFLURAMINES

Fenfluramines (DL-fenfluramine and its analogue dexfenfluramine) are phenylethylamine derivatives (see Fig. 29.2). They are serotonin uptake inhibitors, and have been widely used for the treatment of obesity. Several anecdotal reports associating the use of fenfluramine or dexfenfluramine with PAH appeared in Europe between 1981 and 1992.[10-15] All reported cases of PAH with a history of fenfluramine or dexfenfluramine intake were women, from 20 to 58 years old. The time between exposure until the presentation of early symptoms of pulmonary hypertension, such as dyspnoea on exertion, or hospital admission, varied from 3 months to 8 years.[10-16] In 1993, investigators at a French referral centre performed a 5-year retrospective review of their experience with 73 PAH patients, and found that about 20 per cent of them had been exposed to fenfluramine or dexfenfluramine.[16] There had been a large increase

in the sales of fenfluramine or dexfenfluramine in France after the introduction of dexfenfluramine to the market (1985 to 1992). The role of fenfluramines in development of pulmonary hypertension has been established by an international case–control study conducted in 35 specialized centres in France, Belgium, the Netherlands and the United Kingdom from 1992 to 1994.[17] The International Primary Pulmonary Hypertension Study (IPPHS) found a strong association between appetite suppressants and PAH, with an odds ratio (relative risk estimates) of 6.3 (95% confidence interval: 3.0–13.2). Ninety per cent of patients for whom a defined product could be traced had used a fenfluramine or dexfenfluramine. The risk increased markedly with duration of use (odds ratio of 23.1 for more than 3 months; 95% CI: 6.9–77.7) and decreased after cessation of use.

Despite those IPPHS results, dexfenfluramine was approved for prescription use by the Food and Drug Administration (FDA) for the long-term treatment of obesity. The first cases for fenfluramine-induced PAH were reported in the United States in 1997.[18-20] Several patients took the off-label combination of fenfluramine and phentermine, commonly called 'Fen-Phen'. Fenfluramine and phentermine were approved as individual agents for appetite suppression, with phentermine approved for use in the USA in 1957 and fenfluramine approved in 1973. The total number of prescriptions in the USA exceeded 18 million.[21] Following the discovery of valvular heart disease associated with fenfluramine–phentermine,[21] fenfluramine and dexfenfluramine were removed from the worldwide market in 1997. After their removal, several North American (Surveillance of North American Pulmonary Hypertension (SNAP)[22] and Surveillance of Pulmonary Hypertension in America (SOPHIA)[23]) and European[24,25] observational studies showed the association between the use of fenfluramine or dexfenfluramine and PAH consistent with the IPPHS results.[17] They also provided better understanding of PAH associated with the use of these drugs. Their possible association with female gender was suggested after the observation of a high female to male ratio in series studying these drugs.[25] This is not confirmed, however, in view of the usual sex ratio in PAH at large (1.9:1) and the higher use of appetite suppressants by women.[24]

In the French National Registry which included 17 university hospitals, 674 patients with PAH were enrolled between 2002 and 2003 (i.e. 5 years after fenfluramine derivatives withdrawal). This showed that 9.5 per cent of patients had had a history of anorexigen exposure mainly with fenfluramine derivatives (77 per cent). Two-thirds of them (65 per cent) were exposed for more than 6 months. Delay between last appetite suppressant intake and the first symptoms of pulmonary hypertension was more than 2 years in 76 per cent of them, and 44 per cent more than 5 years.[24] The median duration of fenfluramine exposure was 6 months, with a median of 4.5 years between exposure and onset of symptoms among 109 patients followed up in the French national referral PAH centre.[25]

Regarding survival, Rich et al. reported a poorer survival among patients with fenfluramine-induced PAH compared with patients with idiopathic or heritable PAH (median 1.2 vs. 4.1 years).[26] Analysing a larger sample of patients, Souza et al. did not find a significant difference in survival between patients with fenfluramine-induced PAH compared with patients with idiopathic and heritable PAH, with a median survival of 6.4 years.[25] In both studies, patients were exposed to a similar PAH treatment regimen regardless of their exposure status.

PATHOGENIC MECHANISMS

The pathogenic mechanisms of PAH induced by appetite suppressants are still unknown. Investigators have hypothesized that the anorexigens elevate circulating levels of serotonin (5-hydroxytryptamine, 5-HT) normally stored in platelets – the '5-HT hypothesis of PAH'. Serotonin is a direct pulmonary artery vasoconstrictor, and an active mitogen for smooth muscle. Its levels have been found to be increased in patients with PAH, including patients with a history of anorexigen intake.[27]

Anorexigen phenethylamine derivatives inhibit the uptake and provoke the release of serotonin from platelets.[28] Although aminorex has been shown to release noradrenaline, it also releases serotonin.[28,29] However, several data show that administration of fenfluramines in animals and humans actually lowers blood levels of serotonin and does not increase plasma level of serotonin.[30–32] Rothman et al. tested the hypothesis that the fenfluramines and other appetite suppressants might increase the risk of PAH through interactions with serotonin transporters (SERT) in the lung.[33] They showed that aminorex and fenfluramine are SERT substrates and increase extracellular 5-HT at therapeutically relevant doses. Concentrations of 5-HT in blood and plasma may not be indicative of 5-HT concentrations

in local microenvironments surrounding pulmonary epithelial cells and arterial smooth muscle cells. To explain the hyperplasia of pulmonary artery smooth muscle, Rothman et al. hypothesized that SERT substrates other than 5-HT might also be mitogenic.[33] They suggested that the accumulation of anorexigens in smooth muscle cells after their translocation could trigger mitogenesis through inhibition of K^+ channels. The authors also suggested that the role of SERT in the pathogenesis of PAH is to serve as a 'gateway' for the accumulation and concentration of medications in pulmonary cells.[33] Recent findings support the important role of SERT in the pathogenesis of PAH either idiopathic or anorexigen-induced.[34–36] Anorexigens may have other important actions on the lung circulation. Aminorex, fenfluramine and dexfenfluramine inhibit potassium current in rat pulmonary vascular smooth muscle cells, lowering the membrane potential and leading to contraction.[37]

Since only a minority of patients exposed to fenfluramines develop PAH, the need for other predisposing factors, such as genetic susceptibility, could be postulated, with anorexigens acting as a 'second hit' in those predisposed subjects. BMPR2 mutations have been described in more than 70 per cent of PAH familial cases and in 10–40 per cent of patients with idiopathic apparently sporadic PAH.[38,39] BMPR2 mutations were found among 9 per cent[40] and 18 per cent[25] of patients with fenfluramine-associated PAH. Interestingly, the duration of exposure in patients with BMPR2 mutation was significantly lower than in patients without the mutation. This is in agreement with the 'multiple hit' concept, fenfluramine exposure being a trigger/ risk factor in genetically predisposed individuals. Recently, in a patient with hereditary haemorrhagic telangiectasia and dexfenfluramine-associated PAH, an endoglin germline mutation has been observed.[41] Endoglin and BMPR2 genes encode transmembrane proteins which belong to the TGF-β receptor superfamily which plays an important role in the maturation and development of pulmonary vessels. These data are in favour of a role of the TGF-β signalling pathway in anorexigen-induced PAH.[41] Regarding the role of serotonin receptors in anorexigen-induced PAH, a mutation causing loss of function of the serotonin 5-HT2B receptor was found in 1 out of 10 patients with fenfluramine-induced PAH, whereas no mutations were found among 18 patients with idiopathic PAH.[35] These findings stress the complexity of the pathophysiological mechanisms by which appetite suppressants may cause pulmonary hypertension.

OTHER APPETITE SUPPRESSANTS

Phentermine

Phentermine is an appetite suppressant of the amphetamine and phenethylamine class (see Fig. 29.2). It has often been used in combination with fenfluramine as an appetite suppressant (see above). At the time of writing, no PAH induced by phentermine alone has been reported.

Propylhexedrine

Propylhexedrine is an amphetamine-like appetite suppressant (see Fig. 29.2). One case of PAH, which developed over a period of 8 years of use, was reported in 1984, in a 39-year-old woman.[42] Twelve months after the discontinuation of propylhexedrine, pulmonary symptoms and electrocardiographic signs of the right ventricular hypertrophy were resolved. Propylhexedrine is commonly found in over-the-counter remedies for colds, allergies and allergic rhinitis, and for temporary symptomatic relief of nasal congestion.

Phendimetrazine

Phendimetrazine is an appetite suppressant as potent as amphetamines (see Fig. 29.2). One case of PAH associated with phendimetrazine was reported in 1991, in a 41-year-old woman, exposed 6 months before the onset of her symptoms. Dyspnoea, electrocardiographic abnormalities and pulmonary haemodynamics improved several months after discontinuation of the drug.[43]

Mazindol

Mazindol is a sympathomimetic amine, similar to amphetamine (see Fig. 29.2). We found only a single case report of PAH diagnosed 12 months after discontinuing mazindol that had been taken for 10 weeks.[44]

Phenylpropanolamine

Phenylpropanolamine is a synthetic sympathomimetic amine commonly used as an appetite suppressant (see Fig. 29.2). A case–control study found that it increased the risk of haemorrhagic stroke.[45] Therefore, since 2000, it is no longer marketed in the United States as an appetite suppressant. Between 1998 and 2001, in the SOPHIA study, an association was found between PAH and over-the-counter diet pills containing this drug.[23] In addition, a fatal case of PAH has been reported in a 7.5-year-old boy who had a very high exposure to the substance, which is commonly found in cough or cold remedies.[46] These reports strengthen the hypothesis that phenylpropanolamine could be a risk factor for the development of PAH.

Diethylpropion

Diethylpropion (amfepramone) is structurally related to amphetamines (see Fig. 29.2) and is a sympathomimetic stimulant. The implication of diethylpropion in the development of PAH is not clear because only a small number of patients were exclusively exposed to this drug. Between 1995 and 2003, only two cases of diethylpropion-associated PAH have been reported.[47,48] The first case was observed in the United Kingdom. A 22-year-old woman was prescribed diethylpropion for 3–5 months. Breathlessness on exertion developed when she was 27, and the diagnosis of PAH was made 4 years later.[47] The second case was observed in Belgium. A 27-year-old woman took diethylpropion during two courses of 4 and 5 weeks, respectively 3.5 and 1.5 years before onset of symptoms.[48] Interestingly, a mutation of the BMPR2 was found in the second case of PAH associated with diethylpropion.

Other drugs associated with isolated PAH

BENFLUOREX

Benfluorex is marketed as an adjunctive drug for patients with hypertriglyceridaemia or diabetes who are overweight. This drug, structurally related to the fenfluramine derivative (see Fig. 29.2), is classified by the World Health Organization as an appetite suppressant, and is sometimes used off-licence as a slimming aid.[49] In France, where benfluorex has been on the market since 1976, Boutet et al. reported five cases of PAH related to benfluorex consumption.[50] All patients were overweight diabetic females, aged 50–57 years. The duration of benfluorex exposure ranged from 3 months to 10 years. One patient underwent lung transplantation. Pathological examination of her lungs revealed characteristic lesions of PAH. This small case series does not formally demonstrate a causal relationship between benfluorex use and PAH. However, the pharmacological background and clinical findings support the hypothesis that benfluorex may play a role in this association.[50] In Spain, benfluorex was removed from the market in 2003 because cases of valvular heart disease were observed after its use.[51] One case of this disease has also been reported by Boutet et al.[50] Well-designed studies similar to the IPPHS[17] and the SOPHIA[23] are needed to further investigate fenfluramine cardiovascular-like effects of benfluorex.

PERGOLIDE

Pergolide is an ergot alkaloid-derived dopamine agonist (see Fig. 29.2) prescribed in Parkinson's disease. It alleviates the symptoms of Parkinson's disease by stimulating the dopamine D1 and D2 receptors. It has been prescribed since the late 1980s. Pergolide is known to induce valvulopathies similar to those seen with the fenfluramines. Interestingly, a recent publication reported a case of reversible PAH in a patient after 4 years of pergolide use.[52] Like fenfluramines, pergolide has been shown to cause inhibition of voltage-gated K^+ channels in pulmonary arterial smooth muscle cells, leading to vasoconstriction.[53]

PHENFORMIN

Phenformin is a biguanide (see Fig. 29.2) and has been used as an antidiabetes drug. It was reported to be associated with severe or fatal lactic acidosis, which could induce pulmonary vasoconstriction in animal experiments. In 1973, two cases of pulmonary hypertension, believed to be associated with treatment with phenformin, were reported.[54] This drug was withdrawn from the US market in 1977. To date, no other publication on its use and PAH has been reported.

ORAL CONTRACEPTIVES AND HORMONE REPLACEMENT THERAPY

Kleiger et al. reported six cases of pulmonary hypertension in women with a history of oral contraceptive use.[55] Three had known predisposing causes of pulmonary hypertension. For the other three cases, PAH was diagnosed after they had been exposed to oral contraceptives for about 5 years. There was no evidence of thromboembolism in the lungs or leg veins. In the IPPHS, oral contraceptive use was not associated with a significantly increased risk.[17] One case report suggested the association between a fatal histologically proven PVOD and oral contraceptives use in a 24-year-old woman.[56] Morse et al. described the case of a 64-year-old woman known to be an obligate carrier of familial form of PAH who, after hormone replacement therapy for 3 months, had PAH.[57] She had normal pulmonary function before this treatment. Experts recently considered that it was unlikely that oral contraceptives and hormone replacement therapy could be associated with PAH (an 'unlikely' association was defined as one in which a drug has been studied in epidemiological studies and an association with PAH has not been demonstrated).

LEFLUNOMIDE

Leflunomide is a pyrimidine synthesis inhibitor (see Fig. 29.2). It is a disease-modifying antirheumatic drug that reduces the signs and symptoms of rheumatoid arthritis and slows down disease progression. A 70-year-old woman has been reported to have PAH after 1.5 years of treatment with leflunomide.[58] Symptoms and echocardiographic signs improved after removal of the drug.

INTERFERON-α_2

Interferon-α_2 is an immunomodulator widely used in hepatitis and high-risk melanoma. It is also widely used in treating multiple sclerosis. Jochmann et al. described a 40-year-old woman who developed a severe PAH 30 months after initiation of interferon-α_2 for adjuvant treatment of spreading melanoma.[59] Symptoms and echocardiographic signs were improved after its discontinuation. This association was earlier observed by Al-Zahrani et al.[60]

ALGLUCERASE

Gaucher's disease is an autosomal-recessive disorder that is caused by a deficiency of the enzyme glucocerebrosidase. PAH has been detected in patients with type 1 Gaucher's disease. This may be due at least in part to perivascular infiltration by Gaucher's cells with vascular obliteration and plugging of the capillaries with Gaucher's cells. However, three cases of PAH have been reported to be associated with the use of alglucerase after eliminating the possible involvement of Gaucher's cells.[61,62]

THALIDOMIDE

Thalidomide is an immunosuppressive agent with anti-angiogenic and anti-inflammatory activities (see Fig. 29.2). Before 2006, it was given as monotherapy or in combination therapy against refractory or relapsing multiple myeloma, and since then thalidomide plus dexamethasone is approved as first-line treatment of multiple myeloma. Since 2003, several case reports have suggested a possible association between the use of thalidomide for treatment of multiple myeloma and PAH.[63–65] As thalidomide is known to be associated with an increased risk of venous thromboembolic disease, a pilot study designed to detect clinical and subclinical non-thromboembolic pulmonary hypertension in multiple myeloma patients after thalidomide treatment was conducted among

82 patients.[66] Clinical and echocardiographic evaluation revealed 4 out of 82 patients (5 per cent) with non-thromboembolic pulmonary hypertension between 1 and 6 months after thalidomide administration. After discontinuing thalidomide therapy, pulmonary pressure decreased but without recovering down to the baseline value, possibly because of an irreversible drug effect on pulmonary artery endothelial and smooth muscle cells. Echocardiography may be an interesting tool in monitoring patients treated with thalidomide in order to detect the preclinical stage of the disease. However, one must emphasize the requirement of right heart catheterization to confirm pulmonary hypertension because the diagnostic accuracy of echocardiography is suboptimal.[1]

CYCLOPHOSPHAMIDE

Combination chemotherapy has been reported to be associated with PAH. To our knowledge, two cases of histologically confirmed PAH have been observed.[67,68] In both cases a cyclophosphamide-containing regimen was used to treat neuroblasma and myelomonocytic leukaemia. However, it must be emphasized that cyclophosphamide has been successfully used in the treatment of PAH complicating the course of systemic lupus erythematosus and mixed connective disease.[69,70]

BEVACIZUMAB

Bevacizumab is a recombinant humanized antivascular endothelial growth factor monoclonal antibody. It is approved by the FDA for treatment of several types of cancer: metastatic renal cell carcinoma, second-line treatment of glioblastoma, metastatic HER2-negative breast cancer, first-line treatment of non-small-cell lung cancer, and first- and second-line treatment of metastatic colorectal cancer. In 2008, three cases of fatal pulmonary hypertension were reported when bevacizumab was given to treat recurrent ovarian cancer.[71,72] Pulmonary hypertension occurred during the fourth and fourteenth cycles of therapy,[71] and 18 months after initiation of bevacizumab.[72] The role of bevacizumab in inducing PAH is not clear because in these cases cyclophosphamide was given in combination with it.

DASATINIB

Dasatinib is an oral selective tyrosine kinase inhibitor approved by the FDA for treatment of chronic myelogenous leukaemia after failure of imatinib (unsatisfactory response or toxicity).[73] In a study, pleural effusion was observed in 35 per cent of patients (48/138).[74] When an echocardiogram was done at the onset of pleural effusion, an increase in right ventricular systolic pressure, a non-invasive surrogate marker of pulmonary artery pressure, was observed. Two cases of PAH, confirmed by right heart catheterization, were also reported at 6 and 26 months after the initiation of dasatinib.[75,76] Right ventricular size and systolic function had normalized 6 weeks and 6 months after its discontinuation. Interestingly, imatinib, which is also a tyrosine kinase inhibitor, was reported being effective in treating idiopathic PAH.[77] This stresses the complexity of the pathophysiology in inducing PAH by dasatinib.

Substance abuse and PAH

See also Chapter 12.

METHAMPHETAMINE

As seen above for amphetamine-like appetite suppressants, amphetamine and methamphetamine are central nervous system stimulants (Fig. 29.3). A case report and an epidemiological study have established the relationship between PAH and methamphetamine abuse.[78,79] Rothman *et al.* suggested that the underlying mechanism might involve interactions with SERT in the lung,[33,80] as postulated for aminorex and fenfluramine derivatives.

COCAINE

Cocaine blocks the release and reuptake of catecholamines, including dopamine, and blocks serotonin release and reuptake (see Fig. 29.3). Cocaine- or 'crack'-induced PAH have been described among individuals who have abused these substances for more than 3 years.[81,82] Lung biopsy specimens revealed medial hypertrophy of muscular pulmonary arteries. In the IPPHS, use of illicit drugs (cocaine or intravenous drugs) was associated with an increased point estimate of the relative risk (odds ratio of 2.8; 95% CI: 0.5–15.7), but numbers were too small to draw conclusions.[17] In most studies it is difficult to disentangle the role of the product used (cocaine, heroin, other drugs mixed with talc) from that of emboli associated with intravenous drug use. Finally, cocaine abuse by a mother produced signs of persistent pulmonary hypertension in a newborn.[83]

OTHER SUBSTANCES

In the early 1980s, one small series reported that among 18 boys with a history of chronic glue (toluene) abuse for 6 months or more (see Fig. 29.3), obvious pulmonary hypertension was found on clinical examination in five of them.[84] No lung biopsies were performed.

Herbal preparations

HYPERICUM PERFORATUM

Hypericum perforatum (St John's wort) is a plant whose extract is licensed in continental Europe for the treatment of depression and anxiety.[85] Recently, the SOPHIA study showed an association between use of St John's wort and PAH with an odds ratio of 3.6 (95% CI: 1.0–13.0, versus chronic thromboembolic pulmonary hypertension).[23] This risk factor has been classified in the 'possible' category.[2]

PYRROLIZIDINE ALKALOIDS

Pyrrolizidine is a heterocyclic organic compound that forms the central chemical structure of a variety of alkaloids known collectively as pyrrolizidine alkaloids. They are plant hepatotoxins that can also cause injury to the vasculature of the lungs leading to pulmonary hypertension and right heart failure in animals. They are named differently according to species of plants *Crotalaria* and *Senecio*. Monocrotaline is found in *C. spectabilis* and fulvine in *C. fulva* (see Fig. 29.3).[86] Monocrotaline-induced pulmonary hypertension in rats is a major animal model for the study of pulmonary hypertension. In *S. jacobaea*, several alkaloids are found, in particular seneciphylline (see Fig. 29.3).[86] Leaves and seeds of these plants are used for the preparation of bush teas, which are consumed by indigenous populations for medicinal and other purposes.

To our knowledge, only one case of PAH suspected to be related to the ingestion of pyrrolizidine alkaloids has been reported.[87] After a meticulous investigation, clinicians found that the 66-year-old woman took several herbal remedies in the months prior to hospitalization. One of them, comfrey

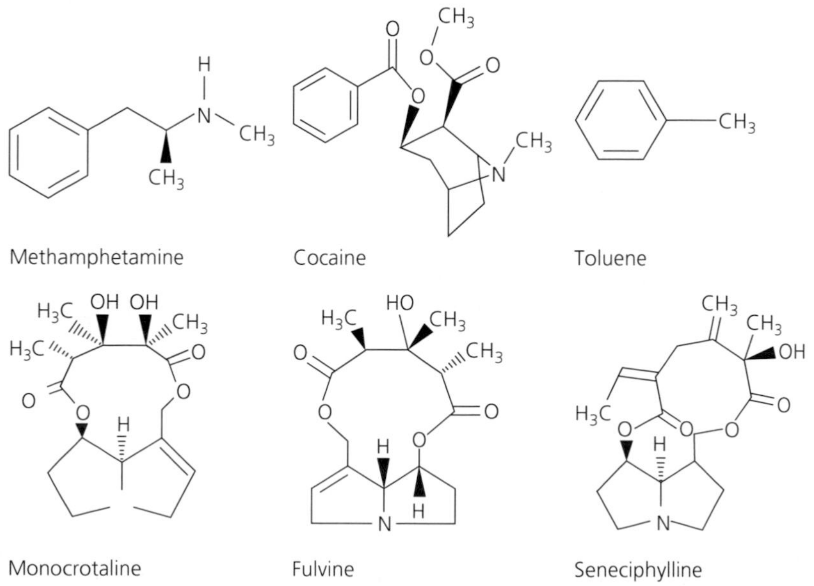

Fig. 29.3 Chemical structures of various substances associated with pulmonary arterial hypertension (PAH).

(*Symphytum officinale*), is known to contain pyrrolizidine alkaloids.

In cases of unexplained pulmonary hypertension and even interstitial lung disease in humans, a history of ingestion of herbal remedies should be sought in a systematic way.

DRUGS ASSOCIATED WITH PULMONARY VENO-OCCLUSIVE DISEASE

Mitomycin-C

Mitomycin-C (Fig. 29.4) is used in chemotherapy for malignant tumours combined with other chemotherapy. Anecdotal reports showed that histologically confirmed PVOD occurred in patients treated for metastatic cervical carcinoma,[88,89] metastatic gastric adenocarcinoma[90] and non-small-cell lung cancer.[91,92] It also causes an interstitial lung disease and a vasculitis with the haemolytic–uraemia syndrome.

Other agents for chemotherapy

Chemotherapy regimens other than those that incorporate mitomycin-C have been shown to be temporally related to histologically confirmed PVOD:

- BCNU (carmustine) (see Fig. 29.5) alone given for treatment of glioma;[93]
- cyclophosphamide and etoposide regimen with autologous bone marrow transplantation for neuroblastoma (see Fig. 29.5);[94]
- bleomycin regimen used for lymphocytic lymphoma;[95]
- MOPP and COPP, or MOPP alone, for Hodgkin's disease.[93,96]

Several physicians have noted that a few patients treated for acute lymphoblastic leukaemia have had histologically confirmed PVOD following chemotherapy.[97–100] PVOD

occurred in a particular context. First, all the cases featured, in total, at least seven drugs before the appearance of PVOD. Second, all the patients had used cyclophosphamide. Finally, all had had an allogeneic bone marrow transplantation after total body irradiation. PVOD has also been noted after treating myelofibrosis with chemotherapy and bone marrow transplantation.[101]

After examining these cases it is very difficult to determine whether chemotherapy is a causal agent. First, we cannot exclude the role of malignant neoplasms in inducing PVOD. For example, Capewell *et al.* described a patient with Hodgkin's lymphoma who developed PVOD at the time of the diagnosis of lymphoma prior to any therapy.[102] Second, we do not know how radiation therapy interacts with chemotherapy to provoke these pulmonary vascular diseases. Kramer *et al.* observed a patient who developed PVOD 6 years after receiving radiation therapy in a mantle distribution for Hodgkin's disease, without using any chemotherapy.[103] In addition, an association between PVOD and bone marrow transplantation, either allogeneic or autologous, has been noted. In most cases, conditioning was carried out with cytotoxic chemotherapy as well as total body irradiation. Finally, some chemotherapy is known to be lung toxic, in particular mitomycin, which can provoke also interstitial pneumonitis. Therefore, a lung biopsy may be needed to classify pulmonary hypertension in that setting but this surgical procedure is often contraindicated because of its very high risk.

We are very far away from understanding the pathophysiological mechanisms of chemotherapy in inducing PVOD. The aetiology is likely to be multifactorial.

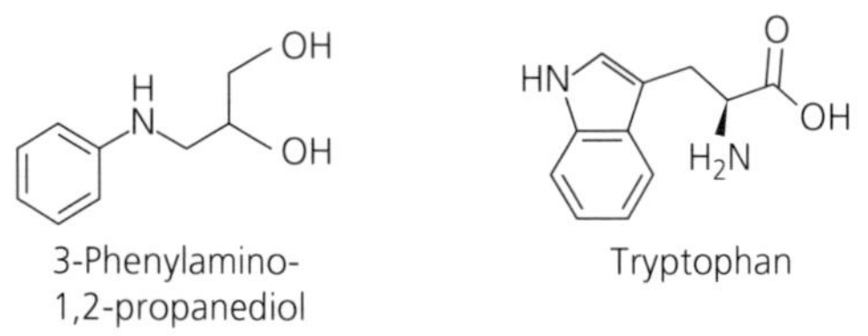

DRUGS ASSOCIATED WITH PERSISTANT PULMONARY HYPERTENSION OF THE NEWBORN

Non-steroidal anti-inflammatory drugs (NSAIDs) and selective serotonin-reuptake inhibitors (SSRIs) may be associated with pulmonary hypertension in the fetus or in the newborn after antenatal or postnatal exposure (these drugs are known to cross the placenta). Among newborns, these drug classes have been reported to be associated with idiopathic PPHN. This syndrome is due to an increased pulmonary vascular resistance that prevents normal pulmonary blood flow and causes a right-to-left shunt through persistent fetal channels (i.e. the patent foramen ovale and patent ductus arteriosus).[104] The prevalence of PPHN is estimated at 2 per 1000 live births.[105] PPHN is suspected when faced with neonatal hypoxaemia,

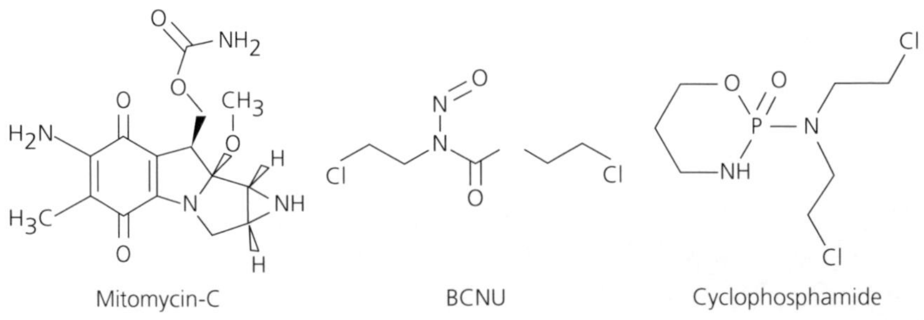

3-Phenylamino-
1,2-propanediol

Tryptophan

Fig. 29.4 Chemical structures of drugs associated with pulmonary veno-occlusive disease (PVOD).

Mitomycin-C

BCNU

Cyclophosphamide

Fig. 29.5 Chemical structures of substances associated with systemic diseases.

which is often refractory and is associated with a high mortality (11 per cent).[105]

Non-steroidal anti-inflammatory drugs

Indomethacin is a non-selective inhibitor of cyclooxygenase which inhibits prostaglandin synthesis. It is used in obstetrics for tocolytic effect in pregnant women with premature labour, and for prophylaxis of patent ductus arteriosus in premature infants. Use of indomethacin for premature labour has been associated with PPHN. The first cases of PPHN in two newborns in relation to prenatal exposure to indomethacin were reported in 1976.[106] In another report, Dalens et al. described five premature newborns whose mothers had received prolonged indomethacin therapy (at least 9 days) at 25 or 32 weeks of gestation.[107] They had clinical and biological symptoms of pulmonary hypertension with persistence of the fetal circulation with a patent ductus arteriosus. Two newborns died and their autopsies showed important reduction of the lumen of pulmonary arterioles due to a thickening of the tunica media. Similarly, in a randomized clinical trial, Besinger et al. reported three cases of PPHN in newborns whose mothers received indomethacin in the long-term treatment (more than 16 days) for preterm labour at 23 to 34 weeks of gestation.[108] Pulmonary hypertension has been observed also among newborn pups from parturient rats that received daily injections of indomethacin.[109] The newborn pups had significantly increased numbers of muscularized peripheral pulmonary vessels. Pulmonary hypertension, however, did not persist to adulthood.

A randomized trial compared short-term (48 hours) indomethacin with ritodrine for preterm labour at ≤ 32 weeks of gestation. Among an indomethacin group composed of 52 patients, no cases of premature closure of the ductus arteriosus or pulmonary hypertension were observed.[110]

According to these findings, the risk of PPHN may be significantly increased with prolonged use of indomethacin, and not with short-term (up to 48 hours) courses of therapy. These results should be interpreted with caution, as the diagnosis of PPHN is difficult to make because anatomical abnormalities have to be considered. As it is a prostaglandin synthetase inhibitor, another well-known adverse effect of indomethacin is constriction of the fetal ductus arteriosus, which can lead to permanent pulmonary hypertension in a premature newborn. Levin et al.[111] and Tarcan et al.[112] reported a case of permanent pulmonary hypertension in a premature newborn after short-term antenatal indomethacin exposure. At autopsy, they described an enlarged pulmonary arterial media with increased smooth muscle cells.[111]

A few case reports have described, among preterm or term newborns, persistent pulmonary hypertension with constricted ductus arteriosus with the use of naproxen, ibuprofen diclofenac and nimesulide, as is seen with indomethacin.[113]

Fluoxetine

Fluoxetine is a selective serotonin reuptake inhibitor (SSRI), as are aminorex and fenfluramines (see Fig. 29.2). However, we are not aware of any report of a relationship between the chronic intake of SSRI antidepressants and PAH in adults. It is estimated that 2–3 per cent of pregnant women receive antidepressive medication of which the SSRI class of drugs is the most commonly utilized.[114] In 1996, Chambers et al. observed in a prospective small cohort an increased risk for perinatal complications among women who took fluoxetine in the third trimester.[115] They suggested a possible association between maternal use of fluoxetine late in the third trimester of pregnancy and the risk of PPHN in the infant. Recent data from a retrospective case–control study conducted between 1998 and 2003 by Chambers et al. support the association between the maternal use of SSRIs after the completion of the 20th week of gestation and PPHN in the offspring: 14 infants with PPHN had been exposed to an SSRI compared with 6 in a control group (adjusted odds ratio of 6.1; 95% CI: 2.2–16.8).[116] Another cohort study based on the Swedish Medical Birth Register was in favour for this association.[117]

Källén et al. observed the risk of PPHN when women took an SSRI at an early stage of pregnancy. Experimental studies in rats support this association, showing that fluoxetine exposure in utero induced pulmonary hypertension in the fetal rat as a result of a developmentally regulated increase in pulmonary vascular smooth muscle proliferation.[118]

Of note, some authors have reported that SSRIs may be a therapeutic agent in adult pulmonary hypertension, as suggested by studies in experimental models of the disease.[119] Data in humans are still preliminary and inconclusive.[120]

DRUGS AND TOXINS ASSOCIATED WITH SYSTEMIC DISEASES

Toxic rapeseed oil

In May of 1981, a large foodborne epidemic occurred in central and north-western Spain. The epidemic was characterized in the initial phase by atypical pneumonia and pronounced eosinophilia in the next few days after exposure. From May to October 1981, it affected over 20 000 people. Over 300 died of this new persistent multisystemic disorder, now known as the 'toxic oil syndrome' (TOS), within 8 months following the onset of the outbreak. Criteria for TOS diagnosis were defined by the Spanish Clinical Commission in August 1981.[121] Epidemiological investigations established that TOS was due to the consumption of adulterated food oil, which was sold principally in unlabelled 5-litre containers by travelling salesmen. This oil was in fact a mixture containing rapeseed oil originating from France. It was denatured with 2 per cent of aniline before being imported to Spain, and was theoretically for industrial use. In Spain, it was denatured and illegally refined to remove the aniline, and finally mixed with other edible oils.

Toxicoepidemiological studies have suggested that 3-phenylamino-1,2-propanediol (DEPAP) or related compounds found in toxic oil may be the causal agent of TOS. However, to date, none of these oil contaminants has produced results in animals or in vitro suggestive of TOS in man.

In the natural history of the disease, most authors agree on the existence of three consecutive phases. The acute phase (first and second month after onset of the disease) was characterized by non-cardiogenic pulmonary oedema with dyspnoea, skin rash, eosinophilia and myalgias. The intermediate phase (third and fourth months) was characterized by severe myalgias, skin oedema, hepatic disease, arthritis and pulmonary hypertension. Pulmonary hypertension was found in about 20 per cent of hospitalized patients at the second to fourth month from onset of the disease.[122] The early chronic phase (from the fourth month to the second year) was marked by sclerodermatous skin changes, polyneuropathy, sicca syndrome and joint contractures. The late chronic phase (from the second year to end of the follow-up) was characterized by muscle cramps, chronic musculoskeletal pain, Raynaud's phenomenon, carpal tunnel syndrome and chronic lung disease.[123]

Alonso-Ruiz et al. followed 332 cases of TOS for up to 8 years and found that 8.1 per cent of them developed pulmonary hypertension; the condition regressed in 74 per cent, and only 2.1 per cent developed a malignant form of pulmonary hypertension at the end of a follow-up period of 8 years.[123] Abaitua Borda et al. showed that mortality during the first 13 years following TOS diagnosis was 5-fold higher among a TOS cohort compared with the Spanish population as a whole (standardized mortality ratio (SMR) of 4.9; 95% CI: 4.4–5.5).[124] No specific cause of death has been shown to be more frequent in the TOS cohort than in the Spanish population, with the exceptions of TOS and pulmonary hypertension. Pulmonary hypertension represented an 18-fold higher risk of death in this cohort than in the general population during this study period (SMR = 18.2; 95% CI: 3.7–53.0). Pathological changes at a later stage (at least 3 years after the onset of the disease) in the evolution of the patients with TOS were characterized by medial hypertrophy, intimal fibrosis and plexiform lesions in pulmonary arteries.[125]

L–tryptophan

L-tryptophan is an essential amino-acid that is normally ingested as a constituent of dietary proteins. L-tryptophan supplements are used by some people for disorders such as premenstrual syndrome, insomnia and depression. In October 1989, three patients in New Mexico had a syndrome in which eosinophilia was accompanied by severe myalgias. This syndrome was associated with L-tryptophan ingestion.[126] It became a newly recognized disease named eosinophilia–myalgia syndrome (EMS) with the following Centers for Disease Control definition: peripheral blood eosinophilia (>109/L) and generalized, disabling myalgias without other recognized causes. The apparent median latency period was 127 days.[126] EMS is a multisystemic disease; other clinical manifestations include pulmonary involvement (interstitial infiltrates and pleural effusions), skin rash and oedema, axonal polyneuropathy, perimyositis, and possible adverse neurocognitive effects.[126] Pulmonary vascular involvement has rarely led to pulmonary hypertension. It has gradually resolved except in a few catastrophic cases. Case–control studies conducted by state health departments confirmed the association of L-tryptophan use with EMS.[126] In November 1989, the FDA recalled all L-tryptophan-containing dietary supplements. After that, the number of new cases dropped

sharply. In August 1990, over 1500 cases of EMS had been reported in the USA. Epidemiological studies clearly linked illness to the ingestion of tryptophan produced by a single manufacturer in Japan, and the time course of the epidemic was most consistent with it being caused by a product dimeric tryptophan contaminant 1,1'-ethylidenebis [L-tryptophan] known as peak E.[126] However, no aetiological agent has been identified. Lung biopsy specimens have been collected among female patients taking L-tryptophan for 1–9 months prior to developing rapidly progressive shortness of breath. They showed both interstitial inflammatory infiltrates with lymphocytes and small numbers of eosinophils, as well as vascular changes with arterial and arteriolar medial hypertrophy being accompanied by mild pulmonary hypertension.[127]

KEY POINTS

- Drugs and toxins related to pulmonary arterial hypertension have been classified by an international group as definite, likely, possible or unlikely. Several drugs and toxins listed in this chapter have not yet been categorized because of insufficient evidence.

- It is difficult to evaluate the roles of drugs and toxins in the occurrence of PAH because we do not know the contributions of patient susceptibility, initial disease (including insidious underlying idiopathic PAH) and other treatments. In oncology, for example, several drugs are given and radiotherapy and bone marrow transplantation are frequent.

- When a suspected drug or toxin is discontinued sufficiently early, a regression of pulmonary hypertension and right ventricular hypertrophy can be observed.

- Newborns can be at risk for persistent pulmonary hypertension when they are exposed in utero to some widely used drugs such as SSRIs or indomethacin in the mother.

- Epidemics of PAH due to drugs or toxins are quite possible in the future because drugs chemically similar to fenfluramine derivatives are still used widely (in particular benfluorex), and certain amphetamine-like appetite suppressants such as propylhexedrine and phenylpropanolamine are commonly found in over-the-counter remedies for colds and allergies.

REFERENCES

1. Chin KM, Rubin LJ. Pulmonary arterial hypertension. *J Am Coll Cardiol* 2008;**51**:1527–38.
2. Simonneau G, Robbins IM, Beghetti M, *et al.* Updated clinical classification of pulmonary hypertension. *J Am Coll Cardiol* 2009;**54**:S43–54.
3. Gurtner HP. Aminorex and pulmonary hypertension: a review. *Cor Vasa* 1985;**27**:160–71.
4. Follath F, Burkart F, Schweizer W. Drug-induced pulmonary hypertension? *Br Med J* 1971;**1**:265–6.

5. Greiser E. [Epidemiological studies on the relation between use of appetite depressants and primary vascular pulmonary hypertension]. *Internist* (Berlin) 1973;**14**:437–42.

6. Olivari MT. Primary pulmonary hypertension. *Am J Med Sci* 1991;**302**:185–98.

7. Loogen F, Worth H, Schwan G, *et al.* Long-term follow-up of pulmonary hypertension in patients with and without anorectic drug intake. *Cor Vasa* 1985;**27**:111–24.

8. Kay JM, Smith P, Heath D. Aminorex and the pulmonary circulation. *Thorax* 1971;**26**:262–70.

9. Gaine SP, Rubin LJ, Kmetzo JJ, *et al.* Recreational use of aminorex and pulmonary hypertension. *Chest* 2000;**118**: 1496–7.

10. Douglas JG, Munro JF, Kitchin AH, *et al.* Pulmonary hypertension and fenfluramine. *Br Med J* 1981;**283**:881–3.

11. Pouwels HM, Smeets JL, Cheriex EC, Wouters EF. Pulmonary hypertension and fenfluramine. *Eur Respir J* 1990;**3**:606–7.

12. McMurray J, Bloomfield P, Miller HC. Irreversible pulmonary hypertension after treatment with fenfluramine. *Br Med J* 1986;**292**:239–40.

13. Fotiadis I, Apostolou T, Koukoulas A, *et al.* Fenfluramine-induced irreversible pulmonary hypertension. *Postgrad Med J* 1991;**67**:776–7.

14. Atanassoff PG, Weiss BM, Schmid ER, Tornic M. Pulmonary hypertension and dexfenfluramine. *Lancet* 1992;**339**:436.

15. Roche N, Labrune S, Braun JM, Huchon GJ. Pulmonary hypertension and dexfenfluramine. *Lancet* 1992;**339**:436–7.

16. Brenot F, Herve P, Petitpretz P, *et al.* Primary pulmonary hypertension and fenfluramine use. *Br Heart J* 1993;**70**: 537–41.

17. Abenhaim L, *et al.*, International Primary Pulmonary Hypertension Study Group. Appetite-suppressant drugs and the risk of primary pulmonary hypertension. *New Engl J Med* 1996;**335**:609–16.

18. Mark EJ, Patalas ED, Chang HT, *et al.* Fatal pulmonary hypertension associated with short-term use of fenfluramine and phentermine. *New Engl J Med* 1997;**337**:602–6.

19. Strother J, Fedullo P, Yi ES, Masliah E. Complex vascular lesions at autopsy in a patient with phentermine-fenfluramine use and rapidly progressing pulmonary hypertension. *Arch Pathol Lab Med* 1999;**123**:539–40.

20. Tomita T, Zhao Q. Autopsy findings of heart and lungs in a patient with primary pulmonary hypertension associated with use of fenfluramine and phentermine. *Chest* 2002;**121**:649–52.

21. Connolly HM, Crary JL, McGoon MD, *et al.* Valvular heart disease associated with fenfluramine-phentermine. *New Engl J Med* 1997;**337**:581–8.

22. Rich S, Rubin L, Walker AM, *et al.* Anorexigens and pulmonary hypertension in the United States: results from the surveillance of North American pulmonary hypertension. *Chest* 2000;**117**:870–4.

23. Walker AM, Langleben D, Korelitz JJ, *et al.* Temporal trends and drug exposures in pulmonary hypertension: an American experience. *Am Heart J* 2006;**152**:521–6.

24. Humbert M, Sitbon O, Chaouat A, *et al.* Pulmonary arterial hypertension in France: results from a national registry. *Am J Respir Crit Care Med* 2006;**173**:1023–30.

25. Souza R, Humbert M, Sztrymf B, *et al.* Pulmonary arterial hypertension associated with fenfluramine exposure: report of 109 cases. *Eur Respir J* 2008;**31**:343–8.

26. Rich S, Shillington A, McLaughlin V. Comparison of survival in patients with pulmonary hypertension associated with fenfluramine to patients with primary pulmonary hypertension. *Am J Cardiol* 2003;**92**:1366–8.

27. Hervé P, Launay JM, Scrobohaci ML, *et al.* Increased plasma serotonin in primary pulmonary hypertension. *Am J Med* 1995;**99**:249–54.

28. Friström S, Airaksinen MM, Halmekoski J. Release of platelet 5-hydroxytryptamine by some anorexics and other sympathomimetics and their acetyl derivatives. *Acta Pharmacol Toxicol* 1977;**41**:218–24.

29. Zheng Y, Russell B, Schmierer D, Laverty R. The effects of aminorex and related compounds on brain monoamines and metabolites in CBA mice. *J Pharm Pharmacol* 1997;**49**: 89–96.

30. Celada P, Martín F, Artigas F. Effects of chronic treatment with dexfenfluramine on serotonin in rat blood, brain and lung tissue. *Life Sci* 1994;**55**:1237–43.

31. Redmon B, Raatz S, Bantle JP. Valvular heart disease associated with fenfluramine–phentermine. *New Engl J Med* 1997;**337**:1773–4.

32. Rothman RB, Redmon JB, Raatz SK, *et al.* Chronic treatment with phentermine combined with fenfluramine lowers plasma serotonin. *Am J Cardiol* 2000;**85**:913–15.

33. Rothman RB, Ayestas MA, Dersch CM, Baumann MH. Aminorex, fenfluramine, and chlorphentermine are serotonin transporter substrates: implications for primary pulmonary hypertension. *Circulation* 1999;**100**:869–75.

34. Eddahibi S, Humbert M, Fadel E, *et al.* Serotonin transporter overexpression is responsible for pulmonary artery smooth muscle hyperplasia in primary pulmonary hypertension. *J Clin Invest* 2001;**108**:1141–50.

35. Blanpain C, Le Poul E, Parma J, *et al.* Serotonin 5-HT2B receptor loss of function mutation in a patient with fenfluramine-associated primary pulmonary hypertension. *Cardiovasc Res* 2003;**60**:518–28.

36. Dempsie Y, Morecroft I, Welsh DJ, *et al.* Converging evidence in support of the serotonin hypothesis of dexfenfluramine-induced pulmonary hypertension with novel transgenic mice. *Circulation* 2008;**117**:2928–37.

37. Weir EK, Reeve HL, Huang JM, *et al.* Anorexic agents aminorex, fenfluramine, and dexfenfluramine inhibit potassium current in rat pulmonary vascular smooth muscle and cause pulmonary vasoconstriction. *Circulation* 1996;**94**:2216–20.

38. Sztrymf B, Coulet F, Girerd B, *et al.* Clinical outcomes of pulmonary arterial hypertension in carriers of BMPR2 mutation. *Am J Respir Crit Care Med* 2008;**177**:1377–83.

39. Machado RD, Eickelberg O, Elliott G, *et al.* Genetics and genomics of pulmonary arterial hypertension. *J Am Coll Cardiol* 2009;**54**:S32–42.

40. Humbert M, Deng Z, Simonneau G, *et al.* BMPR2 germline mutations in pulmonary hypertension associated with fenfluramine derivatives. *Eur Respir J* 2002;**20**:518–23.

41. Chaouat A, Coulet F, Favre C, *et al.* Endoglin germline mutation in a patient with hereditary haemorrhagic telangiectasia and dexfenfluramine associated pulmonary arterial hypertension. *Thorax* 2004;**59**:446–8.

42. Cameron J, Waugh L, Loadsman T, *et al.* Possible association of pulmonary hypertension with an anorectic drug. *Med J Aust* 1984;**140**:595–7.

43. Nall KC, Rubin LJ, Lipskind S, Sennesh JD. Reversible pulmonary hypertension associated with anorexigen use. *Am J Med* 1991;**91**:97–9.

44. Hagiwara M, Tsuchida A, Hyakkoku M, *et al.* Delayed onset of pulmonary hypertension associated with an appetite suppressant, mazindol: a case report. *Jpn Circ J* 2000;**64**:218–21.

45. Kernan WN, Viscoli CM, Brass LM, *et al.* Phenylpropanolamine and the risk of hemorrhagic stroke. *New Engl J Med* 2000; **343**:1826–32.

46. Barst RJ, Abenhaim L. Fatal pulmonary arterial hypertension associated with phenylpropanolamine exposure. *Heart* 2004;**90**:e42.

47. Thomas SH, Butt AY, Corris PA, *et al.* Appetite suppressants and primary pulmonary hypertension in the United Kingdom. *Br Heart J* 1995;**74**:660–3.

48. Abramowicz MJ, Van Haecke P, Demedts M, Delcroix M. Primary pulmonary hypertension after amfepramone (diethylpropion) with BMPR2 mutation. *Eur Respir J* 2003;**22**:560–2.

49. Hidden amphetamines: from smoking cessation to diabetes. *Prescrire Int* 2004;**13**:18–20.

50. Boutet K, Frachon I, Jobic Y, *et al.* Fenfluramine-like cardiovascular side effects of benfluorex. *Eur Respir J* 2009;**33**:684–8.

51. Rafel Ribera J, Casañas Muñoz R, Anguera Ferrando N, *et al.* [Valvular heart disease associated with benfluorex]. *Rev Esp Cardiol* 2003;**56**:215–16.

52. Evrard F, Dupuis M, Muller T, Jacquerye P. [Isolated pulmonary hypertension and pergolide]. *Rev Neurol* 2008;**164**: 278–9.

53. Hong Z, Smith AJ, Archer SL, *et al.* Pergolide is an inhibitor of voltage-gated potassium channels, including Kv1.5, and causes pulmonary vasoconstriction. *Circulation* 2005;**112**: 1494–9.

54. Fahlén M, Bergman H, Helder G, *et al.* Phenformin and pulmonary hypertension. *Br Heart J* 1973;**35**:824–8.

55. Kleiger RE, Boxer M, Ingham RE, Harrison DC. Pulmonary hypertension in patients using oral contraceptives: a report of six cases. *Chest* 1976;**69**:143–7.

56. Townend JN, Roberts DH, Jones EL, Davies MK. Fatal pulmonary venoocclusive disease after use of oral contraceptives. *Am Heart J* 1992;**124**:1643–4.

57. Morse JH, Horn EM, Barst RJ. Hormone replacement therapy: a possible risk factor in carriers of familial primary pulmonary hypertension. *Chest* 1999;**116**:847.

58. Martinez-Taboada VM, Rodriguez-Valverde V, Gonzalez-Vilchez F, Armijo JA. Pulmonary hypertension in a patient with rheumatoid arthritis treated with leflunomide. *Rheumatology* 2004;**43**:1451–3.

59. Jochmann N, Kiecker F, Borges AC, *et al.* Long-term therapy of interferon-alpha induced pulmonary arterial hypertension with different PDE-5 inhibitors: a case report. *Cardiovasc Ultrasound* 2005;**3**:26.

60. Al-Zahrani H, Gupta V, Minden MD, *et al.* Vascular events associated with alpha interferon therapy. *Leuk Lymphoma* 2003;**44**:471–5.

61. Dawson A, Elias DJ, Rubenson D, *et al.* Pulmonary hypertension developing after alglucerase therapy in two patients with type 1 Gaucher disease complicated by the hepatopulmonary syndrome. *Ann Intern Med* 1996;**125**: 901–4.

62. Belmatoug N, *et al.*, Comité d'Evaluation du Traitement de la Maladie de Gaucher. Pulmonary hypertension in type 1 Gaucher's disease. *Lancet* 1998;**352**:240.

63. Younis TH, Alam A, Paplham P, *et al.* Reversible pulmonary hypertension and thalidomide therapy for multiple myeloma. *Br J Haematol* 2003;**121**:191–2.

64. Antonioli E, Nozzoli C, Gianfaldoni G, *et al.* Pulmonary hypertension related to thalidomide therapy in refractory multiple myeloma. *Ann Oncol* 2005;**16**:1849–50.

65. Hattori Y, Shimoda M, Okamoto S, *et al.* Pulmonary hypertension and thalidomide therapy in multiple myeloma. *Br J Haematol* 2005;**128**:885–7.

66. Lafaras C, Mandala E, Verrou E, *et al.* Non-thromboembolic pulmonary hypertension in multiple myeloma, after thalidomide treatment: a pilot study. *Ann Oncol* 2008;**19**: 1765–9.

67. Bentur L, Cullinane C, Wilson P, *et al.* Fatal pulmonary arterial occlusive vascular disease following chemotherapy in a 9-month-old infant. *Hum Pathol* 1991;**22**:1295–8.

68. Vaksmann G, Nelken B, Deshildre A, Rey C. Pulmonary arterial occlusive disease following chemotherapy and bone marrow transplantation for leukaemia. *Eur J Pediatr* 2002;**161**: 247–9.

69. Sanchez O, Sitbon O, Jaïs X, *et al.* Immunosuppressive therapy in connective tissue diseases-associated pulmonary arterial hypertension. *Chest* 2006;**130**:182–9.

70. Jais X, Launay D, Yaici A, *et al.* Immunosuppressive therapy in lupus- and mixed connective tissue disease-associated pulmonary arterial hypertension: a retrospective analysis of twenty-three cases. *Arthritis Rheum* 2008;**58**:521–3.

71. Garcia AA, Hirte H, Fleming G, *et al.* Phase II clinical trial of bevacizumab and low-dose metronomic oral cyclophosphamide in recurrent ovarian cancer: a trial of the California, Chicago, and Princess Margaret Hospital phase II consortia. *J Clin Oncol* 2008;**26**:76–82.

72. Manica Sodhi, Omar Minai, and Hina Sahi. Pulmonary hypertension associated with vascular endothelial growth factor inhibitor. *Chest* 2008;**134**:15001S.

73. Druker BJ, Guilhot F, O'Brien SG, *et al.* Five-year follow-up of patients receiving imatinib for chronic myeloid leukemia. *New Engl J Med* 2006;**355**:2408–17.

74. Quintás-Cardama A, Kantarjian H, O'brien S, *et al.* Pleural effusion in patients with chronic myelogenous leukemia treated with dasatinib after imatinib failure. *J Clin Oncol* 2007;**25**:3908–14.

75. Mattei D, Feola M, Orzan F, *et al.* Reversible dasatinib-induced pulmonary arterial hypertension and right ventricle failure in a previously allografted CML patient. *Bone Marrow Transplant* 2009;**43**:967–8.

76. Rasheed W, Flaim B, Seymour JF. Reversible severe pulmonary hypertension secondary to dasatinib in a patient with chronic myeloid leukemia. *Leuk Res* 2009;33:861–4.

77. Souza R, Sitbon O, Parent F, Simonneau G, Humbert M. Long term imatinib treatment in pulmonary arterial hypertension. *Thorax* 2006;**61**:736.

78. Schaiberger PH, Kennedy TC, Miller FC, *et al.* Pulmonary hypertension associated with long-term inhalation of 'crank' methamphetamine. *Chest* 1993;**104**:614–16.

79. Chin KM, Channick RN, Rubin LJ. Is methamphetamine use associated with idiopathic pulmonary arterial hypertension? *Chest* 2006;**130**:1657–63.

80. Rothman RB, Baumann MH. Methamphetamine and idiopathic pulmonary arterial hypertension: role of the serotonin transporter. *Chest* 2007;**132**:1412–13.

81. Murray RJ, Smialek JE, Golle M, Albin RJ. Pulmonary artery medial hypertrophy in cocaine users without foreign particle microembolization. *Chest* 1989;**96**:1050–3.

82. Russell LA, Spehlmann JC, Clarke M, et al. Pulmonary hypertension in female crack users. *Am Rev Respir Dis* 1992;**145**:A717.

83. Collins E, Hardwick RJ, Jeffery H. Perinatal cocaine intoxication. *Med J Aust* 1989;**150**:331–2,334.

84. Devathasan G, Low D, Teoh PC, et al. Complications of chronic glue (toluene) abuse in adolescents. *Aust NZ J Med* 1984;**14**:39–43.

85. Woelk H. Comparison of St John's wort and imipramine for treating depression: randomised controlled trial. *Br Med J* 2000;**321**:536–9.

86. Kay JM. Dietary pulmonary hypertension. *Thorax* 1994;**49**: S33–8.

87. Györik S, Stricker H. Severe pulmonary hypertension possibly due to pyrrolizidine alkaloids in polyphytotherapy. *Swiss Med Wkly* 2009;**139**:210–11.

88. Joselson R, Warnock M. Pulmonary veno-occlusive disease after chemotherapy. *Hum Pathol* 1983;**14**:88–91.

89. Knight BK, Rose AG. Pulmonary veno-occlusive disease after chemotherapy. *Thorax* 1985;**40**:874–5.

90. Waldhorn RE, Tsou E, Smith FP, Kerwin DM. Pulmonary veno-occlusive disease associated with microangiopathic hemolytic anemia and chemotherapy of gastric adenocarcinoma. *Med Pediatr Oncol* 1984;**12**:394–6.

91. Vansteenkiste JF, Bomans P, Verbeken EK, et al. Fatal pulmonary veno-occlusive disease possibly related to gemcitabine. *Lung Cancer* 2001;**31**:83–5.

92. Gagnadoux F, Capron F, Lebeau B. Pulmonary veno-occlusive disease after neoadjuvant mitomycin chemotherapy and surgery for lung carcinoma. *Lung Cancer* 2002;**36**:213–15.

93. Lombard CM, Churg A, Winokur S. Pulmonary veno-occlusive disease following therapy for malignant neoplasms. *Chest* 1987;**92**:871–6.

94. Trobaugh-Lotrario AD, Greffe B, Deterding R, et al. Pulmonary veno-occlusive disease after autologous bone marrow transplant in a child with stage IV neuroblastoma: case report and literature review. *J Pediatr Hematol Oncol* 2003;**25**:405–9.

95. Rose AG. Pulmonary veno-occlusive disease due to bleomycin therapy for lymphoma: case reports. *S Afr Med J* 1983;**64**:636–8.

96. Swift GL, Gibbs A, Campbell IA, et al. Pulmonary veno-occlusive disease and Hodgkin's lymphoma. *Eur Respir J* 1993;**6**:596–8.

97. Troussard X, Bernaudin JF, Cordonnier C, et al. Pulmonary veno-occlusive disease after bone marrow transplantation. *Thorax* 1984;**39**:956–7.

98. Hackman RC, Madtes DK, Petersen FB, Clark JG. Pulmonary venoocclusive disease following bone marrow transplantation. *Transplantation* 1989;**47**:989–92.

99. Kuga T, Kohda K, Hirayama Y, et al. Pulmonary veno-occlusive disease accompanied by microangiopathic hemolytic anemia 1 year after a second bone marrow transplantation for acute lymphoblastic leukemia. *Int J Hematol* 1996;**64**:143–50.

100. Williams LM, Fussell S, Veith RW, et al. Pulmonary veno-occlusive disease in an adult following bone marrow transplantation: case report and review of the literature. *Chest* 1996;**109**:1388–91.

101. Shankar S, Choi JK, Dermody TS, et al. Pulmonary hypertension complicating bone marrow transplantation for idiopathic myelofibrosis. *J Pediatr Hematol Oncol* 2004;**26**:393–7.

102. Capewell SJ, Wright AJ, Ellis DA. Pulmonary veno-occlusive disease in association with Hodgkin's disease. *Thorax* 1984;**39**:554–5.

103. Kramer MR, Estenne M, Berkman N, et al. Radiation-induced pulmonary veno-occlusive disease. *Chest* 1993;**104**:1282–4.

104. Graves ED, Redmond CR, Arensman RM. Persistent pulmonary hypertension in the neonate. *Chest* 1988;**93**:638–41.

105. Walsh-Sukys MC, Tyson JE, Wright LL, et al. Persistent pulmonary hypertension of the newborn in the era before nitric oxide: practice variation and outcomes. *Pediatrics* 2000;**105**:14–20.

106. Manchester D, Margolis HS, Sheldon RE. Possible association between maternal indomethacin therapy and primary pulmonary hypertension of the newborn. *Am J Obstet Gynecol* 1976;**126**:467–9.

107. Dalens B, Dechelotte P, Gaulme J, Raynaud EJ. [Maternal treatment with indomethacin and severe neonatal pulmonary hypertension]. *Arch Fr Pediatr* 1981;**38**:261–5.

108. Besinger RE, Niebyl JR, Keyes WG, et al. Randomized comparative trial of indomethacin and ritodrine for the long-term treatment of preterm labor. *Am J Obstet Gynecol* 1991;**164**:981–6.

109. Herget J, Hampl V, Povýsilová V, Slavík Z. Long-term effects of prenatal indomethacin administration on the pulmonary circulation in rats. *Eur Respir J* 1995;**8**:209–15.

110. Morales WJ, Smith SG, Angel JL, et al. Efficacy and safety of indomethacin versus ritodrine in the management of preterm labor: a randomized study. *Obstet Gynecol* 1989;**74**:567–72.

111. Levin DL, Fixler DE, Morriss FC, Tyson J. Morphologic analysis of the pulmonary vascular bed in infants exposed *in utero* to prostaglandin synthetase inhibitors. *J Pediatr* 1978;**92**: 478–83.

112. Tarcan A, Gürakan B, Yildirim S, et al. Persistent pulmonary hypertension in a premature newborn after 16 hours of antenatal indomethacin exposure. *J Perinat Med* 2004;**32**:98–9.

113. Silvani P, Camporesi A. Drug-induced pulmonary hypertension in newborns: a review. *Curr Vasc Pharmacol* 2007;**5**:129–33.

114. Belik J. Fetal and neonatal effects of maternal drug treatment for depression. *Semin Perinatol* 2008;**32**:350–4.

115. Chambers CD, Johnson KA, Dick LM, et al. Birth outcomes in pregnant women taking fluoxetine. *New Engl J Med* 1996;**335**:1010–15.

116. Chambers CD, Hernandez-Diaz S, Van Marter LJ, et al. Selective serotonin-reuptake inhibitors and risk of persistent pulmonary hypertension of the newborn. *New Engl J Med* 2006;**354**:579–87.

117. Källén B, Olausson PO. Maternal use of selective serotonin re-uptake inhibitors and persistent pulmonary hypertension of the newborn. *Pharmacoepidemiol Drug Safety* 2008;**17**: 801-6.

118. Fornaro E, Li D, Pan J, Belik J. Prenatal exposure to fluoxetine induces fetal pulmonary hypertension in the rat. *Am J Respir Crit Care Med* 2007;**176**:1035-40.

119. Guignabert C, Raffestin B, Benferhat R, *et al.* Serotonin transporter inhibition prevents and reverses monocrotaline-induced pulmonary hypertension in rats. *Circulation* 2005;**111**:2812-19.

120. Kawut SM, Horn EM, Berekashvili KK, *et al.* Selective serotonin reuptake inhibitor use and outcomes in pulmonary arterial hypertension. *Pulm Pharmacol Ther* 2006; **19**:370-4.

121. Posada de la Paz M, Philen RM, Borda AI. Toxic oil syndrome: the perspective after 20 years. *Epidemiol Rev* 2001;**23**: 231-47.

122. Castro García M, Posada de la Paz M, Díaz de Rojas F, *et al.* Pulmonary hypertension after toxic rapeseed oil ingestion. *J Am Coll Cardiol* 1984;**4**:443.

123. Alonso-Ruiz A, Calabozo M, Perez-Ruiz F, Mancebo L. Toxic oil syndrome: a long-term follow-up of a cohort of 332 patients. *Medicine* 1993;**72**:285-95.

124. Abaitua Borda I, Philen RM, Posada de la Paz M, *et al.* Toxic oil syndrome mortality: the first 13 years. *Int J Epidemiol* 1998;**27**:1057-63.

125. Gómez-Sánchez MA, Mestre de Juan MJ, Gómez-Pajuelo C, *et al.* Pulmonary hypertension due to toxic oil syndrome: a clinicopathologic study. *Chest* 1989;**95**:325-31.

126. Kilbourne EM. Eosinophilia-myalgia syndrome: coming to grips with a new illness. *Epidemiol Rev* 1992;**14**:16-36.

127. Tazelaar HD, Myers JL, Drage CW, *et al.* Pulmonary disease associated with L-tryptophan-induced eosinophilic myalgia syndrome: clinical and pathologic features. *Chest* 1990;**97**:1032-6.

30

Respiratory involvement from herbals

TRACEY K RILEY, KAHOKO TAKI, CHRISTOPHER P HOLSTEGE

INTRODUCTION

Archcological evidence suggests that humans have been using herbal remedies for over 60 000 years.[1] Chinese herbal medicine itself has been used for over 2000 years and has recently integrated numerous plants from South East Asia, India, the Americas and the Middle East with native Chinese plants.[1,2] Herbal products are widely available, relatively inexpensive, and often make alluring but unsubstantiated claims. Herbal medicine appeals to consumers who believe that natural herbal products are preferable to synthetic pharmaceuticals. Exacerbating the problem, herbal remedies are often marketed on the Internet with misleading and unproven claims. Despite repeated warnings, consumers continue to equate 'natural' with safe.

There are numerous products currently on the market that have been associated with pulmonary toxicity. The medical literature has begun to highlight an association of specific herbal use with the development of pneumonitis, acute respiratory distress syndrome, interstitial pneumonia, non-cardiogenic pulmonary oedema, bronchiolitis obliterans (BOOP), pulmonary infiltration with eosinophilia (PIE syndrome), pulmonary hypertension and occupational asthma.[3,4] It is imperative that clinicians consider herbals in their differential diagnosis when evaluating patients presenting with new pulmonary complaints.

SOURCES OF TOXICITY

Currently, numerous different herbal products are available on the market in an unrestricted fashion. Of herbals that are associated with pulmonary toxicity, the majority are derived from China. The use of Chinese herbal products is growing in other countries around the world. Chinese herbals commonly contain multiple ingredients touted to ensure positive results on multiple target areas at the same time, and thus restore homeostasis in the body.[2] For example, sho-saiko-to is likely the most frequently used Chinese medicine to be associated with pulmonary toxicity.[5] Sho-saiko-to-induced pneumonitis was first described by Tsukiyama in 1989.[6] Since then numerous case reports have described Chinese medicine-induced pneumonitis (Box 30.1).

In Japan, blends of herbs or plants are known as 'kampo' drugs and are widely incorporated into medical treatments.[7] In 1996, several deaths from pneumonitis associated with sho-saiko-to, the most popular kampo drug in Japan, were reported.[7] Rarely, if ever, are the causative agents in these blends determined. However, in 1999, Nishimori et al. reported a case of pneumonitis associated with sho-saiko-to and ouren-geodoku-to where the bronchoalveolar lavage (BAL) fluid lymphocyte stimulation test was positive. Since ou-gon was the only ingredient common between the two kampos, it was speculated that it was the cause of the pneumonitis.[8] Ou-gon is an extract of Scuttelaria root, otherwise known as skullcap (scullcap) in Europe and the United States.[4,7]

CLINICAL PRESENTATIONS

The signs and symptoms commonly encountered in patients with pulmonary toxicity due to herbal products include fever, cough, dyspnoea with exertion and hypoxaemia.[19,26] Diffuse pulmonary infiltrates can be seen on chest roentgenograms.[19,26] Chest computed tomography may show diffuse infiltrates and widely disseminated ground-glass opacities.[19,44] Also, patients may demonstrate a restrictive pattern on pulmonary function tests and lymphocytic alveolitis on transbronchial lung biopsy.[7] Lymphocytosis is commonly found on bronchoalveolar lavage with a variable CD4 to CD8 lymphocyte ratio.[26] A lymphocyte stimulation test to potential causative Chinese medicines with lymphocytes from BAL fluid, as well as peripheral blood lymphocytes, may be useful for the diagnosis of Chinese drug-induced pneumonitis.[9]

Box 30.1   Herbals associated with pulmonary toxicity [with references]

Pneumonitis

- Dai-saiko-to [9]
- Saiboku-to [10]
- Rikkunshi-to [11]
- Hangeshashin-to [12]
- Sairei-to [13]
- Saikokeisikankyou-tou [3]
- Bofutsusyo-san [14]
- Saiko-ka-ryukotsu-borei-to [15]
- Sho-seiryu-to [16]
- Bofu-tsusho-san [17, 18]
- Sho-saiko-to [6, 8, 19–24]
- Moku-boui-to [25]
- Bakumon-do-to [25]

- Saiko-keishi-kankyo-to [25]
- Ouren-gedoku-to [8]
- Ou-gon (scullcap) [7, 8]

Adult respiratory distress syndrome

- Sho-saiko-to [4]
- Sai-rei-to [4]
- Otsu-ji-to [4]
- Kamisyoyo-san [26]

Non-cardiogenic pulmonary oedema

- Shosaikoto [27, 28]

Interstitial pneumonia

- Sho-seiryu-to [16]

- Gosha-jinki-gan [29]
- Otsu-ji-to [7]
- Sho-saiko-to [30–36]
- Dai-saiko-to [37]
- Saiboku-to [38]
- Pien-tze-huang [39]

Bronchiolitis obliterans

- *Sauropus androgynus* [40–42]

Pulmonary infiltrate with eosinophilia (PIE) syndrome

- Saiboku-to [43]

MECHANISMS OF TOXICITY

The exact mechanism of toxicity has not been clearly elucidated for many of the herbals purported to induce pulmonary injury. The cause of the pulmonary damage may be from the herb itself, a misidentified plant substituted for the marketed herb, a uniquely genetically susceptible individual, or a contaminant introduced in the manufacturing process. The lungs may be damaged by toxins produced by transformation reactions within the lung parenchyma or from reactive metabolites present in the vascular compartment that were produced elsewhere.[45] A P450 monooxygenase system very similar to the one present in the liver can be found in specific lung cell types. The presence of a system such as this in certain lung cell types translates to the potential for the formation of toxic or even carcinogenic metabolites within the lung parenchyma.[45] Two herbal products associated with pulmonary toxicity, comfrey (*Symphytum* spp) and pennyroyal (*Mentha pulegium*), have been more thoroughly investigated.

Comfrey

The pneumotoxicity associated with comfrey provides an example of how the lungs may be damaged by a metabolite produced elsewhere in the body that travels to the lungs via the bloodstream.[45] The pyrrolizidine alkaloids (PAs) are present in numerous plant species found worldwide (i.e. *Symphytum* spp) and are responsible for significant morbidity and mortality in many mammalian species, including man.[46] The two major sources of human exposure to PAs are through accidental ingestion of contaminated foodstuffs or the deliberate ingestion of pyrrolizidine-containing herbal preparations.[47] Hundreds of PAs have been identified and the type and degree of toxicity vary for different alkaloids and between mammalian species.[48] PAs have been associated with veno-occlusive disease and necrosis of the liver, pulmonary vascular lesions, pulmonary hypertension and cor pulmonale.[46,49] Pneumotoxicity is due to reactive

metabolites generated through hepatic metabolism of the alkaloids as the lung itself is unable to bioactivate pyrrolizidine alkaloids.[49] Lung damage from administration of pyrrolizidine alkaloids has reproducibly occurred in a variety of animal species including rats, dogs, turkeys, pigs, sheep and horses.[49]

The pyrrolizidine alkaloid that has been most investigated for the development of pneumotoxicity is monocrotaline. This alkaloid is present in over 20 plants. In the published research model, it was isolated from the seeds of *Crotalaria spectabilis*. The occurrence of pulmonary arterial hypertension and right ventricular hypertrophy from exposure to monocrotaline, especially in rats, is very similar to the pathology that occurs in humans who develop primary pulmonary hypertension.[49] The occurrence of the pathology in rats is not immediately evident. Vascular remodelling that results in the development of pulmonary hypertension takes days to weeks, long after the exposure to the alkaloid has ended.[49] Also, the dose of monocrotaline administered has an effect on the pathology that develops: low doses of monocrotaline were more pulmonary toxic, while higher doses produce hepatotoxicity when given to rats.[46]

A model for pulmonary injury from a metabolite formed within the lung parenchyma is the pneumotoxicity that develops after exposure to the furan derivative 4-ipomeanol.[45] This compound has been shown to be the major toxic substance isolated from mould-infected (*Fusarium solani*) sweet potatoes (*Ipomoea batatas*).[50] Cattle that ingest mold-infected sweet potatoes develop severe lung injury and death.[45,51] Through the study of 4-ipomeanol toxicity on lung cells, it has been deduced that 4-ipomeanol is metabolized by a P450 system in the non-ciliated pulmonary bronchiolar Clara cells to a highly reactive alkylating metabolite that covalently binds to protein macromolecules at the site of formation.[45,50]

Pennyroyal

Oil of pennyroyal is derived from the plant *Mentha pulegium* and is known for its insect-repellent effects and for being an abortifacient and an emmenagogue. Unfortunately, it has also been found in herbal mint teas mistakenly taken or given as home remedies to treat common ailments in children.[52] The active ingredient in pennyroyal oil is R−(+)-pulegone which is metabolized by the cytochrome P450 system to menthofuran, which is one of several reactive metabolites. Hepatotoxicity, the most common manifestation of a toxic ingestion of pennyroyal oil, is thought to occur due to depletion of hepatic glutathione stores by the reactive metabolites. However, the exact mechanism of oil of pennyroyal toxicity is not clearly elucidated and toxicity is known to occur in extrahepatic tissues. Pulegone and menthofuran were found to cause both hepatotoxicity and pulmonary toxicity in mice.[53] Currently, there are no published minimal toxic levels for pulegone and menthofuran in humans.[52]

A case report published by Bakerink[52] describes respiratory symptoms in an 8-week-old Hispanic male that had mistakenly been given a tea brewed from a home-grown mint plant. The child eventually developed severe hepatic toxicity and ultimately died. Serum obtained during the hospitalization was analysed for the presence of pulegone and menthofuran. The presence of menthofuran but not pulegone was confirmed. Autopsy showed confluent hepatocellular necrosis and bilateral lung consolidation with diffuse alveolar damage and haemorrhage.[52] The lung injury in this case may be due to metabolism of the furan by Clara cells in the lung. The authors conclude that the patient's multiple organ system involvement was likely due to the combination of direct toxicity from pennyroyal oil and shock.

TREATMENT

There is no specific treatment for herbal-induced lung disease. The primary role for the clinician is the recognition that specific herbal products have been associated with pulmonary injury. If signs and symptoms develop, therapy is primarily supportive.

> **KEY POINTS**
>
> - Healthcare providers are being increasingly confronted with the use of herbal medications by their patients.
> - Numerous herbal products, especially from China, are associated with pulmonary injury.
> - It is imperative that clinicians consider herbals in their history taking and differential diagnosis when evaluating patients presenting with new pulmonary complaints.

REFERENCES

1. Bateman J, Chapman RD, Simpson D. Possible toxicity of herbal remedies. *Scot Med J* 1998;**43**(1):7–15.
2. Chan TY, Critchley JA. Usage and adverse effects of Chinese herbal medicines. *Hum Exper Toxicol* 1996;**15**:5–12.
3. Heki U, Fujimura M, Ogawa H, Matsuda T, Kitagawa M. Pneumonitis caused by saikokeisikankyou-tou, an herbal drug. *Intern Med* 1997;**36**:214–17.
4. Sakamoto O, Ichikado K, Kohrogi H, Suga M. Clinical and CT characteristics of Chinese medicine-induced acute respiratory distress syndrome. *Respirology* 2003;**8**:344–50.
5. Izumi T. Respiratory disorders induced by Chinese medicines on the increase. *Intern Med* 1996;**35**:433.
6. Tsukiyama K, Tasaka Y, Nakajima M, et al. [A case of pneumonitis due to sho-saiko-to]. *Nihon Kyobu Shikkan Gakkai Zasshi* 1989;**27**:1556–61.
7. Takeshita K, Saisho Y, Kitamura K, et al. Pneumonitis induced by ou-gon (scullcap). *Intern Med* 2001;**40**:764–8.
8. Nishimori F, Yamazaki K, Jin Y, et al. [Pneumonitis induced by the drug ougon]. *Nihon Kokyuki Gakkai Zasshi* 1999;**37**:396–400.
9. Kawasaki A, Mizushima Y, Kunitani H, Kitagawa M, Kobayashi M. A useful diagnostic method for drug-induced pneumonitis: a case report. *Am J Chin Med* 1994;**22**:329–36.
10. Temaru R, Yamashita N, Matsui S, et al. [A case of drug-induced pneumonitis caused by saiboku-to]. *Nihon Kyobu Shikkan Gakkai Zasshi* 1994;**32**:485–90.
11. Maruyama Y, Maruyama M, Takada T, Haraguchi M, Uno K. [A case of pneumonitis due to rikkunshi-to]. *Nihon Kyobu Shikkan Gakkai Zasshi* 1994;**32**:84–9.
12. Oketani N, Saito H, Ebe T. [Pneumonitis due to hangeshashin-to]. *Nihon Kyobu Shikkan Gakkai Zasshi* 1996;**34**:983–8.
13. Maeno T, Ubukata M, Maeno T, et al. [A case of sairei-to-induced pneumonitis diagnosed by lymphocyte stimulation test of bronchoalveolar lavage fluid]. *Nihon Kyobu Shikkan Gakkai Zasshi* 1997;**35**:1347–51.
14. Hatanaka N, Yamagishi T, Kamemura H, et al. [A case of hepatitis and pneumonitis caused by bofutsusyo-san herbal medicine]. *Nihon Kokyuki Gakkai Zasshi* 2006;**44**:335–9.
15. Tokunaga T. [A case of drug-induced pneumonitis associated with Chinese herbal drugs and valsartan]. *Nihon Kokyuki Gakkai Zasshi* 2005;**43**:406–11.
16. Hata Y, Uehara H. [A case where herbal medicine sho-seiryu-to induced interstitial pneumonitis]. *Nihon Kokyuki Gakkai Zasshi* 2005;**43**:23–31.
17. Suzuki S, Tanaka A, Arai T, Adachi M. [Case of interstitial pneumonitis induced by a Chinese herbal medicine, bofu-tsusho-san]. *Nihon Kokyuki Gakkai Zasshi* 2004;**42**:777–81.
18. Matsushima H, Takayanagi N, Ubukata M, et al. [A case of pneumonitis induced by bofu-tsusho-san]. *Nihon Kokyuki Gakkai Zasshi* 2002;**40**:955–9.
19. Ishizaki T, Sasaki F, Ameshima S, et al. Pneumonitis during interferon and/or herbal drug therapy in patients with chronic active hepatitis. *Eur Respir J* 1996;**9**:2691–6.
20. Hatakeyama S, Tachibana A, Morita M, Suzuki K, Okano H. [Five cases of pneumonitis induced by sho-saiko-to]. *Nihon Kyobu Shikkan Gakkai Zasshi* 1997;**35**:505–10.
21. Sato A, Toyoshima M, Kondo A, et al. [Pneumonitis induced by the herbal medicine sho-saiko-to in Japan]. *Nihon Kyobu Shikkan Gakkai Zasshi* 1997;**35**:391–5.
22. Nakagawa A, Yamaguchi T, Takao T, Amano H. [Five cases of drug-induced pneumonitis due to sho-saiko-to or interferon-alpha or both]. *Nihon Kyobu Shikkan Gakkai Zasshi* 1995;**33**:1361–6.
23. Takada N, Arai S, Kusuhara N, et al. [A case of sho-saiko-to-induced pneumonitis, diagnosed by lymphocyte

stimulation test using bronchoalveolar lavage fluid]. *Nihon Kyobu Shikkan Gakkai Zasshi* 1993;**31**:1163–9.

24. Daibo A, Yoshida Y, Kitazawa, *et al.* [A case of pneumonitis and hepatic injury caused by a herbal drug (sho-saiko-to)]. *Nihon Kyobu Shikkan Gakkai Zasshi* 1992;**30**:1583–8.

25. Suzuki K, Kinebuchi S, Sugiyama K, *et al.* [A case of interstitial pneumonia exacerbated by kampo-induced pneumonitis]. *Nihon Kokyuki Gakkai Zasshi* 2002;**40**:605–11.

26. Shiota Y, Wilson JG, Matsumoto H, *et al.* Adult respiratory distress syndrome induced by a Chinese medicine, kamisyoyo-san. *Intern Med* 1996;**35**:494–6.

27. Yoshida Y. [A non-cardiogenic type of pulmonary edema after administration of Chinese herbal medicine (shosaikoto): case report]. *Nihon Kokyuki Gakkai Zasshi* 2003;**41**:300–3.

28. Miyazaki E, Ando M, Ih K, *et al.* [Pulmonary edema associated with the Chinese medicine shosaikoto]. *Nihon Kokyuki Gakkai Zasshi* 1998;**36**:776–80.

29. Katayama H, Hamada H, Yokoyama A, *et al.* [A case of interstitial pneumonia caused by gosha-jinki-gan]. *Nippon Ronen Igakkai Zasshi* 2004;**41**:675–8.

30. Katou K, Mori K. Autoimmune hepatitis with drug-induced pneumonia due to sho-saiko-to. *Nihon Kokyuki Gakkai Zasshi* 1999;**37**:641–6.

31. Tomioka H, Hashimoto K, Ohnishi H, *et al.* [An autopsy case of interstitial pneumonia probably induced by sho-saiko-to]. *Nihon Kokyuki Gakkai Zasshi* 1999;**37**:1013–18.

32. Kobashi Y, Nakajima M, Niki Y, Matsushima T. [A case of acute eosinophilic pneumonia due to sho-saiko-to]. *Nihon Kyobu Shikkan Gakkai Zasshi* 1997;**35**:1372–7.

33. Wada Y, Kubo M. [Acute lymphoblastic leukemia complicated by type C hepatitis during treatment and further by acute interstitial pneumonia due to sho-saiko-to in 7-year-old]. *Arerugi* 1997;**46**:1148–55.

34. Tojima H, Yamazaki T, Tokudome T. [Two cases of pneumonia caused by sho-saiko-to]. *Nihon Kyobu Shikkan Gakkai Zasshi* 1996;**34**:904–10.

35. Sugiyama H, Nagai M, Kotajima F, *et al.* [A case of interstitial pneumonia with chronic hepatitis C following interferon-alfa and sho-saiko-to therapy]. *Arerugi* 1995;**44**: 711–14.

36. Murakami K, Okajima K, Sakata K, Takatsuki K. [A possible mechanism of interstitial pneumonia during interferon therapy with sho-saiko-to]. *Nihon Kyobu Shikkan Gakkai Zasshi* 1995;**33**:389–94.

37. Matsuda R, Takahashi D, Chiba E, *et al.* [A case of drug-induced hepatitis and interstitial pneumonia caused by a herbal drug, dai-saiko-to]. *Nippon Shokakibyo Gakkai Zasshi* 1997;**94**:787–91.

38. Katsura H, Hashimoto I, Taira M, Kadoriku C, Yamawaki I. [Pneumonia caused by saiboku-to]. *Nihon Kyobu Shikkan Gakkai Zasshi* 1996;**34**:1239–43.

39. Kobayashi Y, Hasegawa T, Sato M, Suzuki E, Arakawa M. [Pneumonia due to the Chinese medicine pien tze huang]. *Nihon Kyobu Shikkan Gakkai Zasshi* 1996;**34**:810–15.

40. Wu CL, Hsu WH, Chiang CD, *et al.* Lung injury related to consuming *Sauropus androgynus* vegetable. *J Toxicol Clin Toxicol* 1997;**35**:241–8.

41. Higenbottam TW. Bronchiolitis obliterans following the ingestion of an Asian shrub leaf. *Thorax* 1997;**52**:S68–72.

42. Lai RS, Chiang AA, Wu MT, *et al.* Outbreak of bronchiolitis obliterans associated with consumption of *Sauropus androgynus* in Taiwan. *Lancet* 1996;**348**:83–5.

43. Soda R, Takahashi K, Tada S, *et al.* [A case of pulmonary infiltration with eosinophilia (PIE) syndrome induced by Saiboku-To (TJ96): detection of ECF activity in lymphocytes stimulated with TJ96]. *Nihon Kyobu Shikkan Gakkai Zasshi* 1992;**30**:662–7.

44. Ikezoe J, Kohno N, Johkoh T, *et al.* Pulmonary abnormalities caused by interferon with or without herbal drug: CT and radiographic findings. *Nippon Igaku Hoshasen Gakkai Zasshi* (Nippon Acta Radiologica) 1995;**55**(3):150–6.

45. Boyd MR. Metabolic activation and lung toxicity: a basis for cell-selective pulmonary damage by foreign chemicals. *Environ Health Persp* 1984;**55**:47–51.

46. Schultze AE, Wagner JG, White SM, Roth RA. Early indications of monocrotaline pyrrole-induced lung injury in rats. *Toxicol Appl Pharmacol* 1991;**109**(1):41–50.

47. Huxtable RJ. Herbal teas and toxins: novel aspects of pyrrolizidine poisoning in the United States. *Perspect Biol Med* 1980;**24**(1):1–14.

48. Yan CC, Huxtable RJ. The effect of the pyrrolizidine alkaloids, monocrotaline and trichodesmine, on tissue pyrrole binding and glutathione metabolism in the rat. *Toxicon* 1995;**33**:627–34.

49. Huxtable RJ. Activation and pulmonary toxicity of pyrrolizidine alkaloids. *Pharmacol Therapeut* 1990;**47**:371–89.

50. Gram TE. Chemically reactive intermediates and pulmonary xenobiotic toxicity. *Pharmacol Rev* 1997;**49**:297–341.

51. Gram TE. Pulmonary toxicity of 4-ipomeanol. *Pharmacol Therapeut* 1989;**43**:291–7.

52. Bakerink JA, Gospe SM, Dimand RJ, Eldridge MW. Multiple organ failure after ingestion of pennyroyal oil from herbal tea in two infants. *Pediatrics* 1996;**98**:944–7.

53. Gordon WP, Forte AJ, McMurtry RJ, Gal J, Nelson SD. Hepatotoxicity and pulmonary toxicity of pennyroyal oil and its constituent terpenes in the mouse. *Toxicol Appl Pharmacol* 1982;**65**:413–24.

Index